ATM Theory
and Application

Wale Akinpelu

Signature Series

ATM Theory and Application

David E. McDysan

Darren L. Spohn

Signature Edition

McGraw-Hill

New York San Francisco Washington, D.C. Auckland Bogotá
Caracas Lisbon London Madrid Mexico City Milan
Montreal New Delhi San Juan Singapore
Sydney Tokyo Toronto

Library of Congress Cataloging-in-Publication Data

McDysan, David E.
    ATM theory and application / David E. McDysan, Darren L. Spohn. —
Signature ed.
        p.   cm.
    Includes index.
    ISBN 0-07-045346-2
    1. Computer networks. 2. Asynchronous transfer mode.
    3.Telecommunication—Traffic. I. Spohn, Darren L. II. Title.
    TK5105.5.M339   1998
    004.6'6—dc21   98-28203
                         CIP

McGraw-Hill

A Division of The McGraw-Hill Companies

3 4 5 6 7 8 9 0   AGM/AGM   9 0 3 2 1 0 9

ISBN 0-07-045346-2

*The sponsoring editor for this book was Steven Elliot, the editing
supervisor was Frank Kotowski, Jr., and the production supervisor was
Tina Cameron. It was set in Century Schoolbook by the authors, and all
computer graphics were generated by the authors.*

Printed and bound by Quebecor/Martinsburg.

AppleTalk, Macintosh, and LocalTalk are trademarks of Apple
Computer; AIX, AS/400, IBM, NetView, and RISC are trademarks of
International Business Machines Corporation; ARCnet is a trademark
of Datapoint; ARPANET was developed by the Department of Defense;
AT&T and UNIX are trademarks of American Telephone & Telegraph;
DEC, DECnet, DNA, VAX, and MicroVax are trademarks of Digital
Equipment Corporation; Ethernet and XNS are trademarks of Xerox
Corporation; MAP is a trademark of General Motors; Windows, Word,
and Excel are trademarks of Microsoft; SNA and IBM are trademarks
of IBM; X-Windows is a trademark of Massachusetts Institute of
Technology.

McGraw-Hill books are available at special quantity discounts to use
as premiums and sales promotions, or for use in corporate training programs. For more information, please write to the Director of Special
Sales, McGraw-Hill, 11 West 19th Street, New York, NY 10011. Or contact your local bookstore.

 This book is printed on recycled, acid-free paper containing a
minimum of 50% recycled, de-inked fiber.

This book is dedicated to my wife, Debbie, and my parents, Lowell and Martha, for their support and encouragement

DEM

This book is dedicated to my wife, Becky, and son, Alexander, for their support and sacrifice of time so that I might complete yet another project.

DLS

Contents

Chapter 13. ATM Adaptation Layers   345

Part 5. ATM Traffic Management, and Congestion Control   585

Chapter 21. The Traffic Contract   587

Chapter 22. ATM Traffic Control and Available Bit Rate (ABR)   609

Part 7. Operations, Network Management, and ATM Layer Management

Chapter 27. Operational Philosophy and Network Management Architectures

Chapter 28. Network Management Protocols and Management Information Bases (MIBs)

Preface

PURPOSE OF THIS BOOK

Why did we decide to update this book? A lot has happened since the publication of the first edition of this book in the fall of 1994. The content-driven surge of the World Wide Web is taking data networking into the next millennium beyond the previous benchmarks set by the ubiquitous voice networks that encircle our planet. Internet service providers have turned to Asynchronous Transfer Mode (ATM) backbones to handle the intense load as traffic volumes double or quadruple each year. Simultaneously, many government and research networks continue to deploy ATM networks. Concurrently, convinced of ATM's inherently lower life-cycle costs, corporations are installing high-performance ATM local area networks, private networks, and Wide Area Networks (WANs) using private ATM networks and public ATM services. The range of products and services being offered now account for nearly a multibillion-dollar business. As further evidence of ATM's move toward ubiquity, every major carrier around the globe now offers ATM service.

With all of this going on, we decided early on that it all would not fit into a single book. As a matter of fact, this book is already quite large! Therefore, we split the material into two volumes: a high level introduction to ATM and a survey of vendor and service providers in the book titled — *Hands-On ATM* — and this update to our best-selling book, *ATM: Theory and Application*. *Hands-On ATM* takes a higher-level, less technically detailed look at the broad range of technologies that now intersect with ATM. We moved most of the ATM device vendor, service provider, and market information into *Hands-On ATM* to make space for additional technical coverage of advanced topics in this expanded edition of *ATM: Theory and Application*. Since *Hands-On ATM* provides only an introduction to the technical details and practical application of ATM, this book provides a in-depth treatment of the background, foundations, standards, implementation considerations, traffic engineering, management, and application of ATM.

This edition adds background information and detailed technical descriptions along with application notes in many areas. Specifically, many new capabilities are now standardized for operation over ATM. In particular, a number of ATM-based protocols now support Local Area Network (LAN) applications and the Internet Protocol (IP). Therefore, Part 2 now provides a detailed tutorial on LANs and IP, in addition to expanded coverage of other communication networking protocols. Furthermore, ATM services now provide full support for bandwidth on demand via Switched Virtual Connections (SVCs). Since this capability derived from X.25, the Integrated Services Digital Network (ISDN) and frame relay, Part 2 also provides expanded coverage on these subjects as background.

Several activities and close cooperation between standards bodies made this happen. Inevitably, the NATO treaty–initiated International Telecommunications Union and the upstart ATM Forum are cooperating. Also, the de facto standards body for the Internet, the Internet Engineering Task Force, has been busy in concert with the ATM Forum crafting methods for internetworking over ATM. Part 3 introduces these players in the standards arena and lays the foundation for the higher-layer protocols. New material added in Part 3 includes text on the new ATM Adaptation Layer (AAL) supporting voice and video, and greatly expanded coverage of the ATM physical layer, signaling standards, and routing protocols.

Meanwhile, major advances occurred in ATM's support for higher-layer applications as described in Part 4. Although the basics of ATM switching and buffering architectures are now stable, a relatively precise definition of ATM device categories can now be discerned from the marketplace. Additionally, significant progress has occurred in the area of ATM-aware Application Programming Interfaces (APIs). ATM is now fulfilling the vision of its designers by offering standardized support for voice and video, along with interworking with legacy data services like frame relay. Furthermore, a frame-based user interface supports ATM operation at low speeds. Part 4 dedicates two entirely new chapters to the important subject of LAN and IP support.

Part 5 updates the descriptions of ATM's hallmark — traffic management — and adds new material covering the Available Bit Rate (ABR) closed-loop flow control technique. Part 6 adds a new chapter covering basic digital communications engineering involved with signal selection, frequency spectra, error rates, along with an introduction to error detecting and correcting codes. Part 6 also contains new material on equivalent bandwidth, TCP/IP performance, and voice and data integration performance.

As is often the case in many communication technologies, network management is often the last subject addressed. Since ATM is now relatively mature, the standards and approaches for managing ATM-based networks have also matured. Therefore, Part 7 contains a large amount of new material on the management of ATM networks. This part also updates the previous material as the result of some finalized industry standards.

When the ancient Chinese said, "may you live in interesting times," they meant it as a curse. Their bureaucratic society abhorred change, preferring the comfort of stability over the uncertainty of progress. Part 8 explores the current era of interesting times in communication and computer networking. This final part contains mostly new material. Starting in the wide area network where efficient use of expensive WAN bandwidth is key, the text objectively studies the efficiency of voice, video, and data over ATM versus packet networks. We study the efficiency of packet transfer over ATM and compare it with new approaches being defined for IP. Now that router manufacturers have learned to make devices that go real fast, they can do so for ATM without requiring the use of advanced hardware. Moving into the local area network, equipment price is the key consideration since bandwidth is much less expensive in the local area when compared with the wide area. In particular, we look at how Gigabit Ethernet will likely offer local area networking price benefits over ATM, while switched Ethernet effectively serves high-performance clients and servers. Nonetheless, users with applications requiring guaranteed quality and reserved bandwidth are selecting ATM in the LAN as well as the WAN. We then compare and contrast connection-oriented and connectionless networking paradigms of ATM versus LANs and IP.

No matter which technology wins out in the marketplace, the concepts of guaranteed bandwidth and quality of service tailored to specific voice, video, and data application needs are certain to remain. At the same time, ATM continues to expand into new frontiers: wireless, satellite, and high-performance access into the home. We invite you to join us in sharing this fast-paced, exciting glimpse into the future, and welcome your feedback!

INTENDED AUDIENCE

This book contains technical background in local area and wide area networking for use as a comprehensive textbook on communications networking. It focuses on ATM protocols, operation, standards, technology, and services for use by the communications manager, network design engineer, practicing professional, or student of data networking. This book was written to tell the story of the origins of ATM alongside complimentary and competing technologies. The text begins with an overview, before proceeding into technical details. The text summarizes all of the key standards and gives the principles of how ATM meets business and user needs. The book provides specific details key to understanding the technology, with technical references identified for the reader who wishes to pursue a particular subject further.

The reader should have some prior knowledge of telecommunications principles, although most of the basic concepts of communication networking are covered in Part 2. The book gives an extensive list of references to other books for readers who want to study or refresh their knowledge in certain background areas. Specific interest groups include telecommunication and computer professionals, Information Services (IS) staff, LAN/WAN designers, network administrators, capacity planners, programmers, and program managers. This book was written with more technical detail than *Hands-On ATM* and serves as a detailed reference for communications networking, the application of ATM, and the competing and complimentary networking technologies. Therefore, not only will the technical professional benefit from this book, but sales and marketing, end users of the technology, and executives will gain a more in-depth view of how ATM technology and services impact their business. This book should help practicing engineers become well versed in the principles and empower them to communicate these principles effectively to their management. Computing and engineering professionals must have a good understanding of current technology to make wise corpo-

rate decisions on their data communication needs, which may certainly involve ATM for many years to come. Thus, the book seeks to address the issues and needs of the local area network, wide area network, and end-user
application communities; highlighting commonalties as well as differences in
these diverse environments.

The book was written in a light, easy reading style in order to retain the
reader's interest during the dry topics while making the learning process
more enjoyable. Some of the material is necessarily quite detailed. We have
tried to draw simple analogies as a memory aid and a means to explain complex concepts. The text also interleaves practical applications with the detailed technical descriptions to improve readability. Befitting its more detailed approach, this book contains some equations or formulas; however,
these can be evaluated using spreadsheets. A word of caution: readers may
find inconsistencies in descriptions of protocols and the ever-evolving ATM
standards because of the changing state of affairs found in standardization
processes. We attempted to address this by providing an appendix including
detailed references as well as information on how to obtain the latest standards. While we tried to keep the text accurate up to the time of publication,
the reader is urged to use the sources provided to confirm information and
obtain the latest published standards.

OVERVIEW OF THIS BOOK

Part 1 considers key business drivers and technology enablers that create the
need for advances in communication networking. Since computers drive so
much of the change in communication networking, the text then summarizes
how the changes in the information networking environment influence communications technologies. ATM borrows many ideas and concepts from other
communication technologies. Hence, the book provides extensive background
on wide area and local area networking technologies in Part 2. Next, Part 3
introduces the standards organizations and the basic terminology of ATM.
The coverage here includes the physical layer, the ATM layer, the common
adaptation layers, and a detailed discussion of signaling and routing. Part 4
then details the current state of ATM technical standards' support for video,
voice, protocol interworking, local area networks, and IP. Part 5 then moves
on to the important subjects of traffic management and congestion control.
Part 6 describes a number of analytical tools useful in communications network design, traffic engineering and network design. Next, Part 7 provides
an in-depth review of the state of ATM network management and performance measurement. Finally, Part 8 compares ATM with competing technologies, such as IP and Gigabit Ethernet. The book concludes with a survey of
opinions on the prognosis for ATM in the major business networking environments — the home, the workgroup, the local backbone, and the carrier
edge and core switching networks.

This book not only reviews highlights of standards, but it also applies the concepts through illustrations, examples, and real-world applications to make the transition from theory to application. This book strives to teach the reader not only what the ATM-based technologies involve, but why each subject is important.

In some ways, the book presents an unconventional point of view. Taking the same tact as our earlier books, we don't assume that the paradigm of a voice-oriented telephony structure will be the future of data communications. Rather, just the opposite. Narrowband ISDN (N-ISDN) will not make the transition to extensive hands-on application, despite a resurgence as the only technology to provide high-performance access to the Web, while Ethernet and the Internet have taken the world by storm. The promise of even higher performance Asymmetric Digital Subscriber Line (ADSL) technologies will eclipse ISDN's brief renaissance. Since the Internet has become the de facto networking protocol standard, we focus a great deal of material on how ATM applies to real-world internets, intranets, and extranets.

This text covers the three critical aspects of business drivers, technology, and practical, hands-on application. It interleaves descriptive material replete with many drawings with application notes and the results of actual networking experience. The book gives examples for network planners, development engineers, network designers, end users, and network managers. The text cites numerous references to complementary texts and sources for for the reader interested in more depth.

ATM: Theory and Application covers the material in eight parts.

Part 1 covers networking basics, including a discussion of the business drivers for ATM, technological enablers, and computer and information networking directions.

Chapter 1 summarizes the business drivers changing computing and communication networking. This includes changing operational and competitive paradigms in the business world which require increased higher performance communications at reduced unit costs. The combined increase in computing and expanding role of communications creates a demand called the accelerating bandwidth principle. Other business drivers include integration savings, flexibility, and economies of scale. Furthermore, a number of consumer and commercial applications drive communication requirements.

Chapter 2 covers the role of ATM technology enablers. This chapter begins with a brief history of voice and data communications as background to the brave new world of data networking. The text summarizes the key technology trends enabling this paradigm shift, including processor enhancements, modernized transmission networks, and decreasing costs.

Chapter 3 sets ATM in the context of the rapidly changing computer and information networking environment. It begins with a review of the evolving corporate communications environment, ranging from the desktop to the local area network and client-server networking domains, to the rapidly growing

world of intranets, internets, and extranets. This chapter also discusses the
role of ATM as an infrastructure for communications networking. This chapter presents a brief analysis of outsourcing and the status of the battle for the
desktop, the LAN, and the WAN.

Part 2 presents a comprehensive background on communications networking and protocols. This includes the basic concepts of multiplexing and
switching, an introduction to layered protocol models, and tutorials on the
major communication networking techniques in use today in the WAN and
LAN environments.

Chapter 4 covers basics of network topologies, circuit types and services,
and asynchronous and synchronous transmission methods. The definitions of
the terms asynchronous and synchronous are covered in detail. This chapter
concludes with a comprehensive review of the principles of multiplexing and
switching, with key points illustrated through detailed examples.

Chapter 5 begins with a brief history of packet switching. It then introduces
the basic protocol layering concepts used throughout the book. The text then
discusses several layered protocol architectures as applied examples of layered protocols. The chapter also presents a discussion of connectionless and
connection-oriented data services.

Chapter 6 then introduces the connection-oriented digital Time Division
Multiplexing (TDM) communication technique widely used in contemporary
voice and private line networks. The text then moves onto an in-depth review
of one of ATM's key ancestors — the Narrowband Integrated Services Digital
Network (N-ISDN) protocol stack. Here the reader is introduced to the concept of multiple planes of layered protocols serving the user, control, and
management functions.

Chapter 7 covers the key connection-oriented packet switching protocols in
use today: X.25 and frame relay. The text gives the origins of each protocol,
their structure, and protocol details. Examples illustrate key points of operation for each protocol. The text separates the description of the user and control plane protocol stacks for frame relay as an introduction to the separation
of function employed in the ATM standards.

Chapter 8 describes the key connectionless packet switching protocols defined for use in communication networks, namely the Internet Protocol and
the Switched Multimegabit Data Service (SMDS). This chapter traces the
historical background of each protocol, details the packet formats, and illustrates key operational aspects through examples. Here the text introduces
IP's Resource reSerVation Protocol (RSVP) as a means for an Internet user to
signal required Quality of Service (QoS) and bandwidth requirements.

Chapter 9 presents a tutorial on bridging and routing as background to
coverage in Part 4 on ATM support for LAN and IP. The text first introduces
basic terminology and concepts. Next, we describe commonly used LAN protocols like Ethernet, Token Ring, and FDDI. The text then introduces the
concepts of routing and addressing.

Part 3 covers the basics of the ATM protocol landscape, providing a structured introduction and reference to all ATM-related terminology, protocols, and standards.

Chapter 10 covers the foundations of B-ISDN and ATM — the bodies and the manner in which they define the standards and specifications, complete with references to the major standards used in this book and how to acquire them. This chapter introduces the overall broadband ISDN protocol reference model in terms of the user, control, and management planes. The layers common to all of these planes are physical, ATM, and ATM Adaptation Layer (AAL).

Chapter 11 provides an introduction to ATM, beginning with a view of its many facets: an interface, a protocol, a technology, an integrated access method, an infrastructure, and a service. We then introduce ATM terminology through high-level illustrations and analogies. The chapter concludes with some illustrations of ATM's key attribute: multiple service qualities.

Chapter 12 details the physical-layer and ATM-layer protocols. The text describes how a single ATM layer operates over a number of physical media. Furthermore, we discuss how various AALs support a broad range of application attributes. It concludes with an introduction to the concepts of the ATM traffic contract, ATM service categories, and Quality of Service (QoS).

Chapter 13 covers the ATM Adaptation Layer (AAL), which provides support for all higher-layer services, such as signaling, circuit emulation, frame relay, and IP.

Chapter 14 introduces the higher layers in the user plane in the WAN, LAN, and the internetwork; control plane signaling and routing; and the management plane. This chapter then introduces the ATM control plane and its AAL and underlying structure. The coverage then moves on to the important subject of addressing.

Chapter 15 covers the important concepts and standards of the ATM control plane, focusing on the signaling protocol itself and how it establishes connections of various types. This chapter also describes the important Private Network-to-Network Interface (PNNI) routing protocol defined by the ATM Forum. Finally, this chapter describes the control plane protocols employed between ATM service provider networks.

Part 4 covers ATM's support for higher-layer protocols in the user plane in the WAN and LAN.

Chapter 16 introduces the basic concepts of ATM switching and buffering designs as background. The text then presents a taxonomy of ATM device types used in service provider and enterprise networks. We describe the basic attributes of each device class and summarize key attributes from the companion text, *Hands-On ATM*, and further research. The text then moves to the ATM desktop, detailing the state of hardware and software support for end-to-end ATM networking. Finally, we present an overview of ATM integrated circuit technology and standards.

Chapter 17 moves on to the support of video, voice, and TDM oriented protocols in the user and control planes. Specifically, the chapter covers the emulation of Time-Division Multiplexed (TDM) circuits and interworking with Narrowband ISDN (N-ISDN). The text also covers the work on Voice and Telephony over ATM (VTOA) performed by the ATM Forum. Finally, the chapter concludes with protocol support for video over ATM and the Residential BroadBand (RBB) initiative of the ATM Forum.

Chapter 18 covers ATM's role in terms of true protocol interworking, use as an access method, or use as a common trunking vehicle. The text then details the two interworking standards defined for operation with frame relay. We then summarize the support for interworking ATM and SMDS. Finally, the chapter describes the cost-effective ATM Data eXchange Interface (DXI) and Frame-based User-Network Interface (FUNI) capabilities.

Chapter 19 moves into the local area network. It begins with an in-depth tutorial of the LAN Emulation (LANE) protocol and its advantages over traditional LAN bridging. The text then describes the proposed Cells In Frames (CIF) protocol. The chapter concludes with a description of multiprotocol encapsulation over ATM.

Chapter 20 describes standards support for carrying multiple protocols over ATM, negotiating maximum packet sizes, and classical IP over ATM for a single logical IP subnetwork. We then introduce the newer subjects of Multi-Protocol over ATM (MPOA) and IP multicast over ATM. The chapter concludes with a review of proprietary approaches for supporting IP over ATM culminating in the MultiProtocol Label Switching (MPLS) standard being developed for the Internet.

Part 5 provides the reader with an application-oriented view of the ATM traffic contract, congestion control, traffic engineering, and design considerations. Complex principles are presented in a manner intended to be more readable and understandable to a wider audience than other current publications.

Chapter 21 focuses on the ATM traffic contract — the service agreement between the user and the network — that defines the Quality of Service (QoS) that the user expects from the network. The other part of the contract is the load generated by the user as defined by a set of traffic parameters for a leaky bucket algorithm. The chapter concludes with a description of the simplifying concept of ATM layer service categories. Standards use a notation defined by a list of related acronyms; for example, the terms CBR, VBR, UBR correspond to constant, variable, and unspecified bit rates, respectively.

Chapter 22 focuses on the aspect of traffic control in ATM, which includes policing, shaping, and selective cell discard. Principles discussed include use of Usage Parameter Control (UPC), priority queuing, Generic Flow Control (GFC), and Connection Admission Control (CAC). This chapter contains an in-depth description of the closed loop flow control technique called Available Bit Rate (ABR).

Chapter 23 covers the important topic of congestion control. The chapter presents a number of solutions and their range of applications, including the levels of congestion that can occur and the types of responses that exist; the standard method of selective cell discard using Cell Loss Priority (CLP); and long-term control by use of resource allocation, network engineering, and management. The chapter compares the advantages and disadvantages of the major closed loop congestion control methods suitable to ATM: window, rate-based, and credit-based.

Part 6 is the technical center of the book, providing practical guidelines and design formulas that afford considerable insight into the benefits and applicability of ATM for particular applications. As such, it is rather detailed and makes extensive use of mathematics. As much as possible, we try to give a "cookbook" style solution that allows the reader to do some basic calculations regarding their application need. The text gives references to other books and periodicals containing further detail.

Chapter 24 covers the basic philosophy of communications engineering. Central to this discipline is the notion of random processes that model real world events like errors, queuing, and loss. The chapter describes how electrical and optical communication signals occupy a specific frequency spectrum that determines overall transmission capacity. In the presence of noise, Shannon's channel capacity theorem defines the highest achievable transmission rate on a specific channel. The text then describes how a communications system may employ error correction or detection coding.

Chapter 25 provides an in-depth overview of traffic engineering philosophy, basic queuing models, and approximate performance analysis for delay and loss in ATM switches and networks. This second edition adds a section on blocking performance and Erlang models applied to ATM SVCs. The text includes an in-depth discussion regarding statistical multiplex gain and equivalent bandwidth. The emphasis is not on mathematical rigor, but rather on usable approximations and illustration of key tradeoffs.

Chapter 26 discusses additional practical design considerations such as the impact of loss and delay on application performance, the tradeoffs between private and public ATM networking, and traffic engineering complexity. The text features an approximate analysis of TCP/IP performance as a function of buffering, the number of sources, and the round-trip delay.

Part 7 provides the reader a view of operations, network management architectures, standards, ATM layer management, and ATM performance measurement.

Chapter 27 begins with the topic of Operations, Administration, Maintenance, and Provisioning (OAM&P) philosophy. The text continues with an assessment of the state of the solution. The chapter concludes with a description and comparison of the various network management architectural approaches to managing ATM networks.

Chapter 28 continues the theme of network management, but at the next level closer to the actual network elements. The text describes the IETF's Simple Network Management Protocol (SNMP) and the ITU-T's Common Management Interface Protocol (CMIP). The chapter concludes with a description of the ATM Forum's Integrated Local Management Interface (ILMI), IETF ATM MIB (AToMMIB), and other ATM-related Management Information Bases (MIBs).

Chapter 29 introduces ATM layer and fault management. The text presents ATM Operations and Maintenance (OAM) cell structures for use in fault detection and identification using the Alarm Indication Signal (AIS) and Remote Defect Indication (RDI) ATM OAM cells. The chapter then defines reference configurations for specifying and measuring Network Performance (NP) and user Quality of Service (QoS). The chapter describes methods to ensure that a network meets specific QoS objectives through the use of performance measurement OAM cells.

Part 8 provides the reader a comparison of technologies that complement and compete with ATM in the wide area and local area network environments. The book concludes with a look at the future of ATM.

Chapter 30 covers technologies that complement and compete with ATM in the WAN. These include circuit switching, frame relay, and IP. The text outlines key requirements for WAN, foremost being efficient use of expensive bandwidth. The chapter presents an objective efficiency analysis of ATM versus its competitors for voice, video, and data. The text highlights a detailed analysis of the more efficient transfer of packet data by frame-based protocols (like IP of frame relay) than ATM over AAL5 and identifies ways to improve performance when using ATM.

Chapter 31 covers technologies that compete with ATM in the local area network. The analysis compares FDDI, 100 Mbps Fast Ethernet, LAN switching, and Gigabit Ethernet with ATM. The text organizes the discussion according to the desktop, workgroup, and backbone LAN environments. It continues with a comparison of alternative and complementary means to deliver end-to-end quality of service (QoS), namely the IETF's Reservation Protocol. This chapter concludes with a comparison of the advantages and disadvantages of connection-oriented and connectionless networking paradigms.

Chapter 32 concludes the book with a view of the future, beginning with a review of how ATM continues to expand into new frontiers: satellite, ADSL, and wireless. The chapter highlights the emerging ATM over Digital Subscriber Loop (xDSL) technology, which promises to extend high-performance, guaranteed quality service to the home office and residence. The chapter also discusses the contention and cooperation going on between IP and ATM, highlighting some of the challenges envisioned in achieving QoS with frame-based protocols. The text then lists some key challenges for ATM technology and service providers. Finally, it concludes with a look at some possible future networking scenarios for which ATM is the best solution currently available.

This book also contains two Appendixes. *Appendix A* lists the major acronyms and abbreviations used in the book. *Appendix B* provides a reference of national and international ATM-related standards and specifications. The back material also contains a glossary of commonly used terms associated with the technologies, architectures, services, and protocols encountered throughout the book, along with a detailed index.

INTRODUCTION

This book takes a new perspective on ATM, focusing on both a theoretical and hands-on, practical point of view. After an initial wave of media hype, ATM has made the transition from theory to application, and is in full production in Internet service provider networks, carrier services, and private corporate and government networks. The coincidence of the exponentially increasing power of computers, the increasing information transfer needs in the reengineering of business, and the vast content of the World Wide Web and Internet continue to drive a revolution in transmission, switching, and routing technologies. These combined trends create an accelerating need for bandwidth. Furthermore, the issues of service quality and bandwidth reservation have come to the forefront. Technology, applications, and businesses can harness the power of ATM today to simultaneously deliver high-performance and differentiated qualities of service.

With books, timing is everything. The same is true with technology. ATM technology, standards, applications, and implementations have finally come of age. Many standards for ATM signaling, addressing, traffic management, network management, the physical layer, the ATM layer, and the AAL (where the true protocol intelligence resides in ATM) have been finalized for some time. Furthermore, a large degree of interoperability exists between various ATM and non-ATM devices. The standards for higher-layer protocols targeted at the local area network and the Internet are now complete. Therefore, this book documents a stable set of ATM-based technology, standards, application guidelines, and network design hints.

This book provides all the information you need in order to determine if and how ATM will become part of your network. You will find that we do not present ATM from a myopic or biased view, but instead offer you a look at competing and complementing technologies such as Fast Ethernet, frame relay, and IP. In fact, ATM offers the capability to integrate these protocols and services into a single architecture that is scalable, flexible, and capable of handling bandwidths from several to hundreds of megabits per second, and even Gigabits per second in the near future.

Many ATM books on the market today are either too technical (tied up in explaining every detail of the standard without relating to the reader what each aspect means to the user or provider) or too telephony-based, with the reader dragged through the older telephony point of view without the benefit of viewing the technology through the eyes of the data and computer communications user. This book attempts to show the many aspects of ATM through the eyes of the network manager focused on business objectives and the data communications user focused on productivity, as well as the ATM-based service designer or provider.

The demand for flexible, on-demand bandwidth for multimedia applications is growing. One multiservice backbone and access technology for voice, data, and video integration, offering single access to a virtual data service, is a must. Data transfer bandwidths for text, video, voice, and imaging traffic are increasing exponentially, with data communications networks based on technologies such as ATM providing the intelligent network. We live in a distributed data world where everyone needs access to everyone else's data. Private lines are quickly becoming the exception as switched public and private data networks span the globe. Computers need to talk to one another the same way people pick up the phone and dial anyone else in the world. Because of this need, the market for high-speed data transport is exploding. The dawning of the age of gigabit-per-second data communication is upon us — witness the explosive growth of the Internet. LANs, MANs, and WANs have already crossed the 100-Mbps barrier and are moving toward gigabit-per-second, intelligent virtual data services.

LANs and the Internet have become an integral element of almost every major corporation. The move toward visually oriented end-user interfaces in computer software packages through the use of Graphical User Interfaces (GUIs) creates a tremendous need for flexible networking capabilities — witness the rapid adoption by business and residential users of the World Wide Web browser user interface. As the number of LANs and Web traffic continue to grow, so does the requirement to interconnect these LANs at native LAN speeds, and thus there is an emerging need for technologies like ATM. The low-speed private line bridge solutions are reaching their limits, further driving the need for higher-speed, cost-effective, flexible data networking. Frame relay provides users with high-speed bandwidth-on-demand services which have displaced many private line networks. IP is the de facto networking standard to the desktop. ATM is poised to integrate aspects of these multiple services over a single intelligent architecture and technology.

Many business bandwidth requirements are exploding: For example, medical institutions are transferring huge imaging files, and filmmakers are transferring digitized video images, which are then stored and manipulated directly on a high-performance computer. An important aspect of these new-age networks is their ability to store and retrieve large image files. In the search for new technology to provide data communications on this scale, packet switching technology has seen a series of refinements that result in higher performance at lower costs. To wit, frame relay supplants X.25 for higher speeds, efficient IP implementations support sophisticated internetworking, and cell-based multiplexing and switching are replacing time-division multiplexing and switching. As ATM-based services gain increasing support from equipment vendors, local exchange carriers, and interexchange carriers, user acceptance will be a key factor in determining the degree to which the technology succeeds.

After reviewing the available technologies and services, many users ask the classical questions: "Which service do I use?" and "Do I need ATM or just want ATM?" This book shows that the answers to these questions are based on many factors, and there may be multiple answers depending upon the specific user application. There is rarely a single solution, and the decision as to technology and service generally comes down to what is best for the application and what is affordable — price versus performance, as well as migration and future expansion considerations. The decision to use ATM, or an ATM-based public network service, is also a complicated one. This book presents the business and technological cases for the use of ATM, explains its use, and offers methods of implementation.

HOW TO USE THIS BOOK FOR COURSES

This book can be used to teach a single-semester course focused on ATM or a two semester course on data communications with a focus in the second semester on the details of ATM. It is by no means an introductory text to data communications, but it can stand alone for some readers with some technical background.

If the subject matter is to be taught over two semesters, it is recommended that the text be broken into two parts. Material for use in a first-semester course on a business introduction to data communications and basic architectures, protocols, technologies, and services could include Parts 1, 2, 3, and 4. Chapters of focus for a second semester course on advanced ATM protocols and technologies would cover Parts 5, 6, 7 and a recap of Part 4, with either selected outside reading or a research assignment.

A single-semester course dedicated to data communications services (circuit switching, frame relay, SMDS, IP, and ATM) focusing on ATM should consider selections from Parts 1, 2, 3, 4 and 5. The student should have a minimum working knowledge of the material contained in Part 3 if this book is used in a single-semester course.

Labs should contain ATM design problems based on the cumulative knowledge gained from the class readings and outside reading assignments (recent technology updates). The exercises should involve multiple end-system and intermediate-system design problems. Because of the fluid nature of emerging ATM standards, students should be encouraged to use the text as a working document, noting any changes as the standards from the sources listed in the appendices are revised and updated. This is your book — write in it! The authors plan to publish additional updated editions as changes in technology and standards warrant. Supplemental documentation, instructional tools, and training courses can be obtained from the authors at extra charge.

AUTHORS' DISCLAIMER

Accurate and timely information as of the date of publication was provided. Some standards used were drafts at the time of writing, and it was assumed that they would become approved standards by the time of publication. At times, the authors present material which is practical for a large-scale design, but must be scaled down for a smaller business communications environment. Many data communications networks will operate, and continue to run, quite well on a dedicated private line network, but eventually the economics of switched technologies and services, even on the smallest scale, are worth investigating. Please excuse the blatant assumption that the user is ready for these advanced technologies — in some cases it may take some time before these technologies can be implemented. Also, please excuse any personal biases which may have crept into the text.

ACKNOWLEDGMENTS

Many people have helped prepare this book. They have provided comments on various drafts, information on products and services, and other value-added services. In particular, we would like to thank our manuscript reviewers — Mr. Sudhir Gupta of DSC Communications; Mr. Alan Brosz, Mr. Terry Caterisano, Mr. Ather Chaudry, Dr. William Chen, Mr. Tim Dwight, Mr. Mike Fontentot, Mr. Herb Frizzell, Sr., Mr. Wedge Greene, Mr. Saib Jarrar, Ms. Claire Lewis, Mr. James Liou, and Dr. Henry Sinnreich of MCI Communications; Mr. Keith Mumford and Mr. Mike McLoughlin of General Datacomm; Dr. Grenville Armitage of Lucent; and Mr. Steve Davies, Mr. Gabriel Perez, and Mr. Michael Shannon of NetSolve. , Inc.

We also thank our other colleagues whom over the last several years have shared their knowledge and expertise. They have helped us develop a greater understanding and appreciation for ATM and data communications.

This book does not reflect any policy, position, or posture of MCI Communications or NetSolve. This work was not funded or supported financially by MCI Communications or NetSolve, or by MCI Communications or NetSolve resources. Ideas and concepts expressed are strictly those of the authors. Information pertaining to specific vendor or service provider products is based upon open literature or submissions freely provided and adapted as required.

Also, special thanks to Mr. Herb Frizzell for the English style and syntax review, to Wedge Greene for providing much of the new material in Part 7 on Network Management, and to Debbie McDysan for the review and editing of the graphics.

Finally, we would like to thank Debbie McDysan and Becky Spohn for their support, encouragement, and sacrifice of their husband's time while we were writing this book.

The combined support and assistance of these people has made this book possible.

David E. McDysan
Darren L. Spohn

1

Drivers and Enablers
of ATM

Asynchronous Transfer Mode (ATM) is now more than a buzzword — it is reality! This book shows that ATM has made the transition from theory to application. Regardless of the type or size of business you manage, fundamental changes in telecommunications warrant a look at ATM. Before delving into the theory and application of ATM and how it relates to your specific environment, the first three chapters explore business, technology, and networking factors that drive and enable ATM. A key business driver is the move of many corporations away from centralized computing into distributed client-server processing. The resulting re-engineering of the corporation creates increased intranet, internet, and extranet communications that accelerates the need for bandwidth. A technology enabling response is the drastic increase in available local, campus, metropolitan, and wide area network bandwidth. Other technology enablers are the advances in electronics and transmission systems. As the text moves through each of these areas of study, the reader should consider what major businesses and end users are doing, and how these experiences relate to his or her own.

The first two chapters introduce drivers and enablers for ATM. A driver is a force or trend that causes an enterprise to act. Example drivers are changing organizational structures, increased communications requirements, budget pressures, opportunity to integrate separate networks, economies of scale, and new applications. An enabler is an development or trend that makes something possible. Examples of enablers are improved transmission technologies, higher performance computers, sophisticated applications, and information distribution architectures like the web. Thus, there are many drivers for which ATM is a solution, as well as many enablers that make ATM possible.

Chapter 3 then introduces the computer and networking directions that have driven users and service providers to apply ATM. This includes routing versus switching, virtual private networking, and end-to-end capabilities. Furthermore, engineering economics play a key role, along with cost minimization across the entire life cycle. Finally, the text summarizes the important subjects of outsourcing and outtasking.

Business Drivers for ATM

This chapter presents the business drivers for residential and corporate communications networks, specifically for Asynchronous Transfer Mode (ATM) and ATM-based services. The text summarizes the key business reasons that make ATM a viable, attractive communications infrastructure alternative for local and wide area networks. First, the text defines the need for business communications spanning the local, metropolitan, and wide area network domains. The chapter continues by analyzing the move from centralized to distributed networks, resulting in an accelerating bandwidth requirement. The text then looks at the requirements for consumers and commercial enterprises. Finally, we look at applications that drive the technology — not only one "killer application" but multiple applications spanning many communities of interest.

1.1 DEFINING COMMUNICATIONS NEEDS

What information services will create the projected demand for data services? What will the application mix be? What human activity or automated system application will require so much more bandwidth on demand? Part of the answer peeks out at us from the experience of the World Wide Web (WWW). Web browsers evolved from interesting demos to slick professional programs that captured the attention of the world in only a few short years. Beginning with simple text pages employing simple graphics, the contemporary web application sports animated graphics, audio, video, and a wide range of other interactive Java applications downloaded in real-time. The tremendous popularity of cable television also shows a possible source of the future traffic. Many studies indicate that adults in the United States spend an average of 4 to 8 hours a day watching television. If one quarter of these 100 million households used video on demand at 1 Mbps per second, then the total nationwide traffic would be over 1 petabit (10^{15}) per second. Theoretically, 1,000 fibers could meet this demand, but the world needs a network to provide it. Although criticized for the 10 to 15 percent cell tax when compared with other techniques optimized for carrying the current mix of traffic on the internet, this book shows how ATM is actually better at carrying this possible future mix of traffic.   But before getting to this exciting possible future, this book first goes back in time to understand the roots of communication and the hard realities of the practical present.

Corporate management prioritizes communications requirements in terms of the business needs of the organization. Business processes and organizations are changing, and along with them application communications patterns and volumes also change. Communications have shifted from centralized to distributed, and from hierarchical to peer-to-peer. Enterprises continue distributing operations nationally and globally, and with them the requirement to communicate distributes as well. One medium that has enabled this communications globalization is the Internet, which almost overnight has changed network design rules. Intranets, internets, and extranet traffic patterns create unpredictability. Correspondingly, traffic patterns for intraLAN, interLAN, and the WAN communication change drastically. Increasingly sophisticated applications accelerate user's appetite for bandwidth and information, which creates increased capacity across the WAN. Furthermore, cost-conscious network designers demand integration of voice, video, and data to the desktop.

Can you identify with the following scenario? An increasingly flatter, leaner organization is emerging, driving the need for each individual to have access to more data faster as well as communicate with other people or information sources. This drives cost-conscious businesses to increase their reliance on less expensive, user controllable, multiple platform communications in a multivendor environment. As faster computers and distributed applications proliferate, so does the need for faster and more efficient data communications. The business becomes increasingly reliant on network communications, and the network becomes a force to be reckoned with. In fact, the network becomes a mission critical aspect of the business. In other words, the network has become the life blood of the corporation. The communications budget swells each year. This happens although the unit cost of bandwidth generally decreases over time, because the enterprise need for bandwidth increases at a faster rate than that costs decline. Possible scenarios creating the need for communications are: growth in LAN interconnectivity between departments, movement of mainframe applications to a distributed client-server architecture, or the emergence of a new class of image-intensive interactive applications. A large number of possible scenarios each leads to the same conclusion: an explosion of data, voice, and video traffic that requires communication throughout the business to meet ever increasing demands for high-performance computing to the desktop and home of every information worker.

How can these increased data, voice, and video requirements, phenomenal growth of user applications, processing power, and demand for connectivity and bandwidth be defined and understood? How can this parallel shift from hierarchical to flat, decentralized networks and organizations be quantified? This chapter provides answers to these questions and more by first reviewing some perspectives on recent multimedia communication growth that create business drivers for ATM.

1.1.1 The Move from Centralized to Distributed Operations

Computer communications networks evolved over the last thirty years from centralized mainframe computing, through the minicomputer era, into the era of the distributed personal computer and workstation-to-server processing. The first data computer communications networks resembled a hierarchical, or star (also called hub and spoke) topology. Access from remote sites homed back to a central location where the mainframe computer resided (usually an IBM host or DEC VAX). Traditionally, these data networks were completely separate from the voice communications networks. Figure 1.1 shows this centralized, hierarchical computer network topology. Note that the star and hierarchy are actually different ways of drawing the same centralized network topology.

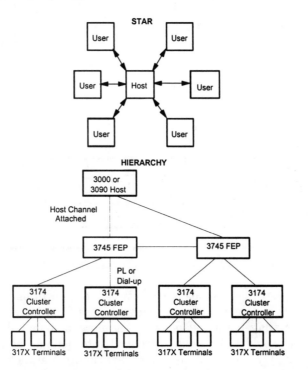

Figure 1.1   Centralized Star and Hierarchical Topology

There is a great parallel happening here between networks and organizations. Networks and entire corporate organizations are making the transition from hierarchical to distributed structures, both requiring greater interconnection and more productivity from each element or individual. We call this move from a hierarchical structure to a distributed structure *flattening*. In networking terminology, flattening means the creation of fewer network elements with greater logical interconnection. In organizations, flattening often refers to the reduction of middle management, which requires greater horizontal communication and interaction within the organization. This trend continues as executives re-engineer the corporation. Organizational and operational changes require re-engineering of the supporting computing and communication network infrastructure.

a. Five Level Hierarchy

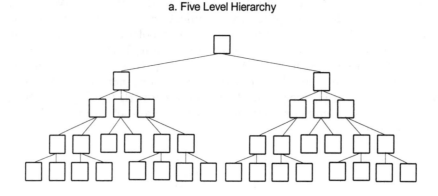

b. Three Level Hierarchy

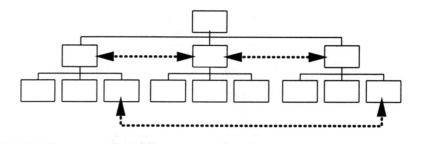

Figure 1.2   Flattening of Network and Organizational Structure

Figure 1.2 illustrates this parallel between organizational and network flattening, where an organization or network goes from a five-tier hierarchical design or management levels (a) to a more distributed three-tier (b) network which provides for nonhierarchical interactions, as shown by the horizontal dashed arrows. Note there are many parallels between the organization and the network models, such as empowering certain individuals (or servers) with the ability to work cross-departmental or cross-company to achieve their goals. Indeed, as discussed in Chapter 3, virtual LAN and Virtual Private Networking (VPN) technology empowers network managers to make changes in response to changing organizational needs. Additionally, networking technology also modifies social behavior by enabling telecommuting and mobile users.

With an ever increasing number of two income households, more enterprises strive to accommodate work at home and telecommuting. The corporate world now realizes that the quality of life and productivity improvements achieved by allowing people to work from home, instead of fighting through hours of rush hour traffic, pay off in the long run. Here too, technology has been an enabler by offering higher and higher modem speeds, a resurgence of ISDN to the home and the promise of standardized, very high speed digital subscriber lines (xDSL) before the twenty-first century.

These evolving corporate infrastructures require communications networking flexibility to respond to ever changing business needs. Reorganizations, mergers, and layoffs frequently place information workers in different physical locations, accessing different network resources. These factors drive network designers to distribute the storage and processing of data to servers supporting groups of clients. The result is that an ever increasing amount of computing occurs via distributed processing using the client-server paradigm. However, as discussed in the section on the wheel of technological reincarnation in Chapter 2, the centralized paradigm may return again with the diskless client workstation and centralized server.

1.1.2 Distributed Processing and the Client-Server Paradigm

Today, most computing occurs in distributed processing and client-server relationships. *Distributed processing* involves the location of network intelligence (i.e., processing) to many network sites, where each site communicates on a peer-to-peer level, rather than through a centralized hierarchy. *Client-server* architectures are the de facto standard today. Client-server computing distributes the actual storage and processing of information among many sites as opposed to storing and processing all information at a single, centralized location. Servers provide the means for multiple clients to share applications and data within a logical, or virtual, workgroup. Servers also allow users to share expensive resources, such as printers, CD-ROM juke boxes, mass storage, high-speed Internet connections, web servers, and centralized databases. This model reflects the increasingly flatter organizational structure and the need for increased connectivity and communications shown earlier.

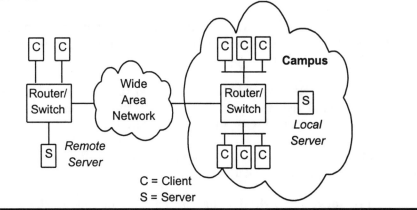

Figure 1.3 Distributed Client-Server Processing Network

Figure 1.3 illustrates a computer communications network supporting the distributed client-server architecture. A server may be locally attached to a router or LAN switch, part of a Virtual LAN (VLAN) but within the same logical local network or remotely accessed across the WAN. The Internet, with its many distributed servers, represents the largest model of distributed processing using client-server communications.

1.1.3 The Need for LAN/MAN/WAN Connectivity

Typically, two scenarios found in corporations and growing businesses drive the need for greater LAN connectivity across the metropolitan or wide area. The first is an increased need for interconnection between distributed computing devices on remote LANs. The second is the logical extension of LAN-speed communications across wider geographic areas; for example, access to remote intranet servers. Geographically dispersed LANs now have a range of connectivity choices — ranging from dedicated circuits to switched wide area and metropolitan area networks to public multimegabit broadband data services. The choice of WAN technology and services is based upon many factors other than cost. In fact, the technology drivers of internetworking LANs across the WAN lead to the hybrid use of private data networks in conjunction with public WAN services in many large corporations.

Computer networking has been defined in many ways. This book uses the following definitions:

Home Area Network (HAN): distance on the order of tens of meters (100 ft); providing connectivity between devices in a personal residence, such as personal computers, televisions, security systems, utility usage metering devices, and telephony.

Local Area Network (LAN): distance on the order of 100 m (350 ft); providing local connectivity, typically within a building, floor, or room.

Campus Area Network (CAN): distance on the order of 1 km (0.6 mi); providing connectivity between buildings in the same general area. FDDI and ATM backbones often serve this need.

Metropolitan Area Network (MAN): distance on the order of 10 km (6.0 mi); providing regional connectivity, typically between campuses over the geographic area associated with a major population center.

Wide Area Network (WAN): distance on the order of 100 to 10,000 km (60 to 6000 mi); providing national connectivity.

Global Area Network (GAN): distance on the order of 1000 to 20,000 km (600 to 12,000 mi) or more; providing connectivity between nations.

Virtual Local Area Network (VLAN): distance varies from a few feet to thousands of kilometers; providing virtual LAN connectivity to geographically diverse users. A VLAN appears as a single, shared physical LAN segment to its users.

1.1.4 Increasing Inter-LAN Traffic

Once an enterprise establishes a LAN, many factors drive the LAN to expand in physical and logical size. The amount of data traffic continues to grow at rates ranging from 100 to 500 percent per year. The capacity of bridges, routers, hubs, gateways, and switches required to transport, switch, and route that data must increase in the same proportion. Many international corporations have already implemented worldwide LAN interconnection using a mix of private and public data services, including the Internet. The business drivers for expanding local area networking usually fall into one or more of the following categories:

- Increased application function, performance, and growth causes an increase in inter-LAN or intra-LAN traffic .
- Users require remote and mobile access to LAN resources.
- Users require faster access to larger volumes of information, and thus higher transmission rates.
- Users require increased intra- and intercompany connectivity in an information driven economy.
- Growth or acquisition causes expansion of the business and communications requirements.
- Organizational structures and right sizing cause changes in traffic patterns.

LAN switching and Gigabit Ethernet create an impedance mismatch between the LAN and expensive WAN access circuits. Microprocessor advances create power hungry desktops and servers. Virtual networks that span the globe create tremendous design challenges. The modernization of WAN infrastructures is playing catch up, while standardization struggles to keep pace.

1.1.5 Bandwidth-on-Demand

Since capitalism reigns over the most of the world, most users seek to minimize costs. Many enterprise users are moving away from dedicated bandwidth terminal to host private line networks to technologies that don't dedicate expensive capacity when it's not needed. A key business driver that drove the move to LANs was the need to reduce the communication cost to each end station, giving it bandwidth only when demanded. Thus, the first interpretation of bandwidth-on-demand arose in the LAN environment, where many users shared a single high bandwidth medium. At any instant only one user was likely to be active and hence had the entire shared medium bandwidth available for their use. Therefore, bandwidth was not dedicated and was available, in a sense, to any user, on demand. This worked well when every user required only a small portion of the shared medium, but became a significant problem as the power of desktop technology increased while the Ethernet capability remained at 10 Mbps. The same problem now occurs for super-servers using 100 Mbps Ethernet.

Another interpretation is analogous to the switched telephone network, where a user places a call (i.e., demand) for bandwidth. The call attempt usually succeeds, failing only with a small blocking probability, and hence is also interpreted as bandwidth-on-demand. The demand for bandwidth has never been greater, and it shows no signs of stopping. User applications require more bandwidth and speed — but only at specific times.

1.1.6 The Accelerating Bandwidth Principle

One measure of computer processing speed is the Millions of Instructions Per Second (MIPS) that it performs. A desktop machine today would have filled a medium-sized office building twenty years ago. In fact, desktop workstations are now available that execute more than a billion instructions per second. Not only are the mainframe MIPS of yesteryear distributed to the desktop today, but the need to interconnect disparate computers is growing as well. Rather than having all the computing power of the network residing in one computer (i.e., in the mainframe in legacy systems), personal computers and servers distribute the total computing power amongst many local and remote devices. Thus, the interconnection of many of these devices becomes the common bus or backplane of old — causing the network to become the computer. New distributed processing applications, like groupware, shared databases, desktop video conferencing, shared workspaces, multimedia, and E-mail accelerate the need to communicate data between desktops and servers. Furthermore, the computing pathways are any-to-any as opposed to the old many-to-one mainframe paradigm. These increases in available desktop computing power and bandwidth, along with the need for any-to-any communications, combine to result in the *accelerating bandwidth principle*. Let's now look at each megatrend in detail to reinforce these key principles.

Applications are processor-intensive. Additionally, as users learn how to apply these applications to increase communications and productivity, bottlenecks form in both the LAN and the WAN. Legacy LAN and WAN bandwidths simply cannot handle this accelerating need for bandwidth. Witness Ethernet, where initially 10 Mbps, even if utilized at only a 40 percent efficiency, initially offered a tremendous amount of bandwidth to local area users.

The cost of WAN bandwidth over time generally decreases, as evidenced by the overall trend of declining tariffs from carrier and private line providers. Note that the prices for frame relay and ATM have either remained flat or actually increased over the last several years. This may be due to the fact that the initial prices were too low. Although there are occasional price increases, the overall trend is driven by decreasing transmission technology and operating costs. However, in comparison to WAN bandwidth, LAN capacity is essentially free.

Client-server application's hunger for more bandwidth along with the distribution of valuable information content across wide areas increases the need for WAN access bandwidth. Show users available bandwidth, and they will find ways to use it. Applications like web browsing, desktop video conferencing, interactive training, and shared database applications have caused LAN and WAN bandwidth requirements to skyrocket! Now Gigabit Ethernet promises to pump data at ten times the speed toward the LAN as 100 Mbps Fast Ethernet does. Thus, the demand for more LAN bandwidth drives the need for more cost-effective use of expensive WAN bandwidth.

As a further illustration of the explosive growth caused by open interconnection and multimedia applications, observe the tremendous growth rate of traffic on the Internet and the World Wide Web (WWW) — by some estimates over 20 percent per month! Furthermore, with the advent of audio and video multicast, bandwidth demand often outstrips Internet capacity during periods of peak usage.

Business methodology, the flow of information, and the nature of the organizational structure drives required connectivity. In a flatter, empowered organization each individual sends, or provides access to, information that was sent upwards in the preceding, hierarchical organization. As witness to this trend, most office workers lament the ever increasing number of email messages they receive and must respond to every day.

The following analysis shows how the combined exponential growth in computing power and the nonlinear growth in intercommunications create an overall demand for data communications with a growth rate greater than exponential, called hyper-exponential by mathematicians. We refer to these phenomena collectively as the *accelerating bandwidth principle*. Before exploring this concept, let's first review Amdahl's law, which states that the average application requires processing cycles, storage, and data communication speeds in roughly equal proportion. For example, an application requiring 1 MIPS also needs 1 megabyte of storage along with 1 Mbps of communications bandwidth. The accelerating bandwidth principle shows that this rule no longer holds true. Instead, the need for communication grows at a faster rate than the growth of processing speed and storage capacity.

Figure 1.4 illustrates the accelerating bandwidth principle. The curve labeled MIPS/desk represents the exponential growth in computing power at the desktop predicted by Moore's law as a doubling of computing power every 18 months. The curve labeled Mbps/MIPS represents the nonlinear growth of the required data communication of approximately 10 percent per year, resulting in a tenfold increase of Amdahl's law of the proportion of bandwidth to processing power over 25 years. The curve labeled Mbps/desk, which is the product of MIPS/desk and Mbps/MIPS, represents the data communications bandwidth predicted by the accelerating bandwidth principle. The growth rate for the Mbps/MIPs is hyper-exponential because the exponent grows at a faster than linear rate due to the combined non-linear increase in interconnection and desktop communications bandwidth. This is precisely the phenomenon of the web driving Internet and corporate data network traffic today. More complete economic models include the increasing cost of operations, including skilled personnel, and network management infrastructure development.

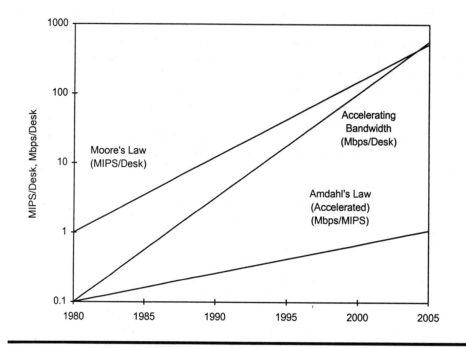

Figure 1.4 Accelerating Bandwidth Principle

1.1.7 A Partial Answer: Compression

The accelerating bandwidth principle points to the inadequacy of Ethernet and the Fiber Distributed Data Interface (FDDI) and identifies the need for true Gigabit-per-second (Gbps) networking in the near future. ATM based LANs already serving high end workgroups and the Gigabit Ethernet effort stand as evidence in support of this principle. One way to offset the increasing need for communications bandwidth is the use of improved compression, which reduces the amount of data requiring transmission.

The high cost of bandwidth has spurred the development of more efficient coding and compression schemes. The most common evidence of this trend is the availability of modems using compression for communication over voice-grade lines that approach the capacity of ISDN as described in Chapter 24. Also note the decrease in video conferencing, North American Television Standard Coding (NTSC), and High Definition Television (HDTV) coding rates over time as summarized in Chapter 17. Acceptable videoconferencing for business runs at DS0 (64 kbps) rates today. Coding of NTSC signals at DS3 (45 Mbps) rates occurred in the late 1980s, and now approaches the DS1 (1.5 Mbps) rate for broadcast programming. The need for 150 Mbps for HDTV transmission has also evaporated due to similar increases in coding efficiencies. Of course, the improvements in coding efficiencies are limited by the actual information content of the signal. In general, these schemes for efficient coding and compression arise when bandwidth is inordinately expensive or a competitive niche occurs that justifies the expense of such coding or compression. However, compression only yields improvement up to the information theoretic limits. We study the effectiveness of this technique for voice and data later in the book.

1.2 BUSINESS DRIVERS FOR THE USE OF ATM

Having reviewed the changes in business communications needs, this section focuses on the key business drivers for the use of ATM. There are four key business drivers when considering ATM:

- Integration savings
- Economies of scale
- Flexibility
- Single enterprise-wide solution

1.2.1 ATM as the Single Network Technology Integrator

Users spend more on service and support than on hardware and software. This statistic shows the decreasing cost of the equipment, as opposed to the increasing cost of bandwidth and support systems required to design, build, administer, and manage networks. Typically, 40 percent of the networking cost goes to WAN bandwidth, 40 percent to network operations and support systems, leaving only 20 percent for the network equipment.

ATM is the first technology that offers a corporation the capability to use a common enterprise-wide protocol and infrastructure for all voice, data, and video communications, desktop-to-desktop. ATM excels when applications require specific quality of service and reserved bandwidth. While most corporate enterprise networks will remain a blend of public network services and private leased lines, many corporations are turning to hybrid outsourced or out-tasked solutions. ATM offers combined switching and routing within the WAN, as well as a backbone technology of choice for high-performance, QoS-aware campus backbones, coexisting with IP and LAN switching. These are the key features that allow ATM to become the common platform for enterprise-wide communications.

Prior to ATM, users required multiple technologies for their voice, data, and in some cases video communication needs as shown in Figure 1.5. Often, enterprises implemented separate networks for voice, video-conferencing, and data communications. Separate devices interface to separate networks to provide each of these services. Many enterprises had separate organizations designing, deploying, and managing these networks. Carriers also used a separate approach: deploying separate networks for different types of voice, data, and video services.

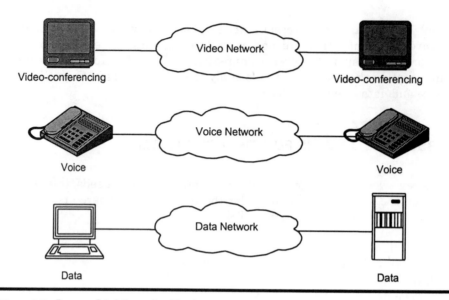

Video-conferencing   Video-conferencing

Voice   Voice

Data   Data

Figure1.5   Legacy Multi-service Environments

ATM changes the situation to that depicted in Figure 1.6, where a user can send voice, video, and data from his or her workstation to peer workstations anywhere across the WAN. ATM provides the infrastructure of a seamless, single protocol for all traffic types that provides logical access to all services (e.g., voice or IP) and interworks with those that have similar characteristics (e.g., Frame Relay and N-ISDN). Part 4 details ATM's solution to these needs.

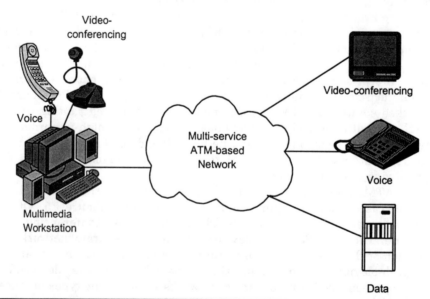

Video-conferencing

Voice

Multimedia Workstation

Multi-service ATM-based Network

Video-conferencing

Voice

Data

Figure 1.6   ATM Multi-service Environment

1.2.2 ATM Offers Economies of Scale

ATM achieves economies of scale in network infrastructure through integration of multiple traffic types. ATM is more efficient than multiple separate networks, because the laws of large numbers make the traffic engineering for larger networks more predictable. ATM is well positioned to be the dominant corporate and service provider of WAN infrastructure of the near future. ATM offers a single, multiservice architecture and universal backbone that integrates narrowband and broadband access technologies, while simultaneously offering a smooth transition path from legacy networks. Thereby, ATM lowers the unit cost of bandwidth, simplifies operations and maintenance support costs, and optimizes overall bandwidth and resource utilization. Many network designers have learned over the past two decades that throwing large amounts of bandwidth at the problem is not always the most cost effective answer. One example is that large private line based networks are moving to more cost effective public network services, like frame relay and ATM. Also, running parallel voice and data networks is inefficient and costly. But, most companies do so today because there is no viable alternative, at least not until recently.

Even though the percentage of non-voice traffic continues to increase, the requirement for voice integration with data over a single access technology remains. A key role for ATM is eliminating the cost of running and managing duplicate infrastructures. Although enterprise ATM networks use ATM for scalable infrastructures, ATM's greatest adoption has occurred with service providers. The record is clear that traditional telephone service carriers as well as many Internet Service Providers (ISPs) increasingly rely on ATM technology in their backbone infrastructure. These service providers are using ATM to achieve a single, common platform backbone technology rather than operating separate networks for voice and various data serviches.

1.2.3 Flexibility of ATM

ATM is a multiplexing and switching technology designed for flexibility, versatility, and scalability at the expense of somewhat decreased efficiency for some traffic types (particularly packet data as analyzed in Chapter 30). For any *single* application it is usually possible to find a better data communications technique, but ATM excels where it is desirable for applications with different performance, quality of service, and business requirements to be performed on the same computer, multiplexer, router, switch, and/or network. This is why some Internet Service Providers are looking at protocols other than ATM for their backbone networks since they need to optimize for the current traffic mix which is predominately packet data. For a wide range of applications, however, the flexibility of ATM can result in a solution that is more cost-effective than several separate, individually optimized technologies. Furthermore, the flexibility of ATM can "future-proof" the network investments since unenvisioned future applications may also be supported by ATM. As shown in Part 8, ATM may be a more efficient solution for an integrated voice, video and data network. Also, ATM enables network managers to flexibly adapt to changing enterprise communication requirements, evolving business environments, and fluctuating traffic patterns and volumes.

1.2.4 ATM as an Enterprise-wide Solution

Clearly, a pivotal success factor for ATM in the LAN is the usage of ATM-aware applications. The cost per user, or LAN workstation port, is, by many studies, less for ATM than for 10 Mbps switched Ethernet when measured on the basis of cost per Megabit per second. Fast Ethernet is clearly ATM's major competitor in the LAN. In the WAN, no technology beats the flexibility and multiple Quality of Service (QoS) levels provided by ATM technology. ATM's penetration in the LAN has been limited, since 100 Mbps switched Ethernet technologies undercut ATM's costs, and Gigabit Ethernet plans to do the same in the LAN backbone. The battle in the LAN requiring users to purchase Network Interface Cards (NICs) in workstations and "forklift upgrade" existing LAN switches and routers makes for a harder sale for ATM in all but new networks or applications which have stringent QoS requirements.

Enterprise-wide ATM can be a dream come true. Visualize an enterprise-wide ATM network. Applications on a workstation run using an ATM aware Application Programming Interface (API) that interfaces to a local ATM network via a high-performance ATM Network Interface Card (NIC). From that point out through the LAN and over the WAN, the network protocol and technology remain ATM. Therefore, a single, enterprise-wide, end-to-end ATM network ensures guaranteed quality of service for multiple traffic types (voice, data, and video). On the other hand, LAN segments must be engineered to low utilization in order to achieve equivalent performance. The ATM WAN could be either a private ATM network or public ATM service.

ATM offers a tighter coupling between the user application and the networking protocol. Protocols like IP across the WAN effectively turn the network into a dumb pipe. ATM places intelligence back into the WAN, making the network smarter and allowing the network to become more like a computer and less like a dumb transport medium.

1.3 APPLICATION DRIVERS FOR ATM

Currently, few applications absolutely require ATM. However, ATM is becoming an increasingly economic alternative. This section outlines these requirements for consumer and commerical applications as well as the the need for increasing bandwidth at acceptable response times for multimedia applications.

1.3.1 Consumer Applications

Consumer service applications requiring high bandwidth, flexibility, and performance include:

- Home-shopping services employing multimedia voice, data, image, and video using on-line databases and catalogs
- Interactive multimedia applications like on-line gaming
- Video-on-demand
- Collaborative applications like shared whiteboard
- Telecommuting
- Electronic commerce
- Multimedia applications to the home (i.e., via xDSL)
- E-mail and multimedia messaging systems

Some applications require not only broadband service but also the capability of a broadcast public service. These include:

- Distance learning and training
- Remote tutoring services
- On-line video libraries for home study
- Video desktop training courses
- Video-on-demand
- High Definition Television (HDTV) distribution

Many of these applications are just now beginning to enjoy wide deployment, driving the need for greater and more flexible bandwidth. A key potential application driver for consumer ATM is to support requirements for consumer services, most notably that of *video-on-demand*. Several cable companies and telecommunications service providers are providing broadcast quality video-on-demand delivery using ATM switches at the head-end of cable distribution networks with sophisticated video servers. ATM is well suited to the ever-changing marketplace of video coding in providing a flexible, multiple-rate, integrated switching vehicle that can handle today's fixed video coding rates (such as MPEG and JPEG carrying a 60 Mbps standard video NTSC signal at a compression ratio of up to 100:1!). Modern video coding schemes operate at variable rates.

1.3.2 Commercial Applications

Meanwhile, there are many commercial public service applications that are pushing the envelope for high and flexible bandwidths such as:

- Seamless LAN/MAN/WAN interconnectivity
- Graphic-intensive industrial engineering applications
- Collaborative and cooperative computing, such as groupware
- Real-time, interactive computer simulations
- Integrated voice, video, data multiplexing, and switching
- Video-conferencing
- Remote access to shared server farms
- Multimedia applications to the desktop
- Medical imaging

- Desktop publishing, electronic publishing, and large file document management
- Remote distributed database access
- Financial modeling and industry reports
- Collapsed backbone campus networks
- Seamless interworking with legacy systems
- Desktop video collaboration (DVC)

Further processing of video programming at the server yields exciting commercial applications, such as desktop video collaboration (DVC), where users share applications and video-conferencing. DVC reduces the amount of travel between meetings and decreases the time to market for many corporations. For example, DVC replaces the situation where three people around the world discuss a blurry fax for a new automobile blueprint on a conference call, with full motion video-conferencing, computer renditions of the proposed automobile, and a shared whiteboard. Much as in the business arena, the rallying cry is for more bandwidth-on-demand (or -call); the hope is that consumers will exclaim "I want my ATM TV!"

1.3.3 Applications' Insatiable Demand for Bandwidth

The above applications offer functionality and benefits ranging from providing consolidation efficiencies and productivity improvements to fundamental changes in people's lives and the way in which enterprises operate. Many applications demonstrate that people increasingly rely on visual or image information, rather than audio or text information, as predicted by the old, but accurate, adage that a picture is worth a thousand words. The increase in telecommuting and conducting business from the home also illustrates the trend to rely on visual or image information. The partnering and buyout of cable firms by information transport and telecommunications providers, as well as the advent of cable modems and technologies such as the digital subscriber line family of technologies (e.g., HDSL, ADSL, VDSL, etc.) using ATM, also portend major changes in the infrastructure for providing interactive multimedia networking to the smaller business location and the home. Indeed, the ADSL Forum predicts that ATM will be the access method from concentrators to various services. Some early ADSL equipment and trials support end-to-end ATM services.

As new technologies emerge, their proponents often look for a single "killer application" for success. More often the case is that many applications or implementations working together make a technology successful in the desktop or the local, campus, metropolitan, and wide area network. We believe this will be true of ATM.

Typically, applications utilize one of two generic types of information: an object of fixed size that must be transferred, or a stream of information characterized by a fixed data rate. Multimedia involves combinations of these basic information transfers. This section illustrates the tradeoffs in response time, throughput, and the number of simultaneous applications supportable by technologies such as ATM through several simple examples.

Figure 1.7 shows the time to transfer an object of a certain size at a particular rate. Along the horizontal axis a number of common information objects are listed as an illustration of the object size in millions of bytes (megabytes or Mbytes). In general, the transfer time decreases as the transfer rate increases. A real-time service requires transfer in the blink of an eye, which is on the order of one twentieth of a second (i.e., 50 ms). The utility of the service in an interactive, near-real-time mode is usually perceived as requiring a transfer time of no more than a few seconds. A non-real-time or batch application may require many seconds up to minutes for transfer of a large object.

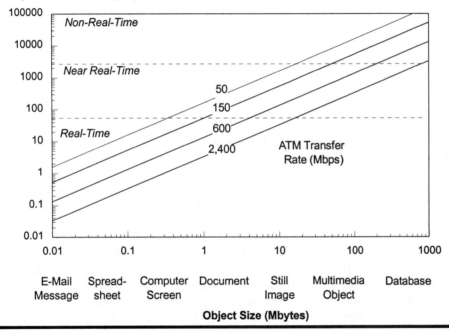

Figure 1.7 Object Transfer Time as a Function of Bandwidth

Now, let's look at applications such as audio and video that use a specified amount of bandwidth. The bandwidth may be a fixed, continuous amount, or an average amount if the network statistically multiplexes the sources. The example considers the simplest case of a fixed bandwidth per application source. Figure 1.8 plots the number of application sessions requiring a specific fixed rate supportable by bandwidths standardized in WAN fiber-optic transmission equipment. In general, as the bandwidth required by each application session increases, the number of simultaneous sessions supported decreases. Of course, allocating more overall bandwidth increases the maximum number of fixed-rate application sessions. Current network technologies support transmitting these applications separately. ATM offers the flexibility to combine a variety of sources in arbitrary proportions as shown by numerous examples throughout this book. This is a key advantage of ATM.

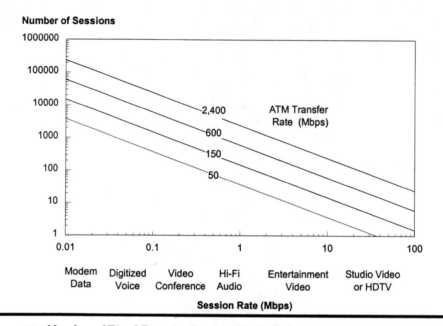

Figure 1.8 Number of Fixed Rate Application Sessions Supported

This acceleration toward sessions demanding greater bandwidth, combined with the advent of ATM, points toward an enabling of virtual reality — a dynamic, interactive visual and audio representation of information rather than just textual or simple, static graphical representation. *Multimedia* is a term often used to represent the combination and presentation of multiple forms of data simultaneously to an end user. The current generation of children, accustomed to sophisticated electronic gaming systems running on 300 MHz Pentium MMX machines, will expect even more sophisticated capabilities in the workplace and their homes in the early twenty-first century.

Figure 1.9 illustrates the bandwidth requirements and response time requirements of multimedia by showing the time required to transfer a high-resolution graphical image in the presence of a number of high-quality video sessions. This example illustrates ATM's flexibility by allowing an application to trade off the number of active fixed rate 10 Mbps high-quality video windows against the required time to transfer a 1 MByte file for representative ATM network bandwidths. Observe how the transfer time of the large file increases as the number of active video sessions increases. The fixed rate application must have continuous bandwidth in order for the video to appear continuous, while the object transfer application can simply take more time to transfer the object as the available bandwidth is decreased due to another, higher priority session being activated.

Time to Transfer 1 Mbyte File (ms)

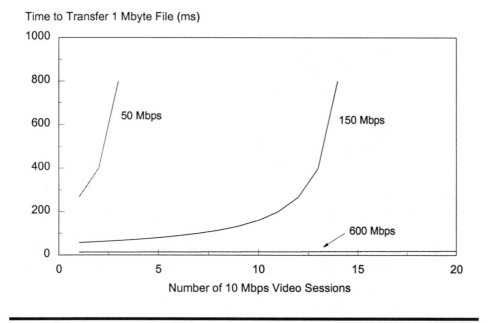

Figure 1.9   Bandwidth Rate Required in Mbps for Multimedia

1.3.4 The Multimedia Wave

Multimedia applications are steadily penetrating the user community. Sound cards, microphones, and cameras are now commonplace in most home and business computers. Internet web pages using Java sport full motion multimedia. Multimedia has also found its way into the commercial market, driven by applications listed in the previous section. Enterprises deploy ATM to support delay-sensitive traffic, typically voice or video. High-speed ATM point-to-multipoint applications include broadcast quality video and video-conferencing applications involving simultaneous transmission of text, data, and video. Applications such as tele-medicine and distance learning that take advantage of the attributes of ATM broadband data technologies are now appearing – to support true voice, data, and video traffic integration with guaranteed Quality of Service (QoS) for each traffic type. ATM supports these applications by flexibly and dynamically allocating bandwidth and connectivity.

Figure 1.10 shows an example of a person's multimedia desktop worksta-tion in Chicago participating in a video and audio conference with four other individuals in New York, London, Paris, and Washington, D.C. Users share a virtual whiteboard, exchange electronic mail (E-mail) interactively, while simultaneously running a web browser to research information on the World Wide Web (WWW). Users insert movie clips into the clipboard so that their peers can play them back on demand. In this example, an automatic translation server (not shown in the figure) could be connected to facilitate communication between the parties speaking different languages.

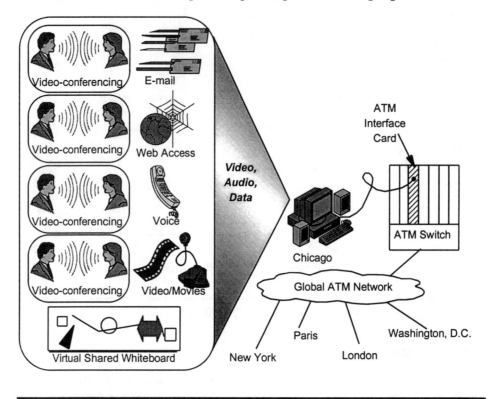

Figure 1.10 Multipoint, Multimedia ATM Application of the Future

Using ATM technology, an ATM interface card in the Chicago workstation combines the video of the built-in monitor camera, the telephone, and text data into a single 25 or 155 Mbps ATM transmission stream to a local ATM switch. The ATM switch then broadcasts the voice, data, and video to switches at all four locations through intervening ATM WANs in the global ATM network. Each of the other sites does the same in return. All attendees have four individual pop-up screens on their workstations so that they can see the other participants, but could share a common whiteboard or database. The conference leader dynamically controls video-conferencing connections. Alternatively, the participants could use a "meet me" or CU-SeeMe type conference. Although this may sound futuristic, applications such as these are emerging in corporations today.

1.4 REVIEW

This chapter first discussed the business needs that drive a change in the communication infrastructure for an enterprise. These include the evolution from centralized to distributed networks, the need for extending LAN and MAN speeds into the WAN, and the ever increasing need for speed. The text defined the accelerating bandwidth principle driven by the ever increasing computing power on every desktop combined with the accelerating communications demand from each user. We then discussed the basic business drivers involving cost reduction through integration, economies of scale, and seamless end-to-end networking. Next, the text looked at the consumer and commercial applications driving the need for ATM. Finally, we presented what may be the greatest benefit of ATM — enabling brave new applications, including ones not yet conceived.

2

Technology Enablers
for ATM

Technology enables innovation. The historical beginnings of voice and data communications define the starting point for the subsequent evolution of present and future communications networking. This chapter begins with an overview of the history of communications technology as background. The text then examines the technologies that serve as enablers to ATM: protocol and processor enhancements, modern digital transmission facilities, and protocol and computing advances. The chapter presents an argument that implementing ATM can provide lower enterprise networking costs in the reduction of transmission and public network service costs, as well as providing a future-proof investment.

2.1 FROM VOICE NETWORKS TO DATA NETWORKS

A dramatic shift in business and consumer reliance on voice communications networks to integrated data communications networks occurred over the past 20 years. Data communications now affects many aspects of our lives on a daily basis. For example, the following applications previously thought of as science fiction are now commonplace: bringing stock market transactions electronically to our fingertips in seconds, enabling hospitals to transmit and receive multiple X-ray images over thousands of miles in minutes [1], delivering hundreds of electronic mail messages to each corporate user on a daily basis, and even providing us with money on practically every street corner through Automatic Teller Machines (the other "ATM!"). What chain of events made communications so integral to our daily life in the last generation? How did the move occur so quickly from voice networks using analog transmission facilities to integrated voice, video and data networks riding digital transmission facilities? With these questions in mind, this chapter begins with a review of the evolution of communications.

2.1.1 A Brief History of Communications

Figure 2.1 depicts a view of the history of communication along the dimensions of analog versus digital encoding and synchronous versus asynchronous timing, or scheduling. The beginnings of spoken analog human communication are over 50,000 years old. Graphic images over 20,000 years old have been found in caves known to be inhabited much earlier. Written records over 5,000 years old from ancient Syrian civilizations scribed in clay tablets record the beginnings of data communication. Digital long distance optical communications began before the birth of Christ when the ancient Greeks used line-of-sight optical communications to relay information using placement of torches on towers at relay stations. The Greeks and Romans also popularized *scheduled* public announcements and speeches as early examples of broadcast communications, as well as individual, *unscheduled* communication in forums and debates.

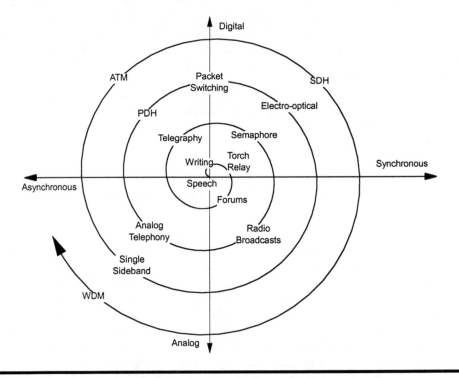

Figure 2.1 Data Communications "Wheel of Reincarnation"

In the seventeenth and eighteenth centuries, optical telegraphy was extensively used in Europe. Later, Samuel F. B. Morse invented electrical telegraphy in 1846, marking the beginning of modern digital electromagnetic communications. Marconi invented radio telegraphy shortly afterwards, enabling communication at sea and access to remote areas. Broadcast analog radio communications of audio signals followed in the late nineteenth and early twentieth centuries. Telephone companies applied this technology to analog voice communication in the same time frame. Television signal broadcasting, with its first scheduled telecast to the public by WGY in Schenectady, NY; in 1928, became commercially viable in the late 1940s with a color broadcast that same year. Then, in the 1950s, the conversion of analog voice to digital signals in the Plesiochronous Digital Hierarchy (PDH) began in large metropolitan areas to make better use of installed cabling. This was followed by the invention of packet switching in the 1960s as an offshoot from research into secure military communication networks. Packet switching includes many technologies covered in detail in the next part, including X.25, frame relay and the Internet Protocol (IP). ATM is a derivative of packet switching. Fiber optic transmission and the concept of synchronous digital transmission as the basis for Narrowband ISDN (N-ISDN) introduced in the early 1980s moved the wheel along another notch. Analog transmission of voice had a brief renaissance using Single SideBand (SSB) technology in the 1980s. ATM next moved the wheel of technology around the circle back into the domain of digital asynchronous communication. The next major leap in technology is Wavelength Division Multiplexing (WDM), which is analog and asynchronous. The speed of digital communication has increased geometrically over time through each of these evolving phases of technology.

This is not a perfect analogy — so be careful if you try to use it to predict the next technology or the rate of change. In fact, the only thing that is certain is that things will change. Sometimes the wheel of reincarnation spins faster than at other times; for example, in the current day we move from Asynchronous Transfer Mode (ATM), an asynchronous digital technology, all the way around to Wave Division Multiplexing (WDM), an asynchronous, analog technology.

2.1.2 Voice as Low-Speed Data

A standard voice-grade channel can be accurately represented by a 64-kbps (or 56-kbps) data stream as described in Chapters 6 and 24. In fact, while voice is typically transmitted at 56-kbps, modern digital encoding techniques now enable a voice channel to be transmitted at speeds as low as 6-kbps as analyzed in Chapter 30.

Telephone companies first used 64-kbps digital representation of voice for engineering economic reasons in large metropolitan areas to multiplex more voice conversations onto bundles of twisted pairs in crowded conduits in the 1950s. Twenty-four voice channels were multiplexed onto a twisted pair in what was known as a T1 repeater system, using a DS1 signal format. Chapter 6 reviews the DS1 signal format in detail. The scarce resource of twisted pairs was now utilized at 2400 percent of its previous capacity, a tremendous enhancement! The fact that these multiplexing methods could be used for the purpose of data communications was utilized only later after carriers upgraded both their transmission and switching networks to an entirely digital infrastructure.

Voice is very sensitive to delay, and somewhat loss-sensitive. Users do not tolerate appreciable delay during a full duplex or half duplex conversation, because it inhibits interaction or results in annoying echo. Nor will voice users accept sentence-flow garbling by the loss of entire syllables. Variations in delay can cause the speaker's voice to become unrecognizable, or render the speech unintelligible. The loss of small portions of syllables or words in voice communications is usually acceptable, however. Satellite delay, which most people have experienced, is a good example of the effects of large delays on interactive conversation. Many current Internet Telephony implementations have similar delay performance; however, when the service is cheap enough, cost conscious users quickly adapt. Certain new technologies such as ATM can handle delay-sensitive voice traffic, while other technologies such as frame relay and IP do not perform as well on lower speed access lines in the presence of packet data as discussed in Chapters 11 and 30.

2.1.3 Voice Network Architecture

The public telephone network naturally evolved into a hierarchy driven by engineering economic considerations. The five-level hierarchy of the public telephone network illustrated in Figure 2.2 minimized cost and achieved traffic engineering economies of scale, which resulted in a corresponding increase in capacity at higher levels in the hierarchy [2]. Customers are connected to the telephone network by a *local* loop, typically provided by a pair of twisted wires at the lowest level of the hierarchy, the Class 5 central office telephone switch. A *twisted pair* is two wires twisted together to minimize impairments in analog transmission such as crosstalk and interference. These twisted pairs are bundled into cables and then split out again at the central office. Generally, if the distance traversed is greater than a few miles, the local loops are aggregated via multiplexing into larger bandwidth for transmission over microwave, copper, or increasingly optical fiber. Telephone groups call these large collections of individual circuits connecting switches *trunk groups*. Indeed, the DS1, DS3 and eventually the SONET multiplex levels were created to efficiently support these trunk groups connecting large voice switches. Their use for data occurred via later invention and innovation rather than initial design.

As shown in Figure 2.2, class 5 end office switches may be directly connected via local tandem switches, or else connect to larger Class 4 tandem switches. Tandem switches connect to other higher levels of the hierarchy via high-usage trunks if a significant community of interest exists. The last choice is called the final route as shown in the figure. Calls first try the high-usage groups. If these are full, then calls progress up the hierarchy on the final route up the hierarchy. Intermediate nodes in the hierarchy may use a high-usage group to reach the destination as shown via dashed lines in the figure. The last route is via the Class 1 switch, which routes the call on the final trunk groups to the Class 5 office serving the destination. Note that all switches still switch only at the same 64-kbps single voice channel level. In general, switches are larger at higher levels in the hierarchy, although a Class 5 switch in a large metropolitan area can be quite large.

Many telephone networks are moving towards a flatter, more distributed network of fully interconnected Class 4 and 5 switches. The number of levels have not yet been reduced to one, but in many cases there are only two or three levels in a hierarchical phone network. Again, this parallels the move of most major businesses from a hierarchical organization to a flatter, distributed organization.

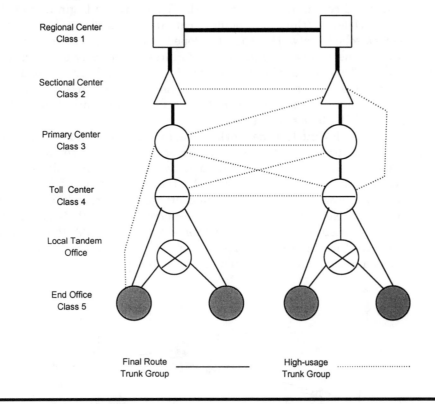

Figure 2.2   Classical Public Telephone Network Hierarchy

Voice channels are time division multiplexed on digital transmission systems. When compared to data, the main difference is that voice calls are circuit switched, whereas data can be either message, packet, cell, or circuit switched as described in Chapter 4. Both voice and data applications require large amounts of bandwidth, but for different reasons. Data applications require large bandwidth to support the peak rate of applications, while voice requires larger aggregate bandwidth to multiplex large numbers of individual voice circuits. Thus, the drive to efficiently interconnect voice switches resulted in a WAN digital transmission technology described in Chapter 6 that enabled contemporary data networks.

2.1.4 Government Funded Networks Enabling the Future

The military has been a key user of data communications networks throughout history. In fact, warfare has been a catalyst for many great (and not so great) inventions throughout time. The telegraph was significant to the Union forces in the American Civil War. Many data processing and early computer systems were developed during World War II, when systems integration was necessary due to complexity. After the war the command and control centers, weapons and sensor systems, voice networks, and the computers which ran these systems needed to be centrally controlled within one interconnected communications network. This was the beginning of the Department of Defense (DoD) telecommunications architecture. In fact, the first packet-switched network was developed with massive amounts of redundancy and the ability to route around failures, a necessity in the case of a nuclear war. This trend of military use of technology lives on today, as shown by live video footage of a guided missile impact on a bunker during the Gulf War, and computer scans of the entire battlefield instantly transmitted back to command HQ to assess the situation.

A major catalyst for modern networking began when the DoD established the Advanced Research Projects Agency NETwork (ARPANET) in 1971 as the first packet-switched network. This data network connected military, and civilian locations as well as universities. In 1983, a majority of ARPANET users, including European and Pacific Rim contingents, were split off to form the Defense Data Network (DDN) — also referred to as MILNET. Some locations in the United States and Europe remained with ARPANET, and are now merged with the DARPA Internet, which provided connectivity to many universities and national telecommunications networks.

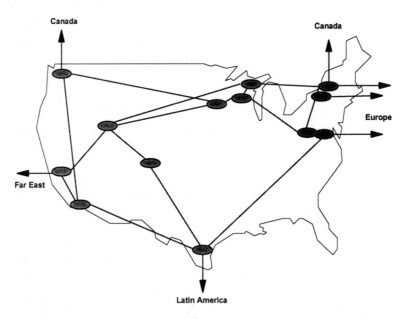

Figure 2.3   NSFNET Backbone for the Internet

The original ARPANET was decommissioned in 1990 and replaced by the NSFnet, which was the ancestor of today's Internet. Many of the advances in computer communications, including routing and networking, were developed through experience on the ARPANET and the NSFnet. The premier example of the largest Internet backbone in the world was the National Science Foundation NETwork (NSFNET) in the early 1990s shown in Figure 2.3. This network had a DS3 backbone. The explosive growth of the Internet, driven by the content accessible over the World Wide Web, continues to drive network designers to build larger, faster, and more feature-rich networks.

The highest performance hybrid, public-private network in existence in 1998 provides a private IP service over a public ATM network. This network is the United States National Science Foundation's very-high-speed Backbone Network Service (vBNS). In April of 1995 MCI turned up this ATM-based OC3 network between five supercomputer centers distributed across the United States. Figure 2.4 illustrates the vBNS network [3]. The OC3 connections are provided by virtual path connections (VPCs) between the five Super Computer Centers (SCCs) over MCI's commercial Hyperstream ATM service. The local ATM switches split the VPCs up into two full meshes of Virtual Channel Connections (VCCs): one for meshing the vBNS routers which utilize the OSPF protocol and a second which acts as a Logical IP Subnet (LIS) planned for use with SVCs in 1997. The NSF plans to use the LIS VCCs for applications like video over ATM. As part of the vBNS effort, engineers collect traffic statistics and report on these regularly [4]. An inexpensive IBM based PC with an OC3 ATM interface card connected to the network via a passive optical splitter is the basis of much of this data. The OC3MON captures the first cell of an AAL5 packet, enabling engineers to analyze all the TCP/IP information [5].

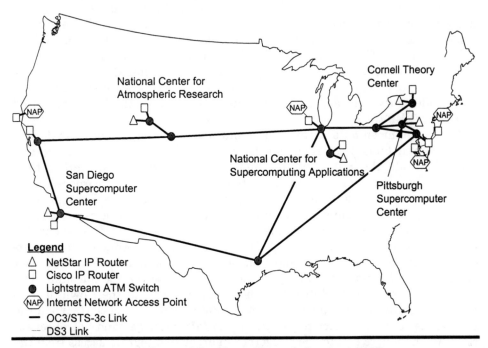

Figure 2.4   NSF's very-high-speed Backbone Network Service (vBNS)

In early 1997 the NSF and MCI announced the upgrade of this network to OC12 trunking and added additional universities with meritorious high bandwidth applications. Basically, the network toplogy looks identical to that above, except the trunks are upgraded to OC-12 (622-Mbps) speeds where FORE ASX 1000 ATM switches replace the Cisco (Lightstream) 2020 switches [4].

2.1.5 Recent History of Data Communications

Host-based networks accessed by local and remote terminals evolved through the use of private networks and packet-switched services. The primary example is the IBM Systems Network Architecture (SNA). This architecture provides the platform for many dumb terminals to communicate with an intelligent host or mainframe in a hierarchical, or star, fashion. This hierarchy developed because collecting expensive intelligence at the host and allowing the terminals to have little resident intelligence was the most cost-effective solution at the time.

Local area networks (LANs) were the next major development in the computer communications networking environment, with the invention of Ethernet by Xerox in 1974. The advent of client-server architectures and distributed processing leads us into the beginning of modern data communications. One challenge corporations face is the flattening of interconnected LAN networks to support the astounding rate of workstation and server proliferation. This flattening began the (r)evolution of the corporate data network. What started as a PC for home use has now become a corporate necessity on every desktop. It was a natural evolutionary choice for the visionary network design engineer to address the users in these islands of information — the distributed LAN interconnecting desktops and the centralized mainframe Management Information Systems (MIS) arena — by creating a common environment using routers to achieve interworking and interconnectivity. In fact, the first router was actually the IBM Front End Processor (FEP).

Just as minicomputers invaded mainframe turf when the cost fell to departmental budget approval levels, so too did PCs, LANs, bridges, and routers invade the minicomputer and mainframe turf. But rather than battling them on even ground, they typically bypassed the corporate MIS budget approval process through low prices, empowering each department and renegade LAN manager to purchase and install these low-cost devices directly. As described in Chapter 9, first bridges then routers enabled interdepartmental connectivity of diverse computing resources in a cost-effective manner. Thus, cost and control were in the hands of the end user more so than ever before. Many users asked themselves the question: why conform to corporate MIS dictates when they could build their own departmental LAN and handle 90 percent of the data processing needs? When the need arose to interconnect these LANs, the MIS manager and entrepreneurial LAN managers had to work together in order to integrate access from the LANs to the mainframes. In addition, WAN interconnectivity often went beyond the scope of a single LAN manager since costs had to be shared across multiple LANs. The router also found its place here as the gateway from the LAN to the WAN as well as the device that interconnected disparate LANs running a common network layer protocol.

Of course, the latest news is how the Internet burst from relative obscurity as a research network to become the hottest thing in networking. Also, fueled by the accelerating bandwidth principle, Gigabit Ethernet promises to relieve the capacity constraint in LANs for many years to come. Subsequent chapters in this book cover these subjects in detail. The text now returns to the roots of the data communications revolution.

2.2 THE DATA REVOLUTION

The 1990s ushered in a broadband data revolution with a rallying cry of "bandwidth-on-demand!" As covered in Chapter 1, the LAN environment enabled a user sharing a high-speed medium with many other users, to have access to the full shared medium bandwidth on demand. Those with a background in circuit switching had another interpretation of bandwidth-on-demand. Their definition is that a user requests bandwidth, and is either granted the full request or is denied access completely (i.e., blocked); this is similar to making a phone call — the call either goes through or you get a busy signal. This section shows how data communications has taken the lead over voice communications in both total volume, growth rate, and the introduction of new services. The discussion then moves to cover how various services best meet application needs in terms of bandwidth, delay, and burstiness. The section concludes with the observation that the data revolution is global, social, and economic in scope.

2.2.1 Data Communications Takes the Lead

Data traffic growth increases at a factor of several hundred percent per year, far outpacing the average growth of voice, at less than 10 percent per year. Unfortunately for carriers, the revenue for data services isn't growing as fast. Of course, many factors may cause these growth rates to change, so this is only a possible outcome. However, it makes the point that the era of emerging data communications dominance over voice communications is near. Web browsing, electronic mail, file transfer, local area network interconnection, interactive applications, and emerging multimedia applications represent just a few examples of how data communications exceeds voice communications in creating the demand for new capabilities. One experiment that brings this into focus is comparing the amount of time you spend on the phone versus surfing the web, working on E-mails, and sending files across the network. E-mail and web browsing overtake voice traffic in terms of sheer volume when large files are exchanged. For example, a typical uncompressed voice call requires approximately 3 Megabytes of data, while many uncompressed files utilize tens to hundreds of thousands of bytes. Look at your own voice and data usage behavior and decide for yourself if the volume of data exceeds that of voice. Ask yourself how inexpensive high performance access would change your behavior.

Data communications not only gives business the competitive edge, it puts them on a leading edge that creates both benefits and risks. Once a business experiences effective data communications, it becomes "hooked" on its own network. The data and computer communications network quickly becomes the lifeblood of the company; for example, E-mail became the standard form of communications, desktop videoconferences reduces travel fares by 50 percent, stock quotes scroll across the screens, accounts payable goes on the web and reduces receivables by 10 days — you get the idea. A company needs voice network communications, but a private voice network can only become so large since it is limited by the company's size. Voice communications traffic can be forecast and has predictable characteristics. On the other hand, difficult to predict data communications traffic characteristics are a different ballgame with a new set of rules. But once a company effectively uses a data network, there is no limit to its potential. Key services and switching technologies representing the emerging data market include:

- The Internet
- Asynchronous Transfer Mode (ATM)
- Frame relay
- Switched high-speed Ethernet
- Asymmetric Digital Subscriber Line (ADSL)

Indeed, the above list defines much of the outline of the remainder of the book. Broadband switched data communication is fast becoming the prevalent market in the 1990s — and at the forefront are ATM and the Internet suite of protocols. This decade may also be called the era of interworking, simply because of the widespread use of these broadband technologies and services and their interaction with LANs, as well as the globalization of IP communications over the Internet and World Wide Web (WWW).

2.2.2 Business Relies on Data Communications

Only now do many business and corporate managers realize the level of dependence on their data communications networks. This is evident by the small amount of personnel resources and funding usually dedicated to data communications in relation to the voice network. But, as Bob Dylan sang, the times are a changin' — studies show that a major portion of most business office budgets are now dedicated to data communications. Obviously, data communications networks are fundamental to the successful operation of most businesses. Computers, terminals, modems, facsimile machines, security systems, and even most telephone systems transit some form of data communications.

Cost also plays a major role in determining network needs, often driven more by the charges for carrier provided services and operating expenses than actual equipment prices. Often services are tax-deductible expenses, whereas the purchase of equipment is a depreciable capital expenditure. Equipment expenses often turn out to be a very small part of the total expenses of operating a network. Ongoing support, especially people costs, is a disproportionately higher cost compared to equipment costs, not to mention the fact that qualified people are much harder to find than good equipment.

Many companies planning to offer data communications services to customers must first demonstrate that these services work internally in their own company. A customer is apt to first ask the question: "Do you use the same service to transmit your own critical traffic?" As is often the case, the company offering the service becomes the test bed for the service before it is sold to a customer, thus becoming in a sense its own best customer.

2.2.3 A New World Order for Data Communications?

The drive is on for corporations to move their communications into global and international integrated data and voice networks. Corporate global enterprise networks continue proliferation at an astonishing rate as international circuit costs decline and countries decentralize their communications infrastructures. In fact, deregulation of international communications services is another driver in reducing the costs of communications. Does this point to a new world order for telecommunications? If it does, one must become part of this new world order of data communications to remain competitive.

Data communications has taken on a global view in many of its facets. This section summarizes some key areas of standardization, global fiber connectivity, and the needs of the multinational corporation.

International standards, of course, are a prerequisite for global connectivity. There has been much progress in this area. The latest generation of digital transmission rates is the same around the world for the first time in history. There is increasing cooperation between standards bodies on an international scale. The entire standard process is changing to better meet the accelerating needs of users; witness the tremendous international success of the ATM Forum driven by users, service providers, and manufacturers of the technology. As shown in Part 3, ATM is standardized on a worldwide basis independent of physical-link speed.

Now the growth in international connectivity requirements exceeds the transoceanic cables installation and satellite launch schedules. In 1998 may experts expect that the data traffic on international cables will exceed that of voice traffic. International switched services are booming, with current product offerings providing a broad range of connectivity options. The entire world is being connected by fiber optic cable systems. Figure 2.5 illustrates the international fiber optic connectivity that is either planned or already in place.

Recent political, technological, economic, and regulatory changes worldwide have spurred international data network interoperability. New interest is occurring in the countries of Mexico and Canada through the North American Free Trade Agreement (NAFTA). Growth has primarily been in the United States, Europe, the Pacific Rim, and Southeast Asia. However, markets in

South America, New Zealand, Australia, Russia, and countries once part of the Eastern Bloc and the now-defunct Soviet Union have been slower to emerge, but are gaining momentum through the development of infrastructures based on state-of-the-art technology. Postal, Telegraphy, and Telephony (PTT) monopolies are realigning with open market competition, making worldwide advanced data communications a reality. Many international businesses are establishing ties and forming agreements with the PTTs for purposes of future network planning. These are the communications markets of the next millennium.

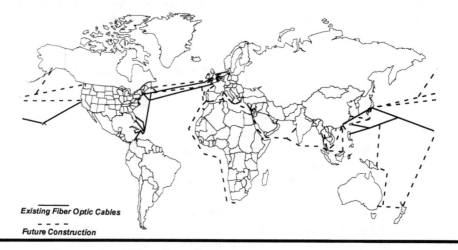

Existing Fiber Optic Cables

Future Construction

Figure 2.5 International Fiber Optic Cable Systems

The 1990s has been the decade of mergers and strategic partnerships. Every day the sun rises on a new international merger or partnership between carriers, hardware vendors, PTTs, governmental agencies, and small companies who fill niche markets. Many joint ventures have sprung up both nationally and internationally. These range from the computer vendors trying to beat out the smaller clone vendors to the large interexchange and international carriers who vie for dominance in the worldwide market.

2.3 WHAT IS ATM? — THE 30,000 FOOT VIEW

Before proceeding any further, we give our first definition of ATM because the book still has a great deal of background to cover. Take your seat in the ATM first class section and stare out the window of our plane soaring high over the communications landscape. Fundamentally, ATM is a technology that simultaneously transmits data, voice, and video traffic over high bandwidth circuits, typically hundreds of megabits per second (Mbps) in 1997 and Gigabits per second (Gbps) before the year 2000. ATM hardware and software platforms form a communications architecture based on the switching and relaying of small units of data, called *cells*. The primary differentiation between ATM-based services and other existing data communications services, such as the Internet Protocol (IP), frame relay, the Switched Multimegabit Data Service (SMDS), and Ethernet, is that ATM is the first technology and protocol structure to effectively integrate voice, data, and video over the same communications channel at any speed.

Now, let's use an everyday example of planes, trains, and automobiles to illustrate ATM's support for multiple service categories. Just as airlines have different service categories (first, business, coach); as trains have sleeper cars, freight cars, and box cars; and as highways have high occupancy vehicle (HOV), emergency, and passing lanes, so too does ATM have methods of handling different types of users, or more accurately application traffic types. As different people have varied class requirements (first class, business, coach), applications also have different class requirements (voice, video, and data; or low delay, and "whatever is available" delay). Applying the notion of service categories to communication networks, think of how corporations want to assign a higher priority to the traffic that carries their executive video conferences, while consumers want to make sure their voice conversation is not interrupted while someone else transfers a file across the Internet.

ATM is well positioned to be the leading technology enabler for communications by the turn of the century. ATM is a standardized multiplexing and switching technology that is designed for flexibility at the expense of efficiency – offering multiple levels of quality of service for integrated voice, video, and data traffic. Proprietary, non-standard systems may deliver some of ATM's features; however, they lock a networking user into a single vendor or single protocol solution — a situation that often is more expensive in the long run. For this reason, much of this book focuses on ATM standards. For these applications the flexibility of ATM can result in a solution that is more cost effective than several separate, individually optimized technologies.

2.4 TRANSMISSION, PROTOCOLS, AND COMPUTING ADVANCES

This section explores technology enablers of ATM including advances in transmission systems, protocol features, and computing power. The principal technology enablers for Asynchronous Transfer Mode (ATM) are:

- Enhancements to protocols that guarantee the effective transmission of information
- Processor advances within servers and personal computers
- High-performance, low-cost digital transmission media
- Cost-effective, high-performance electronics
- Worldwide support for standardization from all sectors of the industry through the ATM Forum

2.4.1 Protocol and Processor Enhancements

A major technology enabler for ATM is the continued growth in desktop processing power, fueled by the continuing decentralization of computing power from the centralized host to the desktop and server, with its associated requirement for more peer-to-peer and client-server networking. Desktop machines not only have the processing power the centralized host once had, they also control information passage employing a wider and more sophisticated range of controlling network and transport protocols, such as TCP/IP and Novell's IPX protocol.

Increased storage in the end stations allows protocols to implement larger retransmission windows and hence increase throughput in congested network environments. RFC 1323 [6] increases TCP window size from 64 kbytes to over 1 Gbyte for this very reason. Increased processing power enables the implementation of more sophisticated flow control and windowing mechanisms. One example of protocol complexity is demonstrated in the sophisticated TCP flow control algorithms in the end station versus the relatively simple Internet Protocol (IP) used in routers.

Older network protocols like X.25 packet switching implemented complex procedures just to ensure that a packet could be reliably sent from node to node, sometimes requiring multiple retransmissions over noisy, error-prone analog links. The simplification of network switching protocols is primarily a result of essentially error-free physical layer communications over digital facilities versus the older error-prone analog facilities. The infrequent occurrence of errors and associated retransmission is then achieved cost-effectively in end systems. Simpler network protocols, such as frame relay, and ATM, rely on the performance of digital fiber-optic transmission, which provides very low error rates, typically less than 10^{-12}. The cost effective availability of plesiochronous digital transmission rates such as DS1, DS3, and synchronous SONET at rates of 150-Mbps, 600-Mbps, and 2.4 Gbps are key technology enablers for ATM services.

2.4.2 Modernization of Transmission Infrastructures

Fiber optics replaced digital microwave transmission in industrialized nations even more rapidly than digital transmission systems replaced analog systems. Satellite communications has evolved as a high-quality digital transmission medium for connectivity to remote areas or as backup to terrestrial facilities. There are even satellites designed specifically to handle ATM traffic, such as those designed by COMSAT [7]. The nationwide and metropolitan area networks of most service providers and ISPs are almost exclusively fiber-based, and many routes rely on SONET automatic restoration facilities. Bandwidth between cities across optical fibers with 44 pairs operating at 10 Gbps will reach into the terabits-per-second (Tbps) range by the year 2000. The prefix "tera" means 10 raised to the 12th power or a thousand-billion. Get used to it; the prefix, "giga," the Greek word meaning 10 raised to the 9th power, will become passé early in the next millennium.

Many carriers are deploying significant amounts of fiber to the curb (FTTC), and even fiber to the home (FTTH). Modern digital and fiber optic transmission communications establish a new baseline for the performance of digital data communications, just as digital transmission made long-distance-calling sound quality comparable to local calls just fifteen short years ago. The impact and benefits of high-performance digital transmission over fiber optics are recurring themes throughout this book. These changes in the modernization of the fiber infrastructure have all accelerated and hence enabled a rapid move toward broadband networking.

2.4.3 Worldwide Industry Support

Most sectors of industry now support and embrace ATM. This began with the telecommunications industry defining initial B-ISDN and ATM standards. The late 1980s saw the development of early, prototype Central Office (CO) type of ATM switches. The traditional customer premises multiplexer and switch vendors then adopted ATM in the early 1990s. Next, router and hub manufacturers began building ATM interfaces for routers and hubs. Now hardware vendors supply ATM interface cards for workstations and servers. The final piece of the puzzle is falling in place with the initiation of operating system and application software development for ATM-based systems. These efforts point toward a strong commitment to the success of ATM because it addresses the user need on an end-to-end basis.

2.4.4 Power to the Desktop

Obviously, one of the biggest trends influencing the success of ATM is the proliferation of computing power to the desktop and its subsequent use in distributed computing. The consumer's personal computer (PC) and corporate workstation processing power (MIPS) (also measured in Millions of FLOating Point operations (MFLOPs)), memory size (Mbytes), and display size (Mpixels) are increasing at an exponential rate. As discussed earlier, organizational re-engineering creates a demand for increased bandwidth to interconnect these PCs and workstations. This increase occurs at a rate faster than the computing performance metrics due to the accelerating bandwidth principle described in Chapter 1.

PCs, such as the Apple computer, were born in "garage shops" and were initially known as toys for games and other amusements. That has changed with the PC and workstation, and their big brother the server, now taking the premier position in the computing industry. However, as software applications were developed and speeds and memory increased as costs dropped, the larger computer manufacturers began to see PCs on the desks of users who previously only had a "dumb" terminal connected to the host. The PC was legitimized by the IBM announcement of their entry into this market in 1983. Since 1983, the PC, renamed "workstation," has been the industry standard for corporate and government microcomputing.

The workstation now provides the user with the device for desktop access to the world of voice, video, and data. Mass storage of information has shrunk to a fraction of its original size and cost. Now the critical element is no longer hardware, but software. Indeed, the cost of software on a modern workstation can easily exceed that of the hardware investment.

2.4.5 Standards and Interoperability

Fast realization and industrywide agreement on a common set of standards have also been prime enablers of ATM. These standards have led to a technology interoperability at all levels of the enterprise network and across many hardware and software platforms. The ATM Forum in particular has set a precedent by fostering cooperation between a wide range of industry segments to develop interoperable specifications from the desktop to the world. Chapter 10 provides a review of the ATM standards players, the processes, results to date, and future plans.

2.5 LOWER COST ATM LOWERS NETWORKING COSTS

ATM provides one of the best price/performance per megabit per second of any public network technology. Better yet, the cost to implement ATM continues to decline. The advent of public carriers offering ATM Switched Virtual Connections (SVCs) providing bandwidth-on-demand makes the economics of ATM even more attractive. Decreasing transmission facility costs also drive down the actual cost of high-speed circuits, if not the price — making ATM even more attractive for large meshed networks. Implementing ATM across the entire enterprise can lower overall network cost by lowering capital and support costs through more efficient use of a single ATM infrastructure. Public service providers now offer WAN ATM services at rates less than dedicated private lines and comparable to hot-selling services like frame relay. Thus, we believe that ATM offers a low risk, high value WAN investment.

2.5.1 Transmission Cost Reduction

Two complementary and competing phenomena are occurring simultaneously. LAN and WAN speeds are converging to provide LAN extension speed connectivity across the WAN. And the cost of WAN and LAN bandwidth is becoming less expensive per bit, per second. Let's explore both.

Bandwidth in the WAN and LAN, when viewed on a cost per megabit-per-second, has steadily decreased in cost over time (years), with LAN costs decreasing much more rapidly than WAN costs. Consider the promise of Wave Division Multiplexing (WDM) in Passive Optical Networks (PONs) proving terabit speeds across a single fiber optic strand, and compare it with the highest commercial-grade fibers carrying 40 Gbps in 1998. And although running fiber to the desktop and home only costs a few cents per feet, it now dominates the overall cost. But here again, creative innovators have found ways to squeeze even more bandwidth out of the existing twisted pairs deployed to every major business and many households through sophisticated Digital Subscriber Line (xDSL) technologies as described in Chapter 32.

Bandwidth in the WAN, as many pundits projected, is not yet "free," but it is becoming less expensive all the time. But these cost decreases are less pronounced than they should be because the transmission and service providers must pass along the cost to upgrade transmission and switching facilities and equipment to the end user. Either way, transmission technology decreases the real cost for transport of high-bandwidth services such as ATM.

2.5.2 Lower Network Lifecycle Costs and Future-Proof Investment

Traditionally, corporations and consumers look for a technology or service that maximizes their investment for at least a 3- to 5-year period. Therefore, a key consideration is the lifetime of a technology. While ATM seems to be a technology leader for the rest of this decade, standards continue to evolve and intervals as short as only a few years now separate generations of equipment. Some public service providers plan to deploy ATM as a core part of their backbone and service offerings, hoping to realize the benefits of cost-effective flexibility and scalability. Flexibility and scalability take on many forms. Being able to increase the number and size of switches, users, and circuits is a key measure of scalability. Support for a distributed architecture where multiple processors share the burden of a single processor failure — translating to higher reliability — is another benefit. The capability to upgrade network elements to faster processors, and upgrade routers, switches, hubs, and workstations to the same standards-based architecture, is also a potential benefit. No technology remains efficient and cost effective forever, but the flexibility and scalability of ATM hardware and software potentially provide a longer life cycle than other MAN and WAN technologies, which "future-proofs" the ATM investment.

Furthermore, purchasing ATM equipment across the enterprise achieves purchasing discounts; since capital equipment purchases for large volumes from a single vendor decreases unit costs. The overall cost per Mbps is greatly reduced, along with the cost of sparing, maintenance, and support staff training. A long term technology like ATM also future-proofs your network, providing a solution that will last many years into the future and lowering the overall life-cycle cost.

2.5.3 Lower Public Service Cost

The demand for public network ATM service is growing as users discover the sobering capital cost of building a private line based ATM WAN, compared with that of a shared public network service. The cost of multiple point-to-point DS3 circuits can be greatly reduced by using a pubic data service that offers a single DS3 ATM access circuit. This benefit is accelerated by the advent of protocol interworking over an ATM service (i.e., frame relay to ATM) and Switched Virtual Connections (SVCs) which allow multiple types of services at different speeds (DS0, FT1, DS3, FT3) to interwork with one another.

Fiscal interests also play a role here, since companies pay deductible expense dollars for public services versus depreciable capital dollars for private equipment. Many corporations more readily justify the expense-oriented approach using a public network service than the capital-oriented approach of purchasing and building a private network. It is interesting to note that over half of the Internet traffic is transported over ATM. Internet service providers have embraced ATM as the transport technology of choice for solving bandwidth bottlenecks in their backbone infrastructure: routers simply can't run fast enough. However, as studied in Part 8, this too is changing as hardware empowered routers rival ATM switches in terms of raw speed.

2.6 REVIEW

Throughout this chapter the evolution from voice to data networks is a recurring theme. History has taken us through various reincarnations of the same types and cycles of technologies. Infrastructures developed for more cost-effective voice networks have evolved to the point where they now support data networks. A revolution in data communications is occurring, led by protocols and services that provide solutions to the new business dependence on data communications. The text then introduced the technologies that serve as enablers to ATM: protocol and processor enhancements, the modernization of digital transmission facilities which provide low bit error rates and SONET/SDH transmission facilities which yield extremely low bit-error rates complemented by very high transmission speeds, along with some significant protocol and computing advances. Finally, technology-oriented business considerations drive some users to choose ATM. The chapter summarizes the inherently lower cost of shared public services, lower life-cycle costs, and future-proofing advantages offered by ATM.

2.7 REFERENCES

[1]   L. Smith, Routing Bandwidth Into the Future," *PC Week*, May 2, 1997.
[2]   G. Kessler, *ISDN, Concepts, Facilities and Services, 2d ed.*, McGraw Hill, 1993.
[3]   J. Jamison, R.Wilder, "vBNS: The Internet Fast Lane for Research and Education," IEEE Communications, January 1997.
[4]   MCI, vBNS Home Page, www.vbns.net.
[5]   "oc3mon happenings," http://www.nlanr.net/NA/Oc3mon/.
[6]   V. Jacobson, R. Braden, D. Borman, "IETF RFC 1323, TCP Extensions for High Performance," May 1992.
[7]   D. M. Chitre and S. Piller, COMSAT's ATM Satellite Services white paper.

3

Computer and Information Networking Directions

This chapter reviews computer and information networking directions. The text begins with a historical review of computer and networking directions through the end of the 1990s and into the twenty-first century, providing insight into the communications revolution sweeping the world. The text then presents the key network trends driving the need for ATM. These new paradigms include the crisis of routing and the birth of switching in the LAN, virtual networking, and the reclassification of internets and extranets. We then take a brief look into the data network of the 2000s. This chapter concludes with a discussion of outsourcing and outtasking trends.

3.1 NETWORKING TRENDS DRIVING THE NEED FOR ATM

Each organization will eventually ask "why do I need ATM?" ATM business drivers take on many forms, from the need for higher speeds, increased flexibility, improved efficiency, and support for multiple traffic types to support entirely new applications. Subsequent sections discuss the following key information and computer networking trends:

- The Middle-Age Crisis of Routing and the Birth of Switching
- Virtual Networking and Virtual LANs
- Intranets and Extranets
- Network Security
- Seamless Interworking
- More Bandwidth for Less Bucks

3.1.1 The Middle-Age Crisis of Routing and the Birth of Switching

Typically, all stations in a LAN connect to a single, shared medium as shown in Figure 3.1a. Users on the shared medium potentially have access to the entire shared medium bandwidth. A problem occurs when more than a few users attempt to use the 10 Mbps Ethernet at the same time, resulting in a usable throughput that is only a fraction of the media speed as analyzed in Chapter 31. Capacity then becomes the limiting factor to the user's applications. The same phenomenon occurs with 100 Mbps Ethernet, Token Ring, FDDI, and DQDB — although some of these protocols achieve network throughputs approaching media speed through more sophisticated resource allocation methods. However, when the users' desktop rate approaches the shared medium speed, there is no choice but to move to the next higher-speed shared medium LAN solution, or to segment the LANs into several LANs with fewer users per segment. Here too, the final step is to dedicate a single segment to each user. Thus, as application demand for more bandwidth increases, LAN administrators must decrease the number of users per LAN segment. The need to manage this environment and enable greater throughput between heavily loaded LAN segments created the market for LAN switches, as shown in Figure 3.1b (although a bridge or router could be used, and a LAN switch is essentially a MAC-layer switch). The shared bus speed inside the LAN switch limits the maximum network throughput. As workstation power increases and each user requires more bandwidth than can be shared with other users on the same segment, our example reduces to a single user per LAN segment connected to a LAN switch, as shown in Figure 3.1c. Typically, LAN switches operate at higher speeds than routers, hence network throughput increases further and the routers can be redeployed to the interface between the LAN and WAN.

As users move from 10 Mbps to 100 Mbps Ethernet as their LAN technology of choice, each user also typically requires more bandwidth to the desktop, having the net effect of reducing the number of users on each Ethernet segment (sometimes down to a single user) despite the higher bandwidth available per segment. A requirement emerges for a device that can provide switching of LAN traffic within the LAN at the MAC layer without complicated routing schemes. Thus, LAN switches begin to dominate the intra-LAN and inter-LAN market, solving the problems of graceful LAN segmentation and growing capacity constraints with a scalable switched solution.

ATM Network Interface Cards (NICs) in workstations and servers, working in conjunction with ATM workgroup switches, act as a common logical interface technology that can scale each workstation and server network interface from 25 Mbps to 155 Mbps without requiring changes in software to support a new shared-medium solution, as shown in Figure 3.1d. Because the ATM network switches work cooperatively, network throughput grows as the number of ATM switches increase — readily achieving Gigabit LAN performance.

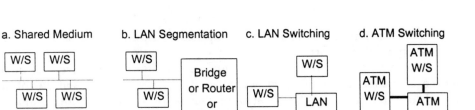

Figure 3.1   Evolution from Shared Medium to High-Performance Switching

Another way to look at this move to LAN switching is from a cost perspective. Intelligent hubs and LAN switches are simpler and thus less expensive per port than routers. Historically, routers handled the more complex multiprotocol routing functions, while hubs handled the less complex aggregation of a single LAN protocol, and in some cases, bridging functions. As studied in Chapter 9, in IP networks, end systems in the same IP subnet communicate at the Media Access Control (MAC) layer without requiring routers to perform the more complex IP routing. A simpler, cheaper technology, bridging, works at the MAC layer, but does not scale well in larger networks.

3.1.2 Virtual LANs (VLANs)

The LANs we have grown to love (or hate) over the last 20 years were designed so that all users of specific resources, like a shared application server, were grouped and assigned on a single physical LAN. First repeaters, and then bridges connected multiple physical LAN segments. As internetworks grew, LAN administrators employed routers to interconnect disparate LANs and address the limited scalability of bridged networks. A trend in this evolution is that each new technology overcame the bottlenecks and limitations of its predecessor. As distributed LAN-based server and application resources proliferate at multiple diverse physical locations, the requirement increases for users not *physically* colocated to share resources — enter Virtual LANs (VLANs). VLANs empower network administrators to transcend the physical LAN connections to their users for the greater nirvana of the *logical* work-group. VLANs create communities between users on different physical LAN media, giving the appearance that they are physically attached to the same physical LAN. VLAN resources still appear local to each user, regardless of whether they share the same physical LAN segment, or all its resources are half a world away. Connectivity is readily achieved over the WAN; however, performance may suffer unless the VLAN is carefully designed. VLANs may support multiple MAC and network protocols rather than requiring the use of the same MAC protocol by every user (such as IP over Ethernet).

VLANs can greatly reduce the administrative burden involved in adds, moves, and changes — the greatest cause of cost, complexity, and downtime for a LAN administrator. One cost saving measure is to only license a limited number of active instances of a particular application; for example, a word processor, spreadsheet, or graphics package. Client workstations then operate only a shell of the original application retrieved from the server on demand. A similar environment is emerging in the Internet, where users with network computers access multiple servers across the Internet, downloading Java programs only when they are required to run, and then releasing them from memory when the user is done with the program. This paradigm also has the side benefit that the user always runs the latest version of the program. Servers typically enable users to share expensive resources such as printers, CD-ROM juke boxes, mass storage, and large databases. Thus, the Internet creates the largest VLAN environment, a scenario that can be replicated within a corporate network.

Figure 3.2 shows an example of a virtual LAN connecting user devices in three different geographic locations. Note that user A, B, and C have virtual access to the server in location 2 as if they all three shared the same physical local Ethernet or FDDI LAN. Many VLAN designs are being built with switching hubs that use some proprietary method of switching between VLAN user groups. The most common methods of implementing VLANs as of press time included the proprietary Inter-Switch Link (ISL), IEEE 802.10, and ATM LAN Emulation (LANE) standards as detailed in Chapter 19.

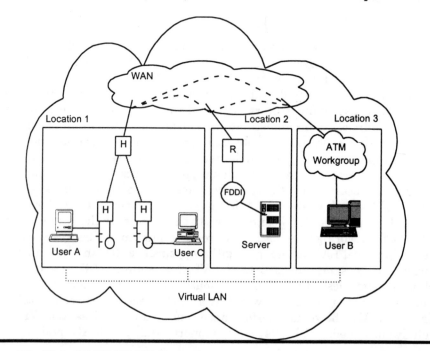

Figure 3.2   Virtual LAN (VLAN) Example

3.1.3 Private and Virtual Private Networks

This section describes and compares two alternative approaches for very high-speed computer networking: dedicated circuits between workstations, servers, routers, and switches (with these systems also performing switching functions) versus shared-access lines connected to network switches which provide virtual, on-demand capacity for any-to-any device communications.

Figure 3.3 illustrates the private network alternative comprised of dedicated circuits connecting customer owned devices. Clients and servers are shown with either dedicated connections to the network or access via routers connected to the network. Each direct connected device has two links to at least two other devices via the network Points of Presence (PoPs) to provide survivability in the event of link failures. When a server or router is not directly connected to the desired destination server or router, the intermediate routers or switches perform switching and routing.

The advantages of this private network approach are full user control and simple, less expensive network technology. The disadvantages are nonproductive redundancy, dedicated link capacity, additional equipment costs to perform switching and routing, and the need to engineer the private line trunks for peak capacity and redundancy.

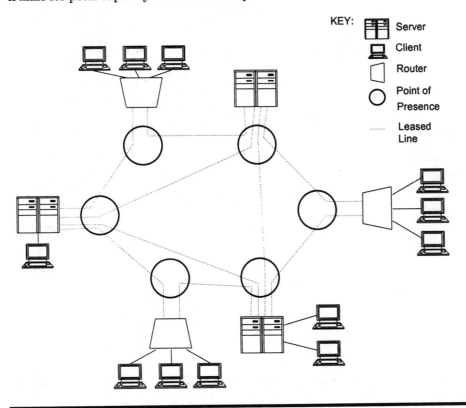

Figure 3.3 Dedicated Circuit Private Network Example

Figure 3.4 illustrates the same set of servers, routers, remote routed/bridged clients, and network PoPs as in the private network, but as part of a shared, virtual ATM network. A separate virtual video network shares the backbone, as shown in the figure. An ATM switch at each network PoP has an access lines to each user. High-speed trunks interconnect the ATM switches. The public ATM network shares these trunks across multiple users by dynamically allocating bandwidth. Each user site requires only a *single* access line and port, thus halving the access cost when compared with the previous private network example. The reliability of access is identical to that of the dedicated network, assuming that the access circuits in the dedicated network take the same route to the network PoP. The ATM switches perform routing and switching, relieving the routers and servers of this task.

In summary, the advantages of the virtual network alternative are: reduced access line charges, the capability to satisfy high peak demands (particularly during low activity intervals for other services), cost impacts proportional to usage (versus cost proportional to peak rate in the dedicated network alternative), and enhanced reliability. Disadvantages are less predictable peak capacity and reduced user control.

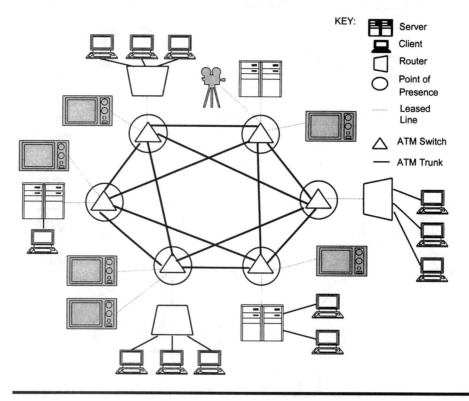

Figure 3.4 Virtual Private Network Example

Virtual private networks are defined as network partitions of shared public network resources between multiple users to form a private network that *appears private to the users;* but is still part of the larger public network. The network assigns shared resources in fair proportion to the bandwidth ordered by customers.

In the virtual private network example above, a single access circuit from each site to the network is sufficient, because multiple virtual connections can be provided from multiple users at a site to their destination on the network. For example, each virtual connection can be allocated a peak rate equal to the access circuit, but have a sum of average rates that is less than or equal to the access circuit speed. Figure 3.5 further demonstrates this concept by showing how users A, B, C, and D at site 1 all have a single physical circuit into their premises ATM device which converts these inputs to four ATM virtual connections (as indicated by the different line styles) and then transmits them over a *single* physical ATM access circuit to the ATM network switch. The ATM network logically switches these individual user virtual connections to the destination premises ATM device where they are delivered to the physical access circuit of the end user as illustrated in the figure. Later, Part 3 of this book details the various types of physical access circuits and ATM virtual connections.

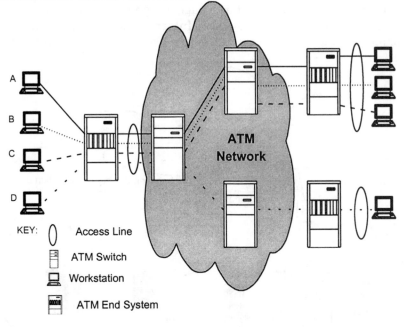

Figure 3.5 Detailed Example of a Virtual Private Network (VPN)

3.1.4 Intranets and Extranets

The Internet has grown phenomenally in the mid 1990s, and shows no signs of slowing down. The Internet continues to be the most commonly used medium for computer communications — sometimes called the information superhighway or the "Infobahn." The estimated number of Internet users is speculated at over 50 million, with a double digit growth rate each month.

Data communications traffic often carries the most private and sensitive information of an enterprise. Therefore, the majority of data traffic is typically intra-enterprise. In contrast, enterprises communicate a large amount of voice traffic over public telephone lines, both on network (intranet) or off the network (extranet). Enterprises now create data and multimedia intranetworks and extranets using private and public network facilities. Some even use the Internet as the transport medium rather than purchasing more expensive dedicated transport facilities. Labeled "intranet within Internet," these communities of users communicate with each other within their company by means of the public Internet. A popular method of doing so is through encrypted IP tunneling. The concept of intranets and extranets is similar to Virtual Private Data Networks (VPDNs), but instead uses the Internet as the transport utility [1]. An "extranet" is created when a corporate intranet is extended to external entities, as when a corporation has a distribution partner with whom they share computing resources. Figure 3.6 illustrates these concepts. In this example location 1 is part of an Intranet for Company ABC, which is part of an extranet with Company XYZ's server at location 2 and high-end web page design group at Company LMN at location 3. The tremendous growth in Internet traffic is an example of the burgeoning demand for data communication over public shared facilities.

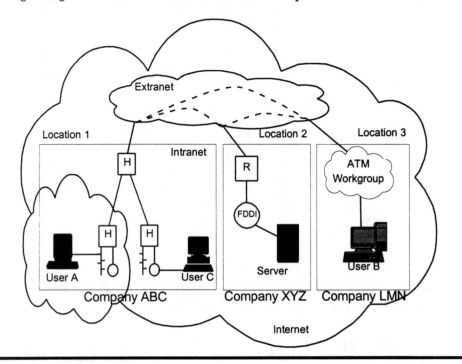

Figure 3.6 Graphical depiction of Internet, Intranet, and Extranet

Companies are designing enterprise and interenterprise (or intercorporate) intranets, internets, and extranets, partitioned by firewalls, guaranteeing bandwidth and delivery using ATM technology. Traffic requirements in the LAN and across the WAN will explode as intranet and extranet users access information via their browsers from multimedia web servers over switched LAN and ATM infrastructures. The challenge for ATM will be to gain market share before the Internet provides a QoS enabled solution (i.e., RSVP or differentiated services) and brings low cost competition to bear.

3.1.5 Internet and Extranet Network Security

When the corporate network of company A is tied to the corporate network of company B, via either private lines or switched services, inter-networking takes place across a secure internet or extranet as shown in Figure 3.7. This is often the case when two companies need to share vital information such as engineering CAD/CAM files, databases, group-ware, and other such applications. This connectivity between two or more corporate, government, or university private networks usually occurs through the use of public network services.

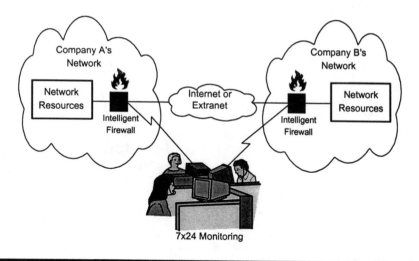

Figure 3.7   Intercorporate Internetworking

If public data services provide good security and partitioning, then the portion of interenterprise traffic sent via carrier services should increase. An enterprise, however, needs to control what communication can occur within, as well as outside, its boundaries. Intrusion monitoring and detection including security screening, and filtering. Passive or active blocking can be done in a connection-oriented environment through signaling, or in a connectionless environment through the use of screening tables or filters in routers. However, implementing too much filtering on a router interface dramatically reduces routed throughput since the router must apply the security rules to every packet. Inter-networking works well as long as security precautions are taken by both the service provider and the end-user networks. It is a common practice today for users to install separate "fire-wall" routers that filter the context and content of packets, circuits, and application protocols to prevent intrusion from the "outside" as well as from the "inside." It is key to keep constant vigilance on these firewalls to make sure the security policy is well enforced. Another trend is to install security servers within the LAN and WAN infrastructure and use physical key-coded security devices on the remote and dial access ports to the network, but placing these "bars on the windows" is just the first step in network perimeter security. Twenty-four hours a day, seven days a week monitoring and proactive management of all security policies is a must. An experienced hacker will always find a way in, no matter how good the security precautions, and someone must be watching round to clock and refute the break-in.

3.1.6 Seamless Interworking

Of course, most enterprises have more than one location and also have more applications than just data. The original premise of Broadband ISDN (B-ISDN) was to provide the capability to serve voice, video, and data using the same technology. This concept has been dubbed "seamless interworking" across the LAN and WAN.

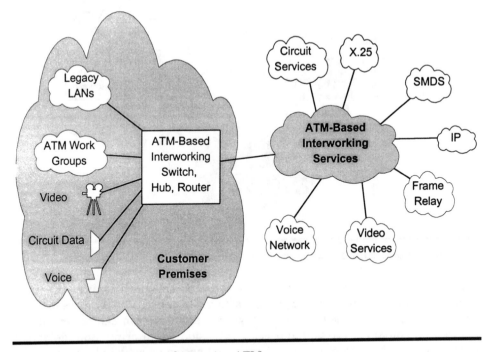

Figure 3.8 Seamless Interworking using ATM

Figure 3.8 illustrates this vision of seamless interworking. Voice, video, and data are converted to ATM at the user site, where they are interconnected to the WAN via ATM. Access to voice and other legacy services is shown in the WAN. ATM provides a smooth migration path from older time division multiplexers through the support of circuit emulation services as illustrated in the figure. ATM also provides direct transport of voice across the WAN. ATM offers the capability to integrate completely separate data and voice networks using different carriers and services into a single unified network. There are other advantages to this approach such as a single network management infrastructure and reduction of access and transmission costs.

The rule of thumb for many years has been that only 10 to 20 percent of the LAN traffic goes over the WAN. Client-server communications over distributed architectures, as well as the Internet as a means of communications, has changed this rule. But the WAN is still "bandwidth-challenged." Current contention-based Ethernet LANs achieve a sustained 30 to 70 percent throughput of the actual LAN speed. ATM can extend the LAN across the WAN with degradation in access time limited essentially by only the speed of light. Chapter 26 describes the tradeoff between bandwidth and latency limits that applications encounter.

This argument of seamless interworking also extends to mobile computing and access via standard telephone lines to these services. Mobile communications and phone links that extend the LAN to the automobile, train, meeting room, or hotel room will become commonplace as business travelers take with them mobile-communications-equipped laptops and personal computing devices. Lower-speed access will perform sophisticated data compression for interconnection into the seamless internetwork.

3.1.7 More Bandwidth for Less Bucks

Let's compare the technology revolution of the 1980s T1 multiplexer technology to ATM technology of the late 1990s. The widespread use of "T1" multiplexers in the 1980s was predicted to be a precursor to a wave of "T3" multiplexer deployment: a prediction that never came true. Understanding the reasons for the T1 multiplexer success and the lack of the adoption of T3 multiplexers is central to placing the potential benefits of ATM in perspective. T1 multiplexers allowed high-performance, relatively low-cost DS1 (colloquially called T1) facilities to be shared between a variety of applications on a quasi-static basis using time division multiplexing (TDM). TDM bandwidth allocation is not well suited to high-performance, bursty data communications because it limits the peak transmission rate, and wastes bandwidth during frequent idle periods typical in data communications. While data communications demand increased dramatically, the demand for TDM-based service did not keep pace with the overall demand for bandwidth. DS3 speeds are over 28 times that of the DS1, but cost only 5 to 10 times more. Most users couldn't justify the economics and restrictions of TDM inherent in the T3 multiplexer paradigm. Instead, network designers saw better choices for public services within the planning horizon, such as frame relay, IP, and ATM.

ATM offers the capability to extend the LAN or MAN across the WAN at speeds comparable to the LAN or MAN (currently 10, 100, and 1000 Mbps) at a lower cost than dedicated or TDM circuits, because the bandwidth and switches are economically shared across many users, as shown in Figure 3.9. Instead of having to funnel the bandwidth of interconnected LANs down to the lower bandwidth provided by the static allocation of TDM connecting sites via DS1s in the DS3 access line, as shown in Figure 3.9a, ATM provides each LAN with the capability to burst at the full LAN access speed across the WAN on the DS3 access line, as shown in Figure 3.9b. This figure shows how TDM LAN interconnection takes much longer and much more bandwidth to transfer data, as shown by the time plots of actual usage on the access lines. Since all users cannot burst simultaneously, ATM accommodates access to peak bandwidth on demand virtually all of the time. Therefore, ATM achieves significant economies of scale by integrating multiple applications on the same physical network.

a. LAN Interconnectivity via TDM Network   b. LAN Interconnectivity via ATM Network

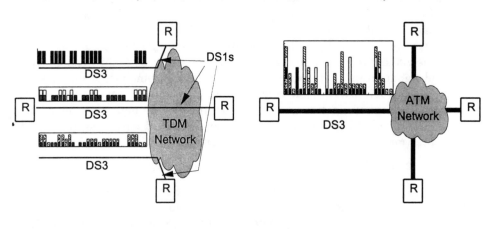

Figure 3.9   ATM Provides More Usable Bandwidth to Data for Less Cost than TDM

3.2 ATM — NETWORK INFRASTRUCTURE OF THE NEXT DECADE

Have you ever wondered what the architecture of tomorrow's data network will look like? This section takes us on a journey to the end of this millennium. Static, predefined private communications networks are migrating to dynamic, virtual networks — networks with ubiquitous access that can interwork past legacy, present, emerging, and future protocols. Virtual, public data networks are adding more and more intelligence, in essence enabling the network to become the computer. The corporation is becoming more and more dependent upon virtual data networking to run the day-to-day business, requiring both partitioning and security. The war on technical obsolescence is under way, and ATM provides us with an excellent technology with which to win the war. Will ATM be the Rosetta Stone for protocol interworking to allow multiple protocols to be deciphered and understood?

Will these networks provide ubiquitous access for all users by public network addressing? Only time will tell.

3.2.1 The Intelligent Data Network

Corporations and governments are moving towards using faster, larger, and more intelligent data communications networks — where the intelligence lies *within* the network, rather than outside it. They are also looking for intelligent network services, rather than simple, traditional private line or circuit switched data services. The term *intelligent network* connotes some level of *value-added* service or function provided by the network. Examples include address translation and interpretation, intelligent routing decisions made within the network by route servers, and protocol conversion. Network intelligence can also mean a service offering based on centralized, intelligent, network-based devices that serve as information servers offering voice, video, and imaging interpretation, routing, and on-line service features. Prime early examples are the interactive database services provided by CompuServe and America On-line. Now much of this content is now available "free" on the web, funded by advertisers. Of course, you often get what you pay for, and therefore there is also a lot of worthless free material out on the web.

Users want networks that are "smarter," not just faster. Users want access to intelligent public data services so that they can better leverage the intelligence within their own networks.   Current and emerging data communication services are just now slowly adding this type of intelligence to their networks — intelligence which in the 1980s resided at the premises. Now, network service providers offer alternative network intelligence that is extended to the user premises where significant intelligence may not be practical.

Thus, a tradeoff exists between intelligent networks and intelligent user equipment. Many factors, driven by global industry standardization and the development of technology, will influence decisions on where the network intelligence will reside. The market is both technology-driven and user-driven. For example, international providers want the network intelligence to reside in international gateway nodes. National carriers want intelligence to reside in carrier Point of Presence (PoP). Local exchange carriers (LECs) or PTTs want intelligence mainly in the serving Central Office (CO). Customer Premises Equipment (CPE) vendors want the intelligence to reside in the CPE. Router manufacturers want the intelligence in their devices, while end-system hardware and software manufacturers strive to make their platforms the seat of intelligence.

The profits of the next century lie in the intelligent functions, not the connectivity function, with all of the aforementioned groups recognizing this fact. The user needs to mix and match all of the above options for the best cost and functional advantage — typically in a "hybrid" networking environment using a mixture of components to meet the needs of the intelligent data communications network of the future. However, the network users and network providers must also work together to ensure that the mix and blend of technology being used meets business goals. This is critical for continued successful business operation and optimization. Relating these responsibilities to the strategic business objectives of the company will guarantee ongoing success.

3.2.2 Meeting the Needs of the Enterprise

Typically, large enterprises have a few large locations that serve as major traffic sources and sinks. Location types include large computer centers, large office complexes with many information workers, campuses requiring high-tech communication, data or image repositories, and large-volume data or image sources. These large locations have a significant community of interest among them; however, the enterprise usually also requires a relatively large number of smaller locations needing at least partial, lower-speed access to this same information. The smaller locations have fewer users, and generally cannot justify higher-cost equipment or networking facilities. Note that cost generally increases as performance, number of features, and flexibility are increased. ATM-based interworking will initially be useful to the largest locations in the largest enterprises. However, the high performance and flexibility of ATM interworking must somehow also meet the need for connectivity to the many smaller locations that also need to be served. This has proved to be true with the success of low-speed ATM access services.

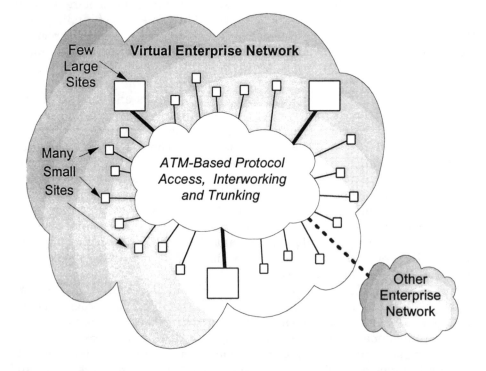

Figure 3.10   Typical Enterprise Network

For low-speed access, efficiency is a more significant concern than the flexibility and higher performance of ATM. This is because the cost per bit per second generally decreases as the public network access speed increases. For example, the approximate ratio of DS1/DS0 and DS3/DS1 tariffs is approximately 10:1, while the speed difference is approximately 25:1. This means that a higher-speed interface can be operated at 40 percent efficiency at the same cost per bit per second. Conversely, the lower-speed interface costs 2.5 times as much per bit per second, and therefore efficiency is more important.

Figure 3.10 illustrates an ATM-based network cloud connecting a few large sites to many small sites. Within the cloud, various standards define how certain protocols can interwork with others like IP and LAN protocols over ATM, true interworking with frame relay, ISDN and private lines, as well as trunking to other networks through a common infrastructure. Typically, large enterprises, such as corporations, governments, and other organizations, deploy networks with many smaller sites and few larger as shown in the figure. Occasionally, some enterprises employ a meshed approach between peer sites.

3.2.3 The War on Technological Obsolescence

Business users are concerned with maximizing their investment in computing and data communications equipment. Rapid advances in computing make a maximum productive lifetime of workstations and servers 3 to 5 years, whereas a networking technology may enjoy a healthy lifespan of 10 to 15 years. Generally, the most expensive computer and communications equipment is justified for only a small set of mission-critical applications. A similar situation exists in the area of data communications.

Currently most intra-enterprise data communication networks are constructed from Customer Premises Equipment (CPE) interconnected by private lines at DS1/E1 speeds or less. The advent of public data network services, such as frame relay and ATM, and their attractive associated pricing, has motivated some customers to migrate a portion of their bursty data from private lines to virtual private networks. Burstiness is defined as the ratio of the peak traffic rate to the average traffic rate. Usage-based billing, a preferred method of payment for users well familiar with their traffic patterns, operates by charging for only the transferred data. Virtual networks offer higher peak rates at affordable levels because there is normally no penalty for idle time, or equivalently, a low average rate.

Frame relay usually can be economically installed in most existing CPE with only a software upgrade. In addition to software upgrades, ATM usually requires new CPE as well as software upgrades.

There is an alternative to ATM called the Frame-based User-Network Interface (FUNI). This HDLC-based protocol encapsulates and/or maps the ATM header functionality between a DTE and a DCE. This means that most CPE can be upgraded via software, at a cost similar to that of frame relay. The FUNI protocol is thus better suited for early, cost-effective, low-speed implementation of ATM. Chapter 17 describes the ATM DXI and FUNI protocols.

Upon study, users find that they can economically justify the performance and flexibility of ATM at their largest locations first, while simultaneously achieving connectivity to their many smaller locations using existing, or less expensive, equipment and access speeds enabled by ATM-based protocol interworking.

3.2.4 ATM as the Rosetta Stone for Interworking?

In 1799, Napoleon's army discovered the Rosetta Stone in Egypt, which provided the key to translation to the undeciphered Egyptian hieroglyphic and demotic writing. The key was that the Rosetta Stone contained the same message in the known Greek language. Can ATM provide a similar role in translating the myriad of existing, sometimes arcane data communication languages into the modern language of data communications? Will ATM-based networks be the Rosetta Stone for networking protocols? Chapter 17 shows how ATM may provide a similar role in the complex world of protocol access, interworking and trunking along a trail that multi-protocol routers have blazed.

Users want a network where any one user can connect with any type of interface and protocol — and talk to any other user on any other network that may have a different interface and protocol. This is the expectation that router-based networks set today. This same interconnection via multiple interfaces and protocols provides a vision of future value-added data network services.

Protocol access, interworking and trunking removes the obstacle of connectivity for smaller locations within the enterprise and slows or circumvents technological obsolescence for existing, or lower-cost, equipment. Furthermore, protocol access, interworking and trunking facilitates more rapid evolution to ATM by allowing a seamless migration on a site-by-site basis, retaining full connectivity at every step, as traffic and application performance requirements grow. Indeed, as traffic grows at large sites, the older, lower-speed ATM equipment can be migrated down to the medium-sized sites and replaced by the latest high-performance equipment, further justifying the need for protocol interworking as a weapon in the war on the high cost of technological obsolescence.

3.2.5 Ubiquitous Access and Addressing

Ubiquitous access to an intelligent data communications network spanning the globe has become a necessity to rapidly growing national and international businesses, as well as the consumer. The Internet is a perfect example. Whether it is the consumer accessing E-mail from home, or the corporate user talking to their manufacturing plant half a world away, users want to access the data network as a large "cloud" and thus be able to talk to any other user connected to that cloud without requiring any knowledge of the internals of the network cloud. A prerequisite to this capability is the assignment of a globally unique address to each user. The public voice network has these characteristics, with several lessons from that domain applicable to data networks. There is also the experience of the Internet, which has different addressing characteristics, that is also a significant factor. If a user cannot reach any other user on the public data network, as is taken for granted in telephony and has become a *fait accompli* on the Internet, then the resulting data service has little utility.

The public phone network uses an addressing, or numbering, scheme called E.164 that basically has a country code part and then a nationally assigned part for each country as detailed in Chapter 14. This is the familiar international dialing plan. Internationally, and usually nationally, this is a geographic numbering plan since the digits taken from left to right hierarchically identify the geographic location of the user. For example in North America the first two digits identify world zone 01, the next three digits identify the area code, the next three digits identify the exchange switch, and the last four digits identify the user access line on the switch. This convention greatly simplifies the routing decisions in telephone networks and has allowed the goal of universal connectivity to be achieved in the telephone network.

The Internet today assigns 32-bit numbers to each user employing an organizational hierarchy. Entire blocks of numbers are assigned to an organization which need not have any geographic meaning whatsoever. A user may move geographically, and it becomes the job of the intelligent network to find him or her using a routing discovery protocol. The organization may structure its block of addresses however it chooses, either geographically, organizationally, or in some other manner. The Internet has also achieved worldwide, nearly ubiquitous access and addressing as well.

3.2.6 As Reliable as the Phone Network

Similar to the expectation of universal connectivity from telephony, data users expect public data networks to be as reliable as the telephone network. Consumers expect to pick up their phone at home and get a dial tone. If there is no dial tone they get angry. Now take this analogy into the data world. Do you ever get a busy signal today when using accessing your E-mail through the LAN at work? No, it is either working or it isn't. However, sometimes when it is working, it is extremely slow. If so, do you find that unacceptable? Wouldn't it be nice if the webtone were as reliable?

Intelligent ATM networks may rely on their fiber transport to be both near error and near outage-free or may detect errors and faults at the ATM level. Redundancy and restoration must be observed at every step in the design, with SONET technology providing this capability in some configurations or ATM-based restoration algorithms in others. Since many applications do not provide error correction or switching to alternate paths, the capability for an ATM network to guarantee nearly error-free transmission and continuous availability is important. These capabilities are key to provide the reliability found in the phone network.

Successful service providers offer services with high availability and low error rates as required by the corporations and government entities that build their enterprise networks on the virtual network. The switch from conventional private lines to broadband switched service is progressing, but there will always be a need for dedicated private lines for specialized applications. The incremental reliability of the public data network, low error rates, and the reduced price of switched data services based on the economies of scale inherent in the carrier frame and cell-based infrastructures make switched data services even more appealing in comparison to dedicated private line services as time progresses — as long as they are extremely reliable.

3.3 OUTSOURCING TRENDS CONTINUE

This study of computer and networking directions would not be complete without a discussion on outsourcing and outtasking trends. The choice between internally designing, building, and managing a network versus outsourcing or outtasking part of it is one that spans every aspect of the business, and is one of the most important decisions for the network manager. Outtasking or outsourcing network needs involves a third party taking over some or all aspects of the corporation's data network.

What does outsourcing and outtasking entail? What issues and considerations are associated with an outsourcing contract? What are the benefits and drawbacks of outsourcing? This section summarizes the primary factors to consider before signing an outsourcing or outtasking agreement. A more detailed analysis of outsourcing and outtasking can be found in Reference 2.

3.3.1 Outsourcing and Outtasking Defined

Outsourcing is contracting one or more outside vendors to design, develop, and implement a solution for a company's communication needs. This may include, but is not limited to, planning, designing, installing, managing, owning, leasing, operating, and controlling a communications network. Full responsibility for some portion of the company's network communications assets will be transferred to, or assumed by, the outsourcing vendor. Typically, outsourcing occurs when a business understands that performing some or all of these functions is not within its core competency. Outtasking is outsourcing selected tasks from this list, either one, some, or all of them. Outtasking allows the outsourcer to maintain some level of control, or perform functions that are within their core competency.

Good strategic planning and innovative techniques must prevail regardless of whether the design is performed internally or is outsourced. First, evaluate the existing available resources. Compare the pros and cons of using existing resources, obtaining new resources, or contracting out the service. If the analysis determines that outsourcing is not justified, then analyze the systems and technologies available that can optimize the in-house operation. If the decision is to outsource, begin by shopping among multiple vendors. Always remember that portions of the organization may be outsourced, while others could be retained in-house. In fact, this is one method of retaining your key people should the outsourcing deal go sour or terminate unexpectedly.

The requirements-gathering process is a critical step before outsourcing can begin. Compare the two major scenarios — *private network* versus *virtual private network over a public network* configurations — before making the decision to outsource. Cost factors, of course, are a major consideration. During the planning stages, a balance must be maintained between designing a network to accommodate internal applications and designing the internal applications to accommodate the network. Network applications and the network itself often grow together after the network is built, but always remember that the network is in place to serve the users and applications. The capacity-planning process is critical throughout the life cycle of the network. Applications are analyzed on an individual and aggregate basis to determine the best network solution. The network design engineers will have various levels of knowledge at their disposal, ranging from projected traffic bandwidths between sites to a complete protocol profile for each application.

3.3.2 Outsourcing Considerations

The decision to outsource is often one of business policy based on many factors, including:

* Corporate resource availability
* What is and what is not your core competency
* Sensitivity of the data
* Return on investment (ROI) analysis
* Skill set and reliability of the vendor
* Cost factors of either owning the business elements or leasing them from the outsourcer
* Retention of control
* Business charter of the company
* Deductible expense versus capital investment

Other factors, such as how much support is required for how long, contract stipulations between user and vendor, and the loyalty to existing company employees, also play a major role in the outsourcing decision. Many business aspects should be considered for possible impact by outsourcing, including:

◊ Resources (staff and existing investments)
◊ Questions on skill sets and reliability of the vendors performing the outsourcing function
◊ Cost savings or eventual loss
◊ Control of the network
◊ Network monitoring and management
◊ Future ability to either continue outsourcing or to bring the business back in-house

As with any business case analysis, all expenses of an outsourcing or outtasking deal should be analyzed in detail for proof of the validity of cost savings versus expenditures. This comes during the process of determining what needs to be outsourced. A business considering outsourcing must consider the following:

- Understand what is your core competency and what needs to be out-sourced
- Compare current employee skill sets with those of the outsourcing company
- Plan what to do with current resources (most importantly people!)
- Choose a reliable outsourcing company
- Understand the monetary impacts before and after outsourcing
- Clearly define levels of control and methods of regaining it should that be required
- Define the extent of outsourcing, for example, protocols, interfaces, and locations
- Determine the duration of the outsourcing contract, ideally at least three to five years
- Fully understand the vendor-user relationship
- Maintain the loyalty of the retained staff, and ensure loyalty of the new staff
- Do not preannounce your intentions until the outsourcing or out-tasking deal is complete

If the ability to manage or control the network internally is lost, and the external vendor source fails to perform, how difficult is it to rebuild the internal networking department? The loss of skilled people may be difficult and costly to regain. When outsourcing is chosen as the alternative, a strategic plan must be implemented with contingencies for each of these possible scenarios. Turning over the network to the vendor in increments until the vendor's capabilities can be judged, called "selective outsourcing", is one answer. Regardless of which solution is chosen, a clear-cut plan must be in place vis-à-vis the vendor for a minimum of two years. Define the vendor's plan for updating technology and workforce. Make sure the outsourcing vendor is able to adapt to your company's business needs, as well as their own, to maintain your future growth and competitiveness. Alternate outsourcing vendors must also remain an option.

3.3.3 Summary of Benefits and Drawbacks

The primary *benefits* of outsourcing are:

- + Uses vendor experience, specialists, and scale
- + Reduces costs
- + Enables you to focus on your core business rather than on build-ing and running networks
- + Taps a good source of quick network resources

  + Augments existing workforce with skilled workers
  + Combines computing and communications departments into single support structure

The *drawbacks* of outsourcing may include:

  – Loss of control
  – Possible loss of resources if networking operations are retained
  – Possible sacrifice of technology flexibility
  – Risk of impact to critical systems if vendor fails

These drawbacks can be countered with good planning, smart management, and proper choice of vendor(s). Unfortunately, these factors are not always controllable.

3.3.4 The Future of Outsourcing

The market for outsourcing continues to grow. Outsourcing and outtasking options are provided by service providers, system integrators, and network management companies. The complexities of outsourcing grow as users move from bridged and multiplexed to routed environments, and from routed networks to combination routed and switched networks. Also, as LANs continue to proliferate, the number of disparate protocols increases drastically, and all of them require interconnectivity. Outsourcing is clearly here to stay, and it is having a major impact on the computer and communications industries. There are obviously many short-term benefits from outsourcing, but what about the long-term effects? Many people who disagree with outsourcing say that the long-term expenses outweigh the short-term gains. This underscores the need to first develop an accurate business case. Currently, most companies that outsource or outtask do so to focus on their core competencies (such as banking or manufacturing) and leave the designing and managing of the network to experts. Either way, outsourcing and outtasking are worth a look for any business.

3.4 REVIEW

This chapter discussed the move from private data networks to virtual private data networks offered by public providers, and the emergence of data internets and extranets reliant on seamless interworking and security. The text then looked at the data network of the next millennium, where ATM could play the role of the Rosetta Stone — unlocking protocol differences and revealing a common infrastructure for many legacy and new access and transport protocols as later chapters reveal. ATM presents some new tactics for use in the war on technological obsolescence. The chapter showed how enterprise-wide ATM implementations provide lower networking costs in the reduction of transmission and public network service costs. Finally, we provided insight into the new trends of outsourcing and outtasking.

3.5 REFERENCES

[1]   P. Ferguson, G. Huston, "What Is a VPN?,"
       http://www.employees.org/~ferguson/vpn.pdf, March 1998.
[2]   D. Spohn, *Data Network Design,* 2nd Edition, McGraw-Hill, 1997.

Digital Communications Network and Protocol Fundamentals

This part describes the topologies, technologies, protocols, and services used in modern digital communications networks. The following chapters cover these subjects as background since ATM borrows successful concepts from prior protocols proven in real world digital communication networks. Chapter 4 provides a comprehensive introduction to these topics by reviewing common network topologies and circuit types, including an in-depth explanation of the principles of multiplexing and switching. We then describe the major physical realizations of multiplexing and switching functions. Chapter 5 introduces the concepts of layered protocols. The text then introduces the layered protocol architectures that influenced the history of networking: IBM's Systems Network Architecture (SNA), the OSI Reference Model (OSIRM), the IEEE's 802.x LAN protocols and the Internet Protocol (IP) suite. Chapter 5 also introduces the key concepts of connection-oriented and connectionless network services. Chapter 6 then defines the basics of the digital transmission hierarchy as the foundation for the Integrated Services Digital Network (ISDN) reference model. Chapter 7 reviews the major connection-oriented services: X.25 packet switching and frame relay. Chapter 8 then surveys key characteristics of commonly used connectionless services, namely the Transmission Control Protocol/Internet Protocol (TCP/IP) suite and the Switched Multimegabit Data Service (SMDS). Finally, Chapter 9 provides a comprehensive summary of local area networks, bridging, and routing.

4

Networks, Circuits, Multiplexing, and Switching

This chapter provides an overview of common network topologies, circuits, and transmission types that form the basis for most network designs. We define the key characteristics of the links or circuits that connect nodes in these topologies, such as the direction of data flow, bit- or byte-oriented transmission, and the broadcast nature of the media. Next, the text reviews the several meanings of synchronous and asynchronous transmission, clarifying the role that Asynchronous Transfer Mode (ATM) plays in the world of networking. Coverage then moves to an explanation of multiplexing techniques — methods of combining and separating units of bandwidth. Finally, the discussion covers the four major methods of switching data: in space, in time, in frequency, by address, or by codes. We point out how that address switching is the foundation of packet switching and ATM. The reader will then have background to understand the concepts underlying common networking protocols, such as frame relay, IP, DQDB, and ATM that — as described in later chapters — are rooted in these multiplexing and switching techniques.

4.1 GENERAL NETWORK TOPOLOGIES

Physical topology defines the interconnection of physical *nodes* by physical transmission *links*. A *node* is a network element, such as an ATM switch, a router, or a multiplexer. We also refer to a node as a *device*, and to a link as a transmission path. A *link* represents a connection between two nodes, either physical or logical. Therefore, a link may be either a physical connection, such as a dedicated private line, or a logical, or virtual, connection such as a Permanent Virtual Connection (PVC).

Logical topology defines connections between two or more logical nodes (or simply interfaces) which may be of either a *point-to-point,* or *point-to-multipoint* configuration in ATM. Furthermore, each connection may be either *unidirectional* or *bidirectional.* A *leaf* is the terminating point of a unidirectional point-to-multipoint topology with originations at the *root.* A spatial point-to-multipoint connection has at most one leaf per physical port, while a logical point-to-multipoint connection may have multiple leaves on a single physical port. When all nodes have a point-to-multipoint connection, then a broadcast logical topology results. Figure 4.1 illustrates each of these logical topologies. Other technologies, such as Ethernet, support a broadcast medium where *all* other stations receive any one station's transmission. Additional protocols and configurations are required to support the broadcast logical topology. Part 3 describes how ATM standards support broadcast via LAN Emulation (LANE), IP Multicast over ATM, Multi-Protocol Label Switching (MPLS) using sets of point-to-multipoint connections.

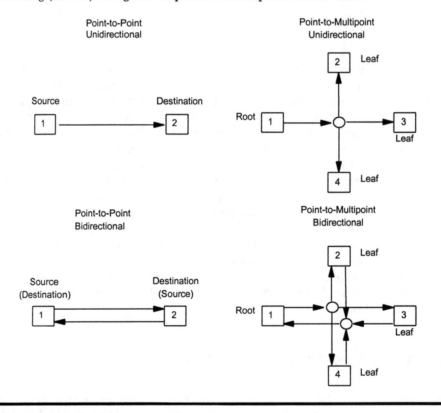

Figure 4.1 Conceptual Illustration of Logical Topologies

The most commonly used physical topologies for computer and data communications networks are: point-to-point, multipoint (or common bus), star, ring (or loop), and mesh. The text provides illustrated examples of each network topology.

4.1.1 Point-to-Point

The point-to-point topology is the simplest, comprised of a single connection between two nodes composed of one or more physical or logical circuits. Figure 4.2 shows three examples of a point-to-point topology. The first example shows a single physical circuit connecting Node A and Node B. The second example depicts a single physical circuit between Node A and Node B carrying multiple logical links. The third example depicts a single connection path between Node A and Node B with multiple physical circuits, each carrying multiple logical links. Typically, network designers employ this configuration when the separate physical circuits traverse diverse routes, in which case any single physical link or circuit failure would not completely disconnect nodes A and B.

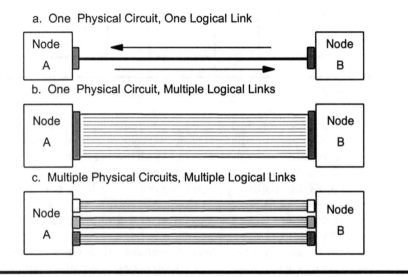

Figure 4.2   Point-to-Point Topology Examples

Point-to-point topologies are the most common method of connectivity in Metropolitan and Wide Area Networks (MANs and WANs). User access to most MAN or WAN network services has some form of point-to-point topology. Examples of the point-to-point topology are private lines, circuit switching, and dedicated or dial-up access lines to packet switched services, frame relay, and ATM.

4.1.2 Multipoint and Broadcast

A common realization of the multipoint topology is a network where all nodes physically connect to (and logically share) a common broadcast medium. Figure 4.3 shows the multipoint topology, where Nodes A through F communicate via a shared physical medium. Sometimes the shared medium is also called a common bus. Most Local Area Networks (LANs) utilize a broadcast (or multipoint) topology. Indeed, the IEEE 802.4 Token Bus, the IEEE 802.3

Ethernet, and the IEEE 802.6 Distributed Queue Dual Bus (DQDB) protocols define different means of logically sharing access to the common physical medium topology, as do other proprietary vendor architectures. Radio and satellite networks also implicitly employ the broadcast topology due to the inherent nature of electromagnetic signal propagation.

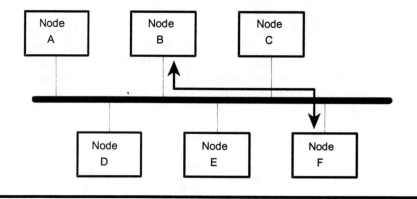

Figure 4.3 Shared Medium (Common Bus) Multipoint (Broadcast) Topology

A multidrop analog line is commonly used for legacy SNA SDLC loop access. In this example, an analog signal is broadcast from a master station (usually a mainframe front end processor) to all slave stations. In the return direction the slaves share the common broadcast medium of the multidrop line. The SNA SDLC polling protocol involves the host polling the slave stations in a round-robin manner, thus preventing any two slaves from transmitting at the same time. See References 1 and 2 for more information on the SNA SDLC protocol.

Other networks, notably the Ethernet protocol, also work on a broadcast medium, but don't provide for orderly coordination for transmission like the SNA SDLC loop does. Instead, these protocols empower stations to transmit whenever they need to as long a another station isn't already sending data. When a collision does occur, a distributed algorithm uses the bandwidth at approximately 50 percent efficiency. Chapter 9 covers the Ethernet and related local area networking protocols.

Figure 4.4 illustrates other conceptual examples of the multipoint topology. Another commonly used multipoint topology is that of broadcast, or multipoint-to-multipoint, which is the case where many other nodes receive one sender's data. Yet another example is that of "incast," or multipoint-to-point, where multiple senders' signals are received at one destination — as in a slave-to-master direction. In this conceptual illustration note that the multipoint-multipoint (i.e., shared medium, or multicast) topology is effectively the combination a full-mesh of multipoint-point topology connections for each of the four nodes. The figure also illustrates emulation of a point-to-multipoint topology via multiple point-to-point links for comparison purposes.

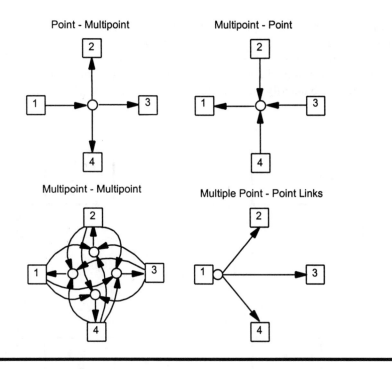

Figure 4.4   Conceptual Illustration of Multipoint Topologies

4.1.3 Star

The star topology developed during the era when mainframes centrally controlled most computer communications. The voice switching world also employs a star topology when multiple remote switching nodes, each serving hundreds to even thousands of telephone subscribers, home-in on a large central switch. This network radiates in a star-like fashion from the central switch through the remote switches to user devices. The central node performs the communication switching and multiplexing functions in the star topology. Nodes communicate with each other through point-to-point or multipoint links radiating from the central node. The difference between this topology and the multipoint topology is that the central node only provides point-to-point connections between any edge node, on either a physical or logically switched basis.

Figure 4.5 shows a star topology, where Node A serves as the center of the star and Nodes B through E communicate via connections switched to and through the central Node A. An example of a star topology is many remote terminal locations, or clients, accessing a centralized server through the central node as illustrated in the figure. The physical star topology is widely used to connect devices to a central hub in LANs, and thus is often called a "hub and spoke" topology. The central hub may logically organize the physical star as a logical bus or ring as is commonly done in LAN wiring hubs. A key benefit of the physical star topology is superior network management of the physical interfaces. For example, if a single interface fails in a physical star topology, then the management system can readily disable it without affecting any other stations. Conversely, in a broadcast topology, a single defective switch can take down the entire shared medium network. As we shall see, many wide area networks also have a star topology, driven basically by the client-server computing paradigm.

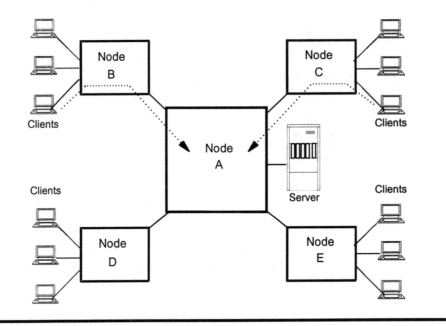

Figure 4.5   Illustration of a Network with a Star Topology

4.1.4 Ring

The ring topology utilizes a shared transmission medium which forms a closed loop. Such networks utilize protocols to share the medium and prevent information from circulating around the closed physical transmission circuit indefinitely. A ring is established, and each device passes information in one direction around the ring.

Figure 4.6 shows a ring network where in step 1 Node A passes information addressed around the ring through Node D in step 2. Node C removes this frame from the ring and then returns a confirmation addressed to Node A in step 3 via Node B, at which point Node A removes this data from the ring in step 4. Note that actual LAN protocols use a somewhat more complicated protocol than that described here. Rings reuse capacity in this example because the destination removes the information from the ring so that other stations can utilize the ring bandwidth. Examples of the ring topology are the IEEE 802.5 Token Ring and the Fiber Distributed Data Interface (FDDI). Although the ring topology looks like a special case of a mesh network, it differs because of the switching action performed at each node. SONET protection rings also use the ring topology, and are also distinguished from a mesh by the difference in nodal switching action from that of a mesh of circuit switches.

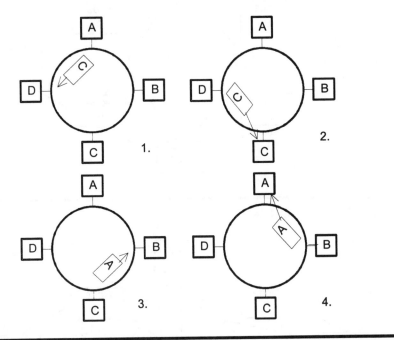

Figure 4.6   Ring or Loop Topology

4.1.5 Mesh

Many switched, bridged, and routed networks employ some form of mesh architecture. Mesh networks have many nodes which are connected by multiple links. If each node is directly connected to every other node, then the network is fully meshed, otherwise the network is only partially meshed. Figure 4.7 shows a partial mesh network where Nodes B, C, D, E, F, and G have a high degree of connectivity by virtue of having at least three links to any other node, while Nodes A and H have only two links to other nodes. Note that Nodes C and G have four links. The number of links connected to a node is that node's degree (of connectivity). For example, Node C has degree 4 while node H has degree 2.

Figure 4.8 shows a *full mesh* network where each node has a link to every other node. Almost every major computer and data communications network uses a partial mesh topology to give alternate routes for backup and traffic loads. Few use a full mesh topology, primarily because of cost factors associated with having a large number of links. This is because a full mesh N-node network has $N(N-1)/2$ links, which is on the order of N^2 for large values of N. For N greater than 4 to 8 nodes, most real world networks employ partial mesh connectivity.

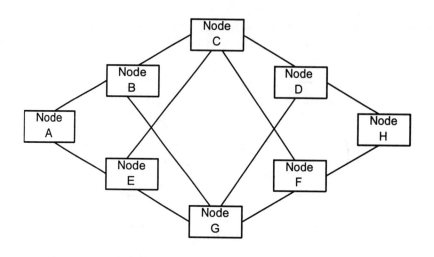

Figure 4.7 Partial Mesh Network

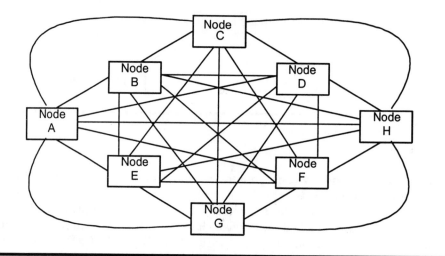

Figure 4.8 Full Mesh Network

4.2 DATA COMMUNICATIONS AND PRIVATE LINES

This section takes a detailed look at the characteristics of connections used in real networks. First, the text introduces the notion of simplex, duplex, and half-duplex communications. The treatment then introduces the concepts of data terminal and communications equipment. We then put these concepts together and apply them to private lines, which form the fundamental component of connectivity for most data communications, multiplexing, and switching architectures. Carriers offer private lines as tariffed services, or as access lines to other data communications services, such as frame relay, ISDN, and ATM.

4.2.1 Simplex, Half-Duplex, and Full-Duplex Transmission

Figure 4.9 illustrates the three types of transmission possible in data communications: simplex, half-duplex, or full-duplex [3].   In the simplex transmission of Figure 4.9*a*, one physical communications channel connects the terminal to the computer, in which case transmission occurs in only one direction.   Examples of simplex communication are a radio broadcast or an alarm system. In the half-duplex transmission example of Figure 4.9*b*, only one physical communications channel connects the terminal and computer; however, a protocol allows communication to occur in both directions, but not simultaneously. Half-duplex communication is also called two-way alternating communication.   An example of half-duplex operation is Citizen Band (CB) radio transmission where a user can either transmit or receive, but not both at the same time on the same channel.   Finally, in the full-duplex example of Figure 4.9*c*, two transmission channels connect the terminal and the computer. In full-duplex operation, communication can take place in both directions simultaneously.

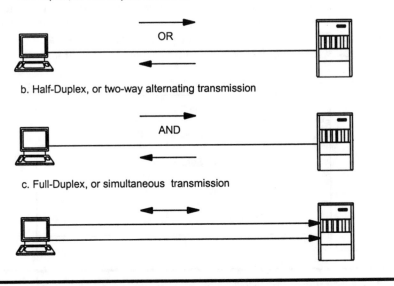

a. Simplex, or one-way transmission

b. Half-Duplex, or two-way alternating transmission

c. Full-Duplex, or simultaneous transmission

Figure 4.9   Illustration of Simplex, Half-Duplex, and Full-Duplex Transmission

4.2.2 DTE-to-DCE Connections

DTE-to-DCE connections provide a local, limited distance physical connection between Data Terminal Equipment (DTE) or Terminal Equipment (TE), such as a computer or PC, and Data Communications Equipment (DCE), such as a modem, designed to connect to a wide area network as illustrated in Figure 4.10. A DCE is equipment that provides functions required to establish, maintain and terminate a connection between a DTE and a wide area network. Typically, the local cabling between the DTE and the DCE is a multistrand cable, but may be coaxial, fiber optic, or twisted pair media. The DCE connects to an access circuit via twisted pair, coaxial cable, or fiber optic media. DTE to DCE and DCE to network communication is a particular example of a point-to-point topology. DTE to DCE communication can operate in any one of the transmission modes defined in the previous section: simple, half-duplex, or full-duplex.

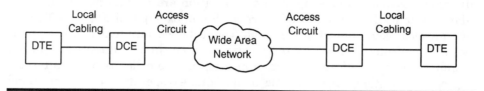

Figure 4.10   Illustration of DTE and DCE Networking

A DTE to DCE connection in a multistrand cable electrically conveys multiple interchange circuits of one of two types: balanced or unbalanced. What's the difference? One wire. Specifically, a balanced interchange circuit uses a pair of wires, while an unbalanced interchange circuit uses a single wire. The practical difference between these techniques is that unbalanced circuits require fewer wires in a cable, but are more susceptible to electrical interference. Hence, they operate over shorter distances than their balanced counterparts, which require twice as many wires in the cable.

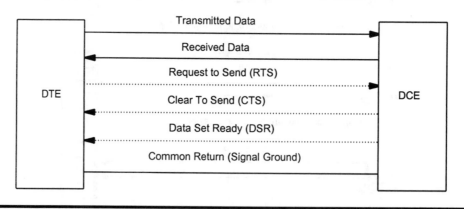

Figure 4.11   Example of DTE-DCE Connection using the RS-232 Interface

A multistrand cable connecting a DTE to a DCE carries multiple interchange circuits, some of which carry data and others which carry control information as shown in the example of Figure 4.11 for the commonly used signals in the RS-232 standard. The widely used RS-232 standard commonly used for serial port connections in personal computer applications employs unbalanced (one-wire) interchange circuits with a common ground as shown in the figure. Each of these circuits appears as a signal associated with a physical pin on the serial port connector. The control circuits (shown by dashed lines in the figure) allow the DTE and DCE to communicate status and request various services from each other. For example, the DCE uses signaling on the control leads between the DTE and DCE at the physical layer to control the direction of transmission in half-duplex communication. When a single pair of wires, a coaxial cable, or a fiber optic cable connects a DTE and DCE, special modulation techniques provide for a functional separation of data and control signals to yield the equivalent of separate interchange circuits for control and data.

4.2.3 Private Lines

A private line, or leased line, is a dedicated physical circuit leased from a carrier for a predetermined period of time, usually in increments of months. A private line may be upgraded by paying extra for a defined quality of service, such that special conditioning is performed to ensure that a better error rate is achieved — resulting in a tremendous improvement in data communications performance. As carriers install all-fiber networks, digital private lines are replacing the old voice-grade analog circuits, at lower prices.

When users purchase leased lines to gain access other services, such as frame relay and ATM, we call this application an *access line*. Users may lease access lines through either the local telephone company or, in an increasing number of locations, through alternative access providers. In some cases, end users own their access facilities; for example, through construction of a fiber optic ring. Generally, access from these alternate sources is less expensive than the local telephone company prices. But, of course, the alternative access carrier usually "cream-skims" the lucrative traffic, leaving the "skimmed milk" (smaller, lower volume, or more remote users) for the LEC to serve.

Private lines in Europe and Pacific Rim countries are still very expensive, and transoceanic fiber access remains limited. A carrier also must make an agreement with the party at the other side of a fiber to offer the transoceanic service. Prices are dropping, but they require significant investment for small amounts of bandwidth (which is often taken for granted in the United States). The high cost of international private lines justifies the expense of sophisticated, statistical multiplexers to utilize the expensive bandwidth as efficiently as possible. ATM offers one alternative to achieve these efficiencies.

Another form of special-purpose private line operating over a four-wire circuit is the High-rate Digital Subscriber Line (HDSL) [4]. HDSLs eliminate the cost of repeaters every 2000 ft as in a standard T1 repeater system, and are not affected by bridge taps (i.e., splices). They need to be within 12,000 ft of the serving central office, which covers over 80 percent of the DS1 customers in the United States. Asymmetric Digital Subscriber Lines (ADSLs) are also becoming available and offer higher speeds and better performance. The goal of the ADSL technology is to deliver a video signal and telephone service over a majority of the existing copper, twisted pairs currently connected to small and medium businesses as well as homes. Part 8 covers the topic of ADSL along with other future roles for ATM.

4.3 DATA TRANSMISSION METHODS

A digital data transmission method is often characterized as being either asynchronous or synchronous. The terms asynchronous and synchronous are used in different contexts and, unfortunately, have entirely different meanings. We first describe definitions different than the ones used in this book; commonly known as asynchronous and synchronous character message transmission [3]. The meaning used in this book is that of Synchronous versus Asynchronous Transfer Mode (STM and ATM). These two entirely different meanings of the same two terms can be confusing. This section presents them together so that you can appreciate the differences and understand the context of each. The traditional meaning of the terms asynchronous and synchronous apply at the character or message level. On the other hand, the same adjectives used in the STM and ATM acronyms define different transmission system paradigms for carrying characters and messages.

4.3.1 Asynchronous and Synchronous Data Transmission

Asynchronous character transmission has no clock either in or associated with the transmitted digital data stream. Instead of a clock signal, start and stop bits delimit characters transmitted as a series of bits numbered 1 through 8 as illustrated in the example of Figure 4.12. There may be a variable amount of time between characters. Analog modem communication employs this method extensively. Chapter 24 explains the notion of baud rate, which is the number of discrete signals transmitted on the communication channel per unit time.

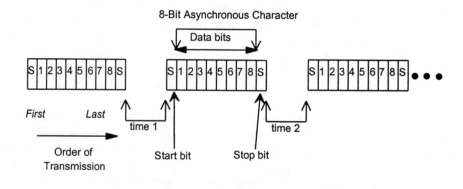

Figure 4.12 Asynchronous Character Transmission

Asynchronous character transmission usually operates at lower speeds ranging from 300 bps up to 28-kbps. Asynchronous interfaces include RS232-C and D, as well as X.21.

On the other hand, synchronous data transmission clocks the bits at a regular rate by a clocking signal either associated with, or derived from, the transmitted digital data stream. The motivation for synchronous signaling is to eliminate the extra time required to send the start and stop bits in asynchronous transmission. Since the start and stop bits are at least one unit time long, they comprise at least 20 percent of the line transmission rate. Therefore, in synchronous transmission, the sender and receiver must have a means to derive a clock within a certain frequency tolerance.

Figure 4.13 shows a typical synchronous data stream from the Binary Synchronous Communications (BSC) protocol employed by IBM in the 1960s for the System 360 [2]. The message begins with a PAD character followed by two synchronization (SYNC) characters and a start-of-header (SOH) character. The header supports functions such as addressing and device control. A Start of Text (STX) character then precedes the textual (data). Textual data cannot contain any control characters in BSC, unless another character — Data Link Escape (DLE) — is sent prior to the STX character. If the data contains a DLE character, then the transmitter inserts an additional DLE character. The receiver strips one DLE character for each pair of DLE characters received. This DLE stuffing operation allows users to send transparent data containing other control characters. The End of Text (ETX) character delimits the message. The Block Control Check performs a simple parity check to detect errors in the data characters. For various reasons, other data link control methods eventually replaced the BSC protocol.

PAD	SYN	SYN	SOH	Header	STX	Data	ETX	BCC

SYN = Synchronization   ETX = End of Text
SOH = Start of Header   BCC = Block Control Check
STX = Start of Text   PAD = Padding Character

Figure 4.13   IBM's Binary Synchronous Communications (BSC) Message Format

On a parallel DTE-to-DCE interface a separate clock interchange circuit (or connector pin) conveys the synchronous timing. Synchronous data interfaces include V.35, RS449/RS-442 balanced, RS232-C, RS232-D, HSSI and X.21. Synchronous data transmission usually runs at higher speeds. For example, the High Speed Serial Interface (HSSI) operates at speeds up to 51.84-Mbps. Also, the Synchronous Optical NETwork (SONET) operates at speeds of up to 40 Gbps.

4.3.2 Asynchronous versus Synchronous Transfer Mode

This section introduces the meaning of the adjectives synchronous and asynchronous applied to transfer modes used in this book. Fundamentally, a transfer mode defines the means for conveying sequences of bits between multiple sources and destinations over a single digital transmission system. In effect, a transfer mode is a means for multiplexing multiple data streams onto a single higher speed bit stream, and then sorting it all out at the destination. Chapter 6 describes Synchronous Transfer Mode (STM), or synchronous time division multiplexing, in detail. Asynchronous Transfer Mode (ATM), or asynchronous time division multiplexing, is a different concept with roots in packet switching (covered in detail in Part 3). The following example provides a high-level introduction to the basic difference between the STM and ATM multiplexing methods.

Figure 4.14 shows an example of STM and ATM. Figure 4.14*a* illustrates an STM stream where each time slot represents a reserved piece of bandwidth dedicated to a single channel, such as a DS0 in a DS1. Each frame contains n dedicated time slots per frame; for example, n is 24 8-bit time slots in a DS1. Overhead fields identify STM frames that often contain operations information as well. For example, the 193rd bit in a DS1 delimits the STM frame. Thus, if a channel is not transmitting data, the bits in the time slot remain reserved without conveying any useful information. If other channels have data to transmit, they must wait until their reserved, assigned time slot occurs in turn again. If time slots are frequently empty, then STM results in low utilization.

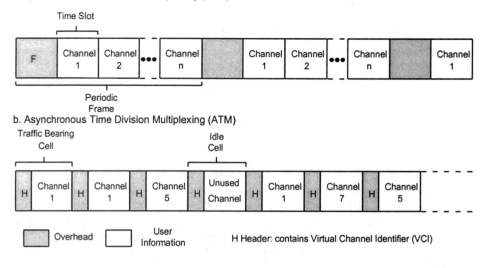

a. Synchronous Time Division Multiplexing (STM)

b. Asynchronous Time Division Multiplexing (ATM)

Figure 4.14 Illustration of STM and ATM Multiplexing

ATM uses a completely different approach. Figure 4.14*b* illustrates the concept. A header field prefixes each fixed-length payload channel, identifying the virtual channel. The combination of the header and payload is a *cell*. The time slots (or cells) are available to *any user* who has data ready to transmit. The traffic need not wait for its next reserved time slot as in STM. If no users are ready to transmit, then ATM sends an empty, or idle, cell. Traffic patterns that are not continuous are usually carried much more efficiently by ATM as compared with STM. The current approach is to carry ATM cells over very high-speed STM transmission networks, such as SONET and SDH. As we shall see, the match between the high transmission speeds of SONET and SDH and the flexibility of ATM is a good one.

4.4 PRINCIPLES OF MULTIPLEXING AND SWITCHING

A close family relationship binds the concepts of multiplexing and switching. Multiplexing defines the means by which multiple streams of information share a common physical transmission medium. Switching, on the other hand, takes information from an input multiplexed information stream and directs this information to other outputs. In other words, a switch takes information from a particular physical link in a specific multiplexing position and connects it to another output physical link, usually in a different multiplexing position. Multiplexing positions are defined by space, time, frequency, address, or code. Since a switch is basically an interconnected network of multiplexers, this section first reviews basic multiplexing methods, and then covers point-to-point and point-to-multipoint switching functions.

4.4.1 Multiplexing Methods Summarized

There are five basic multiplexing methods: space, frequency, time, address, and code. This sequence is also the historical order in which communications networks employed these techniques. Space, frequency, and time division multiplexing all occur at the a physical level. Address switching, or label swapping, and code division multiplexing occur at a logical level. Address switching is the foundation of packet switching, frame relaying, and ATM cell switching.

4.4.2 Space Division Multiplexing (SDM)

An example of space division multiplexing is where multiple, physically separate cables interconnect two pieces of equipment. "Space" implies that there is physical diversity between each channel. The original telephone networks, where a pair of wires connected each end user to communicate, is an example of one of the first uses of space division multiplexing. This approach quickly becomes impractical, as evidenced by old photographs of the sky of major metropolitan cities blackened out by large number of wire pairs strung overhead using space division multiplexing. Early data communications ran a separate cable from every terminal back to the main computer, which is another example of space division multiplexing. When there were only a small number of terminals, this was not too much of a burden. Obviously, since each interconnection requires a separate physical cable, SDM does not scale well to large networks.

4.4.3 Frequency Division Multiplexing (FDM)

As transmission technology matured, engineers discovered how to multiplex many analog conversations onto the same cable, or radio spectrum by modulating each signal by a carrier frequency. Modulation translated the frequency spectrum of the baseband voice signal into a large number of distinct frequency bands. This yielded a marked increase of efficiency and worked reasonably well for analog signals. However, FDM relied on analog electronics that suffered from problems of noise, distortion, and interference between channels that complicated data communications. FDM also made a brief foray into LANs with Wang Laboratories Wang Net. This technology required constant tweaking and maintenance, eventually losing out to Ethernet, Token Ring, and FDDI.

4.4.4 Time Division Multiplexing (TDM)

The next major innovation in multiplexing was motivated by the need in the late 1950s to further increase the multiplexing efficiency in crowded bundles of cables in large cities. This entirely digital technique made use of emerging solid-state electronics. TDM first converts analog voice information to digital information prior to transmission using Pulse Code Modulation (PCM). Although this technique was relatively expensive, it did cost less than replacing existing cables or digging larger tunnels in New York City. Since then, TDM has become the prevalent multiplexing method in all modern telecommunication networks. We now take for granted the fact that the network converts every voice conversation to digital data, transmits it an arbitrary distance, and then converts the digits back to an audible analog signal. The consequence is that the quality of a voice call carried by digital TDM is now essentially independent of distance. This performance results from digital repeaters which decode and retransmit the digital signal at periodic intervals to achieve extremely accurate data transfer. Data communications is more sensitive to noise and errors than digitized voice, but reaps tremendous benefits from the deployment of TDM infrastructure in public networks. In theory TDM may also be applied to analog signals; however, this application was never widely used.

4.4.5 Address or Label Multiplexing

Address, or label, multiplexing was first invented in the era of poor-quality FDM analog transmission. A more common name for address multiplexing is Asynchronous Time Division Multiplexing (ATDM), of which an example appears later in this chapter. Transmission was expensive, and there was a need to share it among many data users. Each "packet" of information was prefixed by an address that each node interpreted. Each node decided whether the packet was received correctly, and if not, arranged to have it resent by the prior node until it was received correctly. SNA, DECNET, and X.25 are early examples of address multiplexing and switching. More recent examples are frame relay, ATM, and the IETF's MultiProtocol Label Swapping (MPLS) standards work. The remainder of this book covers the address multiplexing method in great detail.

4.4.6 Code Division Multiple Access (CDMA)

Code division multiplexing, also called spread spectrum communications, utilizes a unique method to allow multiple users to share the same broadcast communication medium. Also called Code Division Multiple Access (CDMA), this technique works well in environments with high levels of interference. CDMA transmissions are also difficult to detect since the energy transmitted is spread out over a wide frequency passband. This means that others cannot detect the transmission easily, or else the interference between transmitters stays at a very low level.

4.4.7 Point-to-Point Switching Functions

Figure 4.15 illustrates the four basic kinds of point-to-point connection functions that can be performed by a multiplexer or switch.

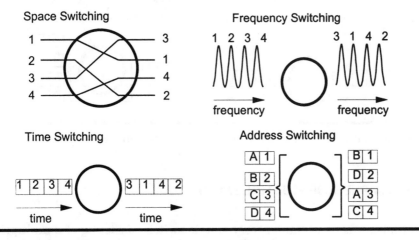

Figure 4.15   Point-to-Point Switching Function Definitions

Space division switching delivers a signal from one physical (i.e., spatial) interface to another physical interface. One example is a copper crosspoint switch. Time division switching changes the order of time slots within a single spatial data stream, organized by the Time Division Multiplexing (TDM) method. Frequency (or wavelength) switching translates signals from one carrier frequency (wavelength) to another. Wavelength Division Multiplexing (WDM) in optical fiber transmission systems uses this method. Finally, address switching changes the address field in data packets, which may be further multiplexed into spatial, time, or frequency signals. This book focuses on this switching method, as applied to packet, frame, and cell switching.

4.4.8 Point-to-Multipoint Switching Functions

Figure 4.16 illustrates the extension of switching from the case of point-to-point to the broadcast, or point-to-multipoint case. A space division broadcast switch replicates a single input signal on two or more outputs. A simple example is a coaxial television signal splitter that delivers the same signal to multiple outputs. TDM broadcast switching fills multiple output time slots with the data from the same input. FDM broadcast switching replicates the same signal on multiple output carrier frequencies. Address broadcast switching fills multiple packets with different addresses with identical information from the same input packet.

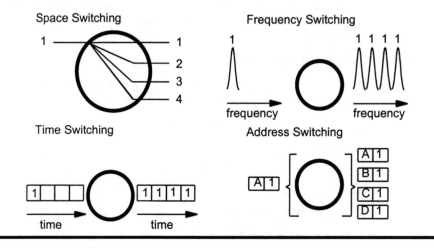

Figure 4.16 Point-to-Multipoint Switching Function Definitions

4.5 EXAMPLES OF MULTIPLEXING

A multiplexer is essentially a very simple switch consisting of a multiplexing function and a demultiplexing function connecting a single trunk port in the network to many access ports connected to individual traffic sources as illustrated in Figure 4.17. The parallelogram symbol with the small end on the side of the single output (called the trunk side) and the large end on the side with multiple interfaces (called the access side) frequently denotes a multiplexer in block diagrams. The symbol graphically illustrates the many-to-one relationship from the many ports on the access side to the smaller number of ports on the trunk side, as well as the one-to-many relationship from the trunk side to the access side.

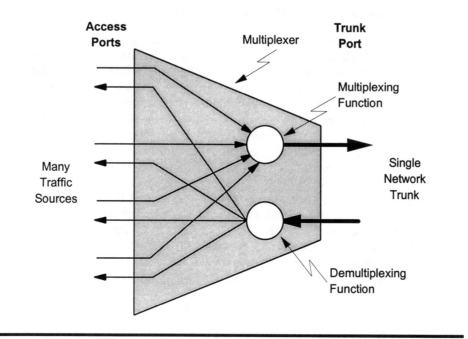

Figure 4.17   Switching Model of a Multiplexer

The multiplexing function shares the single output between many inputs. The demultiplexing function has one input from the network which it distributes to many access outputs. The multiplexing and demultiplexing functions can be implemented by any of the generic switching functions described in the previous section. Usually, the same method is used for both the multiplexing and demultiplexing functions so that the multiplexing method used on each of the interfaces is symmetrical in each direction. Generally, the overall speed or capacity of each port on the access side is less than that on the trunk side. For example, different levels in the Time Division Multiplex (TDM) hierarchy operate at increasingly higher speeds by aggregrating multiple lower-speed TDM signals together. We give more detailed examples for each of the generic methods described in the previous section.

Multiplexers share a physical medium between multiple users at two different sites over a private line with each pair of users requiring some or all of the bandwidth at any given time. Many simple multiplexers statically assigned a fixed amount of capacity to each user. Other multiplexing methods statistically assign bandwidth to users based upon demand to make more efficient use of the transmission facilities which interface to the network. You'll see these called statistical multiplexers in the technical literature. TDM is often used to reduce the effective cost of a private access line or international private line by combining multiple lower-speed users over a single higher-speed facility.

4.5.1 Frequency Division Multiplexing (FDM)

Analog telephone networks made extensive use of Frequency Division Multiplexing (FDM) to aggregate multiple voice channels into larger circuit groups for efficient transport. FDM multiplexes 12 voice-grade, full-duplex channels into a single 48-kHz bandwidth group by translating each voiceband signal's carrier frequency. These groups are then further multiplexed into a mastergroup made up of 24 groups. Multiple mastergroup analog voice signals are then transmitted over analog microwave systems. A lower-frequency analog microwave spectrum was used to frequency division multiplex a DS1 digital data stream in a technique called Data Under Voice (DUV).

Wavelength Division Multiplexing (WDM) on optical fibers is analogous to FDM in coaxial cable and microwave systems. Optical fiber is *transparent* in two windows centered around the wavelengths of 1300 and 1550 nm (10^{-9} m) as shown in the plot of loss versus wavelength in Figure 4.18 [5]. The total bandwidth in these two windows exceeds 30,000 GHz. Assuming 1 bps per Hertz (Hz) would result in a potential bandwidth of over 30 *trillion* bps per fiber!

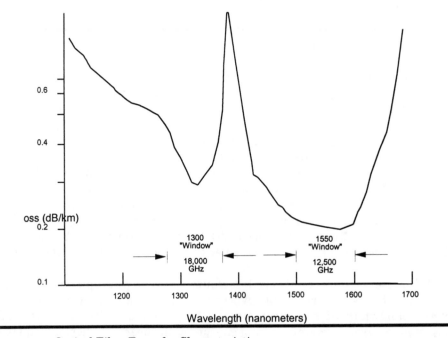

Figure 4.18   Optical Fiber Transfer Characteristic

Recall the basic relationship from college physics $\lambda v = c$; where λ is the wavelength in billionths of a meter (i.e., a nanometer or nm), v is the frequency in billions of cycles per second (gigahertz or GHz), and c is the speed of light in a vacuum (3×10^8 m/s). Applying this formula, the carrier frequency ν at the center of the 1300-nm window is 2300 GHz and 1900 GHz in the 1550-nm window. The available spectrum for signal transmission is 18,000 GHz in the 1300-nm window and 12,500 GHz in the 1550-nm window as shown in the figure. Chapter 24 defines the concepts of optical signals and their frequency spectra. The sharp attenuation peak at 1400-nm is due to residual amounts of water (an OH radical) still present in the glass. Continuing improvements in optical fiber manufacturing will likely make even more optical bandwidth accessible in the future. Commercial long-haul fiber optic transmission is now using between two and eight wavelengths per fiber, in what is called wideband WDM, in these two windows. Implementations of narrowband WDM, supporting up to 100 optical carriers on the same fiber, will likely be deployed prior to the year 2000.

4.5.2 Time Division Multiplexing (TDM)

Time Division Multiplexing (TDM) was originally developed in the public telephone network in the 1950s to reduce costs in metropolitan area networks. It also eliminated FDM filtering and noise problems when multiplexing many signals onto the same transmission medium. In the early 1980s, TDM networks using smart multiplexers began to appear in some private data networks, forming the primary method to share costly data transmission facilities among users. In the last decade, time division multiplexers have matured to form the basis of many corporate data transport networks. The premier example of TDM is DS1 and E1 multiplexing; Chapter 6 describes this for the ISDN Primary Rate Interface (PRI).

4.5.3 Address Multiplexing

A widely used example of address (or label) multiplexing is found in statistical multiplexers. Statistical multiplexing, also called Statistical Time Division Multiplexing (STDM), or Asynchronous Time Division Multiplexing (ATDMs), operates similarly to TDM, except it dynamically assigns the available time slots only to users who need data transmission. Gains of up to 2:1 are achieved for voice transmission by utilizing all available time slots, rather than wasting them on users who are not speaking. Higher or lower statistical multiplex gains can be obtained for data traffic depending upon the burstiness (peak-to-average statistics) of the data traffic. Part 6 covers these tradeoffs in detail. The net effect is a potential increase in overall throughput for users since time slots are not "reserved" or dedicated to individual users — thus dynamic allocation of the unused bandwidth achieves higher overall throughput. Figure 4.19 shows an example of a statistical multiplexer which takes multiple low-speed synchronous user inputs for aggregation into a single 56-kbps synchronous bit stream for transmission. The methods used to interleave the various channels in statistical multiplexers include bit-oriented, character-oriented, packet-oriented, and cell-oriented techniques, each requiring buffering and more overhead and intelligence than basic time division multiplexing.

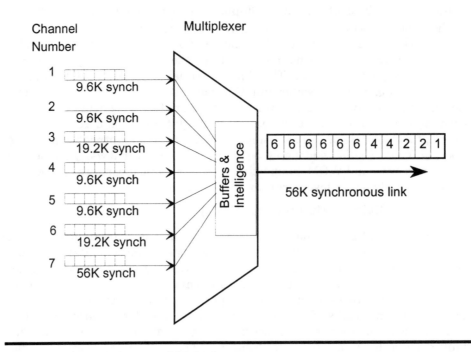

Figure 4.19 Sub-Rate Statistical Multiplexer

Figure 4.19 shows an excerpt from the output of a statistical multiplexed data stream. In a statistical multiplexer, the output bandwidth is *less* than the aggregate input bandwidth. This is done by design, assuming that not all input channels will be transmitting at the same time when each channel is sampled for transmission. Thus, the output synchronous data stream allocates bandwidth only to users who require it. It does not waste time slots by dedicating bandwidth to users who do not require it at the moment. Note in the example that channels 1, 2, 4, and 6 are transmitting, together utilizing 48-kbps of the available 56-kbps trunk bandwidth. Using the same example, if channels 3 (19.2-kbps) and 7 (56-kbps) were to also transmit data at the same instant, the total peak transmission rate of 123.2-kbps exceeds the 56-kbps trunk speed of the multiplexer. The statistical multiplexer overcomes brief intervals of such demand by buffering information and transmitting it when other sources fall idle.

4.5.4 Space Division Multiplexing

Space division multiplexing essentially reduces to the discipline of cable management. This can be facilitated by mechanical patch panels, or increasingly so by automatically controlled optical and electronic patch panels. To a large extent, space division multiplexing is falling out of favor; space division switching or other more efficient types of multiplexing typically replace multiple parallel cables in many network designs.

4.5.5 Code Division Multiplexing

In code division multiplexing, each user bit is modulated by a high-rate "chipping" signal. Each user transmits a unique pseudo-random coded signal at the chip rate for each user data bit. The pseudo-random sequences are chosen so that they are easily generated in hardware. The sequences of ones and zeros cancel each other out for all except the transmitting user. The most commonly used pseudo-noise sequences are generated by maximal length shift registers [6]. These particular sequences have the useful property in that each unique user code is a cyclic shift of another sequence.

Code division multiplexing performs well on channels with high levels of interference, either generated by nature, malicious jammers, or just the interference resulting from simultaneous transmissions by multiple users. Hence, satellite communications, military communications, wireless networks and Personal Communication Systems (PCS) telephones employ code division multiplexing. Some wireless LAN and wireless ATM systems employ code division multiplexing.

4.6 EXAMPLES OF SWITCHING

This section gives an example for each of the major switching techniques: space, time, address, and frequency. The examples chosen define terminology and illustrate concepts as background for material in subsequent chapters. Furthermore, as we shall see fundamental limits of the technique limit the maximum size and highest port speeds of devices built using the space, time, frequency and address multiplexing techniques.

4.6.1 Space Division Switching

Figure 4.20 illustrates a simple two-input, two-output crossbar network, using the crosspoint nodal function. An example connection is shown by the boldface lines and control inputs. Notice that this 2 by 2 switch matrix requires a total of four switching elements. Classical space division switch fabrics are built from electromechanical and electronic elements to provide the crosspoint function. Electrical cross connect switches operate at speeds of up to several Gbps by running multiple circuit traces in parallel.

Future technologies involving optical crosspoint elements with either electronic or optical control are being researched and developed. Optical switches generally have higher port speeds, but probably smaller total port capacities than their electrical counterparts. Electrical and optical space division switches currently scale to capacities on the order of 1 trillion bits per second (Tbps).

Examples of space division switches are: matrix switches, supercomputer High Performance Parallel Interface (HPPI) switches, and 3/3 digital cross connects. Many space division switches employ multiple stages of crosspoint networks to yield larger switch sizes.

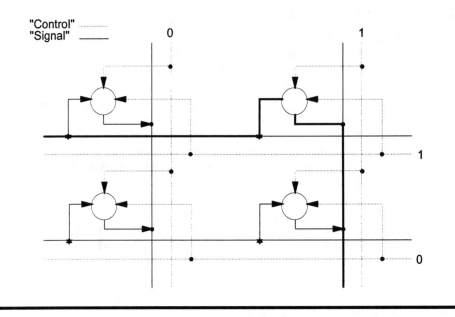

Figure 4.20 Two-Input, Two-Output Crossbar Network

4.6.2 Time Division Switching

The operation of current digital telephone switches may be viewed as an interconnected network of special purpose computers called Time Division Switches (TDS) [7]. Recall that a switch is basically a set of interconnected multiplexers, hence time division switches use Time Division Multiplexing (TDM) as the interface protocol. Our description of time division switching operation references Figure 4.21. Each TDM frame has M time slots. The input time slot m, labeled $I(m)$, is stored in the input sample array $x(t)$ in position m. The output address memory $y(t)$ is scanned sequentially by increasing t from 1 to M each frame time. The contents of the address array $y(t)$ identify the index into the input time slot array x that is to be output during time slot t on the output line. In the example of Figure 4.21, $y(n)$ has the value m, which causes input time slot m to be switched to output time slot n. Note that the input sample array must be double buffered in an actual implementation so that time slot phase can be maintained for inputs and outputs with different frame clock phases.

This TDS function is performed for M time slots, that is, once every frame time. This must occur in less than $\tau = 125$ μs (1/8000) for all slots, $n=1,...,M$. The maximum TDS size is therefore determined by the TDS execution rate, I instructions per second (or equivalently I^{-1} seconds per instruction); then the TDS switch size M must satisfy the inequality

$$M \leq \tau I$$

The TDS is effectively a very special purpose computer designed to operate at very high speeds. For I ranging from 100 to 1000 MIPs, the maximum TDS switch size M ranges from 12,500 to 125,000 which is the range of modern single-stage Time Division Switches (TDS). Larger time division switches can be constructed by interconnecting TDS switches via multiple stage crosspoint type networks [7].

Usually, some time slots are reserved in the input frame in order to be able to update the output address memory. In this way, the update rate of the switch is limited by the usage of some slots for scheduling overhead.

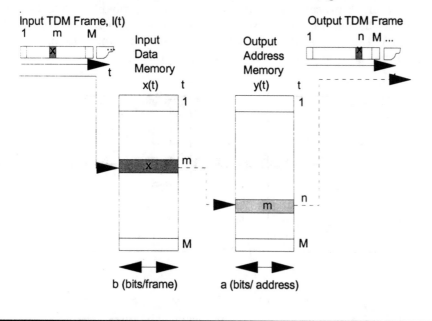

Figure 4.21 Illustration of Time Division Switch Operation

4.6.3 Address Switching

Address switching operates on a data stream in which the data is organized into packets, each with a header and a payload. The header contains address information used in switching decisions at each node to progress the packet towards the destination on a hop-by-hop basis. The address determines which physical output the packet is directed to, along with any translation of the header address. All possible connection topologies can be implemented within this switching architecture: point-point, point-to-multipoint, multipoint-to-point, and multipoint-to-multipoint. We illustrate these topologies in the following example.

Figure 4.22 illustrates four interconnected address switches, each with two inputs and two outputs. Packets (either fixed or variable in length) arrive at the inputs as shown on the left-hand side of the figure with addresses indicated by letters in the header symbolized by the white square prior to each shaded payload. The payload shading is carried through the switching operations from left to right to allow the reader to trace the switching result of the address switches visually. The input address indexes into a table using the column labeled In@, which identifies the address for use on output in the column Out@, and the physical output port on which the packet is sent in the column labeled Port. For example, the input packet addressed as *A* is output on port 1 using address *M*. Conceptually, each switch functions as a pair of busses which connect to the output port buffers. Each switch queues packets destined for a particular output port prior to transmission. This buffering reduces the probability of loss when contention occurs for the same output port. Chapter 25 presents an analysis of the loss probability due to buffer overflow for the main types of switch architectures used in real-world ATM switches. At the next switch the same process occurs until the packets are output on the right-hand side of the figure.

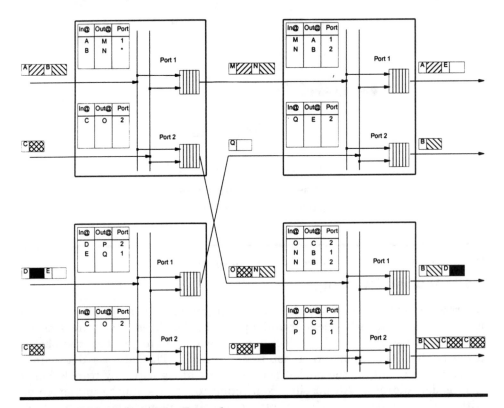

Figure 4.22 Address Switching Example

The packets with header addresses labeled A, D, and E form point-to-point connections. The packets labeled B form point-to-multipoint connections. The packets labeled C form multipoint-to-point connections. Currently, address switching operates at electrical link speeds of up to 2.4 Gbps for both ATM and packet switching systems. Of course, address switching and multiplexing are at the heart of ATM, which subsequent chapters cover in detail. Specifically, Chapter 16 discusses ATM switch fabric designs that scale to total capacities in excess of one trillion bits per second using combinations of address and space switching.

4.6.4 Frequency/Wavelength Switching

A significant amount of research has been conducted recently on all-optical networks [9], [8] using Wavelength Division Multiplexing (WDM). The basic concept is a shared medium, all-photonic network interconnecting a number of optical nodes or "end systems" as shown in Figure 4.23.

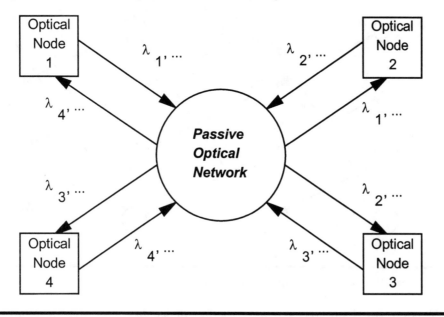

Figure 4. 23   Illustration of Optical WDM Network

The optical end system nodes transmit on at least one wavelength and receive on at least one wavelength. The wavelength for transmission and reception may be tunable, currently in a time frame on the order of milliseconds, with an objective of microseconds. The end systems may also be capable of receiving on more than one wavelength. The wavelengths indicated by the subscripts on the character λ are used in the next example of a multiple-hop optical network.

If the end system cannot receive all of the other wavelengths transmitted by other nodes, then the network must provide some means to provide full interconnectivity. One early method proposed and implemented was that of multiple-hop interconnections. In a multiple-hop system, each end system also performs a routing function. If an end system receives a packet that is not destined for it, it forwards it on its transmit wavelength. Eventually the packet reaches the destination, as shown in the trellis drawing of Figure 4.24. In this example, each node transmits and receives on only one wavelength.

For example, node 1 transmits on wavelength λ_1 and receives on λ_4. For example, in order for station 1 to transmit to station 4, it first sends on wavelength λ_1 which node 2 receives. Node 2 examines the packet header, determines that it is not the destination, and retransmits the packet on wavelength λ_2. Node 3 receives the packet, examines the packet header, and forwards it on λ_3, which the destination node 4 receives after taking three hops across the network.

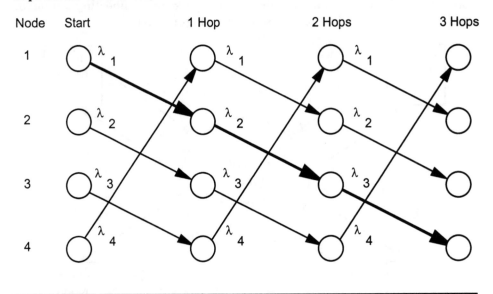

Figure 4.24   Illustration of Multiple-Hop WDM Network Trellis

This multiple-hop process makes inefficient use of the processing power of each node, especially when the number of nodes is large; therefore research focused on single-hop designs. In these designs the tunable transmitter and receiver are often employed. There is a need for some means to allocate and share the bandwidth in the optical network. Connection-oriented signaling has been tried, but the user feedback is that packet switching, and not circuit switching, is required because the connection setup time is unacceptable [8]. Fixed allocation of bandwidth in a time-slotted manner is also not desirable. Dynamic scheduling and collision avoidance hold promise for a solution to the demands of very high-speed networking [9].

4.7 REVIEW

This chapter began with a discussion of the five major network topologies: point-to-point, multipoint, star, ring, and mesh. The treatment then moved to a discussion of the relationship between DTE and DCE connections. The multiple uses of the terms asynchronous and synchronous in data communications was clarified. The text compared the commonly used definition of asynchronous data in modem communications with the meaning of asynchronous in ATM. We showed that ATM is a specialized form of address switching derived from packet switching described by several examples in this chapter. The major principles of multiplexing and switching were also introduced, followed by a number of examples illustrating the major digital communication technologies in use today.

4.8 REFERENCES

[1]   J. Ranade, G. Sackett, Introduction to SNA Networking, McGraw Hill, 1989.
[2]   R. Cypser, Communications Architecture for Distributed Systems, Addison-Wesley, 1978.
[3]   G. Held, R. Sarch, Data Communications, McGraw-Hill, 1995.
[4]   K. Miller, "A Reprieve from WAN's Long Last Mile," Data Communications, July 1993.
[5]   S. Personick, Fiber Optics Technology and Applications, Plenum, 1985.
[6]   J. Proakis, Digital Communications, McGraw-Hill, 1983.
[7]   B. Keiser, E. Strange, Digital Telephony and Network Integration, Van Nostrand Reinhold, 1985.
[8]   P. Green, "An All-Optical Computer Network: Lessons Learned," IEEE Network, March 1992.
[9]   Jajszczyk, Mouftah, "Photonic Fast Packet Switching," IEEE Communications, February 1993.

5

Basic Protocol Concepts

Protocols shape our everyday lives. A *protocol* is similar to a language, conveying meaning and understanding through some form of communication. Computer communication protocols are sets of rules governing the exchange of messages that define the way machines communicate and behave. Similar to the requirement of a shared language for intelligent human discourse, in order for one computer to talk to another, each must be able to understand the other's protocol. Protocols play an important role in data communications; without them islands of users would be unable to communicate.

This chapter begins with an introduction to packet-switching, articulating the reasons for this new suite of protocols, defining some basic principles, and discussing how changes in the telecommunication environment impacted the evolution of packet-switching protocols. Next, the text explains the key concept of layered models, where lower layer protocols provide services to the next higher layer. The concept of layered protocols is largely due to the OSI model; however, the specific protocols of OSI are not widely implemented. We then summarize how protocol suites collect sets of layered protocols into a single group; for example, the well known seven layer Open Systems Interconnect Reference Model (OSIRM), IBM's System Network Architecture (SNA), the IEEE 802 series of local area networking specifications, and the Internet Protocol (IP) suite. Instead, the computer communication community predominantly adopted the LAN and IP protocol suites. Furthermore, the concept of layering enabled the entire industry of multiprotocol routing. The chapter concludes with a comparison of the OSI Connection-Oriented Network Services (CONS) and ConnectionLess Network Services (CLNS) as an introduction to the radically different world of packet-switching.

5.1   A BRIEF HISTORY OF PACKET-SWITCHING

Packet-switched networks have evolved for over 30 years and form the basis of many advanced data communications networks today. Packet-switching initially provided the network environment needed to handle bursty, terminal-to-host data traffic over noisy analog telephone network facilities. Packet-switching has been widely implemented, especially in Europe, where it constitutes the majority of public and private data services.

5.1.1   Early Reasons for Packet-Switching

Paul Baran and his research team at the RAND Corporation invented the concept of packet-switching in the early 1960s as a secure, reliable means of transmitting military communications. The challenge was to enable the United States military communications system to survive a nuclear attack. The solution was to segment a longer message into many smaller pieces and wrap routing and protocol information around these pieces, resulting in data "packets." The routing and control information ensured the correct and accurate delivery and eventual reassembly of the original message at the end-user destination. Early systems had packets with a fixed maximum size assigned, typically 128 or 256 bytes. Through the use of multiple independent packets, the entire message could be transmitted over multiple paths and diverse facilities to a receiver which reassembled the original message.

The next step in packet switch history was taken when the Advanced Research Projects Agency (ARPA) of the United States Department of Defense (DoD) implemented packet-switching to handle computer communications requirements, thus forming the basis for the network called the ARPANET. This was also the first time that layered protocols were used, as well as a meshed backbone topology. Packet-switching was chosen as the method to implement WAN computer communications, which mainly consisted of connecting large computing centers. Soon after ARPANET, many commercial companies also developed packet-based networks. Today, most of Europe's nonfiber transmission networks are tied together with reliable private and public packet-switched networks.

The early days of computing also saw the development of new interfaces and data communication protocols by each major computer manufacturer. Large computer manufacturers, like IBM and DEC, developed protocols that were standardized, but only across their own product line. This tactic often locked a user into a single, proprietary protocol. Indeed, a key objective of the Open System Interconnection (OSI) standardization effort was enabling standard computer communication interfaces and protocols in a multiple vendor environment.

Early packet-switching systems targeted terminal-to-host communications. The typical transaction involved the user typing a few lines, or even just a few characters, and then sending a transaction to the host. The host would then return a few lines, or possibly an entire screen's worth of data. This terminal-host application was very bursty; that is, the peak transmit rate of each terminal was much greater than its average rate. Packet-switching equipment statistically multiplexed many such bursty users onto a single expensive transmission facility.

As the number of computers, applications, and people using computers increased, the need for interconnection increased, creating the accelerating need for bandwidth discussed in Chapter 1. Similar to the growth in telephony, it quickly became absurd to have a dedicated circuit to connect every pair of computers that needed to communicate. Packet-switching and routing protocols were developed to connect terminals to hosts, and hosts to hosts.

5.1.2 Principles of Packet-Switching

In summary, several factors created the need for packet-switching: the need to create standard interfaces between computing devices, the ability to extend computer communication over noisy analog transmission facilities, a requirement to make more efficient use of expensive transmission bandwidth, and a means to enable the interconnection of a large number of computing devices.

Packet-switching is a special case of the general address multiplexing and switching method described in Chapter 4. Packet-switching provides a service in which blocks of user data are conveyed over a network. User data, such as files, are broken down into blocks of units called "payload" information. Packet-switching adds overhead to the user data payload blocks, resulting in a combination called a *packet*. All of the protocols studied in this book have this characteristic, including: SNA, IEEE 802.x LAN protocols, X.25, IP, frame relay, SMDS, Ethernet, FDDI, Token Ring, and ATM.

The functions implemented using the packet overhead are either link layer, packet layer, or transport layer from the OSI Reference Model (OSIRM) as described later in this chapter. Older protocols, such as X.25 and IP, perform both the link layer and packet layer function. Newer protocols, such as frame relay, SMDS, and ATM, perform only a subset of link layer functions, but with addressing functions that have a networkwide meaning.

Link layer functions always have a means to indicate the boundaries of the packet, perform error detection, provide for multiplexing of multiple logical connections, and provide some basic network management capability. Optional link layer functions are flow control, retransmission, command/response protocol support, and link level establishment procedures.

Packets were designed with Cyclic Redundancy Check (CRC) fields that detected bit errors. Early packet switches retransmitted errored packets on a link-by-link basis. The advent of low bit-error rate fiber-optic transmission media made implementation of such error detection and retransmission cost-effective in the end system, since errors rarely occurred.

Network layer functions always have a means to identify a uniquely addressed network station. Optional network layer functions include retransmission, flow control, prioritized data flows, automatic routing, and network layer connection establishment procedures.

Packet-switching enables statistical multiplexing by allowing multiple logical users to share a single physical network access circuit. Buffers in the packet switches reduce the probability of loss during rare intervals when many users transmit simultaneously. Packet switches control the quality provided to an individual user by allocating bandwidth, allocating buffer space, policing the traffic offered by users, or by flow control. Part 6 covers the application of these methods to ATM.

Packet-switching also extends the concept of statistical multiplexing to an entire network. In order to appreciate the power of packet-switching, compare the full mesh network of dedicated circuits in network Figure 5.1*a* versus that of the packet-switched network in Figure 5.1*b*. The dedicated network has three lines connected to every host, while the packet network has only one connecting the host to the packet switch. A virtual circuit connects every user through the packet-switched network, as shown by the lines within the physical trunks.

The dedicated circuit network has higher overall throughput, but will not scale well. The packet-switched network requires additional complexity in the packet switches, and has lower throughput, but reduces circuit transmission costs by over 40 percent in this simple example with the nodes placed on the corners of a square. Sharing of network resources allows savings over the cost of many dedicated, low-speed communications channels, each of which is often underutilized the majority of the time. Virtual circuits are a concept that carries through into both frame relay and ATM networking.

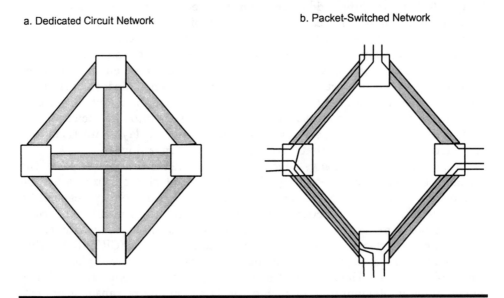

a. Dedicated Circuit Network   b. Packet-Switched Network

Figure 5.1   The Power of Packet-Switching

Packet-switching employs queuing to control loss and resolve contention at the expense of added, variable delay. The packet may take longer to get there with packet-switching, but the chances of loss are lower during periods of network congestion, assuming a reasonable buffer size. Packet data protocols employ two types of flow and congestion control: implicit and explicit congestion notification.

Implicit congestion notification usually involves a layer 4 transport protocol, such as the Transmission Control Protocol (TCP), in either the network device or the user premises equipment. These protocols adaptively alter the rate of packets sent into the network by estimating loss and delay.

Explicit congestion notification occurs when the protocol notifies the sender and/or receiver of congestion in the network. If the sender or receiver reacts to the explicit indication of congestion quickly enough, it avoids loss entirely. Part 5 covers the subject of implicit and explicit flow and congestion control in detail.

5.1.3 Darwin's Theory and Packet-Switching Evolution

Darwin spent nearly as much time to arrive at his theory of evolution as it took for engineers to conceive of implement packet-switching throughout the world of data communications. The basic tenets of Darwin's theory of evolution is natural selection, survival of the fittest, and the need to adapt to a changing environment. In the communications jungle, packet-switching has all of these attributes.

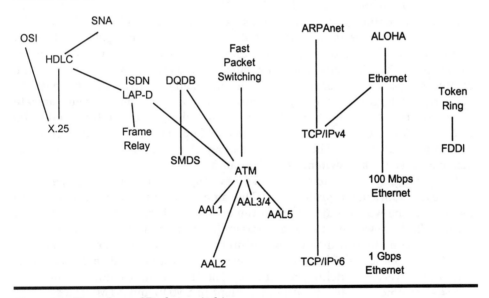

Figure 5.2 Genealogy of Packet-switching

This section takes the reader through a brief summary of the genealogy of packet-switching with reference to Figure 5.2. The genesis of packet-switching began with two proprietary computer communication architectures: IBM's Systems Network Architecture (SNA) and DEC's Digital Network Architecture (DNA). Standards bodies refined the Synchronous Data Link Control (SDLC) protocol from SNA, resulting in the High-level Data Link Control (HDLC) protocol — which begat X.25 and Link Access Procedure D (LAP-D) within ISDN. Frame relay evolved as basically a leaner, meaner

LAP-D protocol. OSI adopted the X.25 protocol as the first link and packet layer standard. Combining the conventions from preceding protocols and the concepts of hardware-oriented Fast Packet Switching (FPS) resulted in the Distributed Queue Dual Bus (DQDB) protocol, which is the basis of the Switched Multimegabit Data Service (SMDS) and, finally, ATM. On top of ATM a number of ATM Adaptation Layers (AALs) support not only data, but voice and video as well, as described in Chapter 17.

Around the same time that the International Standards Organization was developing the Open Systems Interconnect (OSI) protocol suite, the U.S. Advanced Research Projects Agency (ARPA) was working on a network with universities and industry that eventually resulted in the suite of applications and higher level protocols based upon the highly successful Internet Protocol (IP) version 4. Ethernet also sprung up at this time as a result of an experiments packet radio communication network in Hawaii called ALOHA. Ethernet then evolved into 100-Mbps fast Ethernet, and recently increased its speed to over one billion bits per second, called gigabit Ethernet. TCP/IP will also add a much larger address space in version 6, as well as a means to group application flows with similar performance requirements and allocate bandwidth. Token Ring was also developed shortly after Ethernet, and has evolved into the higher speed FDDI.

Packet-switching plays an increasingly important role in the rapidly changing environment of distributed processing of the 1990s. Several environmental factors drive the direction of data communications evolution. There is an accelerating need for more bandwidth driven by increasing computing power, increasing need for interconnectivity, and the need to support ever larger networks where any user or application can communicate with any other. The low error rate of modern fiber-optic, satellite, and radio communications enables more cost-effective implementation of higher-speed data communications. The same technology that increases computer power also increases packet-switching performance.

This changing environment creates new opportunities for new species of data communications protocols. The improved quality of transmission facilities alone was a major force in the evolution of frame relay, SMDS, and ATM. These newer protocols are streamlined in that they do not perform error correction by retransmission within the network. The fixed slot and cell size of SMDS and ATM have also enabled cost-effective hardware implementation of powerful switching machines. The increasing capabilities of high-speed electronics are an essential ingredient in SMDS and ATM devices.

Now that we've surveyed the history, background, and directions of packet-switching, let's take a more detailed look at some of the underlying concepts and terminology involved in layered protocols.

5.2 BASIC PROTOCOL LAYERING CONCEPTS

Webster's New World Dictionary defines a *protocol* as "a set of rules governing the communications and the transfer of data between machines, as in computer systems." As a means to divide and conquer complex protocols, system designers arrange the communications and data transfer between such machines into logical layers that pass messages between themselves. A convention created by the OSI reference models refers to the messages passed between such layers as an interface. Each layer has a specific interface to the layer above it and the layer below it, with two exceptions: the lowest layer interfaces directly with the physical transmission medium, and the highest layer interfaces directly with the end user application. The study of computer communication networking covers primarily the lowest three layers: physical, data link, and network. Devices implementing these protocols may realize these layers in either software, hardware, or a combination of the two. Typically, hardware implementations achieve much higher speeds and lower delays than software ones do; however, with the ever increasing performance of microprocessors this distinction blurs at times. Figure 5.3 illustrates the basic concept of protocol layering that is relevant to the protocols described in this book. Let's now look at these protocol-layering concepts in more detail.

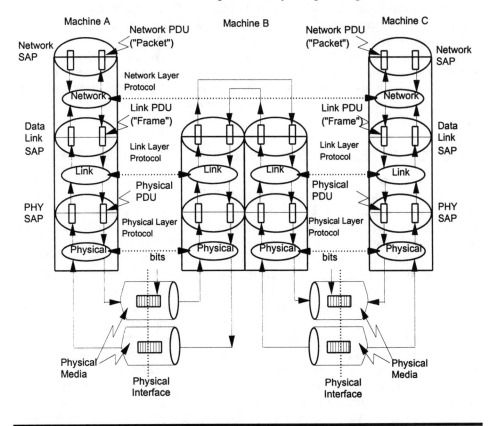

Figure 5.3   Protocol Model of the Physical, Link, and Network Layers

The term *interface* is used in two ways in different standards bodies. First, primarily in the CCITT/ITU view, physical interfaces provide the physical connection between different types of hardware, with protocols providing rules, conventions, and the intelligence to pass data over these interfaces between peer protocol layers. In summary, the CCITT/ITU view is that bits flow over physical interfaces as shown at the bottom of the figure. Secondly, in the OSI view, interfaces exist between protocol layers within end and intermediate systems as indicated in the figure. In this view, *Protocol Data Units* (PDUs), or *messages*, pass over protocol interfaces. OSI standards also call the interfaces between layers Service Access Points (SAPs); because they are the points where the higher-layer protocol accesses the service provided by the lower-layer protocol. ATM protocol models use the OSI concept of SAPs extensively. Stated another way, physical interfaces provide the path for data flow between machines, while protocols manage that data flow across this path using SAPs (or protocol interfaces) between the layers within the machines traversed by PDUs across a network. Obviously, compatibility of both the protocol and physical interfaces is essential in the implementation of communications networks involving machines manufactured by different vendors. Indeed, the original motivation for the OSI reference model was to drive towards a multi-vendor standard that would enable competition with IBM's proprietary System Network Architecture (SNA).

The concepts behind the use of multiple protocol layers are important. The concepts of physical, data link, and network layer protocols can now be defined on a high level. We hearken back to the OSI physical, data link, and network layers when discussing the roots of many other protocols.

5.3   OPEN SYSTEMS INTERCONNECTION REFERENCE MODEL

The Open Systems Interconnection Reference Model (OSIRM) effort strove to define the functions and protocols necessary for any computer system to connect to any other computer system, regardless of the manufacturer. The International Organization for Standardization (ISO) created the OSIRM in the ISO Technical Subcommittee 97 (TC97) starting in 1977, with Subcommittee 16 (SC16) officially documenting the protocol architecture in 1983 as ISO standard 7498. Figure 5.4 depicts the basic OSIRM showing end system (A), intermediate system (B), and end system (C) and the protocol stack within each. The layers are represented starting from the bottom at the first layer, which has a physical interface to the adjacent node, to the topmost seventh layer, which usually resides on the user end device (workstation) or host that interacts with or contains the user applications. Each of these seven layers represents one or more protocols that define the functional operation of communications between user and network elements. All protocol communications between layers are "peer-to-peer" — depicted as horizontal arrows between the layers. Standards span all seven layers of the model, as summarized below. Although OSI has standardized many of these protocols, only a few are in widespread use. The layering concept, however, has been widely adopted by every major computer and communications standards body and most proprietary implementations as well.

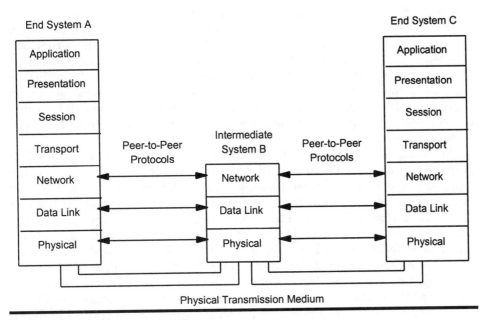

Figure 5.4 OSI Reference Model

Figure 5.5 illustrates the basic elements common to every layer of the OSI reference model. This is the portion of the OSIRM that has become widely used to categorize computer and communications protocols according to characteristics contained in this generic model. Often the correspondence is not exact or one-to-one; for example, ATM is often described as embodying characteristics of *both* the data link and network layers.

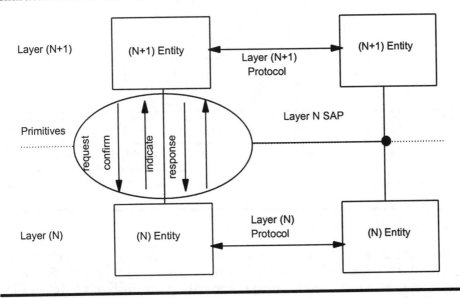

Figure 5.5 Illustration of Layered Protocol Model

Referring to Figure 5.5, a layer (N+1) entity communicates with a peer layer (N+1) entity by way of a service supported at layer (N) through a Service Access Point (SAP). The layer (N) SAP provides the primitives between layer (N) and (N+1) of request, indicate, confirm, and response. Parameters are associated with each primitive. Protocol Data Units (PDUs) are passed down from layer (N+1) to layer (N) using the request primitive, while PDUs from layer (N) are passed up from layer (N) to layer (N+1) using the indicate primitive. Control and error information utilize the confirm and response primitives.

This book utilizes the shorthand notation illustrated in Figure 5.6 to graphically express the concept of protocol layering. This simple syntax represents the PDU structure passing between layers via stacks that only contain the name or acronym of the protocol. Separate portions of the text define the bit-by-bit Protocol Data Unit (PDU) structures. Starting at the left-hand side, Node A takes data at layer (N+1), which is connected to Node B by a layer (N–1) protocol. On the link between Nodes A and B we illustrate the resultant enveloping of the layer headers (HDR) and trailers (TRLR) that are carried by the layer (N–1) protocol. Node B performs a transformation from layer (N) to the correspondingly layered, different protocols called layer (N)' and layer (N–1)'. The resultant action of these protocol entities is shown by the layer (N–1)' PDU on the link between nodes B and C.

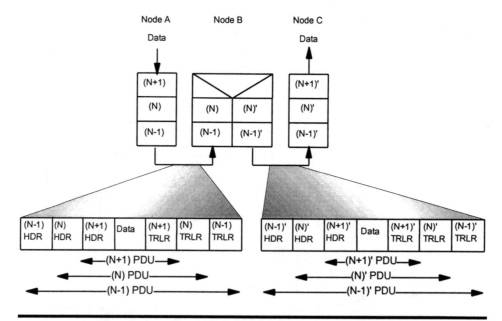

Figure 5.6 Shorthand Protocol Model Notation

Since the model of Figure 5.6 is somewhat abstract, let's take a closer look at a real world example to better illustrate the concept. Figure 5.7 illustrates an example of a workstation connected via an IEEE 802.5 Token Ring (TR) LAN to a bridge, which is connected via an 802.3 Ethernet LAN to a server. Both the workstation and the server are using the Internet Protocol (IP) at the network layer. Over the Token Ring physical layer connection, the workstation and the bridge use the Token Ring link layer. Hence, the protocol data unit on the Token Ring LAN begins and ends with an Token Ring header and trailer that envelop an IP header (since IP has no trailer) and also the workstation's data. The bridge takes in the Token Ring link layer and converts this to the IEEE 802.3 Ethernet header and trailer, and also converts this to the 10-Mbps rate running over unshielded twisted pair using the 10-Base T standard. The resulting PDU sent over the wires to the server via the bridge is illustrated in the lower right hand corner of the figure. An 802.3 Ethernet header and trailer envelop the same IP header and user data. Communication from the server back to the workstation basically reverses this process. This example serves as a brief introduction only. The curious reader can find more details on the IP protocol in Chapter 8 and the IEEE 802.3 Ethernet and IEEE 802.5 Token Ring protocols in Chapter 9.

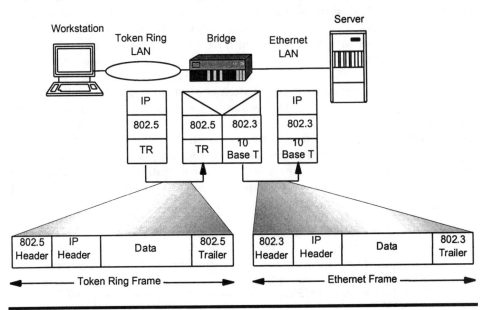

Figure 5.7   Example of Layered Protocol Model and Resulting Protocol Data Units

5.4 LAYERS OF THE OSI REFERENCE MODEL

We now cover each layer of the OSIRM in more detail. The OSIRM outlines a layered approach to data transmission: seven layers, with each successively higher layer providing a value-added service to the layer above it. Data flows down from layer 7 (application layer) at the originating end system to layer 1 (physical layer), where it is transmitted across a network of intermediate nodes over interconnecting physical medium, and back up to the layer 7 of the destination end system. Not all seven levels need be used. The specific

OSI protocols for each of the seven layers have not been widely adopted in practice, particularly at the application, presentation, and session layers. The following sections summarize the generic functions of all seven layers starting with the physical layer, which is the one closest to the physical transmission medium.

5.4.1 Physical Layer

The first layer encountered is the physical layer (L1) which provides for the transparent transmission of a bit stream across the physical connection between network elements. The intelligence managing the data stream and protocols residing above the physical layer are transparently conveyed by the physical layer.

The physical layer connections are either point-to-point or multipoint. The physical layer operates in either a simplex, half-duplex, or full-duplex mode as described in Chapter 4. Simplex means that transmission is in one direction only. Half-duplex involves the use of physical layer signaling to change the direction of simplex transmission to support bi-directional communication, but at any one point in time data flows only in one direction. Full-duplex means that transmission occurs in both directions simultaneously. Furthermore, the bit stream may be transmitted serially or in parallel.

The physical layer includes specification of electrical voltages and currents, optical pulse shapes and levels, mechanical connector specifications, basic signaling through connections, and signaling conventions. The physical layer can also activate or deactivate the transmission medium as well as communicate status through protocol primitives with the layer 2 data link layer. The physical medium can either be an electrical or optical cable, or a satellite or radio transmission channel. Examples of commonly encountered physical layer specifications are: EIA-RS-232-C, EIA-RS-449, and the HSSI interface.

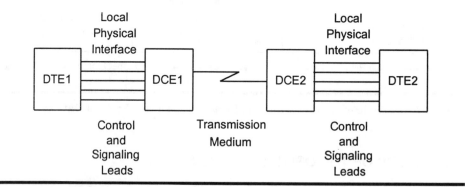

Figure 5.8 DTE to DTE Communications

The terms *Data Termination Equipment* (DTE) and *Data Communication Equipment* (DCE) refer to the hardware on either side of a communications channel interface. DTE equipment is typically a computer or terminal which acts as an end point for transmitted and received data via a physical interface to a DCE. DCE equipment is typically a modem or communication device, which has a different physical interface than that of the DTE. One commonly used type of DCE is a Channel Service Unit/Data Service Unit (CSU/DSU); it converts the DTE/DCE interface to a telephony-based interface.

Figure 5.8 shows a common end-to-end network configuration where DTE1 talks to DCE1, which in turn formats the transmission for transfer over the network to DCE2, which then interfaces to DTE2. Some devices can be configured to act as either a DTE or a DCE.

5.4.2 Data Link Layer

The data link layer is the second layer (L2) in the seven-layer OSIRM, and the second layer in most other computer architecture models as well. The primary function of the data link layer is to establish a reliable protocol interface across the physical layer (L1) on behalf of the network layer (L3). This means that the link layer performs error detection and in some cases, error correction. Toward this end, the data link control functions establish a peer-to-peer relationship across each physical link between machines. The data link layer entities exchange clearly delimited protocol data units, which are commonly called *frames*. The data link layer may use a limited form of addressing such that multiple data link layer protocol interfaces can be multiplexed across a single physical layer interface. There may be a flow control function to control the flow of frames such that a fast sender does not overrun a slow receiver.

Computer communications via local area networks utilize special functions of the data link layer called the Medium Access Control (MAC) and Logical Link Control (LLC) layers. The MAC layer protocols form the basis of LAN and MAN standards used by the IEEE 802.X LAN protocol suite introduced later in this chapter, which includes Ethernet, Token Ring, and Token Bus. Examples of link layer protocol standards include ISO 7776, , ISDN LAP-D, ISO HDLC, and MAC-layer protocols such as the ISO 9314-2 FDDI Token Ring MAC.

Some of the new services, such as frame relay and ATM, can be viewed as using only the first two layers of the OSIRM. They rely heavily on reducing the link layer services to increase speeds at lower costs because of the resulting protocol simplification. A key difference between frame relay and ATM is that the addresses can take on an end-to-end significance, whereas in the OSIRM link layer addresses are only significant between nodes.

5.4.3 Network Layer

The third layer (L3) encountered in the OSIRM is the network layer. The principal function of the network layer is to provide reliable, in-sequence delivery of protocol data between transport layer entities. In order to do this, the network layer must have an end-to-end addressing capability. A unique network-layer address is assigned to each network-layer protocol entity. A network-layer protocol may communicate with its peer over a route of intermediate machines with physical, data link, and network layers. The determination of this route is called the *routing function*. Chapter 9 covers the important subject routing protocols and their operation. Network layer PDUs are often called *packets*.

The network layer may also perform end-to-end flow control and the segmentation and reassembly of data. The network layer is the most protocol-intensive portion of packet networks. Some examples of protocols used in the network layer are the ITU X.25 and X.75 packet level and gateway protocols, the Internet Protocol (IP), CCITT/ITU-T Q.931, Q.933, Q.2931, and the OSI CLNP.

The network layer is also used to define data call establishment procedures for packet and cell-switched networks in ISDN and B-ISDN. For example, ATM signaling utilizes a layer 3 protocol for call setup and disconnection. SMDS also employs a layer 3 protocol to provide an end-to-end datagram service using E.164 (i.e., telephone numbers) for addressing. We cover each of these concepts in subsequent chapters.

5.4.4 Transport Layer

The fourth layer (L4) encountered is the transport layer. The principal function of the transport layer is to interconnect session layer entities. Historically, it was also called the host-to-host layer. Principal functions that it performs are segmentation, re-assembly, and multiplexing over a single network-layer interface. The transport layer allows a session-layer entity to request a class of service, which must be mapped onto appropriate network-layer capabilities. Frequently, the transport layer manages end-to-end flow control. The transport layer may often perform error detection and correction as well. This has become increasingly important since it provides a higher-level error correction and retransmission protocol for services that usually don't provide reliable delivery, such as frame relay, SMDS, IP, and ATM. Often, frame relay users ask what happens when frames are lost. The answer is that the transport layer retransmits these lost packets.

One example of the transport layer includes the ITU X.224 OSI transport protocol TP4. Another widely used example of a transport type of protocol is the Internet Transmission Control Protocol (TCP).

5.4.5 Session Layer

The fifth layer (L5) encountered is the session layer. The session layer is essentially the user's interface to the network, which may have some data transformations performed by the presentation layer. Sessions usually provide connections between a user, such as a terminal or LAN workstation, and a central processor or host. So-called peer-to-peer session-layer protocols can directly connect user applications. Session-layer protocols are usually rather complex, involving negotiation of parameters and exchange of information about the end user applications. The session layer employs addresses or names that are meaningful to end users. Other session-layer functions include flow control, dialog management, control over the direction of data transfer, and transaction support.

Some examples of the session layer are terminal-to-mainframe log-on procedures, transfer of user information, and the setup of information and resource allocations. The ISO standard for the session layer is the ISO 8327/ITU X.225 connection-oriented session protocol.

5.4.6 Presentation Layer

The sixth layer (L6) is the presentation layer; which determines how data is presented to the user. Official standards are now complete for this layer. Many vendors have also implemented proprietary solutions. One reason for these proprietary solutions is that the use of the presentation layer is predominantly equipment dependent. Some examples of presentation-layer protocols are video and text display formats, data code conversion between software programs, and peripheral management and control, using protocols such as ITU X.410 and ITU X.226 OSI connection-oriented protocol.

5.4.7 Application Layer

The seventh and final layer (L7) is the application layer. This layer manages the program or device generating the data to the network. More importantly, this layer provides the actual interface to the end user. The application layer is an "equipment-dependent" protocol, and lends itself to proprietary vendor interpretation. Examples of standardized application-layer protocols include ITU X.400, X.420 X.500 – X.520 directory management, ISO 8613/ITU T.411-419 Office Document Architecture (ODA), and ISO 10026 distributed Transaction Processing (TP).

5.4.8 Mapping of Generic Devices to OSI Layers

In the mainframe and minicomputer era, the bottom three layers (network, data link, and physical) were implemented on different pieces of equipment than the next three higher layers (presentation, session, and transport). The first three layers were implemented on a Front End Processor (FEP), while the higher three layers were implemented on a host. Current customer premises devices such as bridges, routers, and hubs usually manipulate the protocols of the first three layers: network, data link, and physical. They can often connect dissimilar protocols and interfaces. Many implementations of user software that cover the top three non-application layers (presentation, session, and transport) operate as a single program.

5.5 LAYERED DATA COMMUNICATION ARCHITECTURES

In addition to the OSIRM, several other data communication protocol architectures shaped and standardized the computer networking industry in the late twentieth century. These include: IBM's SNA, the IEEE 802.X LAN standards, the ITU-T's ISDN, and the Internet Protocol (IP) suite. This section introduces these important architectures.

5.5.1 IBM's Systems Network Architecture (SNA)

The introduction by IBM of Systems Network Architecture (SNA) in 1974 signaled the beginning of a vendor proprietary architecture which remained prominent in the computing industry into the 1990s. SNA architectures and protocols are still widely used by many businesses and institutions. Many users have multiprotocol environments where designers either separate networks for their IP/IPX traffic and SNA traffic, or encapsulate (or tunnel) their SNA traffic inside another network protocol (such as frame relay). This technique is not without problems. However, since SNA protocol timers assume a private line interconnection environment, the variable delay sometimes encountered with tunneling over another protocol may cause timeouts resulting in session losses. A common solution to this problem is protocol spoofing. This technique involves the device attached to the SNA network sending acknowledgments to the timeout sensitive SNA device prior to receiving the distant acknowledgment. SNA was IBM's method of creating a computing empire through standardization that centered on the mainframe and (distributed) front-end processors. The problem that the huge IBM corporation faced in standardizing data communications among its own products was formidable. By providing a hierarchy of network-access methods, IBM created a network that accommodated a wide variety of users, protocols, and applications, while retaining ultimate control at the mainframe host and front-end processors. The move from centralized to distributed processing has had pronounced effects on SNA. IBM dubbed its latest evolution of SNA Advanced Peer-to-Peer Networking (APPN) using Advanced Peer-to-Peer Communications (APPC) in an attempt to preserve homogeneous SNA networks. However, many SNA users now look to less elegant (and less expensive) — yet functional — solutions.

The SNA architecture layers are shown in Figure 5.9 where an SNA terminal is connected via a Front End Processor (FEP) to a mainframe computer [1]. Since SNA preceded the OSIRM, most of the names for the layers differ, except for the physical and link layers. The SNA layered stack is divided into two main components: node-by-node transmission services and end-to-end services. Every node implements the transmission services, while only end systems implement the end-to-end services as half-sessions as shown in the figure. The common transmission services encompass the physical, data link, and path control layers. The physical and data link layers define functions similar to the OSIRM, with serial data links employing the SDLC protocol and channel attachments between Front End Processors (FEPs) and mainframes employing the System 370 protocol. The path control layer provides connectivity between half sessions based upon the addresses of the source and destination Network Addressable Units (NAUs). Hence, a key function of path control is to determine the node-by-node route from the source to the destination. SNA employs a hierarchical structure to enable scaling to large networks by grouping nodes into sub-areas. Path control utilizes precomputed explicit routes to route packets between sessions. Since the philosophy in SNA is per-session flow control, the path control layer basically provides only an indication back to higher layers about network congestion.

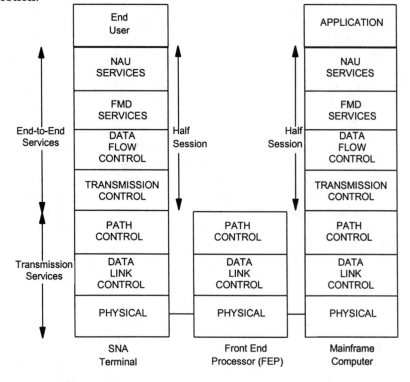

Figure 5.9 IBM's Systems Network Architecture (SNA) Protocol Structure

Fundamental to SNA is the concept of a *session*, where each system in an end-to-end connection implements half of the protocol. The transmission control layer establishes, maintains, and terminates sessions between Logical Units (LU) of various types. The transmission control layer also performs some routing functions, for example within a mainframe where multiple simultaneous sessions exist. The data flow control layer involves the processing of Request-Response Units exchanged over sessions. It provides functions controlling priority, grouping of packets into larger logical sets (such as all the lines on a screen), and controls the operation over half-duplex links. Moving up the protocol stack, Function Manager Directory (FMD) services and Network Addressable Unit (NAU) services complete the SNA protocol stack. FMD services provide user-to-user as well as presentation services, including data compression and character set conversion. NAU services cover the concepts of System Services Control Points (SSCPs), Physical Units (PU), and Logical Units (LUs). The NAU services are the interface to the end user and applications.

5.5.2 IEEE 802.X Series (LAN/MAN/WAN)

The Institute of Electrical and Electronics Engineers (IEEE) established the 802 working group to standardize local and metropolitan area networks (LANs and MANs). These standards have become so important that the International Standards Organization (ISO), the Internation Electrotechnical Committee (IEC), and American National Standards Institute (ANSI) also publish these standards. Generically called the IEEE 802.X series of standards, these important specifications cover the physical and link layers shown in Figure 5.10. The physical specification, commonly abbreviated as PHY, covers the interface to a variety of physical media, such as twisted pair, coax, fiber, and radio frequencies. Above the PHY layer, the data link layer embodies two sublayers: a Medium Access Control (MAC) sublayer, and a Logical Link Control (LLC) sublayer. The LLC layer provides one or more Service Access Points (SAPs) to higher-layer protocols as indicated in the figure. A commonly encountered example of a SAP is the LAN driver in personal computer software. We cover this important series of standards in greater detail in Chapter 9.

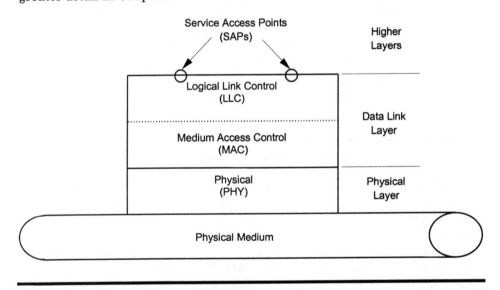

Figure 5.10   IEEE 802.x Protocol Architecture

The logical link layer operates the same way for all LAN architectures, but not for the 802.6 MAN architecture, which is a completely different beast, as discussed in Chapter 8. The medium access control layer and physical layer operate differently for each of the local and metropolitan area network architectures. The LLC layer defines common procedures for call establishment, data transfer, and call termination through three types of services: connection-oriented, unacknowledged connectionless, and acknowledged connection-oriented.

Figure 5.11 depicts the IEEE 802.X protocol suite. The IEEE 802.2 standard [2] defines the LLC sublayer. Also note that the 802.1d standard defines LAN bridging at the MAC level, a topic covered in depth in Chapter 9 as an introduction to LAN Emulation (LANE) over ATM. Two major LAN architectures defined in the 802.X standards are commonly used: Ethernet and Token Ring, also defined in Chapter 9. Ethernet is by far the most commonly used LAN protocol. One major MAN protocol defined by the IEEE was implemented: the Distributed Queue Dual Bus (DQDB) as defined in IEEE 802.6. Chapter 8 covers this protocol as it is used in the Switched Multimegabit Data Service (SMDS). Chapter 31 covers the integrated services and 100 Mbps demand priority LAN standards in the treatment comparing the guaranteed bandwidth and quality of service of other implementations with those of ATM.

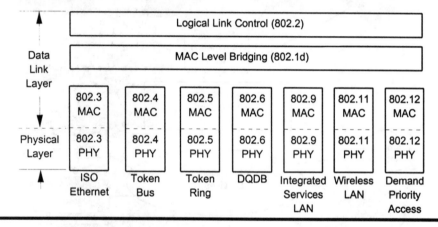

Figure 5.11   IEEE 802.x Protocol Suite

The IEEE 802.3 standard and other specifications form what is called the Ethernet Standard. The first Ethernet products appeared in 1981, and now sales for Ethernet outpace all other 802 protocols combined. Ethernet users contend for a shared medium after first sensing for a carrier transmission by other users. The original interface specified 10 Mbps over twisted pair or coaxial cable physical media. Now, fast Ethernet speeds are available at 100 Mbps and at 1 Gbps.

IBM invented the Token Ring architecture in its development labs in Zurich, Switzerland. The first Token Ring products appeared in 1986. The IEEE 802.5 standard defines the protocol which involves passing a "token" between stations to control access to the shared medium. Token Ring initially competed with Ethernet as a popular LAN standard, but gave up significant market share to the economical Ethernet alternative in the 1990s. Today most estimates place Token Ring as less than 15 percent of LANs in use, compared with Ethernet use exceeding 80 percent.

5.5.3 Integrated Services Digital Network Protocol Architecture

The ITU began work on the Integrated Services Digital Network (ISDN) standards in 1972, with the first documents published in 1984. ISDN's initial goal was aimed at converting the entire telecommunications transmission and switching architecture to a digital architecture, providing end-user-to-end-user digital service for voice, data, and video over a single physical access circuit. Narrowband ISDN standards are the root of Broadband ISDN (B-ISDN) standards, of which ATM is the key protocol layer as described in Chapter 10.

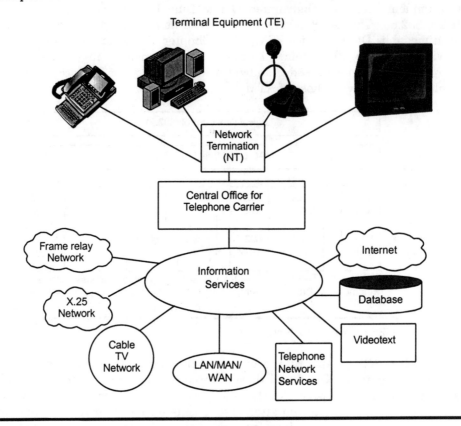

Figure 5.12 ISDN Information Services Concept

The basic concept of ISDN is of multiple types of Terminal Equipment (TE), such as phones and computers, connecting through an ISDN network termination point (called an NT) into the central office environment that provides access to a range of information services as shown in Figure 5.12. While all seven protocol layers are the same as the OSIRM, the physical, data link, and network layers define the lower-layer functions, also called bearer services. These layers define physical connectivity and transmission as defined in ITU-T Recommendations I.430, I.431, and I.432; data link management, flow, error, and synchronization control as defined in ITU-T Q.921(LAP-D); and network addressing, congestion control, end-to-end call establishment, routing or relaying, and switching as defined in Recommendations Q.931/I.451, Q.932/I.452, and Q.933/I.453. The transport, session, presentation, and application layers define the higher-layer functions, including the teleservices which define services such as messaging, telephone, and telex. Standards for these layers are host-to-host, as well as application-specific.

The ISDN architecture was the first to introduce the concepts of multiple layered protocol planes. Figure 5.13 illustrates the user, control, and management planes of ISDN. Note how addition of this dimension to the overall protocol architecture provides for specialization of particular nodes, as well as operation at multiple protocol layers simultaneously. The ITU-T applied this same multiple plane model to the B-ISDN specifications as covered in the next part.

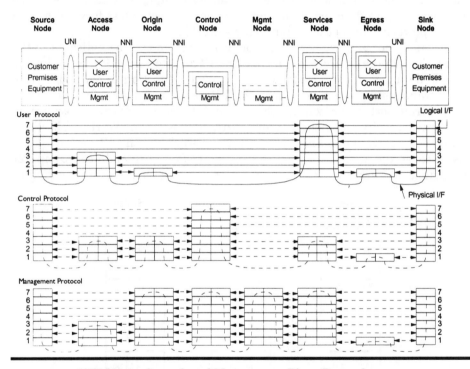

Figure 5.13   ISDN User, Control, and Management Plane Protocols

The user protocol (or bearer service) is layer 1 for circuit-mode (shown in Figure 5.13), layer 2 for frame-mode, and layer 3 for packet-mode services. Teleservices and value-added services operate at higher layers. Intermediate nodes may provide only physical connectivity. User Network Interfaces (UNI) and Network-to-Network Interfaces (NNI) will be explained later. Conceptually, another application runs the control, or signaling, plane. The purpose of the control plane protocols is to establish, configure, and release the user plane (bearer) capabilities. Finally, the management plane is responsible for monitoring the status, configuring the parameters, and measuring the performance of the user and control planes. We cover the protocol structure and functional building blocks of ISDN in the next chapter.

5.5.4 Internet Protocol (IP) Architecture

The Internet Protocol uses fewer layers than the OSI reference model, as seen by examining Figure 5.14. Basically, an Internet is composed of end systems (synonymously called *hosts*) interconnected by routers. The hosts run networked end-user applications over a small set of transport protocol services, such as the hypertext transfer protocol (http) commonly used on the World Wide Web. The transport layer interfaces to the internet layer, which runs over a comprehensive range of data link layer protocols. Intermediate routers provide a connectionless datagram forwarding service to the transport layer. As we describe in the next section, connectionless forwarding is a simple, powerful concept useful in data networking.

Routers do not perform any transport layer functions — as indicated in the figure. Unlike the OSI layered model, the IP architecture does not require a reliable data link layer. Instead, the important function of error detection and retransmission is performed by the Transmission Control Protocol (TCP) at the transport layer within the IP architecture. Possibly because of inherent simplicity and elegance, and partly because of the World Wide Web's emergence as the dominant force in internetworking, the IP protocol suite has become the defacto worldwide standard for most end-user computer equipment networking. Chapter 8 covers the Internet protocol suite in detail.

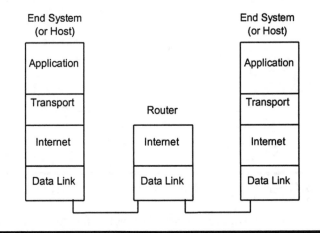

Figure 5.14   Internet Protocol (IP) Layered Architecture

5.6 NETWORK SERVICE PARADIGMS

The OSI reference model categorizes data network services in one of two paradigms: connection-oriented or connectionless. Connection-oriented network services (CONS) involve establishing a connection between physical or logical end points *prior* to the transfer of data. Examples of CONS are Frame Relay, TCP, and ATM. Connectionless network services (CLNS), on the other hand, provide end-to-end logical connectivity *without* establishing any connection before data transfer. Examples of CLNS are IP and LAN protocols. As we shall see repeatedly, the consequences of ATM being connection-oriented while IP an LAN protocol are connectionless are far reaching indeed. Historically, wide area networks employed connection-oriented services, while local area networks used connectionless services. ATM, along with its supporting cast of adaptation layers and higher layer protocols, supports *both* connection-oriented and connectionless services.

5.6.1 Connection-Oriented Network Services (CONS)

Connection-oriented services require establishment of a connection between the origin and destination before transferring data. The connection is established as a single path of one or multiple links through intermediate nodes in a network. Once established, all data travels over the same pre-established path through the network. The fact that data arrives at the destination in the same order as sent by the origin is fundamental to connection-oriented services.

If network management or provisioning actions establish the connection and leave it up indefinitely, then we call the result a Permanent Virtual Connection (PVC). If control signaling of any type dynamically establishes and takes down the connection, then it is called a Switched Virtual Connection (SVC).

A PVC connection may be established by physical wiring, equipment configuration commands, service provider provisioning procedures, or combinations of these actions. These actions may take several minutes to several weeks, depending upon exactly what is required. Once the PVC is established data may be transferred over it. Usually PVCs are established for long periods of time. Examples of physical PVCs are analog private lines, DTE-to-DCE connections, and digital private lines. Examples of logical PVCs are the X.25 PVC, the frame relay PVC, and the ATM PVC.

In the case of an SVC service, only the access line and address for the origin and each destination point are provisioned beforehand. The use of a control signaling protocol plays a central role in SVC services. Via the signaling protocol the origin requests that the network make a connection to a destination. The network determines the physical (and logical) location of the destination and attempts to establish the connection through intermediate node(s) to the destination. The success or failure of the attempt is indicated back to the originator. There may also be a progress indication to the originator, alerting for the destination, or other handshaking elements of the signaling protocol as well. Often the destination utilizes signaling to either accept or reject the call. In the case of a failed attempt, the signaling protocol usually informs the originator of the reason that the attempt failed. Once the connection is established, data can then be transferred. Networks employ SVCs to efficiently share resources by providing dynamic connections and disconnections in response to signaling protocol instructions generated by end users. End users favor the use of SVCs as they can dynamically allocate expensive bandwidth resources without a prior reservation.

Probably, the simplest way to explain an SVC is to compare it to a traditional telephone call. After ordering the service and receiving address assignments, the communications device "picks up the phone" and "requests" a connection to a destination address. The network either establishes the call or rejects it with a busy signal. After call establishment, the connected devices send data until one of the parties takes the call down. There is a direct analogy between establishing and taking down an SVC connection-oriented service and a normal telephone call, as illustrated in Table 5.1.

Table 5.1   Comparison of General Signaling Terminology to a Telephone Call

General Signaling Protocol	Voice Telephone Call
Provision Access/Address	Order Service from Phone Company
Handshaking	Obtain Dial Tone
Origin Request	Dial the Destination Number
Successful Attempt Indication	Ringing Tone
Unsuccessful Attempt Indication	Busy Tone
Destination Acceptance	Answering the Phone
Data Transfer	Talking on the Phone
Disconnect Request	Hanging up the Phone

5.6.2 Connectionless Network Services (CLNS)

As the name implies, connectionless services never establish connections of any kind. Instead, network nodes examine the address field in every packet header to determine the destination. Network nodes provide connectionless service by forwarding packets along a path toward the destination. Each node selects an outgoing link on a hop-by-hop basis. Typically, the nodes run a distributed routing protocol that consistently determines the forwarding tables to result in optimized, loop-free end-to-end paths. Therefore, unlike connection-oriented service, packets do not take a predetermined path through the network. Thus, connectionless services avoid the overhead of call establishment and management incurred by connection-oriented services. The origin node initiates the forwarding process with each intermediate node repeating it until the packet reaches the destination node. The destination node then delivers the packet to its local interface. Pretty simple, right?

Yes, and no. The magic in the above simple description is the routing protocol that consistently determines the next hop at the origin and each intermediate node. Chapter 9 explains more details about different types of routing protocols, but they all achieve the same purpose as stated in the previous sentence: routing protocols determine the contents of the next hop forwarding table such that packets with the same destination address take the same path through the network. A bad routing protocol could create next hop entries that cause endless loops where a packet never arrives at the destination, but instead loops around the network indefinitely. On the other hand, a good routing protocol automatically chooses a path through the network optimized to a specific criterion like minimum cost. Note that if the routing protocol changes the next hop forwarding table in the middle of data transfer between end systems (for example, if a physical circuit fails), then packets may arrive at the destination in a different order than sent by the origin.

Chapter 9 details common network node implementations of connectionless services, namely bridges and routers. As we shall see, some aspects of connectionless services are truly "plug and play," while others require address configuration, subnet mask definitions, and setting of other parameters. The connectionless paradigm requires that each network node (e.g., a router) process each packet independently. Common per-packet processing functions required in real networks include: filtering (out) certain packets based upon addresses and other fields, queuing different packet flows for prioritized service, and data link layer conversions. Often routers implement this complex processing in software, which limits throughput. Practically, using filtering to implement a firewall limits router throughput significantly. However, new hardware based routers promise to open up this bottleneck as described in Chapter 31.

Connectionless services do not guarantee packet delivery; therefore applications rely on higher-level protocols (e.g., TCP) to perform the end-to-end error detection/correction. Additionally, higher-layer protocols must also perform flow control (e.g., TCP or RSVP), since connectionless services typically operate on a best effort basis without any notion of bandwidth allocation.

5.6.3 Connection-Oriented vs. Connectionless Services Analogy

One simple analogy for understanding the difference between CONS and CLNS is that of placing a telephone call compared with sending a telegraph message. To make a phone call, you pick up the phone and dial the number of the destination telephone. The network makes a connection from your house, through one or more telephone switches to the destination switch and rings the phone. Once the called party answers, the telephone network keeps the connection active until one of the parties hangs up.

Now here is a CLNS example. Consider sending a telegraph message in the nineteenth century. A person visits the telegraph office and recites a message, giving the destination address as a city and country. The telegraph operator picks a next hop telegraph station, and keys in the entire message to that telegraph office. Since the originating telegraph operator does not know the status of telegraph lines being up or down except for those lines connected to his own station, he must rely on the other operators to forward the message towards the destination. If there is a path to the destination, then the persistent telegraph operators in this example eventually relay the message to the final destination, even if some telegraph lines on the most direct path are down. This example is not as dated as it may seem — Internet E-mail systems use basically the same method proven over a century ago by telegraph networks to reliably forward and deliver messages.

5.7 REVIEW

This chapter began with an introduction to packet-switching and frame relaying by exploring: their reasons for creation, basic principles, and their history. This chapter then introduced the concept of protocol layering and the notation used throughout the book. Next, a brief description covered the Open System Interconnection Reference Model (OSIRM) and its seven layers: physical, data link, network, transport, session, presentation, and application. The text gave examples of standards and real-world equipment implementations for each layer. Finally, the chapter concluded with definitions of connection-oriented and connectionless network services, giving some foreshadowing as to the manner in which ATM serves both requirements.

5.8 REFERENCES

[1]   R. Cypser, *Communications Architecture for Distributed Systems*, Addison-Wessley, 1978.
[2]   ISO/IEC, *Information Technology — Telecommunications and information exchange between systems — Local and metropolitan area networks — Specific requirements — Part 2: Logical link control*, Standard 8802-2, 1994.

6

Time Division Multiplexing and the Narrowband Integrated Services Digital Network (N-ISDN)

This chapter begins with an overview of circuit switched network services, which was the predecessor to the signaling and connection control aspects of the N-ISDN control plane. This chapter then introduces the basics of Time Division Multiplexing (TDM) used in the North American, European, and Japanese plesiochronous and synchronous digital transmission hierarchies. Begun as a means to more economically transport voice signals, TDM forms the underlying fabric of most wide area communications networks today. We then examine the services provided using the TDM networking paradigm in the Narrowband-Integrated Services Digital Network (N-ISDN).

6.1 CIRCUIT SWITCHING

The following section reviews the long history of circuit switching and how it continues to exert a strong influence on the design of modern communications networks.

6.1.1 History of Circuit Switching

Circuit switching originated in the public telephone network. The first telephone networks had a dedicated electrical circuit from each person to every other person that desired communication, which is essentially a PVC service. This type of connectivity makes sense if you talk to very few people and very few people talk to you. Now let's move forward to the modern day, where the typical person makes calls to hundreds of different destinations for friends and family, business or pleasure. It is unrealistic to think that in this modern age each of these call origination and destination points would have its own dedicated circuit to all others, since it would be much too expensive and difficult to administer.

Historically, early telephone networks dedicated a circuit to each pair of callers until the maze of wires overhead on telephone poles began to block out the sun in urban areas. The next step toward switching was human telephone operators who manually connected parties wishing to communicate using patch cords on a switchboard. Callers identified the called party by telling the operator the name of the person they wished to speak with. This design relieved the problem greatly since now all the wires from each user went back to a central operator station instead of between every pair of users. However, once the number of users grew beyond what a single operator could handle, multiple operators had to communicate in order to route the call through several manually connected switchboards to the final destination. Interestingly, the reason Almon B. Strowger invented the first electrome-chanical circuit switch in 1889 [1] wasn't motivated by engineering efficiency, but basic capitalism. As the story goes, Strowger was actually an undertaker by trade in a moderately sized town which had two undertakers. Unfortu-nately for Strowger, his competitor's wife was the switchboard operator for the town. As the telephone increased in popularity, when anyone died, their relatives called the telephone operator to request funeral services. The operator in this town routed the requests to her husband, of course, and not to Strowger. Seeing his business falling off dramatically, Strowger conceived of the electromechanical telephone switch and the rotary dial telephone so that customers could contact him directly. As a result, Strowger ended up in an entirely different, but highly successful business. Now, we take for granted the ease of picking up the phone virtually anywhere in the world and dialing any other person in the world.

6.1.2 The Basic Circuit Switching Paradigm

Figure 6.1 illustrates the basic process involved in establishing a circuit switched connection between two users, A and B. In the example, user A employs a signaling protocol to request a circuit to the called party, B. An example of a signaling protocol is the dialed digits from a telephone set. The network determines a path and propagates the circuit request to the switch connected to user B, and relays the circuit establishment request, for example, ringing the telephone. Telephone networks typically ring the called party within several seconds after the calling party enters the last digit of the phone number. Typically, telephone engineers refer to this time interval as post dial delay. Call setup delay in data communications over the voice network adds an additional 10 to 30 seconds for modem training. While the network alerts the called party, it provides a progress indication to the calling party A, for example, via a ringing tone. If the called party, B, accepts the request, then the network sets up a circuit and dedicates it to the parties until they release the connection. The parties then engage in an information transfer phase as indicated in the figure. Once the parties complete a higher layer protocol exchange, for example, the wrong number scenario indicated in the figure, either party may initiate a disconnect just by hanging up the phone. The network provides an indication to the other party of disconnection, typically silence, and eventually dial tone or an error message if the other party doesn't hang up the phone, until the other party confirms disconnection by hanging up the phone.

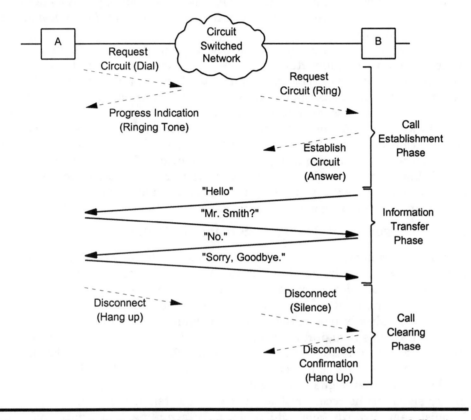

Figure 6.1   Circuit Switching Call Establishment, Information Transfer, and Clearing

Of course, the circuit switching protocol handles a number of exceptional conditions, such as network busy, called party busy, call waiting, and many others. Note that if the same two parties hang up and call each other back later, the telephone network establishes another circuit in the same manner, but not necessarily over the same exact path as before. In this manner, telephone networks share common resources (i.e., circuits and trunk ports) between many users in non-overlapping time intervals, achieving an economy of scale since the average telephone call between people lasts only 3 to 5 minutes. More recently, telephone user behavior has changed when using dial-up access to the Internet, where the holding time may be on the order of hours instead of minutes. This change in traffic patterns presents a challenge to the engineers designing the local exchange switch and the circuits connected to the Internet service provider.

Today, computers "talk" to each other in the same manner using a modem to convert data to signals compatible with the analog access lines connected to the circuit switched voice network. The word *modem* is a contraction for modulator/demodulator, a process that converts digital data into waveforms for transmission over analog access lines. Chapter 26 describes the operation of modems in more detail. The modem presents a DCE interface toward the computer equipment, and outputs a signal that is compatible with a standard phone line. See References 2 and 3 for more information on low-speed modem communications.

The computer application establishes a point-to-point telephone call whenever data communication is needed, for example, when starting up a web browser, and leaves the connection up to allow data transmission. Either the end user disconnects the circuit when he or she are done, or the application automatically disconnects, for example, after a period of inactivity. Since circuit switching is a form of connection-oriented service, the entire circuit bandwidth is dedicated for the duration of the call. Circuit switching is ideal for applications that require constant bandwidth but can tolerate call establishment and disconnection times on the order of many seconds. For example, most users accept waiting 10 to 30 seconds to connect to an Internet service provider for an hour long session of surfing the web.

6.1.3 Digitized Voice Transmission and Switching

Bell Labs engineers faced a decision in the 1950s of either augmenting bundles of twisted pairs run in conduits under the streets in large metropolitan areas or multiplexing more voice conversations onto the existing bundles using a new digital technique. The engineering economic decision drove them to deploy a radically new entirely digital technique. It converted the analog voiceband signals to digital information prior to transmission as illustrated in Figure 6.2. Nyquist derived a theorem in 1924 proving that the digital samples of an analog signal must be taken at a rate no less than *twice* the bandwidth of that signal to enable accurate reproduction of the original analog signal at the receiver. Thirty years later, telephone engineers put the Nyquist sampling theorem into practice by sampling a standard 4000-Hz bandwidth voice channel at 8000 samples per second. Employing 8 (or 7) bits per sample yields the standard 64-kbps (or 56-kbps) digital data stream used in modern digital Time Division Multiplexing (TDM) transmission and switching systems for each direction of a voice channel. Engineers call the digitized coding of each analog voice sample Pulse Code Modulation (PCM). The numerical encoding of each PCM sample uses a nonlinear companding (COMpression/exPANDING) scheme to improve the signal-to-noise ratio by providing greater granularity for larger amplitude values. Unfortunately, the methods used for representing PCM samples differ in networks around the world, with the µ-Law (pronounced "mew-law") standard used in North America and an A-Law method used elsewhere. In fact, while many networks still transmit voice at 56- or 64-kbps, more sophisticated modern digital encoding techniques now enable transmission of a voice channel at speeds as low as 8-kbps (with some loss in quality, of course) when bandwidth is expensive or scarce. International networks and voice over limited bandwidth applications are the primary applications for these low bit-rate techniques.

The transmission system then multiplexes 24 such digitized and sampled voice channel (called a Digital Stream 0 (DS0) in North America) onto a single twisted pair using a T1 repeater signal according to a Digital Stream 1 (DS1) signal format. The DS1 transmission rate of 1.544-Mbps derives from multiplying the DS0 rate of 64-kbps by 24 (i.e., 1.536-Mbps), plus an 8-kbps framing and signaling channel. We review the DS1 signal format later in this chapter since it forms the basis of the narrowband ISDN primary rate interface. This design decision resulted in an improvement of over 2400 percent in utilization of the scarce twisted pair resource— a tremendous gain in efficiency! International standards adopted a similar multiplexing technique to make better use of existing twisted pair plant, but multiplexed together thirty-two 64-kbps channels instead of twenty-four in a standard called E1 operating at 2.048-Mbps. We also cover this format later in the chapter in the section on N-ISDN.

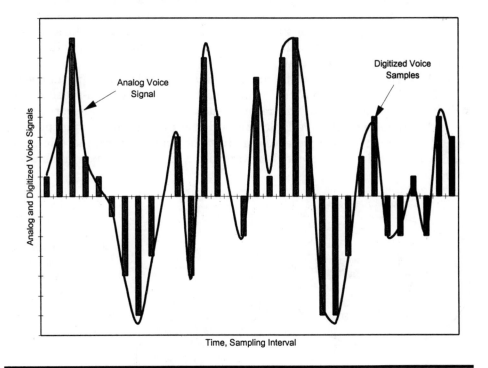

Figure 6.2 Illustration of a Digitized, Sampled Voiceband Signal

6.1.4 Digital Data Circuit Switching

If the access line from the customer is digital instead of analog and the network has digital switches, then another set of digital data circuit switching services apply that eliminate the need for analog to digital conversion in a modem. Of course, data circuit switching arrived only after carriers replaced older analog telephone switches with updated entirely digital versions. Now the availability of digital data service is limited primarily to whether the user's access line is digital or analog. Since Time Division Multiplexing (TDM) uses 8000 samples per second per DS0 channel, a difference arises from the fact that 56-kbps uses only 7 bits per sample, while 64-kbps uses all eight. The 56-kbps rate resulted from the historical use by the North American telephone network of 1 bit per sample for robbed bit signaling.

Switched 56-kbps, or simply switched 56, is a service offered in both the private and public networking environments. Often a Channel Service Unit/Data Service Unit (CSU/DSU) device attaches via a dedicated digital access line to a carrier's switched 56-kbps service. The DSU side presents a standard DCE interface to the computer equipment, as described in Chapter 4. Users employ data circuit switching as a backup for private line services or for on-demand applications. The price of circuit switched data services is close to that of voice service, since that is basically what it is! This pricing makes it a cost-effective option to leased-line services if usage is less than several hours per day, or if multiple destinations require dynamic connectivity. The data communications user, however, needs up to three logical types of communication for one call: the data circuit, a signaling capability, and optionally a management capability.

Many carriers now offer data circuit switching services ranging in speeds from 56/64-kbps up to 1.5-Mbps. Applications which use high-speed circuit switching as an ideal solution are ones such as bulk data transport and/or those that require all the available bandwidth at predetermined time periods. Circuit switching can provide cost reductions and improve the quality of service in contrast to dedicated private lines, depending upon application characteristics.

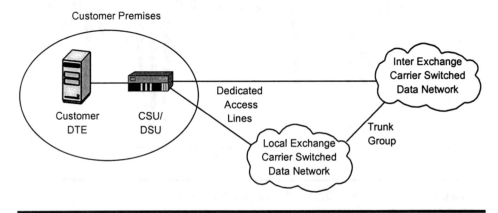

Figure 6.3   Switched Services Interfaces

The interface for switched services can be directly from the CPE to the IntereXchange Carrier (IXC) Point of Presence (PoP), or via a Local Exchange Carrier's (LEC) Switched data service as depicted in Figure 6.3. Dedicated access lines connect the customer equipment to the LEC and IXC carrier services. A trunk group of many circuits connect the LEC and the IXC, achieving an economy of scale by sharing these circuits between many users. Many of these trunk groups carry both digitized voice and data calls. A common use of this configuration is use of the LEC switched service as a backup to the dedicated access line to the IXC circuit switched data service.

Many users implement circuit switched data services as a backup for private lines as illustrated above, or else for occasional, on-demand transfer of data between sites. The typical traffic is long duration, relatively constant bandwidth data, generated by applications such as batch file transfers, database backups, or digital video conferences.

Most carriers offer circuit switched data services that operate at the 56/64-kbps level, but many now also offer services operating at higher speeds, such as nxDS0, DS1, and even DS3 (45-Mbps). Some carriers also offer non-contiguous and contiguous fractional DS1 or nxDS0 reconfigurable or switched services. Reconfigurable services often utilize a computer terminal to rearrange digital cross connects to provide a version of nxDS0 switching. Depending upon whether the control system utilizes semi-automated network management or signaling based control, the circuit establishment times in these services ranges from seconds to minutes. The N-ISDN based version of this service is called the MultiRate Circuit Mode Bearer Service (MRCMBS) which supports switched nxDS0. Videoconferencing is an example application that employs MRCMBS to combine multiple 56/64-kbps circuits to form a single high-speed videoconference channel at higher speeds. Some examples of switched DS1 service traffic include video, imaging, large file transfers, and data center disaster recovery.

6.2 PRIVATE LINE NETWORKS

Users have three physical layer networking options: private line networks, switched networks, and hybrid designs incorporating a mix of both. This section covers some practical aspects of private line networking and concludes with a comparison of private line and circuit switched networks.

6.2.1 Private (Leased) Line Characteristics

Private lines are the simplest form of point-to-point communications. Private lines, also called leased lines, are dedicated circuits between two user locations. Since a service provider dedicates bandwidth to a private line connecting ports on its network, the customer pays a fixed monthly fee dependent upon the distance traversed and the bit rate ordered. Usually, private line tariffs also have a non-recurring installation fee. In return, the service provider guarantees the private line bandwidth effectively 24 hours a day, 7 days a week.

Leased lines come in several grades and speeds. The most basic traditional service available consists of either analog or digital leased lines. Carriers offer digital leased lines at speeds such as 9600-bps, 19.2-kbps, 56/64-kbps,

Fractional-T1, T1 (1.544-Mbps) and higher speeds of the TDM hierarchy defined in the next section. Analog lines require a modem for digital to analog conversion, while digital private lines require a DCE (commonly called a channel service unit/digital service unit (CSU/DSU)) for line conditioning, framing, and formatting. Most local exchange, interexchange, alternate access providers, and international carriers offer digital private line services.

6.2.2 Private Line Networking

Figure 6.4 depicts an network of three user DTEs connected via private lines. User A has a dedicated 56-kbps circuit to user B, as well as a dedicated T1 (1.544-Mbps) circuit to user C. Users B and C have a dedicated 1.544-Mbps circuit between them. Users generally lease a private line when they require continuous access to the entire bandwidth between two sites. The user devices are either voice Private Branch Exchanges (PBXs), T1 multiplexers, routers, or other data communications networking equipment. The key advantage of private lines is that a customer has complete control over the allocation of bandwidth on the private line circuits interconnecting these devices. This is also the primary disadvantage in that the customer must purchase, maintain, and operate these devices in order to make efficient use of the private line bandwidth. Up until the 1980s, most voice networks were private line based, primarily made up of dedicated trunks interconnecting PBXs. The situation changed once carriers introduced cost-effective intelligent voice network services. Now most corporate voice networks use these carrier services. Data networking appears to be moving along a similar trend toward shared carrier provided public data services. In the early 1990s virtually all corporate data networks were private line based. Now, many corporate users have moved from private lines to embrace frame relay, the Internet and ATM for their data networking needs.

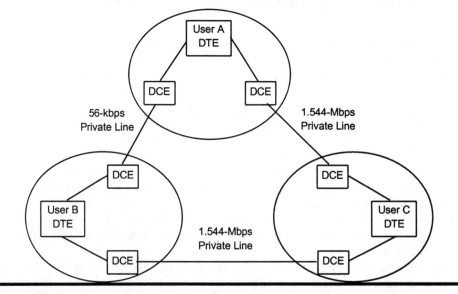

Figure 6.4 Example Private Line Network

While private lines provide dedicated bandwidth, carriers don't guarantee 100 percent availability. Sometimes, a carrier provides for recovery of private line failures via digital cross connects, transmission protection switching or SONET rings. However, in many cases a private line comprises several segments across multiple carriers. For example, a long-haul private line typically has access circuits provided by local exchange carriers on each end and a long distance segment in the middle provided by an inter-exchange carrier. If the private line or any of its associated transmission equipment fails (e.g., a fiber cut) the end users cannot communicate unless the user DTEs have some method of routing, reconnecting, or dialing around the failure. Thus, the user must decide what level of availability is needed for communications between sites.

One option for assuring high availability in private line networks is Automatic Line Protection Switching (ALPS) technology. Basically, ALPS employs diversely routed circuits at the same speed between a pair of sites, automatically detecting a failure on the primary circuit and switching to the standby circuit. Figure 6.5 depicts the use of an ALPS system between user A and user C for the private line network previously described. Now two 1.544-Mbps trunks connect ALPS equipment at sites A and C. If one T1 fails, the ALPS system automatically switches traffic to the alternate 1.544-Mbps trunk, typically within 50 ms or less. Note that the second T1 line (i.e., the standby) cannot carry traffic in this 1:1 protection scheme — it serves only as backup. Newer devices have some 1:N redundancy, which means that 1 standby circuit protects N working circuits. ALPS can be applied to the backbone only, to access circuits, or on an end-to-end basis.

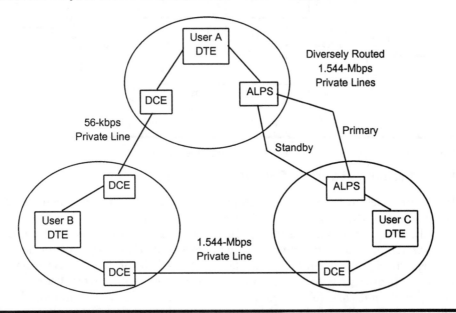

Figure 6.5 Leased Line with Automatic Line Protection Switching (ALPS)

6.2.3 Inverse Multiplexers (I-Muxes)

Early digital private line networks offered service at only 56/64-kbps or DS1/E1 (i.e., 1.5 and 2.0-Mbps) rates. The gap in speed and price between these speeds created a market for inverse multiplexers, commonly called I-Muxes, that provided intermediate speed connectivity by combining multiple lower-speed circuits. As illustrated in Figure 6.6, an I-Mux provides a single high-speed DTE-DCE interface by combining n lower-speed circuits, typically private lines. Inverse multiplexers come in two major categories: nx56/64-kbps, and nxDS1/E1. The inverse multiplexer provides a DCE interface to the DTE operating at a rate of approximately 56/64-kbps or DS1/E1 times n, the number of circuits connecting the I-Muxes. These devices automatically change the DTE-DCE bit rate in response to circuit activations or deactivations. The I-Muxes also account for the differences in delay between the interconnecting circuits. The actual serial bit rate provided to the DTE is slightly less than the nx56/64-kbps or nxDS1/E1 rate because of overhead used to synchronize the lower-speed circuits. Some I-Muxes also support circuit switched data interconnections in addition to private line connections. A bonding standard defines how nxDS0 I-Muxes interoperate. Higher speed nxDS1/E1 I-Muxes utilize a proprietary protocol, and are hence incompatible among vendors. Chapter 12 describes an ATM Forum standard defined for Inverse Multiplexing over ATM (IMA) at the nxDS1/E1 rates over dedicated private lines.

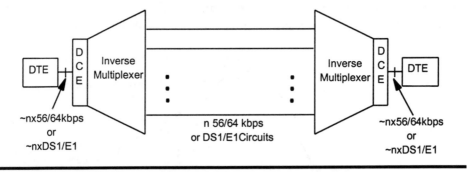

Figure 6.6 Illustration of Inverse Multiplexer Operation

6.2.4 Permanent versus Switched Circuits

We come back to the basic concept of permanent and switched circuits repeatedly in this text, so we begin with a simple example here as an introduction to the basic tradeoffs involved. Figure 6.7 shows a simplified comparison of two communications networks connecting eight users, labeled A through H, which could be LANs, MANs, PBXs or hosts. Network (a) shows dedicated private line circuits connecting each and every user to each other, while network (b) shows circuit switched access to a common, shared network with only a dedicated access line to the network for each user. In network (a) each user has seven access lines into the network for dedicated circuits connecting to a distant access line for each possible destination.

Circuit switching transfers data or voice information at the physical layer. In other words, circuit switching is transparent to higher-layer protocols, which means that the network does not process the information content. The example in the circuit switched network (b) shows User A talking to User H, and User D talking with User E. Any user can communicate with any other user, although not simultaneously, just like in the telephone network.

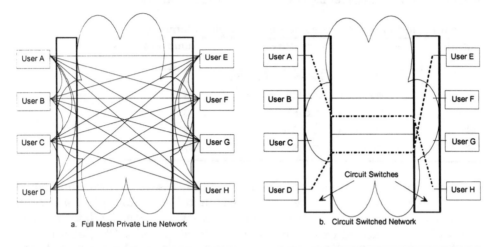

Figure 6.7   Full-Mesh Private Line Networking versus Circuit Switching

The signaling protocols employed by narrowband ISDN, X.25, frame relay and ATM all use the same basic circuit switching paradigm described in the previous section, except that the end device is usually some kind of computer instead of a telephone. The X.25, frame relay and ATM protocols all have a Switched Virtual Connection (SVC) capability similar to the telephone call described above. Furthermore, they also have a Permanent Virtual Connection (PVC) capability analogous to a dedicated pair of wires between each pair of end users wishing to communicate.

6.3 DIGITAL TIME DIVISION MULTIPLEXING (TDM)

Public network carriers first developed plesiochronous digital transmission for economical, high-quality transmission of voice signals. Later on, the carriers used these same transmission systems to offer private line data services. In the 1990s, North American carriers began deployment of the Synchronous Optical NETwork (SONET) while the rest of the world deployed Synchronous Digital Hierarchy (SDH) based transmission systems. The SONET and SDH systems provide higher speeds, standardized interfaces, automatic restoration, and superior transmission quality. This section reviews some basics of these important TDM technologies.

6.3.1 Plesiochronous Digital Hierarchy

The plesiochronous (which means nearly synchronous) digital hierarchy was developed in the 1950s by Bell Labs to carry digitized voice over twisted wire more efficiently in major urban areas. This evolved first as the North American Digital Hierarchy, depicted in Table 6.1. The convention assigns a level to each Digital Stream (DS) format in the hierarchy.

Table 6.1   North American Plesiochronous Digital Hierarchy

Signal Name	Rate	Structure	Number of DS0s
DS0	64 kbps	Individual Time Slot	1
DS1	1.544 Mbps	24xDS0	24
DS1c	3.152 Mbps	2xDS1	48
DS2	6.312 Mbps	2xDS1c	96
DS3	44.736 Mbps	7xDS2	672

Plesiochronous transmission systems multiplex several lower-numbered digital streams into the higher-numbered digital streams within a certain frequency tolerance. No fixed relationship exists between the data between levels of the hierarchy, except at the lowest level, called a DS0, at a rate of 64-kbps. Figure 6.8 illustrates a convention commonly used in North America to label multiplexing between the various levels of the hierarchy depicted in Table 6.1. For example, an M1C multiplexer converts two DS1s into a DS1c signal. A M12 mutiplexer takes two DS1c signals and multiplexes these into a DS2 signal. Finally, an M13 multiplexer takes 7 DS2 signals and combines these into a single DS3 signal. Hence, an M13 multiplexer converts 28 DS1s into a DS3 signal, but uses the M1C and M12 intermediate multiplexing stages to do so.

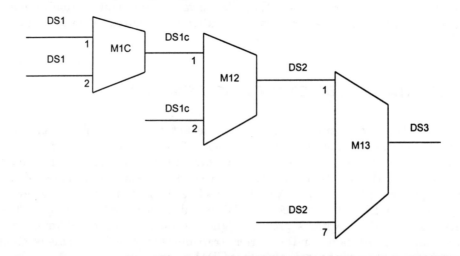

Figure 6.8   North American Plesiochronous Digital Hierarchy Multiplexing Plan

Bell Labs also defined a transmission repeater system over a four-wire twisted pair and called it T1. Many trade press articles and even some technical books use the term "T1" to colloquially refer to a DS1 signal. There is actually no such thing as a "T3" signal, even though many people use this term colloquially when referencing a DS3 signal. The actual interfaces for DS1 and DS3 are called the DSX1 and DSX3 interfaces, respectively, in ANSI standards. The DSX1 is a four-wire interface, while the DSX3 interface is a dual coaxial cable interface.

Europe and Japan developed different plesiochronous digital hierarchies as summarized in Table 6.2 [4]. All of these hierarchies have the property that multiplexing is done in successive levels to move between successively higher speeds. Furthermore, the speed of each level is asynchronous with respect to the others within a certain frequency tolerance.

An important consequence of these digital hierarchies on data communications is that only a discrete set of fixed rates is available, namely nxDS0 (where $1 \leq n \leq 24$ in North America and Japan and $1 \leq n \leq 30$ in Europe), and then the next levels in the respective multiplex hierarchies. The next section on N-ISDN defines the details of the DS0 to DS1 and E1 mappings. Indeed, one of the early ATM proposals [5] emphasized the capability to provide a wide range of very granular speeds as a key advantage of ATM over TDM.

Table 6.2   Summary of International Plesiochronous Digital Hierarchies

Digital Multiplexing Level	Number of Voice Channels	Bit Rate (Mbps) North America	Europe	Japan
0	1	0.064	0.064	0.064
1	24	1.544		1.544
	30		2.048	
	48	3.152		
2	96	6.312		6.312
	120		8.448	7.876
3	480		34.368	32.064
	672	44.376		
	1344	91.053		
	1440			97.728
4	1920		139.268	
	4032	274.176		
	5760			397.200
5	7680		565.148	

6.3.2 SONET and the Synchronous Digital Hierarchy (SDH)

The Bellcore driven North American standards defined a Synchronous Optical NETwork (SONET) while the CCITT/ITU developed a closely related international Synchronous Digital Hierarchy (SDH) in the late 1980s. These standards were the next step in the evolution of Time Division Multiplexing (TDM). SONET/SDH have two key benefits over Plesiochronous Digital Hierarchy (PDH): rates of higher speeds are defined, and direct multiplexing is possible without intermediate multiplexing stages. Direct multiplexing employs pointers in the TDM overhead that directly identifies the position of the payload. Furthermore, the fiber optic transmission signal transfers a very accurate clock rate along the transmission paths all the way to end systems, synchronizing the entire transmission network to a single, highly accurate clock frequency source.

Another key advance of SONET and SDH was the definition of a layered architecture illustrated in Figure 6.9 that defines three levels of transmission spans. This model allowed transmission system manufacturers to develop interoperable products with compatible functions. The SONET framing structure defines overhead operating at each of these levels to estimate error rates, communicate alarm conditions, and provide maintenance support. Devices at the same SONET level communicate this overhead information as indicated by the arrows in the figure. The path layer covers end-to-end transmission, where ATM switches operate as indicated in the figure. This text refers to a transmission path using this definition from SONET. Next, comes the maintenance span, or line layer, which comprises a series of regenerators (or repeaters). An example of a line layer device is a SONET cross connect. The section regenerator operates between repeaters. Finally, the photonic layer involves sending binary data via optical pulses generated by lasers or Light Emitting Diodes (LEDs).

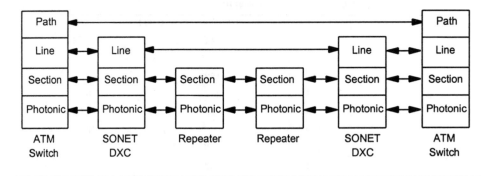

Figure 6.9 Example of SONET Architecture Layers

SONET standards designate signal formats as Synchronous Transfer Signals (STS) at N times the basic STS-1 (51.84-Mbps) building block rate by the term STS-N. SONET designates signals at speeds less than the STS-1 rate as Virtual Tributaries (VTs). The optical characteristics of the signal that carries SONET payloads is called the Optical Carrier (OC-N). An STS-N signal can be carried on any OC-M, as long as M≥N. The standard SONET STS and VT rates are summarized below.

The CCITT/ITU developed a similar synchronous multiplex hierarchy with the same advantages using a basic building block called the Synchronous Transfer Module (STM-1) with a rate of 155.52-Mbps, which is exactly equal to SONET's STS-3 rate to promote interoperability between the different standards. The SDH standards also define a set of lower-speed signals, called Virtual Containers (VCs). Therefore, a direct mapping between the SONET STS-3N rates and the CCITT/ITU STM-N rates exists. An STM-1 frame is equivalent to an STS-3c frame in structure. The overhead byte definitions differ between SONET and the SDH so that direct interconnection is currently not possible. Also, some incompatibilities between SONET and SDH remain in the definition and use of overhead information. Table 6.3 shows the SONET speed hierarchy by OC-level and STS-level as it aligns with the international SDH STM levels and the bit rates of each.

Table 6.3   SONET STS-N/OC-N and SDH STM-M Speed Hierarchy

SONET Level	SDH Level	Bit Rate (Mbps)
STS-1	-	51.84
STS-3, OC-3	STM-1	155.52
STS-12, OC-12	STM-4	622.08
STS-24	-	1,244.16
STS-48, OC-48	STM-16	2,488.32
STS-192, OC-192	STM-64	9,953.28

Table 6.4   SONET/SDH Digital Hierarchy Payload and Overhead Rates

North American SONET	CCITT/ITU SDH	Payload Carried (Mbps)	Payload + Overhead (Mbps)
VT1.5	VC11	1.544	1.728
VT2.0	VC12	2.048	2.304
VT3.0		3.152	3.392
VT6.0	VC2	6.312	6.848
DS3		44.736	50.112
	VC3	34.368, 44.736	48.960
E4	VC4	139.264	139.264
ATM on STS-1		49.536	50.112
ATM on STS-3c	ATM on STM-1	149.760	150.336
ATM on STS-12c	ATM on STM-4	599.040	601.344

Table 6.4 illustrates a similar mapping to that of Table 6.3, comparing the mapping of the North American and CCITT/ITU PDH rates to the corresponding SONET Virtual Tributary (VT) and SDH Virtual Container (VC) rates and terminology. Note that the common 1.5, 2, 6 and 44 Mbps rates are mapped consistently. The other common rates are 155 and 622 Mbps; which is the focus of ATM standardization activity. The rates indicated in the table include the actual payload plus multiplexing overhead, including path

overhead. The table includes ATM carried payload rates for commonly used ATM mappings over SONET/SDH for comparison purposes.

6.3.3 Basic SONET Frame Format

Figure 6.10 illustrates the SONET STS-1 frame format. Notice that the frame is made up of multiple overhead elements (section, line, and path) and a Synchronous Payload Envelope (SPE). The frame size for an STS-1 SPE is 9 rows × 90 columns (1 byte per column) for a total of 810 bytes, comprised of a 783-byte frame (excluding the 27 bytes section and line overhead). The total STS-1 frame of 810 bytes transmitted each 125 μs results in the basic STS-1 rate of 51.84-Mbps. The STS-N SPE format essentially replicates the STS-1 format N times.

STS - 1 Frame

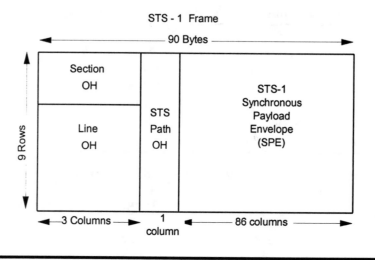

Figure 6.10   STS-1 Frame Format

Figure 6.11 illustrates the mapping of VT1.5s into an STS-1 SONET Synchronous Payload Envelope (SPE). Each VT1.5 uses 27 bytes to carry 24 bytes of a DS1 payload. The first column of 9 bytes is the STS-1 path overhead. The next 28 columns are bytes 1 through 9 of the (28) VT1.5 payloads, followed by a column of stuff bytes. Similarly, columns 31 through 58 are bytes 10 through 18 of the (28) VT1.5 payloads, followed by a column of stuff bytes. The last 28 columns are bytes 19 through 27 of the VT1.5 payloads.

Figure 6.12 shows the format of an individual VT1.5. Note that there are 27 bytes that are transferred every 125 μs in the SPE as defined above, but that only 24 bytes are necessary to carry the user data. User data byte 25 is included to be able to carry the DS1 framing bit transparently. The other two bytes provide a pointer so that the VT can "float" within its allocated bytes and thus allow for timing to be transferred, and the provision of VT level path overhead. The SONET overhead results in a mapping that is less than 100 percent efficient. In fact, VT1.5 multiplexing for support of individual DS0s is approximately 85.8 percent efficient (i.e., 24×28/87/9) in terms of SPE bandwidth utilization. As shown in Chapter 17, ATM support for transport of DS0s using structured mode circuit emulation is approximately 87.4 percent efficient over SONET.

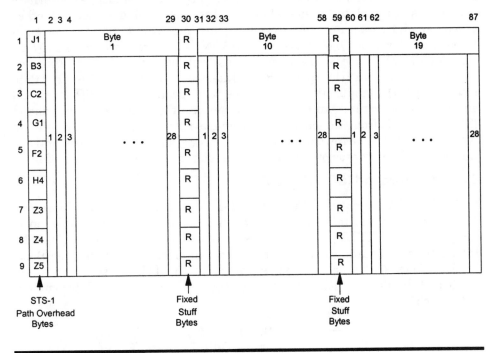

Figure 6.11 VT1.5 Mapping within STS-1 Frame

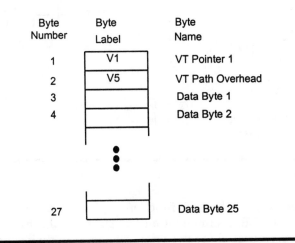

Figure 6.12 Illustration of SONET VT1.5 Format

6.4 BASICS AND HISTORY OF NARROWBAND ISDN (N-ISDN)

N-ISDN is the phoenix of the late 1990s. Rising out of the ashes of the its slowly smoldering adoption in the preceding decade, it takes its place as a switched alternative to the dedicated 56-kbps analog local loop, an alternate dial access protocol to services like the Internet, and a variety of other consumer and corporate uses. We now turn our attention to the original Integrated Services Digital Network (ISDN) standards, where the frame relay and ATM protocols discussed in this book have their roots. First, we summarize the N-ISDN Basic Rate Interface (BRI) and Primary Rate Interface (PRI) configurations. Next, the text covers basic N-ISDN protocol and framing structure. In order to differentiate the ATM-based Broadband ISDN (B-ISDN) from the earlier narrowband ISDN, the standards refer to these protocols as Narrowband ISDN (N-ISDN), a term we use consistently in this book to avoid confusion. See Reference 6 for more N-ISDN information.

6.4.1 Narrowband ISDN Basics

N-ISDN builds upon the Time Division Multiplexing (TDM) hierarchy developed for digital telephony. Although most N-ISDN standardization is complete, the CCITT/ITU-T continues to define new standards for the N-ISDN. Two standards exist for the physical interface to N-ISDN: the Basic Rate Interface (BRI), or basic access, as defined in ITU-T Recommendation I.430, and the Primary Rate Interface (PRI), as defined in ITU-T Recommendation I.431.

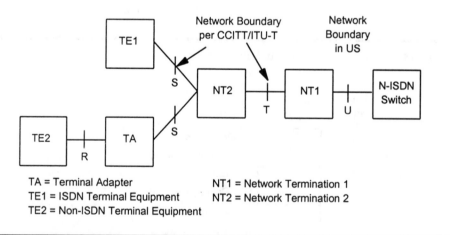

Figure 6.13   N-ISDN Reference Configuration

Figure 6.13 illustrates the N-ISDN functional groupings and reference points (as defined in ITU-T Recommendation I.411). Both the BRI and PRI standards define the electrical characteristics, signaling, coding, and frame formats of N-ISDN communications across the user access interface (S/T) reference point. The physical layer provides transmission capability, activation, and deactivation of Terminal Equipment (TE) and Network Termination (NT) devices, Data (D)-channel access for TEs, Bearer (B) channels for TEs, maintenance functions, and channel status indications. ITU-T Recommendation I.412 defines the basic infrastructure for these physical implementations as well as the detailed definition for the S and T reference points, TEs, and NTs.   The TA manufacturer defines the R reference to non-ISDN terminal equipment. The CCITT/ITU-T defines two possible network interface points at the S and T reference points where a carrier always places equipment on the customers premises. In the United States (US), no formal network boundary exists; however, ANSI standards define this as the U reference point as indicated in the Figure.

The BRI and PRI N-ISDN interfaces provide a set of Bearer (B) Channels and a Data (D) channel as described above. As illustrated in Figure 6.14 the B-Channels provide a layer one service to the N-ISDN terminal equipment, while the D-Channel supports a layer three signaling protocol for control of the B-channels.   Optionally, end user N-ISDN equipment may transfer limited amounts of packet data over the D-channel. Note that the NT1 device operates at only layer one for the D-channel. As we shall see, B-ISDN utilizes a similar concept when labeling the signaling protocol the control plane and the bearer capabilites the user plane.

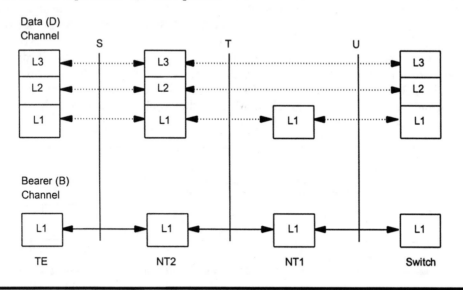

Figure 6.14   Illustration of N-ISDN B-Channel and D-Channel Services

6.4.2 BRI and PRI Service and Protocol Structures

The N-ISDN Basic Rate Interface (BRI) and Primary Rate Interface (PRI) service configurations are defined as follows:

- **Basic Rate Interface (BRI):** provides two 64-kbps Bearer (B) channels for the carriage of user data and one 16-kbps control, messaging, and network management data (D) channel. Documentation commonly refers to the BRI as a 2B+D interface. The BRI was intended for customer access devices such as N-ISDN voice, data, and videophone. Many Internet service providers now use the BRI through the local telephone company to provide high performance access to the Internet at speeds up to 128-kbps.

- **Primary Rate Interface (PRI):** provides twenty-three 64-kbps Bearer (B) channels and one 64-kbps Data (D) signaling channel in North America, commonly referred to as a 23B+D interface. Internationally, 30 B channels are provided in a 30B+D configuration. The PRI was intended for use by higher bandwidth or shared customer devices such as the Private Branch eXchange (PBX), routers, or T1 multiplexers.

The N-ISDN PRI provides a single 1.544-Mbps DS1 or a 2.048-Mbps E1 data rate channel over a full duplex synchronous point-to-point channel using the standard Time Division Multiplexing (TDM) hierarchy introduced earlier in this chapter. CCITT Recommendations G.703 and G.704 define the electrical characteristics and frame formats of the PRI interface, respectively. The 1.544-Mbps rate is accomplished by sending 8000 frames per second with each frame containing 193 bits. Twenty-four DS0 channels of 64-kbps each comprise the DS1 stream. Figure 6.15 shows the format of the DS1 PRI interface. Note that the 193rd framing bit is defined by DS1 standards for error rate estimation and maintenance signaling. A DS1 PRI contains at least 24 B-channels. The 24th DS0 time slot either contains the signaling (D) channel, or a 24th Bearer (B) channel if another D channel controls this DS1.

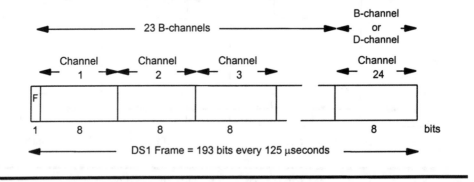

Figure 6.15   1.544-Mbps North American N-ISDN PRI Frame Structure

The CCITT/ITU-T E1-based PRI interface differs somewhat from the DS1 interface as shown in Figure 6.16. The E1 PRI has 30 B-channels, one 64-kbps D-channel in time slot 16, and a channel reserved for physical layer signaling, framing, and synchronization in time slot 0. A primary attribute distinguishing N-ISDN service from telephony is the concept of common channel signaling, or out-of-band signaling using the D-channel. The D-channel and B-channels may share the same physical interface as indicated in the above illustrations, or the D-channel on one interface may control the B-channels on several physical interfaces.

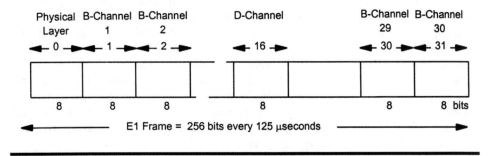

Figure 6.16   CCITT/ITU-T 2.048-Mbps N-ISDN PRI Frame Structure

Since PRIs run at higher speeds, they support additional bearer capabilities called *H-channels*. Two types are defined: H_0-channel signals that have a bit rate of 384-kbps and H_1-channels defined for DS1 and E1 PRIs. H_{11}-channels have a bit rate of 1536-kbps in the United States and Japanese DS1 PRIs, while H_{12}-channels operate at 1920-kbps on E1 PRIs in Europe and other parts of the world. The H_0 channel uses any six time slots in the same DS1 or E1; that is, the time slots need not be contiguous. The H_{11} channel uses all 24 time slots in a DS1, which means that the D signaling channel must be on a separate physical interface.

Standards also define a capability to establish an n×64-kbps bearer service, where n ranges from 1 to 24 (or 30 at the European channel rate) via N-ISDN signaling. The nxDS0 service uses n contiguous time slots or a bit map specified set of DS0 time slots in the DS1 or E1 frame. Standards call this capability the MultiRate Circuit Mode Bearer Service (MRCMBS). Also, N-ISDN signaling can establish a Frame Mode Bearer Service (FMBS) or a switched X.25 connection as discussed in the next Chapter.

6.4.3 ISDN D-Channel Signaling

CCITT Recommendation Q.931 defines a set of message types containing a number of Information Elements used to establish and release N-ISDN bearer connections. Figure 6.17 illustrates an example sequence of such messages involved during the establishment, data transfer and release phase of a N-ISDN call. Starting in the upper left hand corner of the figure, terminal equipment A places a call to B via the N-ISDN network using the SETUP message. The network responds with a CALL PROCEEDING message and relays the call request to the destination, issuing a SETUP message to B. In the example, B alerts the called user, and informs the network of this action via the ALERTING message, which the network relays to A. Eventually, B answers the call with the CONNECT message, which the network acknowledges using the CONNECT ACKNOWLEDGE message. The N-ISDN network also relays the connect indication back to the calling party A, indicating the response using the CONNECT message. Usually, the

CONNECT message identifies the time slot(s) assigned to the bearer connection. Once A confirms the completed call using the CONNECT ACKNOWLEDGE message, A and B may transfer data over the bearer channel established by the above signaling protocol for an indefinite period of time. Either party may release the connection by sending the DISCONNECT message to the network. In our example, A initiates call clearing, which the network propagates to user B in the form of the DISCONNECT message. The normal response to a DISCONNECT message is the RELEASE message followed by a RELEASE COMPLETE message as indicated in the Figure. The information element content of the N-ISDN signaling messages (as well as some additional messages not included in this simple example) emulate all of the traditional functions available in telephony, along with support for a number of additional advanced features.

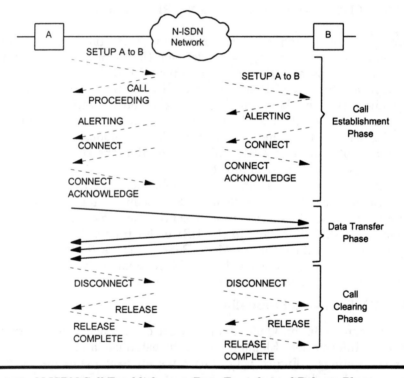

Figure 6.17   N-ISDN Call Establishment, Data Transfer and Release Phases

6.5 REVIEW

This chapter introduced the basic concepts of circuit switching and private line based networking, providing design recommendations throughout. Study of these subjects is important since the circuit switching paradigm described here appears again in X.25, frame relay and ATM. Next, the text described the digital Time Division Multiplexing (TDM) hierarchy and its evolution to the North American Synchronous Optical NETwork (SONET) and the international Synchronous Digital Hierarchy (SDH). Originally designed to provide more cost-effective telephone calls, TDM now provides the foundation for the high-performance digital data communications central to ATM. Upon this foundation, the standards bodies constructed the Narrowband Integrated Services Digital Network (N-ISDN) protocol model. As studied in Part 3, the ATM-based Broadband ISDN (B-ISDN) protocol adopts the concepts of separate user, control, and management protocols from N-ISDN.

6.6 REFERENCES

[1]   D. Bear, *Principles of Telecommunication — Traffic Engineering*, Peter Petringus, Ltd, 1976.

[2]   G. Held, R. Sarch, *Data Communications*, McGraw-Hill, 1995.

[3]   R. Dayton, *Telecommunications*, McGraw-Hill, 1991.

[4]   B. Keiser, E. Strange, *Digital Telephony*, Van Nostrand Reinhold, 1985.

[5]   J. Turner, "Design of an Integrated Services Packet Network," *IEEE Transactions on Communications*, November 1986.

[6]   G. Kessler, *ISDN, Concepts, Facilities and Services, 2d ed.*, McGraw-Hill, 1993.

7

Connection-Oriented Protocols — X.25 and Frame Relay

This chapter presents an overview of packet -switching and frame forwarding protocols that drive the major public and private connection-oriented data services, namely X.25 packet -switching and frame relay. The text describes aspects of each protocol, beginning with the origins of the protocol, followed by an overview of the packet formats and protocol functions. The coverage also illustrates the operation of the protocol through an illustrated example. The traffic and congestion control aspects of the frame relay and X.25 protocols are also surveyed. We also touch on the aspects of the protocol as supported in public services. Since ATM inherits many concepts from X.25 and frame relay, we cover the relevant subjects in sufficient detail to make this book a standalone reference for the reader without a background knowledge of these important protocols. This chapter also introduces the separation of the user and control planes central to B-ISDN and ATM in the treatment on frame relay.

7.1 PACKET -SWITCHING

The CCITT standardized X.25 as the earliest public data network protocol in 1974, continued with refinements and corrections over the years, and most recently updated the standard in 1996 [1]. This section provides an overview of key features and terminology developed for X.25 that are still used in frame relay and ATM today. X.25 packet -switching provides the network environment needed to handle intermittent terminal-to-host data traffic. The typical packet -switching application involves a user inputting keyboard data ranging from a few characters to a few lines, then forwarding the information to a host. Typically, the host then responds with a set of data ranging from many lines to a full screen display. An interval of user "think time" separates these interchanges, resulting in traffic that has a much higher peak transmission rate than the average transmission rate. The data communications industry uses the term *burstiness* to describe the ratio of peak to average transmission rates, derived from the experience with X.25 and SNA networking.

Human nature changes slower than technology. As evidence of this fact, note that this same concept of bursty communication applies today in the World Wide Web (WWW) environment. The basic paradigm for web surfing involves a user inputting a set of data ranging from a single mouse click to filling out data in a form and submitting it to the web server. The server then responds by transmitting an updated web page. Sometimes the user input kicks off the playback of an audio or video clip, or initiates a file transfer. What has changed from the days of X.25 to the web-fueled content of the Internet today is the user's power to unleash bandwidth hungry and QoS aware applications.

7.1.1 Origins of X.25

In the beginning there were proprietary protocols; then the CCITT standardized upon the first international physical, link, and packet layer protocol — X.25. The CCITT developed the X.25 packet-switching standard, along with a number of other X-series standards, to provide a reliable means of data transport for computer communications over the noisy, unreliable analog-grade transmission medium prevalent in the 1970s. By the 1980s, X.25 networks connected the entire planet. X.25 packet -switching still serves many user communities in public and private networks.

7.1.2 Protocol Structure

The CCITT set of X-series standards for the physical, link, and packet layer protocols shown in Figure 7.1 are known collectively as X.25 and were adopted as part of the OSI Reference Model (OSIRM). These standards define the protocol, services, facilities, packet-switching options, and user interfaces for public packet -switched networks.

The physical layer is defined by the X.21 and X.21bis standards. X.21 specifies an interface between Data Terminal Equipment (DTE) and Data Circuit terminating Equipment (DCE). X.21 also specifies a simple circuit switching protocol that operates at the physical layer implemented in the Nordic countries.

The data link layer standard is based upon the High-Level Data Link Control (HDLC) ISO standard [2]. X.25 modified this and initially called it a Link Access Procedure (LAP), subsequently revising it again to align with changes in HDLC resulting in the Link Access Procedure — Balanced (LAPB).

The packet layer standard is called the X.25 Packet Layer Protocol (PLP). The packet layer defines Permanent Virtual Circuit (PVC) and switched Virtual Call (VC) message formats and protocols. As we study later, the concept of PVCs and switched VCs from X.25 is also used in frame relay and ATM.

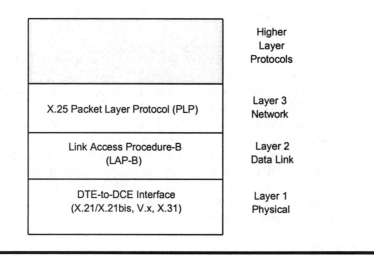

Figure 7.1   X.25 Packet -switching Compared to the OSI Reference Model

The suite of X-series standards covers the entire OSIRM protocol stack, including X.400 (E-mail) messaging and X.500 directory services [3]. The standards related to the network, data link and physical layers include: X.121, X.21, X.25, X.28, X.29, X.3, and X.32. Recommendation X.121 defines the network layer numbering plan for X.25.

CCITT Recommendations X.3, X.28, and X.29 define the method for asynchronous DTEs to interface with X.25 networks via a Packet Assembler/Diassembler (PAD) function. A PAD takes strings of asynchronous characters from the DTE and assembles these into an X.25 packet. The PAD takes data from X.25 packets and delivers asynchronous characters to the DTE. Recommendation X.3 defines PAD parameters, such as terminal characteristics, line length, break key actions, and speed. Recommendation X.28 defines the terminal-to-PAD interface, while Recommendation X.29 defines the procedures between a PAD and a remote packet mode DTE on another PAD. Recommendation X.21 defines a dedicated physical interface, and Recommendation X.32 defines a synchronous dial-up capability. X.25 also supports the V-series of modem physical interfaces, as well as Recommendations X.31 and X.32 for semi-permanent ISDN connections.

7.1.3 Networking Context

Figure 7.2 shows how the X.25 protocol layers operate in a network context interconnecting two end systems; for example, a terminal and a host. In the example, two interconnected intermediate X.25 switches transfer packets between a terminal and a host. The X.25 link layer and network layer protocols define procedures for the establishment of multiple virtual calls over a single physical interface circuit interconnecting terminals and hosts to an X.25 network. Once an X.25 virtual circuit is established, it usually traverses the same physical path between end systems. Each node operates at the physical, link, and network layers as shown in the figure. X.25 packet switches store each packet and then forward it to the next node using a link layer protocol. The transmitting switch deletes the packet from memory only after its link level peer acknowledges receipt.

Some aspects of the operation of the network layer occur only on an end-to-end basis (e.g., packet layer flow control) as indicated by the dashed arrow connecting the end systems in the figure. Of course, X.25 switches use the packet layer address to determine the forwarding path as indicated by the dashed line traversing the layers in the figure. Also, note that the internal interface between X.25 packet switches could be some other protocol, such as frame relay. End systems (e.g., terminals and hosts) also operate at layers 4 through 7 (i.e., transport through application) using either OSI-compatible protocols or other protocol suites, such as SNA or TCP/IP. Now we take a more detailed look at the X.25 protocol involved in layer 2, the link layer, and then layer 3, the network layer.

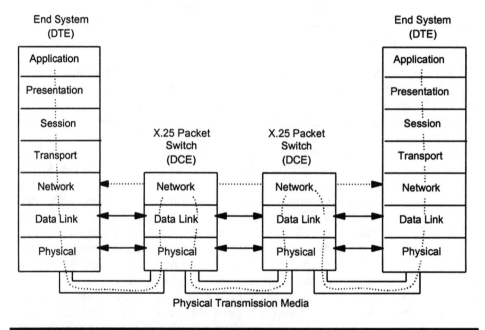

Figure 7.2 X.25 Packet -switching Networking Context

7.1.4 SDLC, HDLC, and X.25's Link Layer Protocol

This section covers the origins and details of X.25's link layer protocol. We begin with IBM's SDLC protocol and move onto the ISO's enhancements, resulting in HDLC. This section concludes with an overview of Link Access Procedures (LAP) defined by the ITU-T for X.25, ISDN and frame relay.

7.1.4.1 Synchronous Data Link Control (SDLC)

In 1973 IBM produced the first bit-oriented data communications protocol, called Synchronous Data Link Control (SDLC). Previous protocols (e.g., IBM's Bisynch) were all character oriented. SDLC, as well as subsequent bit-oriented protocols, allowed computers to transfer arbitrary binary sequences commonly encountered in programs and databases. Also, messages no longer needed to be precisely aligned on an 8-bit character boundary. The International Standards Organization (ISO) adopted this de facto standard and extended it into the widely used High-Level Data Link Control (HDLC) protocol. The present version of IBM's SDLC primarily uses the unbalanced normal response mode of HDLC together with a few proprietary commands and responses for support of polling in loop or ring topologies. SDLC operates independently on each communications link, and can operate in either multipoint or point-to-point, switched or dedicated circuit, and full- or half-duplex operation.

SDLC is steadily replacing the BSC protocol described in Chapter 4. Some improvements of SDLC over BSC include: the ability to send acknowledgments, addressing, block checking, and polling within every frame rather than in a separate sequence; the capability of handling long propagation delays; no restrictions to half-duplex; not susceptible to missed or duplicated blocks; topology independent; and character code transparency.

7.1.4.2 High-Level Data Link Control (HDLC) Terminology

The HDLC protocol is not only the most popular protocol for data link control implementations at layer 2, but also forms the basis for ISDN and frame relay protocols and services. HDLC is an international standard, defined in the document jointly produced by the ISO and the International Electrotechnical Commision (IEC) as ISO/IEC 13239 [2]. HDLC is a bit-oriented synchronous protocol passing variable length frames over either a point-to-point or multipoint network topology. HDLC operates over either dedicated or switched facilities. HDLC operates in either simplex, half-duplex, or full-duplex modes. X.25 uses the HDLC protocol for both PVCs and switched virtual calls.

There are two types of HDLC control links operating over point-to-point circuits for the HDLC connection mode: balanced and unbalanced. Both control link types work on either switched or nonswitched (i.e., dedicated) facilities. In a *balanced link*, each station is responsible for organization and transmission of the information flow, as well as the use of acknowledgments for error recovery, as illustrated in Figure 7.3. The HDLC standards name the stations as combined (primary/secondary) as indicated in the figure.

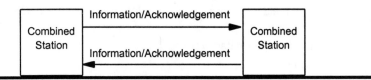

Figure 7.3 HDLC Balanced Control Link Operation

For connection-oriented HDLC, *unbalanced links* involve a primary station and a secondary station as shown in Figure 7.4. In the unbalanced link, the primary/control station polls the secondary/tributary station, which responds with information frames. The primary station then acknowledges receipt of frames from the secondary station. Information flow from the primary to the secondary station occurs within the polled information flow. The unbalanced link emulates the IBM SDLC protocol. The unbalanced link also defines procedures for operation over multipoint lines.

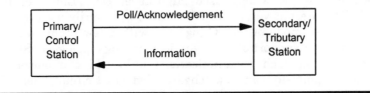

Figure 7.4 HDLC Unbalanced Control Link Operation

HDLC defines three operational modes for data transfer. The Asynchronous Balanced Mode (ABM) applies to the balanced link configuration described above. The other two types apply only to the unbalanced link configuration: the Normal Response Mode (NRM) requires that the secondary station wait for a poll command from the primary station prior to transferring data, and also Asynchronous Response Mode (ARM), which allows a secondary station to transmit data to the primary station if it detects an idle channel.

7.1.4.3 HDLC Frame Formats

The basic one-octet control field HDLC frame format shown in Figure 7.5 is used for both information exchange and link level control. Two flag fields always encapsulate a frame; however, the closing flag for one frame may be reused as the opening flag for the subsequent frame. The HDLC frame format supports several control field formats. An address field provides the address of the secondary station (but is not needed for point-to-point configurations). The information field contains the data being transmitted, and the frame check sequence (FCS) performs error detection for the entire frame. Also included in this frame is a *control field* to identify one of three types of frames available.

8	8	8	Variable length	16	8	bits
F	Address Field	Control	Information	FCS	F	

⬆ leftmost bit transmitted first

F = Flag = '01111110'

FCS = Frame Check Sequence

Figure 7.5 HDLC Frame Format

The Flag (F) sequence is a zero followed by six ones and another zero. Flags delimit the beginning and end of an HDLC frame. A key function of the data link layer is to encode the occurrence of the flag sequence within user data as a different sequence using *bit stuffing* as follows. If the link layer detects a sequence of five consecutive ones in the user data, then it inserts a zero immediately after the fifth one in the transmitted bit stream. The receiving link layer removes these inserted zeroes by looking for sequences of five ones followed by a "stuffed" zero bit. Thus, if an HDLC flag bit pattern, '01111110' is present in the user data, the link layer transmits this as '011111010.' Unfortunately, HDLC's bit stuffing mechanism can be fooled by bit errors on the physical medium, as we shall see later in Chapter 24. Therefore, many higher layer protocols also keep a length count to detect errors caused by bit errors corrupting the HDLC bit stuffing procedure. To determine the boundaries of HDLC frames, the receiver need only check the incoming bit stream for a zero followed by six ones.

The address field of the LAPB frame is primarily used on multidrop lines. The address field also indicates the direction of transmission and differentiates between commands and responses as we detail in the next section.

The sender computes the two-octet Frame Check Sequence (FCS), and the receiver uses the FCS to check the received HDLC frame to determine if any bit errors occurred during transmission. The following generator polynomial specifies the FCS:

$$G(x) = x^{16} + x^{12} + x^5 + 1.$$

The FCS of HDLC is capable of detecting up to three random bit errors or a burst of sixteen bit errors as analyzed in Chapter 24.

The HDLC standard supports control field frame formats of length equal to 8-, 16-, 32-, or 64- bit lengths negotiated at link establishment time. The control field of the X.25 LAPB frame is identical to the corresponding HDLC frame with the one-octet length called basic mode, or the optional modes for a two-octet length called extended mode, and the four octet length called super mode. Figure 7.6 shows the 8-bit version of the three HDLC control field formats: information, supervisory, and unnumbered frame. Figure 7.7 shows the 16-bit version of the information, supervisory, and unnumbered frames. The unnumbered frame control field format is only 8 bits long for all control field formats. Note how the first two bits of the control field uniquely identify the type of frame: *information, supervisory,* or *unnumbered.*

The *information* frame transports user data between DTE and DCE. Within this frame, the N(S) and N(R) fields designate the sequence number of the last frame sent and the expected sequence number of the next frame received, respectively. Information frames always code the Poll (P) bit to a value of one as indicated in the figure. In supervisory and unnumbered formats, the Poll/Final (P/F) bit indicates commands and responses. For example, the DTE (or DCE) sets the P/F bit to one to solicit (i.e., poll) a response from the DCE (or DTE). When the DCE (or DTE) responds, it sets the P/F bit to zero to indicate that its response is complete (i.e., final).

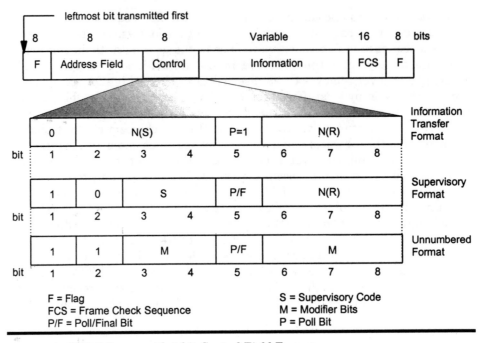

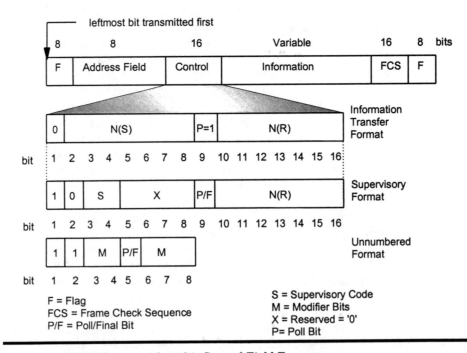

Figure 7.6 HDLC Frame with 8-bit Control Field Formats

Figure 7.7 HDLC Frame with 16-bit Control Field Formats

The *supervisory* frame uses the Supervisory (S) code bits to acknowledge the receipt of, request retransmission, or request temporary suspension of information frame transfer. It performs these functions using the P/F bit in the following command and response pairs: Receive Ready (RR), Receive Not Ready (RNR), REJect (REJ), and Selective REJect (SREJ).

The *unnumbered* frame uses the Modifier (M) bits of the unnumbered format to provide the means for the DTE and DCE to set up and acknowledge the HDLC mode, and to terminate the data link layer connection. The HDLC standard defines a variety of control messages to set up the HDLC mode discussed in the previous section (e.g., NRM, ARM, and ABM). LAPB uses only the Asynchronous Balanced Mode (ABM).

The basic difference between the one-, two-, four-, and eight-octet control field formats is the length of the send and receive sequence number fields, N(S) and N(R) respectively. The HDLC standard defines the modulus as the maximum decimal value of these sequence number fields as given in Table 7.1. In other words, HDLC stations increment the sequence number modulo the modulus value given in the table. For example, for a one-octet control field, stations increment the sequence numbers modulo 8; specifically, the stations generate the following pattern of sequence numbers: 0, 1, 2, 3, 4, 5, 6, 7, 0, 1, etc. The 16-bit control field initially targeted use over long delay satellite links to increase application throughput. A larger sequence number improves performance because the sender can transmit up to the modulus of the sequence number without receipt of an acknowledgment. The 32- and 64-bit versions of the control field were developed for similar reasons as the bandwidth-delay product increased with higher-speed transmission links, such as those used in modern local area and wide area networks.

Table 7.1   HDLC Control Field Lengths and Sequence Number Modulus

Control Field Length (Octets)	Sequence Number Length (Bits)	Sequence Number Modulus
1	3	8
2	7	128
4	15	32,768
8	31	2,147,483,648

7.1.4.4 Single Link Procedure (SLP)

Point-to-point physical X.25 network access supports either a single link or multiple links. The LAPB Single Link Procedure (SLP) supports data interchange over a single physical circuit between a DTE with address "A" and a DCE with address "B." The coding for the one-octet address field for the address "A" is a binary '1100 0000' and the coding for address "B" is a binary '1000 0000.' Figure 7.8 illustrates the SLP frame structure and usage. As seen from the figure, the address field identifies a frame as either a command or a response since command frames contain the address of the other end, while response frames contain the address of the sender. Information frames are always coded as commands in the address field.

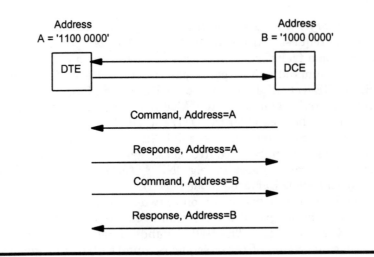

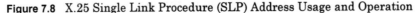

Figure 7.8 X.25 Single Link Procedure (SLP) Address Usage and Operation

7.1.4.5 Multilink Procedure (MLP)

The optional Multilink Procedure (MLP) exists as a an upper sublayer in the data link layer. Multilink operation uses the single link procedures independently over each physical circuit, with the multilink procedure providing the appearance of a single data flow over two or more parallel LAPB data links as illustrated in Figure 7.9. MLP has several applications in real world networks. It allows multiple links to be combined to yield a higher-speed connection; it provides for graceful degradation if any single link should fail; and finally it allows a network designer to gracefully increase or decrease capacity without interrupting service. The X.25 MLP design philosophy appears in inverse multiplexing in the TDM world, an emerging frame relay multilink standard, as well as Inverse Multiplexing over ATM (IMA) as described in Chapter 12.

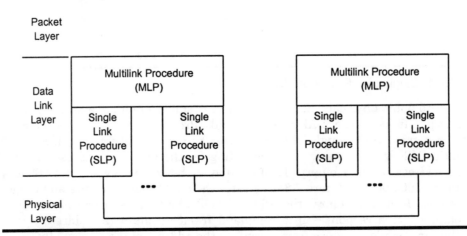

Figure 7.9 X.25 Multilink Procedure (MLP) Protocol Architecture

Figure 7.10 illustrates how MLP operates over independent SLP links using a 16-bit Mutilink Control (MLC) field. MLP uses a convention analogous to SLP for the one-octet LAPB address field to identify commands and responses between DTE and DCE, except that the DTE has an address "C" with binary value "1111 0000" and the DCE has an address "D" with binary value "1110 0000." The use of these different address values assist in diagnostic and maintenance activities since use of incompatible addresses identify a configuration error. A 12-bit Multilink send sequence number used by the transmitter across the SLPs enables the receiver to correctly deliver information frames to the packet layer at the destination. The other bits control the operation of the MLP protocol. If the V bit is set to one, then no sequencing is required, otherwise the receiver performs MLP sequencing. If V is one (no sequencing), the S bit indicates whether a sequence number is included. Even if no (re)sequencing is required, the receiver can use a valid sequence number to detect duplicate or missing multilink frames. The R and C bits control reset and the associated confirmation of multilink sequence number state variables. MLP employs a windowing procedure where a maximum number of multilink frames may be sent without acknowledgment from the associated SLP.

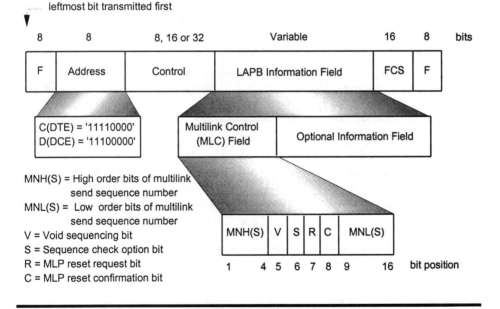

Figure 7.10   X.25 Multilink Procedure (MLP) Frame Format

7.1.4.6 Comparison of Link Access Procedure (LAP) Protocols

The ITU-T defines three types of Link Access Procedure (LAP) protocols. LAP was the first ISDN protocol and was designed based on the HDLC Set Asynchronous Response Mode (SARM) command used in "unbalanced" connections. This mode formed the basis for Link Access Procedure Balanced (LAPB), an HDLC implementation that uses balanced asynchronous mode with error recovery to form the basis of the X.25 packet -switching protocol.

The LAPB protocol is identical in format to a 8-, 16-, or 32-bit control field HDLC frame as described in the next section. The next extension of HDLC and LAP was Link Access Protocol over D-channel (LAPD) standardized by the ITU-T in Recommendations Q.920 and Q.921 as the Digital Subscriber Signaling System number 1 (DSS1) data link layer. This implementation of HDLC uses either the basic or extended asynchronous "balanced" mode configuration and provides the basis for both ISDN and frame relay services. In the late 1980s, the ITU-T removed the sequence numbering, windowing, and retransmission functions for a frame relaying service in the Q.922 standard, resulting in a protocol dubbed LAPF that combined the address and control fields. Note that the removal of the control field in frame relay eliminated the sequence numbers used to implement retransmission of lost frames in LAPB and LAPD. Instead, higher level protocols, for example, TCP, must perform error detection and retransmission when operating over frame relay. As we shall see, a close family relationship exists between other link access procedures employed by X.25 (LAPB), ISDN (LAPD) and frame relay (LAPF) as illustrated in Figure 7.11.

Flag 1 Byte	Address 1 Byte	Control 1, 2 or 4 Bytes	Information ◀▶	FCS 2 bytes	Flag 1 Byte

LAPB

Flag 1 Byte	Address 2 Bytes	Control 1 or 2 Bytes	Information ◀▶	FCS 2 bytes	Flag 1 Byte

LAPD

Flag 1 Byte	Address+Control 2, 3 or 4 Bytes	Information ◀———————▶	FCS 2 bytes	Flag 1 Byte

LAPF

Figure 7.11 Comparison of X.25, ISDN, and Frame Relay Information-Frame Formats

7.1.5 Packet Layer Format and Protocol

Each X.25 packet transferred across the DTE/DCE interface exists within a basic LAPB frame as shown in Figure 7.12. Note that the X.25 layer 3 packet, including packet header and packet data, forms the user data (or information) field of the layer 2 LAPB frame.

An X.25 packet has a header and a user datafield as shown in Figure 7.12. The Qualifier (Q) bit allows a transport layer protocol to separate control data from user data. The D bit is used in delivery confirmation during X.25

switched virtual call setup. The next two bits indicate the packet type, with 01 indicating a data packet with three-octet header. A four-octet header is also standardized. The X.25 packet layer address has a 4-bit group number and an 8-bit logical channel number, together forming a 12-bit Logical Channel Number (LCN). Channel zero is reserved, and therefore there can be up to 2^{16} minus 1, or 4095 logical channels on a physical circuit carrying the X.25 protocol.

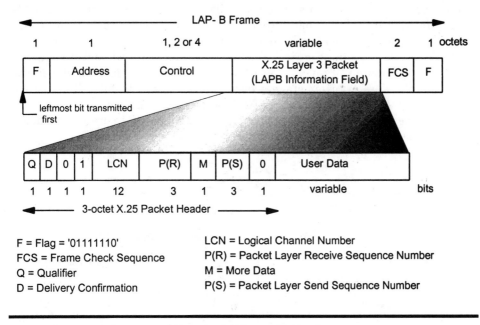

Figure 7.12   LAPB Frame and X.25 Packet Layer Payload

By convention, Logical Channel Numbers (LCNs) are assigned to each of the incoming and outgoing virtual calls for each DCE and DTE, respectively, as well as to all PVCs. Out of the 4095 logical channel numbers available per physical circuit, PVCs are assigned the lowest numbers, followed by one-way incoming virtual calls, then two-way incoming and outgoing calls, with the highest numbers reserved for one-way outgoing virtual calls. Note that LCNs hold only local significance to a specific physical port, but must be mapped to a remote LCN for each virtual call. Packet networks use search algorithms to resolve collisions and assign LCNs to each virtual call.

The packet layer uses the receive and send sequence numbers (P(R) and P(S)) to support a packet layer flow control protocol described later in this section. The More (M) bit supports segmentation and reassembly by identifying the first and intermediate packet segments with a value of 1, with the last segment having a value of zero.

There are two types of services defined in the X.25 standard: virtual circuit and datagram. Virtual circuits assure sequence integrity in the delivery of user data, established either administratively as a PVC, or as a switched Virtual Call (VC) through call control procedures. PVCs are permanently established between a source and destination DTE pair. Datagrams do not require call control procedures and are sent in either a best-effort mode, or request explicit receipt notification.

7.1.6 Control Functions

Call control packets defined in X.25 support switched Virtual Calls (VCs). X.25 VCs use either the X.121 or the E.164 addressing format. VCs act much like telephone calls where a source must first connect to the destination node before transferring data. Therefore, one source can connect to different destinations at different times, as compared with a PVC that is always connected to the same destination. Figure 7.13 shows a typical control packet sequence for the establishment of an X.25 VC, followed by the call clearing procedure. Note that the data transfer stage occurs only after successful call establishment. An issue with VCs is that the X.25 network may become so busy that it blocks connection attempts. Applications that cannot tolerate occasional periods of call blocking should use dedicated PVCs.

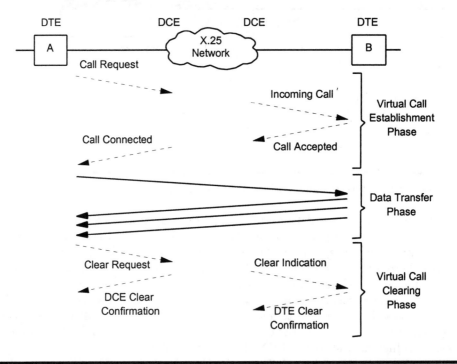

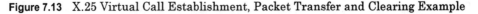

Figure 7.13   X.25 Virtual Call Establishment, Packet Transfer and Clearing Example

7.1.7 Control Packet Formats

Recommendation X.25 defines control packets for virtual call setup and termination. Figure 7.14 shows the generic format for an X.25 control packet. The General Format Identifier (GFI) indicates the format of the remainder of the header, the sequence number modulus, and whether the packet type is either a call setup, clearing, flow control, interrupt, reset, restart, registration, diagnostic, or data. The Logical Channel Group Number has local significance as described earlier. The Packet Type Identifier (PTI) differentiates the various control functions performed by the X.25 protocol, such as virtual call control, flow control, error recovery and diagnostics. For an information packet, the packet type contains the sequence number fields.

Control packets perform many functions, including: call request and incoming call packets; call accepted and call connected; clear request and clear indication; DTE and DCE clear confirmation, data, interrupt, interrupt confirmation, receive ready (RR), receive not ready (RNR), reset confirmation, and restart confirmation; reset request and reset indication; restart request and restart indication; diagnostic; DTE Reject (REJ), registration request, and registration confirmation. Figure 7.14 shows the packet format for call request and incoming calls. The facilities fields support value added X.25 functions like one-way calling, throughput negotiation, closed user groups, charging control and fast-select procedures.

	8	7	6	5	4	3	2	1	bit
Octet 1	General Format Identifier				Logical Channel Group Number (LCGN)				
Octet 2	Logical Channel Number (LCN)								
Octet 3	Packet Type Identifier								
Octet 4	Calling DTE Address Length				Called DTE Address Length				
	DTE Address Field								
	Facilities Fields								
	User Data Fields								

Figure 7.14   X.25 Packet Format for Call Request and Incoming Call Messages

A quick explanation of X.25 control packets is in order. Clear packets terminate user-to-user sessions (DTE-to-DTE). Interrupt packets allow users to bypass the normal flow control procedure. RR and RNR packets manage flow control on the local interface between the DTE and the DCE. These packets work in a manner similar to their HDLC counterparts, providing a level of flow control above the normal HDLC functions over the LCNs. Reset packets set the packet level sequence numbers back to zero for specific PVCs or VCs during normal data transfer. Restart packets clear all VCs, reset all PVCs, and initializes a physical interface port. Diagnostic packets are used as the catch-all identifier of all error conditions not covered by the other control packets. Reject packets reject a specific packet, implicitly requesting retransmission of all packets after the last received sequence number.

7.1.8 Example of X.25 Operation

The LAPB protocol uses a store-and-forward approach to ensure reliable delivery of packets across noisy, error-prone transmission links. The example of Figure 7.15 illustrates the store and forward approach for recovering from errors between two packet -switching nodes, labeled A and B using the X.25 link layer protocol. Two types of packets are exchanged between the nodes:

LAPB Data (LD) and LAPB ACKnowledgments (ACK). In the example, separate link layer control packets (e.g., Receiver Ready (RR)) carry the link layer acknowledgments; however, in actual operation switches often piggyback acknowledgments onto packets heading in the opposite direction. Each LAPB frame has a pair of link layer sequence numbers: N(R), the receive sequence number, and N(S), the send sequence number. N(S) indicates the sequence number of the transmitted packet. N(R) indicates the value that this receiver expects to see in the send sequence number (N(S)) of its peer. Therefore, the receive sequence number, N(R), acts a cumulative acknowledgment of all link layer frames with sequence numbers up to the value N(R) minus one.

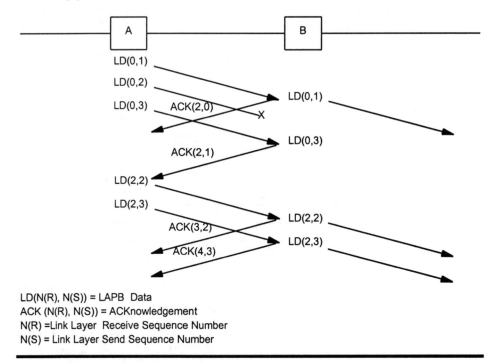

LD(N(R), N(S)) = LAPB Data
ACK (N(R), N(S)) = ACKnowledgement
N(R) =Link Layer Receive Sequence Number
N(S) = Link Layer Send Sequence Number

Figure 7.15   Example of X.25 Link Layer Store and Forward Operation

The example begins with node A sending user data with N(S)=1, which is successfully transferred to node B, which acknowledges its receipt with an ACK containing P(R)=2 indicating that the next expected value of N(S) is 2. Node B now stores packet 1 and attempts to forward this packet to the next node. Meanwhile, node A sent the next packet with N(S)=2; however, node B detected errors and discarded it as indicated by the X in the Figure. B then receives the third packet sent by A with N(S)=3, but it is out of order. B acknowledges this with a link layer RR control packet containing N(R)=2 indicating that it expects the next packet to have a send sequence number of N(S)=2. Node A responds by resending packets with N(S) equal to 2 and 3, which are successfully received and acknowledged by node B. Node B can now attempt to forward these packets to the next node using the same process, but probably with different link layer sequence numbers. The example also illustrates the link layer acknowledgment of the control packets received from B by A.

This simple example illustrates the complexity involved in store and forward packet -switching. Note that not only do acknowledgments occur on every link as shown in this example, but they occur again at the end-to-end network layer as described later in this section. This processing was worthwhile when a significant portion of the packets experienced transmission errors on the noisy facilities, a situation prevalent in the analog and radio transmission facilities of 1970's and 1980's wide area networks. Creative engineers developed even more sophisticated schemes in which only the errored packets were retransmitted using the Selective Reject (SREJ) control protocol. SREJ significantly improves performance over networks such as ATM with long bandwidth-delay products as analyzed in Chapter 26. These schemes became the basis of protocols used in B-ISDN that we study in the next part. However, since the ubiquitous deployment of fiber, many of these error correcting capabilities can now be done at the end points only, instead of at every intermediate node; for example, as is done in frame relay. Furthermore, as we shall see later, other higher-layer protocols also perform error detection and retransmission, which obviates the need to repeat the functions at a lower layer. For example, as described in Chapter 8, the Transmission Control Protocol (TCP) commonly used on the Internet also performs error detection and retransmission.

7.1.9 Traffic and Congestion Control Aspects of X.25

The send and receive sequence numbers in the X.25 packet layer provide flow control between the packet layer source and destination end systems (or DTEs). Figure 7.16 illustrates a simple example of this end-to-end X.25 packet layer flow control between source and destination DTEs connected by an X.25 store and forward network. The packet layer send sequence number, P(S), is a sequential number for the current packet incremented modulo the packet header sequence number modulus. X.25 defines packet layer sequence number modulus values of 8, 128 and 32,768. The destination uses the packet layer receive sequence number, P(R), in the ACKnowledgment to indicate the packet layer send sequence number expected in the next packet from the other end point of that virtual circuit. Therefore, the packet layer receive sequence number, P(R), acts as an acknowledgment for all packets up to P(R) minus one. The reader should note that these packet layer sequence numbers operate on an end-to-end basis, as opposed to a link-by-link basis for LAPB frames.

X.25 defines a separate value of the window size at the transmitter which controls the maximum number of packets it can send without receiving a packet layer acknowledgment from the destination. Similarly, X.25 defines a window size at the receiver which controls how many packets the DTE will accept prior to generating an acknowledgment. Of course, the window size must always be smaller than the packet layer sequence number modulus.

The example of Figure 7.16 employs a window size of 2 at the receiver and a window size of 4 at the transmitter.

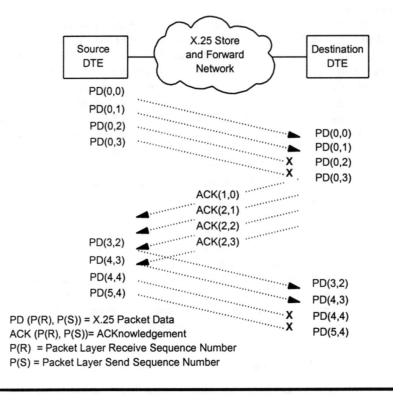

Figure 7.16   Example of End-to-End X.25 Packet Layer Flow Control

As shown in the example Figure 7.16, the source transmits four packets and then waits for the packet level acknowledgment (for example, in a packet level RR packet) before sending additional packets. The destination is slow to acknowledge these packets, and since its window size is only 2, it only acknowledges the first two packets received. The destination indicates that it still expects to receive $P(S)=2$ for the third and fourth packets. The source retransmits the third and fourth packets, along with the fifth and sixth as the process repeats. Normally, the source and destination would coordinate transmit and receive window sizes; however, they don't need to do so. The example also illustrates the packet layer acknowledgment of the control packets between the source and destination.

The generic name for this procedure is a *sliding window* flow control protocol. This procedure allows the transmitter or the receiver to control the maximum rate of transmission over a virtual circuit, and is therefore a form of traffic control. This is still an essential function for a slow receiver (such as a printer) to control a fast transmitter (a computer) in many data communications applications today. The receive sequence number acknowledgment can be "piggybacked" in the packet header for a packet headed in the opposite direction on a virtual circuit, or may be sent in a separate layer 3 control packet.

7.1.10 Service Aspects of X.25

X.25 packet -switching serves many user communities, especially in Europe, where it still constitutes a significant portion of of public and private data services. X.25 traffic levels have remained relatively flat after the introduction of frame relay services absorbed much of the existing public packet - switching communications market growth. Packet -switching remains a popular technology, however, and will likely continue to be used globally well into the twenty first century to reach remote areas of the world.

Recently, due to the ever increasing WAN bandwidth requirements of computing, X25 networking speeds have also increased. Trunk speeds have increased beyond the 300 to 1200 bps access and 56-kbps trunks of the early X.25 networks. Now, many packet switches provide access at 56-kbps with trunks at DS1/E1 speeds.

7.2 FRAME RELAY — OVERVIEW AND USER PLANE

Frame relay led the way in a minimalist trend in data communications, essentially being X.25 on a diet. Proof of frame relay's profound importance in data networking is the fact that public frame relay services have displaced many private line-based data networks. Today, almost every WAN manufacturer and data communications service provider supports the frame relay protocol. This section presents an overview of frame relay, highlighting control and user plane concepts adopted by ATM — imitation is the sincerest form of flattery. Frame relay operates as an interface and as a network service in the user plane as described in this section. The next section describes the control plane operation of frame relay which signals status for permanent connections and establishes switched connections.

7.2.1 Origins of Frame Relay

X.25 packet -switching, proprietary private networks built upon private lines, and legacy networks running HDLC and SDLC dominated the data communications marketplace from 1980 through the early part of the 1990s. In order to keep pace with the increased bandwidth and connectivity requirements of today's applications, users needed a new data communications technology to provide higher throughput at a lower cost. Frame relay responded to this need beginning in the early 1990s to provide higher bandwidth, and more cost-effective transfer of packet data. Frame relay did this by eliminating the overhead of the network layer present in X.25 as well as reducing the complexity of the link layer protocol. The nearly ubiquitous digital fiber-optic transmission facilities made lost packets relatively rare — hence applications networking across a slimmed down frame relay protocol could afford to occasionally retransmit lost packets.

While frame relay has its origins in HDLC and ISDN link layer protocols, its streamlined protocol eliminated the overhead involved in error-correction and flow-control overhead. Higher layer protocols (e.g., TCP) recover lost or corrupted data when operating over frame relay networks.

Not only does frame relay offer the security of private lines; it also provides greater granularity of bandwidth allocation. Since frame relay is a lighter weight protocol, it runs at higher speeds than X.25. Indeed, for software-based implementations that run both frame relay and X.25, the frame relay protocol runs much faster. Frame relay provides an upgrade to existing packet switch technology by supporting speeds ranging from 64-kbps through nxDS1/nxE1 and all the way up to DS3. Frame relay operates at OC3 speeds today, but of course it does so over ATM. The standards bodies and vendors are working on extending native frame relay up to SONET speeds. Frame relay fills a technology gap between X.25 and ATM and at the same time also provides a smooth transition to ATM. In fact, frame relay over ATM and frame relay interworking with ATM are two options now available to facilitate this transition.

Frame relay standards derived from the ISDN Link Access Procedure for the D-channel (LAPD, or as later modifications were called LAPF) as described in ITU-T/CCITT Recommendation Q.921, which led to I.122, I.233, I.370, and Q.922. The signaling procedures in the ISDN control plane specified in ITU-T Recommendation Q.931 led to the frame relay signaling standards defined in Recommendation Q.933 as described in the next section. The corresponding ANSI standards are T1.617 and T1.618, which standardized the service description, protocol, and status signaling for frame relay. The many advantages offered by frame relay also caused the formation of a separate industry group, called the Frame Relay Forum (FRF), whose charter is to develop Implementation Agreements (IAs) to facilitate interoperability between different vendor products and services. Recognizing that frame relay is primarily used today as an access protocol, rather than as a bearer service within ISDN, the ITU-T recently created two new standards. Recommendation X.36 defines the frame relay User-Network Interface (UNI) and Recommendation X.76 defines the Network-Network Interface (NNI). All future ITU-T standards work will follow these recommendations instead of the I-series and Q-series recommendations cited above.

7.2.2 Frame Relay Protocol Structure

The frame relay protocol has two logically separate components: the control plane (C-plane) and the user plane (U-plane). The concept of control and user planes is a practice begun by the ITU-T for ISDN as described in Chapter 6. Figure 7.17 illustrates the structure of the protocols that support frame relay. The user plane of frame relay implements a subset of the OSIRM data link layer functions as specified in ITU-T Recommendation Q.922. Frame relay also has a control plane involved with reporting on the status of PVCs and the establishment of SVCs. Recommendation Q.933 defines the status signaling for PVCs and the call control procedures for

SVCs. The frame relay Q.933 signaling protocol may operate over either the frame relay protocol (Q.922) directly, or be signaled separately via the ISDN protocol (Q.921) on a separate TDM channel. The ITU-T signaling standards for ISDN (Q.931), frame relay (Q.933), and B-ISDN (Q.2931) have a common philosophy, message structure, and approach.

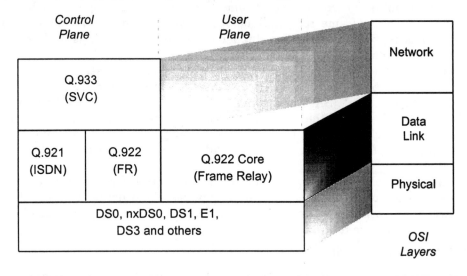

Figure 7.17   Frame Relay Protocol Structure

7.2.3 Frame Relay Networking Context

Figure 7.18 shows how the user plane frame relay protocol layers operate in a network context interconnecting two end systems, for example, two routers of a virtual private network connected by a public frame relay service.

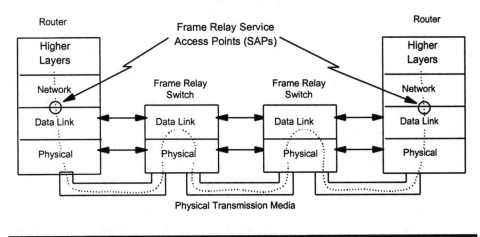

Figure 7.18   Frame Relay Networking Context

In the example, two interconnected intermediate switches relay frames between Service Access Points (SAPs) in the end systems. Multiple permanent or switched virtual connections (PVCs or SVCs) may exist on a single physical access circuit. As shown in the figure, frame relay operates at the link layer only, which reduces complexity and usually allows for higher speed operation, but at a certain price. Frame relay specifies in-sequence delivery of frames. Since frame relay has no sequencing and no retransmission to detect misordered or lost frames, higher layer protocols, such as TCP, must perform retransmission and sequencing.

ITU-T Recommendation Q.922 [4] defines the terms and basic concepts of a frame mode bearer service, called Link Access Procedure F (LAPF). The Service Access Point (SAP) is the logical-level data link interface between the data link layer and the network layer as indicated in Figure 7.18. Each SAP corresponds to a data link connection end point, which are uniquely determined via Data Link Connection Identifiers (DLCIs) in the frame header.

The frame relay protocol also operates in the control plane, which actually runs separate protocols between frame relay users and networks, as well as between networks. The next section covers the frame relay control plane after this section takes an in-depth look at the user plane.

7.2.4 Frame Format

The Link Access Procedure for frame relaying (LAPF) format used by frame relay services is a derivative of the ISDN Link Access Protocol D-channel (LAPD) framing structure. Figure 7.19 shows the basic frame relay frame structure with a two-octet header field from ITU-T Recommendation Q.922 [4]. The first and last one-octet flag fields, labeled F, are HDLC flags. HDLC bit stuffing, identical to that described in the previous section on X.25, is performed to avoid mistaking user data for an HDLC flag. Although HDLC supports a user data field of up to 8188 octets, frame relay standards specify smaller frame sizes. For example, FRF.1 states that all implementations must support a maximum frame size of at least 1600 octets for transport of encapsulated Ethernet traffic. The Frame Check Sequence (FCS) field is two octets long as defined in the HDLC standard.

The reader should note that the X.25 and FR standards use a different notation for the order of bit transmission. In the X.25 standards, bit transmission order was from left to right as indicated by the arrows in the figures and the bit position numbering in the previous section. In frame relay, the bits are grouped into octets, which are transmitted in ascending (i.e., left to right) order. For each octet, bit 1, which is the least significant bit, is transmitted first and bit 8, which is the most significant bit, is transmitted last. There is no right or wrong way to indicate bit order transmission, just as different countries around the world select different sides of the road to drive on. A detailed review of the contents of the frame relay header field follows.

As shown in Figure 7.19, the frame relay header contains an upper and lower Data Link Connection Identifier (DLCI) field, together forming a 10-bit

DLCI that identifies up to 1024 virtual circuits per interface in the two-octet header format. The DLCI only has *local significance* on a physical interface. This means that the DLCIs may differ on the interfaces at each end of the point-to-point frame relay virtual connection. On any interface, each user CPE device has a separate DLCI for each destination. This limits the size of a fully meshed frame relay network to approximately 1000 nodes. Larger frame relay networks require the three- or four-octet header fields, or else a hierarchical, partial mesh topology.

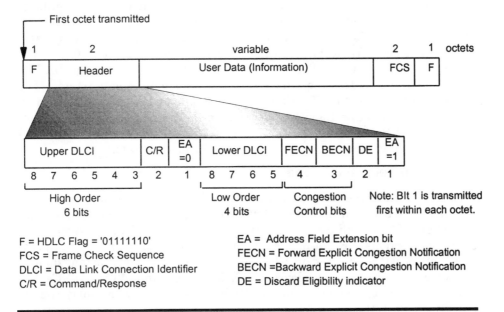

Figure 7.19 Q.922 Frame Mode Bearer Service Frame Structure

The standards require that networks transparently convey the Command/Response (C/R) between frame relay users. Hence, the C/R bit can be employed by user applications.

The Forward Explicit Congestion Notification (FECN) and Backward Explicit Congestion Notification (BECN) bits indicate to the receiver and sender, respectively, the presence of congestion in the network. Specifically, the network sets the FECN indicator in frames traversing the network from sender to receiver which encounter congestion. Receiver-based flow control protocols, such as DECnet, use FECN to initiate congestion avoidance procedures. The network sets the BECN indicator in frames traversing the network from receiver to sender for congestion occurring in the sender to receiver direction. That is, the network sets the BECN indicator in frames traveling in the opposite direction on the same virtual connection to those in which it sets the FECN indicator. Therefore, BECN aids senders who can then dynamically change their source transmission rate.

The Discard Eligibility (DE) bit, when set to 1, indicates that during congestion conditions the network should discard this frame in preference to other frames with a higher priority, for example, those with DE bit set at 0. Note that networks are not constrained to discard only frame with DE set to 1 during periods of congestion. Either the user or the network may set the DE bit. The network sets the DE bit when the received frame rate exceeds the Committed Information Rate (CIR) specified for a particular virtual connection. Users rarely set the DE bit, as there is rarely an advantage to doing so.

The address field extension (EA) bit is the first bit transmitted in each octet of the frame relay address field. When set to 0, it indicates that another octet of the address field follows. When set to 1, it indicates that the current octet is the last octet of the address field. As an example, Figure 7.19 illustrates the basic two-octet frame format where the first EA bit is set to 0, and the second EA bit is set to 1. Recommendation Q.922 defines how the EA bits extend the DLCI addressing range to three- and four-octet formats. The Frame Relay Service Specific Convergence Sublayer (FR-SSCS) defined in ITU-T Recommendation I.365.1 is identical to the frame relay frame without FCS, flags and HDLC zero insertion. Chapter 18 summarizes FR-SSCS and presents the three- and four-octet frame relay header formats.

7.2.5 Frame Relay Functions

A primary use of frame relay is as a user interface to a public data service, or as an interface between networks. Frame relay supports Permanent Virtual Connections (PVCs) and Switched Virtual Connections (SVCs). Frame relay virtual connections are either point-to-point or multipoint-to-multipoint (called multicast) as defined in Chapter 4. SVCs use a call establishment protocol similar to that employed by X.25, ISDN and ATM. PVCs are managed by a status signaling protocol. We cover these frame relay control plane functions in the next section.

7.2.6 Example of Frame Relay Operation

Figure 7.20 illustrates a key aspect of the simplification that frame relay provides. End system A on the left-hand side of the figure transmits frames to a frame relay network which relays the frames to a destination end system B. There are no sequence numbers in the frames. The numbers in the example are for illustrative purposes only. The frame relay network simply relays frames in the same order on the interface to destination B. If a frame is corrupted while in transit, the frame relay network simply discards or drops it. A frame relay network may also discard frames during intervals of congestion. In our example, the frame relay network never delivers the fourth frame, either because of errors or discard due to congestion. This simplified relaying only protocol allows vendors to build simple, fast switches; however, it requires more intelligence in the higher layer protocols — such as TCP residing in the end systems — to recover lost frames. Frames can be lost due to transmission errors or congestion. When frame relay operates over modern transmission media, such as fiber optic systems, frames lost due to errors become a rare occurrence. Loss then occurs primarily due to congestion on the trunks within the frame relay network or on the access line to the destination. Since loss occurs infrequently in well managed frame relay networks, the end systems seldom need to invoke error recovery procedures.

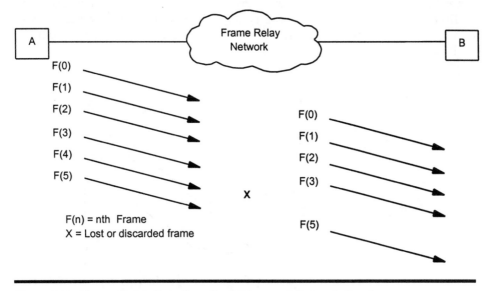

Figure 7.20   Frame Relay Interface, Switching, and Network Service

7.2.7 Traffic and Congestion Control Aspects of Frame Relay

This section covers traffic and congestion control functions implemented in frame relay networks. These functions are simpler than their ATM counterparts and hence provide a good introduction and background to the important concepts of policing, congestion indication, and reaction to congestion as detailed in Part 5 for ATM.

7.2.7.1 Frame Relay Traffic Control

Frame relay defines a Committed Information Rate (CIR) which is the information transfer rate the network commits to transfer under normal conditions [5]. However, the term "committed" is somewhat ambiguous in that it does not specifically state the frame loss or delay objective. Transmission errors can cause frames to be lost, and finite buffer space in networks which implement statistical multiplexing can result in lost frames due to momentary overloads. The frame relay network sets a CIR for each PVC or SVC.

We describe the frame relay traffic control procedures with reference to the diagram depicted in Figure 7.21 which plots the total number of bits transmitted on the vertical axis versus time on the horizontal axis. The maximum number of bits transmitted per unit time is of course limited by the access line rate, AR, as shown by the thick line in the figure. The CIR is the number of bits in a committed burst size, Bc, that can arrive during a measurement interval T such that CIR=Bc/T. The dashed line in the figure has a slope equal to CIR. Frames are sent at the access line rate, indicated by the trace labeled "transmitted frames" in the figure. If the number of bits that arrive during the interval T exceeds Bc, but is less than an excess threshold, Bc+Be, then the frames associated with those bits are marked as Discard Eligible (DE). The bits that arrive during the interval T in excess of Bc+Be are discarded by the access node. To avoid discards at the ingress node for a given access rate, AR, the value of Be must be greater than or equal to AR×T-Bc.

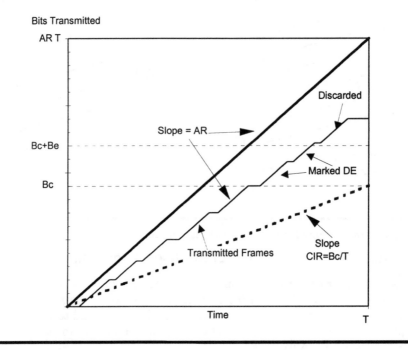

Figure 7.21   Example of Frame Relay Traffic Control Operation

At present, there is no uniform method for setting the interval T. If T is set too small, such that Bc is less than the length of a single frame, then every frame will be marked DE. If T is set too large, the buffer capacity in the FR network required to guarantee CIR may not be practical. Setting T to a value on the order of the maximum round trip delay is a good guideline to achieve good TCP/IP throughput over frame relay. Coast-to-coast round trip delays across the continental United States are typically on the order of one hundred milliseconds across most frame relay networks.

Network providers offer different interpretations of the Committed Information Rate (CIR) and Discard Eligible (DE) functions according to different pricing plans. For example, CIR may be available only in certain increments, or DE traffic may be billed at a substantial discount. Customer network management, performance reporting, and usage-based billing are valued-added services offered by some providers. It pays to shop around carefully.

There are two different philosophies in the setting of the CIR. We call the first full rate allocation, where the sum of the CIRs associated with the PVCs on an interface is no more than the interface speed, and the second is called oversubscription (or overbooking), where the sum of CIRs exceeds the interface rate. The interface may be between the frame relay user and the network, or an internal trunk interface. In full rate allocation, the fact that the sum of the frame relay connection's CIR values across an interface does not exceed the actual bandwidth guarantees predictable, deterministic performance at the expense of lower utilization because all users rarely utilize their entire CIR. In the oversubscription case, loss performance becomes statistical; however, individual PVCs may transmit at higher than the CIR rate than in the regular booking case, since the sum of the PVC CIRs exceeds the interface rate. The oversubscription paradigm relies on the averaging of a larger number of smaller connections in a technique called statistical multiplexing as described in Chapter 25. Typically, oversubscription achieves higher utilization at the expense of a higher loss rate. For properly engineered networks the frame loss level can be controlled through proper parameter settings. Some frame relay networks that use closed loop congestion control algorithms automatically control parameters to reduce loss by allowing less data into the network during congested intervals. However, the setting of the congestion control algorithm parameters is important in achieving acceptable loss and fairness.

7.2.7.2 Frame Relay Congestion Control

The Discard Eligible (DE) bit may be set by either the customer or by the network. DE is rarely set by the customer; rather, the network provider ingress switches set DE for frames exceeding the CIR. If a network node becomes congested, it should first discard the frames with the DE bit set. Beware, however, of the danger of discarding frames marked with DE during long periods of congestion: the applications may react by retransmitting lost frames, intensifying congestion, and possibly resulting in a phenomenon called congestion collapse as described in Chapter 23.

Congestion notification is provided in the frame relay header field by the FECN and BECN bits. Nodes in the frame relay network set the FECN bit when they become congested. The FECN bit informs receiver flow controlled protocols of the congestion situation, or allows the receiver or the egress node to set the BECN bit. The BECN bit is set in FR frames headed in the upstream direction to inform transmitter flow controlled protocols of the congestion situation. An increase in the frequency of FECN and BECN bits received is a good indication of network congestion. At present, little use is being made of this technique by higher layer protocols in end and intermediate systems. One concern is that by the time the FECN/BECN arrives at the controlling end, the congested state may no longer exist in the FR node that sent the FECN/BECN notice. The main concern is that the FECN/BECN notice is delivered to the CPE router, which is not the primary source of flow control. Currently, no technique exists for the CPE router to convey the FECN/BECN message to TCP or the application that could provide flow control.

Figure 7.22 depicts a network of frame relay switches in Houston, Atlanta, Chicago, and Raleigh connecting a host device in Dallas with a LAN in

Charleston via a pair of routers. In this example, the host in Dallas accesses the router via frame relay and is capable of dynamically controlling its transmission rate. In the process of downloading a large volume of files from the host in Dallas to the users on a local area network in Charleston via a PVC (shown as a dashed line), congestion occurs in the output buffers on the Atlanta switch on the physical link to the Raleigh switch as indicated in the figure. The Atlanta switch sets the FECN and BECN bits on all DLCIs traversing the Atlanta-to-Raleigh trunk to notify the senders and receivers of the congestion condition. The Atlanta node sets the FECN bit to one and notifies the router in Charleston using the PVC receiving traffic from Dallas of impending congestion. The Atlanta node also sets the BECN bit to one on frames destined from Charleston to Dallas, informing the router in Dallas using the PVC of the same congestion condition. Either the Dallas user could throttle back, or the Charleston user could flow control the Dallas sender using a higher layer protocol. Either action reduces the frame rate and eventually causes the congestion condition to abate. Once congestion clears, the Atlanta node ceases setting the FECN and BECN bits in frames traversing the queue, and the end systems can transmit more data.

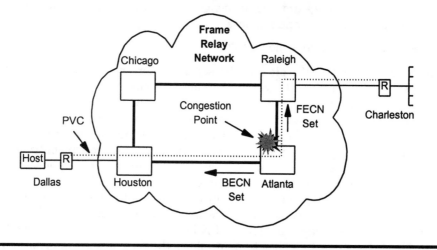

Figure 7.22 Frame Relay Congestion Control Example

Another form of congestion notification defined by ANSI T1.618 is the Consolidated Link Layer Management (CLLM) function. CLLM reserves one of the DLCI addresses on a frame relay interface for transmitting control messages to user devices when there are no frames to transmit, yet a congestion notification still needs to be sent. Although standardized, CLLM is not widely implemented.

7.2.8 Service Aspects of Frame Relay

Frame relay service emerged full force on the public data network market in the 1990s to compete with and erode private line market share. In fact, some industry experts refer to frame relay as a virtual private line service.

Providers of interoperable frame relay hardware and software range from CPE vendors to high-end multiplex and switch vendors.

Most carriers offer Frame Relay (FR) as a public data service with access rates up to DS3 (45 Mbps) speeds connecting a Customer Premises Equipment (CPE) router, bridge, or Frame Relay Access Device (FRAD) into the public frame relay network. Some carriers also provide subrate (i.e., less than 56-kbps) and nxDS1 access speeds. Many carriers also provide dial-up access to frame relay for occasional ad hoc use or backup access. Most carriers offer PVC service, while a few now provide SVC service as well. Carriers interface through FR Network-to-Network Interface (NNI) trunks. However, not many frame relay service provider network interconnections exist today because the interconnection point was a single point of failure. This will likely change as service providers implement a Switched PVC (SPVC) capability that defines a standard for resilient network-to-network interconnection as defined in ITU-T Recommendation X.76 [6] and FRF.10 [7]. This concept is very similar to the Soft PVC concept used in ATM's Private Network-to-Network Interface (PNNI) as discussed in Chapter 9. The FRF.10 specification also defines a means for a FR SVC user to connect to a FR PVC user.

Many service providers have embellished frame relay with additional features, such as prioritization, automatic restoration, closed loop congestion control, and usage-based billing. Several carriers provide international frame relay service. Carriers also offer frame relay in a number of areas in Europe, Asia, and Australia.

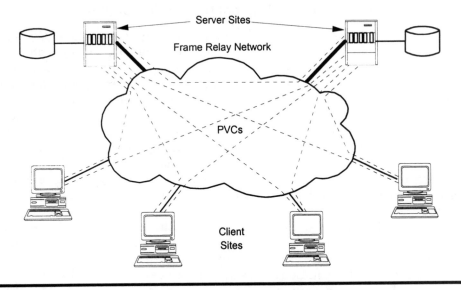

Figure 7.23   Typical Frame Relay Virtual Private Network Client-Server Application

Frame relay provides some basic public network security in that data originating and terminating through an access line is limited to connectivity established by the PVCs for that access line. Indeed, virtual private networks

(VPN) are a key application for frame relay networking. Figure 7.23 illustrates a frame relay network configuration common to many VPNs. At the top of the figure, two larger server sites retain a database for an enterprise with many smaller client sites. The server sites are interconnected via a PVC, and each client site is connected to each server via a pair of PVCs. Thus, any single failure of a server, access line, or network switch leaves the network intact. If the access line to a client site were to fail, then only that site would be disconnected. When used with a public frame relay service that has a per-PVC charge, this type of design is often quite economical when the underlying enterprise paradigm is client-server based.

7.3 FRAME RELAY — CONTROL PLANE

The frame relay control plane functions as a *signaling protocol*. The benefits of using frame relay's signaling capabilities are:

☞ status signaling for permanent virtual connections (PVCs)
☞ dynamic call setup for switched virtual calls (SVCs)

We first cover the relationship of the control plane to the user plane and then cover each of these applications by describing the formats and protocol along with an illustrative example of each.

7.3.1 Frame Relay Control Protocol Networking Context

Figure 7.24 illustrates the context for the frame relay control plane (C-Plane) operating in conjunction with the user plane (U-plane) over a *single* physical interface. Starting in the upper left hand corner on the CPE side of the User-Network Interface (UNI), note that the higher layers interface with both the control and user planes simultaneously. This model corresponds to the real world application where a higher layer protocol, such as a LAN driver, first issues commands to the control plane to establish a frame relay connection through the serving FR network, prior to transfering data over the user plane connection. Next, note that on the right hand side of the figure the network operates at only the core of the data link layer for the user plane as described in the previous section, but also at the layer three control signaling in the control plane.

The shaded portions of the figure indicate the closely related user and network protocols detailed in the frame relay standards discussed in the next section. These standard user side and network side protocols must be present in compatible implementations in each end user device attached to a service provider's frame relay network. Because of these well written standards, excellent interoperability is commonplace in mission-critical, real world applications around the world today. The data link layer provides a core set of services which are the actual frame relay service. On top of this same layer 2 service, the control data link layer adds a reliable, in-sequence delivery service in support of frame relay's layer 3 control signaling protocol used for SVCs. Signaling messages exchanged between a user and a network set up calls in a similar manner to those performed in ISDN and X.25 switched connections as studied earlier.

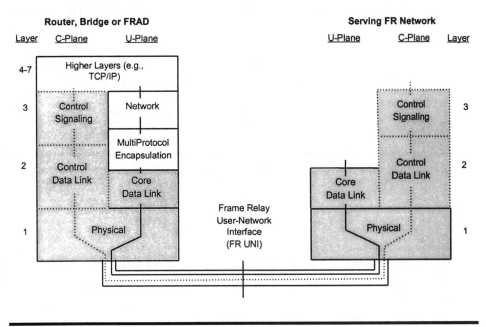

Figure 7.24   Frame Relay Control Protocol Networking Context

7.3.2 Frame Relay Standards and Specifications

A number of standards and specifications define the frame relay protocol at the User-Network Interface (UNI) and the Network-Network Interfaces (NNI). Furthermore, these specifications cover the user plane and the control plane. We start with the Frame Relay Forum (FRF) Implementation Agreements (IA), since these reference the other standards defined by the ITU-T and ANSI. Study of these frame relay standards is important to gain a full understanding of ATM, since the ITU-T signaling standards for ISDN (Q.931), frame relay (Q.933), and B-ISDN (Q.2931) all embrace a common philosophy, message structure, and approach. This is the secret to understanding the evolution of these technologies.

7.3.2.1 Frame Relay Forum Implementation Agreements

Related protocols in frame relay's control plane establish and release Switched Virtual Connections (SVCs), as well as report on the status of Permanent Virtual Connections (PVCs). Figure 7.25 illustrates the various Frame Relay UNI signaling protocols and their context. A frame relay connection can be either a permanent or switched virtual connection (PVC or SVC) as specified in Frame Relay Forum (FRF) implementation agreements FRF.1.1 [8] and FRF.4 [9], respectively. PVC signaling consists of conveying the edge-to-edge status of each frame relay DLCI, while SVCs offer connectivity on-demand based upon the network address and the frame relay traffic parameters. A different protocol interconnects frame relay networks at the NNI level for PVCs as shown in Figure 7.25. The FRF.2.1 implementation agreement defines the NNI for PVCs [10] and FRF.10 [7] defines the NNI for SVCs.

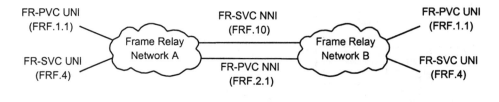

Figure 7.25   Frame Relay Forum Signaling Implementation Agreement Context

Frame Relay Forum (FRF) Implementation Agreements (IAs) are the reference used for interoperability agreements. Usually rather brief, the FRF IAs make extensive references to ITU-T international and ANSI national standards. These documents subset the standards, making modifications or clarifications of the standards as necessary. The next section provides a guide and references for more details on these concepts and important standards.

7.3.2.2 ITU-T and ANSI Frame Relay Standards

Figure 7.26 shows the major standards from the ITU-T (denoted as Q.xxx) and ANSI (denoted as T1.xxx) for the frame relay UNI C-plane and U-plane mapped out according to the protocol functions identified in the shaded area of Figure 7.24. The standards foundation for frame relay resides in ITU-T Recommendation Q.922 [4] at the link layer (layer 2) and Recommendation Q.933 [11] at the network layer (layer 3). The Q.922 standard defines the core frame relaying bearer service (also defined in T1.618 [12]), as well as the reliable data link layer for the Q.933 SVC signaling protocol. Recommendation Q.933 defines the formats, protocols and procedures to dynamically set up and release switched virtual calls (SVCs) for the frame relaying bearer service. Note that standards for SVCs as well as PVCs often appear in the same document. For example, Annex A of Recommendation Q.933 defines status signaling for PVCs, while much of the rest of the document focuses on the call control protocol for FR SVCs. Frame relay standards split the data link layer of the user plane into two areas: core services and user-specified services, such as multiprotocol encapsulation specified in Q.933 Annex E, IETF RFC 1490, FRF.3.1 [13] and T1.617a [14].

Frame relay has two modes of operation: an end user permanently connected to a frame relay switch using the Q.922 link layer, or connected via circuit switched access via Q.931/Q.921 ISDN signaling to a remote frame relay switch. Although not widely used, the second mode of operation is useful for on-demand or backup access to a frame relay network. See T1.617 [15] or References 16 and 17 for more details on switched access to a remote frame switch. Since most frame relay customers connect routers or bridges with frame relay PVCs, they don't require circuit switched access.

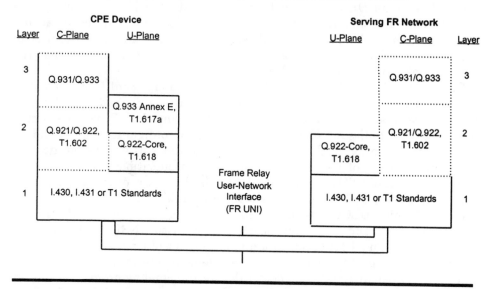

Figure 7.26   Structure of Frame Relay User-to-Network Interface (UNI) Standards

7.3.3 Frame Relay PVC Status Signaling

Figure 7.27 illustrates the basic concept of frame relay status signaling for a CPE device connected to a frame relay switch on the edge of a FR network.

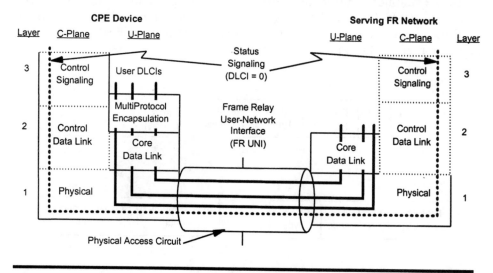

Figure 7.27   Context for Frame Relay Status Signaling

Status signaling allows the end user as well as the network to detect failures between the end points of Permanent Virtual Circuits (PVC). Generally, each physical access line connecting an end user to the serving frame relay switch carries multiple DLCIs. The FR standards define a status signaling protocol running on a particular DLCI (e.g., zero) that reports on the status for all DLCIs carried by the physical FR UNI access line. The periodic exchange of status signaling messages also acts as a heartbeat and fault detection mechanism for the physical access line connecting the user to the network. The status message also contains a field indicating that the network has provisioned a new PVC.

Originally, a group of vendors defined a Local Management Interface (LMI) standard for frame relay. The LMI with extensions defined a protocol to provide a keep-alive signal between the FR CPE and the FR network access port. Also, the LMI simplified initial frame relay configuration by providing automatic notification of changes in PVC connectivity as well as notification of the provisioning of new PVCs. Over time, incompatible standards arose between the LMI, ITU-T and ANSI standards. For example, the LMI extension uses DLCI 1023 for status reporting while T1.617 and ITU-T Q.933 Annex A employ DLCI 0 as shown in the figure. Fortunately, the ITU-T Q.933 Annex A standard is now closely aligned with ANSI T1.617, and the original proprietary LMI specification is falling out of use. Because of basic differences between the various local management interface standards, CPE and networks can automatically detect the status signaling protocol employed by the end user and react appropriately. This removes one configurable item and thus reduces the complexity of turning up a FR PVC for service.

ITU-T Recommendation Q.933 Annex A and ANSI Standard T1.617, Annexes B and D define the modern frame relay status signaling message formats and procedures covered in this chapter. ANSI Standard T1.617 Annex B defines additional status signaling procedures for interfaces carrying both PVCs and SVCs. These specifications define three main areas of PVC management: PVC status signaling, DLCI verification, and the physical interface keep-alive heartbeat.

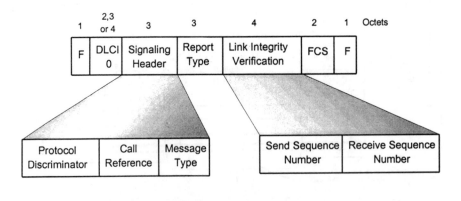

Figure 7.28 Q.933 Annex A STATUS ENQUIRY Signaling Message Content

The PVC status signaling procedures utilize two messages: STATUS ENQUIRY and STATUS. Figure 7.28 shows the Q.933 STATUS ENQUIRY signaling message format and content. The message contains a signaling header common to all signaling messages, comprised of a protocol discriminator, a Call Reference value, and the message type (a code indicating STATUS ENQUIRY in this case). The report type information element indicates a choice of one of the following three options: a full status (FS) report on every provisioned PVC, link integrity verification only, or asynchronous reporting on sets of PVCs whose status has changed.

Figure 7.29 shows the format and content of a frame relay STATUS message from Q.933 Annex A. The STATUS message has the same format as the STATUS ENQUIRY message with the addition of one or more optional PVC status information elements. Of course, the value of message type indicates that this is a STATUS message in response to a STATUS ENQUIRY. One or more optional PVC status information element contains the DLCI of the connection that the status applies to, as defined by two bits: new and active. The new bit is set to 1 for a new PVC, otherwise it is set to 0. The active bit indicates whether the PVC is operational (1) or not (0). The primary benefits of PVC status signaling derive largely from these two simple bits of information as expanded on in the next section.

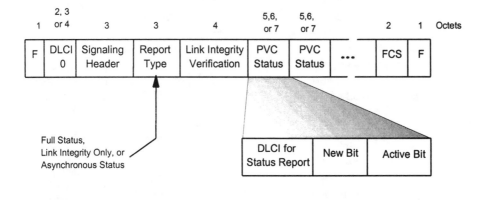

Figure 7.29   Q.933 Annex A STATUS Signaling Message Content

T1.617 Annex D differs slightly from the Q.933 Annex A formats in that it adds a one octet locking shift to codeset 5 information element after the common signaling header to indicate that the link integrity verification and PVC status information elements are for national use (e.g., the United States for ANSI). Status messages contain a link integrity verification field useful for checking continuity as well as acknowledging and detecting lost signaling messages using the send and receive sequence numbers in a manner analogous to the sliding window protocol of X.25.

The maximum number of usable DLCIs on a physical interface is limited by the maximum frame size defined for that interface. This occurs because the maximum frame size dictates how many PVC status information elements fit within a frame used to signal the Full Status (FS) message and thus limits the maximum number of DLCIs on that interface. Recognizing the fact that an NNI will likely need to support more DLCIs than a UNI, the Frame Relay

Forum adopted an event driven procedure documented in the FRF 2.1 implementation agreement. This procedure uses asynchronous status reporting to communicate only changed PVC status along with periodic updates for unchanged PVCs. This procedure overcomes the limit imposed by full status reporting on the maximum number of PVCs supported by an interface.

7.3.4 Frame Relay PVC Status Signaling Example

Figure 7.30 shows an example of frame relay status signaling illustrating access link failure detection and end-to-end PVC failure reporting.

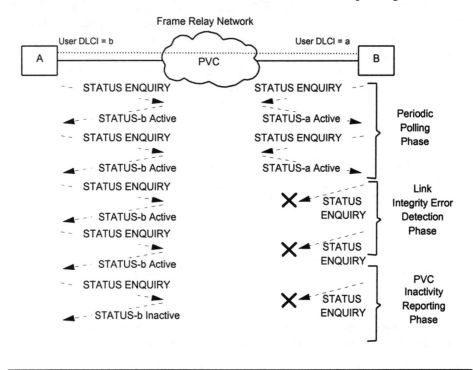

Figure 7.30   Example of Frame Relay Status Signaling Operation

In the example, the frame relay network provides a PVC between CPE devices A and B operating on DLCIs b and a respectively as shown at the top of the figure. Time runs from top to bottom in the figure with the message exchanges between devices A and B and the network shown as labeled arrows. As noticed during the periodic polling phase of messages in the figure, the CPE device periodically sends a STATUS_ENQUIRY message to the network requesting a particular type of status report. The default polling interval is once every ten seconds. In Q.933 Annex A, the default is to only poll for full status once every six message exchanges: the other message exchanges verify link integrity using the send and receive sequence numbers. For brevity, our example assumes that the requested report type is full status for each STATUS ENQUIRY message.   The network responds with a STATUS message reporting that the PVC from A to B is active. Although not shown in the figure, note that this same status message also reports on all other PVCs terminating on the UNI connected to A or B. In general, these status reports differ on A and B's user-network interfaces since at least one PVC will have a different destination, hence also have a different DLCI.

Continuing our example to the next phase, after two cycles of polling status, the physical access line connecting user B to the network fails. In the standards, if the user or network devices miss three consecutive STATUS messages, then they declare the link inactive. For brevity, we assume that this occurs after two missed messages at the expected times. The frame relay network now knows that the end-to-end PVC is down, and reports this fact in the next STATUS message to user A. The network should also report changes in PVC active status for internal failures. Recommendation Q.933 states that the user equipment should cease sending frames if the network reports the PVC as inactive. Once the access line failure clears, the user and network reestablish the enquiry-response status signaling protocol, synchronizing the sequence numbers in the link integrity verification information element, and stand ready to detect subsequent PVC status changes.

7.3.5 Frame Relay Switched Virtual Connections

CPE devices use the Frame Relay Switched Virtual Connection (SVC) signaling protocol (Q.933) to dynamically establish temporary connections for the purposes of data transfer. Frame relay SVCs support both the ITU-T (CCITT) E.164 and X.121 numbering plans for the purposes of identifying interfaces. X.121 supports interworking with X.25 networks, while E.164 is the longer term direction stated in standards. Each service provider administers its own numbering plans using, for example, a carrier prefix code in the E.164 numbering plan. The SVC protocol assigns each end of the connection a unique DLCI used for the duration of the call. SVC call control also provides a mechanism for parameter negotiation (e.g., maximum frame size, traffic parameters, transit delay). The user exchanges SVC call control messages with the network over DLCI 0. SVC control messages require a reliable link layer protocol, as specified in the Frame Relay Forum's FRF.4 Implementation Agreement based upon Recommendation Q.922.

Wide implementation of SVC services has been slow, primarily due to the complexity of SVC management, alignment of standards, and the requirement for strict security and administration. SVCs are outlined along with PVCs in ITU-T Recommendation I.122. Work is in progress in the Frame Relay Forum to specify the means to interoperate FR and ATM SVC services.

7.3.6 Example of Frame Relay SVC Operation

The Frame Relay signaling procedures for switched virtual connections (SVCs) at the UNI are based upon CCITT Recommendation Q.933, which uses Q.922 as a reliable link layer protocol. Q.933 signaling messages are sent on DLCI zero [9]. Figure 7.31 illustrates the sequence of messages

involved in establishing and clearing a Frame Relay SVC. The messages are
similar to those used in ATM since they both have the same ancestor —
narrowband ISDN.   The next section covers the contents and the basic
semantics of these messages.

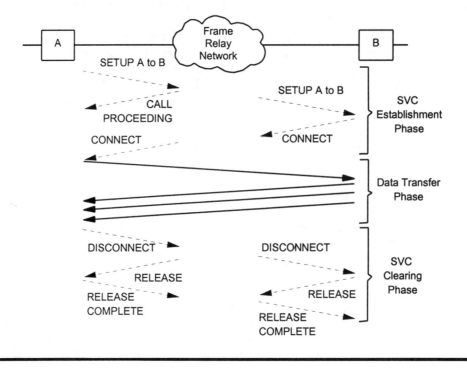

Figure 7.31   Frame Relay SVC Establishment, Data Transfer, and Clearing Example

Note that messages are sent in one of two contexts: from the user to the
network, and from the network to the user. The rules of operation and
allowed message information content differs somewhat in these two contexts
as defined in the standards. In the SVC establishment phase, calling user A
sends a SETUP message to the network indicating B as the called party. If
the network accepts the request, it responds with a CALL PROCEEDING
message, finds a route to the called party and generates a SETUP message
towards B. If B accepts the incoming call request, it responds with a
CONNECT message. Concurrently, the network propagates B's acceptance of
the call back to calling user A via a CONNECT message. Now, A and B may
exchange frames on the dynamically established DLCI until the either party
clears the call by sending a DISCONNECT message. In the example, A
initiates call clearing. A two-phase handshake, involving the RELEASE and
RELEASE COMPLETE messages as shown in the figure, confirms that a call
is indeed cleared. The multiple level handshake in the SVC establishment
and clearing phases is especially important if the network charges for SVCs
based upon call duration. For example, failure to properly release a call could

result in a very large charge to the user since the network would record the event as a long call.

Table 7.2   Frame Relay Signaling Message Information Element Content

	ALERTING*	CALL PROCEEDING	CONNECT	CONNECT ACKNOWLEDGE*	PROGRESS*	SETUP	DISCONNECT	RELEASE	RELEASE COMPLETE	STATUS	STATUS ENQUIRY
Protocol discriminator	M	M	M	M	M	M	M	M	M	M	M
Call reference	M	M	M	M	M	M	M	M	M	M	M
Message type	M	M	M	M	M	M	M	M	M	M	M
Bearer capability						M					
Data link connection identifier	O-1	O-1	O-1			O					
Link layer core parameters			O			O					
Calling party number						O					
Calling party sub-address						O					
Called party number						O					
Called party sub-address						O					
Transit network selection						O					
Cause (code)					O		M	O-2	O-2	M	
Call state										M	
Channel identification *	O-1	O-1	O-1			O					
Progress indicator *	O	O	O		M	O					
Network-specific facilities *						O					
Display *	O	O	O	O	O	O	O	O	O	O	O
End-to-end transit delay *			O			O					
Packet layer binary parameters *			O			O					
Link layer protocol parameters *			O			O					
X.213 priority *			O			O					
Connected number *			O				O	O	O		
Connected sub-address *			O				O	O	O		
Repeat indicator *						O					
Low layer compatibility *			O			O					
High layer compatibility *						O					
User-user *	O		O			O		O	O		

Notes:

1 Mandatory if first message in response to SETUP   M = Mandatory

2 Mandatory if first call clearing message   O = Optional

* Not applicable in FRF.4 implementation agreement   O-x = Optional with Note x

7.3.7 Frame Relay Signaling Message Information Elements

Similar to ISDN signaling messages, frame relay messages contain a number of information elements (IEs), which are either Mandatory (M), or Optional (O) as specified in ITU-T Recommendation Q.933. Table 7.2 depicts the population of the principal frame relay signaling message types with information elements. The Frame Relay Forum FRF.4 implementation agreement further subsets these requirements by not requiring the ALERTING, CONNECT ACKNOWLEDGE or PROGRESS messages as indicated in the Table. Furthermore, the FRF document eliminates a majority of the information elements in interoperable implementations as indicated by asterisks in the table.

As seen from the table, the SETUP message contains by far the largest number of IEs since it must convey all of the information about the connection so that the network or the called user can decide whether to accept the call. The principal information elements are the called party number and subaddress, transit network selection and link layer core parameters (i.e., Bc, Be and T). The calling party number and subaddress are optional. The DLCI information element enables dynamic assignment of the frame relay header values at call setup time. The bearer capability information element indicates transfer of unrestricted digital information using the frame mode service. The cause message indicates the reason for call release or the current status of an active SVC in response to a STATUS ENQUIRY message.

7.3.8 When to Use SVCs Instead of PVCs

A PVC is permanent, as the first letter of the acronym name states; while an SVC is switched, or established on-demand. A user application can establish an SVC only when there is information to send, and then disconnect it once the information is transferred. This can be a considerable economic advantage if the FR SVC service provider charges for the duration of SVC calls. On the other hand, a PVC is always in an established mode, whether there is information to transfer or not. Although this means that there may be periods where unused capacity exists, PVCs avoid the possibility of call blocking that SVCs may have.

Frame relay SVCs offer a more scalable solution than PVCs in networks with a large number of nodes that only require sporadic or ad hoc connectivity: for example, video conferencing, voice, data backup, or file transfer applications. Switched connections also make sense for remote sites which require access to other sites only on an intermittent basis. Furthermore, if the service provider implements address screening or closed user group interlock codes, then SVCs enable intranets and extranets. More service providers will likely interconnect to provide SVC service than those that interconnect to provide FR PVC service in order to serve such markets.

In summary, PVCs are best when replacing a private line network with a virtual private network. A least cost design for many of these PVC-based networks result in a single or multiple star topology as described earlier in this chapter. A full mesh topology for PVCs is, however, often uneconomical for all but the smallest frame relay networks. SVCs avoid the multiple hops through intermediate routers that occur in PVC-based networks; because they dynamically establish a path from source to destination. SVCs are best suited for very large networks with either unknown, or changing connectivity requirements. Furthermore, SVCs can be employed to better track traffic patterns and charge back actual network use to the actual application if the service provider supports usage-based billing. The major issue with SVCs is that as an on-demand service, there is a finite probability of blocking. By definition, PVC networks are non-blocking. If the network blocks an SVC call attempt, then the user cannot transfer any information. If your application cannot tolerate occasional call blocking, then you should design PVCs into the network to meet the need for guaranteed bandwidth that only a PVC can provide. As a final note, hybrid SVC/PVC networks are possible. A node initially connected via SVCs can be changed to a PVC connection if the performance, traffic patterns, and economics warrant such a change.

7.4 REVIEW

This chapter covered the predominant connection-oriented data services: X.25 and frame relay, in terms of the origins, protocol formats, functions, operation, traffic aspects, and commercial service aspects. The text first described the way X.25 performs store and forward packet -switching and windowed flow control to account the relatively poor private line network performance of the 1970s and early 1980s. Next, the coverage moved on to the successor of X.25 and private lines: frame relay. Essentially, frame relay is X.25 on a diet, and lower feature content of frame relay has enabled a ten-fold increase in throughput over classical X.25 packet -switching. Being part of the ITU-T's integrated services digital network extended family, frame relay defines concepts of traffic and congestion control similar to those used in ATM. Furthermore, the operation of switched virtual calls in frame relay is similar to that of ATM. Appreciating the underlying paradigm resulting from years of experience is the knowledge gained from studying technological evolution. This chapter provides the reader with a basic background of the connection-oriented data communication services and concepts that reappear within ATM as described throughout the remainder of the book.

7.5 REFERENCES

[1]   ITU-T, *Interface between Data Terminal Equipment (DTE) and Data Circuit-terminating Equipment (DCE) for terminals operating in the packet mode and connected to public data networks by dedicated circuit*, Recommendation X.25, October, 1996.

[2]   ISO/IEC, *Information-Technology — Telecommunications and information exchange between systems — High-level data link control (HDLC) procedures*, ISO/IEC 13239, 1997-06-15.

[3]   U. Black, *The X Series Recommendations*, McGraw-Hill, 1995.

[4]   CCITT, *ISDN Data Link Layer Specification for Frame Mode Bearer Services, ITU*, Recommendation Q.922, 1992.

[5] CCITT, *Congestion Management for the ISDN Frame Relaying Bearer Service*, Recommendation I.370, October 1991.

[6] ITU-T Study Group 7, "Draft Amendment 1 To Recommendation X.76 (SVC Part)," April 1996.

[7] Frame Relay Forum Technical Committee, *Frame Relay Network-to-Network SVC Implementation Agreement*, FRF.10, September 1996.

[8] Frame Relay Forum Technical Committee, *The Frame Relay Forum User-to-Network Implementation Agreement (UNI)*, FRF 1.1, January 1996.

[9] Frame Relay Forum Technical Committee, *Frame Relay User-to-Network SVC Implementation Agreement*, FRF.4, January 1994.

[10] Frame Relay Forum Technical Committee, *Frame Relay Network-to-Network Interface Implementation Agreement*, FRF.2.1, July 1995.

[11] CCITT, *ISDN Signaling Specification for Frame Mode Bearer Services*, Recommendation Q.933, 1991.

[12] ANSI, *ISDN — Core Aspects of Frame Protocol for Use with Frame Relay Bearer Service*, T1.618-1991, October 1991.

[13] Frame Relay Forum Technical Committee, *Multiprotocol Encapsulation Implementation Agreement*, June 1995.

[14] ANSI, *ISDN — Signaling Specification for Frame Relay Bearer Service for Digital Subscriber Signaling System Number 1 (DSS1) (Protocol encapsulation and PICS)*, T1.617a-1994, January 1994.

[15] ANSI, *ISDN — Signaling Specification for Frame Relay Bearer Service for Digital Subscriber Signaling System Number 1 (DSS1)*, T1.617-1991, 1991.

[16] U. Black, *Frame Relay Networks, Specifications and Implementations*, McGraw-Hill, 1994.

[17] D. Spohn, *Data Network Design*, 2nd Edition, McGraw-Hill, 1996.

8

Connectionless Protocols
— IP and SMDS

This chapter presents an overview of connectionless data services available in public and private data services during the 1990s, namely the Internet Protocol (IP) suite and Switched Multimegabit Data Service (SMDS). We describe the aspects of each protocol, beginning with the historical origins and followed by an overview of the packet formats and protocol functions. Next, the text illustrates the operation of the protocol through examples and then summarizes the traffic and congestion control aspects of the various protocols. Finally, we recount how public services utilize these protocols.

8.1 THE INTERNET PROTOCOL SUITE, TCP/IP

The origins of the Internet Protocol (IP) suite occurred even earlier than X.25. Operating at layer 3, IP supports a number of protocols, most commonly the Transmission Control Protocol (TCP). TCP provides layer-4 type services to applications such as the World Wide Web's hyper-text transfer protocol (http). The emergence of the World Wide Web (WWW) as the predominant user interface to "cyberspace" established IP as the de facto standard for internet-working to the corporate desktop, as well as the home office and the residential user. It is an important set of protocols with some of its influences seen in SMDS and ATM.

8.1.1 Origins of TCP/IP

The U.S. Advanced Research Projects Agency (ARPA) began development of a packet switched network as early as 1969, demonstrating the first packet switching capability in 1972. Named the ARPAnet, this network steadily grew as more universities, government agencies, and research organizations joined the network. Network designers introduced the Transmission Control Protocol/ Internet Protocol (TCP/IP) in 1983, replacing the earlier Network

Control Protocol (NCP) and Interface Message Processor (IMP) Protocol. This is essentially the same TCP/IP standard in use today. Also in 1983, administrators split the ARPAnet into two parts — a military network and a nonmilitary research network that became the origin of today's Internet. In 1986, the National Science Foundation (NSF) constructed a 56-kbps network connecting its six new supercomputer centers, followed shortly thereafter with an upgrade from 56-kbps speeds to 1.5-Mbps DS1 speeds in 1988. The Internet formed its own standards body in 1989, called the Internet Engineering Task Force (IETF), as discussed in Chapter 10. In 1990 the National Science Foundation (NSF) embarked upon a program to upgrade the Internet backbone to DS3 speeds (45-Mbps) for supercomputer interconnection. In 1995 the NSF split the Internet into a backbone for supercomputer communication running at OC-3 speeds (150-Mbps) over ATM and a set of Network Access Points (NAPs) where major backbone Internet Service Providers (ISPs) could interconnect. A significant policy shift also occurred at this time: instead of funding a network for researchers, the NSF gave the funds to research institutions and allowed them to select their own ISPs. True to the spirit of capitalism, many ISPs are now directly interconnected, bypassing the government funded NAPs.

The Internet became a household word in 1995 with the introduction of a user friendly, multimedia browser application accessing the World Wide Web. Having a web page became like advertising — every business had to have it to compete. Now, banks, auto dealerships, retail stores, and even children's television programs beckon users to look at their "web page". On-line information is the new currency, and the WWW is the ATM machine that provides access to that currency. Gone are the good old days when the Internet was the haven of university researchers and government organizations. You want information, to purchase goods, or just play computer games against other players — do it on the web. The web offers the ultimate in interoperability: all you need is Internet access and a web browser, and let the web do all the rest. Let's take an in-depth look at the foundations of this epitome of interoperability in data networking.

8.1.2 TCP/IP Protocol Structure

Figure 8.1 illustrates the layered Internet protocol suite built atop IP. The User Datagram Protocol (UDP), Internet Control Message Protocol (ICMP), routing control protocols, and the Transmission Control Protocol (TCP) interface directly with IP, comprising the transport layer in the Internet architecture. For further details on TCP/IP, see the IETF RFCs referenced in the following sections, or look through one of the many good books that cover IP in much more depth, such as references 1, 2, or 3.

IP is a datagram protocol that is highly resilient to network failures, but does not guarantee sequence delivery. Routers send error and control messages to other routers using the Internet Control Message Protocol (ICMP). ICMP also provides a function in which a user can send a *ping* (echo packet) to verify reachability and round trip delay of an IP-addressed host.

The IP layer also supports the Internet Group Multicast Protocol (IGMP) covered in Chapter 20.

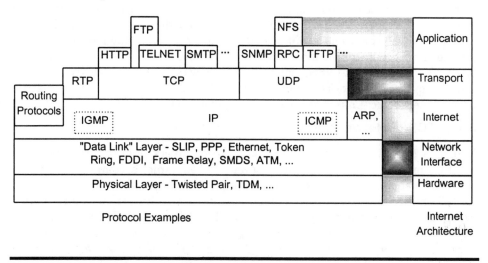

Figure 8.1   Internet Protocol (IP) Suite

The Address Resolution Protocol (ARP) directly interfaces to the data link layer, for example, Ethernet. The purpose of ARP is to map a physical address (e.g., an Ethernet MAC address) to an IP address. This is an important concept used in ATM. Chapter 9 covers routing protocols and ARP after this chapter introduces IP packet format and addressing.

Both TCP and UDP provide the capability for the host to distinguish among multiple applications through port numbers. TCP provides a reliable, sequenced delivery of data to applications. TCP also provides adaptive flow control, segmentation, and reassembly, and prioritized data flows. UDP only provides an unacknowledged datagram capability. The recently defined Real Time Protocol (RTP), RFC 1889, provides real time capabilities in support of multimedia applications [4].

TCP works over IP to provide end-to-end reliable transmission of data across the network. TCP controls the amount of unacknowledged data in transit by dynamically *reducing* either the window size or the segment size. The reverse is also true in that *increased* window or segment size values achieve higher throughput if all intervening network elements have low error rates, support the larger packets, and have sufficient buffering to support larger window sizes.

A number of applications interface to TCP as shown in Figure 8.1. The File Transfer Protocol (FTP) application provides for secure server log-in, directory manipulation, and file transfers. This was an early form of downloading files from the Internet, still used for large file transfers by Internet experts. TELNET provides a remote terminal log-in capability, similar to the old command line interface of terminals to a mainframe. The Hypertext Transfer Protocol (HTTP) supports the popular World Wide Web. The majority of user access now employs this protocol, which runs over TCP as shown in the figure. Other applications operate over UDP. The Simple Network Management Protocol (SNMP) supports configuration setting, data retrieval, and alarm reporting, and is the most commonly used protocol for collecting management data from IP networked devices. The Trivial FTP (TFTP) protocol provides a simplified version of FTP, which is intended to reduce implementation complexity. The Remote Procedure Call (RPC) and Network File Server (NFS) capabilities allow applications to dynamically interact over IP networks. Domain Name Services (DNS) provide a distributed or hierarchical name service running over UDP or TCP.

8.1.3 TCP/IP Networking Context

Figure 8.2 shows how the Internet Protocol layers operate in a network context interconnecting two end systems, for example, a computer workstation and a server. In the example, three interconnected IP routers interface to two end systems. The link layer and internetwork layer protocols define procedures for dynamically discovering the "best" route between the end systems in a process called *routing*, a topic which we cover in some depth in the next chapter. Packets between the same pair of end systems may not always traverse the same physical path, as indicated by the two sets of dashed lines interconnecting the end systems in the figure. This occurs because the routers may dynamically discover a better route and change the forwarding path between transmissions of packets by end systems as illustrated by the dashed line in the figure. Each IP router operates at the physical, link, and network layers as shown in the figure. The transport and application layers use protocols such as TCP, UDP, FTP and HTTP as described in the previous section. Let's take a detailed look at the formats and protocols involved in TCP/IP. Since IP is designed to run over a wide range of link layer protocols, this section describes only those uniquely defined by the IETF. We then principally focus on the specifics of layer 3 — the internetwork layer protocol. Finally, the section concludes with an overview of TCP, which provides services like sequencing, error detection and retransmission.

IP provides a connectionless datagram delivery service to the transport layer. IP does not provide end-to-end reliable delivery, error control, retransmission, or flow control; it relies on higher-layer TCP to provide these functions. A major function of IP concerns the routing protocols, which provide the means for devices to discover the topology of the network, as well as detect changes of state in nodes, links, and hosts. Thus, IP routes packets through available paths and around points of failure. IP has no notion of reserving bandwidth; it only finds the best available path at the moment. Most of the routing algorithms minimize an administratively defined routing "cost." The next chapter provides further details of routing.

Address design within an IP network is a critical task. A network may have a large number of end user and server devices, or "hosts". If every router in a large network needed to know the location of every host attached to every other router, the routing tables could become quite large and cumbersome. A key concept used in routing is that of subsetting the host address portion of the IP address, which is a technique called subnetting. Effectively, subnetting breaks the host address space down into multiple subnetworks by masking the bits in the host address field to create a separate subnetwork for each physical network. This means that a router need only look at a portion of the address, which therefore dramatically reduces network routing table size. The next chapter presents an example of subnetting.

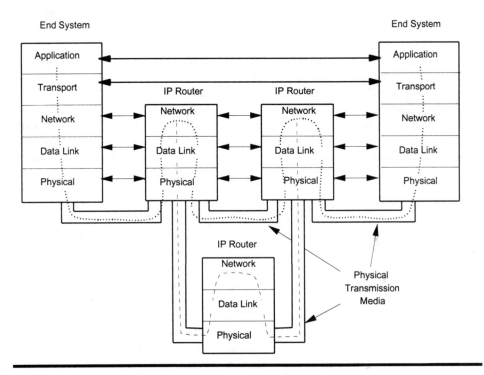

Figure 8.2   Internetworking Protocol (IP) Networking Context

8.1.4 Generic Link Layer Protocols for IP

The earliest generic link layer protocol designed specifically for access to the Internet derived from a proprietary protocol implemented by 3Com in the early 1980s to interconnect computers and routers. Called Serial Line IP (SLIP), the predominant versions of UNIX operating systems implemented this simple protocol. The IETF "standardized" SLIP in 1988 as described in RFC 1055 [5]. Basically, SLIP provides a means for delimiting IP packets on a serial link using specific characters. If the user data contains these specific characters, then SLIP uses a character stuffing mechanism similar to the IBM BSC protocol studied in Chapter 4. Since the protocol is so simple, implementations readily achieve a high degree of interoperability. However, SLIP had several shortcomings: it only supported IP, it performed no error detection or correction, and it required pre-configuration of IP addresses.

Although these shortcomings weren't serious problems in the 1980s, the increasing popularity and sophistication of Internet users required an improved, more feature rich, link layer protocol. Since the current Internet address space was being rapidly consumed by waves of new web users signing up with Internet Service Providers (ISPs), the IETF needed to design a better link layer protocol for access to the Internet. Responding to these needs, the IETF specified the Point-to-Point Protocol (PPP) in 1994, which has now superseded SLIP in most implementations. PPP supports not only IP, but other protocols multiplexed over a shared serial link. PPP now supports the automatic configuration and management of the link layer between dial-up users as well as multiprotocol routers over a wide range of serial interfaces using link and network level negotiation [6]. The basic negotiation procedure involves one party proposing a particular option and parameters, followed by the other party either accepting the proposal, or rejecting it.

The PPP Link Control Protocol (LCP) automatically establishes, configures, and tests a data link connection. The LCP protocol negotiates parameters such as: authentication, maximum packet size, performance monitoring, and header compression. The PPP Network Control Protocol (NCP) is specific to each protocol. Ever wonder how ISPs can sign up thousands of users yet provide the same dial access number to all of them? They do so by using the IP NCP to dynamically assign one of a pre-reserved block of IP addresses to dial-up users as they call in on the common access number. Thus, PPP allows an ISP to efficiently share a limited IP address space (a valuable commodity), since only users who are logged in actually use an IP address. Therefore, an ISP must only have a number of IP addresses equal to the number of dial-up ports in use, and not for the total number of subscribers. This feature of PPP stretches the limited address space of the version 4 Internet protocol (widely used today until replaced by version 6)[3].

PPP assumes that some other protocol delimits the datagrams, for example, HDLC as specified in RFC 1662 [7]. Figure 8.3 illustrates the basic format of an HDLC-framed PPP datagram. Unfortunately, the IETF uses a different convention for indicating the order of bit transmission than that the ISO and ITU HDLC standards use as shown in the figure. In the IETF notation the bits of binary numbers are transmitted from right to left. In other words, bit transmission is from Least Significant Bit (LSB), that is, bit zero, to Most Significant Bit (MSB). Standard HDLC flags delimit the PPP datagram. PPP defines an octet level stuffing method to account for HDLC flags (a binary '01111110') occurring within user data streams. The all-stations HDLC address (all ones) and the control field encoding identify the HDLC frame as Unnumbered Information with the Poll/Final bit set to zero. PPP uses a standard HDLC Frame Check Sequence (FCS) to detect transmission errors, with RFC 1662 giving an example of an efficient software implementation of the FCS function. Inside the PPP header, the one (optionally two) octet protocol field identifies the protocol for the particular PPP datagram. Examples of standard protocol fields are: LCP, padding protocol, PPP Network Level Protocol ID (NLPID) and various NCP protocol identifiers. Optionally, the network layer protocol may insert a variable length pad field to align the datagram to an arbitrary octet boundary to make software implementation more efficient.

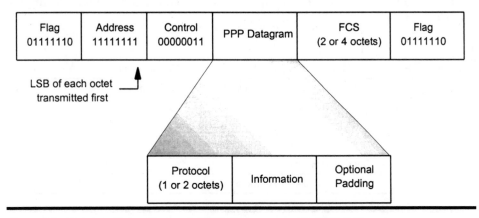

Flag 01111110	Address 11111111	Control 00000011	PPP Datagram	FCS (2 or 4 octets)	Flag 01111110

LSB of each octet transmitted first

Protocol (1 or 2 octets)	Information	Optional Padding

Figure 8.3 Point-to-Point Protocol (PPP) Datagram with HDLC Framing

The authentication feature of PPP allows the network to check and confirm the identity of users attempting to establish a connection. Web browser, operating system, hub, router and server vendors have embraced PPP as the link layer of choice for accessing the Internet. IP also operates over a number of other protocols by treating them all as a link layer protocol, such as frame relay, X.25, SMDS and ATM. We study the operation of IP over ATM as a link layer protocol in Chapter 20.

8.1.5 IP Version 4 (IPv4) Packet Format

Figure 8.4 illustrates the format of the version 4 IP packet [8], [9]. The 4-bit version field specifies the IP protocol version. Each node first checks the version field before processing the datagram. Such use of the version field will be critical in the migration from IP version 4 to IP version 6. Next, the IP Header Length (IP HL) field specifies the datagram header length in units of 32-bit words, the most common length being 5 words, or 20 octets when no options are present. If options are present, then the IP HL field indicates the number of words used for the options, for example, a route trace. The 8-bit Type of Service field contains a 3-bit precedence field, plus 3 separate bits specifying other service attributes, and two unused bits. The precedence field ranges from 0 (i.e., normal priority) through 7 (i.e., network control) indicating eight levels of precedence. The three individual bits request low delay, high throughput, and high reliability. The Total Length field specifies the total IP datagram length for the header plus the user data.

The identification field, flags, and fragment offset fields control fragmentation (or segmentation) and re-assembly of IP datagrams. The Time To Live (TTL) field specifies how many seconds the packet can be forwarded in the network before declaring the packet "dead," and hence disposable. Typically, intermediate nodes or routers decrement the TTL field at each hop. When TTL reaches zero, intermediate nodes discard the packet. Therefore, a packet cannot circulate indefinitely through a complex set of networks. The protocol field identifies the higher-level protocol type (e.g., TCP or UDP), which then specifies the format of the data field. The header checksum ensures integrity of the header fields through a simple bit-parity check that is easy to implement in software.

0	4	8	16	19	24	31
Version	IP HL	Type of Service	Total Length			
Identification			Flags		Fragment Offset	
Time to Live (TTL)		Protocol	Header Checksum			
Source Address						
Destination Address						
Options (0 or more words)					Padding	
Data (0 or more words)						

Figure 8.4 IP version 4 (IPv4) Datagram Format

The one word source and destination IP addresses are required fields discussed in more detail in the next section. The options field can specify security level, source routing, or request a route trace. Two types of source routing option give either a complete list of routers for a complete path, or a list of routers that must be visited in the path. In the route trace, each intermediate router adds its own IP address to the packet header options field (increasing the IP Header Length Field, of course). Optionally, the options can request that each router add a timestamp as well as its IP address when performing a route trace. The data field contains higher-layer protocol information or user data.

8.1.6 Internet Protocol (IP) Addressing

The Internet currently uses a 32-bit global network addressing scheme. Each user, or "host" in Internet parlance, has a unique IP address which is 4 octets in length, represented in the following dotted decimal notation:

XXX.XXX.XXX.XXX,

where XXX ranges from 0 to 255 decimal, corresponding to the range of 00000000 to 11111111 binary. There are 2^{32}, or over 4 billion, possible IP addresses. You would think that this would be enough addresses, but the Internet recently had to drop a previously used strict hierarchical structure that grouped addresses into three classes: A, B, and C. The class determined the maximum network size, measured by the number of hosts. Although network address classes no longer formally exist in the Internet today, the test briefly describes them as an introduction to classless addressing. Table 8.1 illustrates the key properties of the legacy Internet address classes. The IETF also reserved another block of Class D addresses for multicasting. Note the high order bits represent the network address, while the remaining bits represent the hosts in the network.

Table 8.1   Characteristics of Legacy Internet Address Classes

IP Address Class	Bits for Network Address	First Byte Network Address Range	Total Number of Networks	Bits for Host Address	Hosts per Network
A	7	0-127[1]	126	24	16,777,214
B	14	128-191	16,384	16	65,534
C	21	192-223	2,097,152	8	254
D[2]	NA	224-254	NA	NA	NA

Notes:
[1]   Values 0 and 127 are reserved for the all zeroes and all ones addresses
[2]   Values reserved for multicast

A central authority, the Internet Assigned Numbers Authority (IANA), assigns IP addresses to ensure that they are unique. A network administrator then assigns the host addresses within its class A, B or C address space however they wish, as long as the assignment is unique. Reusing IP addresses is a bad idea, since address conflicts arise when networks are interconnected. This assignment scheme worked fine, until the public's attraction to the information content available on the web made the Internet enormously popular. The problem arose because anyone asking for a IP address asked for the largest size they could justify. In the early days of the Internet the address space seemed limitless, so that the administrators generously assigned larger address classes to any user request. This inefficient allocation meant that a lot of IP addresses went unused. Many organizations further exacerbated the problem by requesting Class A and B addresses, and then inefficiently assigned them, leaving large blocks of addresses unused. The Internet community attempted to resolve the crisis a bit late by reallocating unused Class A and C addresses to the many new service providers, who then dynamically allocated addresses to individual users using PPP clamoring to get on the Internet. Now, modern routing protocols treat IP addresses as 32-bit numbers without a class structure.

The Internet called this new scheme Classless Inter-Domain Routing (CIDR) and began deployment in 1993 [10, 11, 12] to improve scaling of the Internet routing system. CIDR generalizes the concept of variable length subnet masks (VLSM) thus eliminating the rigid historical structure of network classes (A, B, and C). Interior (intra-domain) routing protocols supporting CIDR are OSPF, RIP II, Integrated IS-IS, and E-IGRP. Only one exterior (inter-domain) routing protocol, BGP-4, currently supports CIDR.

8.1.7 Next Generation IP — IPv6

It is true that "the Internet is running out of addresses!" The more efficient allocation of IPv4 addresses through CIDR along with the availability of routing protocols that support the variable-length sub-net masks, bought at best a few years time for Internet growth. For example, CIDR is now moving into the Class A address space reserved for expansion (64.0.0.0 through

126.0.0.0) [10]. Anticipating the exhaustion of the IP address space, the IETF issued a call for proposals for a successor to the current internet protocol in 1992 [13]. The twenty-one original proposals were reduced to a short list of seven serious proposals, resulting in the recommendation of RFC 1752 [14] for an IP next generation (IPng) protocol in January, 1995. IPng supersedes IPv4, and is now formally referred to as IPv6. The IETF issued the primary RFCs for IPv6 in December, 1995. These include RFC 1883 for general definitions of IPv6 [15], RFC 1884 for addressing [16], RFC 1885 for ICMPv6 [17], and RFC 1886 for Domain Name Services (DNS) extensions [18]. IPv6 contains the following additions and enhancements over IPv4:

* Expands address field size from 32 to 128 bits
* Simple dynamic auto-configuration capability
* Easier multicast routing with addition of a "scope" field
* Anycast feature, where a host sends a packet to an anycast address which the network delivers to the closest node supporting that function
* Capability to define quality of service for individual traffic flows using the ReSource reserVation Protocol (RSVP)
* Reduction of overhead by making some header fields optional
* More flexible protocol design for future enhancements
* Authentication, data integrity, and confidentiality options
* Easy transition and interoperability with IPv4
* Support for all IPv4 routing algorithms (e.g., OSPF, RIP, BGP, etc.)

The new IPv6 supports all the traditional protocols that IPv4 did, such as: datagram service, FTP file transfers, E-mail, X-windows, Gopher, and of course, the web. Furthermore, IPv6 also supports approximately 340×10^{36} individual addresses. To ease migration, IPv4 addressing is a proper subset of the IPv6 address space.

Figure 8.5 illustrates the version 6 IP packet format [15]. The *Version* field allows routers to examine the first four bits of the packet header to determine the IP version. The IPv4 packet header has the same version field in the first four bits so that routers can support both IPv4 and IPv6 simultaneously for migration purposes. The 4-bit *Priority* field defines eight values (0 through 7) for sources that can be flow controlled during congestion, and eight values (8-15) for sources that cannot respond to congestion (e.g., "real-time" constant bit-rate traffic, such as video). Within each priority group, lower numbered packets are less important than higher numbered ones. The *Flow Label* field is intended for use by protocols such as the Resource ReSerVation Protocol (RSVP) described later in this chapter to guarantee bandwidth and QoS for streams of packets involved in the same flow. The *Payload Length* field indicates the number of bytes following the required 40-byte header. The *Next Header* field identifies the subsequent header extension field. There are six (optional) header extensions: hop-by-hop options, source routing, fragmentation support, destination options, authentication, and security support. The last extension header field identifies the higher-layer protocol type using the same values as IPv4, typically TCP or UDP. The *Hop Limit* field determines the maximum number of nodes a packet may traverse. Nodes decrement by 1 the *Hop Limit* field each time they forward a, analogous to the Time to Live (TTL) field in IPv4, discarding the packet if the value ever reaches zero. The source and destination addresses are both 128 bits in IPv6, four times as large as the address fields used in IPv4. Therefore, the required IPv6 header is a constant 40 bytes; however, the optional extension header fields can make the overall header considerably larger.

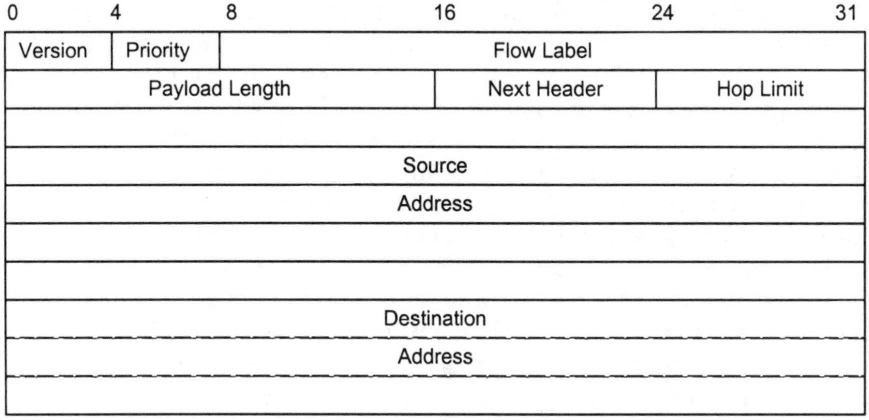

0	4	8	16	24	31
Version	Priority		Flow Label		
Payload Length			Next Header		Hop Limit
Source					
Address					
Destination					
Address					

Figure 8.5 IPv6 Packet Format

8.2 Resource ReSerVation Protocol (RSVP)

Some industry analysts believe that the competition between IP and ATM will be decided on the battleground of Quality of Service (QoS). Although the Internet traditionally offered only best effort service, if its Resource ReSerVation Protocol (RSVP) delivers QoS effectively, then ATM may be relegated to the carrier WAN backbone. Most experts agree that the distinguishing characteristic of ATM is its ability to offer differentiated QoS over a common infrastructure. Speed alone is no longer the overwhelming justification for ATM that it once was. Others believe that RSVP suffers from hype and unrealistically high expectations, similar to accolades lavished upon ATM a few years ago. Others claim that a large effort still remains to make RSVP a commercial reality. This section reviews RSVP in some detail to serve as background for a detailed comparison in Part 8. There the details are explained of how ATM and RSVP cooperatively implement QoS and guaranteed bandwidth.

8.2.1 Background to RSVP and QoS in IP

There is an old saying that "when everything is high priority, there is no priority." Analogously, RSVP over IP only delivers QoS if the underlying layer 2 network delivers QoS. The most straightforward way to assure QoS in the underlying layer 2 network is to ensure that it isn't congested. In the LAN, Switched Ethernet often fills the bill by allocating a dedicated 10 or 100-Mbps segment to each server or client host. In wide area networks subject to congestion, ATM is the only current layer two wide area network that delivers QoS. Therefore, many believe that ATM and RSVP together are the most viable near-term approach for supporting QoS-aware IP based applications.

RSVP grew out of work first published in 1993 by researchers at Xerox and University of Southern California (USC) [19]. The IETF established the integrated services (intserv) working group [20] which resulted in publication of version 1 of the RSVP specification in 1997)after over a dozen drafts in RFC 2205 [21]). Several experimental test beds have run RSVP for the past couple of years, and several vendors are now equipping their routers to implement RSVP [22]. In the applicability statement for RSVP [23], it is interesting to note that the IETF recommends that RSVP be limited to small numbers of multimedia users in intranets only. This caveat corresponds to common industry wisdom that RSVP is not yet ready for prime time. Furthermore, RSVP does not scale well on backbone trunks supporting large numbers of reservations. Another protocol must serve the aggregate backbone bandwidth reservation and QoS guarantee requirement. RSVP also needs some mechanism for deciding which applications get reservations for high-quality resources; otherwise applications (or their human controllers) will indiscriminately ask for the highest quality service available. Recent work in the IETF defines policy servers — which either approve or reject individual RSVP requests. Also, the IETF specified some basic security measures to mitigate the chance of false reservations or service theft in RFCs 2206 and 2207 [24, 25].

OK, you say, now that I have this background, tell me what RSVP is. How will it impact me? And why should I care? RSVP is a resource reservation setup (also called signaling protocol) specifically designed for an integrated services Internet that supports end-to-end QoS. Unlike the connection-oriented services studied so far, RSVP employs receiver-initiated setup of resource reservations for multicast or unicast data flows, in a scalable, robust manner.

8.2.2 The RSVP Protocol

Receiver applications in IP hosts utilize RSVP to indicate to upstream nodes their traffic requirements in terms of bandwidth availability, acceptable jitter, and available buffer space. RSVP is not a routing protocol, but like routing protocols it operates in the background, not in the data forwarding path shown by the solid arrow and shaded boxes in Figure 8.6.

A *packet classifier* in each RSVP capable device utilizes the filter specification to determine the QoS class of incoming data packets, and then selects the route. Each node also utilizes a *packet scheduler*, employing methods such as packet level traffic shaping, priority queuing, and weighted fair queuing to achieve the requested QoS.

The admission control function, resident in RSVP aware nodes along the path between the destination and the source, interprets the flow specification in RSVP control packets and determines whether the node has sufficient resources to support the requested traffic flows. The node may also perform policy control, for example, to ascertain that the requester has the right to make such reservations. If either the admission or control process checks fail, the RSVP process in the node returns an error message to the requesting application. If the node does not accept a request, it must take some action on the offending flow, such as discarding the packets or treating them at a lower priority. Thus, every intermediate node must be capable of prioritized service and selective discard. If a layer 2 network interconnects layer 3 devices (e.g., routers), then it too must be QoS-aware. This is an important point, as many IP networks span multiple link layer types and intermediate networks, making a ubiquitous, end-to-end implementation of RSVP unlikely. One device or link layer in the midst of an end-to-end RSVP flow may invalidate any traffic or quality level guarantees. RSVP capable systems maintain "soft-state" about RSVP traffic flows, that is, the state will time out and be deleted unless it is periodically refreshed.

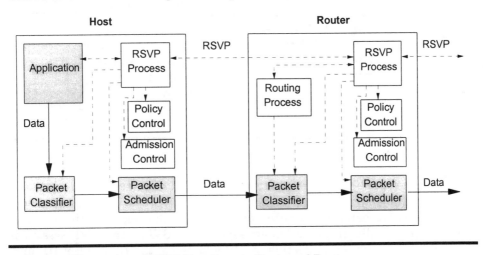

Figure 8.6   Illustration of RSVP Functions in Hosts and Routers

To achieve the performance promised by RSVP, IP equipment vendors will likely require software and/or hardware upgrades. Thus, implementing RSVP network-wide may be an expensive and difficult proposition. In particular, hosts and routers must perform at least some of the following functions [26]:

- Identify different incoming traffic types using various fields, such as source and destination addresses, port number, and protocol types.
- Assign the traffic to the appropriate priority queue and determine scheduling weights based upon the QoS needed by the application.

- Transmit RSVP signaling information to other equipment on the network to communicate desired QoS and traffic parameters for the flow.
- Fragment frames to reduce delay and jitter; to reduce the likelihood of a delay-sensitive, short voice frame getting delayed by a lengthy file transfer data frame.

RSVP builds upon a multicast paradigm which routes traffic flows along point-to-multipoint trees rooted at the source, where unicast is a special case involving a tree with only a single branch. Figure 8.7 illustrates a simple example involving two RSVP capable routers and two senders, both multicasting to each of four receivers (R1 through R4) . Note that R3 and R4 are on the same Ethernet segment, and hence receive all messages to the port on Router B.

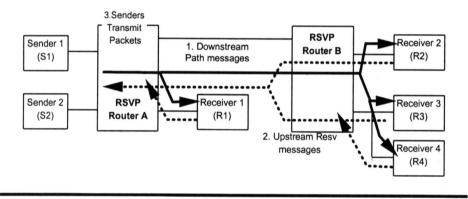

Figure 8.7 Illustration of RSVP Terminology

An example application using this topology could be three end users tuned into a simultaneous video application from S1 of a presenter at a conference whose slides are being transmitted by S2. The following narrative employs this example to illustrate the major aspects of RSVP. An RSVP *session* is defined by the source and destination IP addresses, the IP protocol ID, and, optionally, a destination port in the receiver node's application layer. There are eight sessions in the example shown by the thick solid arrow: S1 transmitting to R1, R2, R3 and R4 plus S2 transmitting to R1, R2, R3 and R4. The dashed arrow symbolizes the 8 sessions involved in communication from each of the 4 receivers back to the two servers. Once an RSVP capable sender and receiver both join a multicast group, senders begin transmitting *RSVP Path* messages downstream (i.e., from sender to receiver) in step 1. Receivers respond to these senders in step 2 with *RSVP Resv* messages containing a flow descriptor. Once the sender receives RSVP Resv packets, it begins transmitting packets to the receivers in step 3. Routers A and B determine whether they accept the request, communicating their decision to the receivers and process the packets for these flows accordingly.

8.2.3 RSVP Flow Specs and Service Classes

A flow descriptor defines a reservation in an RSVP Resv message in terms of a flow spec and a filter spec. A *flowspec* specifies the desired QoS in terms of loss, delay, and jitter. The *filter spec* determines what packets are part of the flow to which the flowspec applies.

An RSVP flowspec includes a service class; a Reserve spec (called an Rspec) that defines the desired QoS; and a Traffic spec (called Tspec) that uses a peak rate (p), a token bucket to define the rate (r) and burst size (b) [27], minimum policed unit (m), and a packet size. The Tspec defines the flow's traffic characteristics for which each IP network node should provide the requested QoS (defined in the Rspec) as part of its admission control procedure.

The IETF currently defines two service classes: Controlled Load in RFC 2211 [28] and Guaranteed Service in RFC 2212 [29], in addition to the de facto Best Effort service of the Internet. Note that the ATM Forum defines five classes, while the ITU-T also defines three categories as described in Part 3. Table 8.2 summarizes the RSVP flowspec's Rspec (i.e., QoS) and Tspec (i.e., traffic) parameters [30] specified for each of the IETF service classes.

Table 8.2   IETF RSVP Service Classes, Rspec and Tspec Parameters

	Guaranteed Service	Controlled Load	Best Effort
Reserve Spec (Rspec)			
Packet Loss	Specified	Not Specified	Not Specified
Packet Delay	Not Specified	Specified	Not Specified
Packet Delay Variation	Not Specified	Not Specified	Not Specified
Traffic spec (Tspec)			
Peak Rate (p)	Optionally Specified	Not Specified	Not Specified
Token Bucket (b,r)	Specified	Specified	Not Specified
Minimum Policed Unit (m)	Specified	Specified	Not Specified
Packet Size	Specified	Specified	MTU Size

A filter spec sets parameters used by the packet classifier to select subsets of packets involved in a particular session (i.e., receiver IP address plus protocol ID). Filter specs may define such subsets in terms of senders using sender IP address, higher-layer protocol identifiers, or, in general, any header fields within the packets involved with the specified session. For example, a filter spec could identify a desktop videoconferencing application running over RTP.

8.2.4 RSVP Reservation Styles

Table 8.3 summarizes the three types of reservation styles defined for RSVP. Reservation sender selection indicates whether the receiver explicitly identifies the sender in Resv messages, or if the reservation applies to all senders (i.e., wildcard). The reservation type specifies whether a distinct reservation is made for each upstream sender, or if the reservation is shared among all upstream senders. The reservation merging column summarizes what merging nodes perform as they propagate reservation requests upstream towards senders. In the Fixed Filter (FF) style there is no merging. The individual reservations for flowspecs for each sender are simply packed into the reservation packet. The Shared Explicit (SE) style takes the union of all senders and chooses the "largest" flowspec. The Wildcard Filter (WF) style is the simplest, because receivers specify no sender information, and the nodes simply choose the "largest" flowspec. The specifications for the individual integrated services describe how one flowspecs is "larger" than another for application in the generic merging process described above. We give examples of the Fixed Filter (FF) and Wilcard Filter (WF) reservation styles below.

Table 8.3   RSVP Reservation Style Attributes and Application

Reservation Style	Sender Selection	Reservation Type	Reservation Merging	Example Application
Fixed Filter (FF)	Explicit	Distinct	Pack all into request	Video conference
Shared Explicit (SE)	Explicit	Shared	Union, pick largest	Audio conference
Wildcard Filter (WF)	Shared	Shared	Pick largest	Audio Lecture

Figure 8.8 illustrates the Fixed Filter (FF) reservation style for the previous example. Starting from the lower right hand corner, router B receives reservations from R3 and R4 as indicated in the Receive column and merges them into a combined request in the Reserve column, while it simply copies R2's request. Router B then combines the two requests in the Reserve column into a single request for sending upstream. Router A receives requests from router B and R1 and makes reservations as indicated in the Receive and Reserve columns, respectively. Finally, router A parses out the individual reservation requests to each of the senders from those packed in the FF message.

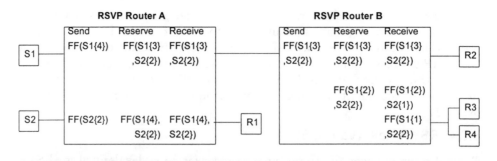

Figure 8.8   Example of Fixed Filter Reservation Style Merging

Figure 8.9 illustrates the Wildcard Filter (FF) reservation style. Starting from the lower right hand corner, router B receives reservations from R3 and R4 in the Receive column and combines them in the Reserve column, while it simply copies R2's request. Router B then combines the two requests in the Reserve column into a single request for transmission upstream. Router A receives requests from router B and R1 and makes reservations as indicated in the Receive and Reserve columns, respectively. Finally, router A sends an identical merged reservation request to each of the senders. Note that the WF method reserved the largest request from all receivers, and did not preserve the distinction between senders as the fixed filter style did in the previous example. The WF method requires shorter messages and less table space, but does not provide the specificity of the FF method.

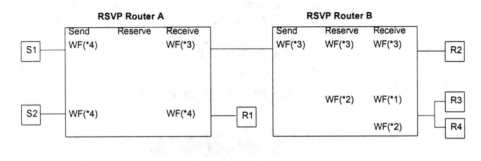

Figure 8.9   Example of Wild Card Filter Reservation Style Merging

Since RSVP only utilizes one pass from receiver back to the sender(s), the protocol defines an optional Advertisement specification (Aspec) to send packets downstream from the senders to the receivers — enabling a report on the expected QoS. An example of an Aspec QoS parameter would be the sum of the latencies along the path. The advertisement is not a guarantee, rather it is a confirmation that upstream senders have admitted the request. The senders may use this information to modify their upstream reservation requests

Changes in the IP Version 6 (IPv6) protocol support an explicit flow ID within the packet header, similar to the VPI/VCI field in ATM cell headers. RSVP associates a flow specification characterizing the flow's traffic pattern with each flowID.

8.2.5 TCP Packet Formats

Now, our coverage of TCP/IP moves up to the transport layer, namely the Transmission Control Protocol (TCP). Figure 8.10 illustrates the version 4 TCP packet format and meanings specified in RFC 793 [31] and RFC 1112 [32]. TCP employs the source and destination port numbers to identify a specific application program running in the source and destination hosts. The 16-bit port number in conjunction with the 32-bit host address comprise the 48-bit *socket* identifier.   Port numbers less than 256 are called "well-known ports" as defined in RFC 1700 and are reserved for standard services. For example, the *Sequence Number* field identifies the position of the sender's byte stream in the data field. The *Acknowledgment Number* field identifies the sequence number of the next byte expected at the receiver. The *Data Offset* field tells how many 32-bit words are in the TCP header. The default header length is five words, as shown in the figure. The code bits field contains six bits: URG, ACK, PSH, RST, SYN and FIN. URG indicates that the Urgent Pointer is used. The ACK bit indicates that the Acknowledgement Number field is valid. The PSH (i.e., push) bit indicates that TCP should deliver the data to the destination port prior to filling an entire software buffer in the destination host. The RST bit indicates that the connection should be reset. TCP also uses the RST bit to reject invalid segments and refuse a connection attempt. TCP employs the SYN bit to establish connections and synchronize sequence numbers. The FIN bit releases a TCP connection. The 16-bit *Window* field identifies the amount of data the application is willing to accept, usually determined by the remaining buffer space in the destination host. The *Checksum* applied across the TCP header and the user data detects errors. The *Urgent Pointer* field specifies the position in the data segment where the urgent data begins — if the URG code bit indicates that this segment contains urgent data. The options field is not mandatory, but provides additional functions, such as a larger window size field as specified in RFC 1323 [33]. Another popular option is selective retransmission as specified in RFC 1106 instead of the default go-back-n protocol — an important protocol difference that dramatically increases throughput on channels with excessive bit errors or loss. This is a topic, studied in Chapter 26. The padding field aligns the TCP header to a 32-bit boundary for more efficient processing in hosts.

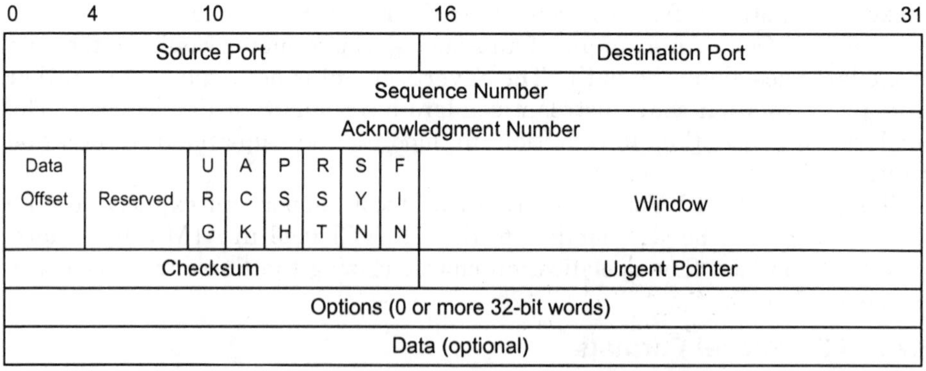

0 4 10									16 31
Source Port									Destination Port
Sequence Number									
Acknowledgment Number									
Data Offset	Reserved	U R G	A C K	P S H	R S T	S Y N	F I N		Window
Checksum									Urgent Pointer
Options (0 or more 32-bit words)									
Data (optional)									

Figure 8.10   TCP Segment Format

TCP is a connection-oriented protocol and therefore has additional, specific messages and a protocol for an application to request a distant connection, as well as a means for a destination to identify that it is ready to receive incoming connection requests.

8.2.6 Example of TCP/IP Operation

Figure 8.11 shows an example of a TCP/IP network transferring data from a workstation client to a server. TCP assumes that the underlying network is a connectionless datagram network (for example, IP) that can deliver packets out of order, or even deliver duplicate packets. TCP handles this by segmentation and reassembly using the sequence number in the TCP header, while IP does this using the fragment control fields in the IP header. Either method, or both, may be used. The client on the left hand side of the figure segments the user data 'ABCD' into four TCP segments. Router R1 initially routes IP datagram A via the X.25 network as shown in the figure. R1 then becomes aware of a direct connection to the destination router, R2, and routes the remaining datagrams (B, C, and D) via the direct route. This routing action causes the datagrams to arrive at the destination server out of order, with datagram A traversing the X.25 network and arriving significantly later. On the right hand side of the figure, the TCP stack running in the server resequences the datagrams and delivers the block of data to the destination socket in the original order. IP performs a similar process using fragmentation and reassembly on an individual packet basis.

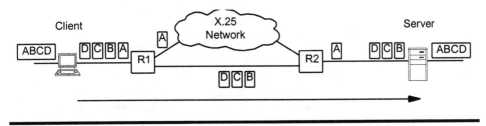

Figure 8.11   Example of Data Transfer using TCP/IP

This operation by TCP/IP of accepting datagrams out of order, and being able to operate over an unreliable underlying network, makes it quite robust. No other standard modern data communication protocol has this attribute.

8.2.7 Traffic and Congestion Control Aspects of TCP

TCP works over IP to achieve end-to-end reliable transmission of data across a network. TCP flow control uses a sliding window flow control protocol, like X.25; however, the window is of a variable size, instead of the fixed window size used by X.25. The tuning and refinement of the TCP dynamic window flow control protocol has been the subject of a great deal of research. The following example is a simple case of the Van Jacobson "Slow Start" TCP algorithm [34], an enhancement to the initial TCP implementation designed to dynamically maximize throughput and prevent congestion collapse. TCP keeps track of a congestion window, which is never greater than the window size reported by the receiver in the TCP packet shown in Figure 8.10.

Figure 8.12 illustrates a simplified example of key concepts in the dynamic TCP window flow control protocol between a workstation and a server. The sender starts with a congestion window size equal to that of one TCP segment

(segment 0). The IP network delivers TCP segment 0 to the destination server, which acknowledges receipt. The sending workstation then increases the congestion window size to two segments. When the destination acknowledges both of these segments, the sender increases the window size to four segments, doubling the window size for each received acknowledgment. Thus, the IP network becomes congested and loses the fifth and sixth segments at this point (the figure indicates the lost packets by 'X's). The sender detects this by starting a timer immediately after sending a segment. If the timer expires before the sender receives a acknowledgment from the receiver, then the sender retransmits the segment. Upon such a retransmission timeout, the sender resets its window size to one segment and begins the above process again. The timeout may be immediate, or typically 500 milliseconds in an attempt to "piggyback" the acknowledgment onto another packet destined for the same socket identifier.

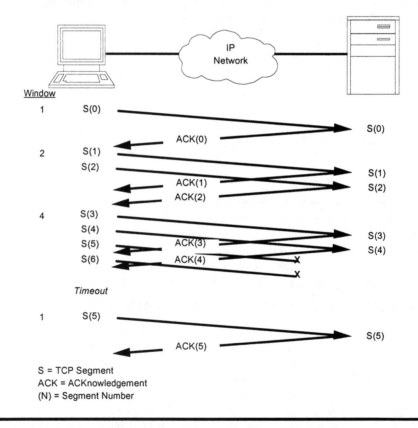

Figure 8.12 Example of TCP Dynamic Windowing Flow Control

Note that TCP actually uses byte counts and not individual segments; however, if all segments are of the same size then our simple example is accurate. The astute reader will observe that TCP is geometrically increasing the window size (i.e., 1, 2, 4, 8, and so forth), so this is not really a slow start at all. TCP congestion control has another function that limits the interval of geometric increase until the window size reaches a threshold of one half the congestion window size achieved before the previous unacknowledged segment. After this point, the TCP increases the congestion window by one segment for each round trip time, instead of doubling it during the geometric growth phase as shown in the above example. This linear phase of TCP window growth is called "congestion avoidance", while the geometric growth phase is called "slow start".

Hence, during steady-state operation when there is a bottleneck in the IP network between the sender and the receiver, the congestion window size at the sender has a sawtooth-like pattern as illustrated in Figure 8.13. In this example, the segment size is always 1 kbytes, and the timeout occurs when the congestion window size reaches 16 kbytes. This type of oscillating window behavior occurs when multiple TCP sources contend for a bottleneck resource in an IP network; for example, a busy trunk connecting routers in the Internet or an access circuit connecting to a popular web server, such as the NASA site when the first pictures of Mars became available. Since the World Wide Web's HTTP protocol runs over TCP, this phenomenon occurs frequently during intervals of congestion on the Internet.

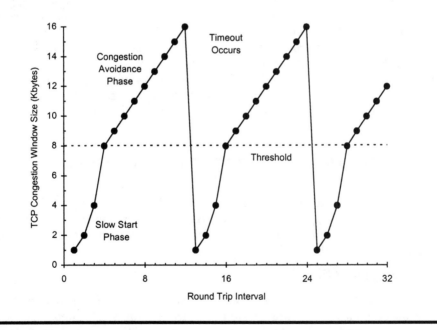

Figure 8.13   Example of TCP Slow Start Congestion Window Size Behavior

The most recent TCP standard, RFC 2001 [34] describes further enhancements to TCP called fast retransmit and fast recovery. TCP generates a duplicate acknowledgment whenever it receives an out of order segment, since it is acknowledging the last byte received in sequence. Fast retransmit uses duplicate acknowledgments to detect a lost segment and attempts to avoid the throughput-robbing reduction of the congestion window size to one after a timeout by retransmitting only the lost segment. Fast recovery follows after the transmitter sends the missing segment, reverting to the linear window size increase algorithm of congestion avoidance even if the transmitter is in the slow start phase.

8.2.8 Service Aspects of TCP/IP

Typically, TCP/IP implementations constitute a router, TCP/IP workstation, server software, and network management. TCP/IP protocol implementations span Windows, UNIX, DOS, VM, and MVS environments. A majority of UNIX users employ TCP/IP for internetworking. Many Network Operating System (NOS) vendors now integrate TCP/IP into their implementations, including Novell Netware and Banyan Vines.

Standards define operation of IP over a number of network, data link, and physical layer services. At the network layer, standards define IP operation over X.25 and SMDS. At the data link layer, other standards define IP operation over PPP, frame relay, Ethernet, and ATM. IP operation over circuit switched and dedicated physical layer facilities is also defined, such as IP over SONET.

Internet Service Providers (ISPs) provide access to TCP/IP users connected via dedicated access lines or dial-up modem pools. Every major country around the world has one or more ISPs. In countries with large volumes of Internet traffic, some ISPs provide backbone transport for other ISPs. Some of these ISPs are directly connected, while many are not. If ISPs are not directly connected, then the IP routing protocols previously described provide the means for packets to traverse multiple networks such that every IP address on the planet is reachable from every other IP address.

8.3 SWITCHED MULTIMEGABIT DATA SERVICE (SMDS)

ATM adopted a number of concepts originally developed for SMDS. As background to the upcoming part on ATM, this section covers the service aspects of SMDS and specifics of the Distributed Queue Dual Bus (DQDB) protocol defined in the IEEE 802.6 standard.

8.3.1 Origins of SMDS

The idea of metropolitan area networks began when the IEEE began work in 1982 on standards for transmission of voice, compressed video, LAN interconnectivity, and bulk-data transfer. It was first presented to the cable television (CATV) community, which didn't tune into the idea. Then Burroughs, National, and Plessey initiated a second effort in 1985 with the slotted ring concept. This effort died when the leveraged buyout of Sperry Univac cut required funding, and again MAN technology waited in the wings. The most recent effort began with a Bell Labs MAN standard proposal developed in parallel with the ex-Burroughs FDDI venture called MST (Multiplexed Slot and Token). This new Bellcore MAN standard became the IEEE Project 802.6.

The IEEE 802.6 standard is based upon the distributed queue dual bus (DQDB) technology [35]. The DQDB architecture (which resembles the Bell Technology Lab's dual coax with reservation MAN architecture, called Fastnet) was invented at the University of Western Australia and hardware was first produced by QPSX LTD (a University of Western Australia and Telecom Australia spin-off).

As SMDS was created as a Metropolitan Area Network (MAN) service by Bellcore, it is in the purest sense a service definition and *not* a protocol. The first realization of SMDS was defined using the DQDB technology, as specified in the IEEE 802.6 standard. The IEEE 802.6 DQDB standard defines connectionless data-transport service using 53-byte slots to provide integrated data, video, and voice services over a MAN, which is typically a geographic area of diameter less than 150 km (90 miles). This cell-switching architecture combines the best of two worlds: connectionless datagram public data-transfer services similar to packet switching, and speeds up to 155-Mbps. The SMDS implementations based upon the IEEE 802.6 standard were the first public services to use ATM-like technology. Although the IEEE 802.6 standard also defines connection-oriented isochronous services, SMDS today supports only a connectionless datagram service primarily targeted for LAN interconnection. Some vendors provide their own proprietary version of isochronous transport.

SMDS utilizes a form of cell switching defined in terms of standards, underlying architectures, services implementation (such as SMDS), and protocols. Cell switching has taken two development paths: connectionless data transport in the form of IEEE 802.6 (DQDB), and connection-oriented Asynchronous Transfer Mode (ATM) with connectionless servers. SMDS services use the IEEE 802.6 DQDB CL (ConnectionLess) service. While SMDS provides LAN/WAN interconnection, a network design offering SMDS service over a DQDB architecture is not limited to a geographical area. While ATM may well become the long-term WAN transport technology of choice, either technology serves the purpose of replacing data networks composed of private lines.

Central-office switch vendors such as Siemens Stromberg-Carlson were the primary players for the first versions of cell switching to hit the telecommunications market: Switched Multimegabit Data Service (SMDS) using the DQDB architecture as access. These switches first made use of DQDB's connectionless service. Versions of SMDS service have been offered by IXCs, LECs, and PTTs worldwide, including MCI Communications, British Telecom, Telecom Ireland, and Deutsche Telekom.

8.3.2 SMDS/IEEE 802.6 Protocol Structure

SMDS and the IEEE 802.6 DQDB protocol have a one-to-one mapping to each other as illustrated in Figure 8.14. The SMDS Interface Protocol (SIP) has Protocol Data Units (PDUs) at levels 2 and 3. The level 2 SIP PDU corresponds to the Distributed Queue Dual Bus (DQDB) Media Access Control (MAC) PDU of the IEEE 802.6 standard. The level 3 SIP PDU is treated as the upper layers in IEEE 802.6. There is also a strong correspondence between these levels and the OSI reference model.

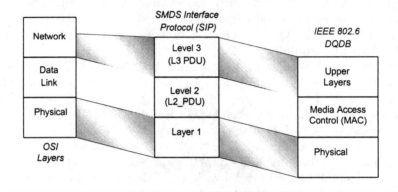
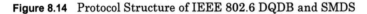

Figure 8.14   Protocol Structure of IEEE 802.6 DQDB and SMDS

8.3.3 SMDS/802.6 Protocol Data Unit (PDU) Formats

Figure 8.15 illustrates the relationship between the user data, the level 3 SMDS PDU, and the level 2 SMDS PDU. The user data field may be up to 9188 octets in length. The level 3 protocol adds a header and trailer fields, padding the overall length to be on a 4-octet boundary. Level 2 performs a segmentation and reassembly function, transporting the level 3 payload in 44-octet segments. The level 2 PDU has a 7-octet header and 2-octet trailer resulting in a 53-octet slot length, the same length as an ATM cell. The level 2 header identifies each slot as being either the Beginning, Continuation, or End Of Message (BOM, COM, or EOM). The cells are transmitted header first.

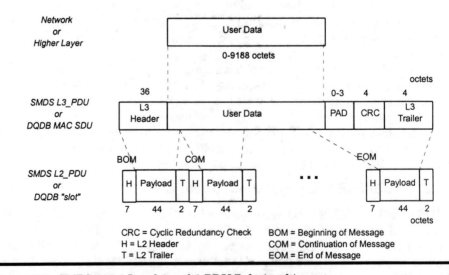

Figure 8.15   SMDS/802.6 Level 2 and 3 PDU Relationships

Figure 8.16 illustrates the details of the SMDS level 3 PDU (L3_PDU) format. The first two octets and last two octets of the SMDS L3_PDU are identical to ATM's AAL3/4 Common Part Convergence Sublayer (CPCS) described in Chapter 13. The SMDS L3_PDU header contains the SMDS Source and Destination Addresses (SA and DA) and a number of other fields. Most of these other fields are included for alignment with the IEEE 802.6 protocol and are not actually used in the SMDS service. When the SMDS level 3 PDU is segmented by level 2, all information needed to switch the cell is carried in either an SSM or BOM slot. This design means that an SMDS switch need only examine the first slot to make a switching decision.

The addressing plan for SMDS and Connectionless Broadband Data Service (CBDS) employs the International Telecommunications Union (ITU) Recommendation E.164, which in the United States is similar to the North American Numbering Plan (NANP) used for telephone service. As the SMDS E.164 address is globally unique, SMDS provides the capability for ubiquitous connectivity. The IEEE 802.6 standard also allows the option for 48-bit IEEE Media Access Control (MAC) addresses to be employed in the DA and SA fields.

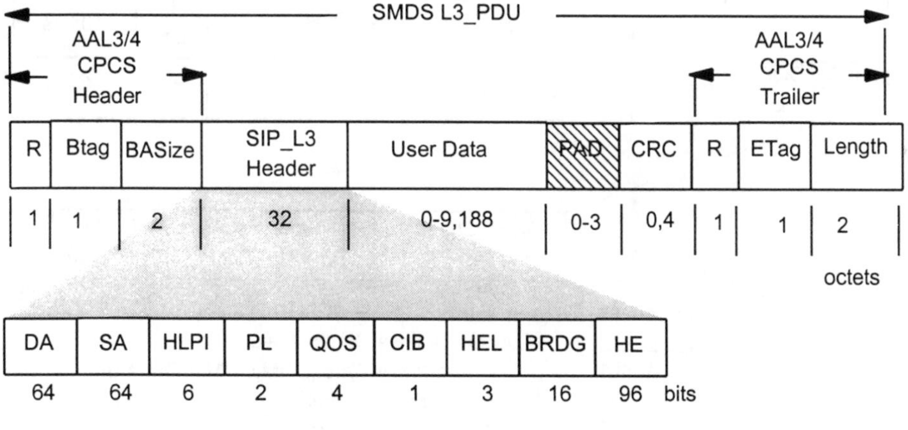

DA = Destination Address
SA = Source Address
HLPI = Higher Layer Protocol Indicator
PL = Pad Length
QOS = Quality Of Service
HEL = Header Extension Length
CIB = CRC Indicator Bit
BRDG = Bridging
HE = Header Extension
R = Reserved
Btag = Beginning Tag
BASize = Buffer Allocation Size
ETag = Ending Tag
Length = SMDS L3_PDU Length

Figure 8.16 SMDS Level 3 PDU Format

Figure 8.17 illustrates the 48-octet SMDS level 2 PDU format encapsulated in a 53-octet DQDB slot. The other four octets of SMDS level 2 overhead in the DQDB payload are used for the SMDS Segment Type (ST), Message IDentifier (MID), payload length, and a Cyclical Redundancy Check (CRC) on the 44-octet payload. The SMDS level 2 overhead and function are identical to the ATM AAL3/4 SAR described in Chapter 13. The ST field identifies either a Single Segment Message (SSM), Beginning Of Message (BOM), Continuation Of Message (COM), or End Of Message (EOM) slot. The MID field associates the BOM with any subsequent COM and EOM segments that make up an SMDS L3_PDU. When an SMDS switch receives an SSM or BOM segment, the destination address determines the outgoing link on which the slots are transmitted.

The DQDB Access Control Field (ACF) and header provide a distributed queue for multiple stations on a bus, provide self-healing of the physical network, provide isochronous support, and control management functions. The next sections describe the distributed queuing and self-healing properties of DQDB. Note that the DQDB ACF and header taken together are exactly five bytes, exactly the same size as the ATM cell header. This choice was made intentionally to make the design of a device that converted between DQDB slots and ATM cells simpler.

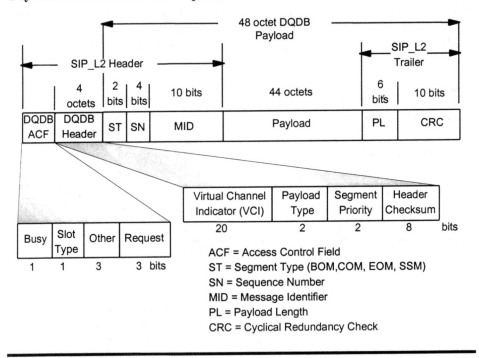

Figure 8.17   DQDB Slot Structure

8.3.4 DQDB and SMDS Operation

This section describes the distributed queuing and self-healing ring properties of the IEEE 802.6 DQDB protocol with reference to Figure 8.18. Two unidirectional buses A and B interconnect a number of nodes, often configured in a physical ring. Even though the physical configuration may be a ring, logical operation is still bus-oriented. Nodes read from both buses, usually passing along any data on to the next node in the bus. Any node may assume the role of Head Of Bus (HOB) or End Of Bus (EOB) according to the rules of the DQDB protocol. The HOB generates 53-octet slots in a framing structure to which the other nodes synchronize. The EOB node simply terminates the bus. If a node fails or is powered down, it is designed so that it passively passes data. Therefore, each node effectively has four ports, two for each bus. Normally one node would be the HOB for both buses, as shown for node C in the figure. However, in the event of a failure of one of the buses connecting a pair of nodes, the DQDB protocol ensures that the nodes on either side of the break become the new HOB within a short period of time. This allows network designers to plan for link failures and ensures that a single link failure will not affect the entire network.

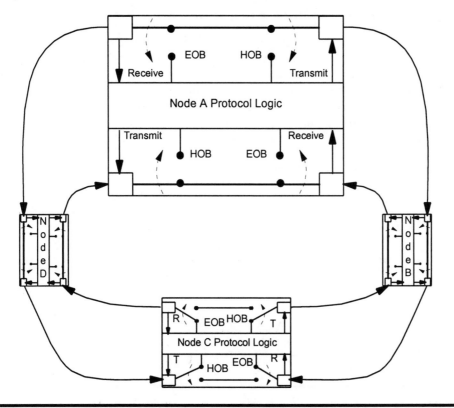

Figure 8.18 Dual DQDB Architecture

The Busy and Request bits in the DQDB Access Control Field (ACF) implement a distributed queue. Each node has two counters: one for requests and the other as a countdown for transmission. If a slot passes through the node with the request bit set, then the node increments the request counter, otherwise the node decrements the request counter. Thus, the request counter reflects how many upstream nodes have slots to send. When a node has data to send, it writes a logical 1 to the request bit in the first slot received which has the request bit equal to 0 and loads the countdown register with the value of the request counter. The countdown timer is decremented each time a slot passes by in the opposite direction. Therefore, when the countdown counter reaches 0, the slot can be sent because all of the upstream nodes have already sent the slots that were reserved in the opposite direction.

This elegantly simple protocol, however, has several problems that complicate the IEEE 802.6 standard. First, the nodes which are closer to the head end of the bus have first access to the request bits, and can dominate the traffic on the bus effectively drowning out users at the end of the bus. Secondly, provisions must be made for stations to join and leave the bus and handle bit errors. The IEEE 802.6 standard defines procedures to handle all of these cases.

8.3.5 Example of SMDS over DQDB Operation

Figure 8.19 illustrates an example of SMDS over DQDB operation. Three DQDB buses, configured as physical rings, are interconnected as shown. A series of slots from a node on the far left are generated with Destination Address (DA) of a node on the far right. The first slot carries the address for the subsequent slots with the same MID. The nodes use the DQDB protocol to queue and transmit the slots corresponding to this datagram. The result is that a reassembled datagram is delivered to the destination. In general, the MID values differ on each DQDB segment. Once a DQDB segment completes transmission of an L3_PDU, the MID numbers can be recycled as indicated in the figure.

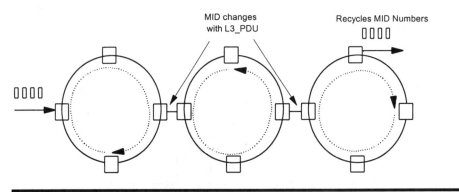

Figure 8.19   Example of SMDS/DQDB Operation

8.3.6 Traffic and Congestion Control Aspects of DQDB and SMDS

There are two aspects of traffic control in SMDS, operating at the DQDB and SMDS service levels. We summarize the important attributes of each aspect in this section.

8.3.6.1 DQDB Access Control

Prearbitrated access enables the service provider to allocate bandwidth to isochronous connections with a requirement for constrained jitter. This allows priority bandwidth to be provided to the services that are most affected by variations in delay – namely video and voice. Video and voice are prime examples of data requiring prearbitrated access to bandwidth because of their intolerance of variation in delay.

Queued arbitrated access is the other access control method. Queued arbitrated access has distance limitations. The longer the distance between the stations, the less effective the priority queuing. This is usually allocated to lower-priority services, or services that can tolerate longer delays and retransmissions more readily than video or voice. This service also assumes that bandwidth balancing is disabled. Bandwidth balancing will be discussed later.

Unfortunately, queuing fairness becomes an issue at utilization rates higher than 80 percent, such as during periods of high network activity on DQDB networks. Typically, users located at the end of the bus receive poorer performance than those at the head of the bus because the transmit tokens are taken before the reservation slot arrives. The standards groups invented bandwidth balancing to give users an "equal share" of the available bandwidth by equally dividing bus bandwidth among all users on the bus.

The network designer should strive to design the subnetworks and CPE that interface with the DQDB bus in a manner to ensure that peak traffic conditions do not cause excessive delay and loss of traffic, especially to end of bus users. This can be accomplished by the effective use of these techniques.

8.3.6.2 SMDS Access Classes

SMDS has an open loop flow control mechanism called Sustained Information Rate (SIR). SMDS SIR is based on the aggregate of all data originating on the SMDS access line regardless of it's destination. Thus, there are no levels of bandwidth management granularity at the individual address level as found in frame relay and ATM. Table 8.4 lists the five access classes defined for DS3 access lines.

The 34-Mbps for Access Class 5 is the maximum throughput achievable on a 45-Mbps DS3 access line after subtracting out Physical Layer Convergence Protocol (PLCP) and L2_PDU overhead. SIR uses a credit manager, or leaky bucket type of rate enforcement method. Basically, no more than M out of N cells may contain non-idle slots or cells. For class 5 M=N, while for the lower numbered classes the relationship is that SIR=M*34/N Mbps. The value of M controls the number of consecutive slots/cells that can be sent at the DS3 rate.

Table 8.4   SMDS Access Classes

Access Classes	SIR (Mbps)
1	4
2	10
3	16
4	25
5	34

Data arriving at a rate higher that the SIR rate is discarded at the originating SMDS switch. Note that access classes 1 through 3 line up with standard LAN speeds, so that traffic from a single LAN cannot experience loss due to the SIR credit manager operation. SMDS has a L3_PDU loss objective of 10^{-4}, which is several orders of magnitude greater than that caused by transmission errors. This is consistent with the character of the SMDS service, emulating that of a LAN but providing MAN or WAN coverage.

8.3.7 Service Aspects of SMDS

Switched Multi-megabit Data Service (SMDS) is a combination packet- and cell-based public data service that supports DS1, E1, E3, and DS3 access speeds through a DQDB Subscriber to Network Interface (SNI) . SMDS is also supported through a Data eXchange Interface (DXI) at speeds ranging from 56-kbps up to and including 45-Mbps. SMDS is a connectionless switched data service with many of the characteristics of Local Area Networks (LANs). SMDS provides the subscriber with the capability to connect diverse LAN protocols and leased lines into a true switched public network solution.

SMDS offers either a point-to-point datagram delivery service, or a point-to-multipoint service, defined by a group multicast address. SMDS service operates on both the E.164 source and destination addresses. At the source access line, the SMDS network screens the source address to authenticate the source addresses subscription. At the source access line, the SMDS network also screens on the destination address to limit the destinations reachable from a particular access line. This provides an excellent access and egress level of security not found in many other data network protocols. At the destination access line, either the SMDS network and/or customer device screens the incoming PDUs to only accept data from specific source SMDS addresses or block data from specific SMDS source addresses. SMDS customers can have ubiquitous connectivity, or they can use these screening tools to achieve tightly controlled closed user groups.

In the early 1990s, experts billed SMDS as the gap filler between frame relay and ATM; however, SMDS now finds its unique position in time diminishing. Central office switch vendors, such as Siemens, AT&T, Fujitsu, and NEC, were primary players for the first version of SMDS cell switching to enter the telecommunications market. Public SMDS services based on the IEEE 802.6 standard architecture have been slow to emerge; however, they were rolled out in the United States and many European countries by 1994. Lack of effective marketing, the requirement to purchase special CSU/DSUs, the tremendous success of frame relay, and early ATM standardization probably had a lot to do with the low penetration rate of SMDS services. Internationally, a close relative of SMDS is the Connectionless Broadband Data Service (CBDS) as defined by the European Telecommunications Standards Institute (ETSI) standards. A North American SMDS Interest Group (SIG) and European SIG formed in the 1990s (to develop detailed implementation specifications to facilitate interoperable implementations of SMDS and CBDS) was disbanded in 1996.

Customers either really like, or really hate, SMDS — there doesn't seem to be much of a middle ground. In a very real sense, SMDS was ahead of its time: modern public data services have not yet met the real need for multicast and group addressing services provided in SMDS. IP routers have not provided either of these functions very well either. Router access control lists provide authentication and screening, but dramatically decrease overall throughput. Over the past few years IP and IPv6 are clearly the network protocol of choice to the desktop. Therefore, SMDS had little chance against IP as the connectionless protocol of choice to the router and the desktop. A key challenge for ATM is to cost effectively implement these mission critical networking demands.

8.4 REVIEW

This chapter covered the predominant connectionless data services offered in the 1990s: IP and SMDS in terms of the origins, protocol formats, functions, operation, traffic aspects, and commercial service aspects. First, we summarized the alternative to the OSI protocol stack embraced by most of the networking world, namely, the de facto standards of Internet Protocol (IP) suite. The text detailed the concepts of resequencing and handling unreliable underlying networks, central to TCP/IP using descriptions and examples. We introduced the recently defined method for IP-based networks to support multiple quality of service classes and guarantee bandwidth through the Resource ReSerVation Prtoocol (RSVP) . The text then described the modern notion of a dynamically adjusted window flow control protocol implemented widely in TCP. Finally, the chapter covered the novel concepts of a distributed queue, self-healing network in the treatment of SMDS and DQDB. We also looked at some of the origins of ATM's cell structure in DQDB's slotted protocol. This chapter provides the reader with a basic background of the connectionless data communication services and concepts. These concepts reappear within ATM and related protocols in the remainder of the book.

8.5 REFERENCES

[1]   D. Comer, *Internetworking with TCP/IP - Second Edition*, Prentice-Hall, 1991.
[2]   U. Black, *TCP/IP and Related Protocols*, McGraw-Hill, 1992.
[3]   A. Tannenbaum, *Computer Communications*, Third Edition," Prentice-Hall, 1996.
[4]   H. Schulzrinne, S. Casner, R. Frederick, V. Jacobson, "RFC 1889 - RTP: A Transport Protocol for Real-Time Applications," IETF, January, 1996.
[5]   J. Romkey, *A Nonstandard for Transmission of IP Datagrams Over Serial Lines: SLIP*, RFC 1055 , IETF, June 1988.
[6]   W. Simpson, *The Point-to-Point Protocol (PPP)*, RFC 1661, IETF, July 1994.
[7]   W. Simpson, *PPP in HDLC-like Framing*, RFC 1662, IETF, July 1994.
[8]   J. Postel, *Internet Protocol*, RFC 791, IETF, September 1981.
[9]   F. Baker, *Requirements for IP Version 4 Routers*, RFC 1812, IETF, June 1995.
[10] Y. Rekhter, *CIDR and Classful Routing*, RFC 1817, IETF, August 1995.

[11]   Y. Rechter, T. Li, *An Architecture for IP Address Allocation with CIDR*, RFC 1518, IETF, September 1993.

[12]   V. Fuller, T. Li, J. Yu, K. Varadhan, *Classless Inter-Domain Routing (CIDR): an Address Assignment and Aggregation Strategy*, RFC 1519, IETF, September, 1993.

[13]   S. Bradner, A. Mankin, *IP: Next Generation (IPng) White Paper Solicitation*, RFC 1550, IETF, December 1993.

[14]   S. Bradner, A. Mankin *The Recommendation for the IP Next Generation Protocol*, RFC 1752, January 1995.

[15]   S. Deering, R. Hinden, *Internet Protocol, Version 6 (IPv6) Specification*, RFC 1883, December 1995.

[16]   R. Hinden, S. Deering, *IP Version 6 Addressing Architecture*, RFC 1884, December 1995.

[17]   A. Conta, S. Deering, *Internet Control Message Protocol (ICMPv6) for the Internet Protocol Version 6 (IPv6)*, RFC 1885, December 1995.

[18]   S. Thomson, C. Huitema, *DNS Extensions to support IP version 6*, RFC 1886, December 1995.

[19]   L. Zhang, S. Deering, D. Estrin, S. Shenker and D. Zappala, "RSVP: A New Resource ReSerVation Protocol," *IEEE Network Magazine*, September 1993.

[20]   R. Braden, D. Clark, S. Shenker, *Integrated Services in the Internet Architecture: an Overview*, RFC 1633, IETF, June 1994.

[21]   R. Braden, Ed., L. Zhang, S. Berson, S. Herzog, S. Jamin, , *Resource ReSerVation Protocol (RSVP) -- Version 1 Functional Specification*, RFC 2205, IETF, September 1997.

[22]   E. Germain, "Fast lanes on the Internet," *Science,* Aug 2, 1996.

[23]   A. Mankin, Ed., F. Baker, B. Braden, S. Bradner, M . O`Dell, A. Romanow, A. Weinrib, L. Zhang, *Resource ReSerVation Protocol (RSVP) -- Version 1 Applicability Statement Some Guidelines on Deployment*, RFC 2208, September 1997.

[24]   F. Baker, J. Krawczyk, A. Sastry, *RSVP Management Information Base using SMIv2*, RFC 2206, IETF, September1997.

[25]   L. Berger, T. O'Malley, *RSVP Extensions for IPSEC Data Flows*, RFC 2207, IETF, September 1997.

[26]   L. Henderson, "Multimedia over IP: A new choice?" *Telephony*, July 22, 1996.

[27]   C. Partridge, *Gigabit Networking*, Addison-Wesley, 1994.

[28]   J. Wroclawski. *Specification of the Controlled-Load Network Element Service*, RFC 2211, IETF, September 1997.

[29]   S. Shenker, C. Partridge, and R Guerin. *Specification of Guaranteed Quality of Service*, RFC 2212, IETF, September 1997.

[30]   J. Wroclawski. *The Use of RSVP with IETF Integrated Services*, RFC 2210, IETF, September 1997.

[31]   J. Postel, *Transmission Control Protocol*, RFC 793, IETF, September 1981.

[32]   R. Braden, *Requirements for Internet hosts - communication layers*, RFC 1122, October 1989.

[33]   V. Jacobson,R. Braden,D. Borman, *TCP Extensions for High Performance*, RFC 1323, May 1992.

[34]   W. Stevens, *TCP Slow Start, Congestion Avoidance, Fast Retransmit, and Fast Recovery Algorithms*, RFC 2001, IETF, January 1997.

[35]   ISO/IEC, *Information processing systems — local and metropolitan area networks — Part 6: Distributed queue dual bus (DQDB) subnetwork of a metropolitan area network (MAN)*, 8802-6, 1993.

9

LANS, Bridging, and Routing

Most major enterprises embraced Local Area Networks (LANs) in the 1980s, and now even some residences have LANs. Network designers then invented bridging to interconnect multiple LANs to provide greater connectivity. Meanwhile, incompatible LAN standards created the need for routers in the environment of diverse interconnected LANs. Since LANs, bridges and routers utilize some unique concepts, we begin by introducing the terminology related to bridging, routing and internetworking used in the remainder of this book. Next, we review LAN protocol standards and bridging in some depth as an introduction to Part 4 regarding the ATM Forum's LAN Emulation (LANE). Finally, we cover the subjects of address resolution and routing as background for the discussion of ATM's routing protocol, the Private Network-Network Interface (PNNI) in Chapter 15 as well as ATM protocol support for the Internet Protocol in Chapter 20. Chapter 31 also references this background material in the LAN networking technology comparison.

9.1 BRIDGING, ROUTING AND INTERNETWORKING

This section introduces the basic terminology of local area networks and how bridges connect LANs. We also introduce the closely related subjects of routing and internetworking. The section concludes with a discussion of some of the key issues regarding address assignment, address resolution, route selection and scalability.

9.1.1 Basic Terminology

Let's first review some basic LAN and internetworking terminology of bridging and routing with reference to Figure 9.1, based upon RFC 1932 [1], which defines the framework for the IETF's work on IP over ATM.

A *Host* (also called an End System) delivers and receives IP packets to and from other hosts. A host does not relay packets. Examples of hosts are workstations, personal computers and servers.

A *Router* (also called an Intermediate System) also delivers and receives IP packets, but also relays IP packets between end and intermediate systems. Internet Service Providers (ISPs) and large enterprises often utilize routers made by companies like Cisco, 3Com, Ascend, and Bay Networks.

All members of an *IP Subnet* can directly transmit packets to each other. There may be repeaters, hubs, bridges, or switches between the physical interfaces of IP subnet members. Ethernet or Token Ring LANs are examples of an IP subnet. However, multiple Ethernets bridged together may also be a subnet. The assignment of IP addresses and subnet masks determines the specific subnet boundaries described in more detail in this Chapter.

Bridging makes two or more physically disjoint media appear as a single bridged IP subnet. Bridging implementations occur at the medium access control (MAC) level or via a proxy Address Resolution Protocol (ARP).

A *Broadcast Subnet* allows any system to transmit the same packet to all other systems in the subnet. An Ethernet LAN is an example of a Broadcast Subnet.

A *Multicast-Capable Subnet* provides a facility that enables a system to send packets to a subset of the subnet members. For example, a full mesh of ATM point-to-multipoint connections provides this capability.

A *Non-Broadcast Multiple Access (NBMA) Subnet* does not support a convenient multi-destination connectionless delivery capability that broadcast and multicast capable subnetworks do. A set of point-to-point ATM VCCs is an example of an NBMA subnet.

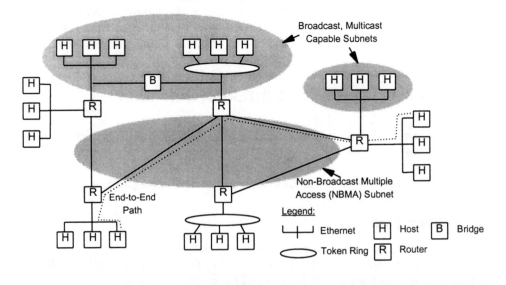

Figure 9.1 Basic LAN and Internetworking Terminology

An *internetwork* is a concatenation of networks, often employing different media and lower level encapsulations, that form an integrated larger network supporting communication between hosts. Figure 9.1 illustrates a relatively small internet when judged in comparison with *the* Internet, which comprises over 40,000 networks.

The term *network* may refer to a piece of Ethernet cable or to a collection of many devices internetworked across a large geographic area sharing a coordinated addressing scheme.

Vendors of intermediate and end systems strive to implement an efficient process for deciding what to do with a received packet. Possible decisions are local delivery, or forwarding the packet onto another external interface. *IP Forwarding* is the process of deciding what to do with a received IP packet. IP forwarding may also require replacement or modification of the media layer encapsulation when transitioning between different LAN media.

IP routing involves the exchange of topology information that enables systems to make IP forwarding decisions that cause packets to advance along an end-to-end path toward a destination. Sometimes routing is also called a topology distribution protocol. We cover this important topic later in this chapter.

An *End-to-End path* is an arbitrary number of routers and subnets over which two hosts communicate; for example, as illustrated by the dashed line in Figure 9.1. Routers implement this path by the process of IP forwarding at each node as determined by an IP routing algorithm.

An *IP address resolution protocol (ARP)* provides a quasi-static mapping between an IP address and the media address on the local subnet.

Scalability refers to the ability of routing and address resolution protocols to support a large number of subnets, as well as handle the dynamics of a large internetwork. In networks with large numbers of interconnected subnets, routers must exchange large amounts of topology data and store the resultant forwarding information in high-speed memory. Furthermore, as network size increases, so does the likelihood that some network elements are changing state, thus creating the need to update routing topology information. Hence, scalability impacts the required processing power and storage for the routing protocols within routers.

9.1.2 Address Assignment and Resolution

A key requirement in any communications network is that unique addresses be assigned to each of the entities that want to communicate. This is the case in the telephone network where every phone in the world has a unique number. It is also found in the 48-bit IEEE 802.3 Media Access Control (MAC) assignments built into every Ethernet interface. Administrators must assure that every user or "host" in the Internet has a unique IP address.

Assuring that the address assignments are unique and that they efficiently administer an address space presents some challenges. Not only must addresses be handed out, but a means for users to return addresses and request additional blocks of addresses is also required. Furthermore, if there is more than one administrative authority, then the scope of assignments allocated to each administration must be clearly defined. It is sometimes difficult to predict the demand for addresses. For example, telephone companies periodically realign area codes because the demand differs from the forecast of just a few years ago. If an administrative authority hands out blocks of addresses too freely, then the network can run out of unique addresses well before the limit determined by the number of bits in the addresses: this has occurred with many of the IP address blocks already assigned.

Once you have your own address and the address of someone that you wish to communicate with, how do you resolve the address of the desired destination into information about how to get it there? First, consider the following simple analogy. Let's say that you have spoken to an individual on the telephone for the first time and have agreed to meet him or her at a party to which you both have been invited by the same host. Once you arrive at the party, you can find the individual (resolve the address) in one of two ways: you can jump up on stage, grab the microphone, and broadcast your presence — or you can locate the host and ask to be introduced to the individual. Broadcast (i.e., the microphone) is commonly used in shared-medium LANs to resolve addresses. A problem arises, however, when the volume of broadcast traffic approaches the level of user traffic: like when it's difficult to hear at a crowded party. The analogy to having someone who has the information (the party's host) resolve the address (match the name with the individual) is like that of an Address Resolution Protocol (ARP) server used in every many ATM designs to support LAN and internetworking protocols. The analogy of you jumping up on stage and announcing your intentions is the broadcast ARP protocol commonly used on LANs described at the end of this Chapter.

9.1.3 Routing, Restoration, and Reconfiguration

After resolving the destination address, there is the issue of what is the best way to reach that address through a network of nodes. A commonly used solution to this problem assigns each link in a network a *cost*, and then employs a routing algorithm to find the least-cost route. This cost may be economic, or may reflect some other information about the link, such as the delay or throughput. One analogy is as follows: you are traveling from your house to the grocery store. One route takes more gas yet takes less time, and the other route takes less gas yet takes more time (e.g., because of construction delays). Each route has a different set of costs (time and money). You choose the cost measure most important to you — that is, do you want to minimize time spent or gas used — and you take that route choice. Routing protocols employ a similar decision criteria and cost metric.

Routing algorithms exchange information about the topology, that is, the links that are connected and their associated costs, in one or two generic methods [2, 3]. The first is a *distance vector* algorithm where neighbor nodes periodically exchange vectors of the distance to every destination subnetwork. This process eventually converges on the optimal solution. The second is where each router learns the entire *link state* topology of the entire network. Currently this is done by flooding only the changes to the link state topology through the network. Flooding involves copying the message from one node to other nodes in the network in a tree-like fashion such that only one copy of the message is received by every node. The link state approach is more complex, but converges much more rapidly than the distance vector approach. Convergence is the rate in which the knowledge of network topology at every node goes from an unstable state to a stable state. When the topology of the network changes due to a link or node failure, or the addition of a new node or link, or because of a link metric change, the flooding protocol must disseminate this information to every other node. The *convergence time* is the interval required to update all nodes in the network about the topology change.

The distance vector method was used in the initial data communication networks such as the ARPAnet and is used by the Internet's Routing Information Protocol (RIP). A key advantage of the distance vector protocol is its simplicity. A key disadvantage of the distance vector protocol is that the topology information message grows larger with the network, and the time for it to propagate through the network increases as the network grows. Another disadvantage of the distance vector algorithm is that it uses hop count instead of a weighted link metric. Minimum hop routing leads to some pathological route choices in certain network topologies. Convergence times on the order of minutes are common in distance vector algorithm implementations.

The link state advertisement method is a more recent development designed to address the scalability issues of the distance vector technique. A fundamental tenet of the approach is to reliably flood an advertisement throughout the network whenever the state of a link changes. Examples of link state change are adding a new link, deleting a link, and an unexpected link failure. Thus, each node obtains complete knowledge of the network topology in the convergence time t (usually several seconds). After any change, each node computes the least-cost routes to every destination using an algorithm such as the Dijkstra algorithm [2]. Since every node has the same topology database, they all compute consistent next hop forwarding tables. Examples of link state routing protocols are the Internet's Open Shortest Path First (OSPF) [3] and the OSI IS-IS Routing Protocol [2], where IS stands for Intermediate System. Key advantages of these protocols are reduction of topology update information and decreased convergence times when compared with distance vector approaches. A key disadvantage is the increased complexity of these methods and consequently increased difficulty in achieving interoperability between different vendor implementations.

Routing is a complicated subject, and the above descriptions serve as only a brief introduction sufficient for our purposes. This chapter provides more detail after first covering LAN standards and bridging protocols. Readers interested in even more detail should consult more detailed descriptions, such as Reference 2 for OSI-based routing and References 4 and 5 for IP-based routing.

9.2 IEEE LOCAL AREA NETWORKING (LAN) STANDARDS

This section summarizes the most popular LAN standards in terms of protocol layering concepts, example deployment scenarios, and then a review of each of the major LAN protocols. The coverage begins with the Logical Link Control (LLC) and then moves on through the Media Access Control (MAC) specifications for Ethernet, Token Ring, Fast Ethernet, and Demand Priority Ethernet.

9.2.1 Layered LAN Protocol Model

LAN protocols implement the protocol stack shown in Figure 9.2. The Logical Link Control (LLC) and Media Access Control (MAC) sublayers of the IEEE 802.X standards map to the link layer, while the actual physical medium (e.g., twisted pair) that interconnects stations on a LAN maps to the physical layer. The MAC layer manages communications across the physical medium, defines frame assembly and disassembly, performs error detection, and addressing functions. Figure 9.2 shows how the LLC layer interfaces with the network layer protocols through one or more Service Access Points (SAPs). These SAPs provide a means for multiplexing within a single host over a single MAC layer address as illustrated in the figure. As we shall see later, address fields within the LLC portion of the data link frame specify the source and destination SAP.

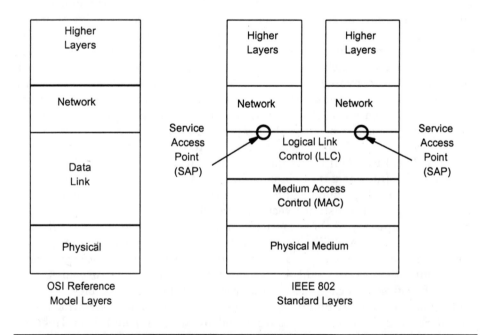

Figure 9.2 LAN Protocol Standards Layered Model

9.2.2 Typical LLC and MAC Sublayer Implementations

The IEEE 802.2 standard defines the logical link control layer, while the IEEE 802.3 through 802.12 standards define various aspects of MAC layer protocols. Figure 9.3 shows some examples of the physical relationship between LLC and MAC interface points in devices connected to the two Ethernet networks and Token Ring network in the center of the figure. Starting in the upper left hand corner, a host runs multiple applications, each with a separate LLC, and interfaces to an Ethernet LAN via a single MAC address. In the upper middle portion of the figure, a bridge forwards MAC frames based upon information obtained by a bridging protocol between the Ethernets and the Token Ring. In the upper right hand corner of the figure, a Network Interface Unit (NIU) provides a single MAC address to the same Ethernet LAN, but supports a terminal and workstation each with separate LLC addresses. In the lower right hand corner a router interconnects an Ethernet, the Token Ring, and a wide area network. The router has multiple MAC addresses, but forwards packets based upon information in the LLC portion of the packet header. Routers also run a routing protocol over the network layer connecting LLC layer entities. Routers forward packets based upon the network layer header contained inside the LLC PDU. In the lower left hand corner of the figure a workstation interfaces via the Token Ring MAC interface. Now that we've seen the basic layering concepts and have some context in terms of example implementations, let's take a look at some further details of the LLC and MAC sublayers.

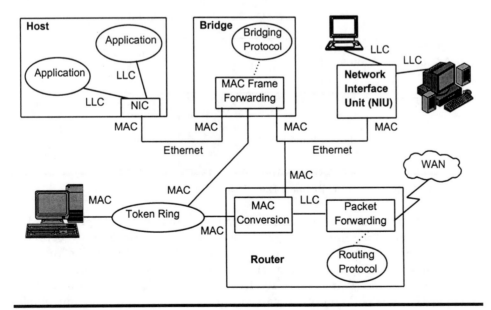

Figure 9.3 Illustration of LLC and MAC Physical Interface Points

9.2.3 The Logical Link Control (LLC) Sublayer

The IEEE Standard 802.2 [6] defines the LLC protocol that hides the differences between various MAC sublayer implementations from the network layer protocol. This allows systems on very different types of LANs, for example Token Ring and Ethernet, to communicate. LLC provides services to the network layer that are either connection-oriented or connectionless. Connection-oriented service uses peer-to-peer communications and provides acknowledgments, flow control, and error recovery. There are three classes of services provided in the LLC: unreliable datagram, acknowledged datagram service, and connection-oriented service.

When the LLC layer receives user data, it places it in the information field and adds a header to form a LLC Protocol Data Unit (PDU) as shown in Figure 9.4. Destination and Source Service Access Point (DSAP and SSAP) address fields along with a control field precede a variable length information field in the LLC PDU. The 802.2 standard defines the least significant bit of the DSAP address to identify a group address. The least significant bit of the SSAP address field is a Command/Response (C/R) bit for use by LLC services. Implementations may use only the high-order 6 bits of the DSAP and SSAP fields, limiting the total number of SAPs to 64.

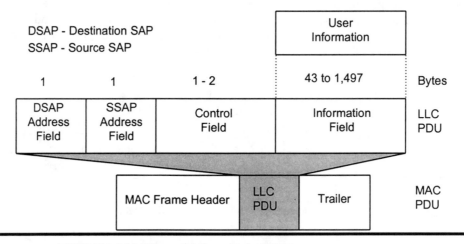

Figure 9.4   IEEE 802.2 LLC Protocol Data Unit (PDU)

The 802.2 standard defines SAP values for a number of ISO protocols, but none for some other important protocols such as IP. Since many potential combinations of LAN types and protocols exist, the IEEE extended the LLC control header using a SubNetwork Access Point (SNAP) structure. Normally, the LLC DSAP and SSAP are 1 byte, and the LLC control 1 or 2 bytes. When both the DSAP and SSAP values equal x'AA', and the LLC control field equals x'03', then the extended LLC header adds two fields — a 3-byte Organizationally Unique Identifier (OUI) for defining an organization that assigns a 2-byte Protocol Identifier (PID). One example would be an Ethernet frame carrying an IP datagram, where the OUI and PID values would be x'000000' and x'0800', respectively. Figure 9.5 illustrates the SubNetwork Access Point (SNAP) structure. RFC 1340 lists the assigned LLC1/SNAP header types. As we describe in Chapter 19, the multiprotocol encapsulation over ATM defined in IETF RFC 1577 employs this extended LLC/SNAP header.

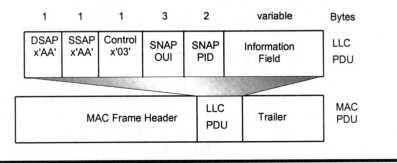

Figure 9.5   IEEE Extended LLC/SNAP Header

9.2.4 The Medium Access Control (MAC) Sublayer

The Medium Access Control (MAC) sublayer manages and controls communications across the physical medium, assembles and disassembles frames, and performs error detection and addressing functions. Table 9.1 summarizes some key attributes of the IEEE 802 series of standards for MAC sublayer protocols.

Table 9.1   Important Attributes of MAC Sublayer Standards

IEEE Number	Standard Title	MAC Speed (Mbps)	Physical Media Supported	Maximum Payload Size (Bytes)
802.3	Ethernet, Fast Ethernet	1, 10, 100	Coax, 2W UTP, STP, Fiber	1500
802.4	Token Bus	1, 5, 10	Coax	8191
802.5	Token Ring	1, 4, 16	STP	5000[1]
802.6	Distributed Queue Dual Bus (DQDB)	34, 44, 155	Coax, Fiber	8192
802.9a	IsoEthernet	16	2W UTP, STP	
802.12	100VG-AnyLAN	100	4W UTP, STP, Fiber	

Notes:
2W UTP   2 Wire Unshielded Twisted Pair
4W UTP   4 Wire Unshielded Twisted Pair
STP   Shielded Twisted Pair
[1] Computed for 4 Mbps media speed and a token-holding time of 10 ms.

The MAC layer adds a header and trailer to the LLC PDU prior to transmission across the physical medium. Each MAC layer PDU contains a unique 6-octet (48-bit) MAC address for the source and destination stations. The IEEE assigns 24-bit identifiers called "Organizationally Unique Identifiers" (OUIs) to each enterprise manufacturing Ethernet interfaces. The manufacturer uniquely assigns the remaining 24 bits, resulting in a unique 48-bit address known as the physical address, hardware address, or MAC address. Thus, a unique address identifies every LAN Network Interface Card (NIC). Each station on an LAN compares the destination MAC address in a received frame to determine if the frame should be passed up to the LLC entity addressed by the DSAP field in the LLC PDU. We now examine the most common MAC protocols: Ethernet, Token Ring, Fast Ethernet, and Demand Priority Ethernet.

9.2.5 Ethernet and the CSMA/CD 802.3 MAC Sublayer

Dr. Robert M. Metcalfe invented Ethernet in the late 1970s at the Xerox Palo Alto Research Center. It was based upon some concepts used in the ALOHA system deployed to implement radio communication in the Hawaiian islands [5]. See Reference 7 for the original 1976 drawing by Dr. Metcalfe of the Ethernet concept and a very readable tutorial on Ethernet. Following successful trials of the Xerox Ethernet, a multivendor consortium of DEC, Intel, and Xerox (abbreviated DIX) published the first Ethernet specifications in 1980, with the first Ethernet products appearing shortly thereafter in 1981. The original Ethernet specification covered operation at 10-Mbps over a 50 ohm coaxial cable, also called thick Ethernet. Indeed, the term "Ether" refers to the broadcast media of an electrically terminated coaxial cable with "vampire" taps connecting each station to the cable. In a classical case of vendor-driven standards development, this proprietary standard provided the basis for the IEEE 802.3 committee. The IEEE 802.3 standard, adopted by the ISO, IEC and ANSI [8] defined a Media Access Control (MAC) protocol called Carrier Sense Multiple Access with Collision Detection (CSMA/CD) at several speeds over a variety of different media as indicated in Table 9.1.

An analogy helpful in understanding CSMA/CD is a Citizen's Band (CB) radio channel. Users first listen to see if anyone is already talking (Carrier Sense) before speaking. Everyone has an equal chance to seize the radio channel (Multiple Access), whether they have anything useful to say or not. Once a user begins speaking, no one interrupts until the transmission completes, indicated by the speaker saying "over." Two people may begin transmitting simultaneously, in which case everyone gets a garbled signal (Collision Detection). The CBer's then back off and repeat the process, with the contending users hopefully waiting different amounts of time to transmit again. As long as the channel isn't too full, collisions don't occur too frequently and useful communication results. Ethernet CSMA/CD works in a very similar manner to this CB channel example, but with some specific refinements to make the collision resolution process work smoothly. Now let's take a more in-depth look at the real CSMA/CD.

CSMA/CD is a fancy name for a relatively simple protocol (analogous to the shared CB radio channel discussed above) that allows stations to transmit and receive data across a multiple access medium, called a "segment" or "collision domain," shared by two or more stations as illustrated in Figure 9.6. Note that LAN switches support a support a single user per segment decreasing the number of contending users to two: the LAN switch port and the host LAN interface card. We use the notation of a single line with multiple taps depicted in this figure throughout this text to indicate an Ethernet segment.

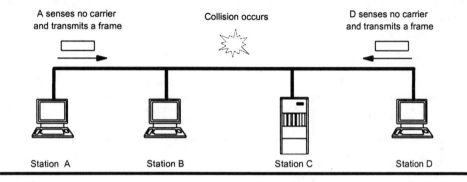

Figure 9.6   Shared Medium Ethernet Configuration and CSMA/CD Example

When a station has data to send, it first senses the channel for a carrier transmitted by any other station. If it detects that the channel is idle, the station transmits an 802.3 frame in serial bit form, as stations A and B do in the example of Figure 9.6. If the transmitting station receives it own data frame without any errors as determined by the frame check sequence, then it knows that the frame was delivered successfully. However, due to the finite propagation delay of electromagnetic signals, the transmissions of two stations may still collide at other points on the shared medium if they begin transmitting at nearly the same time as illustrated in the figure. Stations detect such collisions via means such as an increase in received signal power or a garbled signal. In order to ensure that all stations detect the collision event, each station that detects a collision immediately sends out a short burst of data to garble the signal received by other stations. Eventually, the transmitting station also detects the collision and stops transmitting. Following a collision event the end stations initiate an exponential back-off algorithm to randomize the contending station's next attempt to retransmit the lost frame using contention slots of fixed duration. The design of the back-off algorithm normally results in the resolution of collisions within hundreds of microseconds. The 802.3 CSMA/CD system can achieve throughput approaching fifty percent under heavy load conditions for reasons that we study in Chapter 26.

Figure 9.7 shows the IEEE 802.3 CSMA/CD MAC PDU frame (a) compared to a DIX Ethernet frame (b). The IEEE uses the convention where the order of octet and bit transmission is from left to right as shown in the figure. The 802.3 frame differs from the Ethernet frame only in the interpretation of the two bytes after the address fields, unfortunately making the protocols incompatible. The 7-byte preamble field contains the bit pattern '10101010' which allows all stations to synchronize their clocks to the transmitter. The Starting Frame Delimiter (SFD) contains the "start-of-frame" character, '10101011' to identify the start of the frame. The next two 6-byte (48-bit) fields identify the MAC-layer destination and source addresses formed from the IEEE assigned 24-bit OUI and the 24-bit unique address assigned by the manufacturer of the Ethernet interface card. The 802.3 standard defines some unique addresses. An address of all ones is reserved for the broadcast address. The high-order bit of the address field is zero for ordinary addresses or one for group addresses, which allow a set of stations to communicate on a single address.

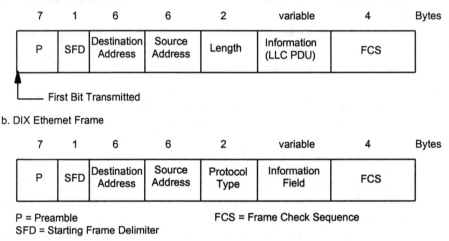

Figure 9.7   (a) IEEE 802.3 CSMA/CD and (b) DIX Ethernet MAC PDU Frames

Next, in the 802.3 frame, the 2-byte length field gives the number of bytes present in the data field, ranging from 0 to 1500. Incompatibly, the DIX Ethernet frame instead uses this field as a protocol type. The information field is the LLC PDU in the 802.3 frame, and the information field is the DIX Ethernet payload. The 802.3 standard defines a pad field in the information field so that all frames are at least 64 bytes long. This minimum frame length prevents a station from completing transmission of a short frame before the last bit reaches the end of the cable, where it may still collide with another frame. Operating at 10-Mbps with a 64-byte minimum frame size constrains the Ethernet cable to be less than 2500 meters [5]. Finally, the frame check sequence uses a standard CRC-32 code as we study further in Chapter 24.

9.2.6 Token Ring

The IBM labs in Zurich, Switzerland, developed the Token Ring protocol in the late 1960s, with the first commercial products appearing in 1986. The IEEE adopted Token Ring as the 802.5 MAC Standard, which also works with the IEEE 802.2 logical link control layer.

A Token Ring is actually a closed loop of unidirectional point-to-point physical links connecting ring interfaces on individual stations as shown in Figure 9.8. This text uses the notation of a circle to indicate a Token Ring.

As shown in the figure, the ring interfaces operates in one of three modes: transmit, listen/modify, or powered-down. In the listen/modify mode, the ring interface inserts a 1-bit delay so that each station can receive, and possibly modify, the bit before transmitting the bit on to the next station as shown for stations C and D in the figure. Since stations actively participate in the transmission of bits, the standard requires that they electrically pass the bit when they are in a powered downed mode, for example, using a relay as illustrated for station B in the figure. A unique 3-byte pattern, called a *token*, circulates around the ring when all stations are idle. Thus, a token ring must have enough delay to contain an entire token, sometimes requiring the insertion of additional delay via one station dynamically designated as the ring monitor.

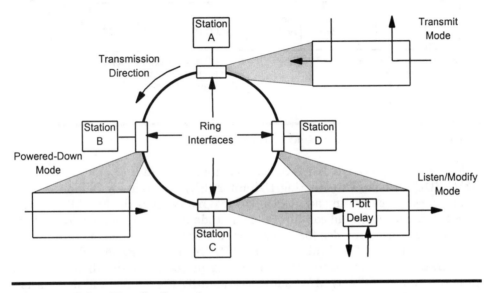

Figure 9.8 Token Ring Configuration

As the token circulates the ring, each station can seize the token, temporarily remove it from the ring, and send one or more data frames to the next station on the ring as shown for station A in Figure 9.8. The data frame then circulates around the ring to every station. The destination station's ring interface copies the frame and forwards it to the next station on the ring. The originating station receives its own data frame, and can verify correct reception through the frame check sequence. The originating station then removes the frame from the ring. The station holding the token can transmit data for a maximum token holding time, which has a default value of 9.1 ms, which constrains the maximum MAC frame size to approximately 4,500 bytes on a 4-Mbps token ring or 18,200 bytes on a 16-Mbps token ring. Once the station completes data transmission (or reaches the holding time limit) it stops transmitting data frames and sends the token to the next station. Hence, token ring LANs have greater minimum delay than Ethernets because a station must wait for the token prior to transmitting any data. On the other hand, since only the station holding the token can transmit data, no collisions occur as in CSMA/CD. Therefore, under heavy load conditions, the Token Ring protocol throughput approaches one hundred percent efficiency.

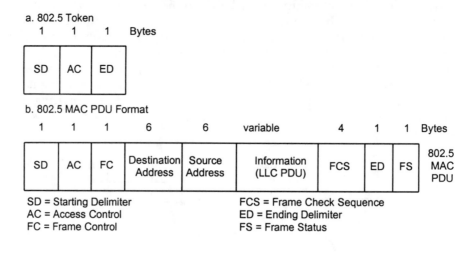

Figure 9.9   IEEE 802.5 Token Ring MAC PDU

Figure 9.9a shows the IEEE 802.5 Token and Figure 9.9b shows the 802.5 MAC PDU format The Starting Delimiter (SD) and Ending Delimiter (ED) fields mark the beginning and end of the token as well as the MAC PDU. The Access Control (AC) field contains a bit that identifies the frame as the token. The AC field also provides for a multiple level priority scheme according to a complex protocol. The Frame Control (FC) field distinguishes data frames from frames used to control the token ring protocol. As in all IEEE MAC protocols, 6-byte (48-bit) MAC addresses identify the source and destination stations. The Frame Check Sequence (FCS) field uses a CRC-32 code to detect bit errors. Stations that detect an error through the FCS, set a bit in the Ending Delimiter field, which also serves to mark the last frame in a sequence of frames sent by a single station. The destination can acknowledge receipt of a data frame using bits in the Frame Status (FS) byte.

Every Token Ring must also have a monitor station that performs various functions related to ring maintenance. The monitor ensures that there is always a token, and also detects orphan frames created by a station crashing in the middle of a transmission, which means that the frame is not removed and would circulate indefinitely. Like many real-world Ethernet networks, Token Ring implementations typically are a physical star homed on a hub in a wiring closet. This physical design also allows the hub to automatically repair a broken ring due to failures in the wiring.

9.2.7 100-Mbps Fast Ethernet

Ethernet is the most popular LAN technology in use today. However, as workstation and server technology advanced, 10-Mbps simply wasn't fast enough. Furthermore, many LAN mangers attempted to utilize complex designs of bridges and routers to meet the capacity demand. Since FDDI ended up being a complex, and expensive, protocol, the IEEE reconvened the 802.3 committee in 1992 to upgrade the popular 802.3 LAN standard to 100-Mbps. Two competing proposals emerged: a conservative proposal to simply increase the speed of the current 802.3 protocol, and another proposal to rework the entire protocol to give it new features that resulted in the establishment of the 802.12 committee. As often happens in standards bodies, both proposals won out. The first, covered in this section, resulted in an updated 802.3 specification in 1996 [8] that added specifications for 100-Mbps, commonly called Fast Ethernet or 100Base-T [9] which quickly became the predominant 100-Mbps LAN standard. The second, covered in the next section, resulted in the 802.12 standard called 100VG-AnyLAN.

Basically, Fast Ethernet speeds up the existing CSMA/CD media access control mechanism from 10-Mbps to 100-Mbps. All the frame formats, procedures and protocols remain pretty much the same. This means that applications designed for 10-Mbps Ethernet can run essentially unchanged over 100-Mbps Fast Ethernet. One important change is that the shared media topology was eliminated in favor of exclusive use of a star topology. The designers chose to use twisted pair copper media at distances up 100 m and fiber optic cables for distances up to 2000 m.

The 802.3 standard update defines three physical media: 100BASE-T4, 100BASE-TX, and 100BASE-FX. The 100BASE-T4 standard uses four unshielded twisted pairs from each device back to a hub, a situation commonly available in telephone grade office telephone wiring. Avoiding the need to rewire an existing office is an increasingly important practical consideration for network designers. Since the cost of LAN technology continues to decrease with respect to labor costs, the cost of rewiring can easily exceed equipment costs. The standards achieve higher effective throughput over the shielded twisted pair and especially fiber optic media since transmission is full-duplex. The 100BASE-TX standard uses two data grade twisted pairs. The 100BASE-FX standard uses a pair of optical fiber cables defined by the same ANSI standard used for FDDI. User equipment (i.e., a DTE) may interface to any of these three physical media through a 40-pin Media Independent Interface (MII), or directly via the 4-pair, 2-pair set of wires or pair of optical fibers.

However, Fast Ethernet still has the variable delay and collision limited throughput that the old slow Ethernet did. Furthermore, the central hub based design limits effective network diameter to approximately 200 meters, less than ten percent of the old Ethernet. But at only two to three times the price of 10-Mbps Ethernet, the 100-Mbps Fast Ethernet is a good value for a ten-fold increase in performance for a LAN designer who doesn't need quality of service features.

9.2.8 100VG-AnyLAN

The IEEE 802.12 standard [10] defines the competing proposal for high-speed Internet, also called 100VG-AnyLAN in the literature [9]. As the name implies, it supports operation with any (existing) LAN protocol, namely Ethernet or Token Ring, but not both at the same time with new stations supporting the 802.12 standard. A key new feature of 100VG-AnyLAN is the demand priority scheme implemented in a hierarchy of LAN repeaters as illustrated in Figure 9.10.   The level 1, or root, repeater may have several lower levels of repeaters connected via cascade ports. Local ports connect 100VG-AnyLAN end node devices to any of the repeaters as shown in the figure.

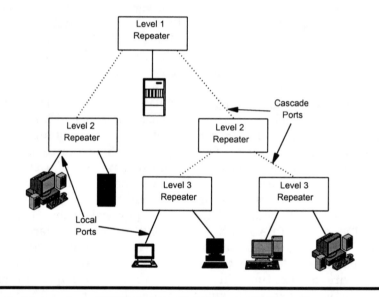

Figure 9.10   Illustration of IEEE 802.12 100VG-AnyLAN Hierarchy

   Like Fast Ethernet, 100VG-AnyLAN avoids any shared media and uses switching hubs exclusively. The demand priority protocol utilizes a round robin polling scheme where individual stations request priority for each MAC frame from the repeater. The 802.12 standard gives prioritized service to frames sensitive to variations in delay over other frames requiring only best effort service across the hierarchy by repeaters selectively granting permission to transmit in response to prioritized end station requests. Furthermore, since the access control algorithm is deterministic, no collisions occur and throughput of the LAN can approach 100-Mbps. Thus, 100VG-AnyLAN overcomes the main disadvantages of the traditional CSMA/CD Ethernet protocol: lack of prioritized service, and collision limited throughput.
   The 802.12 standard requires that 100VG-AnyLAN hubs support either the traditional 10-Mbps Ethernet frame format, or the standard Token Ring frame format (but not both) on the same LAN without bridges. Therefore, the IEEE 100VG-AnyLAN standard provides an easy migration path from

existing 10-Mbps Ethernet and 4/16-Mbps Token Ring networks. 100VG-AnyLAN can operate over four-pair UTP (up to 100 m), two-pair STP (up to 200 m), and fiber-optic cable (up to 2000 m). So far, 100 VG-AnyLAN has seen little commercial adoption since the marketplace has adopted switched 100-Mbps Ethernet.

9.3 FIBER DISTRIBUTED DATA INTERFACE (FDDI)

This section covers the base Fiber Distributed Data Interface (FDDI) capabilities, as well as the additional support for isochronous traffic in FDDI-II.

9.3.1 Basic Fiber Distributed Data Interface (FDDI)

The American National Standards Institute (ANSI) issued the first Fiber Distributed Data Interface (FDDI) standard in 1987 [11] targeting high-performance LANs and campus backbones. Therefore, FDDI is the senior citizen in the class of other high-performance LAN standards. With age come maturity and stability, which many experts acknowledge as a virtue of FDDI [9]. Unfortunately, FDDI's station management protocol was too complex, making the resulting chip sets and implementations expensive in comparison with other alternatives [5]. Also, since FDDI only achieved limited penetration in the marketplace, it never benefited from the volume production manufacture of integrated circuits. We define the basic concepts, operation and capabilities of FDDI mainly as background for a comparison analysis with ATM and other competing LAN/MAN technologies in Part 8.

Similar to Token Ring, FDDI also utilizes a ring topology made up of physical point-to-point fiber optic connections attaching a circular arrangement of stations. However, FDDI uses two counter-rotating rings instead of one for greater reliability as illustrated in Figure 9.11. As indicated in the figure, dual attached stations (also called class A stations) connect to both rings. Data flows in one direction around primary ring, with the secondary ring providing an alternate path that the dual attached stations use to form a longer ring in the event of a failure in the primary ring. See Reference 12 for a good tutorial on FDDI and further details.

FDDI standards define physical interfaces for single-mode fiber using lasers, but most implementations use multimode fiber with cheaper (and safer) Light-Emitting Diode (LED) transmitters. A special type of dual attached station, commonly called an FDDI concentrator [9], supports simpler (and cheaper) single attached stations such as workstations and high-performance servers as shown on the right hand side of the figure. As shown on the left hand side of the figure, single attached stations (also called class B stations) may connect to only the primary FDDI ring, but could be disconnected in the event of a primary ring failure. FDDI protocols also operate over shielded and unshielded twisted-pair copper at distances up to 100 meters in a standard defined by the Unshielded Twisted Pair Forum (UTF) called the Copper Distributed Data Interface (CDDI). Typically, FDDI networks use a star-wiring arrangement, where a patch panel or hub allows for easier cable management as well as the addition and deletion of stations.

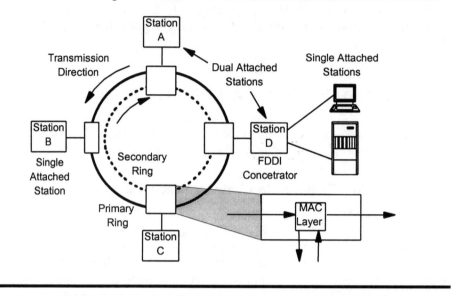

Figure 9.11   FDDI Counter-rotating Ring and Station Attachment Configuration

Figure 9.12 shows a conceptual model of the FDDI protocol stack [13]. As shown in the figure, each station has a Physical Medium Dependent (PMD) sublayer that defines optical levels and signal requirements for interfaces to the fiber optic cables via Medium Interface Connectors (MICs). The PHYsical protocol (PHY) layer defines framing, clocking, and data encoding/decoding procedures.In a manner similar to token ring, powered down dual attached stations convey the optical signal so that both the primary and secondary rings remain intact. Active stations implement a Medium Access Control (MAC) sublayer that defines token passing, data framing, and interface to the IEEE 802.2 Logical Link Control Layer (LLC) as indicated in the figure. The Station ManagemenT (SMT) function interacts with the PMD, PHY and MAC sublayers as shown on the right hand side of the figure to provide management services. SMT functions include provision of unique identifiers, fault isolation, station configuration, status reporting, and ring configuration. SMT also provides the means of inserting and removing stations from the FDDI ring.

Basic FDDI operation is similar to that of 802.5 Token Ring, with one important exception. As in Token Ring, an FDDI a station must first seize the token before transmitting. However, since an FDDI ring may contain a maximum of up to 1000 stations with up to 2 km between each station or a total ring circumference of 200 km, the delay involved could be tens of milliseconds in waiting for the frame to circulate around the ring so that the transmitting station can remove it and re-insert the token. If FDDI operated as token ring does and required other stations to wait for the token prior to transmission, throughput would be dramatically reduced. Therefore, the

FDDI standard allows the transmitting station to insert the token immediately after completing transmission of its frame(s). Hence, subsequent stations can transmit their frames before the sending station removes its frame from the ring. This modification to the passing protocol allows FDDI to achieve near 100 percent transmission efficiency under heavy loads in a manner similar to token ring. Unfortunately, FDDI suffers from the same issue of increased latency due to the token rotation time as the Token Ring protocol does. Furthermore, due to the shared media design of FDDI, the bandwidth available to each user decreases as more stations are added to the ring. Also, the variable delay caused by waiting for the token makes FDDI inappropriate for multi-media and voice applications. Realizing this, the FDDI designers defined a follow-on standard called FDDI-II to support real-time services such as voice and video as summarized in the next section.

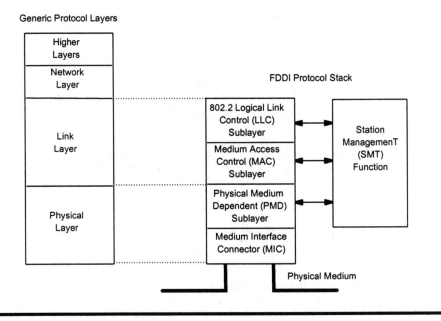

Figure 9.12   FDDI Protocol Stack

The ANSI X3.239 standard [14] defines the FDDI MAC protocol. Note that the FDDI token and MAC frame shown in Figure 9.13 is similar to that used in Token Ring. The FDDI physical layer utilizes a scheme called 4 out of 5 encoding, where every four bits of the MAC layer protocol are actual encoded as five baud on the physical medium to reduce the cost of FDDI equipment. In other words, each FDDI symbol either contains four bits of MAC information, or performs special functions on the FDDI ring. The token and MAC frame both begin with a Preamble field containing at least 4 symbols. Other stations recover clocking from the preamble since the line coding does not provide automatic timing recovery as it does on Ethernet. Next, a two symbol Starting Delimiter (SD) field precedes a two symbol Frame Control (FC) field. The preamble, SD, and FC fields make up a Start of Frame

Sequence (SFS) as indicated in the figure. In the token, the FC field has the token bit set to one, while the FDDI MAC frame has a value of zero in the token bit. The token then concludes with a two symbol Ending Delimiter (ED) field. The FDDI MAC PDU then contains a four (or twelve) symbol Destination Address (DA) and Source Address (SA) field. Next, an optional Routing Information (RI) field comprised of zero to sixty symbols in multiples of four symbols precedes and optional information field. An eight symbol (32-bit) Frame Check Sequence (FCS) provides error detection for the fields indicated in the figure. An End of Frame Sequence (EFS) brackets the FDDI MAC frame with a two symbol Ending Delimiter (ED) and a Frame Status (FS) field of three or more symbols. The maximum total FDDI MAC frame length is 9,000 symbols, or 4,500 bytes.

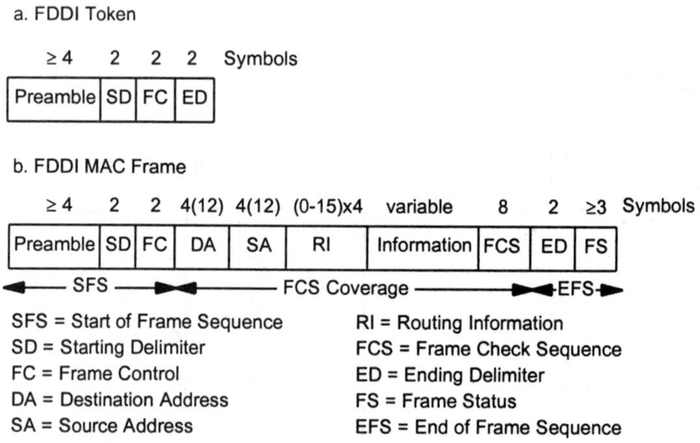

a. FDDI Token

≥ 4	2	2	2	Symbols
Preamble	SD	FC	ED	

b. FDDI MAC Frame

≥ 4	2	2	4(12)	4(12)	(0-15)x4	variable	8	2	≥3	Symbols
Preamble	SD	FC	DA	SA	RI	Information	FCS	ED	FS	

◄——— SFS ———►◄——————— FCS Coverage ———————►◄EFS►

SFS = Start of Frame Sequence   RI = Routing Information
SD = Starting Delimiter   FCS = Frame Check Sequence
FC = Frame Control   ED = Ending Delimiter
DA = Destination Address   FS = Frame Status
SA = Source Address   EFS = End of Frame Sequence

Figure 9.13   FDDI Token and MAC Frame Formats

9.3.2 Hybrid Ring Control (FDDI-II)

Recognizing the problems with basic FDDI in supporting multimedia traffic, ANSI issued a second version of the FDDI protocol, called FDDI-II or Hybrid Ring Control (HRC) [15] in 1992. This effort also resulted in a modified Medium Access Control layer (MAC-2) specification in 1994 [14]. The aim of FDDI-II was to transport multiplexed asynchronous packet data along with isochronous circuit-switched data over FDDI LANs and MANs. Figure 9.14 shows the FDDI-II protocol structure. As part of the set of optional HRC layers, a new hybrid multiplexer layer sits between the physical and MAC sublayers as shown in the shaded portion of the figure. The standard uses either the existing FDDI MAC layer or an enhanced version called MAC-2. The optional HRC capability also adds an Isochronous MAC (IMAC) sublayer supporting one or more Circuit Switching multiplexers (CS-MUX) . The MAC and MAC-2 layers support conventional data applications via a set of asynchronous services using the IEEE 802.2 standard Logical Link Control (LLC) sublayer as shown in the figure. The new IMAC sublayer supports circuit type services, such as constant bit rate digitized voice and video. FDDI-II uses a deterministic multiplexing scheme to split the 100-Mbps bandwidth up into sixteen synchronous frames operating at 6.144 Mbps, which is 96 times the standard 64-kbps telephony rate we studied in Chapter 6. The choice of this unit of bandwidth supports exactly four standard North American DS1 circuits (4×24) or three standard International E1 circuits (3×32). For isochronous traffic, the node-to-node delay reduces to 125 µs plus propagation delay, while the lower priority asynchronous traffic operates as in basic FDDI similar to token ring. The hybrid multiplexer layer strips off the header and passes the frame to the appropriate MAC layer. Enhanced station management functions support the hybrid operation of isochronous and asynchronous data on the same physical ring.

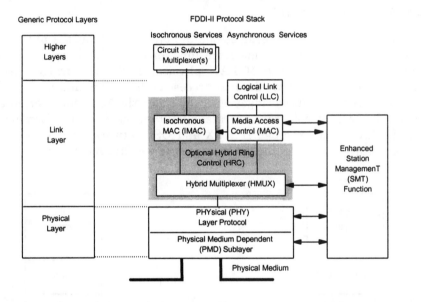

Figure 9.14 FDDI-II Protocol Structure

Theoretically, FDDI-II allows interconnection of PBX equipment to the LAN and MAN, combining voice and data communications on the same fiber. Practically, however, few vendors implement FDDI-II since other technologies that support multiple qualities of service like 100VG-AnyLAN and low-speed LAN ATM network interface cards cost much less [9].

9.4 BRIDGING CONCEPTS, SYSTEMS, AND PROTOCOLS

Bridging performs several critical functions in local area networks. Initially, bridges provided connectivity between LANs of the same media type; for example, Ethernet to Ethernet, or Token Ring to Token Ring. Many network designers deployed bridges to scale local area networks beyond the limits of a single LAN collision domain to support greater distances, more hosts, or larger aggregate bandwidth. Inevitably, the generation of incompatible LAN standards created the need for more sophisticated bridges that could translate between different LAN types as enterprises continued interconnecting local area networks. Enterprises first employed bridges to link LANs across the hallway, and then across entire continents in the 1980s. As we shall see, the design of bridged networks achieves plug and play operation to a greater extent than routed networks. But as the complexity of large bridged networks increased, a need arose for automatic topology discovery and reconfiguration. This section provides a summary of the key concepts, configurations, and protocols involved in bridging.

9.4.1 Bridging Context

As depicted in Figure 9.15, bridges operate on user data at the physical and MAC sublayer. Bridges provide more functions than LAN repeaters, which simply extend the distance of LAN physical medium. Bridges use less processing than a router which must process both the link and network layer portions of packets. Since bridges operate at only the MAC sublayer, multiple network layer protocols can operate simultaneously on the same set of bridged LANs. For example, IP, IPX and OSI network layer protocols can operate simultaneously on the same set of bridged Ethernets.

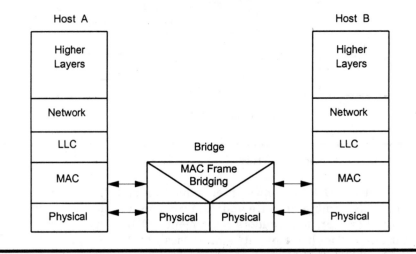

Figure 9.15 MAC Bridging Context for Forwarding User MAC Frames

Communication between bridges using a bridging protocol occurs at both the logical link control (LLC) and the media access control (MAC) sublayers as depicted in Figure 9.16. Similar to the notion of separate planes of operation for user and control data defined in ISDN, the bridging protocol operates separately from the forwarding of user MAC frames depicted in the figure above. Bridging protocols automatically determine the forwarding port for the MAC frame bridging function. Examples of bridging protocols described later in this section are the spanning tree protocol and the source routing protocol.

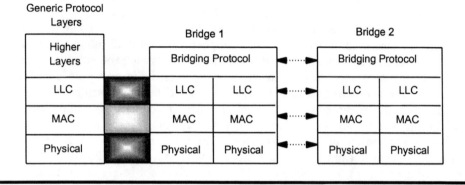

Figure 9.16 Context for Bridging Protocol Communication between Bridges

9.4.2 A Taxonomy of Bridges

Bridges pass traffic from one LAN segment to another based solely upon the destination MAC address. If the destination address of the frame received by the bridge is not local to the bridge, the frame is obviously destined for another LAN. Different types of bridges use different methods to determine the port on which to send the frame. There are four major types of bridges: transparent (or spanning tree), encapsulating, translating, and source routed. This section briefly summarizes the key attributes of each type of bridge, and then explores the spanning tree and source routing algorithms in greater detail. For a very readable, detailed treatment of bridging, see reference 2. We conclude with some observations on the topic of bridge design.

When operating in *transparent* mode, bridges at both ends of a transmission support the same physical media and link layer (MAC-level) protocols from the IEEE 802.X suite, however the transmission speeds may differ. Figure 9.17a and 9.17b show examples of transparent bridging between two local Ethernet LANs and two local Token Ring LANs. Transparent bridges utilize the spanning tree protocol described in the next section to automatically determine the bridged network topology and reliably forward frames based upon the MAC address.

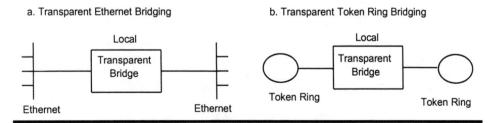

Figure 9.17 Transparent Bridging

Sometimes, network designers require bridges to interconnect dissimilar LANs; for example, Ethernet and Token Ring as shown in Figure 9.18. Unfortunately, the IEEE committees developing LAN standards did not agree on the same conventions, thus creating the need for MAC level conversion. Although this may seem trivial, it's not: for example, even the bit order differs on these media! Bridges must support a *translation* mode to interconnect dissimilar physical LAN media and MAC sublayer protocols. A number of issues must be resolved for conversion between LAN protocols, including: changing bit transmission order, reformatting, checksum recalculation, and control field interpretation [5]. Translation bridges cannot overcome different maximum frame sizes on different LAN types. Instead, the network administrator must ensure that users employ a maximum network layer packet size that doesn't exceed the most constraining LAN. For example, the maximum Ethernet frame size is 1500 bytes, while a 4-Mbps token ring has a maximum frame size of 4000 bytes. In this case, all hosts should use a maximum packet size of less than 1500 bytes.

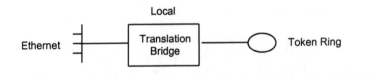

Figure 9.18 Translation Bridging

When operating in *encapsulation* mode, bridges at both ends of the transmission use the same physical and MAC-level LAN protocols, but the network between the bridges may be a different physical medium and/or MAC-level protocol. Encapsulating bridges place MAC frames from the originator within another MAC layer envelope and forward the encapsulated frame to another bridge, which then deencapsulates the MAC frame for delivery to the destination host.

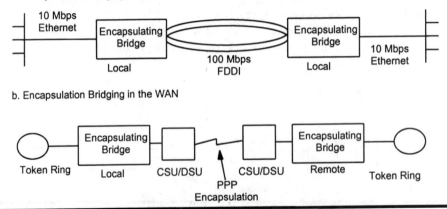

Figure 9.19 Encapsulation Bridging

Figure 9.19 illustrates two typical examples of encapsulation bridging in local and wide area networking. The first example (a) illustrates two local 10-Mbps Ethernet LANs bridged via a campus area 100-Mbps FDDI network. Users frequently employ encapsulation bridges to interconnect multiple Ethernet segments via a high speed FDDI backbone. In the second example (b), a serial WAN link connects two encapsulation bridges using CSU/DSUs supporting bridging between two Token Ring LANs. In this example, the encapsulation bridges place the Token Ring frame in a PPP data link layer frame for transfer over the serial point-to-point WAN circuit. The bridges strip off the PPP envelope and deliver the unmodified Token Ring frame to the other bridge.

Figure 9.20 illustrates an application of *source route bridging* between source and destination Token Ring LANs through three intermediate source route bridges connected by a transit LAN and a WAN serial link. As we study later in this section, source route bridging automatically distributes network topology information so that the source can determine the hop-by-hop path to the destination through specific intermediate bridges. The explorer packets used for this topology discovery add additional traffic to the network, but bridges and hosts cache the topology information for subsequent use. The name source route bridging derives from the fact that the source determines the entire route.

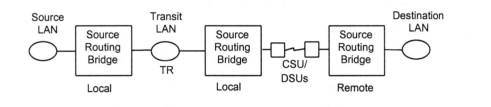

Figure 9.20   Source Route Bridging

9.4.3 Spanning Tree Protocol

Early bridge implementations simply broadcast MAC frames on every port except the one the frame was received on. Soon, network designers found that this basic form of bridging had several problems [2]. The broadcast bridges worked fine as long as the LAN physical network topology had only a single path between any bridges. Unfortunately, if a single link or bridge failed, then the bridged network was down. The IEEE's 802.1d Spanning Tree learning bridge Protocol (STP) [16] solved this problem and delivered reliable, automatically configuring network bridging. The spanning tree protocol dynamically discovers network topology changes and modifies the forwarding tables to automatically recover from failures and revert to an optimized configuration as the topology changes.

The 802.1d spanning tree protocol provides reliable networking by utilizing an algorithm that determines a loop-free topology within a short time interval. The STP algorithm runs continuously to react to link failures as well as automatically add and delete stations and LANs. The resulting path through the bridged network looks like a tree rooted at the bridge with the lowest numerical MAC address. All other bridges forward packets up the tree to this root bridge; which then forwards packets back down the tree to the destination leaf, or until the destination is reached where that bridge simply transmits the frame onto the destination LAN.

Figure 9.21 illustrates a simple example of this property of the spanning tree algorithm. Figure 9.21a illustrates the physical topology of seven LANs, labeled A through G, interconnected by five bridges, labeled B1 through B5. The physical topology has multiple paths between LANs B through G, which the spanning tree algorithm resolves to a single logical topology of a tree rooted in the port on LAN A in Bridge 1 (B1) as shown in Figure 9.21b. Note that traffic in such a bridged network often does not take the most direct path. For example, LAN frames between LANs C and F will not flow through B5, but instead flows through B2, up to the root B1 and back down to B4. Designers can control which links the STP bridge chooses when multiple parallel paths connect bridges by setting administrative costs in the STP algorithm. Also, network designers can choose the MAC address utilized by the bridges to control the resulting topology to a certain extent. Therefore, users should employ these techniques to carefully design STP bridged networks in WANs to minimize traffic flowing across the WAN to the root of the spanning tree, otherwise, since the STP protocol will send all traffic to the lowest numerical address on the bridged network.

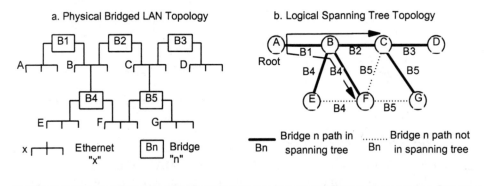

Figure 9.21   Example of IEEE 802.1 Spanning Tree Protocol Operation

9.4.4 Source Routing Protocol

Although STP bridges offer the convenience of plug and play operation, they inefficiently utilize link and bridge port capacity as illustrated in the previous example. The token group of IEEE 802.5, with support from IBM, responded to this challenge by designing the Source Routing Protocol (SRP). Each SRP LAN station specifies the end-to-end routing for each frame for bridging between token LANs. Hence, source route bridges utilize bandwidth more efficiently than spanning tree bridging; however, they require more configuration before they will operate. Stations utilize the SRP for frames destined to stations on other LANs by setting the multicast bit in the source address. This convention works since no station should ever transmit from a multicast address. Each frame contains a complete set of routing information that describes the sequence of bridges and LANs the frame must traverse from the source to the destination station. LAN stations obtain the information to compute this optimal path from explorer packets broadcast periodically throughout the bridged network.

SRP utilizes an addressing scheme in the routing information field illustrated in Figure 9.22, which shows the IEEE 802.5 (Token Ring) MAC PDU fields. The Routing Information (RI) field starts with a 2-byte control header, comprised of a type field to distinguish explorer packets from source routed packets, a length field that limits the source route to 14 hops, a direction bit indicating left-to-right or right-to-left route scanning, and a largest frame size indicator. The remainder of the RI field contains a sequence of 12-bit token ring and 4-bit bridge numbers, defining the hops from source to destination. In real implementations each bridge uses a pair of ring-bridge number pairs, so in practice the maximum number of bridges traversed on a source route is seven. Note that the ATM Forum's Private Network-Network Interface (PNNI) routing protocol also utilizes a source routing paradigm in the interest of efficiently utilizing link and port capacity as covered in Chapter 15.

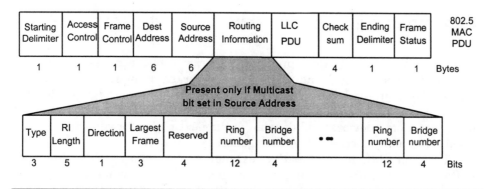

Figure 9.22   IEEE 802.5 Token Ring MAC Frame and Routing Information Field

9.4.5 Bridge Network Design

Careful future planning is required when deploying a bridged network solution. The network engineer who employs a bridged network may find that their design resembles a bridge designed to accommodate a horse and carriage that must now carry automobile and truck traffic. Bridges are best used in networks with few nodes which have a limited geographic extent. Bridging device capacities and speeds vary, supporting low-speed serial links up to DS1 or DS3 across the WAN, and 100-Mbps FDDI or 155-Mbps ATM in the LAN. Higher speeds are needed to support high-speed LANs connected to the bridge, such as 100-Mbps Ethernet and 16-Mbps Token Ring. Bridges provide either local, remote, or both local and remote support.

Although simplicity along with true plug and play operation are major advantages for bridging, there are some major disadvantages as well. Until a transparent bridge learns the destination LAN, it broadcasts packets on all outgoing LAN ports. When destinations are unreachable or have problems, applications resend data, further intensifying traffic overload conditions. Also, the spanning tree bridge uses LAN bandwidth inefficiently by sending all traffic up a tree to a root node, and back out to destinations. Discovery packets used by source route bridges to determine the network topology add additional network overhead traffic. The amount of memory in bridges for storing MAC addresses is also a limiting factor to the size of bridged networks.

Higher layer protocols, such as Netbeui, also generate significant amounts of broadcast traffic, which bridges forward to every host and re-propagate throughout the network. These phenomena and others can create broadcast storms, a problem that increases with the size of the bridged network and number of attached users. Broadcast storms can bring a network to its knees. To minimize these problems, smart bridging techniques provide some level of traffic isolation by segmenting the bridged network into domains that restrict broadcast to a limited area. This containment method, coupled with a ceiling on the amount of multicast traffic, provides some control over broadcast storms.

9.5 ROUTING CONCEPTS, SYSTEMS, AND PROTOCOLS

Unlike most bridges, routers provide connectivity between like and unlike devices attached to local and wide area networks. Routers operate at the network layer protocol, but usually also support link layer bridging. Routers have a common summarized view of the entire network topology, not just locally connected devices, and determine the next hop to forward a packet on based upon many factors. The first generation of routers appeared at MIT, Stanford, and CMU in 1983 as the next step in the evolution following their ARPAnet predecessors three years earlier. Routers emerged into the marketplace over the last decade as the hottest thing since multiplexers, with much more intelligence than bridges or multiplexers. The distinction blurs today since many so-called bridges and hubs have been enhanced to also perform routing functions.

Larger networks usually implement some form of routing protocol that automatically discovers neighbors, distributes topology, and computes optimized routes. This section begins by defining generic router functions and operation. The text then surveys only the modern link-state routing protocols, and not the older historical protocols like RIP. We then cover how larger networks scale through subnetting as an introduction to hierarchical routing in the ATM PNNI protocol. Finally, the text introduces the subject of address resolution in LANs as background to similar concepts discussed in Part 5 regarding ATM's support for LAN emulation, IP over ATM, and MultiProtocol routing Over ATM (MPOA).

9.5.1 Packet Forwarding and Routing Protocol Functions

Routers operate on the user data stream at the physical, link, and network layers to provide a connectionless service between end systems (or hosts) as shown in Figure 9.23. In contrast to bridges, routers operate on the fields in packets at the network level, instead of the MAC sublayer. Subsequent discussion calls this operation on the user data stream a *packet forwarding* function. Of course, systems must employ the same network, transport and application layer protocols when communicating via routers as they would with bridges.

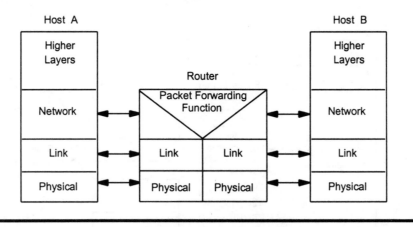

Figure 9.23   Router Packet Forwarding Function Protocol Context

Analogous to the separate planes of operation in bridging protocols, routers communicate via routing protocols as illustrated in Figure 9.24. A dashed line indicates the routing protocol communication between peer entities to distinguish from the packet forwarding peer communication illustrated by solids lines in the previous figure. Routing protocols automatically determine the forwarding port for the packet forwarding function. Examples of routing protocols described later in this section are the topology distribution (i.e., flooding) and the least cost path determination algorithms.

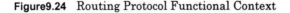

Figure9.24   Routing Protocol Functional Context

Basically, routers implement several interrelated functions as illustrated in Figure 9.25. Starting on the left-hand side of the figure, routers interface to a variety of LAN or WAN media, encapsulating and converting between link layer protocols as necessary. Thus, routers naturally interconnect Ethernet, Token Ring, FDDI and even ATM networks. Routers employ a packet forwarding function, often implemented in hardware in high performance machines. The packet forwarding function contains a lookup table which identifies the physical interface of the next hop toward the destination based upon the high order bits contained in the packet address.

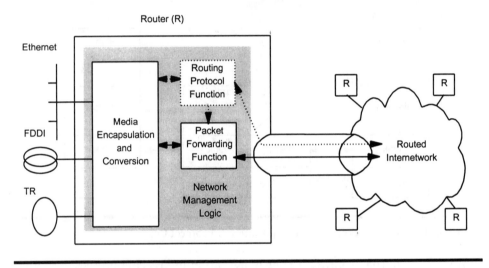

Figure 9.25   Logical Decomposition of Router Functions

Routers employ routing protocols to dynamically obtain knowledge of the location of address prefixes across the entire routed internetwork. The routing protocol engine fills in the next hop forwarding table for the packet forwarding function. Routing protocols determine the next hop based upon specific criteria, such as least-cost, minimum delay, minimum distance, or least-congestion conditions. These routing protocols discover network topology changes and provide rerouting by updating the forwarding tables. Routers employ routing protocols to continually monitor the state of the links that interconnect routers in a network, or the links with other networks. Routers often limit the number of hops traversed by a packet through use of a "time to live" type of algorithm, for example, the one described for IP in Chapter 8. Routers employ large addressing schemes, typically 4 bytes worth in IP and even more with Novell IPX or OSI CLNP. Routers also support large packet sizes. Internal router bus speeds are also much higher than that of bridges, typically in excess of a gigabit per second. The other major advantage of routers is their ability to perform these functions primarily through the use of software, which makes future revision and support for upgrades much easier. A corresponding disadvantage is that the more complex software implementations of routers may have less throughput than simpler implementations of bridges.

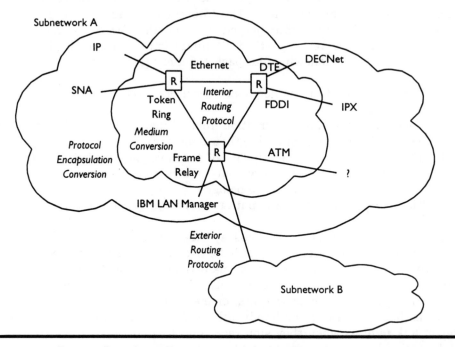

Figure 9.26   Routing Interfaces, Functions, and Architecture

As the name indicates, multiprotocol routers support multiple protocols simultaneously. Figure 9.26 illustrates the range of interfaces and scope of routing. Physical or virtual circuits interconnect routers which are pre-configured with various bits of information needed by the routing protocol, such as address assignments to ports, subnet masks, and routing hierarchy information. As shown in the figure, routers employ an interior routing protocol within a single subnetwork, for example, the Routing Information Protocol (RIP), or the Open Shortest Path First (OSPF). Interconnected networks usually employ an exterior routing protocol, such as the Border Gateway Protocol (BGP). For example, ISPs use BGP to exchange summarized IP address reachability. Static routing, where an administrator manually configures the forwarding tables, should be used with great care since the routing protocol propagates any routing configuration errors throughout the network.

When routers support multiple protocols, they route them to another port on the router which converts it to the corresponding link layer protocol and encapsulation used on that port. Note that the network layer protocol used by the source and destination must be the same; the router just hides the link layer and protocol encapsulation differences. For example, a router encapsulates IP traffic into frame relay frames, SMDS slots, or ATM cells for transmission over a public network service. Most routers must first reassemble ATM cells into packets before processing them.

9.5.2 Link-State Routing Protocols Defined

The class of routing protocols that use the link-state paradigm replaced the earlier distance vector algorithm. The Internet first began using link state routing in 1979. The link-state method overcame the slow convergence times of the prior distance vector method. Routers implementing the link state protocols perform the following four basic functions [5]:

1. They say hello to their neighbors, learn addresses and collect routing "cost" information
2. They collect the state information from all their links and place these in advertisements transmitted to their neighbors
3. They reliably and efficiently "flood" the link state packets throughout the network such that every router quickly converges to an accurate view of the entire network's topology
4. They compute the least cost path to every other router in the network

Let's look into each of these steps a little further with reference to the simple example in Figures 9.27 and 9.28. Neighboring routers run a "Hello" protocol once they boot up, or once a link activates as shown in Figure 9.27a. The hello protocol messages contain routing "cost" information, which may be economic information about the link, or may reflect some other information such as distance, latency or bandwidth of a particular link. Routers also detect link failures when they stop receiving the periodic "heartbeat" of the Hello protocol messages from a neighbor.

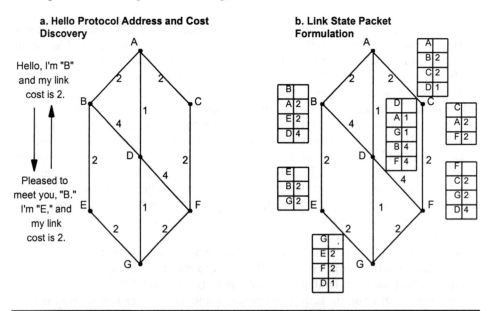

Figure 9.27   Examples of Hello Protocol and Link State Packets

Each node in the network assembles its view of the current link states into a packet and forwards it to each of its neighbors as shown in Figure 9.27b. The link state packet identifies the source router in the first line, and the destination routers and link costs in each of the subsequent lines in the figure. Routers send these link state packets upon startup, or whenever a significant change occurs in the network, such as a link failing or coming back on line.

Intermediate nodes forward link state packets on to other nodes as illustrated in Figure 9.28a for router "A" in a procedure called *flooding*. Flooding involves replicating the link state packets in an efficient and reliable manner such that each node quickly obtains an identical copy of the entire network link state topology. Note that some nodes receive multiple copies of the link state packets. Additional fields in the link state packets that record a sequence number and an aging count eliminate duplicate packets and handle other error scenarios. Routers acknowledge flooded link state packets to ensure reliable topology distribution.

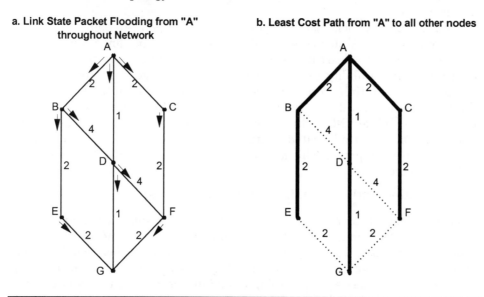

a. Link State Packet Flooding from "A" throughout Network

b. Least Cost Path from "A" to all other nodes

Figure 9.28 Examples of Flooding and Least Cost Route Determination

Finally, each router computes the least cost path to every other router in the network. Figure 9.28b illustrates the result of this calculation for router "A" in the solid, bold lines. The net effect of the routing calculation, using the Djikstra algorithm, for example, is a minimum spanning tree rooted at each node as shown by the upside-down tree rooted in node "A" in the figure. This concept embodies the essence of routing — we encounter it several times within this book. When a link or node fails, the new node or link is added, or a link or node is deleted, then the above procedure repeats. The time for this entire process to complete for every node in the network is called the *convergence time*. Current link state routing algorithms converge within a matter of seconds in moderately sized networks. Rapid convergence to a common topology database in every network node is critical to achieving consistent end-to-end routing decisions. If nodes have different topology databases, then they may create routing loops. As we saw in Chapter 8, IP explicitly handles the possibility of routing loops through the Time To Live (TTL) field in the IP packet header, which prevents a packet from circulating indefinitely.

Three major implementations of link state routing protocols dominate the market: the OSI's Intermediate System to Intermediate System (IS-IS) Routing Protocol, the Internet's Open Shortest Path First (OSPF) protocol [17], and Novell's NLSP. A popular implementation of link state routing is the OSPF, which uses the Dijkstra, or Shortest Path First (SPF), algorithm for determining routing. All costs for links are designated on the outbound router port, so that costs may be different in each direction (unlike the simple example above). OSPF also supports a limited form of hierarchical routing by sectioning the network into independent, lower-level areas interconnected by a backbone area.

OSPF routing supports three types of networks: point-to-point, broadcast, and NonBroadcast MultiAccess (NBMA). Point-to-point links join a single pair of routers. Broadcast networks attach more than two routers, with each router having the ability to broadcast a single message to multiple routers. Nonbroadcast multiple access networks, such as ATM, interconnect more than two routers but do not have broadcast capability. OSPF only supports IP networks, unlike IS-IS, which supports multiple protocols simultaneously. OSPF also supports bifurcated routing: that is, the ability to split packets between two equal paths. This is also commonly referred to as "load sharing," or "load balancing."

9.5.3 Routing and Logical IP Subnetworks (LISs)

A critical concept in routing in large networks is summarization of host addresses, called subnetting. If every router in a network needed to have a routing table entry for every host, then the routing tables would become unmanageably large. Furthermore, routing table size is not the most critical constraint — the message processing to update the routing tables practically limits routing table size before physical storage does.

How does an IP host determine when routing is necessary? For example, how do two hosts on the same LAN know that they can directly transmit packets to one another without using routing? The answer in general is that when the two hosts are not on the same (bridged) LAN, then routing is needed. Historically, IP preceded LANs. Therefore, IP adopted the conventions of a subnetwork bit mask that constrains address assignments to allow hosts to determine whether routing was required based solely upon the source and destination address. The subnet mask convention dictates that IP hosts are in the same Logical IP Subnet (LIS) if a certain number of high order bits of their IP addresses match. A station determines if two IP addresses are on the same subnet by bit-wise ANDing the subnet mask with each address and comparing the results. If both addresses ANDed with the subnet mask result in the same value, then they are on the same subnet. As described in Chapter 8, Classless Inter-Domain Routing (CIDR) generalized the concept of subnet masks even further by allowing them to be of variable length for different subnetworks within the same administrative domain.

Subnet masks use the same IP address format where a certain number of the high order bits all have the value of binary one. In dotted decimal notation it means that the four decimal values of a subnet mask have one of the values listed in Table 9.2. The table also lists the number of consecutive ones in the subnet mask along with the number of networks and hosts allowed in the one byte subnet mask. See Appendices C and D of [18] for more details on subnet mask values and the associated IP address values. For example, the old IP class A, B and C addresses had implicit subnet masks as follows:

Class A	255.0.0.0
Class B	255.255.0.0
Class C	255.255.255.0

Table 9.2   Valid Subnet Mask Decimal Values and CIDR Length fields

Subnet Mask Number	Consecutive Ones	Number of Networks	Number of Hosts								
255	8	254	0	1	1	1	1	1	1	1	1
254	7	126	0	1	1	1	1	1	1	1	0
252	6	62	2	1	1	1	1	1	1	0	0
248	5	30	6	1	1	1	1	1	0	0	0
240	4	14	14	1	1	1	1	0	0	0	0
224	3	6	30	1	1	1	0	0	0	0	0
192	2	2	62	1	1	0	0	0	0	0	0
128	1	1	126	1	0	0	0	0	0	0	0
0	0	1	254	0	0	0	0	0	0	0	0

The old class-based IP address structure suffered from a degenerate variant of Goldilock's syndrome in the Three Bears fairy tale; Class A with over 16 million hosts was much too big for all but the largest networks, and class B with 65,000 addresses was also too big for most networks, yet Class C with only 254 hosts was too small for most networks. It was hard to find a "just right" answer. Although the 32-bit address enabled 2 billion networks, the inefficiency of assigning addresses in only these three sizes threatened to exhaust the Class B address space in 1993. CIDR usage of a variable length subnet mask allowed Internet administrators to split up the IP address space more efficiently and keep the Internet growing until the next IP version (IPv6) arrives. CIDR also defines the concept of "supernetting", where multiple Class C style addresses are combined into a single route advertisement.

Let's look at a simple example — the Alamo Trader's Market in Texas. This network has an old style class C address range of 198.62.193.1 to 198.62.193.

254. Our example network has four routers, one at the headquarters in Austin and three remote sites at Dallas, Houston and San Antonio as illustrated in Figure 9.29. Each site requires up to 10 hosts per router; therefore, we can use the subnet mask of 255.255.255.240 since it allows up to 14 hosts per network (i.e., site) as shown in Table 9.2. We see in the example how most sites like Dallas only use three hosts today, but can expand to support up to 13 total hosts (198.62.193.34 to .46) with one reserved (.47). This choice allows the network administrator to add more sites in the future. The 14 network addresses available under the subnet mask 255.255.255.240 are (see [18] Appendix C):

198.62.193.16
198.62.193.32
198.62.193.48
198.62.193.64
198.62.193.80
198.62.193.96
198.62.193.112
198.62.193.128
198.62.193.144
198.62.193.160
198.62.193.176
198.62.193.192
198.62.193.208
198.62.193.224

The network administrator assigns the address 198.62.193.16 to the headquarters subnetwork in Austin, 198.62.193.32 to the Dallas subnetwork, 198.62.193.48 to the Houston subnetwork, and 198.62.193.64 to the San Antonio subnetwork as shown in Figure 9.29. The administrator assigns addresses to hosts and servers at each of the locations. Note that we leave out the common Class C address prefix 198.62.193 for clarity in the following. Starting in the Dallas 198.62.193.32 subnetwork, the administrator assigns addresses, .34, .35, and .36, to the three hosts (Ken, Sue, and Julie) attached to the router port on the Ethernet segment with address .33. Note that the administrator can add up to 10 more addresses within this subnet (.37 through .46). Moving on to the next subnet, Houston, the network administrator assigns .49 to the router interface on the Ethernet segment and .50, .51, and .52 to the hosts named Bill, Kelly, and Joe, respectively. Finally, she assigns addresses within the San Antonio subnet to the router interface on the Ethernet segment (.65), Rodney (.66), Kim (.67), and Steve (.68). The same address assignment process also applies at the Austin location.

The network administrator in our example now must assign IP addresses and subnet masks to the WAN links in the example of Figure 9.29. Planning for growth of her network, the administrator chooses a different subnet mask for the point-to-point links, since the 255.255.255.240 subnet mask reserves 14 addresses per WAN link, when in fact only two are needed. This is possible because the routers run a version of OSPF that supports CIDR's variable length subnet masks. Therefore, the administrator assigns all three WAN links the same subnet mask of 255.255.255.252, which allows exactly two hosts per subnet as identified in Table 9.2. The network administrator chooses to use this longer subnet mask to split up the 198.62.193.224 network, leaving room to add other subnetworks for planned expansion to other cities in Texas. Hence, the valid network numbers are .224 through 252 in increments of 4 (see [18] Appendix D). The administrator assigns the network address 198.62.193.224 with a subnet mask of 255.255.255.252 to the link between Austin and Dallas as shown in Figure 9.29. The router port in Austin gets the IP address 198.62.193.225 and the Dallas router gets 198.62.193.226. Note that these address choices avoid the all zeroes and all ones host addresses on the WAN link subnet. The other WAN links are then assigned sequential addresses under the longer 255.255.255.252 subnet mask as shown in the figure.

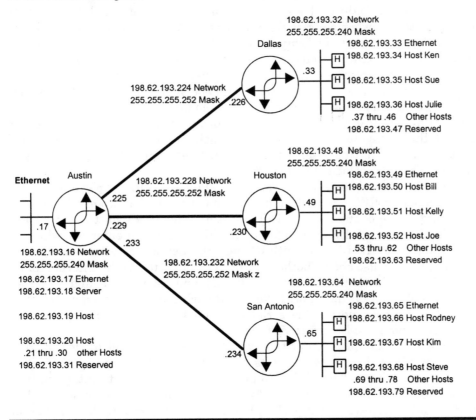

Figure 9.29 Network Example for IP Address Assignment and Subnet Masks

This assignment allows the administrator to add another 9 subnets under the 255.255.255.240 mask and another 11 WAN links out of the 198.62.193.224 network under the 255.255.255.252 mask. If the routers had not used CIDR, then the administrator would need to assign a separate network to each WAN link, wasting three network address blocks in the process, and thus limiting potential expansion to 6 additional subnets and/or WAN links. Patting herself on the back for such a forward looking address assignment and clever use of variable length subnet masks, the LAN administrator of our example heads out of the office for a well deserved, ice cold Lone Star beer at her favorite watering hole on Sixth Street in Austin.

9.5.4 Address Resolution Protocol (ARP)

Another concept from local area networking used in several ATM address protocols is that of address resolution. In modern networks, most hosts attach to local area networks with Network Interface Cards (NICs) that only understand MAC level addresses. When a host wishes to send a packet to another host using a network level (e.g., IP) address, then the sending host must first determine the MAC level address of either the destination host or the next hop router that can progress the packet towards the destination. Note that most real-world applications involve an additional step of first obtaining the network address from a name, such as a Domain Name Server (DNS). In the interest of brevity, we omit this step in the following examples. From the subnetting discussion in the previous chapter, a host knows whether the destination is in the same subnet based upon the result of bit-wise ANDing the destination address with the subnet mask. If the destination is in the same subnet, then all that remains is to determine the MAC address of the destination host. If the destination address is on another subnet, then the source must forward the packet to the pre-configured default router address. In this case, the source must determine the MAC address of the default router attached to its subnet.

The IETF defined the *Address Resolution Protocol* (ARP) to perform exactly this function in RFC 826 [19]. Let's see how it works for two examples with reference to the configuration of Figure 9.30. Here we depict several hosts (labeled H1 through H3) connected to an Ethernet with Router R1, which in turn connects to the Internet with IP and Ethernet addresses as indicated on the figure. We also show another router, R2, connected to the Internet and a local Ethernet with two servers labeled S1 and S2 with IP and Ethernet address indicated on the figure. Note that each router has a separate set of addresses for each port. The serial ports on the routers connected to the Internet do not have Ethernet MAC addresses.

In our first example, Host 1 (H1) wishes to send a packet to Host 3 (H3). Examining H3's IP address using its subnet mask, H1 determines that H3 is on the same subnet. Therefore, H3 prepares an ARP packet containing H3's IP address (A.3) and sends it to the "all Ethernet" broadcast address. H1, H2, and H3 all receive the ARP packet, but only H3 responds with its Ethernet address (E3) in the ARP response packet. Now H1 can place its IP packet addressed to H3 inside an Ethernet MAC frame addressed to E3, and send the MAC frame over the Ethernet. In order to avoid performing this procedure for every packet, hosts cache the resolved Ethernet addresses for recently used IP addresses. Eventually, the ARP cache times out and the above process must be repeated. Also, since H3 will typically need to respond to H1's packet, the ARP protocol includes the following clever way to avoid another broadcast message. The ARP packet sent by H1 includes H1's IP address (A.1) and Ethernet MAC address (E1), so that not only can H3 copy this into its ARP table, but so can every other station on the Ethernet. A station may also send an ARP request for itself in a process called *gratuitous ARP* to flush out the cache of other stations and use the current MAC address associated with a particular IP address.

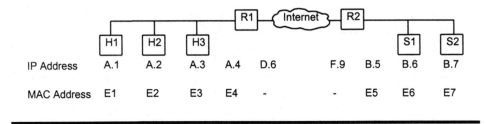

	H1	H2	H3					S1	S2
IP Address	A.1	A.2	A.3	A.4	D.6	F.9	B.5	B.6	B.7
MAC Address	E1	E2	E3	E4	-	-	E5	E6	E7

Figure 9.30 Address Resolution Protocol (ARP) Example Configuration

In our second example, Host H1 wishes to send an IP packet to Server S2 at IP address B.7. Comparing S2's IP address to its own ANDed with the subnet mask, Host H1 determines that S2 is not on the same subnet. Therefore, H1 prepares to send the packet to the default router IP address pre-configured by the network administrator in its storage, namely address A.4 for Router R1. For the first packet sent to R1, Host H1 must ARP for the Ethernet address of A.4 using the above procedure. Once Host H1 receives the ARP response from R1, it updates its ARP cache to indicate that Ethernet address E4 corresponds to the IP address of the "default router", A.4. The default router gives, in effect, a target for all destination IP addresses on a subnet different than the sender's. Next, Host H1 takes the packet addressed to S2 (with IP address B.7), places it in a MAC frame with address E4 (i.e., the MAC address of the default router R1) and transmits it on the Ethernet. R1 receives this packet and examines the destination IP address. From the routing protocol that R1 has been running with the Internet, it determines that the next hop is on the port with IP address D.6. The Internet routes the packet and eventually delivers it to the port with on Router R2 IP address F.9. Router R2 compares the destination IP address (B.7) against its internal forwarding table and determines that the interface with IP address B.5 is on the same subnet as the destination address. If this is the first packet destined for IP address B.7, then router R2 must send an ARP packet on the Ethernet to determine the MAC address. Once Router R2 stores the mapping of IP address B.7 to Ethernet MAC address E7 in its ARP cache, it can forward the MAC frame on to S2.

Although these examples may seem somewhat complicated, computers repeatedly perform these procedures very rapidly without difficulty. In fact, if you compare ARP with the possibility of manually configuring all of the mappings between IP addresses and MAC addresses, the inherent simplicity of the concept becomes obvious. The above examples illustrate the minimum set of addresses that must be configured in every IP host, namely: the stations own IP address, the default router's IP address, and the DNS address.

9.6 BRIDGING AND ROUTING SYSTEMS DESIGN

A great sage once wrote, "bridge when you can, but route when you must." The reason this is true is that bridges offer true plug and play operation, while routed networks require at least some configuration of hosts and routers before they will work. Thus, installing a bridged network is simpler, but only scales to a limited network size. Some devices blur the distinction between routing and bridging by implementing more sophisticated, proprietary bridging protocols which perform some network layer functions.

While routing is more complex, it also provides more features and advantages over bridging, but at a price. Routers *dynamically* reroute traffic over, for example, the least-cost path. Routers reduce the danger of broadcast storms by terminating broadcast sources, such as Netbeui or Banyan Vines. Routers allow a network designer to build a hierarchical addressing scheme that scales to very large networks, which the construction of the global Internet proves. Routers also provide filtering capabilities similar to those in bridges to restrict access to known users, and can also be programmed through filters to block out specific higher layer protocols in a process commonly called a "firewall." Routers have the additional flexibility to define virtual networks within a larger network definition. Routers using IP solve packet-size incompatibility problems by fragmenting larger packets into smaller ones and reassembling them. However, this solution should be used with care since it significantly impacts performance due to the additional software processing required.

However, routers do have a few disadvantages. Routing algorithms typically require more system memory resources than bridges, and addressing schemes that require specialized skills to design and manage. Also, true routers cost somewhat more than simple bridges and hubs.. Modern routing algorithms and implementations (i.e., IS-IS, OSPF, and BGP) are comparable to bridging (such as STP) in the amount of bandwidth overhead required for topology updates. Many router vendors have implemented multiple processors within the network interface card and faster platforms and processors (such as RISC machines) to eliminate throughput problems caused by increased traffic loads of routing protocols. Table 9.3 shows a comparison of bridge and router uses and capabilities.

It is a good idea to *bridge* when you desire simplicity, have the same LAN medium type across the entire network, have a small centralized LAN with a simple topology, or need to transport protocols that cannot be routed, such as NetBIOS and DEC LAT. Select *routing* when you want to route traffic based upon network parameters like least-cost route; have multiple MAC protocol environments; have large, dynamic networks with complex topologies; want optimized dynamic routing around failed links over paths that run in parallel; or have network and subnetworking requirements.

Although many workstations, PCs, and servers have built-in bridging and routing functions, beware of hidden implications. While this packaging seems to offer the cost and management advantages of using only a single device, such products often suffer from limited support, scalability, upgradability, and manageability. Choosing the right device that will grow with your network pays back in benefits, such as lower upgrade costs with minimal operational impact. One option is to purchase a full router rather than a bridge — you may not need routing today, but as your network grows you may be able to upgrade without having to replace the LAN device. Avoiding the operational impact of downtime and addressing changes may be well worth the additional cost of a later upgrade. Beware though that the router may be one generation of technology behind before you finally exploit its routing capability. It may be less expensive in the long run to purchase a router with port expansion, rather than taking the network down and installing a larger or more feature-rich router later. Port sizing and traffic growth patterns typically dictate the size of the router required. Paying careful attention to network design can help you make the right hardware and software design decisions. For further information on the tradeoffs between bridging, routing, and overall network design, see References 2, 5, and 18.

Table 9.3   Comparison of Bridging with Routing

Function	Bridging	Routing
Network addressing	No	Yes
Packet handling	Interpret packet at MAC layer only	Interpret packet at link and network layers
Packet forwarding efficiency	Poor for spanning tree, good for source routing	Good for least cost (OSPF), moderate for distance vector (RIP)
Configuration required	None, except for source routing	Some always required, can be quite complex
Priority schemes	No	Yes
Security	Based on hardware isolation of LAN segments	Based on processor intensive filtering of each packet

9.7 REVIEW

The chapter began with an overview of terminology commonly used in computer communication networks. Next, we surveyed important terminology and concepts at the data link layer critical to LANs, such as Logical Link Control (LLC) and Medium Access Control (MAC). The text then covered the most commonly used LAN media: Ethernet, Token Ring and the Fiber Distributed Data Interface (FDDI). The coverage continued on to the first technique used to connect LANs — bridging. Finally, we covered the pivotal concept of routing, including examples of shortest path routing, subnetworking, and address resolution. Chapter 15 references this material when covering ATM's routing protocol: the Private Network to Network Interface (PNNI). Armed with this knowledge, the reader now has the background to understand the descriptions in Part 4 of how ATM supports LAN Emulation (LANE), IP over ATM, and MultiProtocol routing Over ATM (MPOA).

9.8 REFERENCES

[1]   R. Cole, D. Shur, C. Villamizar, *IP over ATM: A Framework Document*, RFC 1932, IETF, April 1996.

[2]   R. Perlman, *Interconnections*, Addison-Wesley, 1992.

[3]   U. Black, *TCP/IP and Related Protocols*, McGraw-Hill, 1992

[4]   D. L. Comer, *Interworking with TCP/IP - Volume I Principles, Protocols and Architecture*, Prentice Hall, 1991

[5]   A. Tannenbaum, *Computer Communications, Third Edition"* Prentice-Hall, 1996.

[6]   ISO/IEC 8802-2, ANSI/IEEE Std 802.2, *Information processing systems — local and metropolitan area networks — Part 2: Logical link control, Second edition*, ISO/IEC, 1994.

[7]   C. Spurgeon, "Ethernet Quick Reference Guides," http://152.15.16.15/ELET3281/ethernet-home.html, 1998.

[8]   ISO/IEC 8802-3, ANSI/IEEE Std 802.3, *Information processing systems — local and metropolitan area networks — Part 3: Carrier sense multiple access with collision detection (CSMA/CD) access method and physical layer specifications, Fifth edition*, ISO/IEC, 1996.

[9]   S. Saunders, *The McGraw-Hill High-Speed LANs Handbook*, McGraw-Hill, 1996.

[10]   IEEE Std 802.12-1995, *Demand Priority Access Method, Physical Layer and Repeater Specification for 100 Mb/s Operation*, IEEE, 1995.

[11]   ANSI, *American National Standard for information systems — fiber distributed data interface (FDDI) — token ring media access control (MAC)*, X3.139-1987, 1987.

[12]   FDDI Consortium, "FDDI Tutorial," http://www.iol.unh.edu/training/fddi/htmls/index.html, 1998.

[13]   E. Taylor, *Internetworking Handbook*, McGraw-Hill, 1998.

[14]   ANSI, *American National Standard for information systems — fiber distributed data interface (FDDI) — Token Ring Media Access Control-2 (MAC-2)*, X3.239 - 1994.

[15]   ANSI, *American National Standard for Information Systems — Fiber Distributed Data Interface (FDDI) — Hybrid Ring Control (HRC)*, X3.186-1992.

[16]   IEEE, *IEEE Standards for Local and Metropolitan Area Networks: Media Access Control (MAC) Bridges*, Std 802.1d-1990, March 1991.

[17]   J. Moy, *OSPF Version 2*, RFC 1247, July 1991.

[18]   D. Spohn, *"Data Network Design, 2nd Edition,"* McGraw-Hill, 1997.

[19]   D. Plummer, *Ethernet Address Resolution Protocol: Or converting network protocol addresses to 48.bit Ethernet address for transmission on Ethernet hardware*, RFC826, November 1982.

B-ISDN/ATM
Standards, Basics,
Protocol, and Structure

This part introduces the standards bodies, basic concepts, foundational protocols and their structure for Broadband-ISDN and Asynchronous Transfer Mode (ATM). Chapter 10 introduces the creators of ATM — the organizations that define the B-ISDN and ATM standards and specifications. We describe the major standards bodies and summarize key accomplishments. Appendix B provides a list of all standards documents developed to date along with a brief abstract of each one. The text frequently cross references chapters which cover particular aspects of these standards. Finally, Chapter 10 introduces the B-ISDN protocol model used throughout the remainder of the book. This model employs user, control, and management planes in addition to the concepts of protocol layering already discussed in Part 2 that structures the standards and interoperation of various aspects of ATM and B-ISDN. Next, Chapter 11 provides a high-level introduction to ATM. Descriptive analogies and several examples illustrate the concept of addressing and switching fixed length cells. A multimedia example illustrates the operation of differentiated QoS in support of different applications. Chapter 12 then covers the lowest two layers of the protocol reference model: the PHYsical layer, and the ATM layer that introduces the cell structure. Moving up the protocol stack, Chapter 13 next covers the ATM Adaptation Layer (AAL) that provides support for higher layer services, such as circuit emulation, voice, video, and data packets. Chapter 14 then surveys the vast landscape of higher layer protocols running on top of the various AAL protocols; articulating the separation between the signaling protocols of the control plane. This chapter also introduces the management plane, but leaves the details for Part 7. Chapter 15 concludes Part 3 by covering the specifics of the control plane, specifically, signaling standards and routing protocols used in ATM networks. Later on, Part 7 covers the management plane in detail.

10

ATM and B-ISDN Standards and Specifications

This chapter covers ATM and B-ISDN standards and specifications. Beginning with a summary of the major groups of standards and specification bodies currently active in the areas of B-ISDN and ATM, the chapter then moves to a discussion of the players involved and the process followed in the production of standards and specifications. The exponential rate of technology development and technological advance presents an ever increasing challenge to the players, testing the ability of the standards and specifications process to respond. We then compare and contrast the basic standards model of B-ISDN with the Open Systems Interconnect (OSI) and N-ISDN models. The major standards and specifications produced to date are listed and briefly summarized in appendices for each major standards body identified in this chapter. Finally, the text summarizes the current direction of standards and specifications activities, identifying some open issues and topics that readers should closely follow in the future.

10.1 STANDARDS AND SPECIFICATION BODIES

The two classes of standardization and specification bodies actively involved in B-ISDN and ATM are: formal standards bodies, and industry forums. The formal international standards body is the International Telecommunications Union-Telecommunications (ITU-T) standardization sector, formerly called the International Telegraph and Telephone Consultative Committee (CCITT). The official B-ISDN/ATM standards organization in the United States is the American National Standards Institute (ANSI). The formal B-ISDN/ATM standards organization in Europe is the European Telecommunications Standards Institute (ETSI).

There are three major industry forums currently active in the B-ISDN/ATM specification area: the ATM Forum, the Internet Engineering Task Force (IETF), and the Frame Relay Forum. The SMDS Interest Group (SIG) produced an updated document on ATM access to SMDS service, but is currently in a quiescent state.

The ITU-T defined the concept of Broadband-ISDN (B-ISDN) in the late 1980s, choosing ATM technology as the basis for future standards. Recall from Chapter 6 that this same organization also created the concept of the Integrated Services Digital Network (ISDN) in the early 1970s, which is now called Narrowband-ISDN (N-ISDN). Originally, experts anticipated that the standards for B-ISDN would take many years to develop. However, many of the formal standards bodies adopted new procedures in an effort to accelerate the pace of standardization required by the demand from industry and users. In fact, formal standards bodies such as the ITU-T and ANSI sometimes take inputs from the industry forums such as the ATM Forum, communicating via liaison letters and representatives that attend meetings of these different groups. For example, since the publication of the first edition of this book, the ITU-T and the ATM Forum have instituted a formal liaison process.

A new style of "jump starting" the standards process in the 1990s is through the formation of industry forums. A recent example of this phenomenon is the ADSL Forum [1] started in late 1994 specifying a leading edge technology involving ATM. Chapter 32 discusses the role of ATM in this important new networking technology. While they are not formally sanctioned standards committees, these industry forums are independent groups formed by vendors, users, and industry experts who have a vested interest in ensuring interoperability between different vendor products and carrier networks. This process works because those players without large market share want to compete with the industry leader(s); while the dominant market players must participate or else risk becoming obsolete. History shows that this process allows the industry to document further implementation details without reverting to the slower standards process. These specifications, also called implementation agreements, select an interoperable subset of requirements from standards documents, clarify ambiguities, or in some cases, specify certain aspects in advance of the formal standard. Sometimes these forums provide valuable contributions to the formal standards organizations (often as already implemented, and hence proven and not theoretical approaches), thus speeding along the acceptance of an interface, protocol, or other aspect of B-ISDN and ATM technology. Usually, more than just writing and publishing a standard is necessary for success. The multiple vendor and provider agreements developed in these industry forums are often essential to a standard's eventual success in the marketplace. Let's take a look at these standards bodies and industry forums in more detail. This knowledge often plays an important role when selecting equipment or a service provider.

10.1.1 International Telecommunications Union (ITU)

The United Nations founded the International Telecommunications Union (ITU) in 1948 to produce telegraphy and telephone technical, operating, and tariff issue recommendations. Formerly known as the Consultative Committee International Telegraph and Telephone (CCITT), the Telecommunications standardization sector, referred to as the ITU-T, produce recommendations on B-ISDN in which ATM is but one component in an overall set of services.

Up until 1988, the ITU-T published approved recommendations once every four years in the form of a set of books. These books were often referred to by the color of their covers — red, yellow, blue, etc. The ITU-T published the first B-ISDN/ATM standards in the 1988 blue books. After 1988, the ITU-T adopted an accelerated standards process, and published subsequent recommendations when completed, rather than waiting until the next four year anniversary.

During a study period, which is now typically two years instead of four, the ITU-T assigns questions to a study group. These groups organize into lower level committees and produce working documents and draft recommendations. Study group 1, for example, covers B-ISDN services aspects; study group 11, on the other hand, works on signaling protocols, while study group 13 defines ATM-related functions. For details to obtain further information about the ITU-T contact:

International Telecommunication Union (ITU)
Place des Nations
CH-1211 Geneva 20
Switzerland
Voice: +41 22 730 5111
Fax: +41 22 733 7256
Email:itumail@itu.int
X.400S=itumail; A=400net; P=itu; C=ch

Another way to get ITU-T documents is over the web at http://www.itu.ch. The ITU-T began offering this service in June 1995 allowing registered subscribers to download the standards. Many individual standards can also be ordered from this web page. The service supports both Microsoft Word and Postcript formats. A one year subscription is 3,200 Swiss Francs (approximately $2,200 US) and allows a user to read or download all of the standards. You may also purchase individual documents over the web.

The ITU-T has been busy advancing the state of internationally approved standards, particularly in the complex yet pivotal area of signaling (which they spell as "signalling"). Furthermore, the record shows that the ATM Forum and ITU-T have been working together more closely than ever — again, particularly in the area of signaling, but also in other areas such as traffic management. A formal liaison arrangement between the ATM Forum and the ITU-T initiated in 1995 facilitated much of this close cooperation.

The ITU-T defines ATM as a component of an overarching vision called Broadband ISDN (B-ISDN). The ITU-T organizes recommendations covering various aspects of B-ISDN in a set of alphabetically identified series. The following series are relevant to ATM (B-ISDN) as summarized in Appendix B:

- Series E — Telephone network and N-ISDN
- Series F — Non-telephone telecommunication services
- Series G — Transmission systems and media
- Series H — Transmission of non-telephone signals
- Series I — Integrated services digital network
- Series M — Maintenance: international transmission systems, telephone circuits, telegraphy, facsimile and leased circuits
- Series Q — Switching and signalling

Starting in 1988, the CCITT published Recommendations I.113 and I.121 that defined the vocabulary, terms, principles, and basic objectives for broadband aspects of ISDN, called B-ISDN. The CCITT revised and approved these recommendations in November 1990, publishing them in 1991. Also in 1991, the ITU-T issued eleven more recommendations: I.150, I.211, I.311, I.321, I.327, I.361, I.362, I.363, I.413, I.432, and I.610, further detailing the functions, service aspects, protocol layer functions, Operations, and Maintenance (OAM), and user-to-network and network-to-network interfaces. In 1992, the ITU-T approved the following additional recommendations: I.364, I.371, and I.580. In 1993, the ITU-T revised I.113 and I.321, issued a new Recommendations I.356, added a new section of I.363 for AAL5, and also approved I.365, and I.555. The ITU-T has approved a large number of ATM-related recommendations to date as summarized in Appendix B (which presents a tabular summary of each approved standard, giving the title and date followed by a brief description for many of the standards).

10.1.2 American National Standards Institute (ANSI)

The American National Standards Institute (ANSI) structures its work into several committees, which cover a broad spectrum of areas. The ANSI committee T1 is primarily involved in the standardization of B-ISDN and ATM for the United States. ANSI develops these standards in close coordination with the ITU-T, addressing characteristics of technology unique to North America. Specific ANSI T1 subcommittees cover particular aspects of standardization, such as T1E1 covering the physical interface aspects of ATM, T1M1 covering the maintenance aspects of ATM, T1A1 covering performance aspects, and T1S1 covering the network, service, signaling, ATM layer, interfaces and the AAL aspects of ATM. ANSI committee T1 recommends positions on technical issues that the United States Department of State presents at ITU-T meetings. For details to obtain further information about ANSI standards contact:

American National Standards Institute (ANSI)
11 West 42nd Street

New York, New York 10036
Telephone: +01 212 642 4900
Fax: +01 212 398 0023
http://web.ansi.org/

ANSI has approved many ATM-related standards to date as summarized in Appendix B (which presents a tabular summary of each approved standard, giving the title and date followed by a brief description of most standards).

10.1.3 European Telecommunications Standards Institute (ETSI)

The European Telecommunications Standards Institute (ETSI) addresses the standardization of European telecommunications. While the ITU-T develops recommendations for worldwide use, regional bodies, such as ETSI in Europe and ANSI in America, generate more detailed specifications adapted to the unique historical, technical, and regulatory situation of each region. For details to obtain further information about ETSI standards contact:

European Telecommunications Standards Institute
Route des Lucioles
F-06921 Sophia Antipolis Cedex
France
Tel: +33 (0)4 92 94 42 00
Fax: +33 (0)4 93 65 47 16
X.400: c=fr, a=atlas, p=etsi, s=SECRETARIAT
email:webmaster@etsi.fr
http://www.etsi.org/

ETSI has approved a number of ATM-related standards to date as summarized in Appendix B (which presents a tabular listing of the approved standards giving the title and date).

10.1.4 ATM Forum

Four companies — Northern Telecom (now called Nortel), SPRINT, SUN Microsystems, and Digital Equipment Corporation (DEC) — founded the ATM Forum in October 1991. The ATM Forum opened membership to the public in January 1992, for a price. There are currently three categories of membership: principal, auditing, and user. Only principal members can participate in technical and marketing committee meetings. Auditing members receive copies of the technical and marketing committee documents, but cannot participate in the committee meetings. Only user members may participate in Enterprise Network Roundtable meetings. The technical committee produces specifications and is organized into a number of technical "subject matter expert" subcommittees, which define the document naming and numbering convention. As of December 1997, the technical subcommittees are:

- Control Signaling (CS)

- LAN Emulation (LANE) and MultiProtocol Over ATM (MPOA)
- Network Management (NM)
- PHYsical layer (PHY)
- Residential Broadband (RBB)
- Routing and Addressing (RA)
   - Private Network-Network Interface (PNNI)
   - Broadband InterCarrier Interface (B-ICI)
- Service Aspects and Applications (SAA)
- Testing (TEST)
- Traffic Management (TM)
- Voice Telephony Over ATM (VTOA)
- Wireless ATM (WATM)

These subcommittees meet six times per year with attendees paying a per meeting fee in addition to a $10,000 annual fee per member company. They also conduct their business over email and a restricted members-only web page. During the first five years of operation, the ATM Forum produced over 60 specifications; garnering a reputation as a fast paced group covering a broad scope of topics. ATM Forum membership swelled during this period to the point as of December, 1997, that the ATM Forum had approximately 220 principal members, 460 auditing members, and 190 end-user members — up from less than 20 principal members at the end of 1992.

The adage that a large organization cannot move as quickly as a smaller one proved true once again with the ATM Forum's announcement of the "The Anchorage Accord" [2] in the fall of 1996. To clear up some of the confusion in implementers' minds about which standards they need to adhere to when building ATM products, "The Anchorage Accord" defined two groups of specifications: foundation and expanded features. Foundation specifications include traffic management, signaling, Private Network-to-Network Interface (PNNI), network management, and the existing User-Network Interface (UNI) 3.1 specification. The expanded feature category includes: LAN Emulation (LANE), MultiProtocol Over ATM (MPOA), Audio-visual Multimedia Service (AMS), Application Programming Interface (API) and Voice Telephony Over ATM (VTOA). Parts 2 and 3 cover each of these specifications in detail. The McGraw-Hill CD of the "The Anchorage Accord" specifications [3] is a handy reference for the practicing ATM professional.

The new accord also defined mechanisms for determining when an existing standard required updating. The Forum agreed to freeze foundation specifications until one of these mechanisms. triggered new work. An example of a trigger mechanism is an activity in another standards body that affects interoperability with an ATM Forum specification. Since ATM Forum specifications often come out before standards from the International Telecommunications Union (ITU) and other standards bodies, subsequent agreements and compromises in the final standards documents often require updates to ATM Forum specifications.

Another trigger mechanism is the need to maintain internal consistency. For example, new traffic management specifications, such as Available Bit

Rate (ABR) and signaling capabilities in the 4.0 series of specifications, mandate updates in other documents, such as the Broadband Inter-Carrier Interface (B- ICI) version 2.1, LANE, MPOA, network management, and API specifications.

The Market Awareness and Education (MA&E) committee produces tutorials, presentations, press releases, newsletters, and other informative material. They publish an informative newsletter, called "53 Bytes," on the ATM Forum's home page. Currently, this committee has branches in the North American, Europen, and Asia-Pacific regions. The user group formed in August 1993 had the goal of collecting higher level requirements and providing these to the technical and MA&E committees. See the ATM Forum web page for more details on these committees. The address, phone number, email address and web page to obtain information about the ATM Forum are:

The ATM Forum
2570 West El Camino Real, Suite 304
Mountain View, CA 94040-1313
+1.650.949.6700 or +1.415.949.6700 Phone
+1.650.949.6705 or +1.415.949.6705 Fax
info@atmforum.com
http://www.atmforum.com

The ATM Forum has approved many ATM-related specifications to date. Appendix B presents a tabular summary of each approved specification, giving the title and date followed by a brief description of each specification. All recent ATM Forum specifications are available on-line at the ATM Forum web site in Adobe Acrobat format.

10.1.5 Internet Engineering Task Force (IETF)

Commensurate with the mandate for global interoperability of the Internet is the need for standards, and lots of them. The United States Defense Advance Research Projects Agency (DARPA) formed the Internet Activities Board (IAB) in 1983. By 1989 the Internet had grown so large that the IAB reorganized, delegating the work of developing specifications for interoperability to an Internet Engineering Task Force (IETF), split into eight areas, each with an area director. The initial objective of the IAB/IETF was to define the necessary specifications required for interoperable implementations using the Internet Protocol (IP) suite. Specifications are drafted in documents called Request For Comments (RFC). These RFCs pass through a draft stage and a proposed stage prior to becoming an approved standard. Another possible outcome of an RFC is archival as an experimental RFC. The IETF also publishes informational RFCs. As a housekeeping matter, the IETF archives out-of-date RFCs as historical standards. The archiving of all approved (as well as historical or experimental RFCs) serves as a storehouse of protocol and networking knowledge available to the world — accessible, of course, over the web. By the spring of 1998, the IETF had produced over 2,300 RFCs since 1969.

There are a number of good introductory books on the Internet, for example: References 4 to 8. The best way to get more information on the IETF and the Internet Protocol (IP) is from the IETF home page at http://www.ietf.org/home.html. You can also contact them via snail mail or telephone at the secretariat's office as follows:

IETF Secretariat
c/o Corporation for National Research Initiatives
1895 Preston White Drive, Suite 100
Reston, VA 20191-5434, USA
+1 703 620 8990 (voice)
+1 703 758-5913 (fax)

The IETF has issued many ATM-related RFCs to date as summarized in Appendix B (which presents a tabular summary of each approved RFC, giving the title and date followed by a a brief description of each RFC).

10.1.6 Frame Relay Forum

Many aspects of frame relay are similar to those of ATM. They are both connection-oriented protocols, involve N-ISDN-based signaling, and require similar network management functions. Indeed, the ATM Forum and the Frame Relay Forum corroborated closely in the production of Frame Relay/ATM interworking specifications. Chapter 18 summarizes the status of the interworking of frame relay and ATM networks.

10.1.7 SMDS Interest Group

The SMDS Interest Group (SIG) is also working closely with the ATM Forum to specify access to SMDS service over an ATM UNI interface. The technical committee of the SIG was dormant at the time of publication. Chapter 18 summarizes the status of the interworking of SMDS and ATM networks.

10.2 CREATING STANDARDS — THE PLAYERS

Perhaps the single most important driving factor to achieve successful standards and industry specifications is responsiveness to real user needs. "Real" means what users actually need (and will pay for), not what engineers think is technically elegant. Standards created for things with no real user need rarely succeed. Some of the most important questions a user can present to a vendor are "Does it conform to industry standards, which ones, and how?" Standards play a critical role in an age where national and international interoperability is a requirement for successful communications networking. People from many nationalities, diverse cultures and differing value systems must find a consensus (or a meeting of the minds) in interna-tional standards bodies. That these groups of earnest people are able to produce such highly successful, interoperable results testifies to the power of human cooperation.

10.2.1 Vendors

Standards present a dilemma to vendors: on the one hand they must consider the standards, while on the other hand they must develop something proprietary (often leading-edge) to differentiate their products from the rest of the competitive pack. Being the market leader and conforming to standards are often contradictory objectives to achieve simultaneously. The proprietary feature may increase the cost but add value, or remove some non-critical portion of the standard to achieve lower cost. Usually, vendors are very active in the standards process. In the emerging era of ATM, successful vendors frequently lead standards development. Vendors who remain completely proprietary, or try to dictate the standards only to complement their own offerings, are sometimes confronted by users unwilling to risk future business plans on vendor proprietary systems.

Vendors can also drive standards, either by de facto industry standardization, through formal standards bodies, or industry forums. De facto standardization occurs when a vendor is either an entrepreneur or the dominant supplier in the industry and wants to associate the new technology with their name, such as IBM with SNA or Microsoft with Windows. De facto standards in high-technology areas, however, do not last forever. Sometimes the dominant vendor is not the only one in the market with a product, but the market share, quality, or technology of their product makes them the de facto standard around which other vendors must design. B-ISDN and ATM are still too new for de facto standards to have had a significant effect.

10.2.2 Users

Generally, users benefit when they purchase equipment conforming to industry standards rather than proprietary solutions. They can competitively shop for products and services with the assurance of some level of interoperability. A certain comfort level exists in knowing that the equipment a company stakes its business communications on has the ability to interface with equipment from other vendors. Standards, remember, are of paramount importance in the context of international interconnectivity. Also, users play a key role in helping develop standards since the use of standard equipment (as well as vendor acceptance) may well determine the success or failure of the standard. Ubiquitous deployment is often required for success. Vendors say: "We will provide it when customers sign up." Customers say: "We will sign up when it is universally available at the right price, unless we see something else better and less expensive." For example, consider the plight of N-ISDN in North America, which did not become widely available until the web created an urgent demand for high-performance data communication. N-ISDN is more widely available and has less competition in other countries, especially in Europe, and has seen greater success. Users usually do not play a very active part in the standardization and specification process. Instead they signal their approval with their purchases — in other words, they vote with their money.

10.2.3 Network Service Providers

Network service providers (e.g., carriers, LECs, and ISPs) also actively participate in the standard-making process. In a very real sense they are also key users of vendor equipment. Service providers often select vendors that adhere to industry standards but that still provide some (usually nonstandard) capability for differentiation. This approach does not lock a service provider into one vendor's proprietary implementation, and it allows for a multiple vendor environment. Providers must not only make multiple vendor implementations interoperate within their networks, but they must also interface with other networks. Additionally, they must also ensure the availability of industry standard interfaces to provide value-added services to users. Many service providers still utilize single vendor networks since full implementation of standards is not complete. Furthermore, an example where standards are still incomplete is notably in the management and administrative areas, requiring proprietary solutions to meet essential business needs.

10.3 CREATING STANDARDS — THE PROCESS

This section reviews the general standards and specification process illustrated in Figure 10.1. The process begins with the organization defining a plan to work on a certain area. Technical meetings progress the work through a series of written contributions, debates, and drafting sessions. The result is usually a document drafted and updated by the editor in response to contributions and agreements achieved in the meetings. The group reviews the drafts of this document, often progressing through several stages of voting and approval — eventually resulting in a final standard or specification. Business and political agendas often influence the standards process: sometimes the industry accepts a de facto standard, while at other times the standards bodies form a compromise and adopt incompatible standards performing the same function. If your company has a vested interest in the outcome of the standard, they should have input and involvement in its creation. Of course, the final measure of success for any standard is the degree of user acceptance and the volume of interoperable implementations produced by the industry as indicated at the bottom of the figure.

10.3.1 Work Plan

Most standards and specifications groups must first agree on a work plan. An exception is a vendor bringing in a fully documented proposal to a standards body. A work plan defines the topics addressed, a charter for the activity, an organization for performing the work, and usually a high-level set of objectives. User input and involvement most likely occurs at this stage, either indirectly or sometimes through direct participation. This is the time that vendors and service providers often voice their high-level requirements. The work plan for updating an existing standard usually includes some changes resulting from user feedback or interoperability issues that have arisen. Often, the work plan sets an approximate schedule for completion of the standard or specification.

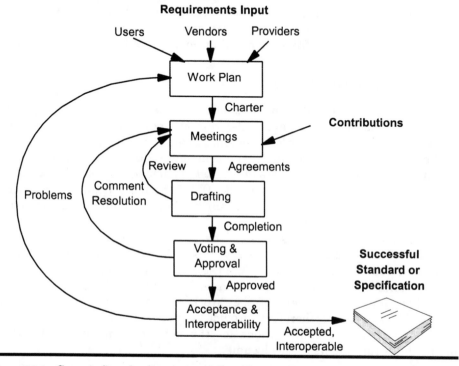

Figure 10.1   Generic Standardization and Specification Process

10.3.2 Meetings and Contributions

The majority of the work occurs at technical meetings, which last from several days to a week or more. Typically, participants submit written contributions in advance of meeting face-to-face. Contributions take on many flavors, ranging from proposing specific text and drawings for the standard, presenting background information, articulating arguments for or against a particular approach, or serving as liaisons with other standards or specification bodies. Usually, smaller subcommittees discuss these detailed contributions, except when a contribution proposes something of interest to the entire standards body. If the contribution proposes adding or changing text to a baseline document, then the subcommittee employs a process to determine whether the proposal is accepted, amended, or rejected. Formal standards bodies usually attempt to achieve consensus before agreeing to include a contribution's input, while other industry forums employ a straw vote method to accept or reject proposals.

The ITU-T, ANSI, ETSI, ATM Forum and the IETF are all large committees, with hundreds to thousands of members attending each meeting. These large committee meetings normally begin and end with a plenary session where representatives from all the subcommittees attend. After the plenary meeting, multiple subcommittee meetings usually occur in parallel. The subcommittees are granted some autonomy. However, they usually review major changes or key decisions in the plenary session. Meetings also are used to resolve issues that arise from the drafting, review, voting, or approval process described below.

10.3.3 Drafting and Review

Key individuals in the development of a standard or specification are the editors. Standards would never exist without the efforts of these dedicated individuals. Many standards documents explicitly give credit to the editor and acknowledge key individuals who contributed to the overall standard or specification. This inspires extra effort and participation. As indicated in the reference section of each chapter, the IETF explicitly acknowledges these individuals as authors of an RFC. Many other documents list the contributors in the introductory sections.

The editor drafts text based upon the contributions, as amended in the meeting. The working group usually trusts the editor to research related standards and specifications and align the document accordingly. A key part of any standards or specification technical activity is the ongoing review, correction, and improvement of the "working" document or baseline text. "Working" documents therefore provide a major input to meetings and become the basis for contributions for the next meeting that will further define the requirements in the document.

10.3.4 Voting and Approval

Once a particular document has reached a "draft" status, the committee usually distributes it for a preliminary vote. Comments that members believe must be addressed in order to approve the document as a standard or specification are often addressed via a comment resolution process at meetings, resulting in more drafting for the editor and the subcommittee. The voting step of the process differs in various bodies in the number of members required to approve a change. If complete concurrence is the objective, then the process can be quite lengthy; if only a majority vote is required, then progress may be more rapid, but can possibly increase the risk. Since human beings work on the standards, the occasional instance of human error is inevitable. After completing the comment resolution process, the standard or specification then goes to a final ballot. Again, depending upon the rules governing the standards or specification body, anything ranging from a simple majority to unanimous approval is necessary for the body to release the document as an approved standard or specification. Often a supervisory board reviews the proposed standard for consistency with the format, style, scope, and quality required by that body in the final approval stage.

10.3.5 User Acceptance and Interoperability

Some users have business problems today that can only be solved by proprietary implementations *prior* to the development of standards. Waiting for an approved standard could put these users out of business. Therefore, the user is caught in the dilemma of adopting an emerging standard now, or else waiting for it to mature. Users primarily determine the success of standards by creating the demand for specific capabilities, and even technology, by purchasing various implementations from vendors and carriers supporting that standard.

The key technical measure of a standard, or specification's success is whether implementations from multiple vendors or carriers interoperate according to the details of the documentation. Remember too that specifications also point out where systems will *not* interwork. The documents should specify a minimum subset of interfaces, function, and protocol to achieve this goal. In support of interoperability, additional documentation, testing, and industry interoperability forums may be required, such as those established for the Internet (IETF), FDDI (ISO), and N-ISDN (CCITT).

If users do not accept a standard, or if significant interoperability issues arise, then this feedback is provided back into the standards process for future consideration. Acceptance by the vendor community also plays a key role in the success or failure of standards — if no implementation of the standard is built, no user can buy it!

10.3.6 Business and Politics

Standards organizations and industry forums have had increased participation and scope in recent years. With this increased number of people working on a plethora of problems there comes the inevitable burden of bureaucracy. Service providers, vendors, and to some extent, users view the chance to participate in the standard-setting process as an opportunity to express and impress their views upon the industry. This is a double-edged sword: while participation is necessary, biases are brought to the committees which can tie up decision making and bog down the process for years in making standards. One example in B-ISDN is Generic Flow Control (GFC), where the attempt to achieve agreement on a shared medium solution for ATM failed for this very reason. The impact of this type of situation depends on whether the committee operates on a complete consensus basis or some form of majority rule. All too often a consensus-based approach ends up being a compromise with multiple, incompatible options stated in the standard. There is then a need to further subset a standard as an interoperability specification to reduce the number of choices, and to translate the ambiguities of the standard into specific equipment requirements.

Standards can also have omissions or "holes" left to vendor interpretation because they simply weren't conceived originally as issues. These "holes" may exist because no agreement could be reached on how the requirement should be standardized, or merely because of oversight. Typically, standards identify known "holes" as items "for further study" (sometimes denoted as ffs) just to point out that there is an awareness of a need for a function or element that isn't yet standardized.

Some vendors play a game of supporting their proprietary solution to make it a standard before their competitor's proprietary solution becomes a standard. This alone can delay and draw out a standards process for many months or even years. While standards organizations take their time to publish standards, some vendors try to take the lead and build equipment designed around a draft standard and then promise compliance with the standard once it is finally published. This is savvy marketing if they guess right because they can be well ahead of the pack. However, if they miss the mark, a significant investment could be lost in retooling equipment and rewriting software to meet the final standard.

10.4 B-ISDN PROTOCOL MODEL AND ARCHITECTURE

This section describes the B-ISDN protocol model and structure. We then present the standards vision of how B-ISDN interconnects with N-ISDN, SS7, and OSI.

10.4.1 B-ISDN Protocol Reference Model

The protocol model for the ITU-T's Broadband Integrated Services Digital Network (B-ISDN) builds upon the foundation of ATM. Figure 10.2 depicts this model from ITU-T Recommendation I.321, which provides a structure for the remaining recommendations. The introduction to subsequent portions of text often use a shaded version of the model in Figure 10.2 to illustrate the topics covered in the following part, chapter or section. The top of the cube labels the planes, which stretch over the front and side of the cube. The user plane and control plane span all the layers, from the higher layers, down through the AALs (which can be null), to the ATM layer and the physical layer. Therefore the physical layer, ATM layer, and AALs are the foundation for B-ISDN. The user and control planes make use of common ATM and physical layer protocols, and even utilize some of the same AAL protocols. However, service specific components of the AALs and the higher layers differ according to function. Therefore, ATM provides a common foundation over a variety of physical media for a range of higher layer protocols serving voice, video and data.

ITU-T Recommendation I.321 further decomposes the management plane into layer management and plane management. As shown in the figure, layer management interfaces with each layer in the control and user planes. Plane management has no layered structure and is currently only an abstract concept with little standardization at this point. It can be viewed as a catchall for items that do not fit into the other portions of this model, such as the role of overall system management.

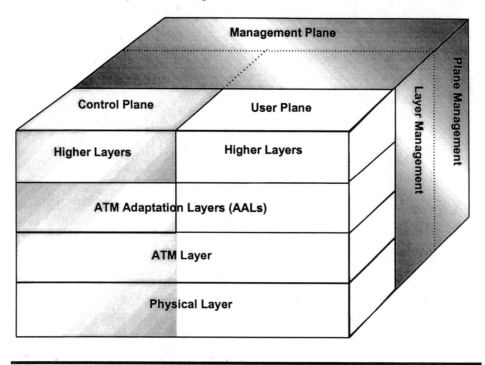

Figure 10.2   B-ISDN Protocol Model

10.4.2 B-ISDN Architecture

Figure 10.3 depicts the vision of how B-ISDN should interconnect with N-ISDN, SS7, and OSI as defined in CCITT Recommendation I.327, which complements the N-ISDN architecture defined in Recommendation I.324. Signaling System 7 (SS7) is the signaling protocol that connects switches within a telephone or N-ISDN network. As shown in the figure, SS7, N-ISDN, and B-ISDN are lower level capabilities that interconnect Terminal Equipment (TE) or service providers through local functional capabilities. SS7 provides out-of-band interexchange signaling capabilities for telephony and N-ISDN, while N-ISDN provides signaling capabilities for TDM-based services, X.25, and frame relay. B-ISDN provides bearer services of various types and signaling. All of these services support higher layer capabilities. Initially, B-ISDN covered interfaces of speeds greater than 34-Mbps, and hence the choice of the adjective broadband. However, the subsequent standardization of 1.5- and 2-Mbps physical interfaces for B-ISDN blurs the distinction with other protocols such as frame relay on speed alone.

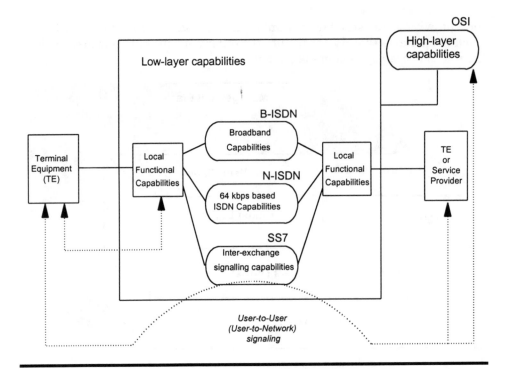

Figure 10.3 Integrated B-ISDN, N-ISDN, SS7, and OSI Architecture

10.5 CURRENT DIRECTIONS AND OPEN ISSUES

This section concludes with a discussion identifying critical success factors, new approaches, standardization and specification work that is in progress, along with the enumeration of some open issues.

10.5.1 Measures of Success and New Approaches

Standards-conscious players should always keep the mission in mind, regardless of the dangers present. Those bodies that keep the user's application foremost in mind — while dealing with the problems and dangers of the standards process — will live long and prosper, or at least live longer than those who concentrate primarily on the development of standards.

User trial communities and university test beds are another method employed by these forums to help speed up the testing and acceptance of new technologies. The Internet protocols are an example of the work by university and academia to successfully lead standards development. Indeed, the notion of "deployment propagation" created by the IETF is that a proposed standard must first demonstrate interoperability before final ratification begins catching on in other standards bodies [9]. Standards development seems to have improved by the adoption of this free market approach, just as the same free market of ideas has stimulated the world.

10.5.2 Future Work Plans

The ITU-T, ANSI, and ETSI continue to refine and expand upon the set of standards introduced earlier. As described, the ITU-T leads the way for ANSI and ETSI to add the geographically specific details for ATM-related standards. Significant activity is occurring in the areas of the control and management planes. Much of the work on traffic management is now complete, but much work still remains in the definition of value-added signaling protocol extensions. Chapter 15 summarizes the extensive new additional features proposed for the signaling protocol. These bodies are also putting the finishing touches on ATM network management as covered in Part 7.

The ATM Forum tracks its work plan on its web page in a section called the "spec watch." Some highlights of the upcoming work by the various technical subcommittees are:

- Control Signaling/Routing Addressing Group
  - PNNI Augmented Routing (PAR)
  - PNNI 2.0 (including B-QSIG PNNI interworking)
  - Interworking ATM Networks
  - ATM Addressing Guidelines
  - Closed User Group Support, Third Party Connection, Security
- Physical Layer and RBB Groups
  - 50 Mbps over Plastic Optical Fiber (POF) Specification
- LAN Emulation/MPOA Group
  - LANE v2.0 LEC MIB
  - LANE v2.0 Server-to-server Interface
  - Multi-Protocol Over ATM v1.0 MIB
- Network Management Group
  - Enterprise/Carrier Management Interface (M4) Requirements & Logical MIB SVC Function NE View V2.0
  - Enterprise/Carrier Network Management (M4) SNMP MIB
  - Carrier Interface (M5) Requirements & CMIP MIB
  - Management System Network Interface Security Requirements & Logical MIB
  - ATM Access Function Specification Requirements & Logical MIB
  - M4 Requirements & Logical MIB Network View v2.0
- Physical Layer Group
  - nxDS0 Interface
  - 2.5 Gbps Interface
  - 1-2.5 Gbps Interface
  - 10 Gbps Interface
  - Plastic Optical Fiber (POF) Jack Connector
  - Utopia Level 3
  RBB (Residential Broadband)Specification
- Service Aspects & Applications
  - API Semantic Doc 2.0 for UNI 4.0

- - Native ATM Services DLPI Addendum v1.0
  - VBR MPEG-2 Over ATM
  - Broadband Multimedia Services
  - H.323 Media Transport over ATM
- Testing
  - Conformance Abstract Test Suite for Signaling (UNI 3.1) for the User Side
  - Performance Testing Specification
  - PICS for Signaling (UNI v3.1) - User Side
  - Conformance Abstract Test Suite for LANE 1.0 Server
  - Conformance Abstract Test Suite for UNI 3.0/3.1 ILMI Registration (User Side & Network Side)
  - ATM Test Access Function (ATAF) Spec
  - UNI Signaling Performance Test Suite
  - Interoperability Test Suite for PNNI v1.0
  - Interoperability Test Suite for LANE v1.0
  - Traffic Management 5.0
- Voice and Telephony over ATM
  - ATM Trunking Using AAL2 for Narrowband
  - Low Speed CES
- Wireless ATM (WATM) Specification 1.0

As you can see from the above list, the ATM Forum plans to produce a significant number of new and revised specifications. We elaborate on these planned work areas of the ATM Forum at the appropriate time later in this book.

The IETF consolidated its activities regarding ATM into the Internetworking Over NBMA (ION) group. See Chapter 9 for the definition of Non-Broadcast Multiple Access Networks (NBMA), such as ATM and frame relay. Work also continues in the IETF Integrated services (Intserv) group on interworking the IPv6 Resource ReSerVation Protocol (RSVP) with ATM signaling and traffic management procedures.

10.5.3 Unaddressed Areas

Currently, a few areas still remain unaddressed in any standard, specification, or documentation effort. Several reasons account for this. A common reason is that product differentiation is allowable in certain areas, and standardization is not really required for interoperability. Details of an implementation are often in this category. In other areas, it is a matter of prioritization of the standards and specification work to be performed. Certain areas are simply not as important or as urgent as others. Usually, if these areas are acknowledged by a standards body, they are briefly defined and then identified as areas designated "for further study" (ffs).

10.5.4 Predicting the Future

One critical aspect often overlooked that influences standards acceptance is the development of services and applications. An example of this is the relatively slow adoption of the N-ISDN Basic Rate Interface (BRI) in the United States, where even after the technology was fully developed, applications and demand were scarce. Were it not for the demand created by leading-edge users requiring high-speed access to the web immediately, history would have recorded BRI as a failure. Now, however, BRI finds its window of opportunity closing as technologies twenty years younger (e.g., ADSL) promise to deliver ten times the bandwidth over the same twisted pair of wires. This example shows the interplay between technologies that makes the telecommunications industry so fascinating. Originally, N-ISDN was basically a telephone company plan to upgrade, digitize, and put the latest technology into the utilities' networks for maximum efficiencies when they were primarily in the voice business — and the end user was an afterthought and not significant in the standardization equation. Little wonder, then, that end users did not perceive any immediate value in the technology upgrades touted by the telephone companies.

Does the same fate that made N-ISDN obsolete now await B-ISDN? We predict that it will not for several reasons. First, many of the same players from N-ISDN have a vested interest in the success of B-ISDN and will probably not repeat the same mistakes. Furthermore, the ATM Forum and IETF are bringing the data communications and telecommunication vendors, providers, and users together in a synergistic new process for developing standards. It promises to serve telecommunications well in the twenty-first century.

10.6 REVIEW

This chapter identified the organizations taking an active role in standardizing and specifying ATM and B-ISDN equipment and services. The CCITT, now called the ITU-T, started this accelerated standards process in 1988. The process and level of detail have come a long way since then, shown by the number of standards and specifications now produced on B-ISDN and ATM. In fact, there are so many standards that we relegated the listings and summary descriptions to Appendix B in this edition. The text then covered the role of the various players in the standards process: user, vendors, and service providers. The chapter then described the standards development process using a flowchart with narrative for each major step along the way. We identified the differences in approach between the various organizations. As an introduction to the remainder of the rest of the book, the text presented an overview of the three-dimensional, layered B-ISDN protocol model. We also presented the ITU-T view of how B-ISDN relates to the traditional N-ISDN, SS7, and OSI protocols introduced in Part 2. This chapter concluded with a discussion covering current directions for future work and identified some open issues still requiring resolution.

10.7 REFERENCES

[1]   ADSL Forum Home Page, http://www.adsl.com/.
[2]   G. Dobrowski, "Accord Anchors ATM Industry," ATM Forum 53 Bytes, Vol 4, No.2, 1996.
[3]   ATM Forum, *The Anchorage Accord - The Official Interoperability Specification*, McGraw-Hill, 1997.
[4]   U. Black, *TCP/IP and Related Protocols*, McGraw-Hill, 1992.
[5]   D. L. Comer, *Interworking with TCP/IP - Volume I Principles, Protocols and Architecture*, Prentice Hall, 1991.
[6]   D. Dern, *The Internet Guide for New Users*, McGraw-Hill, 1993.
[7]   P. Simoneau, *Hands-on TCP/IP*, McGraw-Hill, 1997.
[8]   S. Feit, *TCP/IP*, McGraw-Hill, 1996.
[9]   G. Kowack, "Internet Governance and the Emergence of Global Civil Society," *IEEE Communications*, May 1997.

11

Introduction to ATM

Although ATM is a technology, most users experience ATM through the use of both ATM equipment and services. This chapter introduces the reader to the basic principles and concepts of ATM. First, we look at ATM through many different glasses — viewing it as an interface, a protocol, a technology, integrated access, a scalable infrastructure, and a service. The exposition continues by giving a chef's menu of ATM terminology, taking one course at a time, starting with the user-network interface, continuing with the service categories and traffic parameters, and then finishing with a wide range of options. Next the text explains the operation of ATM and quality of service through a series of simple analogies and examples. The coverage then proceeds with the building blocks of ATM — transmission interfaces, virtual paths, and virtual channels. The chapter concludes with a description and background of how the standards bodies selected a compromise 53-octet cell size for ATM between a smaller cell size optimized for voice and a larger cell size optimized for data.

11.1 ATM'S MANY FACES

ATM plays many roles in modern networks. First, it provides a user-to-network (UNI) interface protocol for the simultaneous transfer of voice, video, and data. Secondly, it acts as a signaling protocol for controlling ATM services. Next, ATM multiplexers and switches utilize ATM as a technology to implement large, fast machines. Also, many service providers view ATM as an economical, integrated network access method, and, in addition, as a scalable core network infrastructure. Finally, ATM acts as a common multi-service platform for public networks. In summary, ATM shows the following faces:

- Interface
- Protocol
- Technology
- Economical, Integrated Access

- Scaleable Infrastructure
- End-to-end Service

Figure 11.1 illustrates these concepts in a typical ATM network configuration. Let's now explore each in more detail.

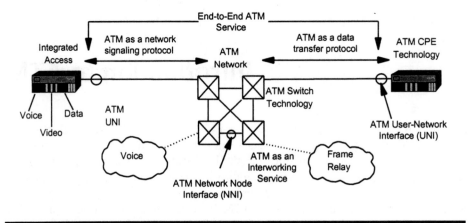

Figure 11.1 ATM's Many Faces

11.1.1 ATM as an Interface

Asynchronous Transfer Mode (ATM) is an interface defined between the user and a network, as well as an interface between networks. Well, what does an interface do? The user-network interface, commonly abbreviated UNI, defines physical characteristics, ATM cell format, signaling, and management processes. Interfaces also provide the boundary between different types of hardware, while protocols provide rules, conventions, and the intelligence to pass voice, video, and data traffic over these interfaces. The network-node interface, commonly called NNI, is essentially a superset of the UNI, addressing aspects unique to interconnection between nodes and connections between networks. ITU-T-defined B-ISDN standards and ATM Forum specifications define a number of variants of UNI and NNI interfaces as detailed in Chapter 12.

Figure 11.2 illustrates these different interface points in a typical configuration of ATM UNI access to a public ATM service. A typical configuration involves a workstation attached to a LAN interface on an ATM access device (e.g., an ATM-capable router) which has an ATM UNI interface. The access device in turn connects to an edge ATM switch in the service provider's network, which in turn connects to an ATM backbone switch. Wide area transmission facilities interconnect the ATM access device with the Edge switch as well as interconnect service provider switches in different geographic locations. A different interface, the ATM Network Node Interface (NNI), interconnects ATM nodes within a public or private network as indicated in the figure.

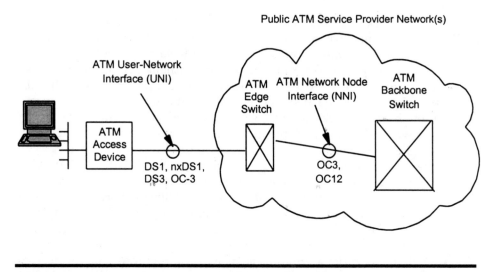

Figure 11.2   ATM as an Interface in a Public Network Context

11.1.2 ATM as a Protocol

ATM is a protocol designed to switch constant, variable, unspecified, and available bit rate traffic over a common transmission medium. The ITU-T's B-ISDN standards address primarily telephone company needs for simultaneously supporting video, voice, and data as an evolutionary sucessor to Narrowband ISDN (N-ISDN). Extensions to this protocol structure by the ATM Forum took the B-ISDN standards as a starting point and addressed end user devices and protocols to enable users to utilize existing networks and devices, such as PBXs, LANs, and network layer data protocols (e.g., IP), in a more cost-effective and efficient manner. Furthermore, ATM interworks with legacy protocols such as frame relay, SMDS, and IP. Additionally, the ATM Forum's LAN Emulation (LANE) protocol enables seamless interworking between ATM-powered devices and legacy LANs employing Token Ring, Ethernet, or FDDI. Additionally, the ATM Forum has specified a sophisticated private ATM network NNI in the Private Network-to-Network Interface (PNNI) specification that supports automatic configuration, optimized routing, bandwidth allocation, hierarchical scalability, and multiple qualities of service. The latest ATM Forum Multi-Protocol Over ATM (MPOA) specification supports any network layer protocol over a high-performance ATM infrastructure.

11.1.3 ATM as a Technology

In a most basic sense ATM is a technology, defined by protocols standardized by the ITU-T, ANSI, ETSI, and the ATM Forum introduced in the previous chapter. ATM is a cell-based switching and multiplexing technology designed to be a general-purpose, connection-oriented transfer mode for a wide range of services. ATM technology comprises hardware and software conforming to ATM protocol standards, which provides multiplexing, cross-connect, and switching function in a network. ATM technology takes the form of a network interface card, router, multiplexer, cross-connect, or intelligent switch in Customer Premises Equipment (CPE). Today, ATM prevails in the switch market and as a WAN interface on traditional data communications products like routers and hubs, although ATM end systems (NICs) and applications (APIs) are beginning to appear in significant quantities. Carriers use ATM switches of different sizes and capabilities on the edge of their networks as well as in the backbone. Since the ATM standard is well defined and stable, it provides an insurance policy against technological obsolesence.

The late 1980s ushered in early prototype Central Office (CO) ATM switches. The traditional customer premises multiplexer and switch vendors then adopted ATM in the early 1990s. The mid-1990s ushered in the next generation of CO switches, some modified from experience gained from trials and Customer Premises Equipment (CPE) designs. CO switches continued to evolve rapidly in the mid-1990s, becoming larger, faster, and more capable than preceding generations. Simultaneously, router and hub manufacturers began building ATM interfaces for their existing models, as well as including ATM capabilities in their latest designs.

Computer vendors soon started building ATM Network Interface Cards (NICs) for workstations and personal computers, and the loaded per port or per workstation cost of ATM to the NIC has fallen to within $100 of that of a 100-Mbps Ethernet connection. The final piece of the puzzle is falling in place with the initiation of operating system and Application Program Interface (API) software development for ATM-based systems, offering the capability for a true end-to-end homogeneous ATM network. These efforts point toward a strong commitment to the success of ATM because they address the users' need for consistent and predictable quality of service on an end-to-end basis.

11.1.4 ATM as Economical, Integrated Access

Virtually all carriers provide some form of public ATM service, enabling users to capitalize on a basic advantage of ATM — integrated physical and service access which reduces cost. The development of ATM nxDS1/E1 inverse multiplexers and CPE access products ranging from DS1 to OC-N speeds has extended many of ATM's benefits to users who currently employ separate TDM-based networks. Now the ATM Forum's Frame-based UNI (FUNI) specification defines operation of ATM at nx64-kbps speeds.

Figure 11.3 illustrates how ATM provides integrated access for voice, data, and video applications to a wide range of network services over a single access line. ATM delivers equipment, bandwidth, and operational savings when supporting all data, voice, and video requirements over a *single* access line. As shown in the figure, the configuration involves an ATM-based, multi-service access device connecting voice, video and data applications over a single access line to a multi-service edge switch in a service provider network. Multiple services economically share the single physical access circuit, allowing tremendous savings in network interface equipment, eliminating the need for multiple local loops, and also reducing wide area network service costs. When used for integrated access, the provider's edge switch performs circuit emulation to split off the TDM traffic destined for the telephone and private line networks as detailed in Chapter 17. Simultaneously, customer devices access the Internet, and a public frame relay or ATM data service, using protocol interworking as defined in Chapters 18 through 20. Furthermore, video applications may access video content networks, or interwork with legacy private line-based video services as indicated in the figure. Of course, other combinations of integrated access connections than those shown in the figure between customer applications or devices with public or private network services are also possible.

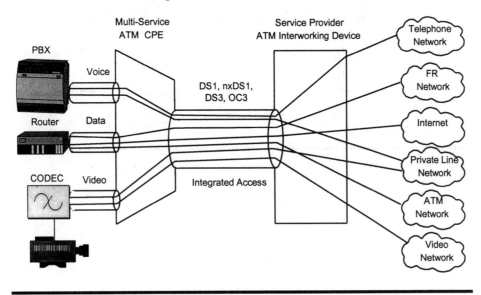

Figure 11.3 Example of the ATM Integrated Access Concept

11.1.5 ATM as a Scalable Infrastructure

ATM technology also has advantages over legacy technologies such as IP, FR, or SMDS in a network infrastructure. ATM hardware and associated software together provide an efficient backbone technology for an advanced communications network. In fact, ATM-based architectures offer the best integrated platform for voice, video, and data. ATM also provides a highly scalable infrastructure, from the campus environment to the central office. Scalability occurs along the dimensions of interface speed, port density, switch size, network size, multiple application support, and addressing.

ATM also provides greater bandwidth granularity and flexibility in designing network topologies. As an illustration of this fact, consider the example network shown in Figure 11.4. In Figure 11.4a, the headquarters (HQ) site router (R) communicates with routers at sites A, B, C, and D over dedicated DS3 private lines. Each private line has an average utilization of 20 percent, but can burst up to the full DS3 line speed of approximately 45-Mbps. With this private line solution, the HQ site requires four DS3

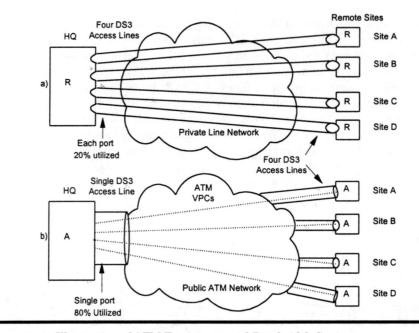

Figure 11.4   Illustration of ATM Equipment and Bandwidth Savings

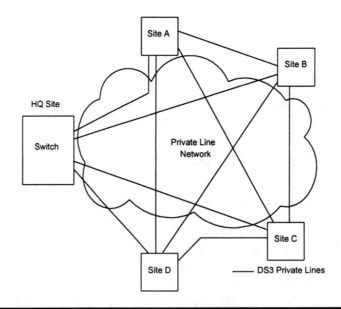

Figure 11.5   Peer-to-Peer Communications Private Line Example

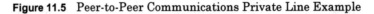

network (router) access ports, four local loops, and four wide area (i.e., interLATA) circuits.

Alternatively, for the network using ATM capable CPE devices (indicated by the character A) shown in Figure 11.4b, the HQ site provisions only a single DS3 physical access port to a public ATM network. A logical Virtual Path Connection (VPC) connects the HQ ATM access device to remote ATM access devices at sites A, B, C, and D. Thus, the HQ site saves the cost of three DS3 ports, three DS3 local loops, and three dedicated DS3 wide area circuits. Usually, these private line costs shown in Figure 11.4a exceed those of a public ATM service. An example later in this section demonstrates these economic considerations further.

Even greater savings accrue if all users communicate peer-to-peer instead of in a hierarchical manner through a central site. If each site communicates directly with the others, each site requires four DS3 connections as illustrated in the private line example in Figure 11.5. This means that each site has four DS3 ports, four access lines and four DS3 private lines.

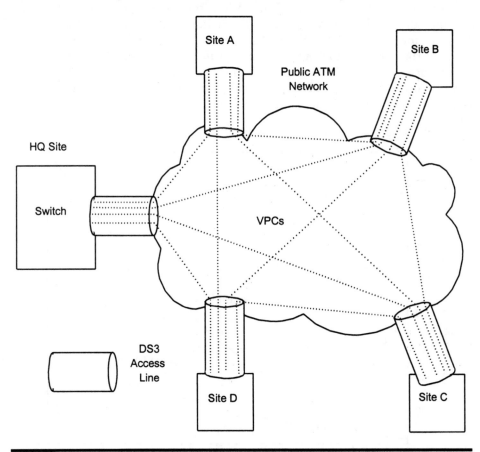

Figure 11.6 Peer-to-Peer Communications for Shared ATM Access Example

On the other hand, the shared network ATM design shown in Figure 11.6 requires only a single DS3 access port at each site for access to the public ATM switch, thus reducing the number of ports and local loops by a count of 15. Each site is then connected to every other site using a logical Virtual Path Connection (VPC) across the public ATM network. This design also has the by-product of reducing processing power and memory requirements in the network access devices by focusing traffic on a single WAN port. Depending upon the service provided by the public ATM network, each VPC could burst up to its full line rate. If all VPCs have this capability, then the service provider can statistically oversubscribe the VPCs when multiplexing together the traffic of many customers on shared facilities since the end user transmission rates are constrained by the access line speeds. For example, if the VPCs connecting the headquarters site to sites A and B are both running at 20-Mbps, then the VPCs to sites C and D must be idle. Public network switches use traffic control to partition bandwidth and buffer resources between customers to insure guaranteed QoS and fair service during periods of congestion. Typically, carrier pricing for logical ATM connections in the Variable Bit Rate (VBR) service category cost less than the equivalent dedicated bandwidth private line service. On the other hand, the price for a public ATM service Constant Bit Rate (CBR) service should be about the same as that of an equivalent private line service. The advantage to users with ATM over private lines in this case is that ATM CBR service has much finer bandwidth granularity than the rigid TDM hierarchy as well as the fact that the same access line supports both CBR and VBR services simultaneously.

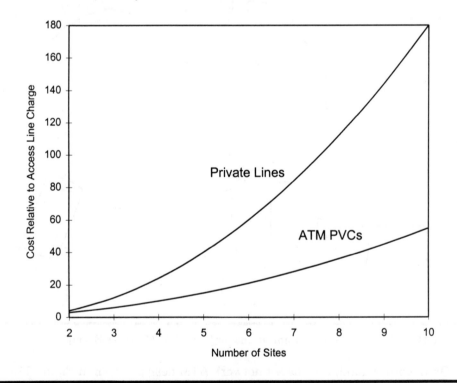

Figure 11.7   ATM versus Private Line Meshing Efficiencies

Figure 11.7 shows the results of an example calculation of the ATM VPC versus private line costs relative to the access line cost (for the above configurations) as a function of the number of sites in the network. In the example, an n node full mesh private line network requires n(n-1) access lines and ((n)(n-1))/2 WAN private line connections. The ratio of the WAN private line charge (W) to the access line charge (A) varies with distance, and was set equal to 2 for the example.   By comparison, the number of ATM access circuits required is just n; however, the ATM design still requires ((n)(n-1))/2 VPCs to retain full mesh connectivity. Typically, the charge for the equivalent ATM VPC for VBR service varies by carrier as a function the Sustainable Cell Rte (SCR). This analysis assumes that the ratio of the VPC charge to the private line charge is 75 percent in this example.   This graph therefore illustrates the advantage of ATM as the number of sites requiring meshed connectivity increases. In addition, as shown in Chapter 25, the statistical multiplexing properties inherent in ATM also improve backbone link utilization, further improving the price/performance ratio of ATM versus private line based networks.

ATM architecture also enables dynamic and flexible adds, moves, and changes to the network topology. Once a site has a port and access line connecting to the public ATM network, adding VPCs and VCCs to new remote sites is much easier than ordering and installing a new dedicated private line circuit. Typically, carriers provision new logical ATM connections within minutes to no more than a few days. On the other hand, new private line connections often take several weeks to provision and may require coordination between multiple service providers.

11.1.6 ATM as a Service

ATM handles both connection-oriented traffic (either directly (cell-based) or through adaptation layers) as well as connectionless traffic through the use of servers and adaptation layers. ATM virtual connections operate according to one of the following service categories:

- Constant Bit Rate (CBR)
- Variable Bit Rate (VBR), in either real-time (rt) or non-real-time (nrt) modes
- Unspecified Bit Rate (UBR)
- Available Bit Rate (ABR)

These service categories support a wide range of applications. ATM also applies to the LAN and private network technologies as specified by the ATM Forum.   ATM handles connection-oriented traffic directly or through adaptation layers.   ATM uses servers to support connectionless traffic through the use of adaptation layers. ATM virtual connections operate according to a traffic contract and a specific service category which guarantees a particular Quality of Service (QoS). Each ATM cell contains addressing information that establishes a virtual connection for each ATM switching or multiplexing node. ATM provides either Permanent or Switched Virtual

Connections (PVCs or SVCs) to set up a sequence of such nodal connections. All cells are then transferred, in sequence, over this virtual connection. ATM, as it name implies, is asynchronous because the transmitted cells need not be periodic as time slots of data are in Synchronous Transfer Mode (STM), as described in Chapter 4.

ATM standardizes on one network architecture for multiplexing and switching and operates over a multiplicity of physical layers. ATM also supports multiple Quality of Service (QoS) classes for applications requiring different delay and loss performance. Thus, the synergistic vision of ATM is that a network constructed — using ATM and ATM Application Layers (AALs) switching and multiplexing principles — supports a wide range of services, including:

★ Voice
★ Packet data (SMDS, IP, FR)
★ Video
★ Imaging
★ Circuit emulation

ATM provides bandwidth-on-demand through the use of SVCs, and also supports LAN-like, best effort access to available bandwidth.

11.2 ATM TERMINOLOGY - A HIGH-LEVEL INTRODUCTION

This section introduces some key concepts and terminology of ATM through simple analogies and examples. Subsequent chapters in this part then build upon this foundation by providing more detail and references for further reading.

11.2.1 The ATM Menu

A user defines an ATM connection by selecting from a set of options, just as when ordering from a menu. Some people say that ATM has so many options that it is like going into a restaurant where you can order anything you want; as long as you can specify its precise molecular structure. Although ATM does indeed have a jargon all its own, one of the main objectives of this book is to explain how to order from the ATM restaurant. Here is the menu and its service options offered by the chefs at the ATM Forum and the ITU-T. The following dialogue takes you through a somewhat whimsical conversation with a waiter in ordering an ATM service. It serves as a means of introducing a number of new terms. If you don't fully grasp the terms in this little story, don't worry. Each of them will be covered in more detail and summarized later.

⇒ UNI access port speed: over 30 entrees to choose from!
⇒ Virtual Path or Virtual Channel Connection (VPC or VCC): Choose a set selection or meal, or order a la carte.

⇒ PVC or SVC: Do you want a guaranteed reservation at Chez Nois or just an on demand meal at McDonald's? With PVCs you'll work with a tuxedoed waiter (a network management system) or else order at the drive-through window (i.e., use signaling messages).

 ⇒ Specify your source and destination addresses: Who are the host and guest of honor?

 ⇒ If you're ordering an SVC, please select from the options on the menu; that is, fill out your sushi order, give it to me and I'll fill it at once.

⇒ Select point-to-point or point-to-multipoint: Will it just be the two of you, or will more people be joining your party?

⇒ Unidirectional or bidirectional: Do you know and want to tell me what you want, or would you like to discuss your options?

⇒ For each direction of service, please select one service class:

 ⇒ Constant Bit Rate (CBR): Dedicated, most expensive service

 ⇒ Variable Bit Rate (VBR), real-time and non-real-time: Bring me my food when it is ready! Or bring the entrees when you see me finish my appetizer.

 ⇒ Available Bit Rate (ABR): Shared waiters, and you get your food and drink delivered to your table when a waiter is available.

 ⇒ Unspecified Bit Rate (UBR): You get whatever service is available, or possibly no service at all. I'll give you the lowest price for this service however.

 ⇒ Would you care for tagging (i.e., cell loss priority)?

  ⇒ For all services, please select your:

   ⇒ Peak Cell Rate (PCR)

  ⇒ For the VBR services, also please choose:

   ⇒ Sustainable Cell Rate (SCR)

   ⇒ Maximum Burst Size (MBS)

  ⇒ For ABR, the choices are:

   ⇒ Minimum Cell Rate (MCR)

   ⇒ Initial Cell Rate (ICR)

   ⇒ Your closed-loop congestion control parameters? If you don't know this, would you like me to suggest a predefined set?

⇒ And finally, would you like to select from a number of other options, such as no pickles, hold the lettuce, etc.?

As you can see from the above list, there are a number of selections and choices that must be made to order up an ATM SVC or PVC. The following sections delve into the physical components of an ATM connection, making some other simple analogies to further explain some of the above terms. Chapter 12 then presents a more formal summary of the relevant physical and ATM layer standards.

11.2.2 A Simple Transportation Analogy

Let's begin with the simple example in Figure 11.8. Imagine each user node as a city surrounded by mountains and the ocean. From the mountain passes leaving the cities, highways lead out to the other cities. In this analogy, the mountain passes are the physical access circuits, the highways are Virtual Paths (VPs), and the highway lanes are virtual channels. Two VPs of different quality connect cities A and C: VP 1 takes the seaside route, while VP 2 takes the overland route through the mountains. VPs 3 and 4 take another overland route and pass through City B. This analogy isn't perfect, since the counterpart of ATM cells would actually be the vehicles traveling on the highway. On real highways, vehicles come in all shapes and sizes, while ATM cells are always the same size — 53 bytes. Hopefully, this simple example gives the reader a mental picture to act as an aid in remembering the new terminology.

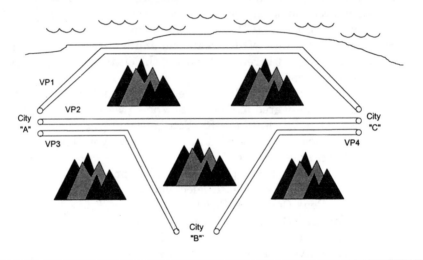

Figure 11.8 Simple Physical Interface, ATM Virtual Path and Channel Analogy

11.2.3 VP and VC Post-Office Analogy

Let's use, for example, a post office that delivers packages of various sizes and weights. A large package weighing several pounds, in which a company includes many smaller letters (each individually addressed), is analogous to a VP. Individually addressed and stamped letters are analogous to VCs. VPs make sense only when the two locations (origination and destination) have a large volume of mail (information) to exchange. VCs make sense when small, ad hoc, intermittent correspondence is required, such as for remote offices which may use overnight delivery for a single letter when the correspondence is urgent.

11.2.4 TDM Switching and Digital Cross Connect Analogy

This capability to switch down to a virtual channel level is similar to the operation of a Private or Public Branch Exchange (PBX) or telephone switch in the world of telephony. A PBX or switch operates on each individual channel within a trunk group (path). Figure 11.9 illustrates this analogy. ATM devices which perform VC connections are commonly called VC switches because of this analogy with telephone switches. Transmission networks use a TDM cross-connect, which is basically a space division switch, or effectively an electronic patch panel. ATM devices which connect VPs are often called VP cross-connects in the industry literature because of their analogy with the transmission network.

These analogies are useful to understand ATM for those familiar with TDM/STM and telephony, but should not be taken literally. There is little reason for an ATM cell switching machine to restrict switching to only VCs and cross-connection to only VPs, although some manufacturers do so. Another use of VP and VC switching is to build larger networks using the hierarchy of virtual paths and channels defined in ATM.

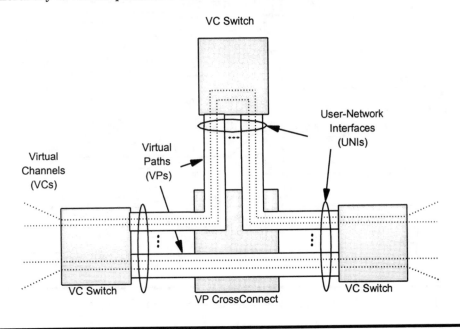

Figure 11.9 TDM Switch and Cross-Connect Analogy

11.2.5 ATM Switching and Label Swapping

ATM is a special case of a multiplexing and switching technique called address switching or label swapping introduced in Chapter 4. Figure 11.10 depicts a simple illustration of the operation of label swapping for several ATM CPE devices connected to an ATM switch [1]. The following example traces the following three point-to-point virtual connections in the figure: D to E, A to C, and B to F.

In the upper left hand corner a continuous video source provides input to a *packetizing* function, which has logical destination VPI/VCI address *D* on the transmit side of a CPE device. The packetizing function takes the bit stream from the video coder-decoder (CODEC) and breaks it up into fixed-length cells made up of a header and a payload field (indicated by the shading). The video source is multiplexed with the packetized output of a continuous DS3 bit stream destined for logical address *A* onto a shared SONET access line. Also, another ATM CPE device high-speed computer inputs packetized data addressed to *B*. These sources are shown *time division multiplexed* over two SONET transmission paths going into a translation function on the input to an ATM switch.

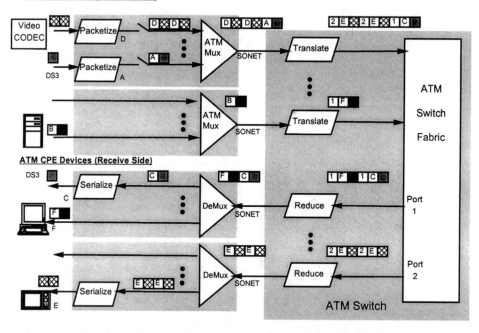

Figure 11.10   Asynchronous Transfer Mode Label Swapping Example

The ATM switch *translates* the logical addresses into physical outgoing switch port addresses and logical VPI/VCI addresses as shown in the upper right hand side of the figure. Thus, the translation function prefixes a modified, longer header to every ATM cell entering the switch fabric as shown in the figure. Most real ATM switches use an internal cell size greater than 53 bytes within their switching matrix (also called a fabric) to accommodate such physical routing information in the expanded cell. Specifically, the switch translates the following addresses and port numbers: DS3 source address *A* into address *C* destined for physical port 1, the video source with address D into address E destined for physical port 2, and the computer source with destination address B to address F destined for physical port 1. The ATM switch fabric then utilizes the physical destination address field to deliver the ATM cells to the appropriate physical switch port and associated transmission link. Chapter 4 gave a simple example of how an ATM switch performs this function using self-routing, while Chapter 16 describes other commonly encountered implementations.

At the output of the ATM switch, the *reduce* function removes the expanded cell header, reducing the cell size back to the standard 53 bytes. The ATM switch port then time division multiplexes the logically addressed ATM cells onto the outgoing transmission circuits for delivery to the receive side of other ATM CPE devices. The ATM CPE devices demultiplex the ATM cells to the appropriate devices based upon the addresses in the cell header as indicated in the figure. The Continuous Bit Rate (CBR) connections (i.e., video and the DS3) then have the logical addresses removed, and are reclocked to the information sink via the *serialize* function. ATM-capable devices, such as the workstation in the example with ATM address F, receive ATM cells directly.

11.3 INTRODUCTION TO ATM CELLS AND CONNECTIONS

This section introduces the ATM cell and defines some basic reference configurations that utilize UNI and NNI interfaces. This section defines the basic concepts of physical connections between ATM devices and the logical capabilities of ATM at the Virtual Path (VP) and Virtual Channel (VC) connection level. Several examples extend these definitions to a network context. Chapter 12 goes into more detail in these areas, defining each bit in the ATM cell header and its detailed meaning with reference to the relevant standards. The intent of this section is to provide a high-level overview of the basic concepts of ATM detailed in the remainder of the book.

11.3.1 ATM User-Network Interface (UNI)

Basically, an ATM UNI is two ports connected by a pair of wires (or optical fibers) as depicted in Figure 11.11. On the left hand side of the figure, a physical port on the CPE ATM device (e.g., a workgroup, LAN backbone, edge switch, ATM multiplexer, ATM access device, or router) interfaces to a full duplex physical circuit. The access circuit, typically at DS1, DS3, or OC-3 or higher transmission speeds in the WAN, or a connection via twisted pair, or optical fiber in the LAN and MAN, connects the end user ATM device to a port on an ATM switch. The ATM switch may be part of a private network, or within a service provider's public ATM network. Standards call this collection of two ports interconnected by physical transmission, along with the format, protocols, and procedures operating over the physical medium, a user-to-network interface (UNI).

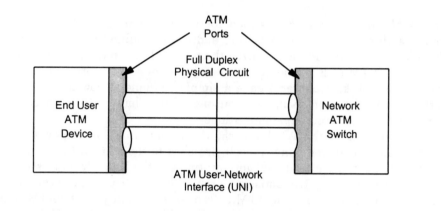

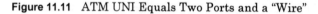

Figure 11.11 ATM UNI Equals Two Ports and a "Wire"

11.3.2 ATM UNI and NNI Defined

Standards give different names to the particular context of a physical interface connecting ports on devices. Figure 11.12 illustrates the commonly used terminology for ATM reference configurations at the UNI and the NNI.

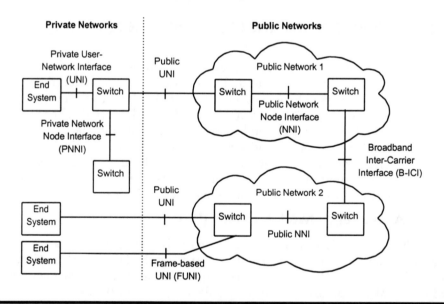

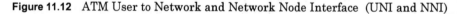

Figure 11.12 ATM User to Network and Network Node Interface (UNI and NNI)

The ATM UNI exists between the user equipment or End System (ES) and switches, also called Intermediate Systems (IS). The figure illustrates the ATM Forum terminology and context for private and public UNIs. The ATM Forum's Frame-based UNI (FUNI) protocol allows legacy devices to interface with ATM networks after only a software upgrade as described in Chapter 17. Standards call the connections between ATM switches either a Network Node Interface, or a Network-Network Interface — employing the same NNI acronym to indicate this particular connection of ports. Chapter 15 details the ATM Forum's Private Network Network Interface (PNNI) specification, as well as the interface between carrier networks, called the Broadband Intercarrier Interface (B-ICI).

11.3.3 Overview of the ATM Cell

The unit of transmission, multiplexing and switching in ATM is the fixed-length *cell*. Standards bodies chose a constant-length ATM cell to simplify the design of electronics in ATM switches and multiplexers, because the hardware manipulation of variable-length packets is more complex than processing a fixed-length cell. Indeed, for this very reason many packet switching devices allocate hardware buffers for each packet equal to the size of the maximum packet to simplify the hardware design. As the price of memory declines and the density of logic increases, the advantage of a short fixed-length cell in simplifying hardware design decreases.

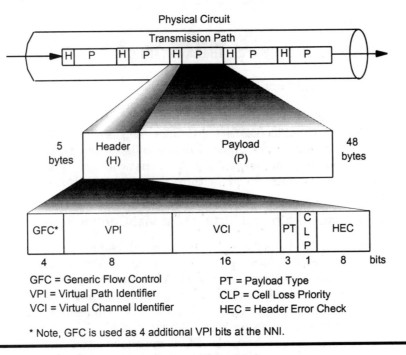

Figure 11.13 ATM Cell Format - Header and Payload

ATM standards define a fixed-size cell with a length of 53 octets (or bytes) made up of a 5-octet header (H) and a 48-octet payload (P), as shown in Figure 11.13. Chapter 12 defines the details of the ATM cell — this section provides a brief introduction to the examples used in the remainder of the chapter. An ATM device transmits the bits from the cells over the physical transmission path in a continuous stream.

ATM networks switch and multiplex all information using these fixed-length cells. The cell header identifies the destination port through the label swapping technique as well as the cell type and priority. As we shall see, the VPI and VCI have significance only on a single interface since each switch translates the VPI/VCI values from input port to output port along the path of an end-to-end connection. Therefore, in general, the VPI/VCI values on the input and output ports of the same switch differ. The sequence of VPI/VCI mappings in the switches along the path makes up the end-to-end connection. The ITU-T and the ATM Forum reserve VCI values 0 through 31 for specific functions detailed in the next chapter. The Generic Flow Control (GFC) field allows a multiplexer to control the rate of an ATM terminal. While B-ISDN standards define GFC, no ATM Forum implementation standards or agreements actually describe how to use or implement it. The format of the ATM cell at the Network-Node Interface (NNI) eliminates the GFC field and instead uses the 4 bits to increase the VPI field to 12 bits as compared to 8 bits at the User-Network Interface (UNI).

The Payload Type (PT) indicates whether the cell contains:

- User data
- VCC-level OAM information
- Explicit Forward Congestion Indication (EFCI)
- AAL information
- Resource management (RM) information

The Cell Loss Priority (CLP) bit indicates the relative priority of the cell similar to Discard Eligible (DE) bit in frame relay service. Lower-priority cells may be discarded before higher-priority cells by the Usage Parameter Control (UPC) at the ingress to the ATM network, if cells violate the predetermined user contract, or within the network during periods of congestion.

11.3.4 ATM Virtual Path and Virtual Channel Connections

Three major concepts in ATM are: the physical transmission circuit, the logical Virtual Path (VP), and, optionally, the logical Virtual Channel (VC). These form the basic building blocks of ATM networking. Figure 11.14 graphically depicts the relationship between the physical circuit and the logical Virtual Path (VP) and Virtual Channel (VC) connections carried. A physical circuit supports one or more virtual paths. A virtual path may support one or more virtual channels. Thus, multiple virtual channels can be trunked over a single virtual path. ATM switching and multiplexing operates at either the virtual path or virtual channel level.

ATM users view VPs (or VCs) as an end-to-end logical connection from the source endpoint (node A) to the exit endpoint (node B) as shown in Figure 11.15a. Standards call this an end-to-end VP (or VC) . Think of a PVC VP (or VC) as a static logical highway from one location (user access port) to another. SVCs are dynamically established by signaling, similar to telephone calls, as detailed in Chapter 15. A VP (or VC) PVC has characteristics similar to those of a dedicated private line: the destination never changes, the network allocates a certain bandwidth at a specified quality of

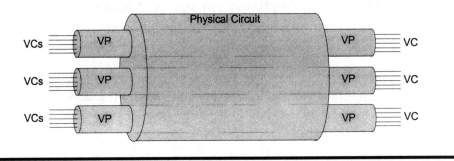

Figure 11.14 Physical Circuit, Virtual Path (VP) and Virtual Channel (VC)

service. Sequences of cells transferred over an ATM VP (or VC) arrive in exactly the same order at the destination.

From the network's point of view the end-to-end VP (or VC) actually involves switching based upon the VPI (and for VCs the VCI value as well) at each connecting point within the network, as shown in Figure 11.15b. This network implements connecting points and the associated switching within the network transparently to the end users. Standards utilize the term *VP* (or *VC*) *link* for the smaller segments between connecting points that make up the overall end-to-end connection, as identified in the figure.

a. At the End Points a VP (or VC) appears as a single-pipe to devices A and B

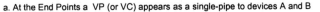

b. To the network, a VP (or VC) appears as a set of links linked together at Connecting Points

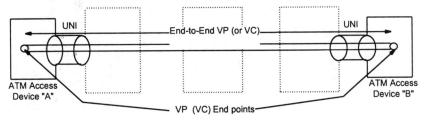

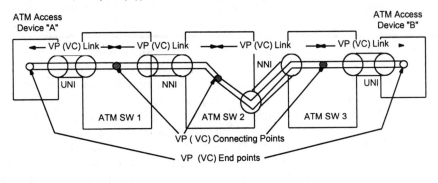

Figure 11.15 Basic VP and VC Terminology

11.3.5 ATM Cell Switching Using VPI and VCI Values

More formally, a VP or a VC Connection (VPC or VCC) operates according to the following definitions:

Virtual Path Connections (VPCs) switch based only upon the value of the Virtual Path Identifier (VPI) in the ATM cell header. The users of the VPC may assign the VCCs within that VPI transparently since they follow the same route.

Virtual Channel Connections (VCCs) switch based upon the combined VPI and Virtual Channel Identifier (VCI) values in the ATM cell header.

Intermediate ATM switches may modify both the VPI and VCI values at connecting points to forward cells along VC links that comprise an end-to-end VC. Note that VPI and VCI values must be unique on any particular physical interface on an ATM device. Thus, ATM switches use VPIs and VCIs independently on each UNI or NNI. In other words, the VPI and VCI have local significance only.

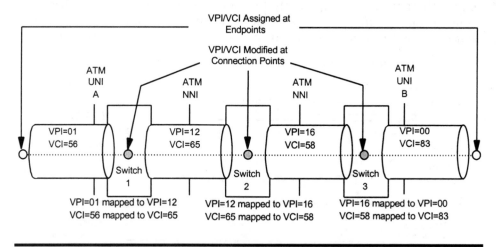

Figure 11.16 VPI/VCI Switching in a Link and End-to-End Context

Figure 11.16 illustrates this concept through a simple example where each switch maps VPI and VCIs to different VPIs and VCIs at each connecting point. In this example, all three switches have a single physical interface between them. At ATM UNI A, the input device to switch 1 provides a video channel over Virtual Path 1 (VPI 1) and Virtual Channel 56 (VCI 56). Switch 1 then maps VCI 56 on the UNI to VCI 65 on the NNI connecting to switch 2, and VPI 1 on the UNI to VPI 12 on the NNI. In our example this mapping is symmetric in both directions, although real switches may do the mapping independently in each direction. In the middle of the figure, switch 2 then maps the VPI and VCI values on the NNI to switch 3 to different virtual path and channel identifiers (i.e., VPI 16 and VCI 58). Thus, the network ties together the VPIs and VCIs used on each individual link to deliver the end-to-end connection to the end points. This is similar to frame relay, where Data Link Connection Identifiers (DLCIs) address a virtual channel at each end of a link. Finally, switch 3 translates VCI 58 on VPI 16 on its NNI from switch 2 to VCI 83 on VPI 0 on ATM UNI B. This example illustrates how the destination VPI and VCI need not be the same at the endpoints of a connection.

11.3.6 ATM Virtual Connection Networking

A user connects to an ATM network through the use of an ATM UNI access port and circuit. As described in Chapter 6, transmission standards define operation of the UNI as the Transmission Path (TP) layer in the SONET and SDH hierarchies. The logical connection between user ports is called a VP and VC, where one or more VCs reside within each VP. Each VP and VC can be unidirectional or bidirectional, point-to-point or point-to-multipoint, and carries a service category and traffic parameters from the service menu assigned to it. At each VP endpoint a VPI acts as a locally significant address. At VC endpoints a combination of a VPI and a VCI acts as a locally significant address.

One concept used consistently throughout the book is that of the network "cloud," which represents a variety of network environments. Cases generically represented by a cloud include, but are not limited to, service provider networks, the local exchange environment, and even multiple intervening networks that interconnect end users. Typically, our book employs the cloud symbol when the network transparently interconnects end users.

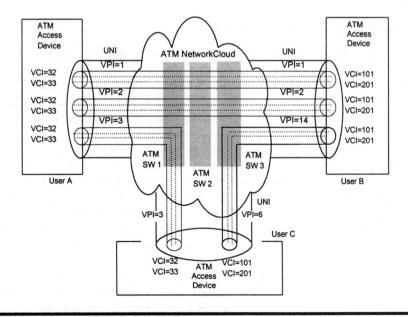

Figure 11.17 Network Cloud Example

Figure 11.17 illustrates a sample ATM network environment with three users, A, B and C, each accessing the ATM network cloud via a single UNI. This network illustrates the ATM terminology details of the simple analogy employed earlier in this chapter of cities interconnected by highways, with the lanes representing channels. In the figure, each VP connects with another site. The "pipes" between ATM access devices depict VPCs. VCCs within each of these VPs, as depicted with dashed lines, are identified by VPIs and VCIs as indicated in the figure. Note that the VPI and VCI values can be different on each end of the connection, since the intermediate switches can change the values. In general, it's wise to keep the VPI/VCI values identical on each end; however, in some operational situations, such as those resulting from network migrations or changes, this is not always possible. Do not worry though since ATM has the flexibility to support these cases. This example illustrates the concept that the VPI and the VCI are addresses significant only to the local interface.

11.4 USING ATM SERVICE CATEGORIES

Now that we've covered the three major building blocks of ATM: the physical interfaces, the Virtual Path (VP), and the Virtual Channel (VC), let's move on to an example of how an end user would employ these in a multimedia application. The coverage then provides several analogies illustrating the need for multiple service categories, as well as basic mechanisms for implementing them.

11.4.1 Multimedia Application Example

ATM switches take a user's data, voice, and video, chops this information up into fixed-length cells, and multiplexes them into a single stream of cells transmitted across the User-Network Interface (UNI) physical medium. An example of a multimedia application is that of one team of information workers needing to collaborate on an important project with another team at another site.

Figure 11.18 illustrates the role of ATM in this real-life example, with one team sitting around an ATM workstation equipped with ATM interface card, sound board with microphone, and video camera. A cost-effective 25-Mbps ATM connection over existing building wiring connects the workstation to a local ATM switch, which in turn attaches to a public ATM-based wide area network service provider via a 45-Mbps ATM UNI. The ATM network connects to another ATM switch that serves a workstation serving the other team.

One team places a multimedia call to the other, transmits the data from several documents and drawing programs, and begins an interactive meeting over the network. The call involves three connections: one for voice, one for video, and a third for data interchange. The teams interact through separate data, voice, and video virtual connections, all integrated by their workstations in real time. Using an interactive white board application, one team looks through the document at its workstation, interactively questioning the other team about certain points in the work. Let's break down this scenario into its working ATM components.

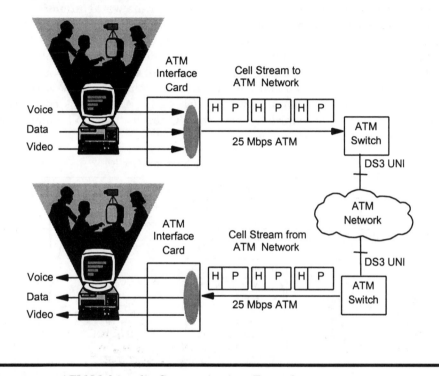

Figure 11.18 ATM Multimedia Communications Example

Video and voice are delay-sensitive; that is, the information cannot be delayed for more than a blink of the eye. Furthermore, delay cannot have significant variations for voice and video traffic, either. Disruption in the video image or distortion of voice destroys the interactive, near-real-life quality of this multimedia application. Voice traffic may be sent at a constant rate, while newer video coding schemes send at a variable rate in proportion to the amount of image movement. Transferring a video segment from a high-action motion picture, such as dogfights between jet fighters in the movie *Top Gun*, requires a much higher transmission rate than the session where the two teams make eye contact and exchange simple gestures.

Data, such as a file transfer using TCP/IP, on the other hand, isn't nearly as delay-sensitive as voice or video traffic. Data traffic, however, while not delay-sensitive, is very sensitive to loss. Therefore, ATM must discriminate between voice, video, and data traffic, giving voice and video traffic priority with guaranteed, bounded delay, while simultaneously assuring that data traffic has very low loss.

When using ATM, longer packets cannot delay shorter packets as in other packet-switched implementations because ATM chops long packets up into many cells. This enables ATM to carry Constant Bit Rate (CBR) traffic such as voice and real-time Variable Bit Rate (rt-VBR) video in conjunction with non-real-time Variable Bit Rate (nrt-VBR) data traffic, potentially having very long packets within the same network in a standard way.

Examining this example further, the software in the workstations establish three separate virtual connections for data, voice, and video, each VC with a different service category. The workstation software assigns nrt-VBR to data, CBR to voice , and rt-VBR to video. Figure 11.19 shows how the multiplexing function inside the ATM interface card combines all three types of traffic over a single ATM UNI with Virtual Channels (VCs) being assigned to the data (VCI = 51), voice (VCI = 52), and video (VCI = 53). All of these connections utilize VPI=0, which standards reserve for VC connections. VP connections occur only on non-zero VPI values.

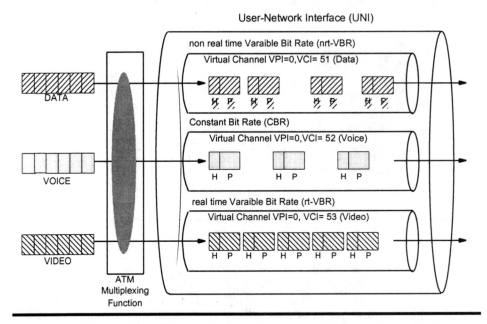

Figure 11.19   Illustration of Virtual Channels Supporting Multimedia Application

The ATM interface card simultaneously transmits data, voice, and video data traffic. The ATM multiplexing function inside the card contains a segmentation, or "chopper," function which slices and dices the separate voice, video and data streams into 48-octet pieces, as shown in Figure 11.20. In the next step, a "postman" addresses the payload by prefixing it with the VPI, VCI, and the remaining fields of the 5-octet cell header. The "postman" assigns VCI 51 for data, VCI 52 for voice, and VCI 53 for video, all on VPI = 0. The result of this operation is a stream of 53-octet ATM cells from each source: voice, video, and data delivered by the postman. Since each source independently generates these cells, contention may arise for cell slot times on the interface connected to the network.

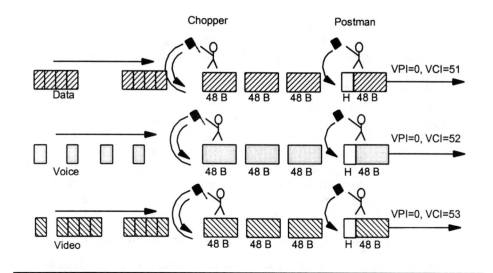

Figure 11.20 ATM Multiplexing Function – the "Chopper" and the "Postman"

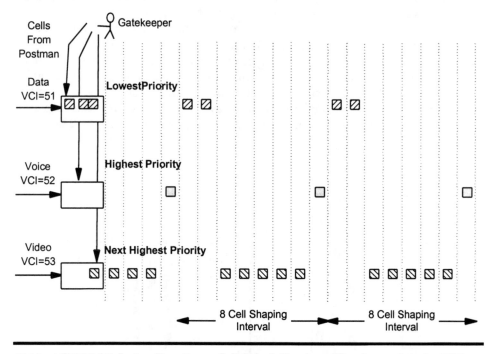

11.21 ATM Multiplexing Function – Cell Scheduling by a "Gatekeeper" at the UNI

Figure 11.21 shows an example of how the ATM-powered workstations send the combined voice, video, and text data over the UNI towards the local switch. A "gatekeeper" in the interface card in her workstation sends the cells received from the postman in a specific proportion over intervals of eight cells (about 80 μs at the DS3's 45-Mbps physical transmission rate), normally allowing one voice cell, then five video cells, and finally what is left — two data cells — to be transmitted. This shaping action corresponds to about 4-Mbps for high-fidelity audio, 24-Mbps for video, and 9-Mbps for data. All sources (data, voice, and video) contend for the bandwidth within each eight cell time shaping interval.

The gatekeeper gives priority to cell transmit time opportunities among the various streams. *Voice* cells have the highest priority, *video* cells have the second priority, and *data* cells the lowest priority. The gatekeeper sends the highest priority cells first in the shaping interval, retaining cells in a buffer for each service category in case all of the cell slot times are full during the shaping interval. If the buffer fills up, then the gatekeeper discards any cells received from the postman for that service category. Real implementations employ a much larger shaping interval than used in this example to provide greater bandwidth allocation granularity.

11.4.2 ATM and Packet Switching Prioritization Compared

Why are long, variable-length packets detrimental to short, fixed-length pulse-code modulation (PCM) type voice samples on a packet-switched network? Why are cells important? We use a simple example based upon an analogy suggested by Goralski [2] to illustrate these points.

In the United States today, railroads are still used for moving large, bulky loads of freight. Railroad engines pull variable-length trains of boxcars, up to 200 or so, on special "networks" of railroad tracks. Highways were built as another special kind of network, but this time they are optimal for variable-length and variable-speed cars optimized for carrying people. As long as the networks are separate — and the trains stay on the tracks and the cars stay on the roads — the networks both work fine, as shown in Figure 11.22a.

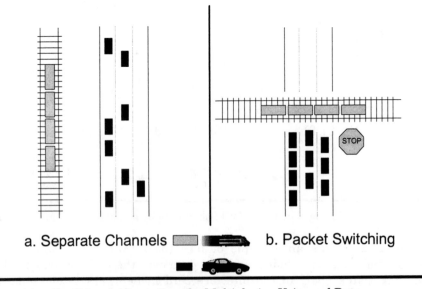

Figure 11.22 Traditional Alternatives for Multiplexing Voice and Data

The problem arises however when the highway crosses the railroad tracks as illustrated in Figure 11.22b. At a railroad crossing, the train clearly takes priority, and once the train starts across the highway, the cars must all queue until the entire train clears the intersection. This delays the cars, possibly for a long time if the train has a large number of boxcars. Only after the train clears the crossing may the queued-up cars cross the intersection. The number of queued-up cars may be quite large, further exacerbating the delay.

In this analogy, the train is a large, variable-length packet, and the cars are the smaller, but more numerous, fixed-length voice PCM samples. The entirely separate railroad tracks and highways are the separate voice and data Time-Division Multiplexed (TDM) networks commonly in use today. The crossing situation represents an attempt to mix the data and delay-sensitive voice samples on a single packet network. Figure 11.23a and b illustrates the separate TDM and integrated packet data networks corresponding to transportation analogy of Figure 11.22. An example of an intergrated packet switched network is voice over IP carried by a router using packet switching.

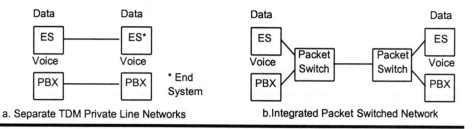

a. Separate TDM Private Line Networks   b.Integrated Packet Switched Network

Figure 11.23 Analogy Applied to Real World Voice and Data Network Integration

Clearly, scenario b causes a problem if the train is too slow. The backup of voice traffic will cause the receiver to think that something is wrong if the absence of "cars" arriving persists too long. Exactly how long it persists, however, depends on the length and speed of the train.

There are actually two potential solutions to this problem. First, if the engine pulling the train runs much *faster* than the cars, then the delay for cars waiting for the train to pass is correspondingly *shorter*. That is, if the train runs twice as fast, the wait is half as long. If it runs ten times as fast, the wait is one tenth as long, and so on. The point is that if the train runs fast enough, the time that the cars have to wait at the crossing becomes rather insignificant. Figure 11.24a illustrates this solution. In packet switching, this corresponds to using very high circuit speeds to connect routers, a technique commonly call Packet Over SONET (POS) in the industry literature.

There is a drawback to this approach, however. Data requirements have grown by leaps and bounds in the past, and the trend is upward. Files today are larger than the entire DOS partition (32-MB limit with DOS 3.2) not long ago. It is not unusual to have 16 MB of RAM on a PC today — larger than the most common hard drive size of 10 years ago (10 MB). In other words, the data packet trains keep getting longer and longer! Rebuilding the railroad to stay ahead of the trend may be a self-defeating proposition in the long run. In other words, faster railroads may actually encourage longer trains. Applying this reasoning back to packet switching, the network engineer may not be able to afford a fast enough circuit to utilize the fast packet switching of approach a.

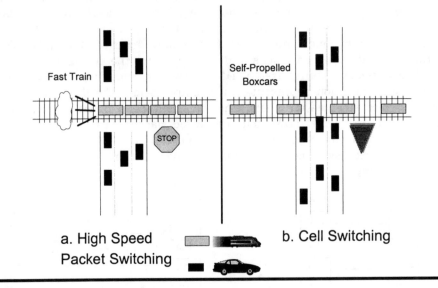

Figure 11.24 Alternative Solutions for Mixing Voice and Data

There is another solution. What if each boxcar were a self-propelled unit about the same size as a car? Now if a "train" starts across the intersection and a car arrives, there is only a very short delay before the boxcar is through the crossing and the car may pass. This is much simpler than the previous solution and requires no reengineering of the railroad or the highway. Figure 11.24b illustrates this approach. This is the ATM solution, with the boxcars and cars becoming cells and ATM switches replacing the packet switches of the last example with the capability prioritize voice cells (cars) over data cells (train cars) . Of course, a key engineering decision is to decide upon what the cell size should be. The next section covers this important topic.

11.5 RATIONALE FOR THE CHOICE OF ATM CELL SIZE

When it came to deciding on a standard cell size in the CCITT, a debate raged between a 32-octet versus a 64-octet payload size for valid technical reasons. The final decision on the 48-byte payload size was actually a compromise between these two positions. The choice of the 5-octet header size was a separate tradeoff between a 3-octet header and an 8-octet header between increased function and efficiency.

The debate centered over the basic tradeoff between packet data transport efficiency and voice packetization delay versus cell payload size as illustrated in Figure 11.25. The figure shows packet data efficiency for a 5-octet cell header. Voice packetization delay is the amount of time required to fill the cell payload at a rate of 64-kbps, that is, the delay waiting to fill the cell with digitized voice samples. Ideally, high efficiency and low packetization delay are both desirable, but cannot be achieved simultaneously as seen from the figure. Better efficiency occurs at large cell sizes at the expense of increased packetization delay. In order to carry voice over ATM and interwork with two-wire analog telephone sets, if the total round-trip delay exceeds 50 ms, then the network must employ echo cancellation. During this time the echo cancellation delay objective was more stringently set at 15 ms. Hence, a cell size of 32 octets avoided the need for echo cancellation. Looking at the figure, a cell payload size of 32 octets results in a best case data packet efficiency (for a very long packet) of approximately 85 percent. Increasing the cell payload size to 64 octets would have increased the data efficiency to 95 percent. Thus, the ITU-T adopted the fixed-length 48-octet cell payload as a compromise between a long cell size for more efficient transfer of delay-insensitive data traffic (64 octets) and smaller cell sizes for delay-sensitive voice traffic (32 octets). A consequence of this compromise is that voice over ATM connections in a local geographic area require echo cancellation, whereas voice over TDM connections do not. Thus, voice over ATM in local geographic areas starts at an economic disadvantage with respect to traditional digitized voice over TDM. Chapter 17 explores the operation of voice over ATM and the role echo cancellation in more depth.

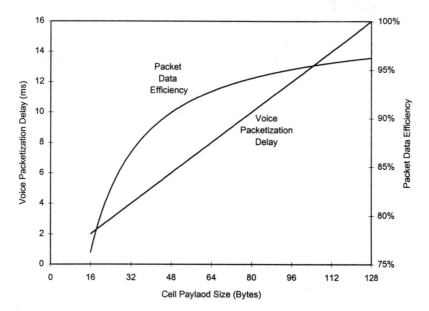

Figure 11.25   Voice Packetization Delay versus Cell Size Tradeoff

11.6 REVIEW

This chapter introduced the reader to the foundational elements of ATM. These include ATM's multifaceted nature, acting as: an interface, a protocol, a technology, economical integrated access, a scalable infrastructure, and a service. We summarized each of these aspects to set the stage for the rest of the book. The text then introduced the reader to the basic building blocks of ATM: the physical transmission interfaces, the virtual path (VP), and the virtual channel (VC) through several analogies. After a brief stop in the ATM café to peruse the menu of options available from ATM's diverse cuisine, the presentation moved on to define ATM's basic building block — the cell. Through several graphical examples, the text defined the method of constructing cells by assigning VPI and VCI addresses to a header field which prefixes the user payload. The chapter then provided several examples illustrating ATM switching of VPs and VCs using Virtual Path and Virtual Channel Identifiers (VPIs and VCIs) on a node-by-node and an end-to-end basis. The coverage then moved on to a multimedia example of multiple service categories for voice, video and data. We then presented a transportation analogy to illustrate the importance of priority and different quality of service requirements for voice and data. Finally, the chapter concluded with a brief ATM history, describing the tradeoffs involved in the decision process of the standards bodies resulting in the 53-byte cell size used today.

11.7 REFERENCES

[1]   D. McDysan, "Performance Analysis of Queuing System Models for Resource Allocation in Distributed Computer Networks," D.Sc. Dissertation, George Washington University, 1989.

[2]   W. Goralski, *Introduction to ATM Networking*, McGraw-Hill, 1995.

12

Physical and ATM Layers

This chapter explores the foundation of the B-ISDN protocol stack: the physical and ATM layers. We start at the bottom with the physical (PHY) layer, and then move to the ATM layer, which defines virtual paths and virtual channels. This chapter provides details of the concepts introduced in the last chapter.

The primary layers of the B-ISDN protocol reference model are: the PHYsical layer and the ATM layer cell structure and functions. The ATM Adaptation Layer (AAL) provides support for higher layer services such as signaling, circuit emulation, and frame relay as defined in the next chapter. The PHY layer corresponds to OSI Reference Model (OSIRM) layer 1, the ATM layer and part of the AAL correspond to OSIRM layer 2, and the higher layers in the B-ISDN stack correspond to OSIRM layers 3 and above.

First the description covers the broad range of physical interfaces and media currently specified for the transmission of ATM cells. A detailed discussion of definitions and concepts of the ATM layer, defining the cell structure for both the User-Network Interface (UNI) and the Network-Node Interface (NNI), follows. At a lower level, the text provides a description of the meanings of the cell header fields, payload types, and generic functions.

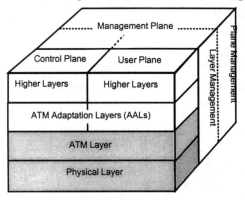

The remainder of the book uses figures like the one shown on the left depicting the B-ISDN protocol model from ITU-T Recommendation I.321 (see Chapter 10). The shaded portion indicates the subject matter covered in that section. This figure, for instance, shows what this chapter covers: the physical layer and the ATM layer. But first, let's look at how the bottom three layers work together to provide services to the higher layers.

12.1 THE PLANE-LAYER TRUTH — AN OVERVIEW

This section introduces the basic B-ISDN layered structure, looking at the next level of detail in the physical, ATM and AAL layers from several points of view. This overview provides an outline for the remainder of this chapter as well as more extensive coverage of the AAL layer in Chapter 13.

12.1.1 Physical, ATM, and AAL Layer Functions

Unfolding the front and right sides of the B-ISDN protocol cube yields the two-dimensional layered model shown in Figure 12.1, which lists the functions of the four B-ISDN/ATM layers along with the sublayer structure of the AAL and Physical (PHY) layer, as defined by ITU-T Recommendation I.321 [1].

Layer Name		Functions Performed	
Higher Layers		Higher Layer Functions	L A Y E R
A A L	Convergence Sublayer (CS)	Service Specific (SS)	L A Y E R
		Common Part (CP)	
	SAR Sublayer	Segmentation And Reassembly	
ATM		Generic Flow Control Call Header Generation/Extraction Cell VCI/VPI Translation Cell Multiplexing/Demultiplexing Cell Rate Decoupling (Unassigned Cells)	M A N A G E
P H Y S I C A L	Transmission Convergence (TC) Sublayer	Cell Rate Decoupling (Idle Cells) Cell Delineation Transmission Frame Adaptation Transmission Frame Generation/ Recovery	E M E N T
	Physical Medium Dependent (PMD)	Bit Timing Physical Medium	

Figure 12.1 B-ISDN/ATM Layer and Sublayer Model

Starting from the bottom, the physical layer has two sublayers: Transmission Convergence (TC) and Physical Medium Dependent (PMD). The PMD sublayer interfaces with the actual electrical or optical transmission medium, detecting the signals, transferring bit timing, and passing the bit stream to and from the TC sublayer. The TC sublayer extracts and inserts ATM cells within either a Plesiochronous or Synchronous (PDH or SDH) Time-Division Multiplexed (TDM) frame and passes these to and from the ATM layer, respectively. The ATM layer performs multiplexing, switching, and control actions based upon information in the ATM cell header and passes cells to, and accepts cells from, the AAL. The generic AAL has two sublayers: Segmentation and Reassembly (SAR) and Convergence Sublayer (CS). The CS is further broken down into Common Part (CP) and Service-Specific (SS) components. Not all AALs follow this model; for example, AAL2 does not have CS and SAR sublayers. Instead, AAL2 efficiently multiplexes short packets from multiple sources into a single cell. AALs pass Protocol Data Units (PDUs) to and accept PDUs from higher layers. These PDUs may either be of variable or fixed length. Chapter 13 details the AALs currently standardized for operation over the common ATM layer.

12.1.2 B-ISDN Protocol Layer Structure

The physical layer corresponds to layer 1 in the OSI model. Most experts concede that the ATM layer and AAL correspond to parts of OSI layer 2, but other experts assert that the Virtual Path Identifier (VPI) and Virtual Channel Identifier (VCI) fields of the ATM cell header have a network-wide connotation similar to OSI layer 3 [2]. A precise alignment with the OSI layers is really not necessary; because the OSI layers are only a conceptual model. Use any model that best suits your networking point of view. As we shall see, B-ISDN and ATM protocols and interfaces make extensive use of the OSI concepts of layering and sublayering, even if a precise mapping to the OSI reference model isn't always possible or required.

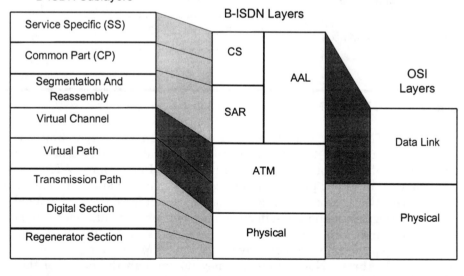

Figure 12.2 B-ISDN Layers and Sublayers and OSI Layers

Figure 12.2 illustrates the mapping of the B-ISDN PHY, ATM, and AAL sublayers to the OSI layers employed in this text. This book uses a protocol model comprised of multiple planes, instead of attempting to fit all of the

ATM-related protocols into the monolithic OSI reference model as some texts attempt to do. Of course, as seen from the I.321 B-ISDN protocol model, the physical and ATM layers are the foundation for the user, control, and management planes.

12.1.3 Hardware and Software Implementations of B-ISDN Layers

The number of standardized protocols for each layer, and whether their target implementation is in hardware or software, tells us a great deal about ATM. Figure 12.3 depicts the number of instances of standard protocols at each layer by rectangles in the center of the figure. The arrows on the right-hand side illustrate the fact that ATM implementations move from being hardware-intensive at the lower layers (PHY and ATM layers) to software-intensive at the higher layers (AALs and higher layers). The figure shows the single ATM layer at the center as the pivotal protocol, shown at the tip of the inverse pyramid in the illustration. The singular instance of the ATM cell structure operates over a large number of physical media described in the next section.

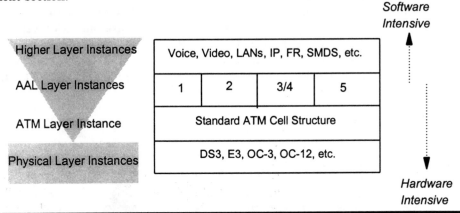

Figure 12.3   ATM Protocol Model Hardware-to-Software Progression

Atop the ATM layer, only four AALs support an ever-expanding set of higher-layer functions. Chapter 14 introduces the detailed coverage of this suite of higher-layer protocols presented in Part 4. In summary, ATM allows machines with different physical interfaces to transport data, independently of the higher-layer protocols, using a common, well-defined protocol amenable to a high-performance and cost-effective hardware implementation to support guaranteed bandwidth and service quality. This flexibility of a single, multipurpose protocol is a key advantage of ATM-based equipment and service architectures. Now, in our journey up through the layers of the B-ISDN/ATM protocol model, we start with the physical layer.

12.2 PHYSICAL (PHY) LAYER

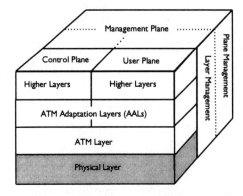

This section covers the key aspects of the PHY layer. The PHY layer provides for transmission of ATM cells over an electrical or optical physical transmission medium connecting ATM devices. The PHY layer has two sublayers: the Physical Medium Dependent (PMD) sublayer and the Transmission Convergence (TC) sublayer. The PMD sublayer provides for the actual transmission of the bits in the ATM cells. The TC sublayer transforms the flow of cells into a steady flow of bits and bytes for transmission over the physical medium, such as a DS1/E1, DS3/E3, or OC3/STM-1 private line access circuit into the WAN, or twisted-pair cabling within an office or residence.

12.2.1 Physical Medium Dependent Sublayer

The PMD sublayer interfaces with the TC sublayer via a serial bit stream. Table 12.1 summarizes each of the standardized interfaces in terms of its name, physical medium, interface speed, user bit rate, and standardizing group(s). The PMD sublayer clocks the bits transmitted over a variety of physical media at the line rate indicated in Table 12.1. Multimode fiber allows transmitters to use inexpensive Light Emitting Diodes (LEDs), while single mode requires use of more expensive lasers. As seen from the table, shielded twisted-pair physical media generally support higher bit rates than unshielded twisted-pair. The user bit rate column indicates the actual bandwidth available to transmit 53-byte ATM cells after removing physical layer overhead. Multiple standards bodies define the physical layer in support of ATM: ANSI, ITU-T, ETSI, and the ATM Forum as indicated in the table.

12.2.2 Transmission Convergence (TC) Sublayer

The TC sublayer maps ATM cells to and from the TDM bit stream provided by the PMD sublayer. The TC sublayer delivers cells, including the 5-byte cell header, to the ATM layer at speeds up to the user bit rate indicated in Table 12.1. The user bit rate is the cell rate times the cell size of 424 bits (53 bytes). The difference between the interface speed and the user bit rate is due to physical-layer overhead. On transmit, TC maps the cells into the physical layer frame format. On reception, it delineates ATM cells in the received bit stream. Generating the HEC on transmit and using it to correct and detect errors on receive are also important TC functions.

Table 12.1   Physical Layer Interfaces, Media, and Bit Rates

Interface Name/ Description	Physical Medium	Line Rate (Mbps)	User Bit Rate (Mbps)	Standardizing Group(s)
DS1	Twisted pair	1.544	1.536	ITU, ANSI, ATMF
E1	Coaxial cable	2.048	1.920	ITU, ETSI, ATMF
nxDS1 IMA	Unshielded twisted pair	$n \times 1.544$	$n \times 1.488^*$	ATMF
nxE1 IMA	Coaxial cable	$n \times 2.048$	$n \times 1.860^*$	ATMF
J2	Coaxial cable	6.312	6.144	ITU, ATMF
Token Ring– based	(Un)shielded twisted pair	32	25.6	ATMF
E3	Coaxial cable	34.368	33.92	ITU, ETSI, ATMF
DS3	Coaxial cable	44.736	40.704, 44.21[†]	ITU, ANSI, ATMF
Midrange PHY	Unshielded twisted pair	51.84, 25.92, 12.96	49.536, 24.768, 12.384[‡]	ATMF
STS-1	Single/multimode fiber	51.84	49.536	ANSI
FDDI-based	Multimode fiber	125	98.15	ATMF
E4	Fiber, coaxial cable	139.264	138.24	ITU, ETSI
STS-3c	Single/multimode fiber	155.52	149.76	ITU, ANSI, ATMF
STM-1	Fiber, coaxial cable	155.52	149.76	ITU, ETSI, ATMF
155.52 Mbps	Unshielded twisted pair	155.52	149.76	ATMF
Fiber channel– based	Multimode fiber, Shielded twisted pair	194.4	155.52	ATMF
STS-12c	Single/multimode fiber	622.08	599.04	ITU, ANSI, ATMF
STM-4	Fiber, coax	622.08	599.04	ITU, ETSI
STS-48c	Single-mode fiber	2,488.32	2,377.728	ATMF

[*]The user bit rate for nxDS1 and nxE1 inverse multiplexing over ATM (IMA) physical interfaces specified by the ATM Forum assumes the default value of one overhead cell for every 32 cells at the DS1 or El level. The IMA standard supports up to eight DS1/E1s.

[†]The two user bit rates for DS3 are for the older method called Physical Layer Convergence Protocol (PLCP), taken from the 802.6 Distributed Queue Dual Bus (DQDB) standard, and the new method that employs cell delineation.

[‡]The lower bit-rate values for the midrange PHY are for longer cable runs.

The following sections cover three examples of TC mapping of ATM cells: direct mapping to a DS1 payload, direct mapping to a DS1 payload, and the PLCP mapping to a DS3. The section then covers the use of the Header Error Check (HEC) and why it is so important. Another important function that TC performs is cell rate decoupling by sending idle (or unassigned) cells when the ATM layer has not provided a cell. This critical function allows the ATM layer to operate with a wide range of different speed physical interfaces. An example later in this section illustrates cell rate decoupling using unassigned or idle cells.

12.2.3 Examples of TC Mapping

This section gives examples of direct and Physical Layer Convergence Protocol (PLCP) mapping by the Transmission Convergence (TC) sublayer to and from the physical layer bit stream provided by the PMD sublayer.

12.2.3.1 DS1 Direct Mapping

Figure 12.4 illustrates direct cell delineation mapping by the TC sublayer for the DS1 physical interface. Note that the cell boundaries need not align with octet boundaries defined for DS1, for example, as defined in ISDN. Most TC layer specifications are similar in form to this standard. For further details, see the ATM Forum DS1 UNI physical layer specification [3] or the ANSI T1.646 standard [4].

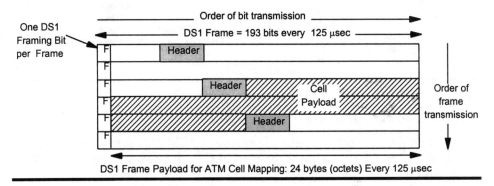

Figure 12.4   DS1 Transmission Convergence Sublayer Mapping

12.2.3.2 SONET STS-3c Direct Mapping

Figure 12.5 illustrates the direct mapping of ATM cells into the SONET STS-3c (155.52-Mbps) Synchronous Payload Envelope (SPE) defined in Chapter 6. Note that ATM cells continuously fill the STS-3c payload since an integer number of 53-octet cells do not fit in an STS-3c frame.

The ATM User-Network Interface (UNI) 3.1 specification [5] eliminates a number of SONET overhead functions to reduce the complexity, and hence cost of SONET-based ATM interfaces. For example, the UNI eliminates the requirement for the processing intensive SONET Data Communications Channel (DCC). The TC sublayer uses the HEC field to delineate cells from within the SONET payload. The user data rate is computed as 9 rows times 260 columns of bytes at 8,000 SONET frames per second, or 149.76-Mbps. For further details see the ATM Forum UNI 3.1 specification or ANSI standard T1.646 [4]. For a more readable discussion of ATM mapping into the SONET payload, see [6]. The mapping over STS-12c is similar in nature. The difference between the North American SONET format and the international SDH format exists in the TDM overhead bytes.

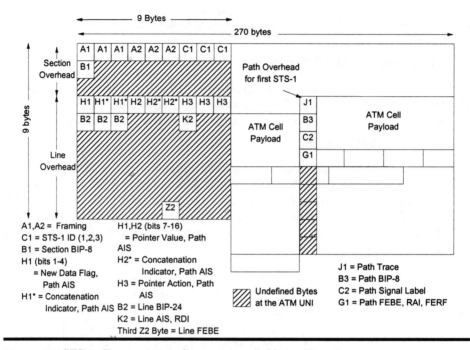

Figure 12.5 STS-3c Transmission Convergence Sublayer Mapping

12.2.3.3 DS3 PLCP Mapping

Figure 12.6 illustrates the original mapping of ATM cells into a DS3 using the Physical Layer Convergence Protocol (PLCP) defined in the IEEE 802.6 standard [7].

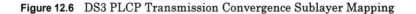

Figure 12.6 DS3 PLCP Transmission Convergence Sublayer Mapping

PLCP encloses the ATM cells within a 125-µs frame which rides inside the standard DS3 M-frame. The PLCP mapping transfers 8 kHz timing across the DS3 interface, but achieves a cell transfer rate of only 40.704-Mbps, utilizing only about 90 percent of the DS3's approximately 44.21-Mbps payload rate. The BIP-8 indicator covers the Path OverHead (POH) and associated ATM cells of the previous PLCP frame.

The more recent ANSI T1.646 standard [4] defines a direct mapping of ATM cells into the 44.21-Mbps DS3 payload similar to that defined earlier for DS1 and STS-3c. With direct mapping, timing transfer occurs by tying the 44.736-Mbps DS3 line rate to a Stratum 1 reference clock. The DS3 PLCP standard is falling out of use because it is approximately 10 percent less efficient than the direct method.

12.2.4 TC Header Error Check (HEC) Functions

The Header Error Check (HEC) is a 1-byte code applied to the 5-byte ATM cell header capable of correcting any single-bit error in the header. It also detects many patterns of multiple-bit errors. The TC sublayer generates the HEC on transmit and uses the received HEC field to determine if the received header has any errors. If the receiver may either correct or discard the cell if HEC detects a single bit error; but must discard the cell if HEC detects more than one error. Chapter 24 presents an analysis of the undetected error rate to help users decide on the HEC correction or detection option for their particular application. Since the header tells the ATM layer what to do with the cell, it is very important that it not have errors; if it did, the cell might be delivered to the wrong user or inadvertently invoke a function in the ATM layer.

The TC also uses the HEC to locate cells when they are directly mapped into a TDM payload; for example, as in the DS1 and STS-3c mappings described in earlier sections. This method works because HEC infrequently matches random data in the cell payloads when the 5 bytes being checked are not part of a valid cell header. Thus, almost all ATM standards employ the HEC to locate cell boundaries in the received bit stream. One notable exception is the North American standard for DS3, which uses a separate Physical Layer Convergence Protocol (PLCP) for cell delineation as described in the previous section. Once the TC sublayer locates several consecutive cell headers in the bit stream received from the PMD sublayer through the use of the HEC, then the TC knows to expect the next cell 53 bytes later. Standards call this process *HEC-based cell delineation*. Thus, the fixed length of ATM cells aids in detecting valid cells reliably.

12.2.5 TC Cell Rate Decoupling

The TC sublayer performs a cell rate decoupling, or speed matching, function, as well. Physical media that have synchronous cell time slots (e.g., DS3, SONET, SDH, STP and the Fiber Channel-based method) require this function, while asynchronous media such as the FDDI PMD do not. As we shall see in the next section, special codings of the ATM cell header indicate whether a cell is either *idle* or *unassigned*. All other cells are *assigned* and correspond to cells generated by the ATM layer. Figure 12.7 illustrates an example of cell rate decoupling between a transmitting and receiving ATM device. Starting at the left-hand side of the figure, the transmitting ATM device multiplexes multiple cell streams together, queuing them if a time slot is not immediately available on the physical medium. If the queue is empty when the time to fill the next cell time slot arrives as determined by the physical layer, then the TC sublayer in the transmitter inserts either an unassigned cell or an idle cell, indicated by the solid black filled cells in the figure. The receiving device extracts unassigned or idle cells and distributes the other, assigned cells to the destinations as determined from the VPI and VCI values as shown on the right hand side of the figure. Also, note how the act of multiplexing and switching changes the inter-cell spacing in the output cell streams, as the reader can see from the example.

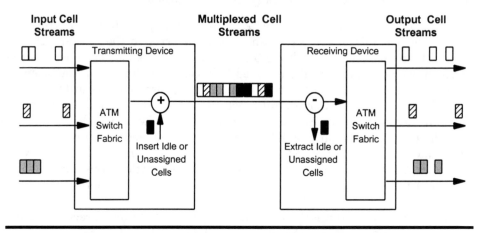

Figure 12.7   Cell Rate Decoupling Using either Idle or Unassigned Cells

ITU-T Recommendation I.321 originally placed the rate decoupling function in the TC sublayer of the PHY layer using idle cells, while the ATM Forum placed it in the ATM layer using unassigned cells. The ITU-T model viewed the ATM layer as independent of whether or not the physical medium has synchronous time slots. Since the ITU-T initially standardized on idle cells, while the ATM Forum chose unassigned cells for cell-rate decoupling, these methods were incompatible since the coding of the cell header to indicate idle or unassigned cells differed. In 1996, ITU-T Recommendation I.361 aligned itself with the ATM Forum's usage of the unassigned cell for rate decoupling. Therefore, the difference in usage of idle versus unassigned cells occurs only in older equipment.

12.2.6 Inverse Multiplexing over ATM (IMA)

The ATM Forum's Inverse Multiplexing over ATM (IMA) specification [8] standardizes access at nxDS1 and nxE1 rates. IMA combines multiple DS1 or E1 access circuits into a single aggregate transmission path with approximately n times the bandwidth of a single DS1 or E1 circuit. Hence, IMA provides a capability similar to proprietary nxDS1 or nxE1 TDM multiplexers, but in a standard manner. Figure 12.8 illustrates the operation of an IMA system. Cells arriving from the sending ATM layer originate at the upper left-hand side of the figure. The IMA transmitter multiplexes these cells in a round-robin manner onto the three physical links in the IMA group, interleaving IMA Control Protocol (ICP) cells for control and synchronization purposes. IMA cells (ICP or filler) are specially coded OAM cells that IAM does not convey to the ATM layer. IMA transparently carries standard ATM OAM cells as detailed in Part 7. The shading of cells in the figure indicates the link selected by the IMA transmitter. At the receiver, the cells on each of the physical links may arrive after different delays on each of the physical connections, for example, due to different circuit lengths. The maximum differential delay required by the IMA specification is 25 ms, or approximately a difference of 2,500 miles in the physical circuit propagation delay when operating over optical fiber.

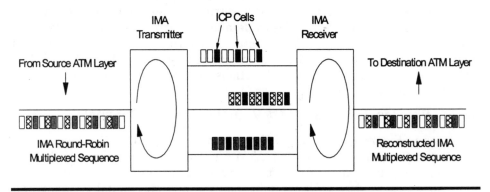

Figure 12.8 Illustration of Inverse Multiplexing over ATM (IMA)

On the right hand side of the figure, the IMA receiver employs the ICP cells to realign the user cells before multiplexing them back into an accurately reproduced version of the original high-speed ATM cell stream, which it then delivers to the destination ATM layer. The net effect of IMA is that the end equipment sees an ATM cell stream operating at approximately an nxDS1 or nxE1 aggregate rate. At the default rate of one ICP cell every 32 cell times, the IMA consumes approximately 3 percent overhead.

Many ATM switch, access multiplexer, and concentrator vendors now offer support for IMA. IMA offers a cost-effective way to garner greater than DS1 or E1 bandwidth across the WAN, without having to lease a full DS3 or E3 access line. Furthermore, IMA allows users to add bandwidth in DS1 or E1 increments, typically up to 8 DS1s or E1s, at which point a user can cost justify the purchase a more expensive DS3/E3 or SONET/SDH access line. Carriers began offering nxDS1 IMA services in the United States in 1997. This service is especially useful for international connectivity where bandwidth is even more expensive than in the United States.

12.3 ATM LAYER — PROTOCOL MODEL

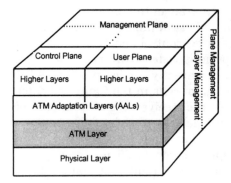

This section moves up one more layer to the focal point of B-ISDN protocol model, namely the Asynchronous Transfer Mode (ATM) layer. First, the text covers the relationship of the ATM layer to the physical layer and its division into Virtual Path (VP) and Virtual Channel (VC) levels in detail. This is a key concept, which is the reason Chapter 11 presented a brief introduction backed up by several analogies and examples. This section provides more detailed examples portraying the role of end and intermediate systems in a real-world setting rather than just the formal standards model. These examples explain how intermediate and end systems use the ATM layer VP and VC functions in terms of the layered protocol model. The examples show how intermediate systems perform ATM VP or VC switching or cross-connection functions, and how end systems exchange cells with the ATM Adaptation Layer (AAL).

12.3.1 Physical Links and ATM Virtual Paths and Channels

As shown in Figure 12.9 from ITU-T Recommendation I.311 [9], a key concept is the construction of end-to-end ATM Virtual Path and Virtual Channel Connections (VPCs and VCCs) from one or more VP or VC links. The physical layer has three levels: regenerator section, digital section, and transmission path as shown in the figure. The ATM layer uses only the transmission path of the physical layer, which is effectively the TDM payload that connects ATM devices. Generically, an ATM device may be either an endpoint or a connecting point for a VP or VC. A Virtual Path Connection (VPC) or a Virtual Channel Connection (VCC) exists only between endpoints as shown in the figure. A VP link or a VC link exists between an endpoint and a connecting point or between connecting points as also indicated in the figure. A VPC or VCC is an ordered list of VP or VC links, respectively. The control plane establishes VPCs and VCCs by provisioning, in the case of a Permanent Virtual Connection (PVC), or by signaling, in the case of Switched Virtual Connections (SVCs). The following sections define these terms more precisely.

12.3.1.1 VC level

The *Virtual Channel Identifier (VCI)* in the cell header identifies a single VC on a particular Virtual Path (VP). A VC connecting point switches based upon the combination of the VPI and VCI. A *VC link* is a unidirectional flow of ATM cells with the same VPI and VCI between a VC connecting point and either a VC endpoint or another VC connecting point. *A Virtual Channel Connection (VCC)* is a concatenated list of VC links traversing adjacent VC switching ATM nodes. A VCC defines a *unidirectional* flow of ATM cells from one user to one or more other users. A point-to-point bi-directional VCC is actually a pair of point-to-point unidirectional VCCs relaying cells in opposite directions between the same endpoints.

A network must preserve cell sequence integrity for a VCC; that is, the cells must be delivered in the same order in which they were sent. This means that ATM devices deliver cells to intermediate connecting points and the destination endpoint in the same order transmitted by the originating endpoint. Each VCC has an associated Quality of Service (QoS).

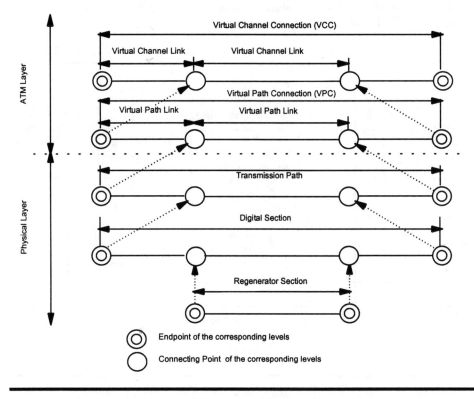

Figure 12.9 Physical Layer, Virtual Paths, and Virtual Channels

12.3.1.2 VP Level

Virtual Paths (VPs) define an aggregate bundle of VCs between VP endpoints. A *Virtual Path Identifier (VPI)* in the cell header identifies a bundle of one or more VCs. A *VP link* provides unidirectional transfer of cells with the same VPI between VP endpoints or VP connecting points. A VP connecting point switches based upon the VPI only — it ignores the VCI field. A Virtual Path Connection (VPC) is a concatenated list of VP links between adjacent VP switching nodes. A VPC defines a unidirectional flow of ATM cells from one user to one or more other users. A point-to-point bidirectional VPC is actually a pair of point-to-point unidirectional VPCs relaying cells in opposite directions between the same endpoints.

Standards do not require a network to preserve cell sequence integrity for a VPC; however, the cell sequence integrity requirement of a VCC still applies. Each VPC has an associated Quality of Service (QoS). If a VPC contains VCCs with different QoS classes, then the VPC assumes the QoS of the VCC with the highest QoS class.

12.3.2 Intermediate Systems (IS) and End Systems (ES)

As a more concrete illustration of connecting and endpoint functions, Figure 12.10 looks inside an Intermediate System (IS) and End System (ES) based upon the ANSI T1.627-defined protocol model [10].

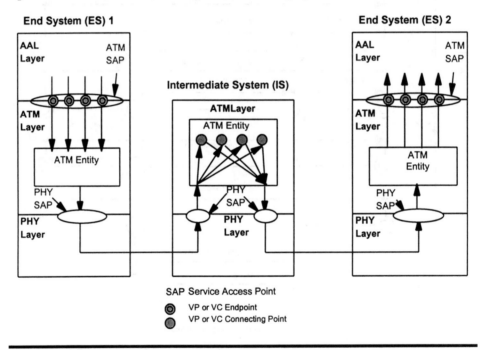

Figure 12.10   Intermediate System (IS)/End System (ES) VP/VC Functions

This example depicts an intermediate network node (IS) connecting two ATM CPE devices (ESs). The ATM Adaptation Layer in End System 1 (ES1) generates cell payloads for a number of connection endpoints at the boundary between the AAL and the ATM layers, which is the ATM Service Access Point (SAP) as indicated in the figure. The ATM entity in ES1 multiplexes these virtual connections and passes ATM cells to the PHYsical (PHY) layer across the PHY SAP. End System 1 (ES1) passes the bit stream over the physical layer to the Intermediate System (IS) over the physical interface. The IS demultiplexes the ATM connections and applies each to a connecting point in the IS ATM entity. The protocol model does not formally specify this action; however, we show it in the example for clarity. The connecting point in the IS ATM entity translates the VPI and/or the VCI depending on whether it is a VP or VC connecting point, determines the outgoing physical interface, and performs other ATM layer functions defined later in this chapter. The IS ATM entity then multiplexes these onto the outgoing PHY-SAP for transfer to the destination — End System 2. The PHY layer in ES2 delivers these to its ATM entity via the PHY-SAP. The ES2 ATM entity demultiplexes the cells and delivers them to the endpoint of the corresponding VP or VC via the ATM-SAP for processing by higher layers.

12.3.3 VP and VC Switching and Cross-Connection

This section provides a specific example of VP and VC endpoints and connecting points in intermediate and end systems for VP links, VPCs, VC links, and VCCs. Figure 12.11 depicts two end systems (or CPE) and an intermediate system (or switch). The endpoint and connecting points use the terminology and notation from Figure 12.9. The physical interface, virtual path, and virtual channel are shown as a nested set of *pipes* using the convention introduced in Chapter 11 from ITU-T Recommendation I.311. The transmission path PHY layer carries Virtual Paths (VPs) and Virtual Channels (VCs) which are either unidirectional, or bidirectional. This example shows end systems (or CPE) with both VP and VC endpoints. The left-hand-side end system, or CPE, originates a VP with VPI x and two VCs with VCI values equal to a and c.

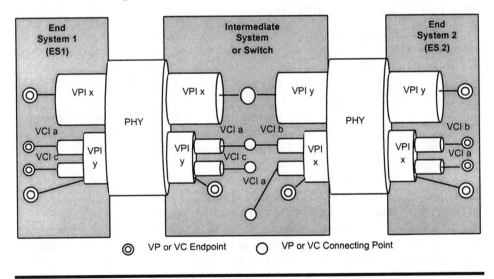

Figure 12.11 VP Link, VPC, VC Link, and VCC Example

The intermediate system (or switch) contains VP and VC switching functions as shown in Figure 12.11. The intermediate system VP switching function translates the VPI from x to y, since VPI x is already in use on the physical interface to the destination end system. The switch automatically connects all VCIs within VPI x to VPI y since by the use of the VP cross-connect function for this pair of VPI values. This simultaneous switching of a large number of VCs within a VP is the principal reason for the standardization of VPs. If ATM employed only a single level of interface level addressing in the cell header, this function would be impossible. VC switching operates within VPs as illustrated by the other VC connection terminating within the switch. Now looking at the VCs on VPI y, the VC switching function in the switch translates the received VCI a to an outgoing VCI b on VPI x for delivery to the destination. VCI c from VPI y is switched to some other destination (not shown in the figure) using a connecting point. Similarly, the switch takes VCI a coming from another physical interface and/or VPI and places it within VPI x for delivery to End System 2.

12.4 ATM LAYER AND CELL — DEFINITION

Let's take a detailed look inside the ATM cell header and ascertain the meaning of each field. First, we define the User-Network and Network-Network Interfaces (UNI and NNI), followed by a summary of the ITU-T Recommendation I.361 [11], ANSI, and ATM Forum definitions of the cell structure at the ATM UNI and NNI. Next, this section describes the functions of the ATM layer in terms of the specific bit-level codings and meanings of the ATM cell header fields.

12.4.1 ATM UNI and NNI Defined

Figure 12.12a gives the ITU-T and ANSI oriented B-ISDN view defined in I.413 [12] for the User-Network Interface (UNI) and the Network Node Interface (NNI). ATM UNIs interconnect user equipment, called Broadband Terminal Equipment (B-TE), to either a Terminal Adapter (TA) or Network Termination (NT) device prior to interfacing to an ATM network. These standards assign letters to the interfaces between each of these functional blocks. Similar to the ISDN reference model described in Chapter 6, the ANSI standards define the U reference point while the international standards do not.

Figure 12.12b shows the ATM Forum terminology of private and public UNIs corresponding to the ITU-T reference point terminology above it. The ATM UNI is either a private ATM UNI, which occurs at the R or S reference points in ITU-T Recommendation I.413 and ANSI T1.624 [13], or a public ATM UNI, which occurs at reference points T or U as shown in the figure. Normally, we refer to the Network-Node Interface (NNI) defined in ITU-T Recommendation I.113 as the standard interface between networks; however, it will most likely also be the interface used between nodes within a network. The ATM Forum distinguishes between an NNI used for private networks and public networks as shown in the figure. This book uses primarily the ATM Forum terminology.

12.4.2 Detailed ATM Cell Structure at the UNI and NNI

Two standardized coding schemes exist for cell structure: the User-Network Interface (UNI) and the Network-Node, or Network-Network Interface (NNI). The UNI is the interface between the user, or CPE, and the network switch. The NNI is the interface between switches or between networks. We introduce the UNI and NNI coding schemes and then detail the format and meaning of each field as defined in this section. ITU-T Recommendation I.361 [11] is the basis of these definitions, with further clarifications and implementation details given in the ANSI T1.627 [10] standard and the ATM Forum UNI and Broadband Inter-Carrier Interface (B-ICI) specifications.

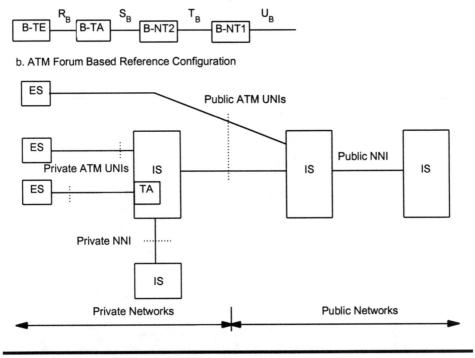

Figure 12.12 ATM UNI and NNI Reference Configurations

12.4.2.1 ATM UNI Cell Structure

Figure 12.13 illustrates the format of the 53-byte ATM cell at the User-Network Interface (UNI). The cell header contains an address significant only to the local interface in two parts: an 8-bit Virtual Path Identifier (VPI) and a 16-bit Virtual Channel Identifier (VCI). The cell header also contains a 4-bit Generic Flow Control (GFC), 3-bit Payload Type (PT), and a 1-bit Cell Loss Priority (CLP) indicator. An 8-bit Header Error Check (HEC) field protects the entire header from errors. Later in this section details are given as to the meaning of each header field. A fundamental concept of ATM is that switching occurs based upon the VPI/VCI fields of *each* cell. Switching done on the VPI only is called a Virtual Path Connection (VPC), while switching done on both the VPI/VCI values is called a Virtual Channel Connection (VCC).

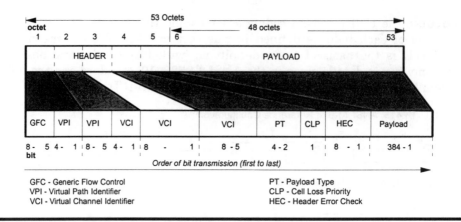

GFC - Generic Flow Control
VPI - Virtual Path Identifier
VCI - Virtual Channel Identifier

PT - Payload Type
CLP - Cell Loss Priority
HEC - Header Error Check

Figure 12.13 ATM User-Network Interface (UNI) Cell Structure

12.4.2.2 ATM NNI Cell Structure

Figure 12.14 illustrates the format of the 53-byte ATM cell at the Network-Node Interface (NNI). The format is identical to the UNI format with two exceptions. First, there is no Generic Flow Control (GFC) field. Secondly, the NNI uses the 4 bits used for the GFC at the UNI to increase the VPI field to 12 bits at the NNI as compared to 8 bits at the UNI. SVCs involve a significantly different control plane for the NNI as covered in Chapter 15.

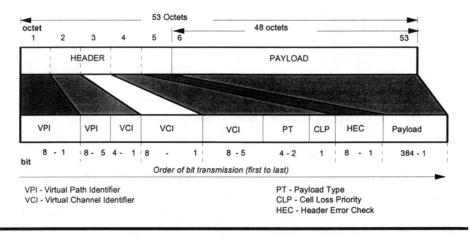

VPI - Virtual Path Identifier
VCI - Virtual Channel Identifier

PT - Payload Type
CLP - Cell Loss Priority
HEC - Header Error Check

Figure 12.14 ATM Network Node Interface (NNI) Cell Structure

12.4.2.3 Definition of ATM Cell Header Fields

This section provides a brief description of each header field.

Generic Flow Control (GFC) is a 4-bit field that supports simple multiplexing implementations. The current ATM Forum implementation agreements define only the *uncontrolled* mode, where the 4-bit GFC field is always coded as zeroes. If too many non-zero GFC values are received, layer management should be notified. Chapter 22 describes the controlled mode of GFC operation as standardized in I.361.

Cell Loss Priority (CLP) is a 1-bit field that indicates the loss priority of an individual cell. Either the end user or the network may set this bit.

Payload Type (PT) is a 3-bit field that discriminates between a cell payload carrying user information or one carrying management information. A subsequent section details the coding of the payload type field.

The Header Error Control (HEC) field provides error checking of the header for use by the Transmission Convergence (TC) sublayer of the PHYsical layer as defined earlier in this chapter.

12.4.3 ATM Layer Functions

This section details the key functions of the ATM layer. The ATM Layer provides many functions, including:

- Cell Construction
- Cell Reception and Header Validation
- Cell Relaying, Forwarding, and Copying Using the VPI/VCI
- Cell Multiplexing and Demultiplexing Using the VPI/VCI
- Cell Payload Type Discrimination
- Interpretation of Pre-defined Reserved Header Values
- Cell Loss Priority (CLP) bit Processing
- Support for Multiple QoS Classes
- Usage Parameter Control (UPC)
- Explicit Forward Congestion Indication (EFCI)
- Generic Flow Control (GFC)
- Connection Assignment and Removal

We already covered the cell construction, reception and header validation, along with several examples of relaying, forwarding, and copying. Subsequent sections cover descriptions of payload type discrimination, interpretation of predefined header values, and cell loss priority processing. Part 5 covers the key topics of support for multiple QoS classes, UPC, EFCI, GFC, and connection assignment and removal.

12.4.4 Relaying and Multiplexing Using the VPI/VCI

As shown through several earlier examples, the heart of ATM is the use of the VPI and VCI for relaying or switching. ATM also effectively performs multiplexing and demultiplexing of multiple logical connections with different quality requirements using the fixed length ATM cell.

Standards reserve VPI=0 for VCCs. Therefore, the number of bits allocated in the ATM cell header to the VPI limit each physical UNI to no more than $2^8-1 = 255$ virtual paths and each physical NNI to no more than $2^{12}-1 = 4095$ virtual paths. Each virtual path can support no more than 2^{16}

= 65,536 virtual channels on the UNI or the NNI. Even though each ATM access circuit can contain a combination of up to 255 VPCs and 65,536 VCCs per VP in theory, service providers and equipment manufacturers typically support less. Check with your equipment vendor or service provider regarding the specific VPI and VCI values supported. Generally, higher speed circuits support for more VPCs and VCCs than lower speed ones do.

Although the UNI and NNI cell formats specify 8 and 12 bits for the VPI, respectively, and 16 bits for the VCI on both interfaces, real ATM systems typically support a smaller number of the lower-order bits in the VPI and VCI. Ranges of VPI/VCI bits supported by interconnected devices must be identical for interoperability. One way to handle VPI/VCI interoperability is to use the ATM Forum's Integrated Local Management Interface (ILMI), which allows each system to query the other about the number of bits supported, thus guaranteeing interoperability.

12.4.5 Meaning of Preassigned Reserved Header Values

A key function of the ATM Layer is the identification and processing of preassigned, reserved header values. Table 12.2 shows the preassigned (also called predefined) header field values for the UNI defined in ITU-T Recommendation and I.361. The 4-bit GFC field applies to all of these values. The ITU-T reserves the first 16 VCIs for future assignment as preassigned, reserved header value functions. Other portions of the book cover the use of these specific header values as indicated below.

Table 12.2   Preassigned, Reserved ATM Cell Header Values

Usage	VPI	VCI		PT	CLP
Unassigned Cell	00000000	00000000	00000000	XXX	0
Invalid	00000000	00000000	00000000	XXX	1
Physical Layer OAM Cell	00000000	00000000	00000000	100	1
Reserved for PHY layer	00000000	00000000	00000000	PPP	1
Meta-signalling (I.311)	XXXXXXXX	00000000	00000001	0AA	C
General broadcast signaling	XXXXXXXX	00000000	00000010	0AA	C
Point-point signaling	XXXXXXXX	00000000	00000101	0AA	C
Segment OAM F4 Cell	YYYYYYYY	00000000	00000011	0A0	A
End-to-end OAM F4 Cell	YYYYYYYY	00000000	00000100	0A0	A
Reserved for VP Functions	YYYYYYYY	00000000	00000111	0AA	A
Segment OAM F5 Cell	YYYYYYYY	ZZZZZZZZ	ZZZZZZZZ	100	A
End-to-End OAM F5 Cell	YYYYYYYY	ZZZZZZZZ	ZZZZZZZZ	101	A
VC RM Cell	YYYYYYYY	ZZZZZZZZ	ZZZZZZZZ	110	A
VP RM Cell	YYYYYYYY	00000000	00000110	110	A

X = "Don't Care"	A = Use by appropriate function
Y = Any VPI value	C = Originator set CLP
Z = Any non-zero VCI	P = Reserved for PHY Layer

ITU-T Recommendation I.432 [14] defines specific values and meanings for the physical layer OAM cells. Chapter 15 defines the metasignaling, general broadcast, and point-to-point signaling functions. Chapters 28 and 29 detail the OAM cell flow formats and functions. Chapter 22 details the use of the Resource Management (RM) cell in the Available Bit Rate (ABR) service category. VC RM cells are invalid on VCI values reserved for other functions, such as signaling or OAM. We described the use of the unassigned and idle cell types earlier in this chapter. The NNI has an additional 4 bits in the VPI field. The NNI preassigned, reserved header fields have not been completely standardized. The current version of the ATM Forum B-ICI specification only requires support for the F4 OAM flows, point-to-point signaling, invalid patterns, and unassigned cells.

The binary coding of the VCI values in Table 12.2 identifies the currently defined VCCs out of the range 0-15 reserved by the ITU-T. Furthermore, the ATM Forum reserves the next 16 VCI values (16–31) on every VPI in support of other protocols. Table 12.3 summarizes these reserved VCI values.

Table 12.3   VCI Values Reserved by the ITU-T and ATM Forum

VCI Value	Purpose
1	Meta-Signaling
2	General broadcast signaling
3	Segment F4 OAM Cell
4	End-to-end F4 OAM Cell
5	Signaling Channel
6	VP Resource Management Cell
7	Reserved for Future VP Functions
16	Integrated Local management Interface (ILMI)
17	LAN Emulation Configuration Server (LECS)
18	Private Network-Network Interface (PNNI) routing channel

12.4.6 Meaning of the Payload Type (PT) Field

Table 12.4 depicts Payload Type (PT) encoding from ITU-T Recommendation I.361. Observe that the rightmost bit is an AAL indication bit (currently used by AAL5 to identify the last cell in a packet). The middle bit indicates upstream congestion, and the first bit discriminates between data and operations cells. Payload types carrying user information may indicate congestion by the Explicit Forward Congestion Indication (EFCI) bit. Also, user cells may indicate whether the cell contains an indication to the AAL protocol. OAM and resource management cells cannot indicate congestion or AAL function. The management information payload type for the F5 flow indicates whether the cell is either a segment or end-to-end Operations Administration and Maintenance (OAM) cell for a VCC. A specific PT coding indicates the presence of a Resource Management (RM) cell. Chapter 13 covers the usage of the AAL_indicate bit by AAL5. Part 5 covers the use of

EFCI and resource management cells. Chapters 28 and 29 detail OAM cell usage.

Table 12.4   ATM Payload Type (PT) Encoding and Meaning

PT Coding	Payload Type (PT) Meaning
000	User Data Cell, EFCI = 0, AAL_indicate = 0
001	User Data Cell, EFCI = 0, AAL_indicate = 1
010	User Data Cell, EFCI = 1, AAL_indicate = 0
011	User Data Cell, EFCI = 1, AAL_indicate = 1
100	OAM F5 segment associated cell
101	OAM F5 end-to-end associated cell
110	Resource management cell
111	Reserved for future VC functions

EFCI = Explicit Forward Congestion Indication
AAL_indicate = ATM-layer-user-to-ATM-layer-user indication

12.4.7 Meaning of the Cell Loss Priority (CLP) Field

A value of 0 in the Cell Loss Priority (CLP) field means that the cell is of the highest priority — or in other words, the network is least likely to discard CLP=0 cells in the event of congestion. A value of 1 in the CLP field means that this cell has low priority — or in other words, the network may selectively discard CLP=1 cells during congested intervals in order to maintain a low loss rate for the high-priority CLP=0 cells. The value of CLP may be set by the user or by the network as a result of a policing action. Part 5 details the uses of the CLP bit in traffic and congestion control.

12.5 ATM-LAYER QOS, TRAFFIC PARAMETERS AND SERVICE CATEGORIES

As you read about ATM, you will encounter the claim that ATM is the best communications networking technology to guarantee Quality of Service (QoS) and reserve bandwidth. The next few sections introduce QoS, traffic parameters, and ATM layer service categories, leaving the details to Part 5.

12.5.1 Introduction to ATM-Layer Quality of Service (QoS)

The most commonly used ATM QoS parameters are:

* Cell Transfer Delay (CTD)
* Cell Delay Variation (CDV)
* Cell Loss Ratio (CLR) (defined for CLP = 0 and CLP = 1 cells)
* Cell Error Ratio (CER)
* Cell Misinsertion Rate (CMR)

For all applications, the CER and the CMR must be extremely small, on the order of one in a billion or less. Therefore, the principal QoS parameters are delay (CTD), variation in delay (CDV), and loss ratio (CLR). To a large extent, human sensory perceptions determine the acceptable values of these major QoS parameters, while data communication protocol dynamics define the rest. Let's explore these QoS drivers in more detail.

Characteristics of the human body's nervous system and sensory perceptions drive many QoS requirements for delay, loss, and delay variation for voice and video applications. The blink of an eye is approximately one-fiftieth of a second, or 20 ms. Video broadcast and recording systems utilize frame rates of between 25 and 30 video frames per second. When frames are played back at this rate, in conjunction with the image persistence provided by television displays, the human eye-brain perceives this as continuous motion. When lost or errored cells disrupt a few frames in succession, the human eye-brain detects the discontinuities in motion, which are subjectively objectionable. The human ear is also sensitive to such differences in delay on a similar time scale. This shows up in telephony, where the reception of a speaker's echo becomes objectionable within less than the blink of an eye; indeed, 50 ms is the round-trip delay where standards require echo cancellation. Delay variation also affects the perception of audio and video. Audio is the most sensitive, since the human ear perceives even small delay variations as changes in pitch. The human ear-brain combination is less sensitive to short dropouts in received speech, being able to accept loss rates on the order of 0.5 percent.

Data applications determine other aspects of QoS. Many data protocols respond to delay and loss through retransmission strategies to provide guaranteed delivery. One example is the Transmission Control Protocol (TCP), widely used to flow control the transfer of World Wide Web text, image, sound, and video files. Data applications are extremely sensitive to loss because they respond by retransmitting information. A user perceives this as increased delay if, for example, retransmissions due to loss extend the time required to transfer a large file carrying a video or audio image to a user surfing the web. Since most data networks exhibit significant delay variation (especially over the Internet), applications insert a substantial delay before starting playback. This technique works fine for one-way communication, but impedes two-way, interactive communication if the round-trip delay exceeds 300 ms. Using satellite communications for a voice conversation illustrates the problem with long delays: since the listener can't tell if the speaker has stopped, or merely paused, and simultaneous conversation frequently occurs. Most data communication protocols remain relatively insensitive to delay variations, unless the delay varies by values on the order of a large fraction of a second which causes a retransmission time-out.

Table 12.5 lists the major causes of these QoS impairments [15]. Note that the offered traffic load and functions performed by the switch primarily determine the delay, variation in delay, and loss parameters. Error statistics largely involve cell errors, but they also include misinsertion events when a cell header is corrupted by several errors so that the cell erroneously appears as valid. Of course, the more switching nodes a cell traverses, the more the quality degrades. All QoS parameters accrue in approximately a linear fashion except for delay variation, which grows at a rate no less than the square root of the number of nodes traversed as discussed in Chapter 26.

Table 12.5   Mapping of Network Impairments to ATM QoS Parameters

Impairment	CTD	CDV	CLR	CER	CMR
Propagation delay	✓				
Switch queuing architecture	✓	✓	✓		
Switch buffer capacity	✓	✓	✓		
Switch resource allocation/admission control	✓	✓	✓		
Variations in traffic load	✓	✓	✓		✓
Switch and link failures			✓		
Media bit error rate and error burst statistics			✓	✓	✓
Number of switching nodes traversed	✓	✓	✓	✓	✓

12.5.2 Introduction to the ATM Traffic Contract

The ATM traffic contract is an agreement between a user and a network, where the network guarantees a specified QoS — *if and only if* the user's cell flow conforms to a negotiated set of traffic parameters. The traffic descriptor is a list of parameters which captures intrinsic source traffic characteristics.

Figure 12.15 illustrates the following traffic contract parameters [15] in a worst-case, bursty traffic scenario.

◎ PCR, expressed in units of cells per second, defines the fastest rate at which a user can send cells to the network.
◎ CDVT, expressed in units of seconds, constrains the number of cells the user can send at the physical medium rate as indicated in the figure.
◎ SCR, expressed in units of cells per second, defines maximum sustainable, average rate that a user can send cells to the network.
◎ MBS is the maximum number of cells that the user can send at the peak rate in a burst, within the sustainable rate.

We cover this model in greater depth in Chapter 21, giving exact expressions and illustrating these concepts with further examples. This introduction serves to define the concepts until then.

ATM standards utilize a *leaky bucket* algorithm to determine conformance of an arriving cell stream to the above traffic parameters in an action called Usage Parameter Control (UPC). Standards also refer to this technique as policing. A leaky bucket algorithm in the network checks conformance of a cell flow from the user by conceptually pouring a cup of fluid for each cell into two buckets, one with a depth proportional to CDVT leaking at a rate corresponding to the PCR, and the second with depth defined by MBS leaking at a rate corresponding to the SCR. If the addition of any cup of cell fluid would cause a bucket to overflow, then the cell arrival is *nonconforming*, and its fluid is not added to the bucket. The ATM Forum gives a formal definition of the leaky bucket algorithm, called the Generic Cell Rate Algorithm (GCRA), in Reference 15.

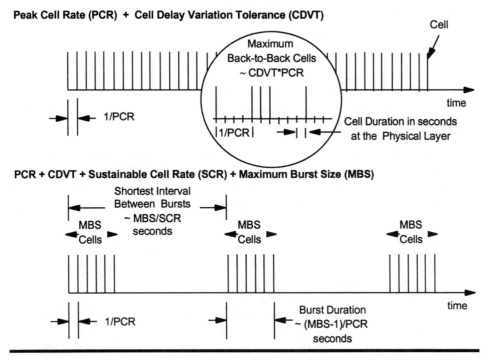

Figure 12.15 Illustration of Principal ATM Traffic Parameters

An increasing amount of user equipment and switches also implement traffic shaping according to the above parameters. Traffic shaping ensures that the cell stream generated by an ATM device for a particular connection conforms to the traffic contract so that the network doesn't discard violating cells. Traffic shaping implementations operate at either the peak rate only, or at the combined peak and sustainable rates subject to a maximum burst size. Shaping is essential for user equipment in a network which does not support tagging using the CLP bit. Also, shaping is very useful when interconnecting networks to ensure that traffic exiting one network conforms to the traffic contract with the next network down the line.

12.5.3 Introduction To ATM Layer Service Categories

We've used the term ATM service category several times, but now we define what all those darn acronyms mean. They are hard to keep straight, so this section seeks to explain each in lay terms, with Chapter 21 providing the rigorous definition of each service category, or in ITU-T parlance, *transfer capability*. Each service category definition includes terms that define the traffic contract parameters and QoS characteristics. The ATM Forum Traffic Management 4.0 specification [15] (abbreviated TM 4.0) defines the following ATM-layer service categories:

- **CBR**   Constant Bit Rate
- **rt-VBR**   Real-time Variable Bit Rate
- **nrt-VBR**   Non-real-time Variable Bit Rate
- **UBR**   Unspecified Bit Rate
- **ABR**   Available Bit Rate

The ATM Forum defines attributes and application characteristics for each of these service categories [16] as summarized below. Chapter 21 addresses this subject in detail.

- The CBR service category supports real-time applications requiring a fixed amount of bandwidth defined by the PCR. CBR supports tightly constrained CTD and CDV for applications that cannot tolerate variations in delay. Example applications are voice, constant-bit-rate video, and Circuit Emulation Services (CES), as covered in Chapter 17.

- The rt-VBR service category supports time-sensitive applications, which also require constrained delay and delay variation requirements, but which transmit at a time varying rate constrained to a PCR and an "average" rate defined by the SCR and MBS. The three parameters PCR, SCR, and MBS define a traffic contract in terms of the worst-case source traffic pattern illustrated in Figure 12.15 for which the network guarantees a specified QoS. Such bursty, delay-variation-sensitive sources, conforming to the traffic contract, such as voice and variable-bit-rate video, may be statistically multiplexed as analyzed in Chapter 25.

- The nrt-VBR service category supports applications that have no constraints on delay and delay variation, but which still have variable-rate, bursty traffic characteristics. The traffic contract is the same as that for rt-VBR. Applications include packet data transfers, terminal sessions, and file transfers. Networks may statistically multiplex these VBR sources effectively.

- The ABR service category works in cooperation with sources that can change their transmission rate in response to rate-based network feedback used in the context of closed-loop flow control. The aim of ABR service is to dynamically provide access to bandwidth currently not in use by other service categories to users who can adjust their transmission rate. In exchange for this cooperation by the user, the network provides a service with very low loss. Applications specify a maximum transmit-rate bandwidth (PCR) and the minimum required rate, called the Minimum Cell Rate (MCR). ABR service does not provide bounded delay variation; hence real-time applications are not good candidates for ABR. Example applications for ABR are LAN interconnection, high-performance file transfers, database archival, non-time-sensitive traffic and web browsing. Chapter 22 covers the subject of ABR in detail.

- The ATM Forum also calls the UBR service category a "best effort" service, which does not require tightly constrained delay and delay variation and provides no specific quality of service or guaranteed throughput. This traffic is therefore "at risk" since the network provides no performance guarantees for UBR traffic. Most LANs and IP implementations provide a "best effort" service today. The Internet and local area networks are examples of this "best effort" delivery performance. Example applications are LAN emulation, IP over ATM, and non-mission-critical traffic.

Table 12.6 summarizes the attributes of these ATM-layer service categories from the ATM Forum's Traffic 4.0 specification. Each ATM service category uses one or more of the traffic contract parameters defined in the previous section. ABR uses these parameters, and more as detailed in Chapter 22. Furthermore, most service categories require guarantees on either loss, delay variation, or bandwidth. The one exception is UBR which, being a best effort service, requires no guarantees whatsoever. Finally, only ABR utilizes feedback control.

Table 12.6   Service Category Attributes and Guarantees

Service Category	Traffic Descriptor	Guarantees			Feedback Control
		Loss (CLR)	Delay Variance (CDV)	Bandwidth	
CBR	PCR	Yes	Yes	Yes	No
rt-VBR	PCR, SCR, MBS	Yes	Yes	Yes	No
nrt-VBR	PCR ,SCR, MBS	Yes	No	Yes	No
ABR	PCR, MCR, and behavior parameters	Yes	No	Yes	Yes
UBR	PCR	No	No	No	No

Table 12.7 summarizes the above discussion to show the suitability of the above service categories for a number of commonly used applications [16]. In general, CBR service applies best to applications that currently require the dedicated bandwidth and minimal delay variation that TDM networking provides. The nrt-VBR service category best serves traditional packet switched data applications, such as FR, IP, and LAN traffic. As noted above, the ATM Forum specified the rt-VBR service category for variable bit rate video transport. The ABR service category best serves LAN and WAN data transport through its sophisticated flow control mechanism. Finally, although UBR may not be the best for any application, it may well be the cheapest.

Table 12.7   Suitability of ATM Forum Service Categories for Various Applications

Applications	CBR	rt-VBR	nrt-VBR	ABR	UBR
Critical data	Good	Fair	Best	Fair	No
LAN interconnect	Fair	Fair	Good	Best	Good
WAN data transport	Fair	Fair	Good	Best	Good
Circuit emulation	Best	Good	No	No	No
Telephony, videoconferencing	Best	TBD	TBD	No	No
Compressed audio	Fair	Best	Good	Good	Fair
Video distribution	Best	Good	Fair	No	No
Interactive multimedia	Best	Best	Good	Good	Fair

Note: TBD = to be determined.

12.6 REVIEW

This chapter covered the foundations of B-ISDN: the PHYsical (PHY) layer and the ATM layer. The chapter started with an overview of how these layers fit into the overall B-ISDN protocol model, and then moved on to investigate the sublayer structure of the PHY and ATM layers. The PHY layer broke down into Physical Medium Dependent (PMD) and Transmission Convergence (TC) sublayers. The text presented examples of how the PMD supports different physical media and interface rates and how the TC sublayer effectively makes the PHY layer appear as a pipe that transfers cells at a maximum rate to the ATM layer. The exposition then continued with the ATM layer protocol model, including an explanation of Virtual Path (VP) and Virtual Channel (VC) links, connections, and concepts complete with network examples from several points of view. Next, the text dissected the ATM cell itself, clearly defining every field in the header, and reviewing some of the basic functions. The discussion referenced detailed treatments of particular subjects in other parts of this book and in various standards. The chapter concluded with an introduction to the important ATM layer concepts of Quality of Service (QoS), the traffic contract, and ATM layer service categories.

12.7 REFERENCES

[1]   ITU-T, *B-ISDN Protocol Reference Model and Its Application*, Recommendation I.321, 1991.

[2]   A. Tannenbaum, *Computer Communications, Third Edition*, Prentice-Hall, 1996.

[3]   ATM Forum, *DS1 Physical Layer Specification – Version 1.0*, af-phy-0016 September 1994.

[4]   ANSI, *ANSI Standard for Telecommunications – Broadband ISDN – Physical Layer Specifications for User-Network Interfaces Including DS1/ATM*, T1.646-1995, May 12, 1995.
[5]   ATM Forum, *User-Network Interface Specification, Version 3.1*, af-uni-0010.002, September 1994.
[6]   W. Goralski, *Introduction to ATM Networking*, McGraw-Hill, 1995.
[7]   ISO/IEC, *Information processing systems — local and metropolitan area networks — Part 6: Distributed queue dual bus (DQDB) subnetwork of a metropolitan area network (MAN)*, , 8802-6, 1993.
[8]   ATM Forum, *Inverse Multiplexing for ATM (IMA) Specification Version 1.0*, af-phy-0086.000, July 1997.
[9]   ITU-T, *B-ISDN General Network Aspects*, Recommendation I.311, August 1996.
[10] ANSI, *BISDN ATM Functionality and Specification*, T1.627-1993, 1993.
[11] ITU-T, *B-ISDN ATM Layer Specification*, Recommendation I.361, November 1995.
[12] ITU-T, *B-ISDN user-network interface*, Recommendation I.413, March 1993.
[13] ANSI, *BISDN UNI: Rates and Formats Specification*, T1.624-1993, 1993.
[14] ITU-T, *B-ISDN user-network interface – Physical layer specification: 155 520 kbit/s and 622 080 kbit/s operation*, Recommendation I.432, August 1996.
[15] ATM Forum, *ATM Forum Traffic Management Specification, Version 4.0*, af-tm-0056.000, April 1996.
[16] L. Lambarelli, *ATM Service Categories: The Benefits to the User*, http://www.atmforum.com/atmforum/service_categories.html.

ATM Adaptation Layers

The ATM Adaptation Layer (AAL) layer provides support for higher-layer services such as signaling, circuit emulation, voice, and video. AALs also support packet-based services, such as IP, LANs, and frame relay. First, this chapter introduces the initial ITU-T notion of AAL service classes and applies this to real world applications. Next, the text covers the AAL generic layered structure. The chapter then delves into each ATM Adaptation Layer, describing the basic characteristics, formats, and procedures. The text gives several examples to further clarify the operation of each AAL. The remaining chapters in this part, as well as the coverage in Part 4 regarding the higher layers of the user and control plane, rest upon this foundation of AALs.

13.1 ATM ADAPTATION LAYER (AAL) — PROTOCOL MODEL

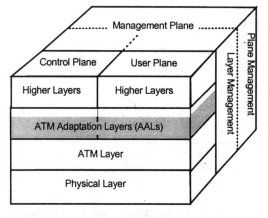

The I.363 series of ITU-T Recommendations define the next higher layer of the B-ISDN protocol stack — the AAL. First, this section covers the attributes for AAL service classes A through D and their use in representative applications. Next, the text describes the generic AAL protocol model comprised of a Segmentation And Reassembly (SAR) sublayer, along with Common Part and Service Specific Convergence Sublayers (CPCS and SSCS). The chapter then describes the Common Part (CP) format and protocol for each standardized AAL, illustrated by example applications.

13.1.1 AAL Service Attributes Classified

In 1993, ITU-T Recommendation I.362 [1] defined the basic principles and classification of AAL functions. Although no longer in force, many concepts initially defined in I.362 are still used in the ATM Forum documents and other standards. The attributes of the service class are: the timing relationship required between the source and destination, whether the bit rate is constant or variable, and whether the connection mode is connection-oriented or connectionless. The AAL service class is a separate concept from the ATM layer's service category and Quality of Service (QoS) introduced in Chapter 12 and detailed in Part 5. Chapter 15 describes how the service class (or bearer capability), service category, and QoS class (or optionally, explicit QoS parameters) can all be signaled separately in a SVC call setup message. Figure 13.1 depicts key attributes and example applications for the four AAL service classes, labeled A through D envisioned in I.362:

- Class A — Constant Bit Rate (CBR) service with end-to-end timing, connection-oriented
- Class B — Variable Bit Rate (VBR) service with end-to-end timing, connection-oriented
- Class C —VBR service with no timing required, connection-oriented
- Class D —VBR service with no timing required, connectionless

Attribute	Service Class			
	Class A	Class B	Class C	Class D
Timing relation between source and destination	Required		Not Required	
Bit Rate	Constant	Variable		
Connection Mode	Connection-Oriented			Connection -less
AAL(s)	AAL1	AAL2	AAL3/4 or AAL5	AAL3/4 or AAL5
Example Applications	DS1, E1, nx64 kbps emulation	Packet Video, Audio	Frame Relay, X.25	IP, SMDS

Figure 13.1 ATM ITU ATM/B-ISDN Service Classes

Some standards — for example, classical IP over ATM — specify in detail how the higher-layer protocol sets the closely related, yet different parameters of AAL service class, ATM service category and QoS in signaling messages. The mapping of service classes to AALs is now complete with the adoption of AAL2. Later, this section maps the AAL(s) to the attributes for each service class and gives example applications.

The ITU-T initially defined AAL1 through AAL4 to directly map to the AAL service classes A through D. The history of AAL development for VBR services changed this simple concept. Initially, the ITU-T targeted AAL3 for connection-oriented services and AAL4 for connectionless services. However, as the experts defined the details, they realized that AAL3 and AAL4 were common enough in structure and function to be combined into one, called AAL3/4, which the IEEE 802.6 standard adopted and applied to SMDS. More recently, the computer industry conceived of AAL5 in response to perceived complexity and implementation difficulties in AAL3/4. When initially proposed in 1991, AAL5 was called the Simple Efficient Adaptation Layer (SEAL) for these reasons [2]. The ATM Forum, ANSI, and the ITU-T adopted AAL5 in a relatively short time compared to the usual standards process. Since then, AAL5 has become the predominant AAL of choice in the majority of data communications equipment. As we detail in Part 4, AAL5 carries the Internet Protocol (IP), LAN frames, signaling messages, frame relay, and video. The ATM Data eXchange Interface (DXI) and ATM Forum Frame-based UNI (FUNI) also use AAL5. The ITU-T initially defined AAL1 to interwork directly with legacy TDM networks and N-ISDN. Subsequent specifications generated by ANSI and ETSI for domestic TDM rates along with ATM Forum circuit emulation specifications detailed interoperability agreements and conventions in support of CBR applications. The standards bodies completed the base AAL2 standard in 1997 to support variable bit rate voice and video applications more efficiently than the constant bit rate AAL1. The ITU-T I.363.x series of recommendations state the standards for all AALs.

Table 13.1 identifies applications that utilize particular combinations of AALs and the ATM layer service category (from Chapter 12). As seen from the blank spaces in the table, not all combinations of AAL and ATM layer service category are relevant. The next chapter introduces the protocols listed in this table with detailed coverage provided in Part 4. AAL1 interworks with Time Division Multiplexing (TDM) protocols using circuit emulation. AAL2 provides the infrastructure to support variable bit rate voice and video. Standards currently define AAL3/4 for ATM access to Switched Multimegabit Data Service (SMDS) as well as an option in ATM DXI and FUNI. Most data-oriented applications utilize AAL5, as seen from the table, including an early specification for video on demand defined by the ATM Forum in advance of AAL2 standardization. Early LAN and internetworking protocols (i.e., LAN Emulation (LANE) and classical IP over ATM) utilized only the Unspecified Bit Rate (UBR), also known as the best effort, service category. The ATM Forum's LANE specification version 2.0 and the Private Network-Network Interface (PNNI) specification provide access to all ATM service categories as indicated in the table. A native ATM Application Programming Interface (API) and basic signaling procedures provide access to any combination of AAL and ATM service category. However, most implementations currently use the standard combinations listed in Table 13.1, although a few proprietary protocols directly access the ATM layer.

Table 13.1   Applications of AAL and ATM Layer Service Category

| AAL | ATM Layer Service Category | | | | |
	CBR	rt-VBR	nrt-VBR	ABR	UBR
AAL1	Circuit emulation, N-ISDN, voice over ATM				
AAL2		Variable bit rate voice and video			
AAL3/4			ATM/SMDS		
AAL5	LANE 2.0, PNNI 1.0	Video on demand, LANE 2.0, PNNI 1.0	Frame relay, ATM/SMDS, LANE 2.0, PNNI 1.0	LANE 2.0, PNNI 1.0	Classical IP over ATM, LANE 1.0/2,0, PNNI 1.0, MPOA
Null	PNNI 1.0	PNNI 1.0	PNNI 1.0	PNNI 1.0	PNNI 1.0

13.1.2 AAL Protocol Structure Defined

The B-ISDN protocol model adapts the services provided by the ATM layer to those required by the higher layers through the AAL. Figure 13.2 depicts the structure and logical interfaces for AAL1, AAL3/4 and AAL5 as defined in the I.363 series of recommendations. AAL2 has a different structure as described in section 13.3.1. At the top of Figure 13.2, an AAL Service Access Point (SAP) provides services to higher layers by passing primitives (e.g., request, indicate, response, and confirm) concerning the AAL Protocol Data Units (AAL-PDUs). Subsequent sections summarize the resulting transfer of PDUs between sublayers and across the AAL-SAP from a functional point of view. See the standards referenced in the following sections for details on the protocol primitives for each AAL.

Standards further subdivide the Common Part (CP) for AAL1, AAL3/4 and AAL5 into the Convergence Sublayer (CS) and the Segmentation and Reassembly (SAR) sublayer as shown at the bottom of the figure. The CS layer contains a Service-Specific (SS) and Common Part (CP) sublayer as indicated in the figure. The SSCS may be null, which means it does nothing, or as the name implies, provides particular services to the higher-layer AAL user. Chapter 15 covers the signaling SSCS while Chapter 18 covers the FR-SSCS. The CPCS must always be implemented along with the SAR sublayer. These layers pass primitives regarding their respective PDUs among themselves as labeled in the figure, resulting in the passing of SAR-PDU primitives (which is the ATM cell payload) to and from the ATM layer via the ATM-SAP.

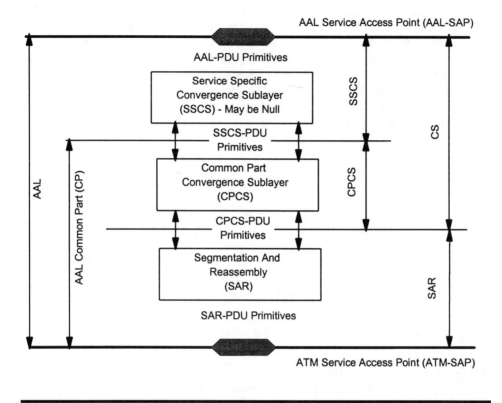

Figure 13.2   Generic AAL Protocol Sublayer Model

The layered AAL protocol model presented above is rather abstract, so let's look at it from another point of view. Let's zero-in on the cell and PDU level in the receiver side of an ATM device.   Figure 13.3 depicts the ATM layer at the bottom and moves up to the AAL SAR sublayer, and up through the AAL CS sublayers to the higher-layer protocol.   Notice how this model follows the general layered methodology found in data communications architectures introduced in Chapter 5. The PDU at each layer has a header, and optionally a trailer, that conveys information for use at the particular sublayer. Starting at the bottom of the figure, the ATM layer passes payloads from cells with valid headers up to the AAL SAR sublayer across the ATM-SAP for a particular Virtual Channel Connection (VCC).   All currently defined AALs operate at the VCC level.   The SAR sublayer interprets the header (and optional trailer) for AAL SAR-PDUs constructed from the cell payloads.   In general, the SAR-PDU and cell payload boundaries are not aligned.   If the SAR layer successfully reassembles an entire AAL CPCS-PDU, then it passes it up to the CPCS layer, which has its own header and optional trailer.   The CPCS layer then extracts the AAL SDU.   If the AAL-SSCS layer is null, then the AAL-SDU is passed directly to the AAL user across the AAL SAP.

If the SSCS sublayer is non-null, then the CPCS sublayer passes its payload up to the SSCS sublayer.   The SSCS sublayer finally extracts the

AAL SDU using its header and optional trailer. In some AAL definitions, the SSCS layer may derive multiple AAL Interface Data Unit (IDUs) from a single SSCS-PDU for transfer across the AAL-SAP interface to the higher layer protocol. If the SSCS is null, then the IDU is exactly the same as the CPCS payload, also called the AAL Service Data Unit (AAL-SDU). The process of receiving AAL-IDUs from the higher-layer protocol and processing them down through the AAL convergence sublayer and segmentation is the reverse of the above description.

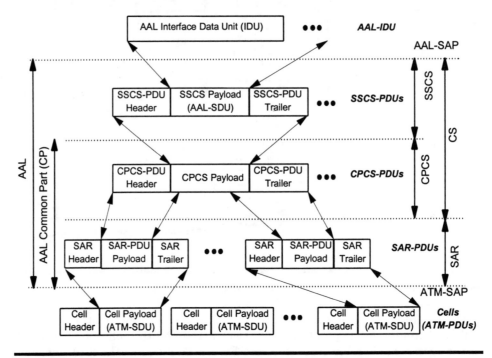

Figure 13.3   AAL Protocol Data Unit (PDU) Model

The AAL-IDU (usually equivalent to the AAL-SDU) provides the all important interface to the AAL that makes the ATM cell level capabilities accessible to voice, video, and packet data applications as detailed in Part 4. The IDU, for example, is the API in the Windows operating system for packet-layer transmission. In the case of AAL1 emulating a TDM circuit, the IDU is a sequence of bits with particular timing attributes.

13.1.3 Key AAL Attributes

Why did the standards bodies define so many AALs when everything ends up packed into the 48-byte payload of ATM cells anyway? The answer lies in differences between key attributes supported by each AAL as summarized in Table 13.2. First, the various AALs have vastly different PDU lengths as seen at the AAL-SAP. AAL1 and AAL2 have shorter PDU lengths to support real-time services such as circuit emulation, voice, and video. On the other hand, AAL3/4 and AAL5 support traditional packet data services by carrying anything ranging from a minimal 1-octet size up to a jumbo size of 65,535 octets. Another key AAL attribute is support for multiple logical channels over a single ATM VCC. AAL2 does this to support packetized voice and video to reduce packetization delay and improve packet fill efficiency. AAL3/4 does this to make more efficient use of VCCs and reduce delay variation encountered by individual packets. Finally, some AALs support a User-User Indication (UUI) information transfer capability. Currently, standards don't exist for these fields; however, they offer an extensibility that other AALs don't.

13.2 ATM Adaptation Layer (AAL) Attributes

Attribute	AAL1	AAL2	AAL3/4	AAL5
AAL-PDU Size Range (Octets)	46-47	1-64	1-65,535	1-65,535
Multiple logical channels per VCC	No	Yes	Yes	No
User-User Indication (UUI)	No	Yes	No	Yes

The following sections define the sublayers and characteristics for each of the currently standardized AALs:

- AAL1 — constant bit rate, real-time traffic
- AAL2 — variable bit rate, real-time traffic
- AAL3/4 — variable bit rate data traffic
- AAL5 — lightweight variable bit rate traffic

Each section then provides one or more examples illustrating the application of the ATM Adaptation Layer to a real world application.

13.2 ATM ADAPTATION LAYER 1 (AAL1)

As stated in ITU-T Recommendation I.363.1 [3], the AAL1 protocol specifies the means to:

☞ Transfer service data units received from a source at a constant source bit rate and then deliver them at the same bit rate to the destination
☞ Optionally transfer timing information between source and destination
☞ Optionally transfer TDM structure information between source and destination
☞ Optionally perform Forward Error Correction (FEC) on the transferred data
☞ Optionally indicate the status of lost or erroneous information

13.2.1 AAL1 Segmentation And Reassemby (SAR) Sublayer

The AAL1 SAR sublayer provides the following services:

☞ Map between the 47 octet CS-PDU and the 48-octet SAR-PDU using a 1-octet SAR-PDU header
☞ Indicate the existence of CS function using a bit in the SAR-PDU header
☞ Generate sequence numbering for SAR-PDUs at the source and validate received sequence numbers at the destination before passing them to the CS sublayer
☞ Perform error detection and correction on the Sequence Number (SN) field

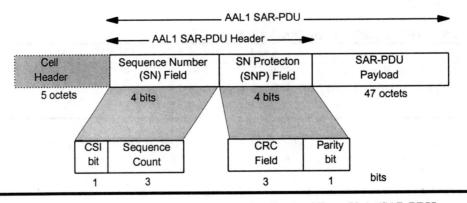

Figure 13.4 AAL1 Segmentation And Reassembly Protocol Data Unit (SAR-PDU)

Figure 13.4 depicts the SAR-PDU for AAL1. Since the AAL 1 SAR uses one octet, 47 octets remain for user data in the SAR-PDU payload. The AAL1 SAR header has two major fields: the Sequence Number (SN) and the Sequence Number Protection (SNP) field as indicated in the figure. Within the SN field, the origin increments the 3-bit sequence count sequentially. The receiver checks for missing or out-of-sequence SAR-PDUs, generating a signal alarm when this occurs. AAL1 CS protocols utilize the Convergence Sublayer Indication (CSI) bit for specific functions described later in this section. The 3-bit CRC field computes a checksum across the 4-bit SN field. As further protection against errors, the parity bit represents even parity across the first 7 bits in the 1-octet SAR-PDU header.

The sequence number is critical to proper operation of AAL1 since an out-of-sequence or missing SAR-PDU disrupts at least 47 octets of the emulated circuit's bit stream. Standards define a detailed procedure to correct many problems due to bit errors in the sequence number field, or to accurately detect uncorrected errors. The state machine of Figure 13.5 illustrates operation at the receiver. While in the correction mode the receiver corrects single-bit errors in the SAR header using the CRC. If after CRC correction the parity check fails, then the receiver switches to detection mode since an uncorrectable multiple bit error has occurred. The receiver stays in detection mode until no error is detected and the sequence number is sequential again (i.e., valid).

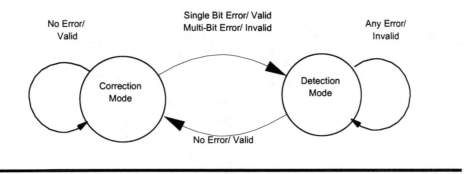

Figure 13.5 AAL1 Sequence Number Protection Operating Modes

13.2.2 AAL1 Convergence Sublayer Functions

The AAL1 Convergence Sublayer (CS) defines the following functions in support of the transport of TDM circuits, video signals, voiceband signals, and high-quality audio signals:

- Blocking and deblocking of user information to and from 47-octet SAR-PDU payloads
- Handling cell delay variation for delivery of AAL-SDUs at a constant bit rate
- Partial fill of the SAR-PDU payload to reduce packetization delay
- Optional processing of the sequence count and its error check status provided by the SAR sublayer to detect lost and misinserted cells
- Synchronous recovery of source clock frequency at the destination end using the Synchronous Residual Time Stamp (SRTS) method
- Transfer of TDM structure information between source and destination using Structured Data Transfer (SDT)
- Utilize the CS Indication (CSI) bit provided by the SAR sublayer to support specific CS functions, such as SRTS and SDT
- Asynchronous recovery of the TDM clock at the destination using only the received cell stream interarrival times or playback buffer fill
- Optional forward error correction combined with interleaving of the AAL user bit stream to protect against bit errors
- Optionally generate reports on the end-to-end performance deduced by the AAL based on lost and misinserted cells; buffer underflow and overflow; and bit error events

These AAL1 CS functions provide a menu that higher-layer applications use to provide required service features (e.g., timing recovery), deliver end-to-end performance (e.g., loss and delay), and account for anticipated network impairments (e.g., cell loss and delay variation). Table 13.3 illustrates some examples of this selection of AAL1 CS menu items by higher-layer services [3], such as Circuit Emulation Service (CES), voice, and video. The following sections give details for use of the SDT pointer, the unstructured mode, and the clock recovery CS functions.

Table 13.3   Application of AAL1 CS Functions to Transport of Specific Services

AAL1 CS Function	Structured TDM CES	Unstructured TDM CES	Video Signals	Voiceband Signals
CBR rate	nx64 kbps	PDH rate *	MPEG2 rate	64 kbps
Clock recovery	Synch	Synch, asynch	Asynch	Synch
Error correction	Not used	Not used	Used	Not used
Error status	Not used	Not used	Used	Not used
SDT pointer	Used†	Not used	Not used	Not used
Partial cell fill	Optional	Not used	Not used	Not used

Notes:

* For example, 1.544, 2.048, 6.312, 8.448, 34.368, or 44.736 Mbps.

† Pointer not necessary for n = 1 (64 kbps).

13.2.3 Structured Data Transfer (SDT) Convergence Sublayer

The SDT CS utilizes a pointer field in even-numbered SAR-PDUs once every eight SAR-PDUs (i.e., once within the wraparound interval of the 3-bit sequence number field) to communicate the beginning of the structure boundary. Most SAR-PDUs use the non-P format as shown in Figure 13.6a.

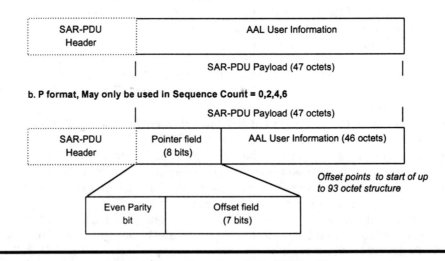

Figure 13.6   AAL1 Structured Data Transfer (SDT) CSI Bit Usage

The SDT CS supports octet structures ranging from 2 to 93 octets at 8 kHz. The CSI bit in the AAL1 SAR PDU indicates when the pointer field is present in an even-sequence-numbered SAR-PDU, called a P-format SAR-PDU as illustrated in Figure 13.6b. This means that CS-PDUs carry only 46 octets of user data when the pointer is present (p-Format), as compared with the 47 octets when the pointer is absent (non-P format). According to ITU-T Recommendation I.363.1, the pointer field should be used at the first opportunity, that is, sequence number zero, as shown in the example below. The SDT CS uses the pointer field to reconstruct the precise 64-kbps time slot alignment at the destination, since the time slot octets generally do not coincide with the CS-PDU boundaries. The 7-bit offset within the pointer field in the CS performs this function by pointing to the first octet of the structure. The octets prior to this octet are filled with data from the structure, from the previous sequence of eight CS-PDUs.

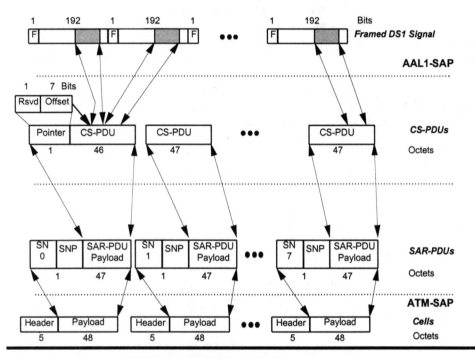

Figure 13.7   Structured Mode AAL1 SAR and CS Example

The Structured Data Transfer (SDT) convergence sublayer uses the pointer field to support transfer of nx64-kbps signals as illustrated in Figure 13.7. Starting at the top of the figure, the SDT CS accepts specific 64-kbps time slots from the AAL1 user (e.g., a N-ISDN video teleconference) as indicated by the shaded portions of the framed DS1 in the figure. Note that the SDT CS does not convey TDM framing information indicated by the fields labeled "F" at the top of the figure. The figure illustrates how the pointer field occurs in only one out of seven SAR-PDUs as described above. The SAR sublayer adds the sequence number, inserts the data from the CS and computes the CRC and parity over the SAR header, passing the 48-octet SAR-PDU to the ATM layer. The ATM layer adds the 5-byte ATM header and outputs the sequence of 53-byte cells as shown at the bottom of the figure. The process at the receiver is analogous to that described above, except the steps are reversed. The receiver then employs its local clock to take the contents from the received CS PDUs and clock out the resulting bit stream to the receiving device, placing it in the appropriate position in the DS1 TDM framing structure.

13.2.4 Unstructured Mode Convergence Sublayer

Figure 13.8 illustrates the unstructured method, which takes data from the AAL1 source, performs CS functions (e.g., asynchronous clocking information or FEC) and passes these data units to the SAR sublayer. The SAR sublayer takes 47-octet CS PDUs, prefixes these segments with a 1-octet SAR header, and passes them to the ATM layer for transmission as the cell payload. At the destination, the receiver's SAR sublayer takes the 48-octet ATM cell payload and examines the SAR-PDU header for errors. If the SAR-PDU header contains errors, the SAR layer then passes an indication to the CS, which may in turn inform the end application. The SAR sublayer passes correctly received SAR-PDU payloads to the CS, if present. The CS then reclocks the received bit stream to the destination AAL user, transparently passing any framing information in the original source signal as indicated at the top of the figure by the fields marked "F" in the AAL1 user bit stream. Unstructured mode may use either a synchronous or asynchronous clock recovery method as described in the next section.

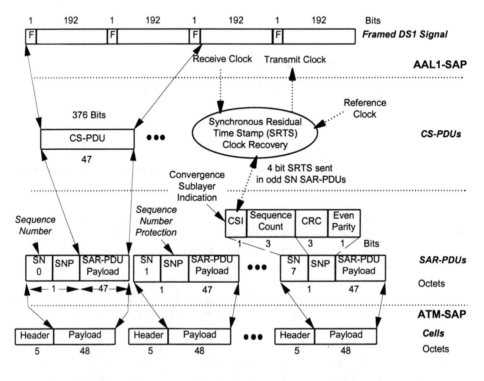

Figure 13.8   Unstructured Mode AAL1 SAR and SRTS CS Example

Figure 13.8 also illustrates the operation of unstructured mode using the Synchronous Residual Time Stamp (SRTS) method, which employs the CSI bit in SAR-PDUs with odd sequence numbers. The next section details the operation of the SRTS and adaptive clock recovery methods.

13.2.5 AAL1 Clock Recovery Methods

ITU-T Recommendation I.363.1 defines two methods for recovery of the source clock at the destination: synchronous and asynchronous. The synchronous method requires that the source and destination have access to the same accurate reference clock, which the standards call synchronous mode. Since N-ISDN is a synchronous TDM network, accurate clocks are usually available in most modern private and public digital telephone systems. Note that SDT CS requires synchronous clock sources at each device which converts between TDM and ATM. The destination SDT CS utilizes the structure pointer along with the structure parameters and its accurate clock to place the received 64-kbps time slots in the framed TDM structure for delivery to the end user. The SDT mode is best suited to network designs that do not require an entire DS1/E1 between sites. For example, SDT works well in designs requiring the interconnection of PBXs in a corporate network. Accurate timing can be obtained from carrier interface circuits connected to the PBXs, from a SONET/SDH ATM UNI, or from the DS3 PLCP layer.

The asynchronous method uses either an adaptive method or the Synchronous Residual Time Stamp (SRTS) method. Asynchronous clock recovery is essential to the support of legacy TDM networks, for example T1 multiplexers, over an ATM network. The adaptive clock recovery method requires no reference clock at all, and hence offers a plug-and-play approach to circuit emulation. Since standards don't exist for adaptive clock recovery, many implementations offer a range of tuning parameters to control clock recovery performance. In adaptive clock recovery, the destination recovers an estimate of the source clock frequency from the inter-cell spacing of the received cell stream or the playback buffer fill level [4]. Adaptive clock recovery effectively controls jitter of the recovered clock signal; however, the long term clock frequency may wander outside of normal TDM clock tolerances. A disadvantage of adaptive clock recovery is sensitivity to time varying cell interarrival times caused by congestion or dynamically established and released CBR connections. In real world networks, adaptive clock recovery usually works quite well.

On the other hand, SRTS requires an accurate reference clock at both the source and the destination devices. Unlike the synchronous method, SRTS aims to accurately transfer the users clock from source to destination, for example a requirement in many legacy T1 multiplexer networks. To do this, SRTS measures the difference between the source data rate and the local accurate clock reference and transmits the difference to the destination in odd-numbered sequence number fields as detailed below. The destination uses this information to compute the clock rate at the destination prior to clocking out the received bit stream to the destination AAL1 user. SRTS works well in a single-carrier network or a private network where accurate clocks tied to the same reference are available. In single carrier ATM networks, the physical access circuit often provides an accurate timing source. SRTS may result in degraded performance in connections that traverse multiple carrier networks, since an accurate time reference may not be available at both the source and the destination. Although the SRTS method delivers less jitter and receive frequency wander than the adaptive method, it requires an accurate reference clock at each end of the connection. Carefully investigate the timing source capabilities of your application to select the clock recovery method best suited to your network.

Figure 13.9 depicts more details on the operation of SRTS CS asynchronous clock recovery. As stated previously, SRTS assumes that both the origin and destination have access to a common clock of frequency fn. The signal (e.g., DS1) has a service clock frequency fs. The objective is to pass sufficient information via the AAL so that the destination can accurately reproduce this clock frequency. At the bottom part of the figure, the network reference clock fn divided by x such that $1 \leq fn/x/fs \leq 2$. The source clock fs divided by N samples the 4-bit counter Ct driven by the network clock fn/x once every $N=3008=47*8*8$ bits generated by the source. The SRTS CS uses the CSI bit in the SAR-PDU to send this sampled 4-bit Residual Time Stamp (RTS) in odd-sequence-numbered SAR-PDUs. ITU-T Recommendation I.363.1, ANSI T1.630, and Bellcore TA-NWT 1113 show how the SRTS method accepts a frequency tolerance for a difference between the source frequency and the reference clock frequency of 200 parts per million (ppm).

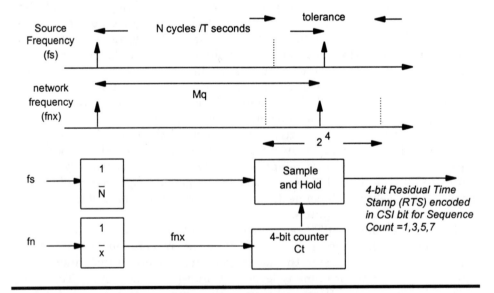

Figure 13.9 AAL1 Synchronous Residual Time Stamp (SRTS) Operation

13.3 ATM ADAPTATION LAYER 2 (AAL2)

AAL2 supports ATM transport of connection-oriented variable bit rate packetized voice and video. Setting a new record for the standards process, the ITU-T approved the basic definition of the AAL2 protocol in Recommendation I.363.2 in September, 1997, after only 9 months [5]. Close cooperation between the ATM Forum's Voice and Telephony Over ATM (VTOA) working group and the ITU-T facilitated this remarkable accomplishment. This section provides an overview of the key methods which the AAL2 protocol employs to minimize delay and improve efficiency for real-time voice and video. Services provided by AAL2 include providing a means for identifying and multiplexing multiple users over a common ATM layer connection, transferring service data units at a variable bit rate, and indication of lost or erroneous information. AAL2 has a number of advantages when transporting voice over ATM [6], including more efficient bandwidth usage due to silence detection and supression as well as idle voice channel deletion. As studied in Chapter 17, earlier standards used either AAL5 or AAL1 for transport of variable bit-rate video or voice in advance of the formal AAL2 standard.

13.3.1 AAL2 Protocol Structure and PDU Formats

Unlike the other AALs covered in this section, AAL2 has no Segmentation and Reassembly Sublayer, but instead employs the structure illustrated in Figure 13.10.

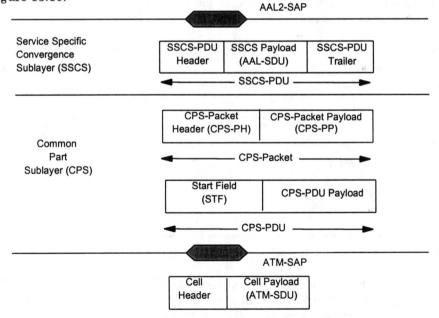

Figure 13.10 AAL2 Protocol Structure

Like all other AALs, AAL2 has an interface to the ATM layer at the ATM-SAP and an interface to the higher-layer user at the AAL2-SAP. The Common Part Sublayer (CPS) has two components as indicated in the figure: a CPS-Packet and a CPS Protocol Data Unit (CPS-PDU). Now that the CPS foundation of the basic AAL2 protocol is complete, standards work on further Service Specific Convergence Sublayers (SSCS) can proceed. For example, the ITU-T draft standard known as I.TRUNK defines trunking of narrow-band ISDN traffic over ATM. Much work still needs to be done in the SSCS for AAL2 to support transfer of timing to support variable bit rate video, such as that defined in the MPEG 2 standard [7]. Fortunately, for support of 64-kbps voice (or even compressed voice), relating the number of octets in the AAL2 payload to a synchronous 8-kHz clock for playback purposes as done in AAL1 structured data transfer mode is a much simpler problem to solve as the ITU-T's I.TRUNK effort plans to do. The more challenging problem for voiceband support over AAL2 is to avoid delay variation in playback caused by statistical fluctuations in voice activity. Such variations in phase could cause disruption to modem or fax transmissions carried over the same AAL2 VCC.

The CPS provides the means to identify AAL users multiplexed over a single ATM VCC, manage assembly and disassembly of the variable length payloads for each user, and interface to the SSCS. The CPS provides an end-to-end service via concatenating a sequence of bi-directional AAL2 channels operating over an ATM Virtual Channel Connection (VCC). Each AAL2 user generates CPS packets with a 3-octet packet header and a variable length payload as illustrated in Figure 13.11.

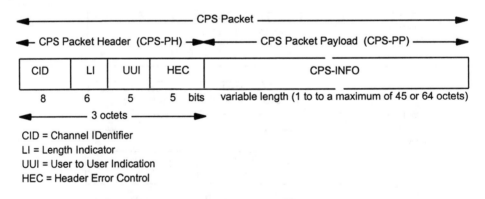

CID = Channel IDentifier
LI = Length Indicator
UUI = User to User Indication
HEC = Header Error Control

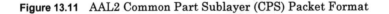

Figure 13.11   AAL2 Common Part Sublayer (CPS) Packet Format

AAL2 uses the 8-bit Channel ID (CID) in the CPS Packet Header (CPS-PH) to multiplex multiple AAL2 users onto a single VCC. The CID field supports up to 248 individual users per VCC, with eight CID values reserved for management procedures and future functions. Next, the 6-bit Length Indicator (LI) field specifies the number of octets (minus one) in the variable length user payload. The maximum length of the user payload is selected as either 45 or 64 octets. Note that selecting the 45 octet value means that exactly one CPS packet fits inside the 48 octet ATM cell payload. The 5-bit User-to-User Indication (UUI) field provides a means for identifying the particular SSCS layer along with support for OAM functions. A 5-bit Header Error Control (HEC) field provides error detection and correction for the CPS-PH. Rationale similar to that described in the last chapter regarding the use of HEC in the ATM cell header drives the need to protect the AAL2 CPS packet header, since undetected errors may affect more than one connection. The CPS packet payload may be up to 64 octets in length.

The CPS sublayer collects CPS packets from the AAL2 users multiplexed onto the same VCC over a specified interval of time, forming CPS-PDUs comprised of 48 octets worth of CPS packets. Figure 13.12 illustrates the CPS-PDU format employed by AAL2. The CPS-PDU employs a one octet Start Field (STF) followed by a 47 octet payload. A 6-bit OffSet Field (OSF) in the start field identifies the starting point of the next CPS packet header within the cell. Note that if more than one CPS packet is present in a cell, then AAL2 uses the Length Indicator (LI) in the CPS packet header to compute the boundary of the next packet. The offset field allows CPS packets to span cells without any wasted payload as the example in the next section demonstrates. Since the start field is critical to the reliable operation of AAL2, a 1 bit Sequence Number (SN) and Parity (P) field provide for error detection and recovery. In order to maintain real-time delivery, the AAL2 protocol times out if no data has been received and inserts a variable length PAD field to fill out the 48-octet ATM cell payload.

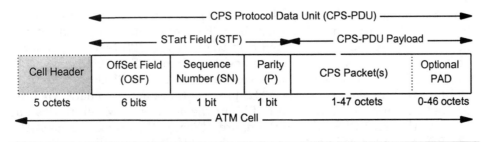

Figure 13.12 AAL2 Common Part Sublayer (CPS) PDU Format

13.3.2 Example of AAL2 Operation

Figure 13.13 illustrates an example where AAL2 multiplexes four real-time, variable bit rate sources (labeled A,B,C, and D) into a single ATM virtual channel connection. Starting at the top of the figure, each source generates 16-octet samples, which pass across the AAL2 Service Access Point to the Common Part Sublayer (CPS), which forms CPS packets by prefixing each sample with a 3-octet CPS Packet Header (CPS-PH) containing a Channel ID (CID) that identifies the source letter in the illustration. The CPS sublayer collects CPS packets over a specified interval of time and forms CPS-PDUs comprised of 48 octets worth of CPS packets using a 1-octet Start Field (STF) as shown in the figure.

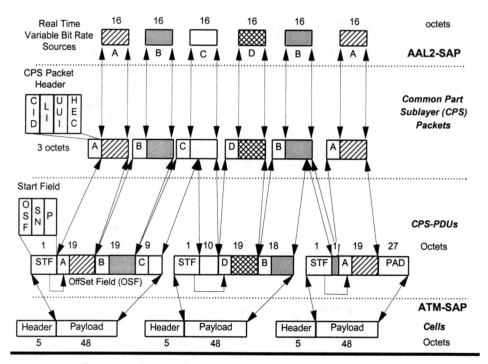

Figure 13.13   Example of AAL2 Operation

In this example, each CPS packet consumes 19 octets in the CPS-PDU. If the sources have been inactive, then the STF offset points to the next octet in the CPS-PDU as shown by the arrow in the leftmost CPS-PDU. In general, the STF offset points to some other position when CPS packets span CPS-PDU boundaries as shown in the second and third CPS-PDUs. In order to maintain real-time delivery, the AAL2 protocol times out if no data has been received and inserts a PAD field to fill out the 48-octet ATM cell payload as shown in the third CPS-PDU in the figure. Finally, the protocol maps AAL2 CPS-PDUs to ATM cell payloads across the ATM-SAP for a Virtual Channel Connection. Thus, AAL2 reduces packetization delay by multiplexing multiple sources together, and controls delay variation by inserting the PAD field if the period of source inactivity exceeds a specific timer threshold. As shown in later chapters, minimal packetization delay is critical for voice over ATM due to echo control problems. Furthermore, control of delay variation is critical for both voice and video over ATM.

13.4 ATM ADAPTATION LAYER 3/4 (AAL3/4)

ITU-T Recommendation I.363.3 [8] combines AAL3 and AAL4 into a single common part, AAL3/4, in support of VBR traffic, both connection-oriented and connectionless. The AAL3/4 protocol conforms to the generic AAL protocol model because it has a Segmentation And Reassembly (SAR) and

Convergence Sublayer (CS). Support for connectionless service is provided at the Service Specific Convergence Sublayer (SSCS) level. The text provides an example of AAL3/4 operation, illustrating the multiplexing function provided by this AAL over a single ATM VCC.

13.4.1 AAL3/4 SAR Sublayer

Figure 13.14 depicts the SAR for AAL3/4. The SAR-PDU encoding and protocol function and format are nearly identical to the L2_PDU from IEEE 802.6 as summarized in Chapter 8. The SAR-PDU has a 2-octet header and trailer. The header contains three fields as shown in the figure. The 2-bit Segment Type (ST) field indicates whether the SAR-PDU is a Beginning Of Message (BOM), a Continuation Of Message (COM), an End Of Message (EOM), or a Single Segment Message (SSM). The sender increments the 2-bit Sequence Number (SN) which the receiver uses to detect lost SAR-PDUs. The numbering and checking begins when the receiver detects a BOM segment. The 10-bit Multiplex IDentification (MID) field allows multiplexing of up to 1024 different CPCS-PDUs over a single ATM VCC. This is a key function which differentiates AAL3/4 from AAL5 which is important when a carrier charges per VCC, motivating users to do their own multiplexing to minimize cost. However, as described in Chapter 19, IETF RFC 1483 defines an analogous means of multiplexing multiple protocols over a single ATM VCC using AAL5. The AAL3/4 SAR function is essentially the same one used in the 802.6 L2 protocol where there the cell header contains effectively no addressing. The transmitter assigns the MID prior to sending a BOM or SSM segment. The MID value in a BOM segment ties together the subsequent COM and EOM portions of a multi-segment message. The SAR-PDU trailer has two fields. The 6-bit Length Indicator (LI) specifies how many of the octets in the SAR-PDU contain CPCS-PDU data. LI has a value of 44 in BOM and COM segments, and may take on a value less than this in EOM and SSM segments. The 10-bit CRC checks the integrity of the segment.

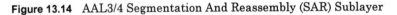

	SAR-PDU

Cell Header	ST	SN	MID	SAR-PDU Payload	LI	CRC
5 octets	2 bits	4 bits	10 bits	44 octets	6 bits	10 bits

ST = Segment Type (BOM, COM, EOM, SSM)
SN = Sequence Number
MID = Multiplex IDentification

LI = Length Indicator
CRC = Cyclic Redundancy Check

Figure 13.14 AAL3/4 Segmentation And Reassembly (SAR) Sublayer

13.4.2 AAL3/4 CPCS Sublayer

Figure 13.15 depicts the CPCS-PDU for AAL3/4. The header has three components as indicated in the figure. The 1-octet Common Part Indicator

(CPI) indicates the number of counting units (bits or octets) for the Buffer Allocation Size (BASize) field. The sender inserts the same value for the 2-octet Beginning Tag (BTag) and the Ending Tag (ETag) so that the receiver can match them as an additional error check. The 2-octet BASize indicates how much buffer space the receiver must reserve to reassemble the CPCS-PDU. A variable-length PAD field ranging between zero and 3 octets makes the CPCS-PDU an integral multiple of 32 bits to make end system processing more efficient.

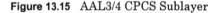

CPI	Btag	BASize	CPCS-PDU Payload	PAD	AL	ETag	Length
1	1	2	1-65,535	0-3	1	1	2

octets

CPI = Common Part Indicator
Btag = Beginning Tag
BASize = Buffer Allocation Size

AL = 32 bit ALignment
ETag = Ending Tag
Length = CPCS-PDU Length

Figure 13.15   AAL3/4 CPCS Sublayer

The trailer also has three fields as shown in Figure 13.15. The 1-octet ALignment field (AL) simply makes the trailer a full 32 bits to simplify the receiver design. The 1-octet ETag must have the same value as the BTag at the receiver for the CPCS-PDU to be valid. The length field encodes the length of the CPCS-PDU field so that the pad portion may be taken out before delivering the payload to the CPCS user.

13.4.3 Example of AAL3/4 Operation

Figure 13.16 depicts an example of the operation of the AAL3/4 SAR and CS sublayers for AAL3/4. Starting from the bottom of the figure, the 48-octet payload of a sequence of cells on the same Virtual Channel Connection (VCC) (i.e., cells having the same Virtual Path Identifier (VPI) and Virtual Channel Identifier (VCI) values) interface with the AAL3/4 SAR sublayer across the ATM-SAP. The 2-bit Segment Type (ST) field indicates that the SAR-PDU is a Beginning Of Message (BOM), a Continuation Of Message (COM), and finally after a number of cells containing COM indications, a cell containing an End Of Message (EOM) segment type as shown in the example.

If the SAR sublayer receives all SAR-PDUs in a message in sequence with correct CRC values, and matching tags, then it passes the reassembled packet up to the CPCS layer. In message mode, AAL3/4 accepts one AAL-IDU at a time and optionally sends multiple AAL-IDUs in a single SSCS-PDU. In streaming mode, the higher-layer protocol may send multiple AAL-IDUs separated in time; the SSCS may deliver these in multiple AAL-IDUs, or reassemble the pieces and deliver only one AAL-IDU [9]. The principal advantages of AAL3/4 are multiplexing of multiple logical connections over a single ATM VCC, additional error-checking fields, and the indication of message length in the first cell for use in efficient buffer allocation in intermediate or destination switches.

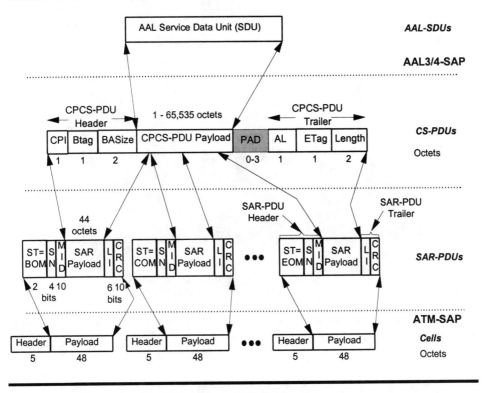

Figure 13.16   Example of AAL3/4 SAR and CS Sublayers

13.4.4 AAL3/4 Multiplexing Example

This section illustrates the operation of AAL3/4 multiplexing through the example shown in Figure 13.16.   This depicts a data communications terminal that has two inputs with two 98-byte (or octet) packets arriving simultaneously destined for a single ATM output port. Two parallel instances of the CPCS sublayer encapsulate the packets with a header and trailer. These then pass to two parallel Segmentation And Reassembly (SAR) processes that segment the CPCS-PDU on two different MIDs, resulting in a BOM, COM, and EOM segment for each input packet.

Because all of this occurred in parallel, the ATM cells resulting from this process are interleaved on output as shown on the right-hand side of the figure. This interleaving over a single VCC is the major additional function of AAL3/4 over AAL5, as seen by comparison with the AAL5 example in section 13.5.4. Also, the multiple levels of error checking in AAL3/4 make the probability of an undetected error very small.

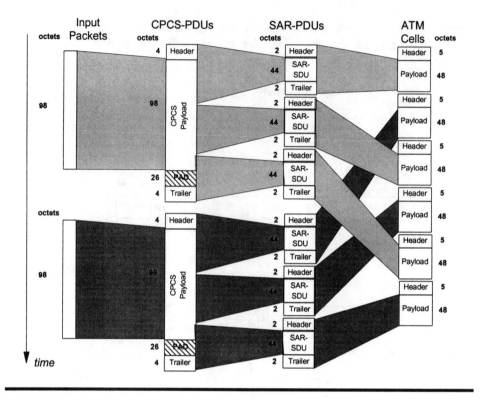

Figure 13.17   Multiplexing Example Using AAL3/4

13.5 ATM ADAPTATION LAYER 5 (AAL5)

If AAL1, AAL2 and AAL3/4 appear complicated, you can now appreciate the motivation for developing a Simple Efficient Adaptation Layer (SEAL). Standards assigned the next available number to this lightweight protocol, which made it AAL5. As of publication time, it was still the last word in AAL supporting packet data, a number of proposals for AAL6 never having achieved sufficient backing for standardization.   The Common Part (CP) AAL5 supports Variable Bit Rate (VBR) traffic, both connection-oriented and connectionless. Support for connectionless or connection-oriented service is provided at the Service Specific Convergence Sublayer (SSCS) level as defined for access to SMDS in Chapter 18. Most other data protocols operate over AAL5, and not AAL3/4; SMDS, the ATM Data eXchange Interface (DXI), and ATM Frame-based UNI (FUNI) being the only exceptions that optionally support AAL3/4. Furthermore, the ATM control plane operates over AAL5 as described in Chapter 15. Part 4 provides examples of video, frame relay, LAN protocols, and IP, all operating over AAL5.   However, despite its simplicity, AAL5 is now coming under criticism for its relatively inefficient operation, especially for the mix of packet sizes typically used on the Internet. Chapter

32 covers this important subject in more detail when the considering the future of ATM.

13.5.1 AAL5 Segmentation and Reassembly (SAR) Sublayer

Figure 13.18 depicts the SAR-PDU for AAL5. The SAR-PDU is simply 48 octets from the CPCS-PDU. The only overhead the SAR sublayer makes use of is the Payload Type code points for *AAL_indicate*. AAL_indicate is zero for all but the last cell in a PDU. A nonzero value of AAL_indicate identifies the last cell of the sequence of cells indicating to the receiver that reassembly can begin. This makes the reassembly design simpler and makes more efficient use of ATM bandwidth.

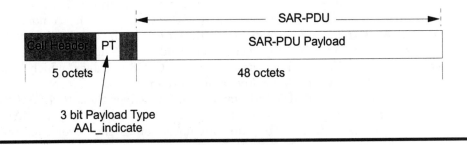

Figure 13.18 AAL5 Segmentation and Reassembly (SAR) Sublayer

13.5.2 AAL5 Common Part Convergence (CPCS) Sublayer

Figure 13.19 depicts the CPCS-PDU for AAL5.

Figure 13.19 AAL5 Common Part Convergence Sublayer (CPCS)

The payload may be any integer number of octets in the range of 1 to $2^{16}-1$ (65,535). The PAD field has a variable length chosen such that the entire CPCS-PDU is an exact multiple of 48 so that it can be directly segmented into cell payloads. The User-to-User (UU) information is conveyed transparently between AAL users by AAL5. The only current function of the Common Part Indicator (CPI) is to align the trailer to a 64-bit boundary, with other functions for further study. The length field identifies the length of the

CPCS-PDU payload so that the receiver can remove the PAD field. Since 16 bits are allocated to the length field, the maximum payload length is $2^{16}-1 = 65,535$ octets. The CRC-32 detects errors in the CPCS-PDU. The CRC-32 is the same one used in IEEE 802.3, IEEE 802.5, FDDI, and Fiber Channel. Chapter 26 compares the undetected error performance of AAL5 and HDLC.

13.5.3 Example of AAL5 Operation

Figure 13.20 depicts an example of the operation of the AAL5 SAR and CPCS sublayers. The relative simplicity of AAL5 with respect to AAL3/4 is readily apparent by comparing this figure with Figure 13.16. Starting from the ATM cell stream on a single VCC at the bottom of the figure, note that the only overhead the SAR sublayer uses is the payload type field in the last cell of a sequence of cells corresponding to a single PDU (i.e., packet). A nonzero value of the AAL_Indicate field identifies the last cell in the sequence of cells indicating that the receiver can begin reassembly. The SAR sublayer reassembles the CPCS-PDU and passes it to the CPCS sublayer, which first uses the CRC-32 field to check for any errors in the received PDU. Normally, CPCS discards corrupted PDUs, but may optionally deliver them to an SSCS. The CPCS removes the PAD and other trailer fields before passing the AAL-SDU across the AAL5-SAP. The transmit operation is the reverse of the above description.

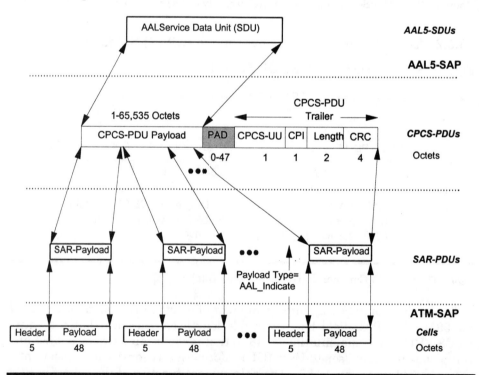

Figure 13.20   AAL5 Common Part SAR and CS Example

The payload may be any integer number of octets in the range of 1 to 2^{16}-1 (65,535). The PAD field has a variable length chosen such that the entire CPCS-PDU is an exact multiple of 48 so that it can be directly segmented into cell payloads. The User-to-User (UU) information is conveyed transparently between AAL users by AAL5. The only current function of the Common Part Indicator (CPI) is to align the trailer to a 64-bit boundary, with other functions for further study. The length field identifies the length of the CPCS-PDU payload so that the receiver can remove the PAD field. Since 16 bits are allocated to the length field, the maximum payload length is 2^{16}-1 = 65,535 octets. The CRC-32 detects errors in the CPCS-PDU. The CRC-32 is the same one used in IEEE 802.3, IEEE 802.5, FDDI, and Fiber Channel. Chapter 26 compares the undetected error performance of AAL5 and HDLC.

13.5.4 AAL5 Multiplexing Example

Figure 13.21 depicts the same example previously used for AAL3/4 to illustrate the major difference in multiplexing operation. The figure depicts a data communications terminal that has two 98-byte packets arriving simultaneously, destined for a single ATM output port, this time using the AAL5 protocol.

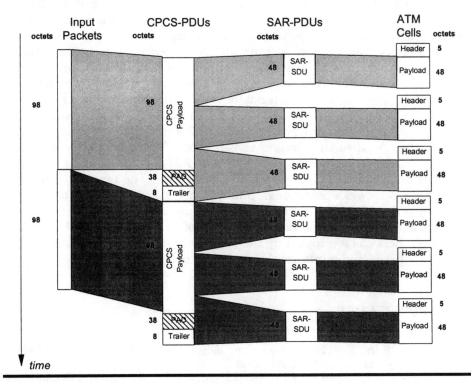

Figure 13.21   Multiplexing Example Using AAL5

On the left-hand side of the figure the two 98-byte packets arrive in close succession. Two parallel instances of the CPCS sublayer add PAD and trailer fields to each packet. Note that in AAL5, the entire packet need not be received before it can begin the SAR function as would be required in AAL3/4 to insert the correct Buffer Allocation Size (BASize) field. Two parallel Segmentation And Reassembly (SAR) processes segment the CPCS-PDU into ATM cells as shown in the middle of the figure. In the example these cell streams share the same VCC, hence the device can send only one at a time. The AAL5 implementation is simpler than AAL3/4, but is unable to keep the link as fully occupied as the additional multiplexing of AAL3/4 could if the packets arrive much faster than the rate at which SAR and ATM cell transmission occurs. Furthermore, the serialization process causes greater delay variation in AAL5 than for AAL3/4 when sharing the same VCC. However, putting delay variation sensitive packet streams on separate VCCs solves this problem when using AAL5.

13.6 REVIEW

This chapter covered the enabling middleware for B-ISDN applications: the Common Part (CP) ATM Adaptation Layers (AALs). The text introduced AALs in terms of the service classes that specify functional attributes of constant or variable bit rate, the need for timing transfer, and the requirement for connection-oriented or connectionless service. We then illustrated this concept by mapping commonly encountered higher-layer protocols to specific AALs and ATM layer service categories. The text then introduced the generic AAL protocol model in terms of the Convergence Sublayer (CS) and the Segmentation And Reassembly (SAR) sublayer, with a further subdivision of the CS sublayer into a Service Specific (SS) and Common Part (CP). The chapter then detailed the formats and protocols for the currently standardized ATM adaptation layers: AAL1, AAL2, AAL3/4, and AAL5. We also provided an example of the operation of each AAL. The text also compared the tradeoff between additional multiplexing capability and complexity involved with AAL3/4 and AAL5 through the use of two examples. Now that we've introduced the theoretical foundation of the physical, ATM and AAL layers, the next chapter moves up the protocol stack to take in an overview of the higher layers and introduce the control plane.

13.7 REFERENCES

[1]   ITU-T, *B-ISDN ATM Adaptation Layer (AAL) Functional Description*, Recommendation I.362, March, 1993.

[2]   T. Lyon, *Simple and Efficient Adaptation Layer (SEAL)*, ANSI T1S1.5/91-292, August 1991.

[3]   ITU-T, *B-ISDN ATM Adaptation Layer (AAL) Specification - Types 1 and 2*, Recommendation I.363.1, , May 1996.

[4]   C. Li, Y. Ofek, "Distributed Source-Destination Synchronization," *IEEE*, 1996.

[5]   M. McLoughlin, J. O'Neil, "A Management Briefing on Adapting Voice for ATM Networks: An AAL2 Tutorial," General Datacomm, http://www.gdc.com/.

[6]   M. McLoughlin, K. Mumford, "Adapting Voice for ATM Network: A Comparison of AAL1 Versus AAL2," General Datacomm, http://www.gdc.com/.

[7]   W. Goralski, *Introduction to ATM Networking*, McGraw-Hill, 1995.

[8]   ITU-T, *B-ISDN ATM Adaptation Layer: Type 3/4 AAL,* Recommendation I.363.3, August 1996.

[9]   ITU-T, *B-ISDN ATM Adaptation Layer (AAL) Specification,* Recommendation I.363, March 1993.

Higher-Layer User, Management, and Control Planes

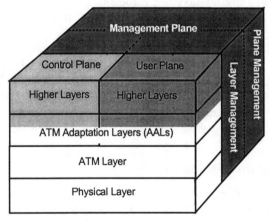

This chapter provides an overview of the higher layers of the user and control planes, along with a brief introduction to the management plane, illustrated by the shaded areas in the B-ISDN cube on the left. This chapter provides an introduction to how these three planes within the B-ISDN architecture work together to fulfill the user's voice, video, and data application needs. The coverage includes the ATM Adaptation Layer (AAL) Service Specific Convergence Sublayer (SSCS) for the user and control planes. The chapter leads off with a summary of higher-layer user plane applications mapped to each AAL to set the stage for the detailed coverage of user plane protocols in Part 4. Of course; a key point is that the principal purpose of the control and management planes is to support the services provided by the user plane, such as LAN emulation, IP over ATM, and Voice and Telephony Over ATM (VTOA). Therefore the text introduces the management plane as a prelude to the detailed coverage provided in Part 7. The text describes the functional decomposition of the management plane into overall plane management and management for the user and control plane layer components. Finally, the chapter moves on to the control plane, which is central in performing the functions needed in a Switched Virtual Connection (SVC) service. The text first provides an overview of the usage and meaning of the various signaling and routing control plane protocols detailed in Chapter 15. We then introduce some key control plane functions, concepts, requirements and goals. The discussion then defines the Control plane Service Specific Convergence Sublayer (SSCS)

functions for B-ISDN signaling, namely the Service Specific Coordination Function (SSCF), and the Service Specific Connection-Oriented Protocol (SSCOP). Next, the text explores the important concept of ATM addressing. Finally, the chapter concludes with an overview of how the ATM Name Service and the Integrated Local Management Interface (ILMI) address registration procedure make the control plane more user friendly.

14.1 HIGHER-LAYER USER PLANE OVERVIEW

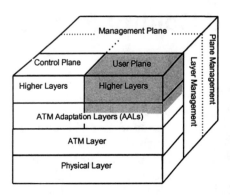

As shown in the shaded portion of the figure to the left, this section covers an overview of the general purpose and function of the service-specific and higher-layer protocols of user plane applications. The treatment references detailed coverage for each user application in Chapters 17 through 20. Chapters 17 and 18 cover higher-layer user protocols for voice, video, and WAN data protocols like frame relay. Chapters 19 and 20 cover LAN and internetworking protocols like IP. First, the text covers the general topics of user plane protocols and functions.

14.1.1 User Plane — Purpose and Function

The B-ISDN protocol cube shows that the user plane spans the PHYsical Layer (PHY), ATM Layer, ATM Adaptation Layer (AAL), and higher layers. The common part AALs and higher layers provide meaningful interfaces and services to end user applications such as voice, video, frame relay, LAN protocols, the Internet Protocol (IP), and Application Programming Interfaces (APIs). The control and management planes exist to support the user plane, in a manner similar to that developed for N-ISDN, as described in Chapter 6. The control plane provides the means to support establishment and release of the following connection types on behalf of the user plane:

* Switched Virtual Connections(SVCs)
* Permanent Virtual Connection (PVC)

SVCs and PVCs can be either point-to-point, or point-to-multipoint, Virtual Path Connections (VPCs) or Virtual Channel Connections (VCCs) as defined in Chapters 11 and 12. A VPC or VCC provides a specified Quality of Service (QoS) with a certain bandwidth defined by traffic parameters in an ATM layer traffic contract, as introduced in Chapter 12 and detailed in Part 5.

14.1.2 User Plane — SSCS Protocols

To date, standards define two Service Specific Convergence Sublayer (SSCS) protocols specifically for the user plane:

* Frame Relay SSCS
* SMDS SSCS

Chapter 18 details both of these protocols. There is no SSCS required for support of IP or circuit emulation over ATM, since the common part of AAL1 and AAL5 directly supports each of these protocols. Some implementations use the Service Specific Connection-Oriented Protocol (SSCOP) defined for signaling later in this chapter to provide an assured data transfer service in the user plane. Standards bodies are developing SSCS protocols for the following user-driven applications:

☺ Voice Trunking (I.TRUNK)
☺ Variable bit rate video over AAL2

More standardization work is required to support the applications listed above and put a real smile on the user's or service provider's face. The standards bodies will likely develop more SSCS protocols to support other applications in the future.

14.1.3 User Plane — Specific Higher-Layer Protocols

The standards and specification bodies listed in Chapter 10 active in the arena of ATM have defined a large number of higher-layer protocols in the user plane to support voice, video, TDM circuits, LAN data protocols, and internetworking protocols over ATM. In fact, Chapter 17 covers ATM support for voice, TDM circuits, and video. Chapter 18 covers WAN data protocols, such as frame relay and SMDS, while Chapter 19 reviews ATM support for LANs. Chapter 20 details the important topic of internetworking over ATM, comparing and contrasting the various standardized approaches defined for supporting IP over ATM. As we shall see, many of these higher-layer user plane applications work hand in hand with control plane protocols to meet user application needs. In particular, a consistent theme is the *emulation* of connectionless data services through address resolution and fast circuit switching in a network of clients and servers designed to support specific end user protocols. Therefore, we introduce the control plane and routing in the next chapter before moving into the higher-layer user plane protocols. As a prelude, this section provides an overview of the higher-layer protocols operating over ATM covered in Part 4.

Figure 14.1 illustrates the SSCS and higher-layer user plane protocols covered in Chapters 17 and 18. Chapter 17 covers support of TDM circuit transport, voice, and video. Circuit emulation along with Voice and Telephony Over ATM (VTOA) specifications from the ATM Forum make exclusive use of AAL1. The video on demand specification used AAL5 in conjunction with the real-time VBR ATM service category, since AAL2 was not standardized at the time of development. After completion of standards for Service Specific Convergence Sublayers (SSCS) operating over the common part AAL2, we expect that subsequent voice and video standards will make use of its unique capabilities as indicated in the figure. Chapter 18 covers higher-layer support for WAN data protocols over ATM, specifically, frame relay, SMDS, the ATM Data eXchange Interface (DXI) and the ATM Frame-based UNI (FUNI). Note that the majority of these protocols utilize AAL5, while only ATM access to SMDS requires AAL3/4. ATM DXI and FUN make support for AAL3/4 optional. The only protocol employing the FR-SSCS sublayer is FR/ATM network interworking, a protocol designed to support trunking of frame relay over ATM as detailed in Chapter 18.

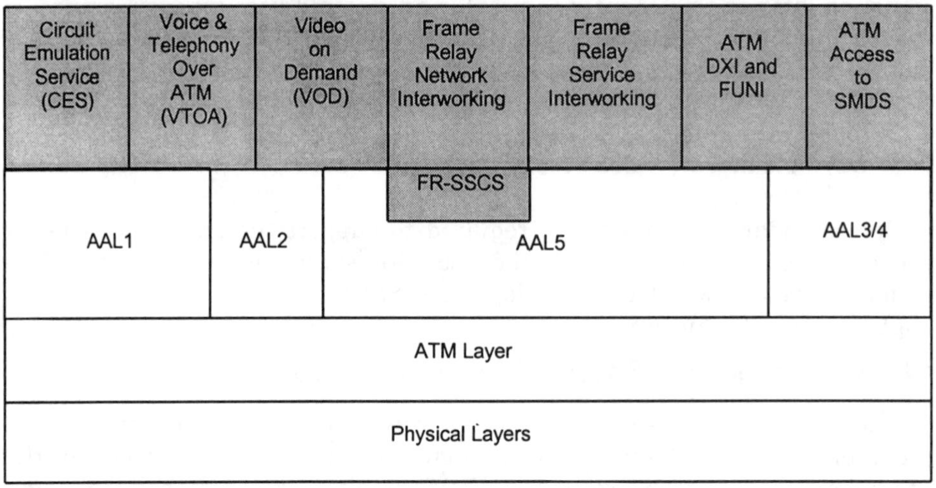

Circuit Emulation Service (CES)	Voice & Telephony Over ATM (VTOA)	Video on Demand (VOD)	Frame Relay Network Interworking	Frame Relay Service Interworking	ATM DXI and FUNI	ATM Access to SMDS
AAL1	AAL2		FR-SSCS			AAL3/4
			AAL5			
ATM Layer						
Physical Layers						

Figure 14.1   User Plane Higher-Layer Protocols for Voice, Video and WAN Data

Figure 14.2 illustrates the higher-layer user plane protocols covered in Chapters 19 and 20 that support LANs and internetworks. Notice that all of these data protocols operate over AAL5. Since data traffic continues growing at a much faster rate than voice traffic, many readers will be interested in more detail on these subjects. This is why we included an extensive coverage on the background of local area networks, bridging, routing and the Internet Protocol (IP) in Chapters 8, 9 and 10 in this edition of the book. Chapter 19 describes LAN Emulation (LANE), multiprotocol encapsulation and the proposed Cells In Frames (CIF) protocol. Multiprotocol encapsulation performs a comparable function to AAL3/4 by multiplexing multiple network layer protocols (e.g., IP, IPX, Appletalk, DECnet, etc.) over a single ATM Virtual Channel Connection (VCC). RFC 1483 defines support for the Internet Protocol (IP) over ATM along with many other protocols. Chapter 20 describes how classical IP subnetworks work over ATM, as well as the protocol that implements a multicast capability over ATM. The treatment covering ATM hardware and software in Chapter 16 covers native ATM Application Programming Interfaces (APIs). The chapter also surveys several proprietary, leading-edge, high-performance, QoS-aware protocols that support IP networking over an ATM infrastructure, such as IP switching, tag switching, and Aggregate Route-based IP Switching (ARIS). Finally, looking toward the future, Chapter 20 summarizes the current status and direction of the IETF's Multi-Protocol Label Switching (MPLS) working group.

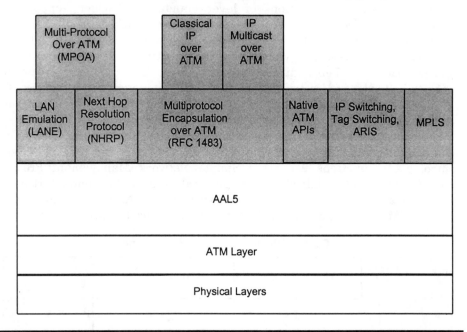

Figure 14.2   User Plane Higher-Layer Protocols for LANs and Internetworking

14.2 MANAGEMENT PLANE

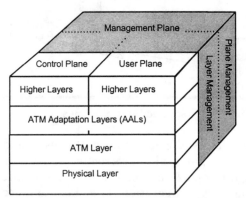

The management plane covers the layer management and plane management functions, as shown in the B-ISDN cube on the left. Layer management interfaces with the Physical and ATM layers, AAL, and higher layers. Plane management is responsible for coordination across layers and planes in support of the user and control planes through layer management facilities. Management ensures that everything works together properly. Typical of new technologies, the management protocols tend to mature later than the technology itself. Since the ATM user plane is now relatively mature and the control plane is stabilizing, detailed definition of the ATM management protocols is well on its way as described in Part 7. This section first discusses layer management, followed by plane management.

14.2.1 Layer Management

Part 7 addresses the functions of layer management which support the Physical and ATM layers, AAL Common Part, SSCS and higher-layer protocol entities in both the control and user planes as depicted in Figure 14.3. This two-dimensional view results from cutting the B-ISDN cube open from the back and then folding it out flat. The ITU-T and the ATM Forum define standards and specifications for these management interfaces for telecommunications equipment using the Telecommunications Management Network (TMN) architecture, often employing the Common Management Information Protocol (CMIP). The IETF and the ATM Forum also define some higher-layer management functions in support of the user plane and control plane functions using the Simple Network Management Protocol (SNMP). Note that standards for the physical, ATM and common part AALs are identical for both the control and user plane. As detailed in Part 7, special ATM cells support Operation And Maintenance (OAM) functions.

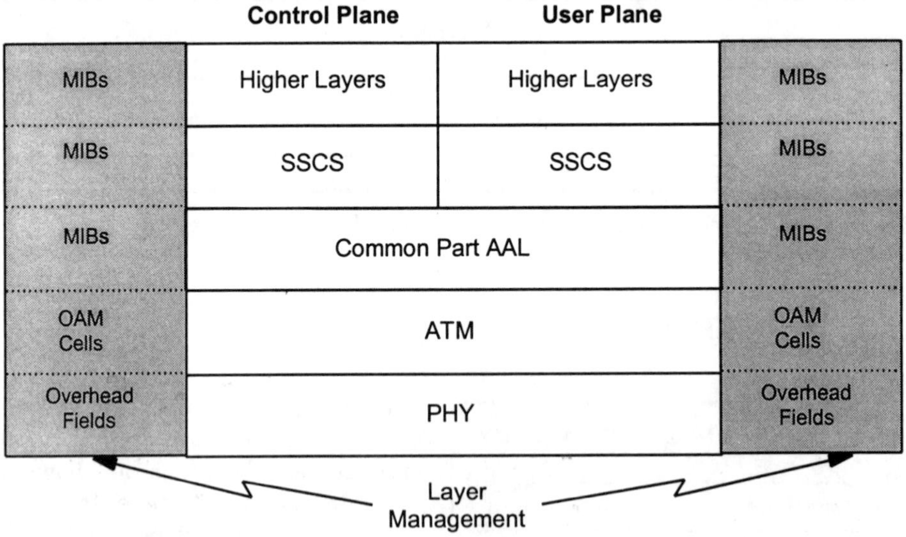

	Control Plane	**User Plane**	
MIBs	Higher Layers	Higher Layers	MIBs
MIBs	SSCS	SSCS	MIBs
MIBs	Common Part AAL		MIBs
OAM Cells	ATM		OAM Cells
Overhead Fields	PHY		Overhead Fields

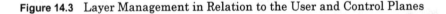

Layer Management

Figure 14.3 Layer Management in Relation to the User and Control Planes

Layer management has the responsibility for monitoring the user and control planes for faults, generating alarms, and taking corrective actions, as well as monitoring for compliance with the performance stated in the traffic contract. Layer management handles the operation and maintenance information functions found within specific layers. These functions include fault management, performance management, and configuration management. PHY-layer management utilizes standardized overhead fields within the physical bit stream in wide area networks. Chapters 28 and 29 describe the standards for ATM-layer fault and performance management. Standardization for management for the AAL and higher layers exists mainly in the definition of object-oriented Management Information Bases (MIBs). Of course, the real value of a management systems comes from a detailed understanding of how the MIB objects relate to actual performance or faults seen by end users. Standards-based Network Management Systems (NMSs) utilize these MIB objects to determine status, detect failures, and automatically configure the managed network element.

14.2.2 Plane Management

Plane management has no defined structure, but instead performs management functions and coordination across all layers and planes in the entire system. The Telecommunication Management Network (TMN) architecture developed by the ITU-T for managing all types of telecommunications networks performs the B-ISDN plane management role.

14.3 CONTROL PLANE OVERVIEW

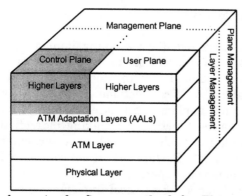

The control plane handles all virtual connection–related functions, most importantly the Switched Virtual Connection (SVC) capability. The control plane also performs the critical functions of addressing and routing. The text summarizes the higher-layer, service-specific AAL portions of the signaling protocol. This remainder of this chapter covers the shaded portions of the B-ISDN cube as shown in the figure on the left. This section introduces the network context for signaling, control plane functions, along with the signaling routing protocols in the control plane detailed in Chapter 15.

14.3.1 Use of Signaling Protocols

In switched ATM networks, users signal switches, which in turn signal other switches, which in some cases signal other networks. Switches and users employ different signaling protocols for each of these contexts. This is shown in Figure 14.4: user-network, inter-switch, and network-to-network.

Users interface to switches and communicate the connection request information via a User-Network Interface (UNI) signaling protocol. Networks interconnect via a more complex network-network (NNI) signaling protocol. Switches employ an interswitch signaling protocol, usually based upon an NNI protocol, frequently employing vendor proprietary extensions. Private and public networks often use different NNI signaling protocols because of different business needs. Private switched networks usually connect to public switched networks via UNI signaling.

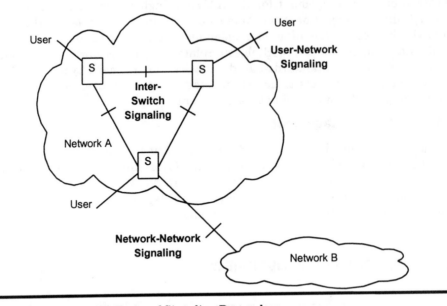

Figure 14.4 Context for Types of Signaling Protocols

Since ATM switches are connection-oriented devices, many implementations utilize these standards-based signaling protocols to establish connections across a network. However, as pointed out in Chapter 20, some approaches use other protocols to establish ATM connections; for example, IP switching and tag switching. Also, some implementations employ network management protocols to emulate signaling functions by making individual ATM cross-connects (i.e., VP or VC links) at each switch along a particular route to build up an end-to-end VPC or VCC. Beware that these centralized network management approaches generally operate at a much slower connection set up rate than distributed signaling protocol implementations. On the other hand, centralized control may provide other features not defined in the distributed signaling and routing protocols. As covered in Part 5, many higher-layer data protocols operating over ATM require large connection establishment rates. For example, in the LAN Emulation protocol, a LAN station sets up an SVC for communication with each link layer LAN address. Although each user typically sets up a few connections to various servers per unit time, the aggregate call rate scales roughly with the number of attached LAN users. The basic ATM signaling architecture responds to this challenge by distributing intelligence to each device, thus maximizing scalabilty of B-ISDN networks by eliminating centralized control.

14.3.2 Control Plane Functions

The control plane provides the means to support the following types of connections on behalf of the user plane:

* Switched Virtual Connections (SVCs)
* Soft Permanent Virtual Connections (SPVCs)

SVCs and SPVCs are either point-to-point, or point-to-multipoint, Virtual Path or Channel Connections (VPCs or VCCs). A switched VPC or VCC of a particular service category provides a specified QoS for specified traffic parameters in an ATM-layer traffic contract. Chapter 12 introduced these concepts while Part 5 covers the details. SVCs allow end users or applications to set up connections on demand, as shown for LAN Emulation, IP over ATM, and Multiprotocol over ATM in Part 4. SPVCs provide a standard means for private network managers or public network service providers to automatically provision semi-permanent connections across multivendor networks using the PNNI protocol described in Chapter 15.

14.3.3 Control Plane Protocols

The shaded area of Figure 14.5 illustrates the ATM control plane protocols detailed in the next Chapter. The specifications for the Service-Specific Connection-Oriented Protocol (SSCOP) provide a guaranteed, reliable packet delivery service to all signaling protocols. This chapter describes SSCOP as background. Chapter 15 covers the signaling protocols at the User-to-Network Interface (UNI) and Network Node Interface (NNI). The ATM Forum has produced three versions of UNI signaling protocols, numbered 3.0, 3.1, and 4.0. ITU-T Recommendation Q.2931 specifies B-ISDN signaling on the ATM UNI. The ATM Forum UNI 4.0 and Q.2931 specifications are closely aligned as described in Chapter 15. The ITU-T's formal name for the ATM UNI signaling protocol is the Digital Subscriber Signaling System 2 (DSS2).

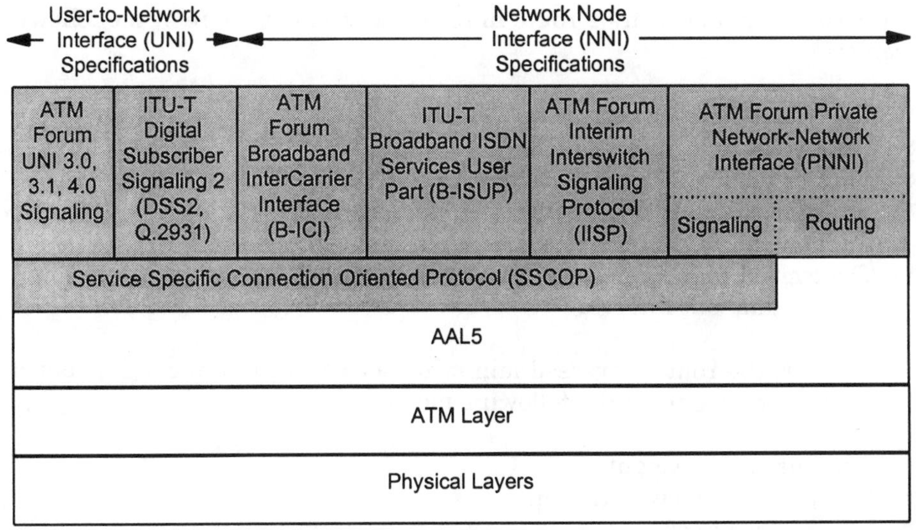

Figure 14.5   Overview of Control Plane Protocol Stack

Next, Chapter 15 covers the Network-Node Interface (NNI) signaling protocols used between switches and between networks. B-ISDN adapts the N-ISDN User Part (ISUP) at the NNI resulting in a protocol called B-ISUP. The ISUP protocol supports N-ISDN connection capabilities between carrier networks. The ATM Forum's adaptation of B-ISUP at the NNI for a Broadband Intercarrier Interface (B-ICI) has two versions, 2.0 and 2.1, aligned with UNI 3.1. The ATM Forum defined an Interim Interswitch Signaling Protocol (IISP) as a simple, multivendor interoperable NNI protocol. Finally, the ATM Forum's PNNI protocol defines not only signaling at the NNI, but a scalable, hierarchical topology distribution and routing protocol.

14.3.4 Basic Routing Requirements

Recall that cells from the same VPC or VCC must follow the same route, defined as the ordered sequence of physical switch ports which the cells traverse from source to destination. As described in Chapters 11 and 12, the VP or VC cross-connects between each pair of ports made by an ATM switch are either VP or VC links. An end-to-end VPC or VCC is an ordered sequence of such VP or VC links.

The control plane establishes an end-to-end VPC or VCC in response to one of the following events:

- A management interface provisions a new SPVC
- Automatic reestablishment of a failed SPVC
- A user makes an SVC connection request via a UNI signaling protocol
- An adjacent network makes an SVC connection request via an NNI signaling protocol

The control plane clears a route in response to one or more of the following events:

- The management interface deletes an SPVC
- Detection of a failure on an SPVC route
- An user requests disconnection of an SVC via a UNI signaling protocol
- An adjacent network requests disconnection of an SVC via an NNI signaling protocol
- The control plane detects a physical link or SSCOP layer failure after a specific time-out interval

Generally, the route traversed minimizes a cost function including, but not limited to, one or more of the following factors:

- Administrative weight
- Delay and/or Delay variation
- Loss performance
- Available Quality of Service (QoS)

☞ Economic expense
☞ Balanced utilization (when multiple links are present between a node-pair)

Many routing algorithms, such as OSPF, optimize routing on a single parameter, typically administrative weight.   The ATM PNNI protocol optimizes across the above list of parameters as described in Chapter 15.

14.3.5 Desirable Routing Attributes

Several functional requirements and goals drive the design of ATM layer routing schemes. Desirable attributes of the routing scheme include at least the following:

* Simplicity
* Automatic determination of an optimized route
* Ease of managing changes in the network
* Ability to easily add or delete links and nodes
* Extensibility of the routing scheme to a large network
* Capability to trace the actual route taken through the network

14.4 CONTROL PLANE STRUCTURE AND AAL

The B-ISDN control plane handles all virtual connection-related functions, most importantly, Switched Virtual Connections (SVCs). The higher-layer and service-specific AAL portions of the signaling protocol are now well standardized. This section summarizes the B-ISDN UNI and NNI signaling protocols and also covers the protocol model for the signaling AAL's Service Specific Convergence Sublayer (SSCS).

14.4.1 Control Plane Architecture and Signaling

Figure 14.6 illustrates the relationships between the major ITU-T signaling standards. The left-hand side of the figure shows the B-ISDN User-Network Interface (UNI) signaling protocol stack. As shown in the center of the figure, the B-ISDN Network Node Interface (NNI) interconnects public networks, but is sometimes used between switches within a single network. The right-hand side of the figure also shows the N-ISDN InterWorking Function (IWF) along with the N-ISDN UNI signaling protocol stack. This section covers the layer 2 signaling protocols, while Chapter 15 covers the layer 3 signaling protocols.

The ITU-T developed standards for the Signaling AAL (SAAL) which the ATM Forum, ETSI, and ANSI subsequently adopted. ITU-T Recommendation Q.2931 [1] (previously called Q.93B) specifies the B-ISDN signaling over the ATM UNI. The two don't interoperate; so beware of equipment touting support for the older, preliminary Q.93B standard or the older predecessor to Q.2110 called Q.SAAL. The Q.2931 standard borrows heavily from both the Q.931 UNI signaling protocol for N-ISDN and the Q.933 UNI signaling protocol for frame relay. The formal name for the ATM UNI signaling protocol is the Digital subscriber Signaling System 2 (DSS2), indicating it as the next evolutionary step after the DSS1 signaling used for N-ISDN.

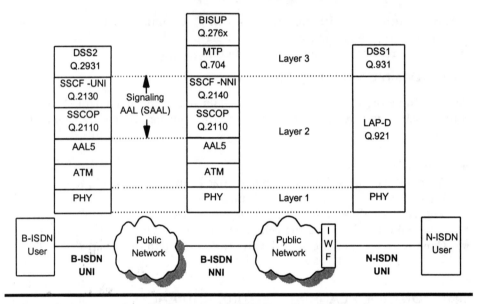

Figure 14.6   Relationship and Context of ITU-T Signaling Standards

ITU-T Recommendation Q.2130 (previously called Q.SAAL.2) specifies the Service Specific Coordination Function (SSCF) for the UNI. ITU-T Recommendation Q.2110 (previously called Q.SAAL.1) specifies the Service Specific Connection-Oriented Protocol (SSCOP) covered in a subsequent section. The standards adapts the N-ISDN User Part (ISUP) concept for supporting B-ISDN UNI signaling between networks using a protocol called B-ISUP. The B-ISUP protocol operates over the Message Transfer Protocol 3 (MTP3), identical to that used in Signaling System 7 (SS7) for out-of-band voice and N-ISDN signaling. This choice of standards will allow B-ISDN network signaling the flexibility to operate over existing signaling networks or work directly over new ATM networks. The series of ITU-T Recommendations Q.2761 through Q.2764 specify the B-ISUP protocol as summarized in Appendix B. ITU-T Recommendation Q.2140 specifies the SSCF at the NNI. The NNI signaling uses the same SSCOP protocol as the UNI.

14.4.2 Types of Signaling Channel Association

Figure 14.7 illustrates the two types of signaling channel configurations: associated and non-associated. Normally, the signaling channel operates on VPI 0, VCI 5 on a physical interface. UNI 3.1 employed a method called *associated* signaling. In general, VCI 5 on any virtual path (specified by the VPI value) controls all VCCs on that VP for associated signaling. UNI 4.0 utilizes *non-associated* signaling, for which Recommendation Q.2931 [1] defines a Virtual Path Connection Identifier (VPCI) that associates a signaling VCC with one or more VPIs on a physical interface as shown in the figure. The ITU-T uses the VPCI instead of the Virtual Path Identifier (VPI) to allow virtual path cross-connects in the access network; while the ATM Forum's UNI 4.0 specification uses the VPCI to support switched virtual paths, proxy signaling, and virtual UNIs as described below. We call the VPI/VPCI(s) controlled by a signaling channel a *logical UNI* in the following sections.

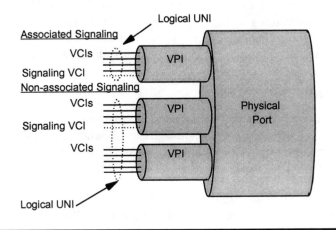

Figure 14.7 Associated and Non-associated Signaling

Figure 14.8 illustrates an example of non-associated signaling. In the example, a VP multiplexer maps the VCI=5 signaling channel to different VPI values, each with a different VPCI into a VC switch. The end user, the VP multiplexer, and the VC switch must be configured with compatible VPCI mappings (in this case physical interface identifier (IFn) plus the user VPI) as indicated in the figure.

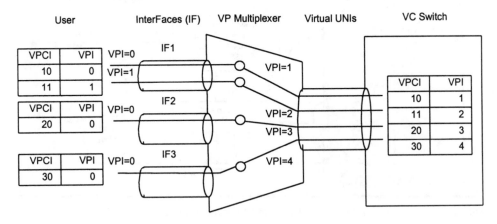

Figure 14.8   Usage of Virtual Path Connection Identifiers (VPCIs) as Virtual UNIs

Looking inside the VC switch, VPCI=11 corresponds to physical interface IF1 and user VPI=0 on the user side of the VP multiplexer which the VC switch sees on VPI=2 on the port from the VP multiplexer. Annex 8 of the ATM Forum UNI 4.0 specification also defines a *virtual UNI* capability where only the VP multiplexer and the VC switch utilize the VPCI. In this case, the end user employs associated signaling as defined in UNI 3.1 where the user does not use the VPCI concept. Annex 2 of the Forum's UNI signaling 4.0 specification also defines a *proxy signaling* capability where a single signaling interface controls multiple VPCIs. This capability allows another signaling user to set up connections on behalf of user devices without a signaling capability.

14.4.3 Layered SSCS Model

Figure 14.9 illustrates the protocol model for the Signaling AAL (SAAL) specified in ITU-T Recommendation Q.2100 [2] and ANSI Standard T1.636 [3]. The Common Part AAL (CP-AAL) is AAL5 as described in Chapter13. The following two protocols comprise the SSCS portion of the SAAL:

&∕ Service Specific Coordination Function (SSCF)
&∕ Service Specific Connection-Oriented Protocol (SSCOP)

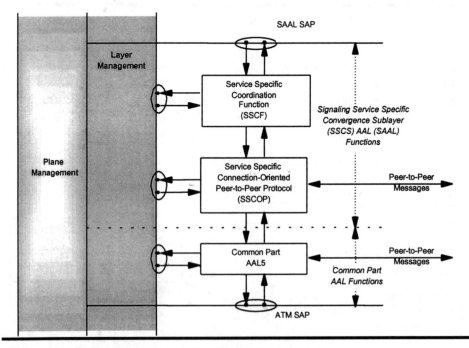

Figure 14.9 Signaling AAL (SAAL) Layered Protocol Model

The SAAL primitives define services at the SAAL Service Access Point (SAP). The CP AAL5 interfaces with the ATM layer at the ATM SAP. A one-to-one correspondence exists between an SAAL SAP and an ATM SAP.

Corresponding layer management functions manage the signaling SSCF and SSCOP protocols and the CP-AAL as separate layers as indicated on the left-hand side of Figure 14.9. Layer management sets parameters in the individual layer protocols; for example, timers and threshold, as well as monitoring their state and performance. For example, layer management may use the state of SSCOP to determine the state of the underlying physical link or virtual connection between two ATM devices. Plane management coordinates across the layer management functions to monitor and maintain the overall end-to-end signaling capability.

14.4.4 Service Specific Coordination Function (SSCF)

The Service Specific Coordination Function (SSCF) provides the following services to the Signaling AAL (SAAL) user:

❖ Independence from the underlying layers
❖ Unacknowledged data transfer mode
❖ Assured data transfer mode
❖ Transparent relay of information
❖ Establishment of connections for assured data transfer mode

The SSCF provides these capabilities primarily by mapping between a simple state machine for the user and the more complex state machine employed by the SSCOP protocol. ITU-T Recommendation Q.2130 [4] defines the SSCF at the UNI, while Recommendation Q.2140 [5] defines the SSCF at the NNI.

14.4.5 Service Specific Connection-Oriented Protocol (SSCOP)

ITU-T Recommendation Q.2110 [6] defines the Service Specific Connection-Oriented Protocol (SSCOP) serving both the UNI and NNI SSCF functions. SSCOP is a sophisticated link layer, peer-to-peer protocol which performs the following functions:

♦ Guaranteed sequence integrity, that is, in sequence message delivery
♦ Error correction via error detection and selective retransmission
♦ Receiver-based flow control of the transmitter
♦ Error reporting to layer management
♦ Keep alive messaging during intervals of no data transfer
♦ Local retrieval of unacknowledged or enqueued messages
♦ Establish, disconnect, and synchronize SSCOP connections
♦ Transfer user data in either an unacknowledged or assured mode
♦ Protocol level error detection
♦ Status reporting between peer entities

SSCOP is a complex protocol, but is specified in the same level of detail as a successful protocol like HDLC. As the name implies, a connection must be established *before* any data transfer occurs. The unacknowledged mode is a simple unacknowledged datagram protocol, similar to the User Datagram

Protocol (UDP) in the IP protocol suite. Much of the complexity of SSCOP occurs in the assured data transfer mode. SSCOP defines the following PDUs in the functional categories indicated in Table 14.1 for use in performing the above functions.

Table 14.1 SSCOP Protocol Data Unit Categories and Descriptions

Functional Category	SSCOP PDU	SSCOP PDU Description
	BGN	Request Initialization
Establishment	BGAK	Request Acknowledgment
	BGREJ	Connection Reject
Release	END	Disconnect Command
	ENDAK	Disconnect Acknowledgment
Resynchronization	RS	Resynchronization Command
	RSAK	Resynchronization Acknowledgment
Recovery	ER	Recovery Command
	ERAK	Recovery Acknowledgment
	SD	Sequenced Connection-mode Data
Assured Data Transfer	POLL	Transmitter State Information with request for Receive State Information
	STAT	Solicited Receiver State Information
	USTAT	Unsolicited Receiver State Information
Unacknowledged Data Transfer	UD	Unnumbered User Data
Management Data Transfer	MD	Unnumbered Management Data

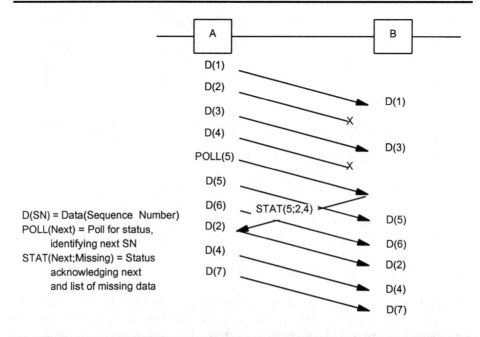

Figure 14.10 Example of SSCOP Retransmission Strategy

Figure 14.10 illustrates an example of the SSCOP selective retransmission strategy. First, the error detection capability of AAL5 reliably determines whether the adjacent signaling node receives a frame successfully. SSCOP requires that the transmitter periodically poll the receiver as a keep-alive action, as well as a means to detect gaps in the sequence of successfully-received frames. The receiver must respond to the poll, and if more than a few poll responses are missed, the transmitter takes down the connection. A key feature is where the receiver identifies that one or more frames are missing in its sequence as illustrated in the figure. The transmitter then only resends the missing frames. Chapter 26 shows how this selective retransmission protocol significantly improves throughput when compared with "Go-Back N" retransmission strategy, such as those employed in X.25 and TCP as described in Part 3. SSCOP PDUs employ a 24-bit sequence number that achieves high throughputs on very high-speed links, such as those typically used in ATM networks.

14.5 CONTROL PLANE ADDRESSING

The control plane performs the critical functions of addressing and routing. The text draws analogies with addressing and routing performed in telephone networks and the Internet where appropriate to leverage the reader's background in these related areas. This section concludes with an overview of the ILMI address registration procedure and the ATM Name Service.

14.5.1 Control Plane Addressing Levels

Two capabilities are critical to a switched network: addressing and routing. *Addressing* occurs at the link level between ATM devices at the VPI/VCI level as previously described, but more importantly at a logical network, or end-to-end level. Addressing also occurs in the association of signaling channels with bearer channels. Since the VPI/VCI is unique only to a physical interface, the higher-level address must be unique across all interconnected networks. Ideally, the address should be unique across all networks in order to provide universal connectivity if the networks later interconnect. Once each entity involved in switching virtual connections has a unique address, there is an even more onerous problem of finding a route from the calling party to the called party. *Routing* solves this problem by one of two means: either static, manual configuration, or dynamic, automatic discovery as described in the next chapter.

14.5.2 ATM Level Addressing

The signaling protocol automatically assigns the VPI/VCI values to SVC calls between ATM addresses corresponding to ATM UNI signaling channels according to a set of rules. This is also called logical UNIs as defined in section 14.4.2. The VPI/VCI values in the case of associated signaling, or in the case of non-associated signaling the VPCI/VCI values, are unique to the signaling channel. SVCs may be either point-to-point or point-to-multipoint. Each ATM UNI signaling channel must have at least one unique ATM address in order to support SVCs. An ATM UNI signaling channel may have more than one ATM address.

14.5.2.1 Point-to-Point Connections

Recall from Chapter 12 that a VCC or VPC is defined in only one direction; that is, it is simplex. A point-to-point duplex SVC (or a SPVC) is actually a pair of simplex VCCs or VPCs: a forward connection from the calling party to the called party, and a backward connection from the called party as illustrated in Figure 14.11. Applications may request different forward and backward traffic parameters and ATM service categories. For example, a file transfer applications might set up and SVC with the forward direction having ten times the bandwidth as the backward direction; since the backward channel is only used for acknowledgments. A video broadcast might specify large forward traffic parameters with zero backward bandwidth.

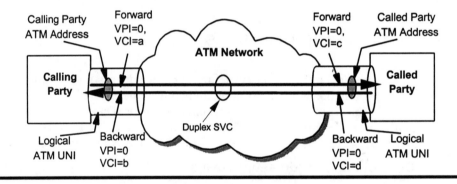

Figure 14.11   Point-to-Point Switched Virtual Connection (SVC)

Thus, the forward and backward VPI (and VCI for a VCC), as well as the ATM address associated with the physical ATM UNI ports at each end of the connection, completely define a point-to-point SVC shown in Figure 14.11. Furthermore, the VPI and VCI assignment may differ for the forward and backward directions of a VPC or VCC at the same end of the connection, as well as being different from the other end of the connection, as illustrated in the figure. In the case of VCCs, the VPI value is often zero. A convention where the VPI (and VCI for a VCC) is identical at the same end of a connection may be used, and is a common implementation method for PVCs because it simplifies operation of ATM networks. Since the SVC procedures dynamically assign the VPI (and VCI for VCCs), the values generally differ for each end of the connection.

14.5.2.2 Point-to-Multipoint Connections

A point-to-multipoint SVC (or SPVC) has one root node and one or more leaf nodes. The VPI (and VCI for VCCs) along with the ATM address associated with the signaling channel of the root node, and the ATM address and VPI and VCI for the signaling channel for each leaf node of the connection define a point-to-multipoint connection, as shown in Figure 14.12. There is essentially only a forward direction in a point-to-multipoint connection, because the network allocates zero bandwidth in the backward direction as specified in the ATM Forum's UNI 3.1 specification. However, the network must provide a backward flow for OAM cells and use by other protocols. As described in Chapter 20, one such application is Aggregate Route-based IP Switching (ARIS) which emulates a best effort broadcast channel by using the backwards direction of a point-to-multipoint VPC. Note that more than one VPI/VCI value and ATM address on a single physical interface may be part of a point-to-multipoint connection. This means that the number of physical ATM UNI ports is always less than or equal to the number of logical leaf endpoints of the point-to-multipoint connection. The implementation of a point-to-multipoint connection should efficiently replicate cells at intermediate switching points within the network as illustrated in the figure. Replication may occur within a public network, or within a local switch. A minimum spanning tree (see Chapter 9) is an efficient method of constructing a point-to-multipoint connection. Both the LAN Emulation and IP Multicast over ATM protocols make extensive use of switched point-to-multipoint ATM connections when emulating broadcast LAN protocols. Other applications, such as video teleconferencing, video broadcasts and simultaneous dissemination of information to multiple users also utilize the point-to-multipoint ATM connection capability.

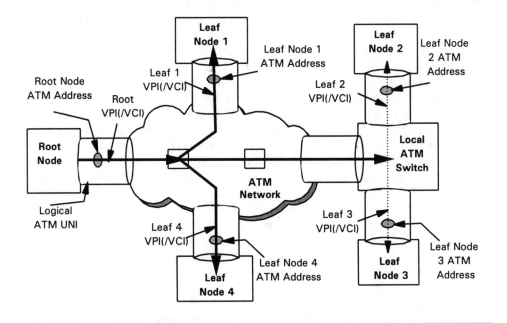

Figure 14.12 Point-to-Multipoint Switched Virtual Connection (SVC)

14.5.3 ATM Addressing Formats

Currently two types of ATM Control Plane (SVC) addressing plans identify an ATM UNI address: a data oriented Network Service Access Point (NSAP) based format defined by the International Standards Organization and the telephony oriented ITU-T E.164 standard. An important contribution of the ATM Forum UNI 3.0 and 3.1 specifications towards the goal of global ATM internetworking was the adoption of an address structure based upon the ISO NSAP syntax. UNI 4.0 continued the use of this addressing structure and clarified several points. On the other hand, the ITU-T initially adopted the use of telephone number-like E.164 addresses as the addressing structure for public ATM (B-ISDN) networks to interwork with legacy telephone and Narrowband ISDN networks. Since E.164 addresses are available only to monopoly carriers, which prevented the assignment of addresses to competing carriers and the private business sector, the ATM Forum chose NSAP-based addresses to provide unique ATM addresses for both private and public networks. Since the E.164 address space was not planned for ATM service support, the ITU-T and the ATM Forum are now standardizing the use of NSAP-based formats for carriers.

There are two fundamental classes of ATM address formats: ATM End System Address (AESA) and native E.164. An AESA uses the ISO defined NSAP address syntax. Figure 14.13a illustrates the NSAP-based AESA format. International (e.g, British Standards Institute) and national (e.g., ANSI) standards bodies assign the Initial Domain Part (IDP) to various organizations, such as carriers, companies and governments for a nominal fee. The remainder of the 20 octet address is called the Domain Specific Part (DSP). T he next section details the AESA formats. The network provider supplies the IDP part obtained from an administrative body as well as part of the DSP. The domain's network administrator defines the remaining octets. The end user part contains at least 7 octets. The NSAP standards define a more rigid structure than adopted by the ATM Forum, which is why we say that the Forum's address structure is *NSAP-based* and not NSAP formatted. The reason the Forum chose a more flexible format was to achieve better scalability through hierarchical assignment of the IDP part of the address as covered in the next Chapter in the section on PNNI.

Figure 14.13b illustrates the ITU-T specified E.164 address format. This is the same format used for international telephone numbers, which begins with a country code (e.g., 01 for North America, 44 for the UK, etc.), followed by a number defined within that country. This plan served voice telecommunications well for over fifty years, but assumed only one monopoly phone company per country. Telecommunications deregulation created multiple carriers within a country, violating the underlying paradigm of the E.164 numbering plan. Furthermore, with the proliferation of fax machines, cellular phones and multiple phones per residence, the E.164 numbering plan has too few digits, necessitating renumbering of area codes and even individual numbers. Unfortunately, this need to change addresses to support continued growth in the telephony sector occurs on an increasingly frequent basis in response to growing demand. Recent ITU-T standards work to evolve the E.164 plan to assign a country code to specific carriers is an attempt to address the emerging global competitive nature of networking. In response to the deficiencies of the E.164 numbering plan, the ATM Forum plans to specify further details regarding the use of NSAP-based addresses by ATM service providers.

a. ISO Network Service Access Point (NSAP) based Address Format

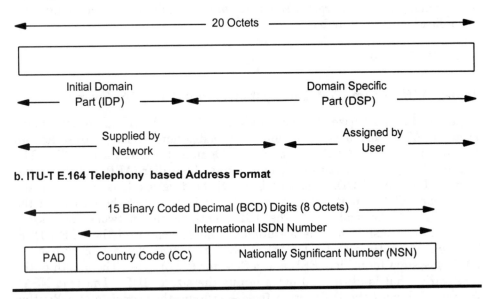

b. ITU-T E.164 Telephony based Address Format

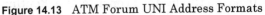

Figure 14.13 ATM Forum UNI Address Formats

The international E.164 number contains up to 15 Binary Coded Decimal (BCD) digits padded with zeroes on the left-hand side to result in a constant length of 15 digits. The ITU-T assigns a Country Code (CC) of between one to three digits as standardized in Recommendation E.163. The remainder of the address is a Nationally Significant Number (NSN). The NSN may be further broken down as a National Destination Code (NDC) and Subscriber Number (SN). The North American Numbering Plan (NANP) is a subset of E.164. Currently, Bellcore administers the NANP with guidance from the Industry Numbering Committee (INC). The NDC currently corresponds to an Numbering Plan Area (NPA) code and switch NXX identifier for voice applications in the United States. For example, the E.164 number +01-214-555-1234 corresponds to a telephone subscriber in Dallas, Texas, in the United States. Further standardization of the NANP by the INC for local number portability is in progress.

NSAP-based numbers can be hard to remember (unless your parents spoke binary code while you were growing up), so the ATM Forum's ATM Names Service [7] provides a means to look up an ATM address based upon a "name," which may be a human readable name, an IP address, or another ATM address. The ANS specification represents an ATM End System Address (AESA) as a string of hexadecimal digits with the "." character separating any pair of digits for readability. An example of an NSAP address is:

39.246f.00.0e7c9c.0312.0001.0001.000012345678.00

The specification represents an E.164 formatted ATM address by a "+" character followed by a string of decimal digits that form an international E.164 number. A "." character separates any set of digits for readability. An example of an E.164 number is:

+01.212.555.1234

14.5.4 ATM Forum ATM End System Address (AESA) Formats

Figure 14.14 summarizes the current version of the NSAP addressing plans from the ATM Forum UNI version 4.0 specification. Each address has an Initial Domain Part (IDP) followed by a Domain Specific Part (DSP). The IDP has two parts: the Authority and Format Identifier (AFI) and the Initial Domain Identifier (IDI). The length of IDP field varies depending upon the particular AESA format. The one byte AFI field identifies the format for the remainder of the address. The IDI identifies the network addressing authority responsible for the assignment and allocation of the AESA DSP. The DSP has a High-Order DSP (HO-DSP) and low-order part comprised of an End System Identifier (ESI) and a Selector (SEL) byte. The length of the DSP varies, but is always 20 bytes minus the size of IDP. The true NSAP format subdivides the DSP into a fixed hierarchy that consists of a Routing Domain (RD), an Area identifier (AREA), and an End System Identifier (ESI). ATM Forum UNI 3.1 combined the RD and AREA fields into a single High-Order DSP (HO-DSP) field in anticipation of specifying a flexible, multi-level hierarchy prefix-based routing protocol. The specific uses of the Higher Order DSP (HO-DSP) are determined by the standards body identified in the IDP. The End System Identifier (ESI) and SELector (SEL) portions of the DSP are identical for all IDI formats as specified in ISO 10589. The ESI must be unique within a given IDP+HO-DSP address prefix. The ESI can also be globally unique, for example, a unique 48-bit IEEE MAC address. Beware that not all MAC addresses are unique since some devices allow the user to assign the MAC address. The SELector (SEL) field is not used by routing, but End Systems (ES) may employ it for local multiplexing.

ITU-T Recommendation X.213 [8] and ISO 8348 [9] specify the format, semantics, syntax and coding of AESAs. The ATM Forum currently defines three AESA formats as indicated in the figure: DCC, ICD, and E.164. The following sections give more details on each of these formats.

14.5.4.1 Data Country Code (DCC) AESA Format

The ITU-T assigns the IDP part using ISO country codes defined in ISO standard 3166 to yield the Data Country Code (DCC) AESA format shown in Figure 14.14a. The ISO member body within each country (e.g., ANSI in the

United States) administers between 2 and 4 of the high-order DSP bytes to allocate the common AFI+DCC prefix to multiple users within the same country. ANSI standard X3.216-1992 refers to this 2-4 byte field as the Country Domain Part (CDP). Thus, only 6 to 8 of the HO-DSP octets are available to the user.

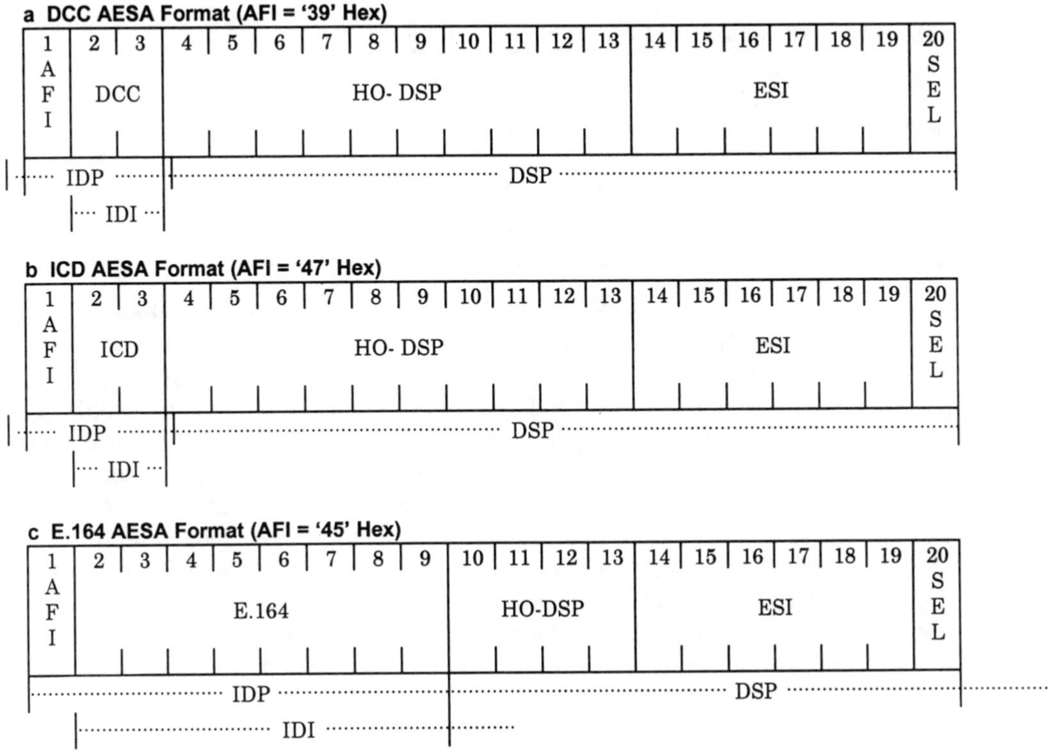

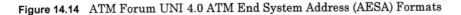

Figure 14.14   ATM Forum UNI 4.0 ATM End System Address (AESA) Formats

14.5.4.2 International Code Designator (ICD) AESA Format

The International Code Designator (ICD) identifies a specific organization as administered by the British Standards Institute as defined in ISO 6523. This address space aims to provide a resource for numbering plans that spans the globe. All 10-bytes of the HO-DSP field are useable by the owner of the ICD code.

14.5.4.3 E.164 Based AESA Format

The E.164 AESA combines the properties of the E.164 addressing scheme with the AESA format. The domain administrator assigns the 4-byte HO-DSP field. Private networks may use the E.164 based AESA format to identify public UNIs. Such private networks take the address prefix from the E.164 number NSAP-formatted number and identify local nodes by using the lower order ESI and SEL address fields.

14.5.5 Group Addresses and Anycast Signaling

Annex 5 of the ATM Forum UNI 4.0 specification [10] defines the concept of a *group address* assigned to more than one UNI port within a network. Group addresses have a unique prefix associated with each AFI to distinguish them from individual addresses as shown in Table 14.2.

Table 14.2 Individual and Group Address AFI Values

AESA Format	Individual AFI Value	Group AFI Value
DCC	'39'Hex	'BD'Hex
ICD	'47'Hex	'C5'Hex
E.164	'45'Hex	'C3'Hex

If a user places a point-to-point call to a group address, then the network routes the call to the port "closest" to the source associated with the specified group address. Since the network sets up the connections to any of several possible destinations, the ATM Forum specification calls this capability *anycast* signaling. Several higher-layer protocols, such as LAN Emulation (LANE) and MultiProtocol Over ATM (MPOA) make use of this anycast and group addressing function to provide multiple servers within a network. Furthermore, the ILMI address registration protocol described later in this section supports dynamic registration of group addresses so that networks dynamically discover new or relocated servers. This procedure also supports the means for the scope of a registering anycast server, which restricts the level of address advertisement, and hence usage by end users of particular anycast servers that may be too far away.

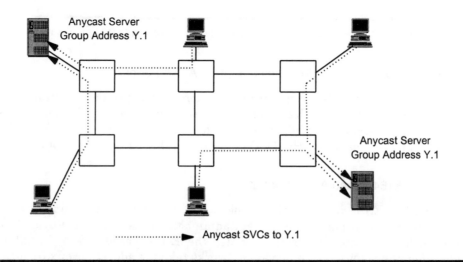

Figure 14.15   Group Address and Anycast Signaling Example

Figure 14.15 illustrates an example application of group addresses and anycast signaling. The network has six nodes, with two servers both assigned group address Y.1, registered with the network using the ILMI address registration procedure described below. The dashed lines indicate the result of the network routing decision for hosts attached to the various nodes resulting from an anycast call to group address Y.1. In this simple example, the scope is the entire network. If one of the servers failed, then the network would route all anycast calls to group address Y.1 to the remaining server. Hence, the definition of group addresses allows ATM network designers to implement resilient servers into the higher-layer user plane protocols studied in Part 4. A final note on terminology: although SMDS also uses the term group address, it does mean the same thing. The SMDS meaning of group address is much more closely tied to the IP multicast described in Chapter 20.

14.5.6 ILMI Address Registration

Address registration using the Integrated Local Management Interface (ILMI) [11] is a key component of automatic configuration of PNNI reachability information when using ATM SVC networking. Basically, address registration allows the network to communicate to the user the valid address prefixes for a particular logical ATM UNI. The user then registers complete addresses by suffixing the address prefix with the ESI and SEL fields. Optionally, the user may register a connection scope along with each address. Thus, ILMI overcomes the need to manually configure large numbers of user addresses. It also enables source authentication, since the originating switch may screen the calling party address information element in the SETUP message against the set of registered addressed prefixes.

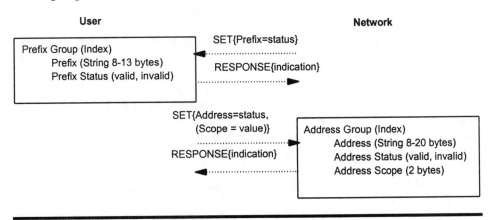

Figure 14.16   Illustration of ILMI MIB Address Registration

Figure 14.16 illustrates the SNMP message flows associated with the address registration portion of the ILMI MIB. The Network Prefix Group resides in the ILMI MIB on the User side of the UNI interface, while the Address Group resides on the Network side of the UNI interface. The

Address Group is not applicable for Native E.164 addressing since the 8-byte network prefix completely specifies the address. All 20-byte AESA formatted addresses have a 13-byte network prefix. Registration occurs at initialization time (i.e., a cold start trap), or whenever either the network or the user needs to add or delete a prefix or address. At initialization, the address and prefix tables on the network and user side are empty. One side registers a prefix or address with the other by first SETting the address value and its status to valid. The other side sends a RESPONSE with either a "No Error" or "Bad Value" parameter to indicate the success or failure, respectively, of the registration attempt. If one side sends a SET message and receives no RESPONSE, then the originator should retransmit the SET request. If one side receives duplicate SET messages, then it returns a RESPONSE indicating no error. If one side receives a SET request which attempts to change the status of an unregistered prefix or address to invalid, then it returns a RESPONSE with a "NoSuchName" error.

The Service Registry MIB information portion of the ILMI provides a general-purpose service registry for locating ATM network services, such as the LAN Emulation Configuration Server (LECS) and the ATM Name Server (ANS). Either of these servers could be assigned a group address and accessed via the anycast signaling procedure described above.

14.5.7 ATM Name Service (ANS)

The ATM Forum adopted the Domain Name System (DNS) concept from the Internet to resolve names into ATM addresses in ATM Name Service (ANS) [7]. ANS supports both NSAP-based and E.164 ATM addresses. In the Internet, a DNS resolves a host name and organization in an E-mail address (i.e., user@host_name.org) or a web site (e.g., http://www.usersite.org) to an IP address. Most human beings find it easier to remember a name rather than a number. There are exceptions among us, such as those capable of rattling off IP addresses and other numeric data more readily than their own children's names; however, you won't likely encounter them at too many cocktail parties.

ANS is a native ATM application defined by the Native ATM Services (NAS) API described in Chapter 16 that employs ATM SVCs for clients to communicate with ANS servers. The protocol also specifies the means for servers to communicate in the processing of providing service to ANS clients. The basic directory services defined in ANS are:

- Domain name–to–ATM address translation using existing top-level domain names
- ATM address–to–domain name translation using a new domain name, ATMA.INT

Figure 14.17 illustrates the operation of the ANS protocol. In the first step, ANS clients either get the ATM address of an ANS server via the Integrated Local Management Interface (ILMI) defined in Chapter 27, or else use a well-known ANS address. In step 2, an ANS client sets up a connection to an ANS server using the SVC procedure, using the address determined in step 1. Once the client establishes a connection with the server, it can send ANS requests in step 3, to which the ANS server responds. The ATM Forum specification advises that the client and server should release the connection if no activity occurs for a long period of time (e.g., minutes) so that other clients may access the ANS server.

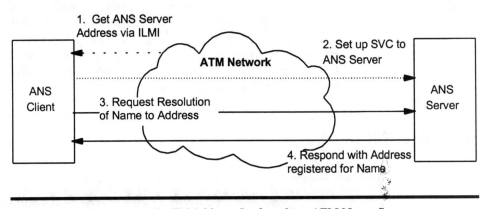

Figure 14.17   Illustration of ATM Address Lookup from ATM Name Server

Although not widely implemented yet, ANS promises to make the ATM control plane much friendlier to end users and applications in a manner similar to the way DNS makes the translation of easy to remember World Wide Web Universal Reference Locator (URL) names into Internet addresses invisible to the user. Furthermore, the ANS can also be used to translate one ATM address into another. This capability could be used to translate a private ATM network address into a public ATM address. Other translations could implement intelligent network services like user mobility, time of day routing, or overflow routing. The next version of the ANS specification will likely include: security extensions, dynamic update capabilities, and prompt notification of changes.

14.6 REVIEW

This chapter defined B-ISDN from the top-down perspective in terms of the user, control, and management planes. The sections on the user plane provided an overview of how a wide range of user plane protocols supporting voice, video, as well as LAN and WAN data protocols utilize the common part AALs defined in the previous chapter. The coverage laid out the presentation of these subjects in Part 4. The text then showed how the control and management planes support the services provided by the user plane. The text described the functional decomposition of the management plane into overall plane management and management of each of the user and control plane layer components.

Finally, the chapter introduced the control plane, which is central in performing the functions needed in a Switched Virtual Connection (SVC) service. The text provided an overview of the usage and meaning of the various signaling and routing control plane protocols detailed in Chapter 15. The text defined the context for the various signaling protocols and articulated the structure of the control plane Service Specific Convergence Sublayer (SSCS) and its constituent components: the Service Specific Coordination Function (SSCF), and the Service Specific Connection-Oriented Protocol (SSCOP). Next, we delved into the important concept of ATM addressing. Finally, the chapter concluded with an overview of group addresses and anycast signaling; the Integrated Local Management Interface (ILMI) address registration procedure; and the ATM Name Service (ANS).

14.7 REFERENCES

[1]   ITU-T, *Broadband Integrated Services Digital Network (B-ISDN) – Digital Subscriber Signalling System No. 2 (DSS 2) – User-Network Interface (UNI) Layer 3 Specification For Basic Call/Connection Control*, Recommendation Q.2931, February 1995.

[2]   ITU-T, B-ISDN *Signalling ATM Adaptation Layer (SAAL) Overview Description,* Q.2100, July 1994.

[3]   ANSI, *Telecommunications - B-ISDN Signaling ATM Adaptation Layer - Overview Description,* T1.636-1994, July 11, 1994.

[4]   ITU-T, B-ISDN *Signalling ATM Adaptation Layer – Service Specific Coordination Function for Support of Signalling at the User Network Interface (SSCF at UNI)*, Recommendation Q.2130, July 1994.

[5]   ITU-T, B-ISDN *Signalling ATM Adaptation Layer – Service Specific Coordination Function for Support of Signalling at the User Network Interface (SSCF at UNI)*, Recommendation Q.2140, February 1995.

[6]   ITU-T, B-ISDN *ATM Adaptation Layer – Service Specific Connection Oriented Protocol (SSCOP)*, Recommendation Q.2110, July 1994.

[7]   ATM Forum, *ATM Name System Specification Version 1.0*, af-saa-0069.000, November 1996.

[8]   ITU-T, *Information Technology – Open Systems Interconnection – Network Service Definition*, Recommendation X.213, November 1995.

[9]   ISO/IEC 8348, *Information Technology - Telecommunications and Information Exchange Between Systems - Network Service Definition*, 1993.

[10]   ATM Forum, *User-Network Interface Signaling Specification*, Version 4.0, af-sig-0061.000, July 1996.

[11]   ATM Forum, *Integrated Local Management InterfaceVersion 4.0*, af-ilmi-0065.000, September 1996.

15

Signaling and Routing in the Control Plane

This chapter covers the B-ISDN control plane, which performs a pivotal role in ATM's Switched Virtual Connection (SVC) service. ATM SVC signaling is a connection-oriented protocol that operates in a manner analogous to a telephone call. However, in ATM computers signal many more parameters than telephone calls do. For example, ATM signaling conveys logical channel and physical network level addresses, quality of parameters, ATM service categories, and traffic parameters. Specifically formatted fields, called Information Elements (IEs), within B-ISDN signaling messages convey these user requests. The text summarizes the definition and usage of signaling messages to help the reader understand the basic functions available from the signaling protocol. When reading this chapter, remember that many ATM-based applications are computer programs issuing B-ISDN signaling messages, and not human beings placing calls across a telephone network.

Next, the text covers the UNI signaling protocols by introducing the key signaling messages and their information elements. Simple examples of point-to-point and point-to-multipoint signaling procedures illustrate how users and networks employ the messages to establish and release connections. The chapter then moves on to signaling protocols used between nodes in a network, as well as protocols used between networks. The treatment addresses the simplest network-network protocol first — the ATM Forum's InterSwitch Signaling Protocol (IISP), before moving on to a more complex protocol, the ATM Forum's Private Network-Network Interface (PNNI). The sophisticated PNNI protocol combines concepts from B-ISDN signaling, local area network source routing, and dynamic Internet style routing to automatically provide guaranteed quality and routing in networks that can scale to global proportions. Finally, the chapter concludes with considerations involved in signaling between carrier networks at the Broadband InterCarrier Interface (B-ICI) using the B-ISDN User Services Part (B-ISUP).

15.1 USER-NETWORK INTERFACE (UNI) SIGNALING

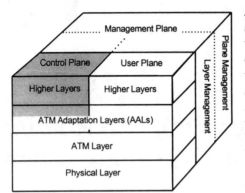

As described in Chapter 5, ATM signaling shares many characteristics with basic telephony, with extensions that add the capabilities to specify bandwidth, quality of service, various end system attributes, different connection topologies, and address formats. First, this section describes the basic functions of the UNI 3.1 and 4.0 UNI signaling protocols. The text presents a comparison of ATM Forum and ITU-T signaling standards with references for readers interested in following up in more detail. Next, we introduce the basic signaling message types and review the role of some of the key information elements in these messages.

15.1.1 Signaling Functions — Q.2931 and UNI 3.1

The ATM Forum UNI version 3.1 signaling specification [1] and a draft version of the ITU-T's Recommendation Q.2931 [2] called Q.93B were not closely aligned. Furthermore, beware of an initial version of the ATM Forum signaling specification, UNI 3.0, which is completely incompatible because the ITU-T changed the layer 2 SSCOP protocol after the ATM Forum issued its 3.0 specification. The UNI 3.1 specification made several corrections and clarifications to UNI 3.0. To maximize interoperability, look for equipment and services that support at least ATM Forum UNI version 3.1 with a migration path to UNI 4.0 signaling. The major signaling functions defined in ITU-T Recommendation Q.2931 are:

- ① Point-to-point connection setup and release
- ① VPI/VCI selection and assignment
- ① Quality of Service (QoS) class request
- ① Identification of calling party
- ① Basic error handling
- ① Communication of specific information in setup request
- ① Subaddress support
- ① Specification of Peak Cell Rate (PCR) traffic parameters
- ① Transit network selection
- ① Interworking with N-ISDN

The ATM Forum UNI version 3.1 specification does not support the following capabilities specified in ITU-T Recommendation Q.2931:

- ✕ No alerting message sent to called party
- ✕ No VPI/VCI selection or negotiation

✂ No overlap sending
✂ No interworking with N-ISDN
✂ No subaddress support
✂ Only a single transit network may be selected

Most of the additional features specified in UNI 3.1 have since been adopted in ITU-T Recommendations as described below. The text describes these features in the context of the latest version of the ATM Forum signaling specification, UNI 4.0. Don't worry if you plan to stay on UNI 3.1 signaling for a while, since the Forum did not create another incompatible set of signaling standards as happened in the change from UNI 3.0 to 3.1.

15.1.2 ATM Forum UNI Signaling 4.0 and the ITU-T Standards

The ATM Forum's UNI Signaling 4.0 specification [3] is based largely upon the ITU-T's Q.2931 signaling standard. In fact, the UNI 4.0 signaling document is written as a set of additions and deletions to the ITU-T's Q.2931 and related Q series recommendations. The major additions and changes from the UNI 3.1 specification that align UNI 4.0 with the Q.2931 standard are:

- Supplementary Service Support
- End-to-end Transit Delay Information Element Support
- Narrowband ISDN (N-ISDN) Interworking support

The current version of the ATM Forum UNI Signaling Specification version 4.0 and the ITU-T Q.2931 standard are closely aligned in the specification of control plane functions. Table 15.1 compares the capabilities defined in the ATM Forum UNI 4.0 signaling specification and those defined in the applicable ITU-T Q Series Recommendation. Note that many ATM Forum UNI 4.0 capabilities are not yet standardized by the ITU-T, while a few ITU-T capabilities are not addressed by the ATM Forum 4.0 specification.

Supplementary services supported in the specification include Direct Dialing In (DDI), Multiple Subscriber Number (MSN), Calling Line Identification Presentation/Restriction (CLIP/CLIR), Connected Line Identification Presentation/Restriction (COLP/COLR), Subaddressing and User to User Signaling (UUS). The UNI 4.0 specification also adds support for the end-to-end Transit Delay Information Element, which allows the calling party to request bounded delay for the connection. Furthermore, the specification adopts a number of ITU-T conventions in support of Narrowband ISDN (N-ISDN) interworking.

However, the ATM Forum continued to add new capabilities in the UNI signaling 4.0 specification, ahead of the slower ITU-T standards adoption process. Specifically, UNI 4.0 signaling adds the following functions:

- Support for Group and Anycast Addresses
- Connection Parameter Negotiation at Setup Time
- Leaf Initiated Join (LIJ) point-to-multipoint SVCs

- User Specified Quality of Service (QoS) parameters
- Switched Virtual Paths
- Support multiple signaling channels over one physical UNI
- Proxy Signaling
- ABR Signaling for point-to-point SVCs

Table 15.1   Comparison of UNI 4.0 and ITU-T UNI Signaling Capabilities

Capability Description	ATM Forum UNI 4.0	ITU-T Recom- mendation
On-Demand (Switched) Connections	Yes	Q.2931
Point-to-point Calls	Yes	Q.2931
N-ISDN Signaling Interworking	No	Q.2931
E.164 Address Support	Yes	Q.2931
NSAP-based Address Support	Yes	Q.2931
Root-initiated Point-to-multipoint Calls	Yes	Q.2971
Leaf-initiated Point-to-multipoint Calls	Yes	No
Signaling of Individual QoS Parameters	Yes	No
ATM Anycast	Yes	No
ABR Signaling for Point-to-point Calls	Yes	No
Generic Identifier Transport	Yes	No
Virtual UNIs	Yes	No
Switched Virtual Path (VP) service	Yes	No
Proxy Signaling	Yes	No
Frame Discard	Yes	No
Traffic Parameter Modification during Active Calls	No	Q.2963.1
Traffic Parameter Negotiation during Call Setup	Yes	Q.2962
Supplementary Services		
Direct Dialing In (DDI)	Yes	Q.2951
Multiple Subscriber Number (MSN)	Yes	Q.2951
Calling Line Identification Presentation (CLIP)	Yes	Q.2951
Calling Line Identification Restriction (CLIR)	Yes	Q.2951
Connected Line Identification Presentation (COLP)	Yes	Q.2951
Connected Line Identification Restriction (COLR)	Yes	Q.2951
Subaddressing (SUB)	Yes	Q.2951
User-user Signaling (UUS)	Yes	Q.2957

An anycast address identifies a particular service, and not a specific node as described in the previous chapter. The UNI 4.0 signaling specification supports a limited form of connection parameter negotiation at call setup time. The user may include in addition to the desired traffic descriptor either the minimum acceptable, or an alternative, traffic descriptor in the SETUP message. The network responds indicating whether it granted either the original traffic descriptor, or the user-specified minimum/alternative.

This response is important to applications, such as video conferencing, that operate best with a preferred bandwidth, but can "step down" to a lower bandwidth in a manner similar to automatic modem speed negotiation dependent upon line quality.

An additional Leaf Initiated Join (LIJ) signaling procedure complements the root originated point-to-multipoint Switched Virtual Connection (SVC) capability of UNI 3.1. The next section describes these capabilities in more detail. Additional signaling information elements allow the end user to explicitly specify Quality of Service (QoS) parameters, such as peak-to-peak Cell Delay Variation (CDV) and Cell Loss Ratio (CLR) in both the forward and backward directions.

UNI signaling 4.0 also supports switched Virtual Path Connection (VPCs). Later in this chapter, we discuss how the PNNI specification utilizes switched VPCs to "tunnel" across other networks. As described in Chapter 14, the specification also defines "virtual UNIs" that allow multiple signaling channels over a single physical UNI. The specification also defines a means to define multiple signaling links on different VPIs on the same physical interface. This capability allows a dumb VP multiplexer to aggregate ATM traffic into an intelligent ATM switch. These features are important to private networks implemented where other protocols tunnel over VPC connections, or in public networks where a single physical interface supports multiple logical customers.

15.1.3 UNI 4.0 Signaling Message Types

The ATM Forum UNI specification version 4.0 uses the message types shown in Table 15.2 for point-to-point and point-to-multipoint connections. The table groups the point-to-point messages according to function: call establishment, call release (or clearing), status, and layer 2 signaling link management. The point-to-multipoint messages support the procedures for adding and dropping root and leaf initiated calls. The next section illustrates the use of many of these messages through call establishment and release examples.

15.1.4 Signaling Message Information Elements

Each of the UNI 4.0 signaling messages has a number of Information Elements (IEs), some of which are Mandatory (M) and others of which are Optional (O) as indicated in Table 15.3. Reference 4 presents a handy reference of the maximum IE size along with a brief description in a tabular format. This table does not include the N-ISDN interworking signaling messages or information elements for brevity. Q.2931 and UNI 4.0 have many information elements in common. An asterisk next to the information element description indicates those IEs unique to UNI 4.0.

Note that all messages related to a particular call attempt each contain a common mandatory information element, the *call reference,* that is unique to a signaling channel. Also, every message must also contain an information element for their type, length, and protocol discriminator (i.e., the set from which these messages are taken). This structure supports variable length messages and the addition of new message types in the future as needed. The following narrative highlights some of the key information elements and their usage.

Table 15.2   UNI 4.0 Signaling Message Types

Point-to-Point Connection Control	Point-to-Multipoint Connection Control
Call Establishment Messages	ADD PARTY
ALERTING	ADD PARTY ACKNOWLEDGE
CALL PROCEEDING	ADD PARTY REJECT
CONNECT	PARTY ALERTING
CONNECT ACKNOWLEDGE	DROP PARTY
SETUP	DROP PARTY ACKNOWLEDGE
Call Clearing Messages	LEAF SETUP REQUEST
RELEASE	LEAF SETUP FAILURE
RELEASE COMPLETE	
Status Messages	
STATUS ENQUIRY	
STATUS (Response)	
NOTIFY	
Signaling Link Management	
RESTART	
RESTART ACKNOWLEDGE	

One insight into ATM signaling comes from analyzing Table 15.3 to determine which message types contain specific information elements. Obviously, the SETUP message contains the majority of the information elements because it conveys the user's request to the network. The key mandatory information elements used in the SETUP message are:

- Called party number
- Broadband Bearer Capability
- ATM traffic descriptor

The called party number may be either an NSAP-based or E.164 ATM address as defined in Chapter 14. The Broadband Bearer Capability specifies the ATM service category introduced in Chapter 12 and detailed in Chapter 21, namely CBR, rt-VBR, nrt-VBR, UBR, or ABR. The ATM traffic descriptor defines the parameters of the traffic contract, such as PCR, SCR and MBS in both the forward and backward directions. The traffic descriptor also determines whether the network tags cells with the CLP bit for inputs exceeding the contract per ITU-T Recommendation Q.2961 [5].

The connection identifier gives the value of the VPI (and VCI) for a switched VPC (or VCC) connection. The user may request a specific value in the SETUP message, or accept a value assigned by the network in an ALERTING, CALL PROCEEDING, or CONNECT message.

Table 15.3   UNI 4.0 Signaling Message Information Element (IE) Content

Information Element (IE) Name	ADD PARTY	ADD PARTY ACK	ADD PARTY REJECT	ALERTING	CALL PROCEEDING	CONNECT	CONNECT ACK	PARTY ALERTING	SETUP	LEAF SETUP REQUEST	LEAF SETUP FAILURE	DROP PARTY	DROP PARTY ACK	RELEASE	RELEASE COMPLETE	RESTART	RESTART ACK	NOTIFY	STATUS	STATUS ENQUIRY
Protocol discriminator	M	M	M	M	M	M	M	M	M	M	M	M	M	M	M	M	M	M	M	M
Call reference	M	M	M	M	M	M	M	M	M	M	M	M	M	M	M	M	M	M	M	M
Message type	M	M	M	M	M	M	M	M	M	M	M	M	M	M	M	M	M	M	M	M
Message Length	M	M	M	M	M	M	M	M	M	M	M	M	M	M	M	M	M	M	M	M
Boradband bearer capability									M											
Endpoint reference	M	M	M	O-1	O-1	O-1		M	O			M	M					O	O	O
Endpoint state																			O	
LIJ call identifier*									O	M										
LIJ parameters*									O											
Leaf sequence number*	O								O	M	M									
Notification indicator	O	O		O	O	O	O	O	O			O		O	O			M		
Connection identifier				O	O-1	O-1			O											
Generic identifier transport	O	O	O	O		O		O	O			O		O	O					
ABR additional parameters*						O			O											
ABR setup parameters*						O			O											
AAL parameters	O	O				O			O											
ATM traffic descriptor						O			M											
Alternative ATM traffic descriptor*									O											
Min. acceptable ATM traffic descriptor*									O											
Broadband high layer information	O								O											
Broadband low layer information	O	O				O			O											
Broadband repeat indicator	M								O											
Broadband sending complete									O											
Calling party number	O								O	M										
Calling party sub-address	O								O	O										
Called party number	M								M	M	O									
Called party sub-address	O								O	O	O									
Connected number		O				O														
Connected sub-address		O				O														
Transit network selection	O								O	O	O									
Connection scope selection*									O											
QoS parameter*									O											
End-to-end transit delay	O	O				O			O											
Extended QoS parameters*						O			O											
Cause (code)			M								M	M	O	O-2	O-2				M	
Call state			O																M	

Notes:

1 Mandatory if included in SETUP

2. Mandatory if first call clearing message

* ATM Forum UNI 4.0 unique Information Element

The cause (code) specified in ITU-T Recommendation I.2610 [6] provides important diagnostic information by indicating the reason for releasing a call. The cause IE must be present in the first message involved in call clearing. For example, the cause indicates whether the destination is busy, if the network is congested, or if the requested traffic contract or service category isn't available.

The key attributes of the other optional parameters not already described are as follows. The root point-to-multipoint call procedures utilize the endpoint reference identifier and endpoint state number information elements. The Leaf Initiated Join (LIJ) call identifier uniquely identifies the point-to-multipoint call at the root. The root uses the LIJ parameters to associate options, such as a screening indication with the call in the SETUP message responding to a LEAF SETUP REQUEST. The leaf sequence number associates the signaling messages involved with a LIJ call. The Available Bit Rate (ABR) parameters detail the requested (and granted) service defined by the ATM Forum as described in Chapter 22. The AAL parameters along with the broadband low- and high-layer information elements convey information between end users about the end systems. Only the end user may employ the called and calling party subaddress information elements to convey additional addressing information across the network. The QoS parameter indicates the QoS class (values 0 through 4) defined in UNI 3.1. The Transit Network Selection (TNS) IE allows an end user to specify the desired network provider. The STATUS message uses the call state IE to indicate the current condition of the referenced call or endpoint in response to a STATUS ENQUIRY message.

15.2 SIGNALING PROCEDURES

Signaling procedures specify the valid sequence of messages exchanged between a user and the network, the rules for verifying consistency of the parameters, and the actions taken to establish and release ATM layer connections. A significant portion of signaling standards and specifications handle error cases, invalid messages, inconsistent parameters, and a number of other unlikely situations. These are all important functions since the signaling protocol must be highly reliable to support user applications. This section gives an example of signaling procedures for the following types of calls using the message types and information elements described above:

* point-to-point connection establishment
* point-to-point connection release
* root initated point-to-multipoint connection establishment
* leaf initated join point-to-multipoint connection establishment

Standards specify signaling protocols in several ways: via narrative text, via state machine tables, or via a semigraphical Specification Definition Language (SDL). The ATM Forum UNI specifications and the ITU-T Q.2931 use the narrative method as well as the SDL technique. For complicated protocols, such as Q.2931, a very large sheet of paper would be needed to draw the resulting state machine in a manner such that a magnifying glass is not required to read it. The SDL allows a complicated state machine to be formally documented on multiple sheets of paper in a tractable manner. In fact, ITU-T Recommendation Q.2931 dedicates almost one hundred pages to SDL diagrams out of a total of 250 pages.

The B-ISDN Q.2931 protocol is based upon the Narrowband ISDN Q.931 and Frame Relay Q.933 protocols. Since a large amount of expertise has been built up over the years on these subjects, the prospects for implementing the rather complex Q.2931 protocol in an interoperable manner are encouraging.

If you're looking for an ATM signaling expert for your network or company, engineers with N-ISDN or frame relay signaling background can help.

15.2.1 Point-to-Point Connection

Figure 15.1 illustrates the point-to-point connection establishment example. This example employs: a calling party with ATM address A on the left, a network shown as a cloud in the middle, and the called party with ATM address B on the right. Time runs from top to bottom in all of the examples.

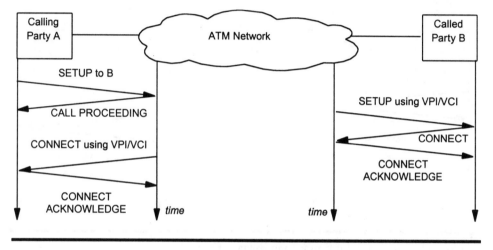

Figure 15.1 Point-to-Point Connection Establishment Example

Starting from the upper left-hand side of Figure 15.1, the calling party initiates the call attempt using a SETUP message indicating B in the called party number IE. Recommendation Q.2931 requires that the network respond to the SETUP message with a CALL PROCEEDING message as indicated in the figure. The network routes the call to the physical interface connected to B and outputs a SETUP message indicating the specified VPI/VCI values in the connection identifier IE. Optionally, the SETUP message may also communicate the identity of the calling party A in the calling party number IE, similar to the calling line ID service in telephony. If the called party chooses to accept the call attempt, it returns the CONNECT message, which the network propagates back to the originator as rapidly as possible in order to keep the call setup time low. Optionally, the called party user may respond with either a CALL PROCEEDING or an ALERTING message prior to sending the CONNECT message; however, unlike the network side, Recommendation Q.2931 does not require the user side to respond to a SETUP with the CALL PROCEEDING or ALERTING message. Both the user and network sides confirm receipt of the CONNECT message via sending the CONNECT ACKNOWLEDGE message as shown in the figure.

Figure 15.2 illustrates the point-to-point connection release example, or in other words the process used to hang up the call. The reference configuration and conventions are the same as in the point-to-point connection establishment example above. Either party may initiate the release process, just as either party may hang up first in a telephone call. This example illustrates the calling party as the one that initiates the disconnect process by sending the RELEASE message. The network then propagates the RELEASE message across the network to the other party B. The network also responds to A with a RELEASE COMPLETE message as indicated in the figure. The other party acknowledges the RELEASE request by returning a RELEASE COMPLETE message. This two-way handshake completes the call release process.

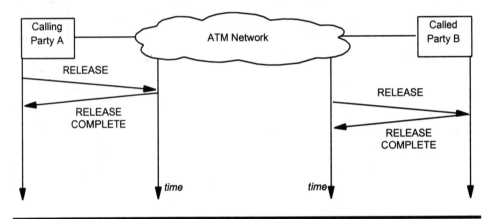

Figure 15.2   Point-to-Point Connection Release Example

15.2.2 Root Initiated Point-to-Multipoint Connection

The ATM Forum UNI 3.1 specification first defined the root initiated point-to-multipoint connection capability in 1994. Subsequently, ITU-T Recommendation Q.2971 [7] standardized this important capability one year later. The UNI 4.0 signaling specification retains this capability and adds the Leaf Initiated Join (LIJ) procedure described in the next section. The root initiated point-to-multipoint connection process is similar to that of three way telephone calling where a conference leader (the root) adds other parties to an existing call. The root initiated point-to-multipoint connection procedure meets the needs of broadcast audio, video and data applications. A full mesh of point-to-multipoint connections emulates a shared media data network as described for the IP multicast over ATM application in Chapter 20.

Figure 15.3 illustrates an example of a root node setting up a point-to-multipoint call from an originator (root) node A to two leaf nodes B and C connected to a local ATM switch on a single ATM UNI, and a third leaf node D connected to a separate ATM UNI. In the example, root node A begins the point-to-multipoint call by sending a SETUP message to the network requesting setup of a point-to-multipoint call identifying leaf node B's ATM address. The network responds with a CALL PROCEEDING message in much the same way as a point-to-point call. The network switches the call attempt to the intended destination and issues a SETUP message to node B identifying the assigned VPI/VCI. Leaf node B then indicates its intention to join the call by returning a CONNECT message that the network in turn acknowledges with a CONNECT ACKNOWLEDGE message. The network informs the calling root node A of a successful addition of party B through a CONNECT and CONNECT ACKNOWLEDGE handshake as shown in the figure.

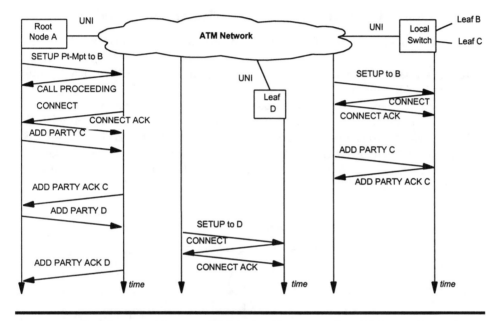

Figure 15.3   Root Initiated Point-to-Multipoint Connection Establishment Example

Continuing on with the same example, the root node requests addition of party C through the ADD PARTY message. The network relays the request to the same ATM UNI used by party B through the ADD PARTY message to inform the local switch of the requested addition. In other words, the network only uses the SETUP message to add the first party on any particular ATM UNI and uses the ADD PARTY message for any subsequent leaf added to an ATM UNI which already has a connected party in the point-to-multipoint call. Party C responds with an ADD PARTY ACKNOWLEDGE C message that the network propagates back to the root node A. Finally, the root node A requests addition of leaf party D through an ADD PARTY message. The network routes this to the UNI connected to party D, and issues a SETUP message since this is the first party connected on this particular ATM UNI. Node D responds with a CONNECT message, which the network responds with a CONNECT ACKNOWLEDGE message. The network communicates the successful addition of leaf party D to the call to the root node A through the ADD PARTY ACKNOWLEDGE D message.

The leaves of the point-to-multipoint call may disconnect from the call by the DROP PARTY message if one or more parties would remain on the call on the same UNI, or by the RELEASE message if the party is the last leaf present on the same UNI. For example, the local switch could disconnect party C by sending a DROP PARTY message, but must send a RELEASE message if party B later disconnected. If the root node initiates disconnection, then it drops each leaf in turn and finally releases the entire connection. Note that the root node is aware of all the parties in the call, since it added each one to the point-to-multipoint connection.

15.2.3 Leaf Initiated Join (LIJ) Point-to-Multipoint Connection

The ATM Forum's UNI 4.0 specification added a Leaf Initiated Join (LIJ) protocol better suited to data applications. For example, IP multicast internetworking allows users to dynamically join (and leave) multicast groups as described in Chapter 20. This procedure is similar to the popular "meet-me" telephony conference bridges used by many businesses and enterprises today. Two new message types perform this function in conjunction with the other message types previously described above: LEAF SETUP REQUEST and LEAF SETUP FAILURE. Figure 15.4 illustrates an example of an LIJ call using the same parties and network configuration employed in the previous section.

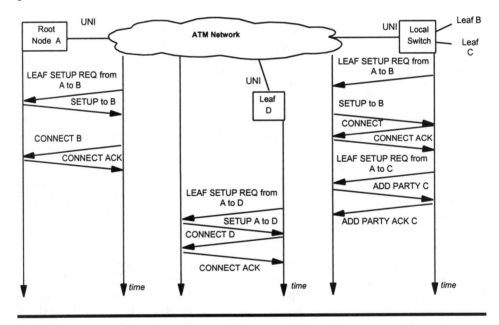

Figure 15.4   Leaf Initiated Join Point-to-Multipoint Example

Starting from the right-hand side, leaf B sends a LEAF SETUP REQUEST identifying A as the root, B (itself) as the calling party and the group address of the point-to-multipoint call B wishes to join. If the network determines that the point-to-multipoint connnection is active, then it sends the LEAF SETUP REQUEST to the root node A. If the network and A both accept the request, then the root node A initiates SETUP of a point-to-multipoint call to B, identifying it as a call that other leaves may join without notifying the root. The specification also supports LIJ calls where the network notifies the root node of every joining leaf. If either the network or the root node A reject B's request, the network sends a LEAF SETUP FAILURE message to B with the cause value indicating the reason for denying the request. The network connects B to the point-to-multipoint call as in the root initiated case, but includes leaf sequence number information from the LEAF SETUP REQUEST message so that B can associate the SETUP message with its original request. Leaf B indicates establishment via the CONNECT message and the network acknowledges in a manner similar to the above examples. Now, when leaf sends a LEAF SETUP REQUEST for the same point-to-multipoint connection group address, the network adds an additional party to the UNI already supporting the point-to-multipoint connection. In the case of party D, the network initiates the SETUP/CONNECT handshake when adding the first party on the point-to-multipoint connection on the UNI. Disconnection from the LIJ point-to-multipoint connection is similar to that for the root initiated case.

Note that the LIJ point-to-multipoint SVC procedure requires that every leaf use the same group address. This can either be a well known address, or an address determined from a higher layer protocol, such as IP multicast as described in Chapter 20.

15.3 NETWORK NODE INTERFACE (NNI) SIGNALING

The remainder of this chapter covers the important topic of Network-Node Interfaces, also known as Network-Network Interfaces, both abbreviated as NNI. These two meanings of the same acronym identify its dual purpose for use between nodes in a single network, as well as interconnection between different networks. The treatment in the remainder of this chapter begins in the private network domain and then moves to the public network domain by covering the following major NNI signaling protocols:

* ATM Forum Interim InterSwitch Signaling Protocol (IISP)
* ATM Forum Private Network-Network Interface (PNNI)
* ATM Forum Broadband Inter-Carrier Interface (B-ICI)
* ITU-T Broadband ISDN Services User Part (B-ISUP)

15.4 INTERIM INTERSWITCH SIGNALING PROTOCOL (IISP)

If you want to build a multi-vendor ATM SVC network, IISP is the simplest way to get started. However, as your networking needs mature and the ATM network grows, the design should move to PNNI.

The ATM Forum recognized the need to produce a standard for a minimum level of interoperability for multi-vendor private ATM networks. Therefore, the Forum rapidly developed and published the Interim InterSwitch

Signaling Protocol (IISP) in late 1994 [8]. The Forum announced that the IISP standard would fill the void until it could complete the PNNI specification. IISP basically extended the UNI 3.0/3.1 protocol to a simple network context.

Figure 15.5 illustrates the conventions that IISP added to the UNI 3.0/3.1 UNI signaling specifications. The IISP physical layer, ATM layer, and traffic management specifications are identical to the UNI 3.0/3.1 specification. IISP employs the UNI cell format, no ILMI, and makes policing optional. IISP specifies a limited set of VCIs ranging from 32 to 255 on VPI 0 to ensure interoperability. The signaling procedures for the user and network sides differ. Simply stated, one may view the network side as the master and the user side as the slave in the signaling protocol. Hence, a key addition of IISP to UNI signaling is identification of each side of a trunk as either the user or network side as depicted in the figure. The functions of the user and network sides are symmetrical in the aspects regarding placement or reception of calls; however, and the assignment of user and network side is arbitrary.

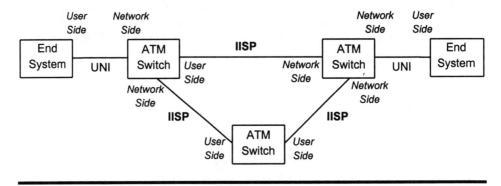

Figure 15.5   Interim Inter-Switch Signaling Protocol (IISP) Context

IISP utilizes the NSAP-based ATM End System Address (AESA) format defined in Chapter 14. IISP defines a simple hop-by-hop routing based upon matching the longest address prefix in a statically configured routing table. Such manual configuration limits the scalability of IISP networks and is difficult to manage and administer. Also, a switch must clear the SVC if it detects a link failure on any IISP interface, a function normally performed only by the network side of a UNI. Support for other features is optional, such as routing over parallel links between nodes, connection admission control for guaranteeing quality of service, and alternate routing.

Furthermore, manual configuration of hop-by-hop routing tables may introduce routing loops, a potential problem the IISP specification identifies, but provides no guidance on how to avoid. Figure 15.6 illustrates a routing loop in a five node network with high-speed, low-cost trunks shown as thick lines and lower-speed higher cost trunks shown as thinner lines each labeled by a lowercase character; r, s, t, v, w; and u and x; respectively. Each node supports end systems with addresses denoted as A.1.S.E, with S taking on values of 1 through 5, and E indicating the end system portion of the NSAP-based address. For example, end system A.1.1.1 attaches to the node with address prefix A.1.1. The figure shows a row from each switch's routing table. The table entries give an ordered list of trunks for routing calls to address prefix A.1.3. For example, when routing a call to an address with prefix A.1.3, switch A.1.1 first tries trunk s, then v, and finally r if both s and v are out of service or busy. Of course, a switch does not route a call back out on the same trunk it received the signaling message on. A routing loop develops between switches A.1.1, A.1.4 and A.1.2 as illustrated in the figure for a call attempt from A.1.1.1 to A.1.3.2 when trunks s and t are either down, or full. The next hop routing table set up in this example chose the expensive trunks (x and u) last, but unfortunately created the potential for a routing loop nonetheless. Beware that the potential for the formation of routing loops always exists with static routing. The only solution is careful engineering, or the ATM Forum's answer — dynamic routing using the PNNI protocol.

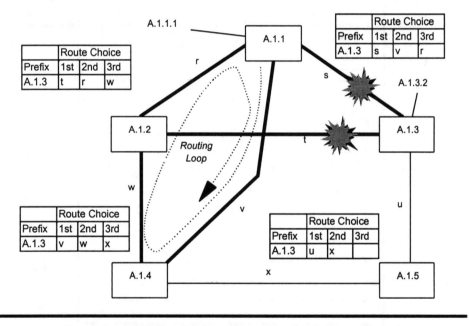

Figure 15.6   Example of IISP Routing Loop formed by Static Route Choices

15.5 THE ATM FORUM'S ROUTING PROTOCOL - PNNI

A key business driver for the Private Network-Node Interface (PNNI) protocol [9] is automatic configuration, also called "plug and play" operation, along with multi-vendor interoperability of ATM hardware and software. Customers today expect these capabilities based upon their IP internetworking experience. Early ATM networks required extensive amounts of manual configuration, which led to errors and connectivity failures. The ATM Forum responded to these challenges by designing the mother of all routing protocols — PNNI.

15.5.1 Architecture and Requirements

The abbreviation PNNI stands for either Private Network-Node Interface or Private Network-to-Network Interface, reflecting its two possible uses for connecting nodes within a network or interconnecting networks. The PNNI protocol specifies two separate, but interrelated, protocols and functions to achieve the goal of controlling the user cell stream between nodes and networks as illustrated in Figure 15.7. The PNNI protocols operate over dedicated links, or else tunnel over Virtual Path Connections, as denoted by the VPI=* notation in the figure.

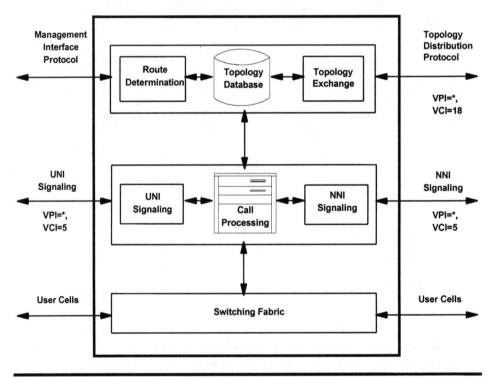

Figure 15.7   PNNI Switching System Architectural Reference Model

A topology distribution protocol defines the methods and messages for disseminating topology information between switches and clusters of switches. PNNI nodes use information exchanged by this protocol to compute optimized paths through the network. PNNI employs a recursive, hierarchical design that scales to huge network sizes. A key feature of the PNNI hierarchy mechanism is its ability to automatically configure itself in networks with an address structure that reflects the topology. PNNI topology and routing is based on the well-known link-state update technique: the same algorithm used in the Open Shortest Path First (OSPF) IP routing protocol.

See Chapter 9 for background on OSPF and the Djikstra shortest path algorithm.

The PNNI signaling protocol uses message flows to establish point-to-point and point-to-multipoint SVCs and PVCs across an ATM network. The PNNI signaling protocol uses the the ATM Forum UNI 4.0 signaling specification as a basis, augmenting it with information elements to support source routing, and the ability to crankback to earlier nodes in order to route around an intermediate node that blocks a call request. The PNNI specification also defines SVC-based Soft Permanent Virtual Paths and Channel Connections (SPVPC and SPVCC) that allow automatic provisioning and restoration of ATM VCC and VPC PVCs. Version 1.0 of the ATM Forum's PNNI specification supports the following functions [9]:

- Supports all UNI 3.1 and some UNI 4.0 capabilities
- Supports hierarchical routing enabling scaling to very large networks
- Supports QoS-based routing
- Supports multiple routing metrics and attributes
- Uses source routed connection setup
- Operates in the presence of partitioned areas
- Automatic topology and address reachability discovery
- Responsive to changes in resource availability
- Uses separate routing protocols within and between peer groups
- Interoperation with external routing domains not using PNNI
- Supports both physical links and tunneling over VPCs as NNIs
- Supports SPVPC/SPVCCs
- Supports anycast signaling and group addresses
- Limits scope of reachability advertisements

15.5.2 Network Addressing Philosophy

Network addressing philosophies fall between two ends of a spectrum: flat and hierarchical [10]. Devices with flat addressing, such as bridged LANs, are aware of every other device in the network. Since the routing table of every node must contain every other address, the memory requirements of networks with flat addressing scale linearly with the number of nodes. Typically, scalability refers to how the nodal memory and processor requirements grow with network growth.

On the opposite end of the spectrum, hierarchical addressing assigns significance to portions of the address. A familiar example is that of the telephone network, which employs E.164 addresses, commonly known as phone numbers. The leftmost digits of an international phone number identify the country in which the addressed device resides (e.g., 01 for North America, 44 for the UK). The identified country defines the remaining digits. Most countries employ a geographically oriented addressing scheme. The reason for this convention was scalability in the telephone switch memory requirements. Each node need only keep track of how to reach other nodes within its own peer hierarchical level: countries, area codes, or exchanges. If the telephone network used flat addressing, then every switch would require hundreds of millions of address entries. Instead, they function with hundreds to thousands of entries, depending upon the telephone switches role in the network. Indeed, without this rigid address hierarchy, a global telephone network could not have been built with the technology available in the 1960s. However, the price of memory has dropped precipitously since the 1960s, consequently changing the drive to minimize memory, while the telephony addressing plan has not changed accordingly. Indeed free telephone services like the 800 service in North America (free at least from the perspective of the calling party) uses a flat addressing plan. Intelligent network servers adjunct to the telephone switches translate these flat addresses into hierarchical telephone numbers for use in routing to the destination.

Data protocols such as IP also utilize the concept of hierarchical address spaces so that network nodes have scalable routing table sizes and processing requirements. Whereas the telephone network addressing is administered by treaty organizations (i.e., the ITU) and national government-appointed regulators, the Internet runs an address registry. The telephone network requires manual administration of the numbering plan, while the Internet offers automatic plug and play operation through the use of sophisticated routing protocols. As described in Chapter 9, routing protocols automatically advertise the relative cost of reaching a particular address prefix via each logical interface. Although there are tens of millions of hosts on the Internet today, the use of hierarchical addressing limits routing table sizes to the order of hundreds of thousands of entries. The Internet primarily utilizes an organizational hierarchy, since address prefixes are assigned to organizations and not countries or geographic areas. However, many organizations have geographic locality, or manage their IP address spaces such that a geographic dimension exists in parallel with the organizational hierarchy. However, no standards dictate this type of address assignment.

The ATM Forum chose the Internet routing protocol as its addressing and routing model over the older, manually maintained telephone network hierarchy — mainly since private network customers insist upon a high degree of automatic configuration.

An undesirable side effect of hierarchical addressing is the generally low utilization of the total available address space. Sparse fill of address space occurs because organizations leave room for growth, or perhaps a network design dictates peer relationships between groups of devices with widely different population sizes. The ATM Forum's choice of the 20-octet NSAP address format for PNNI meets these requirements well, since there is never likely to be a network that approaches a size anywhere near PNNI's theoretical limit of 2^{160} (approximately 10^{48}) nodes. In practice, however, the real number of useable addresses is much less.

The PNNI addressing plan provides an unprecedented level of hierarchy, supporting up to 105 levels. PNNI exploits the flexibility of such a huge address space with the objective of providing an extremely scalable network in the specification of its routing protocols. In contrast to other network routing protocols developed before it, the PNNI specification begins with global scalability as an underlying requirement instead of being an afterthought as now confronts IP in its transition from version 4 to version 6. In the late 1970s, 32 bits of address seemed more than adequate in IPv4. Hindsight resulted in selection of a 16-octet IPv6 address as described in Chapter 8. Note that the IETF has an ICD code reserved, hence IPv6 addressing is actually a subset of the ATM Forum's NSAP-based addressing plan.

15.5.3 A Tale of Two Protocols

Two separate PNNI protocols operate between ATM switching systems connected by either physical or virtual PNNI links: signaling and routing. The signaling protocol sets up the ATM connection along the path determined by the routing protocol. The routing protocol utilizes two types of addresses — topology and end user — in a hierarchical manner. Through exchange of topology information over PNNI links, every node learns about a hierarchically summarized version of the entire network. The distribution of reachability information along with associated metrics, such as administrative cost to reach a particular address prefix over a PNNI link, is similar to that used in the OSPF protocol.

Given that the source node has a summarized, hierarchical view of the entire network and the associated administrative and quality metrics of the candidate paths to the destination, PNNI places the burden of determining the route on the source. The information about the source-to-destination path is computed at the source node and placed in a Designated Transit List (DTL) in the signaling message originated by the source. Intermediate nodes in the path expand the DTL in their domain, and crankback to find alternative paths if a node within their domain blocks the call. Hence, PNNI DTLs are similar to token ring networks which employ source routing. Furthermore, source routing explicitly prevents loops, therefore, a standard route determination protocol isn't necessary, simplifying interoperability.

The PNNI signaling protocol defines extensions of UNI 4.0 signaling through Information Elements (IE) for parameters such as Designated Transit Lists (DTL), Soft PVCs, and crankback indications. The PNNI signaling protocol utilizes the same virtual channel, VCI 5, used for UNI signaling. The VPI value chosen depends on whether the NNI link is physical or virtual.

The PNNI routing protocol operates at the virtual circuit level to route messages from the signaling protocol through the ATM network towards their destination. This protocol operates over AAL5 on VCI 18 of VPI 0 for a dedicated link, or some other non-zero VPI for a PNNI virtual path tunneled across another PVC ATM network. The PNNI routing protocol borrows concepts from some connectionless routing protocols to achieve its goals.

Although PNNI builds upon experience gained from older protocols, its complexity exceeds that of any routing protocol conceived to date. As subsequent sections illustrate, the complexity of PNNI stems from requirements on scalability, support for QoS-based routing, as well as the additional complexities of supporting a connection-oriented service with guaranteed bandwidth — a consideration absent in contemporary routing protocols for connectionless services, such as OSPF and IS-IS. The IETF is working on QoS-aware routing for the Internet; however, it will likely differ from PNNI since the underlying network paradigm is connectionless instead of connection-oriented.

15.5.4 The PNNI Routing Hierarchy and Topology Aggregation

PNNI employs the concept of embedding topological information in hierarchical addressing to summarize routing information. This summarization of address prefixes constrains processing and memory space requirements to grow at lower rates than the number of nodes in the network. At each level of the hierarchy, the PNNI routing protocol defines a uniform network model composed of logical nodes and logical links. PNNI proceeds upward in the hierarchy recursively; that is, the same functions are used again at each successive level. The PNNI model defines:

- Neighbor discovery via a Hello protocol
- Link status determination via a Hello protocol
- Topology database synchronization procedures
- Peer-group determination and peer group-leader election
- Reliable PNNI Topology State Element (PTSE) flooding
- Bootstrapping of the PNNI hierarchy from the lowest level upwards

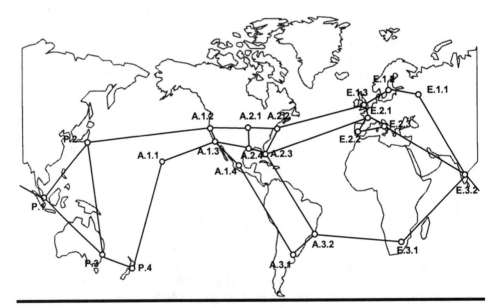

Figure 15.8   Example International Network of Logical Nodes

Figure 15.8 depicts the example of the Antique Trader's Mercantile network used in the remainder of this section to illustrate the concepts of PNNI. The example denotes addresses using a label of the form a.b.c to denote common address prefixes. In these addresses, "a" represents the world region (where P represents the Pacific Rim, A the Americas and E Eurasia), "b" represents the next lower level of hierarchy, and "c" represents the lowest level of hierarchy. At the lowest level (e.g., a single site) the addressing may be flat, for example, using LAN MAC addresses.

15.5.4.1 PNNI Terminology and the Lowest Hierarchical Level

The PNNI routing hierarchy builds from the lowest-level logical nodes and links depicted in Figure 15.9. A lowest-level node may be a physical system, or a network of switches operating a proprietary protocol that supports PNNI for external connectivity. The diagram doesn't depict end systems.

PNNI has its own vocabulary, which we now introduce. A *logical link* may be either a physical link, or a VPC PVC across another ATM network. PNNI defines logical links as either *horizontal links* when they connect logical nodes within a peer, or as *outside links* if they connect peer groups. PNNI also defines *exterior links* that connect nodes to other networks that don't use the PNNI protocol. An example of such connections is a public network or a private network using IISP. In these cases, PNNI simply uses the UNI signaling protocol to dynamically signal connection establishment and release requests with these networks. The PNNI routing protocol terminates at these exterior links.

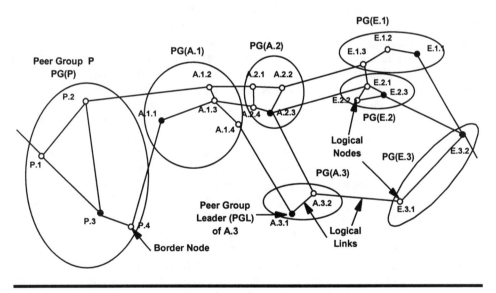

Figure 15.9 Example Lowest Hierarchical Level

A few of the key PNNI terms involve addressing. A 19-octet ATM End System Address (AESA) uniquely identifies each logical node. The 20th octet, the selector byte, has other uses described later. By configuration, lowest-level logical nodes are part of a *Peer Group* (PG) identified by a one octet level indicated and a prefix of, at most, 13 octets of an AESA. Nodes within a peer group share a common address prefix whose length determines the level in the PNNI hierarchy. For example, the nodes with addresses of the form A.1.x are all in the peer group A.1. A peer group leader election process selects one node automatically based upon its configured priority. Border nodes have one or more links crossing a peer group boundary.

15.5.4.2 Dynamically Building the PNNI Hiearchy

When logical nodes (or logical) links transition to the active state, a Hello Protocol executes on VCI 18, called the PNNI *Routing Control Channel* (RCC). Periodically transmitted Hello Packets convey the logical node's AESA, node ID, peer group ID and port ID. The *Hello Protocol* detects link failures via timeouts of unacknowledged packets. Once the nodes acknowledge each other via the Hello Protocol, they begin a database synchronization process. Newly acquainted nodes exchange PNNI Topology State Element (PTSE) headers to determine if they are in synchronization. The protocol resolves mismatches by accepting the latest PTSE. As a result of these information exchanges, neighbor nodes synchronize their databases, and hence learn the overall network topology in an efficient manner.

Once database synchronization completes, nodes flood information about their nodal and link attributes and metrics throughout their peer group. PNNI defines a reliable flooding protocol where every PTSE has a unique identifier containing a timestamp. When a node within a peer group receives a PTSP, it floods the PTSEs not already received on all other links except the link on which it received the PTSP containing the new PTSE. Thus, since PTSPs are acknowledged, the reliable flooding protocol ensures that the topology database of every node in the peer group *converges* to a common state. As in other link state protocols, PNNI sends PTSPs at regular, but not too frequent, intervals. Nodes also send PTSE updates when a significant event occurs, such as a link failure or large change in allocated bandwidth.

Figure 15.10 illustrates how the Hello Protocol performed by border nodes establishes uplinks which dynamically build the PNNI hierarchy. The resulting higher-level peer group elects a leader, flooding summarized topology data within the higher-level peer group. The higher-level logical nodes construct logical horizontal links, sometimes collapsing multiple physical links into a single logical link. For example, the parallel links between A.1.2-A.2.1 and A.1.3-A.2.4 collapse into the one link A.1-A.2 at the next higher-level in the hierarchy. As illustrated in the figure, PG(A) is the parent peer group of all nodes with addresses of the form A.x, who are the child peer groups of A. Higher-level peer group IDs have a common prefix with the IDs of all nodes within their peer group. For example, all nodes in PG(P) have a common address of the form P.x, where x indicates an arbitrary number.

PNNI identifies nodes using a 22-byte node identifier. For nodes present only at the lowest-level, the first octet specifies the level of the node's containing peer group, the second octet uniquely identifies it as a lowest-level node, followed by the node's 20 octet AESA. For *Logical Group Nodes* (LGN) participating at higher levels in the PNNI hierarchy, the first octet is the same, followed by the 14-octet peer group ID and a 6-octet ESI of the end system implementing the LGN function.

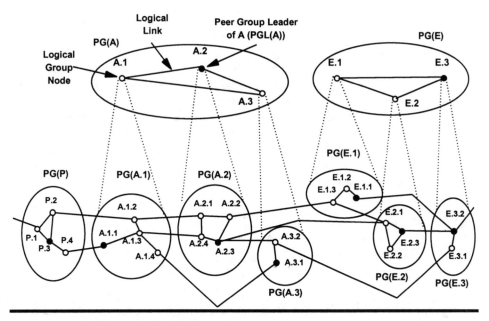

Figure 15.10 Building the PNNI Hierarchy

The PNNI protocol also advertises links in PTSPs. Since ATM link attributes may be asymmetric, and actual resource usage will generally differ in each direction, PNNI identifies links by a combination of a transmitting node ID and a locally assigned port ID. The Hello protocol exchanges these port IDs. PNNI defines additional logic to handle the case when multiple links connect adjacent nodes.

15.5.4.3 Topology Aggregation and Complex Node Representation

All network nodes estimate the current network state from the attributes and metrics contained in the flooded PTSPs advertisements about both links and nodes. Typically, PTSPs include bi-directional information about the transit behavior of particular nodes based upon entry and exit port, taking into account the current internal state. Nodal information often represents an aggregated network and not just a single switch, also called a *summarized peer group* in PNNI parlance.

The complex node representation presents the result of this aggregation in a parent peer group or a lowest-level node. The default representation uses a symmetric star topology centered on a node called the *nucleus* connected to ports around the circumference of the logical node via spokes. Spokes may have either default attributes, or exception attributes. The concatenation of two spokes represents traversal of the symmetric peer group.

Unfortunately, peer groups are frequently asymmetric; therefore, the default representation may hide some key routing information. The complex node representation models such differences by using exceptions to represent particular ports whose connectivity to the nucleus is significantly different from the default. Furthermore, exceptions can represent connectivity between two ports significantly better than that implied by traversing the nucleus. An example of an exception is a high-speed, underutilized connection between border nodes of the peer group.

Figure 15.11a shows the details of Peer Group A.2's available bandwidth for each link. Figure 15.11b shows a possible complex node representation of A.2 using the available bandwidth attribute. The complex node representation need only advertise two values for the link available bandwidth, the default value of 40 and the single exception of 100. Some accuracy is lost in this example, but aggregation achieves a savings of 50 percent in required topology data over the non-hierarchical method of advertising the available bandwidth for every link.

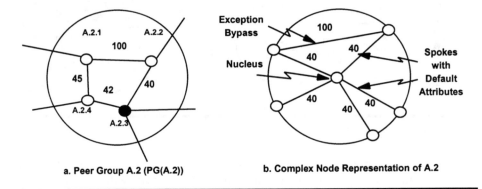

a. Peer Group A.2 (PG(A.2))   b. Complex Node Representation of A.2

Figure 15.11 Complex Node Representation Example

Aggregation reduces complexity and increases scalability by reducing the amount of information required in PTSP flooding. PNNI allows the network designer to trade off improved scalability using aggregation against loss of detailed information necessary for optimized source routing. If you want optimal networking, then minimize aggregation, but if you plan to build a large network, include topology aggregation in the design.

All nodes within a peer group converge to the same topology database as they exchange full topology information using a reliable flooding protocol. Practically, the exchange of full topology information limits peer group size due to processing, storage and transmission of topology data in PTSPs. Complex node aggregation reduces topology information advertised up, across and back down the hierarchy by logical group nodes, as long as the administrator controls the number of advertised exceptions. Although aggregation reduces the volume of topology data, the processing required inserts some additional delay. However, since other parts of the network require less processing time for the smaller volume of aggregated data, the overall convergence time typically decreases.

15.5.4.4 Completing the PNNI Hierarchy

Figure 15.12 completes the hierarchy for this example, recursively employing the previous concepts and protocols. PGLs within each peer group formulate summarized versions of their children using the complex node representation, format these into PTSEs, and flood these in PTSPs with the nodes in their peer groups so that each node in the higher-level peer groups has a common view of the summarized topology.

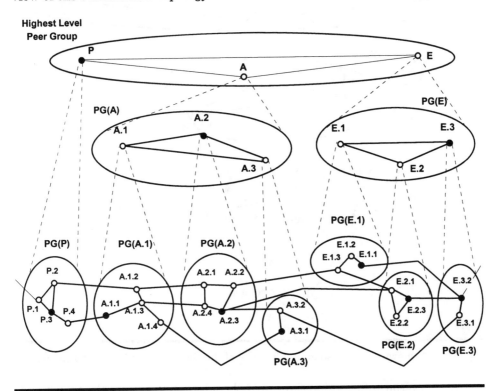

Figure 15.12   Complete PNNI Hierarchy for International Network Example

Furthermore, the LGNs flood the higher-layer topology information within their child peer groups so that lower-level topology databases also contain the same view of the summarized overall network topology. This information is essential for PNNI's source routing paradigm to function. Figure 15.13 illustrates the summarized topology view from node A.1.3's point of view. Uplinks, shown as bold dashed lines in the figure, actually bootstrap the PNNI hierarchy. When border nodes detect that the adjacent node is in a different peer group, they create an uplink and advertise topology informa- tion via PTSEs and advertise this to their respective peer groups. Every member of each peer group is then aware of the uplink. This information enables peer group leaders to establish an RCC via a Switched Virtual Channel Connection (SVCC) at the next level of the hierarchy as shown in the figure. The peer group leaders effectively become logical group nodes at the next level up in the PNNI hierarchy and perform similar functions to logical nodes at the lowest-level. A major difference is that the Hello and topology synchronization protocols operate over the SVCC RCC, and not a physical link.

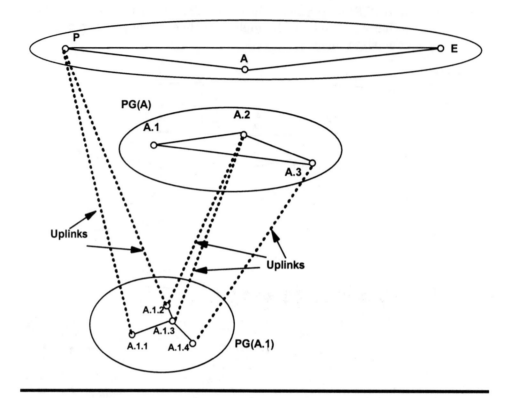

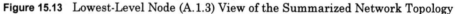

Figure 15.13   Lowest-Level Node (A.1.3) View of the Summarized Network Topology

Note how the information fed down the hierarchy represents an increasingly more summarized view of the remainder of the network. The lowest-level node has a complete view of the topology of its peers, however, information decreases the further up the hierarchy as peer groups are aggregated.

15.5.4.5 Reachability and Scope

In addition to summarized addresses, a number of other elements of reachability information are also carried within PTSP. Routes to external networks, reachable across exterior links, are advertised as external addresses. Peer groups may also include nodes with non-aggregatable addresses, which must also be advertised, as must registered group and anycast addresses. Generally none of these types of information can be summarized, since they fall outside the scope of the default PNNI address hierarchy. Note that the scope of advertisement of the group addresses is a function of how the network administrator maps the scope of a registered node to the corresponding PNNI hierarchy. Use these features only when necessary since they require manual configuration.

15.5.5 Soft Permanent Virtual Connections (SPVCs)

The PNNI protocol also supports Soft Permanent Virtual Path and Channel Connection (SPVPC/SPVCC) establishment. This protocol specifies the means for setting up Permanent Virtual Channel (PVC) Virtual Path Connections (VPC) and Virtual Channel Connections (VCCs) using a combination of the IETF ATM Management Information Base (ATOMMIB) and ATM Forum SPVC MIB [11] parameters at the origin node illustrated in Figure 15.14. The ATOMMIB, defined in RFC 1695 and a forthcoming supplement, keys on the source interface, VPI and VCI indices — which the SPVC MIB also uses to identify the source of either a point-to-point, or a point-to-multipoint SPVCC (SPVPC). The ATOMMIB also defines the traffic descriptors for both directions of the connection.

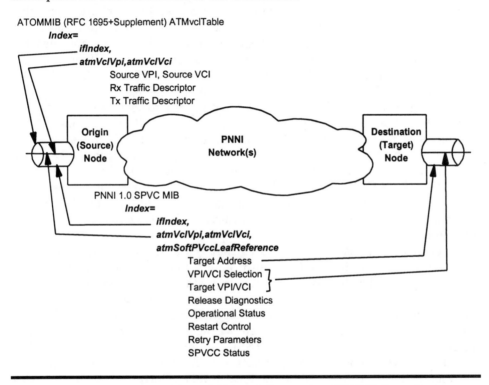

Figure 15.14   Soft Permanent Virtual Channel Connection (SPVCC) Control

The SPVC MIB identifies the NSAP-based address of the destination, or target, interface and whether the destination may select any VPI/VCI, or if particular target VPI/VCI must be used. The SPVC MIB specifies the SPVCCs operational status (i.e., in progress, connected, or retries exhausted) diagnostics giving the cause for an SPVC release and a set of retry control parameters (e.g., interval, maximum number).

Once the SPVCC status is set to active or the SPVCC is restarted by the restart control, the network automatically attempts to establish the end-to-end connection as if it were an SVC. An additional information element in the PNNI signaling message instructs the destination switch to terminate signaling and not forward any signaling to the destination end-system. Furthermore, if the source detects a failed connection, for example, due to an intermediate link failure, it must automatically periodically retry to reconnect the SPVCC/SPVPC. Hence, PNNI soft permanent virtual path and channel connections provide a standard means for automatically provisioning and restoring VPC and VCC PVCs.

15.5.6 Beyond Connectivity to Quality and Bandwidth

PNNI supports two of the fundamental tenets of ATM via its routing and signaling protocols — guaranteed Quality of Service (QoS) and reserved bandwidth on a per connection basis. The routing protocol distributes available bandwidth, cost, and QoS metrics across the network hierarchy. The source node utilizes these metrics to choose the best route to meet the requested required bandwidth and QoS criteria. The source specifies the preferred route in a Designated Transit List (DTL) in the SETUP message.

PNNI 1.0 supports all TM 4.0 service classes: CBR, VBR-rt, VBR-nrt, ABR, and UBR (see Chapters 12 and 21 for details). ATM switches implement Connection Admission Control (CAC) in response to signaling requests to assure that the requested QoS is delivered across all connections. Each switch along a route performs CAC upon receipt of a SETUP message from the preceding switch. The CAC function admits the connection only if QoS can be guaranteed, forwarding the connection request onto the next switch in the DTL. If the CAC decision fails, the switch clears the request back to the preceding switch. CAC is an implementation-dependent function, determined by factors such as the switch architecture, buffer structure, and queuing implementation.

An efficient implementation assures that the source route determined by the origin node specifies a DTL with a high likelihood of success. PNNI meets this objective by utilizing a topology state routing protocol in which nodes flood QoS and reachability information throughout the network in a hierarchical manner. Thus, all nodes obtain knowledge about reachability within the network, the available bandwidth, and QoS across the network. PNNI Topology State Packets (PTSP) contain Type-Length-Value (TLV) encoded PNNI Topology State Elements (PTSE) to pass this information between nodes and different hierarchical levels. Current link state routing protocols, such as OSPF, implement a similar flooding procedure; however, they usually pass only a single parameter between nodes. Flooding of topology information occurs whenever a significant event occurs, such as when a new node joins; a link or node fails; or a a network resource (e.g., available bandwidth) changes dramatically.

PNNI defines two types of link parameters: non-additive link *attributes* applied to a single network link or node in route determination, along with additive link *metrics* which apply to an entire path of nodes and links. The set of non-additive link attributes in PNNI 1.0 are:

∗ Available Cell Rate (ACR): A measure of available bandwidth in cells per second for each traffic class.
∗ Cell Rate Margin (CRM): A measure of the difference between the effective bandwidth allocation per traffic class and the allocation for sus-

tainable cell rate. CRM is the safety margin allocated above the aggregate sustained rate.

The set of additive link metrics in PNNI 1.0 are:

+ Maximum Cell Transfer Delay (MCTD) per traffic class
+ Maximum Cell Delay Variation (MCDV) per traffic class
+ Maximum Cell Loss Ratio (MCLR) for CLP=0 cells, for the CBR and VBR traffic classes
+ Administrative Weight: This value indicates the relative desirability of a network link in a manner analogous to OSPF.
+ Variance Factor (VF): A relative measure of CRM margin normalized by the variance of the aggregate cell rate on the link

15.5.7 Estimating and Refining Source Routes

Recall that the source PNNI node determines a path across the network based upon the requested QoS and its knowledge of the network state obtained from flooded PTSEs. In a dynamically changing network, the source node has only an imperfect approximation to the true network state. This inconsistency occurs because topology information obtained from PTSE flooding is always older than the actual current network state. Furthermore, the topology aggregation process may hide details to improve scalability. Hence, the initial source routed path is at best an estimate. Additionally, each node along the path may perform CAC differently than the source node's estimate. The net result of these considerations is that the source node's best estimate of the ideal path may not result in a successful connection attempt.

15.5.7.1 Generic Connection Admission Control (GCAC)

The PNNI protocol attempts to minimize the probability of failure in the determination of the first source route by defining a Generic CAC (GCAC) algorithm. This algorithm allows the source node to estimate the expected CAC behavior of nodes along candidate paths based upon additive link metrics advertised in PTSPs and the requested QoS of the new connection request. The GCAC algorithm provides a good prediction of a typical node-specific CAC algorithm. Individual nodes may optionally communicate the stringency of their own CAC using an optional complex GCAC calculation by advertising the Cell Rate Margin (CRM) and Variance Factor (VF) metrics.

The GCAC uses the additive metrics subject to the constraints described above. Individual nodes (physical or logical) advertise the values of these parameters based upon their internal structure and current connection status. Note, however, that PNNI 1.0's GCAC algorithm works best for CBR and VBR connections. GCAC for UBR connections simply determines whether nodes on the candidate path advertise support for UBR. For ABR connections, GCAC checks whether the link or node can support any additional ABR connections and also verifies that the Available Cell Rate (ACR) for the ABR traffic class for the node or link resource is greater than the Minimum Cell Rate specified by the connection. Using the GCAC, the source node routing algorithm performs the following functions:

- Eliminates all nodes and links from the topology database that don't meet the end-to-end QoS or bandwidth requirements of the connection request.
- Applies the advertised reachability information against the above reduced topology to compute a set of least cost paths to the destination, for example, using the Djikstra algorithm.
- Removes candidate paths that fail the additive link metrics, such as delay, and do not meet other constraints. If no paths are found, then the source node blocks the connection request. If the routing algorithm finds more than one path, the source node may choose one according to its policy, such as minimum delay, load balancing or least administrative weight.
- The source node encodes the selected route to the destination in a Designated Transit List (DTL) which describes the complete hierarchical route to the destination. The source then inserts the DTL into the SETUP signaling message, which it then forwards to the first node identified in the DTL.

Note that when multiple links connect adjacent nodes, connections traverse only one physical link to preserve cell sequence integrity. For multiple links between adjacent nodes with otherwise equal attributes and metrics, implementations may optionally implement load balancing.

15.5.7.2 Designated Transit Lists (DTLs)

PNNI arranges DTLs in a stack in the SETUP message containing the path elements from the source to the destination. Figure 15.15 illustrates and example for a connection attempt from a source at address P.3.a to a destination at address E.1.1.b. The DTL stack is a list of logical node IDs, along with a pointer indicating the next element, indicated by underlined type in the figure. Beginning at the source node, the DTL articulates each node up to the border node P.4. The border node P.4 removes the top DTL from the stack and sets the pointer to the higher-level peer group A along the path to the destination E. The receiving border node in the peer group A, A.1.1, then constructs a detailed DTL defining how to route the request through peer group A.1 and pushes it onto the top of the DTL stack.

The PNNI signaling protocol forwards the connection request between nodes and peer groups, with border nodes expanding the DTL to traverse the peer group until the request reaches the destination peer group as illustrated in the figure. In the destination peer group E.1, the last element in the border node's DTL expansion is that of the destination AESA in the original SETUP message, which node E.1.1 uses to complete the call.

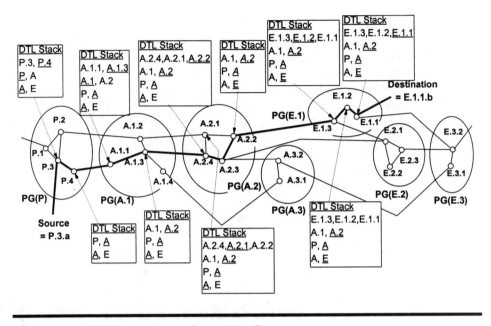

Figure 15.15 DTL Processing in Connection Setup

15.5.7.3 Crankback Procedure

The source launches the SETUP message to a series of logical nodes on the path defined in the DTL. The CAC at each node along the path determines whether to accept the connection request based upon the traffic parameters in the SETUP message and its own current state, which may differ from the information in the topology database at the source node. The connection request may fail because of changed conditions, differences in CAC and GCAC, or due to inaccuracies introduced by the topology aggregation process. PNNI anticipates such events and recovers from them via the concept of *crankback* signaling and routing.

Crankback utilizes the DTL mechanism to return connection attempt failures back to the node that created and inserted the DTL which failed. Only the source node and border nodes create DTLs. They must maintain state information about all outstanding connection requests until the signaling protocol confirms set up, a connection reject is received from the destination, or all possible expansions of the DTL are exhausted. This is why the DTL stack contains all of the information about the expanded path. When a particular node along the source routed path blocks a connection request, it cranks back the request to the last node in the path that performed a DTL operation. This node processes the crankback response by attempting to specify another DTL to the final destination excluding the blocking node or link. This node may also have newer, or more accurate network state information.

Figure 15.16 illustrates the operation of crankback for a call from P.3.a at the lower left to E.1.1.b at the upper right. The call attempt begins as in the previous example, but a failure occurs in the Pacific to America links as indicated by the blots or starbursts on these links in the figure. Node A.1.1 returns a crankback indication in step 3 reporting that it cannot make a connection between peer groups P and E. The network cranks back the call to the node that last made a routing decision, P.3. Node P.3 then selects the next best route via node P.1 and pushes this onto the DTL stack in step 4. The border node P.1 forwards this request to peer group E in step 5 across the link that connects the Pacific region to Europe. Node E.3.2 determines the path to E.1.1.b in Step 6 and forwards the attempt to peer group E.1. Node E.1.1 determines that it is indeed the destination node, and successfully completes the connection via the alternate route in Step 7.

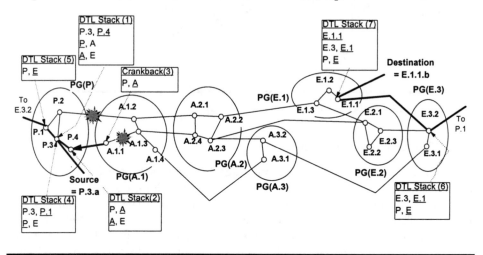

Figure 15.16   Illustration of Crankback Operation

15.5.8 Minimum Interoperable PNNI 1.0 Subset

A complex protocol like PNNI 1.0 must have a definition of a minimum interoperable subset, otherwise manufacturers could choose which portions of PNNI they wish to implement, and the resulting networks would fail to meet the goal of multi-vendor operation. Towards this end, the ATM Forum PNNI 1.0 specification defines base node subsets and options.

Figure 15.17 illustrates the three kinds of nodes PNNI defines for the purposes of interoperability. A minimal function node has only inside PNNI links and forms the mandatory basic function set. A border node has only outside links. A Peer Group Leader/Logical Group Node (PGL/LGN) capable node participates in higher levels of the PNNI hierarchy. Furthermore, PNNI defines a set of optional functions. The following bullets describe these capabilities in greater detail.

Base Functions

Optional Functions

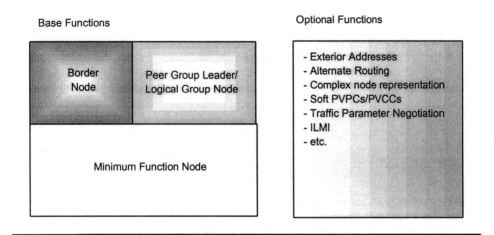

Figure 15.17   Function Sets for PNNI Nodes

Minimum Function Node
- All PNNI signaling functions, except
  - Soft PVPCs/PVCCs
  - Negotiation of ATM traffic descriptors
- At least one PNNI an UNI/NNI link
- Supports only inside links
- Supports version negotiation and Information Group (IG) tags
- Hello protocol - inside over physical links and VPCs
- Database synchronization
- Originates and floods nodal information, horizontal link, and internally reachable address PTSEs
- Votes in PGL elections
- Generates and passes back crankbacks
- Perform default internal address summarization
- DTL processing origination/termination in a PNNI routing domain with hierarchy, including complex node representations

Border Node.
- Supports minimum node function base
- Outside link support
- SVC based RCC
- Supports Hello protocol in all contexts
  - Outside over physical links and VPC's
  - SVC Hello protocol when LGN is peer
  - LGN horizontal link
- Supports incremental database synchronization for LGNs
- Originates uplink PTSE's
- Aggregate uplinks

PGL/LGN Capable Node
- Supports minimum node function base
- Runs complete PGL Election
- Is PGL capable
- Database synchronization deltas for LGNs
- Supports Hello protocol in all contexts
- Link, uplink, and nodal aggregation
- Ability to perform address summarization for internal and exterior addresses

Conditional/Optional Support
- Support exterior reachable addresses
- Alternate routing as a result of crankback
- Originate complex node representation
- Support for Soft PVPC and PVCC (SPVPC and SPVCC)
- Negotiation of ATM traffic descriptors
- ILMI support

15.5.9 The Road Ahead — PNNI 2.0 and PAR

Like many other first generation protocols, PNNI 1.0 isn't perfect and doesn't include all of the features and functions required for every network. Recognizing this fact, the ATM Forum technical committee plans to issue version 2.0 of the PNNI specification in late 1998. The current effort includes detailed specification of procedures in support of point-to-multipoint connections, Closed User Groups (CUGs), interworking with B-QSIG signaling, key distribution for security, rerouting, connection recovery mechanisms, and interworking with frame relay SPVCs.

The point-to-multipoint routing and signaling procedures support both the root and leaf initiated join mechanisms. ISO/IEC 13247 specifies B-QSIG call control based upon the PNNI signalling protocol applied between nodes of private ATM networks that employ static hop-by-hop routing. The CUG capability utilizes interlock codes to indicate that parties may call each other.

Currently, rerouting and connection recovery occur at the edge of the network. For example, if an intermediate link or node fails along the route taken by a PNNI SPVC, the protocol releases the connection back to the edge, which retries the attempt. PNNI 2.0 specifies optional fast connection recovery procedures which restore SPVCs or SVCs at intermediate nodes. These protocols reduce network processing in response to failures. Furthermore, they address a fundamental issue with connection-oriented networks. The ATM and N-ISDN paradigms explicitly select the route through a sequence of switches at call setup time. When a reroute occurs due to a failure, the network may choose a suboptimal path. Once a maintenance action corrects the failure, the connection remains on the suboptimal path. The planned PNNI 2.0 capabilities in support of fast connection recovery and switching at intermediate nodes provides an operational tool to periodically regroom the network to optimize capacity utilization.

PNNI Augmented Routing (PAR) supports flooding of information about non-ATM services, for example, IP, according to an ATM network's PNNI topology. The initial PAR specification supports the IPv4 OSPF or BGP protocols as well as transparent carriage of proprietary routing data. PAR defines a new PTSE type for carriage of this non-ATM-related routing information. Thus, these PAR-capable devices can determine reachability of other protocol addresses across a PNNI ATM network. Routers may also use PAR to automatically identify compatible routing protocols and IP subnet information.

The PAR specification also defines an optional set of protocols called Proxy PAR to allow routers to obtain reachability and routing protocol information without implementing PNNI. Proxy PAR defines a server which acts on behalf of a client seeking PAR related services. By design, PAR and Proxy PAR support Virtual Private Networks (VPNs) through use of a unique VPN ID. PAR also makes use of PNNI's routing scope capability and defines filtering techniques along with subnet masks in support of VPNs.

15.6 Broadband InterCarrier Interface (B-ICI)

The Broadband InterCarrier Interface (B-ICI) supports PVCs as well as SVCs. This section summarizes the key attributes of the ATM Forum B-ICI specifications and the relationship to the ITU-T standards.

15.6.1 B-ICI Version 1.0

The B-ICI version 1.0 from the ATM Forum defines the PVC following services interconnected between carrier networks as illustrated in Figure 15.18:

- ⊳ Cell Relay Service (CRS)
- ⊳ Circuit Emulation Service (CES)
- ⊳ Switched Multimegabit Data Service (SMDS)
- ⊳ Frame Relay Service (FRS).

B-ICI 1.0 defines support for these services between the Local Exchange Carrier (LEC), IntereXchange Carrier (IXC), Independent Local Exchange Carrier (ILEC), and other public ATM network providers. Depending upon regulatory and business arrangements, any one of these carriers may assume any of the roles illustrated in Figure 15.18.

The B-ICI 1.0 specification defines support for both Permanent Virtual Connection (PVC) and SMDS services. B-ICI specifications for support of ATM Switched Virtual Connections (SVCs) were developed after the PVC specifications in versions 2.0 and 2.1 as summarized later in this section. The following paragraphs present highlights of the B-ICI 1.0 specification.

Physical layer specifications at the B-ICI are essentially identical to those defined in Chapter 12 for DS3, OC3/STS-3c, and OC12/STS12c at the UNI.

The specification defines additional operational functions required between carriers to manage transmission paths, usually requiring some degree of bilateral agreement between carriers (i.e., both carriers must agree).

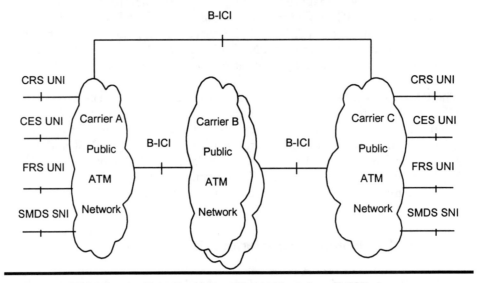

Figure 15.18 Multi-Service Broadband InterCarrier Interface (B-ICI)

The ATM layer specification is the ATM NNI as defined in Chapter 12. B-ICI 1.0 required only Peak Cell Rate (PCR) traffic management, with other traffic control subject to bilateral agreement. The traffic contract as defined in Chapter 12 is now between two carriers instead of between a user and a network. The Quality of Service (QoS) is now a commitment from one carrier to another, and the B-ICI defines reference traffic loads and assumptions as guidance to carriers in resolving this part of the traffic contract. The B-ICI 1.0 specification covers ATM layer management in great detail.

The B-ICI specification not only defines what is required at the B-ICI interface, but also defines the higher-layer functions required in the user plane for InterWorking Functions (IWF) required to support frame relay, SMDS, and circuit emulation since these specifications were incomplete at the time. In fact, the Frame Relay Forum's FR/ATM network interworking specification resulted from work done in the ATM B-ICI committee. Several SMDS specific functions in support of intercarrier SMDS are defined in the ATM B-ICI specification, including carrier selection, group addressing, and routing.

Furthermore, the B-ICI 1.0 specification articulates the general principles for provisioning, network management, and accounting. More work remains to translate these thoughts into more detailed guidelines, interfaces, and protocols.

15.6.2 B-ICI Version 2.0

The Broadband InterCarrier Interface (B-ICI) specification version 2.0 [12] specifies support for UNI 3.1 SVCs across interfaces that connect carrier networks, including support for the following capabilities:

- Bidirectional and unidirectional point-to-point connections
- Unidirectional point-to-multipoint network connections
- Symmetric and asymmetric connections
- Support for UNI 3.1 QoS classes and bearer capabilities
- Support for E.164 ATM End System Addresses (AESA)
- Call rejection due to unavailable ATM resources
- Minimum set of signaling network Operations And Maintenance (OAM) functions (e.g., blocking, testing, and reset)
- Identification of control plane associations between nodes by means of signaling identifiers
- Usage measurement procedures for SVCs

Figure 15.19 illustrates the context for use of the B-ICI specification between carriers. The specification details not only message formats and general procedures, but also specifies exact actions required by originating, intermediate, and terminating switches. Note that the B-ICI specification does not specify the protocol used between switches within a carrier's network, only the protocol used between carrier networks.

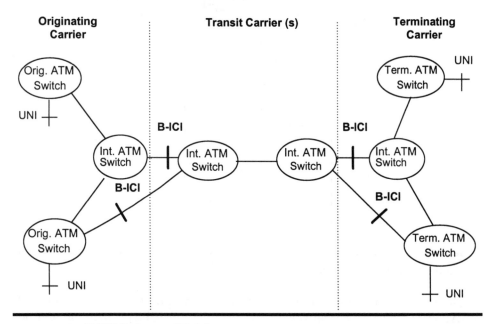

Figure 15.19 B-ICI Reference Model

The ATM Forum chose the ITU-T's B-ISDN Services User Part (BISUP) protocol stack as the basis for the B-ICI 2.0 signaling specification illustrated in Figure 15.20. BICI 2.0 employs a subset of the Message Transfer Part (MTP) protocol, which is used as part of the Signaling System 7 (SS7) protocol in telephone networks today. The Signaling AAL (SAAL) is similar to that utilized at the UNI, with a different Service Specific Convergence Function (SSCF) as described in Chapter 14. The Transaction Capabilities Application Part (TCAP) and Signaling Connection Control Part (SCCP) could provide future support of transaction-based services between switches, and between switches and intelligent network databases, called Service Control Points (SCPs).

15.6.3 B-ICI Version 2.1

An addendum to the 2.0 specification, called B-ICI 2.1, [13] added three key features. A call correlation tag provides a means to identify a call across multiple carriers for billing purposes. The addendum also clarified support for Variable Bit Rate (VBR) connections. The 2.1 version also defined support for all NSAP-based address formats in the connections between carriers. These capabilities defined means to support more end-to-end features of UNI 3.1 SVCs between carriers.

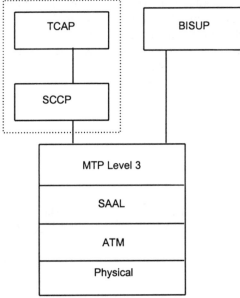

Figure 15.20 BICI 2.0 Protocol Stack

Figure 15.21 illustrates the support added for carriage of ICD and DCC NSAP-based addresses between carriers. The B-ICI specification defines called and calling ATM End System Address (AESA) information elements in addition to called and calling address information elements to support address formats other than E.164 based addresses.

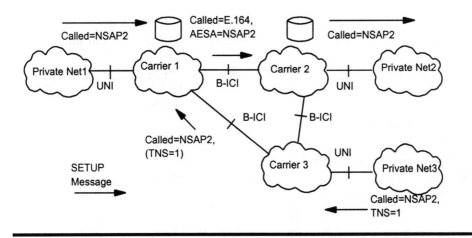

Figure 15.21 B-ICI Support for NSAP-based ATM End System Addresses (AESAs)

Figure 15.21 presents two example calls, one from private network 1 and another from private network 3, both destined for private network 2 with address NSAP2. Carrier 1 translates the call from private network 1 into a called E.164 address for routing to carrier 2 over the B-ICI interface. Carrier 2 takes the NSAP formatted address out of the AESA when it delivers the call to private network 2. The second example of a call from private network 3 is via carrier 3, which, for example, does not perform NSAP to E.164 translation. The user of the private network specifies the transit network via an optional information element called the Transit Network Selection (TNS) field. In the example, the user specifies TNS=1, which instructs Carrier 3 to direct the call to carrier 1. Carrier 1 then translates the NSAP2 address into a routable E.164 address for delivery by carrier 2. Upon reaching the designation UNI defined by the E.164 address provided by carrier 1, carrier 2 delivers the NSAP2 address in the called party information of the SETUP message to private network 2.

15.6.4 B-ISDN User Services Part (BISUP)

The ITU-T specifies the B-ISDN User Services Part (BISUP) protocol for use at the Network-Node Interfaces (NNI), which the ATM Forum uses in the Broadband InterCarrier Interface (B-ICI) specification. The ITU-T BISUP specifications for signaling at the NNI match well with the corresponding ITU-T UNI Signaling standards as summarized in Table 15.4. This table provides a road map that provides some structure to the long list of standards in the Q-Series recommendations summarized in Appendix B.

The ATM Forum based a large part of the B-ICI specification on the ITU-T's BISUP protocol, in a manner analogous to the way the Forum's Signaling 4.0 specification is based largely upon Q.2931. However, B-ICI 2.0 intended to support UNI 3.1 signaling only, while BISUP has already been extended to support Q.2931 capabilities, which are more closely aligned with UNI 4.0.

Therefore, in the area of the specification on signaling interfaces between carriers, look for the ITU-T standards to support UNI 4.0 capabilities since the Forum has announced no plans to update the B-ICI specification.

Table 15.4   Mapping of ITU-T UNI and NNI Signaling Capabilities

Capability Description	ITU-T UNI Standard	ITU-T NNI Standard
On-Demand (Switched) Connections	Q.2931	Q.2761-4
Point-to-point Calls	Q.2931	Q.2761-4
N-ISDN Signaling Interworking	Q.2931	Q.2660
E.164 Address Support	Q.2931	Q.2761-4
NSAP-based Address Support	Q.2931	Q.2726
Root-initiated Point-to-multipoint Calls	Q.2971	Q.2722.1
Traffic Parameter Modification during Active Calls	Q.2963.1	Q.2725.1
Traffic Parameter Negotiation during Call Setup	Q.2962	Q.2725.2
Supplementary Services		
Direct Dialing In (DDI)	Q.2951	Q.2730
Multiple Subscriber Number (MSN)	Q.2951	Q.2730
Calling Line Identification Presentation (CLIP)	Q.2951	Q.2730
Calling Line Identification Restriction (CLIR)	Q.2951	Q.2730
Connected Line Identification Presentation (COLP)	Q.2951	Q.2730
Connected Line Identification Restriction (COLR)	Q.2951	Q.2730
Subaddressing (SUB)	Q.2951	Q.2730
User-user Signaling (UUS)	Q.2957	Q.2730

15.7 REVIEW

This chapter detailed the signaling and routing protocols involved in Switched Virtual Connections (SVC) at the User-Network Interface (UNI) as well as the Network-Network Interface (NNI). The text began with an introduction to ATM UNI SVCs by first describing generic functions defined in ATM Forum and ITU-T documents. Next, the chapter looked at the key signaling messages and the role their information elements provide in providing ATM's rich set of capabilities. The presentation then moved to graphical examples of signaling procedures establishing and releasing point-to-point and point-to-multipoint connections using these messages. The chapter then described the simplest Network-Network Interface (NNI) protocol first, the ATM Forum's InterSwitch Signaling Protocol (IISP). The text then gave an in-depth walkthrough of the most sophisticated routing protocol developed to date — the ATM Forum's Private Network-Network Interface (PNNI). Finally, the chapter concluded by reviewing considerations with signaling protocols employed between carrier networks at the Broadband InterCarrier Interface (B-ICI) using the BIDSN User Services Part (B-ISUP).

15.8 REFERENCES

[1]   ATM Forum, *User-Network Interface Signaling Specification, Version 3.1*, af-uni-0010.002, September 1994.

[2]   ITU-T, *Broadband Integrated Services Digital Network (B-ISDN) – Digital Subscriber Signalling System No. 2 (DSS 2) – User-Network Interface (UNI) Layer 3 Specification For Basic Call/Connection Control*, Recommendation Q.2931, February 1995.

[3]   ATM Forum, *User-Network Interface Signaling Specification, Version 4.0*, af-sig-0061.000, July 1996.

[4]   G. Sackett, C. Metz, *ATM and Multiprotocol Networking*, McGraw-Hill, 1997.

[5]   ITU-T, *Broadband Integrated Services Digital Network (B-ISDN) – Digital Subscriber Signalling System No. 2 (DSS 2) – Additional Traffic Parameters*, Recommendation Q.2961, October 1995.

[6]   ITU-T, *Broadband Integrated Services Digital Network (B-ISDN) – Usage of Cause and Location in B-ISDN User Part and DSS 2*, Recommendation Q.2610, February 1995.

[7]   ITUT-T, *Broadband Integrated Services Digital Network (B-ISDN) – Digital Subscriber Signalling System No. 2 (DSS 2) – User-Network Interface Layer 3 Specification for Point-To-Multipoint Call/Connection Control*, Recommendation Q.2971, October 1995.

[8]   ATM Forum, *Interim Inter-Switch Signaling Protocol, version 1.0, af-pnni-0026.000*, December 1994.

[9]   ATM Forum, *Private Network-Network Interface Specification Version 1.0*, af-pnni-0055.00, March 1996.

[10]   A. Alles, *ATM Interworking,* http://cio.cisco.com/warp/public/614/12.html.

[11]   ATM Forum, *Private Network-Network Interface Specification Version 1.0 Addendum (Soft PVC MIB)*, af-pnni-0066.00, September 1996.

[12]   ATM Forum, *BISDN Inter-Carrier Interface (B-ICI) Specification, Version 2.0*, af-bici-0013.003, December 1995.

[13]   ATM Forum, *Addendum to BISDN Inter Carrier Interface (B-ICI) Specification, v2.0 (B-ICI Specification, v2.1)*, af-bici-0068.000, November 1996.

4

ATM Hardware, Software, and Higher-Layer Protocols

This part begins by reviewing the state of ATM hardware and software and the higher-layer protocols that provide the actual interface to end user applications. Chapter 16 begins by detailing the predominant classes of ATM hardware: Central Office (CO)-based switches; Campus ATM switches; and local devices including routers, switches, hubs, and end systems. As ATM moves to the desktop, we look at the state of ATM interface cards and the enabling native ATM Application Programming Interface (API). Chapter 17 continues with an in-depth look at support for video and voice over ATM. The subjects covered include interworking with narrowband ISDN, circuit emulation, video on demand, and residential broadband services. Next, Chapter 18 covers ATM interworking with the frame relay and SMDS wide area networking data communication protocols. This chapter also describes the cost-effective, low-speed, frame-based ATM access protocols: the ATM Data eXchange Interface (DXI) and the Frame-based UNI (FUNI). Next, Chapter 19 details the ATM Forum's LAN Emulation (LANE) protocol and the multiprotocol over ATM encapsulation technique. Finally, Chapter 20 covers the important topic of ATM's support for network layer protocols, in particular the Internet Protocol. Topics covered include Classical IP over ATM, MultiProtocol Over ATM (MPOA), and IP Multicast over ATM. This final chapter also compares and contrasts other approaches based upon Multi-Protocol Label switching (MPLS), such as IP switching, Aggregate Route-based IP switching (ARIS), and tag switching.

16

ATM Hardware and Software

This chapter first surveys the landscape of ATM hardware devices: switches, concentrators, hubs, routers, and end systems. The treatment divides the myriad of ATM hardware into five major categories: carrier backbone switches, carrier edge switches, enterprise switches, campus/workgroup switches, and ATM-enabled devices. The text then summarizes the status of ATM Network Interface Cards (NIC) hardware which provides ATM connections at speeds ranging from 25-Mbps over twisted pair to 622-Mbps over fiber directly to the desktop computer. Complementing the hardware, we review the emergence of native ATM-capable software, specifically, the generic ATM Application Programming Interface (API) and specific Windows 98 implementation plans. Finally, the chapter reviews the status of standard ATM hardware chip-level interfaces defined by the ATM Forum's UTOPIA standard.

16.1 KEY ATTRIBUTES OF ATM SWITCHES

ATM switches have several distinguishing characteristics. First, a switch architecture is either blocking, virtually non-blocking, or non-blocking. The maximum load at which a switch operates to deliver a specific level of blocking determines the inherent efficiency. Next, the architecture of the switch fabric at the heart of the machine determines the maximum switch size and maximum port speed. Finally, the buffering method has significant implications on the number of service categories and statistical multiplexing gain.

16.1.1 ATM Switch Blocking Performance

Circuit switches initially defined the concept of *blocking*, which traffic engineers then adapted to ATM switches. In circuit switches, if an inlet channel can be connected to any unoccupied outlet, up to the point where all inlets are occupied, then the switch is strictly non-blocking. Circuit switches are often then specified as virtually non-blocking, meaning that a small blocking probability occurs as long as no more than a certain fraction of inlet channels is in use. This fraction of inlets in use when the specified level of blocking occurs determines the efficiency of the switch. For example, if the acceptable blocking level of 1 percent occurs when 90 percent of the inlets are in use, then the switch operates at 90 percent efficiency.

Traffic engineers extended the blocking concept to ATM switches, although ATM switches utilize a completely different paradigm than circuit switches. When connecting an inlet to an outlet, a circuit switch reserves bandwidth, hence completely isolating connections from each other. Generally, this situation is not valid in an ATM switch — cells from virtual path and virtual channel connections (VPCs or VCCs) arriving on different input ports may contend for the same output port. Cell loss can occur, depending upon the statistical nature of this virtual connection traffic as well as how the switch fabric and buffering strategy handle congestion. Although cell loss in an ATM switch is not the analog to call blocking in a circuit switch, you will see the term "blocking" attributed to cell loss performance. Most analyses make the assumption that arriving virtual connection traffic is uniformly and randomly distributed across the outputs. The switch performance is then normally cited as virtually non-blocking (sometimes called non-blocking), meaning that up to a certain input load a very low cell loss ratio occurs. Most vendors surveyed cite non-blocking performance at this load per the traffic patterns specified in Bellcore's GR-1110 specification.

The ATM switch blocking performance is sensitive to the switch architecture and the source traffic assumptions. This is an important practical design consideration, because depending upon traffic characteristics, one switch type may be better than another.

16.1.2 Switch Architectures

Figure 16.1 illustrates several of the more common published switching architectures implemented in current ATM switches. These are also called switch fabrics, or switch matrices. Of course, some switch designs are hybrids employing one larger switch fabric to connect multiple smaller switch fabrics together to yield a larger overall switch. This section describes each architecture in terms of its complexity, maximum overall speed, scalability, ease of support for multicast, blocking level, and other unique attributes.

The single bus shown in Figure 16.1a is the simplest switch type. Basically, ports connect to a single bus, usually implemented via a large number of parallel circuit board traces. The total speed of such a bus usually ranges between 1 and 5 Gbps. Many CPE devices, multiplexers and ATM-enabled devices utilize this architecture. This design introduces some complexity due to the need for bus arbitration, which in combination with the buffering strategy controls the blocking level. Multicast is easily implemented since all output ports "listen" to the common bus.

The multiple bus switch fabric shown in Figure 16.1b extends the single bus concept by providing a broadcast bus for each input port. This design eliminates the need for bus arbitration, but places additional requirements on controlling blocking at the outputs. In general, each bus runs at slightly greater than the port card speed (usually between 500-Mbps and 1 Gbps), and uses multiple circuit traces on a shared backplane that all the other cards plug into. In general, this switch architecture runs n times faster than a single bus, where n is the number of parallel buses. An early version of this switch architecture was called the "knockout" switch [1], because the outputs could receive cells from only a limited number of inputs at the same time in an attempt to make the architecture more scalable. Those cells not selected for reception at the output were "knocked out." For smaller switches, every output can receive from every input simultaneously. Another method employs arbitration to ensure that the output port does not receive too many cells simultaneously; however, this creates a need for input buffering and adds complexity. Multicast is also natural in this switch type since each input is broadcasting to every output.

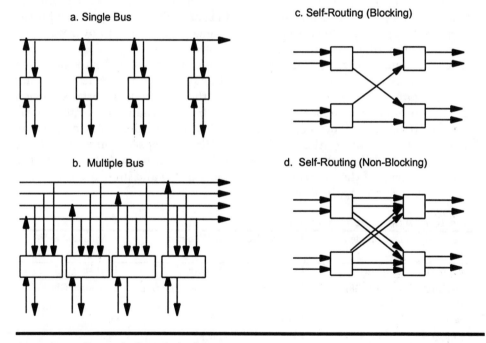

Figure 16.1 Example Switch Architectures

Many single shelf switch designs often used in small carrier edge switches and enterprise switches employ some variation of the multiple bus architecture. The cards sharing the multiple bus backplane may provide connections to other port expansion shelves. Since each bus runs at approximately 500-Mbps to 1 Gbps and a shelf contains 10 to 15 cards, these types of switches typically operate in the range of 5 to 15 Gbps.

The self-routing switch fabrics shown in Figure 16.1c and 16.1d, such as Batcher Banyan networks, employ more complex internal switching elements; however, these can be scaled to larger sizes due to the regular nature of the individual switching elements in Very Large Scale Integration (VLSI) implementations. These types of networks have been the subject of a great deal of research and investigation. See References 2 and 3 for an in-depth review of these switch architectures. These switches generally do not support multicast well, often requiring either a separate copy network [4], or special processing within the internal switching elements. If the self-routing network runs at the same speed as the input ports, then blocking can be quite high. Hence, many of these switch designs run at a higher internal speed than the switch port speed. Self-routing networks generally have some buffering within the switching elements. An augmented self-routing fabric basically runs the internal matrix at a faster speed, or has multiple connections between switching elements as shown in Figure 16.1d. Large multi-shelf switches frequently use self-routing fabric designs and hence scale to sizes much larger than single shelf, multiple bus based designs. Several vendors have announced plans to build such switches with capacities ranging from 100 Gbps to over 1 terabit per second (Tbps). These switches use higher internal port speeds, or else some form of feedback control to overcome the inherently poorer blocking performance of this design.

Table 16.1 summarizes a comparison of the characteristics for these switch fabric architectures. This table indicates the switch fabric capacity and not the aggregate speed of the ports. Unless the switch has port expansion shelves, the sum of the ports speeds may be less than the switch fabric capacity. If the switch does some additional statistical multiplexing prior to connection to the fabric, then the sum of the port speeds may exceed the fabric capacity. When shopping for an ATM switch, be sure to ask for the capacity in terms of the number of ports, their speeds and interface types that you require for your application.

Table 16.1   ATM Switch Characteristics

Characteristic	Single Bus	Multiple Bus	Self-Routing	Augmented Self-Routing
Complexity	Low	Medium	Higher	Higher
Maximum Speed	1-2 Gbps	1-20 Gbps	1 Gbps - 100 Gbps	100-1,000 Gbps
Scalability	Poor	Better	Good	Best
Pt-to-MultiPt Support	Good	Good	Poor	Poor
Blocking Level	Low	Low	Medium	Low-Medium
Unique Attributes	Inexpensive	Inexpensive	Amenable to VLSI	Amenable to VLSI

Space division crosspoint switches described in Chapter 4 can also be used as an ATM switch fabric. A crosspoint switch synchronizes all inputs to a common cell clock, rearranging the connection matrix every cell time. A crosspoint switch performs broadcast by connecting multiple outputs to a single inlet during a cell time.

16.1.3 Switch Buffering Methods

The buffering strategy employed in the switch also plays a key role in the switch blocking (i.e., cell loss) performance. Figure 16.2 illustrates various ATM switch buffering strategies employed in real world switches.

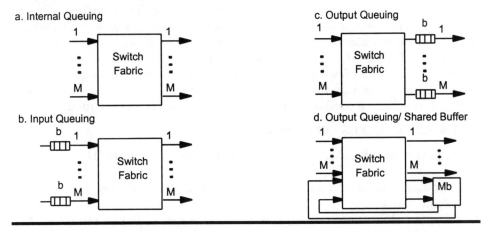

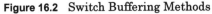

Figure 16.2   Switch Buffering Methods

The following parameters define the notation used in the figure:

M = number of ATM switch ports
b = number of effective cell buffer positions per port

Chapter 4 gave an example of internal queuing in the section on address switching. Figure 16.2a depicts the block diagram view of a switch fabric employing internal queuing. Switch fabrics built with internal queuing, such as a self-routing switch fabric, have the potential to scale to large sizes as indicated in the previous section. On the other hand, internally queued switching fabric designs have difficulty providing other key functions — such as priority queuing, large buffers, and multicast. Switch fabrics with internal queuing exhibit internal blocking. That is, the switch fabric may lose cells destined for the same output.

Input queuing depicted in Figure 16.2b is simple to implement; however, it suffers from a fatal flaw. With input queuing, when the cell at the Head Of Line (HOL) cannot be switched through the fabric, all the cells behind it are delayed. As shown in Chapter 25, input queuing limits switch throughput to only 50 to 60 percent of the switch port speed depending upon whether the input traffic has a constant or variable bit rate. Therefore, input queuing alone is not adequate for many applications, and is usually employed in conjunction with other queuing methods or feedback control. Crosspoint-based ATM switch fabrics always employ input queuing. One way around the HOL blocking problem is to provide an input queue for each switch output port (or even a queue for each virtual connection). In this way, a congested port does not delay cells destined for other uncongested ports (or connections).

As shown in Chapter 25, output queuing depicted in Figure 16.2c is theoretically optimal. A switch fabric that is internally non-blocking, such as a bus or augmented self-routing design, conveys all cells to the output port without loss. Hence, loss only occurs if the output buffers overflow. Therefore, most ATM switches employ at least some output queuing to efficiently utilize the transmission facility. Furthermore, shared output queuing depicted in Figure 16.2d is the best in terms of achieving the maximum throughput with the fewest cell buffer positions. Shared memory switches, however, are limited in size by the speed of the memory chips. Real ATM switches may have a combination of input, output, and internal queuing.

16.1.4 Other Aspects of ATM Switches

Other factors important in comparing switch architectures are [1]:

- Modularity is the minimum increment of ports or fabric capacity a switch design can add.
- Maintainability measures the isolation of a disruption or failure on the remainder of the switch.
- Reliability means that the operation continues in the presence of single or multiple faults.
- Availability measures the fraction of time the switch is operational. Look for switchover time of redundant hardware and software components.
- Complexity, often measured by logic gate counts, chip pin-out, and card pin-out is a measure used when comparing switch implementations.
- Flexibility regards the capability to implement additional functions easily.
- Expandability considers the maximum number of switch ports and increased speeds supportable by the basic switch and buffering architecture.
- Upgradability considers the impact of changing hardware and/or software. CO-based switches often tout in-service upgrades.

16.2 ATM SWITCH TYPES AND CATEGORIES

ATM technology appears in many networking contexts, ranging from the switching backbone of large carriers to small workgroups. Thus, ATM dissolves the fine line between wide area, campus, and local area networking using a broadband, cell-based architecture that scales well in distance and speed, as well as network size. Most ATM switch vendors target their hardware for four different environments corresponding to the following ATM switch types:

☞ Core or Carrier Backbone (also called a Central Office switch)

☞ Carrier Edge
☞ Enterprise Backbone (also labeled CPE access switches)
☞ LAN/Campus Backbone or Workgroup

This section first briefly introduces these switch types, working down from the carrier backbone switch, through the edge and into the enterprise switch, across the campus backbone arriving at the workgroup. A subsequent sections explores the end user workstation hardware and software . Figure 16.3 illustrates a commonly used hierarchical view of these ATM switch types. The carrier backbone switch resides in a Central Office (CO) and forms the core of a public ATM service network. Edge switches located in carrier Points of Presence (PoPs) feed into the core backbone switch network. The carrier backbone and edge switches may also interwork with voice and other data networks, such as frame relay and the Internet, as shown in the figure.

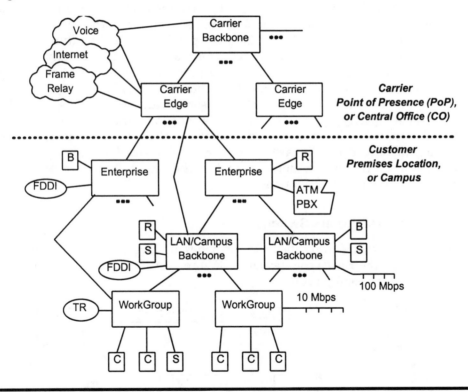

Figure 16.3   Central Office (CO), Enterprise, LAN and Workgroup Switching Roles

Moving out of the carrier domain onto the customer premises, the enterprise ATM switch connects other local switches and devices to the Wide Area Network (WAN) via an ATM User-to-Network Interface (UNI) (e.g., DS3, OC-3).   Enterprise switches interconnect LAN backbone switches, high-performance legacy LANs (such as FDDI), ATM-capable PBXs, and large

Routers (R), Servers (S) and Bridges (B). Enterprise ATM switches often provide access to public data and voice services. The LAN, or Campus, backbone plays a key role in large customer local area networks, often replacing FDDI backbone networks to interconnect LANs, bridges, routers and servers. The LAN backbone switch also interconnects the ATM workgroup switches directly supporting native ATM Clients (C), Routers (R) and Servers (S) in virtual networks. ATM workgroup switches may also support legacy LANs, such as Token Ring (TR) or Ethernet. The hierarchy isn't a strict one, as indicated in the figure where workgroup switches connect to enterprise switches and even the WAN. Note that LAN backbones may also connect directly to the WAN.

Now, let's look at each of these categories in more detail. Often a particular vendor switch product may fit into one or more of these roles. Although the following sections place particular vendor products in a specific role for purposes of presentation, note that a particular product may play more than one role in a real network design. We obtained the capacities and port speed support listed in the tables in the following sections from the survey in our book titled Hands On ATM [5], the web, public domain literature, or the vendors themselves. Where indicated, the maximum port capacity results from selecting the card(s) with the largest aggregate speed and fully populating all slots in the switch. This bounds the usable capacity. If your design uses a mixture of cards including some with lower speeds, then the total usable capacity may be less. All of the capacities or capabilities listed in the table may not be immediately available. Consult the vendors before making any decision. Be aware that many vendors have plans not yet announced in the public domain, and hence not reflected in the data tabulated in the following sections.

16.2.1 ATM Backbone Switches

Generally, backbone (also called core) switches are larger and more industrial strength than their edge counterparts. Figure 16.4 illustrates a typical ATM backbone switch configuration.

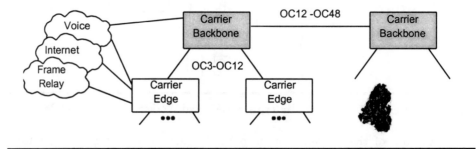

Figure 16.4   Role of the ATM Backbone Switch in a Carrier Network

Carriers deploy ATM WAN backbone switches to form the core of their backbone switching and transport fabric, as they offer the highest-speed trunks and largest switch capacities available in the marketplace. The central office (CO) environment where core backbone and edge switches reside dictates DC power and the capability to scale to a large number of high-speed ports and trunks. Backbone switches often tout throughput in excess of 40 Gbps up to 1 Tbps, primarily supporting native ATM UNI and NNI interfaces. Most backbone switches support growth from a minimum capacity to a larger capacity in a modular manner. Some backbone switches support interfaces to voice and N-ISDN networks as indicated in the figure. These switches support redundant DC power feeds and have hot standby switch fabrics and controllers.

Typically, backbone switches employ multiple interconnected shelves arranged in multiple cabinets. Literally, these are large switches, often occupying four or more equipment racks. They often support SONET/SDH automatic protection switching to work hand in hand with automated restoration in the modern, synchronous digital transmission network universally deployed by carriers. Backbone switches also often have multiple processors in support of management functions and Switched Virtual Connections (SVCs). These switches often support the ITU-T defined CMIP management interface as defined in Part 7.

Table 16.2 lists some representative backbone switch vendor products, their minimum and maximum fabric capacities, the sum of the port speeds for the maximum configuration, along with support for ATM interfaces and other protocols.

Table 16.2   Representative Backbone Switch Vendor Characteristics

Vendor	Model Identification	Minimum Switch Capacity (Gbps)	Maximum Switch Capacity (Gbps)	Maximum Port Capacity (Gbps)	ATM Interfaces					
					DS1/E1	DS3/E3	OC3/STM-1	OC12/STM-4	OC48/STM-16	Voice
Ascend	GX550	25	100	72			Y	Y	Y	
Fujitsu	FETEX-150		160				Y	Y		
NEC	M20	10	160			Y	Y	Y	Y	
Nortel	Concorde	10	80			Y	Y	Y		
Siemens	36190	5	>1,000		Y	Y	Y	Y		Y

16.2.2 ATM Edge Switches

When enterprise ATM switches interface to a carrier's ATM service, they typically do so through a carrier grade edge switch. Figure 16.5 illustrates the roles that an ATM edge switch plays in a carrier network. Carrier networks often deploy many edge switches at smaller Point-of-Presence (PoP) access points to minimize backhaul to the larger backbone switch sites. Edge switches have capacities between 5 and 20 Gbps. Typically, these switches also have a smaller maximum port speed than backbone switches do. Some edge switches support growth from a minimum capacity to a larger capacity in a modular manner. Edge switch designs target high reliability by making common components redundant with automatic switchover.

Edge switches often fit within a single shelf, often employing a single shelf fabric design which dictates a high port density per card unless the product employs expansion shelves. Some designs support multiple interconnected shelves to increase port fan-out, for example, as is done in the Newbridge 36170 and the Cisco BPX/AXIS products. Other designs build up a larger node out of multiple smaller switching systems connected via standard high-speed interfaces. These switches support redundant DC power feeds and have hot standby switch fabrics and controllers. Many edge switches support a variety of other interfaces besides ATM, such as frame relay, HDLC (e.g., for access to the Internet), circuit emulation, and voice. Several vendors call these types of devices multiservice switches. Many edge switches have their own management systems targeting the carrier environment, or support other management systems using the Simple Network Management Protocol (SNMP). Some edge switches also provide software for usage-based billing.

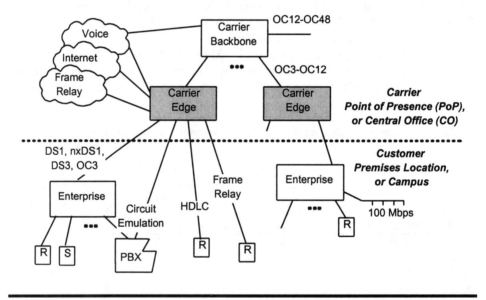

Figure 16.5 Role of an ATM Edge Switch in a Carrier Network

Table 16.3 lists some representative edge switch vendor products, their minimum and maximum fabric capacities, the sum of the port speeds for the maximum configuration, and support for ATM and other protocol interfaces. Looking at the column labeled other interfaces shows why some vendors dub their products multiservice platforms.

Table 16.3   Representative Edge Switch Vendor Characteristics

Vendor	Model Identifier	Minimum Switch Capacity (Gbps)	Maximum Switch Capacity (Gbps)	Maximum Port Capacity (Gbps)	ATM Interfaces					Other Interfaces				
					DS1/E1	nxDS1/E1 IMA	DS3/E3	OC3/STM-1	OC12/STM-4	Frame Relay	Serial	Circuit Emulation	Voice	Ethernet
Ascend	CBX 500	2.5	5	8.4	Y		Y	Y	Y					
Cisco	BPX/AXIS	-	20	14.4	Y	Y	Y	Y	Y	Y	Y	Y		
Fore	TNX-210	-	2.5	2.4	Y		Y	Y	Y			Y		
Fore	TNX-1100	2.5	10	9.6	Y		Y	Y	Y			Y		
GDC	APEX	1.6	6.4	4.2	Y		Y	Y		Y	Y	Y	Y	Y
GDC	Strobos	12.8	25.6	19.2			Y	Y	Y					
N.E.T.	Promina 2000	-	3	2.4	Y		Y	Y						
N.E.T.	Promina 4000	0.6	10	9.6			Y	Y	Y					
Nortel	Vector	2.5	5	4.8	Y		Y	Y	Y			Y		
Newbridge	36170	0.8	12.8	27	Y	Y	Y	Y	Y	Y	Y	Y		

16.2.3 Enterprise ATM Switches

Enterprise ATM switches interface the customer premises to the wide area network (WAN). Enterprise ATM switches typically support [6]:

- Local networking
- Basic user-to-network interface connections
- LAN Emulation
- Multiple QoS classes and traffic management
- Switched Virtual Connection (SVC) signaling
- Available bit rate support
- WAN connectivity

Figure 16.6 illustrates a typical ATM enterprise switch configuration. The principal characteristic of an ATM enterprise switch is the WAN interfaces and support for a wide range of local interfaces, such as native LAN, MAN, HDLC, video, and support for voice equipment. These devices are the Swiss army knife of ATM switches, supporting a multitude of interface types.

Enterprise switches often provide sophisticated traffic management capabilities in support this multiplicity of interface types. Furthermore, enterprise switches may also provide connectivity for both LAN backbone and workgroup ATM switches, as well as high-speed LANs, such as FDDI and 100-Mbps Ethernet as indicated in the figure. They may also support dual homing to more than one carrier edge switch as illustrated in the figure. Some also provide protocol conversion, LAN emulation, and virtual networking. Enterprise switches often have smaller port capacity than LAN backbone switches, focusing on the application of efficiently multiplexing multiple traffic types across the WAN. Typically, these devices range in switching capacity from 1 to 5 Gbps and are designed for the customer premises, as opposed to the central office environment. Since much of the traffic may be switched locally, enterprise switches may be as larger, or even larger than carrier edge switches.

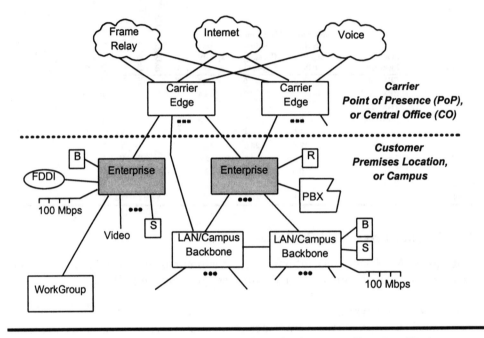

Figure 16.6   Role of an ATM Enterprise Switch in the Customer Premises Environment

The capacity bottleneck frequently exists in the access line bandwidth between the enterprise and the carrier edge switch, not within the customer building or campus.   Here too, sophisticated traffic management aims to make the most efficient use of expensive WAN bandwidth.   For example, enterprise switches shape the traffic from non-ATM sources to meet the traffic contract enforced by the carrier edge switch.

Enterprise switches run on either AC (i.e., wall outlet) or DC power. Many enterprise switches also serve a dual role as carrier edge switches, personalized by the purchaser's choice of interface cards and redundant common equipment.   Enterprise switches often have a management system that controls a network of switches at different locations.   Many enterprise switches support the Simple Network Management Protocol (SNMP).

Table 16.4 lists some representative enterprise switch vendor products, their minimum and maximum fabric capacities, the sum of the port speeds for the maximum configuration and the types of ATM and other interfaces supported. Similar to their edge counterparts, many of these devices support a broad range of non-ATM services. What differentiates an enterprise switch from an edge switch is primarily support for local interfaces.

Table 16.4   Representative Enterprise Switch Vendor Characteristics

				ATM Interfaces					Other Interfaces								
Vendor	Model Identifier	Maximum Switch Capacity (Gbps)	Maximum Port Capacity (Gbps)	DS1/E1	25 Mbps	DS3/E3	OC3/STM-1	OC12/STM-4	Frame Relay	Serial	Circuit Emulation	Voice	Video	Token Ring	FDDI	Fast Ethernet	Ethernet
Cisco	IGX	1.2		Y	Y				Y	Y	Y						
Cisco	Lightstream 1010	5	2.4		Y	Y	Y	Y	Y	Y	Y						Y
Fore	ASX-200	2.5	2.4	Y	Y	Y	Y				Y						Y
Fore	ASX-1000	10	9.6	Y	Y	Y	Y				Y						Y
GDC	DV2, IMX	6.4	5.0	Y	Y	Y			Y	Y	Y	Y	Y				Y
Hitachi	(AN) 1000	20	19.2	Y	Y	Y											
IBM	2220 Nways	12.8		Y	Y	Y			Y	Y	Y	Y					
IBM	2230 Nways	5		Y	Y	Y											
Newbridge	36140/36144	2	1.65		Y	Y			Y		Y						Y
Newbridge	36150	2.5	2.4	Y	Y	Y					Y			Y	Y		Y
Nortel	Passport	1.6		Y	Y	Y			Y	Y	Y	Y		Y	Y		Y
Xylan	Omniswitch	13.2	9.6		Y	Y	Y							Y	Y	Y	Y

16.2.4 Campus Backbone and Workgroup ATM Switches

ATM campus backbone and workgroup switches provide hierarchical connectivity for workgroup ATM switches with higher speed uplinks of 155-Mbps at a minimum, and typically at rates of 622-Mbps. Campus backbone and workgroup switches typically support native ATM workgroup user connectivity as well as local networking.

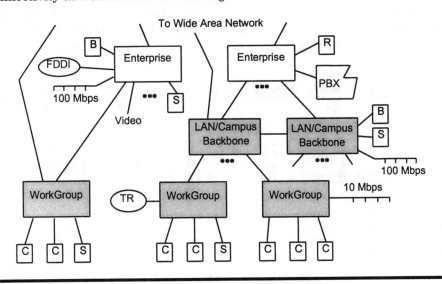

Figure 16.7   Role of a ATM Campus Backbone or Workgroup Switch

Figure 16.7 illustrates a typical campus backbone and workgroup switch configuration providing connections for local devices at ATM backbone speeds within a confined geographic area, such as a campus environment. The

LAN/campus backbone often aggregates traffic into one or more uplinks to an enterprise switch and commonly has only local interfaces since the Enterprise switch has WAN interfaces. However, a LAN backbone switch may also connect directly to the WAN as shown in the figure.

Usually, these devices reside in wiring closets, near the main distribution frame (MDF), or the data center area. Typically, they are AC-powered. They often provide some level of redundancy, especially in processor cards and power supplies since many users access the switch. These single shelf devices range in switching capacity from approximately 1.0 Gbps to over 50 Gbps. The size of a LAN backbone switch approaches that of carrier backbone switches to support large amounts of locally switched traffic.

The most critical application for ATM Desktop Area Networks (DANs) are high-performance ATM workgroups where the high bandwidth, multiple classes of service, and flexible bandwidth allocation require end-to-end ATM capabilities. Workgroup ATM switches provide direct connection to ATM workstations equipped with 25- or 155-Mbps Network Interface Cards (NICs). Workgroup switches usually support high-speed devices, such as super servers, super computers, or high-end multimedia workstations. Workgroup ATM switches are typically low-cost devices that also support legacy LAN interfaces (i.e., Ethernet and Token Ring). A common configuration multiplexes many lower-speed legacy LAN or lower-speed ATM interfaces onto a single, higher-speed ATM uplink connected to a LAN campus backbone switch as shown in Figure 16.7. As shown in the figure, a workgroup may also connect directly to an enterprise switch or the WAN. Workgroup switches often have less capacity than most ATM LAN and WAN switches. These switches typically run on AC (i.e., wall outlet) power and make up for smaller port capacity and less processing power by being modular or stackable.

Table 16.5   Representative Campus and Workgroup Switch Vendor Characteristics

Vendor	Model Identifier	Maximum Switch Capacity (Gbps)	Maximum Port Capacity (Gbps)	ATM Interfaces							Other Interfaces				
				DS1/E1	25 Mbps	100 Mbps	DS3/E3	OC3/STM-1	OC12/STM-4	OC48	Token Ring	FDDI	Gigabit Ethernet	Fast Ethernet	Ethernet
3Com	CoreBuilder 7000	5					Y	Y	Y					Y	Y
3Com	CoreBuilder 9000	70	67.2	Y		Y	Y	Y	Y	Y			Y	Y	
Cabletron	Smartswitch 9000	75.6	33.6				Y	Y	Y						
Bay Networks	Centillion 100	10						Y			Y				Y
Cabletron	Smartcell ZX-250	2.5					Y	Y	Y						
Cabletron	SFCS-1000	10		Y	Y	Y	Y	Y	Y						
Cisco	Catalyst 5000	1.2	2.5	Y	Y		Y	Y	Y				Y	Y	Y
Fore	LE155	2.5	2.4					Y	Y						
Fore	ASX-200WG	2.5	5.1		Y			Y							
IBM	8285 Nways	4.2			Y			Y							
Madge Networks	LANNET 740	2.6		Y				Y	Y						
Madge Networks	LANNET 530/540		0.3					Y			Y				Y
Newbridge	CS1000/3000	1.6/6.4					Y	Y	Y						
Xylan	Omniswitch	13.2	9.6		Y			Y	Y		Y	Y		Y	Y

Table 16.5 lists some representative campus backbone and workgroup switch vendor products, their minimum and maximum fabric capacities, the sum of the port speeds for the maximum configuration, as well as support for ATM interfaces and other protocols. Many of the OC3/STM-1 rate connections run over multimode fiber or shielded twisted pair and hence cannot connect directly to the WAN at these speeds.

16.3 ATM ENABLED ENTERPRISE AND LAN BACKBONE DEVICES

This section reviews the various types of other ATM-enabled equipment used primarily in the local area customer premises. These include routers, hubs, bridges, service access multiplexers, and CSU/DSUs. In some cases these devices serve as the backbone of the LAN or enterprise, and in other cases they directly interface to the WAN. Every major router, bridge, and hub manufacturer interfaces to ATM networks with at least their high-end products. Simultaneously, it appears that everyone is moving into the other's area of functionality. Indeed, many products have a split personality: hubs performing switching, bridges implementing routing, and routers doing bridging. Other devices serve specialized roles acting as simple multiplexers to interconnect to the WAN, LAN bridges with ATM interfaces, and ATM CSU/DSUs that convert serial bit streams into cells using AAL3/4 or AAL5. This chapter reviews ATM capable hardware in each of the following categories:

- Routers
- Bridges
- Hubs
- Multiplexers
- CSU/DSUs

16.3.1 ATM Capable Routers, Bridges, and Hubs

When bridges, hubs and routers first came along, they dealt primarily with local area networking. Now the functions of both have begun to merge through the use of increased processor speeds and technologies (i.e., Reduced Instruction Set Computer (RISC) processors and Application Specific Integrated Circuits (ASIC)). These advances yield reduced costs and enable devices to support both local and wide area networking at speeds ranging from 9.6-kbps legacy analog modems up to gigabit per second speeds over ATM and SONET interfaces.

Current high-end routers forward millions of IP packets per second with aggregate forwarding capacities in the gigabits per second range. Most LAN backbone class routers have support for LAN emulation, as well as ATM UNI and NNI protocol support. The current generation of ATM-capable routers convert between packets and ATM cells on high-performance interface cards. Although the initial versions of these cards were rather expensive, the prices are rapidly dropping. Until 1997, if you wanted to run a router at OC3 speeds, you bought an ATM interface card. Recently, hardware optimized for packet switching allows IP to run directly over SONET. Chapter 32 compares these approaches in greater detail. Therefore, ATM-capable routers accept multiple protocols and either route them to non-ATM ports, or send them to ATM ports for conversion from packets to ATM cells for transport over a switched ATM network.

An ATM-capable bridge takes a bridgeable protocol, such as Ethernet or Token Ring, and connects it over an ATM network. This makes it appear to the user devices as if they were on the same shared LAN medium. Many use the ATM Forum defined LAN Emulation protocol (detailed in Chapter 19) to achieve this appearance. A bridging device encapsulates a bridged protocol, such as Ethernet, and emulates the encapsulated protocol's bridging functions. These include self-learning and self-healing capabilities.

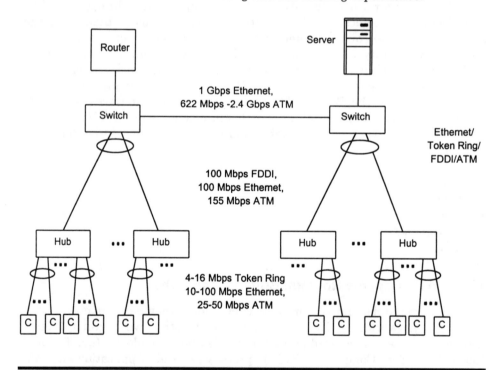

Figure 16.8   Hierarchical LAN Hub Architecture, Interfaces, and Protocols

Figure 16.8 illustrates the wiring collection, segmentation, and network management function a hub typically performs. Usually many Ethernet or Token Ring twisted pair lines, and in some cases FDDI or ATM over twisted pair to individual workstations, run to a hub located in a wiring closet. Hubs allow administrators to assign individual users to a resource (e.g., an

Ethernet segment), shown as an ellipse in the figure, via network management commands. LAN designers often interconnect lower-level hubs in a hierarchy to higher-level switches, sometimes via higher-speed protocols such as FDDI and ATM over optical fiber and high-grade twisted pairs as illustrated in the figure. As more and more traffic aggregates in a hierarchy, higher speeds come into play. Hubs employed in such a hierarchical manner concentrate access for many individual users to a shared resource, such as a server or router as shown in the figure. The highest-level hubs are candidates for a collapsed backbone architecture based on high-speed ATM or Gigabit Ethernet.

Table 16.6 lists some representative router, bridge and hub vendor products, and their support for ATM and other protocol interfaces. As indicated in the figure, several devices may perform more than one role.

Table 16.6   Representative ATM-capable Routers, Bridges and Hub Vendors

| Vendor | Model Identifier | Bridge, Router, Hub | ATM Interfaces | | | | | ISDN | ATM DXI | Frame Relay | Serial | Token Ring | FDDI | Gigabit Ethernet | Fast Ethernet | Ethernet |
			25 Mbps	DS3/E3	100 Mbps	OC3/STM-1	OC12/STM-4									
3Com	NetBuilder II	B/R				Y		Y	Y	Y	Y	Y	Y		Y	Y
Ascend	GRF IP Switch	R				Y	Y						Y		Y	Y
Bay Networks	BCN	R		Y		Y			Y	Y	Y	Y	Y		Y	Y
Cisco	4000 Series	R		Y		Y		Y			Y	Y	Y		Y	Y
Cisco	7000 Series	R		Y	Y	Y		Y			Y	Y	Y		Y	Y
Cisco	12000 Series	R				Y	Y							Y		
IBM	2210 Nways	R	Y					Y		Y	Y	Y				Y
IBM	8281 Nways	B			Y							Y				Y
Newbridge	Vivid Ridge	B/R				Y							Y		Y	Y
Xyplex	Network 9000	R/H				Y									Y	Y

16.3.2 Multiservice ATM Multiplexers and ATM CSU/DSUs

A multiservice ATM multiplexer, also called a concentrator, takes multiple interfaces on input and concentrates them into a single ATM WAN trunk as illustrated in Figure 16.9. Although an ATM multiplexer fits in the same position as an enterprise switch, it generally doesn't perform much switching. However, some ATM multiplexers do perform local switching. Since a multiplexer performs fewer functions than a switch, it is generally less expensive. In essence, an ATM multiplexer allows a user to share an access line to the WAN by exploiting the multiple quality of service and bandwidth management capabilities of ATM. Some multiplexers are simple, taking in only one or two types of inputs for combination on a single output. Others take many different types of inputs and combine these onto a single high-speed ATM output link as illustrated in the figure.

Some carriers use a similar device to serve the Network Termination (NT) function defined in the B-ISDN reference model in Chapter 12. An NT device provides an ATM interface towards the end user as well as an ATM interface toward the network. The NT provides the carrier consistent physical and ATM layer to the demarcation point on the customer's premises. Also, an NT device may perform value added functions such as traffic management.

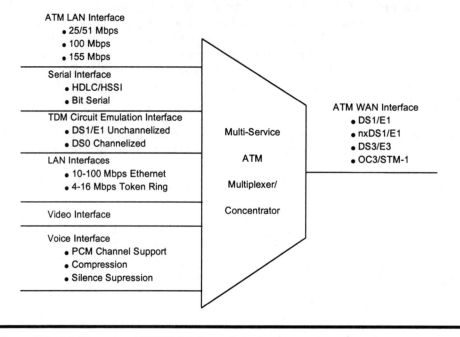

Figure 16.9   Multiservice ATM Multiplexer Interfaces

An ATM Channel Service Unit/ Data Service Unit (CSU/DSU) performs the conversion from a serial DTE/DCE interface utilizing the frame-based ATM DXI or FUNI protocol into a stream of ATM cells for transmission over an ATM UNI interface as illustrated in Figure 16.10.

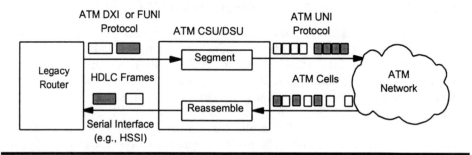

Figure 16.10   ATM CSU/DSU Interfaces and Function

Chapter 18 details the ATM Data eXchange Interface (DXI) and Frame-based UNI (FUNI) protocols. The key functions performed are segmentation of frames into cells for transmission, and reassembly of cells into frames upon reception. The commonly implemented DTE/DCE interface is the High-Speed

Serial Interface (HSSI) operating at up to 51-Mbps. ATM CSU/DSUs were one of the earliest ATM CPE devices since they allowed legacy devices to utilize ATM, for example, routers with serial interfaces. Now that many data devices have ready-made ATM interfaces, end users rarely need ATM CSU/DSUs.

Table 16.7 lists some representative multiservice multiplexer and ATM CSU/DSU vendor products, their capacities, and support for the types of ATM and other protocol interfaces.

Table 16.7   Representative ATM Multiservice Multiplexer and CSU/DSU Vendors

Vendor	Model Identifier	Maximum Capacity (Gbps)	DS1/E1	nxDS1/E1 IMA	HSSI	DS3/E3	100 Mbps	OC3/STM-1	OC12/STM-4	Frame Relay	ATM DXI	HDLC	Circuit Emulation	Voice	Ethernet
			ATM Interfaces							Other Interfaces					
3Com	AccessBuilder 9600	1.6	Y	Y	Y			Y		Y	Y	Y	Y		Y
ADC Kentrox	AAC-1		Y							Y	Y	Y			
ADC Kentrox	AAC-3	0.5		Y	Y	Y		Y		Y	Y	Y	Y		Y
ADC Kentrox	DataSMART T3 ADSU					Y					Y				
ADC Kentrox	CellSMART 201/202/203		Y	Y								Y			Y
Digital Link	DL3200					Y					Y				
Digital Link	DL7100		Y	Y		Y				Y	Y	Y	Y		
Fore	CellPath 90		Y							Y	Y	Y			Y
Fore	CellPath 300		Y			Y		Y				Y			
GDC	MAC1	1.6	Y			Y	Y	Y		Y	Y	Y	Y	Y	Y
Sentient Networks	Ultimate 1000			Y		Y				Y		Y	Y		
Yurie Systems	LDR 5		Y							Y		Y			
Yurie Systems	LDR 50		Y			Y	Y	Y		Y		Y	Y	Y	Y
Yurie Systems	LDR 200	1.2	Y			Y	Y	Y		Y		Y	Y	Y	Y

16.4 END SYSTEM ATM HARDWARE AND SOFTWARE

Users realize the greatest benefits from ATM when using it between end systems. This section covers ATM implications for End System (ES) hardware, operating systems, and application software.

16.4.1 ATM Network Interface Cards (NICs)

ATM end system hardware runs directly on a Network Interface Card (NIC) inserted in the user's computer attached to one of the ATM switch types described above. The key aspects of an ATM NIC are:

- Computer bus type, interface speed and physical media
- Operating System (OS) support
- Features and functions
- Price/performance ratio
- Ease of Installation and Support

Figure 16.11 illustrates the role that an ATM NIC typically plays in real-world applications. Starting on the left hand side, a NIC card attaches to the same computer bus that the CPU, memory, and other peripherals use. The NIC software interfaces with the computer bus and driver code running on the host workstation/computer CPU. The NIC hardware performs the interface to the physical media as well as support for the ATM and AAL protocols. The NIC driver software running on the host interfaces with the operating system. The operating system interfaces with applications, and may have a native ATM API. Typically, the NIC driver software on the host supports LAN Emulation for interconnection with legacy LAN devices as described below. The LANE interface looks like a normal LAN driver interface to the operating system, hence existing applications run without change over ATM LANE.

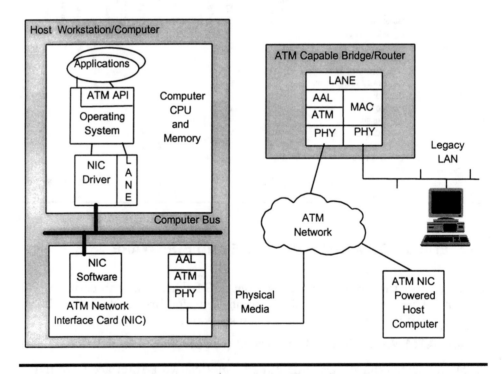

Figure 16.11   ATM Networking Interface Card (NIC) Context and Application

In order to interface with legacy LAN devices, the ATM NIC powered workstation must connect to an ATM-capable bridge, router or hub as illustrated in the right hand side of Figure 16.11. The bridge/router takes the LANE protocol running over ATM and AAL5 and converts this to the Medium Access Control (MAC) protocol for interface to the legacy LAN device. Note that the ATM NIC-powered workstation may directly communicate to other similar ATM-capable workstations directly over a workgroup ATM network without going through a bridge/router. Such desktop-to-desktop ATM networking eliminates the possibility of quality sensitive traffic traversing a legacy network that doesn't support the end-system-specified QoS parameters.

16.4.1.1 Bus, Media and Interface Speed

Commonly supported computer bus architectures are:

- Motorola VME
- IBM Compatible Industry Standard Architecture (ISA) and Extended ISA (EISA)
- IBM PS/2 Micro Channel Architecture (MCA)
- SUN S-bus
- Peripheral Computer Interface (PCI) in Pentium and Macintosh
- GIO used in Silicon Graphics

An important consideration in choosing a NIC card is whether you plan to use your existing wiring and cabling, or plan to install new cabling. Commonly encountered cabling options are:

- Unshielded Twisted Pair (UTP)
- Shielded Twisted Pair (STP)
- Single or Multimode Fiber (SMF or MMF)

Most private networks utilize multimode fiber since it is cheaper and easier to splice and attach connectors to. Most ATM adapter cards (i.e., NIC) operate over the following combinations of speeds and media:

- 25-Mbps over STP, UTP
- 51-Mbps over STP, UTP
- 155-Mbps over UTP, MMF, or SMF

16.4.1.2 Operating Systems Support

Commonly supported operating systems are:

- Sun Microsystems Solaris
- Sun Microsystems SunOS
- Microsoft/IBM DOS
- Microsoft Windows 3.x, Windows 95, Windows NT
- IBM OS-2
- IBM AIX
- Silicon Graphics IRIX
- Novell NetWare
- Apple MacOS

16.4.1.3 Software Features and Functions

Most adapters provide similar higher-layer protocol capabilities, such as ATM Forum UNI V3.0/3.1 signaling, LAN Emulation, and the IETF's Classical IP over ATM. Some cards also support new standards, such as Multiprotocol over ATM (MPOA), and IP multicast over ATM. See Chapters 19 and 20 for definitions of these protocols.

16.4.1.4 ATM NIC Card Vendor Comparison

Table 16.8 presents a representative list of the leading NIC vendors and which of the key characteristics described earlier their products support. We used public domain information and the vendor web pages in the compilation of this table.

Table 16.8   Representative ATM NIC Vendor Characteristics

Vendor	Product Name	Computer Bus						Operating System										Speed				Software					
		PCI	ISA/EISA	S-bus	Microchannel	VME	GIO	Solaris	SunOS	DOS/Windows	Windows 95	Windows NT	MacOS	OS-2	AIX	IRIX	Netware	25 Mbps	100 Mbps	155 Mbps	622 Mbps	UNI 3.0/3.1	LANE	MPOA	Classical IP/ATM	Winsock	XTI
3Com	ATMLink	Y										Y								Y		Y	Y				
Adaptec	ANA-5xxx	Y	Y	Y				Y	Y	Y	Y	Y	Y	Y	Y				Y	Y		Y	Y				
Efficient	ENI	Y	Y	Y				Y				Y	Y							Y		Y	Y				
Fore	ForeRunner 200E	Y	Y	Y	Y	Y	Y	Y	Y			Y			Y	Y				Y		Y	Y		Y		
Fore	ForeRunner LE	Y										Y	Y						Y	Y		Y	Y		Y		
Fore	ForeRunner HE 622	Y																			Y	Y	Y	Y	Y	Y	Y
IBM	Turboways	Y	Y	Y	Y			Y	Y	Y	Y	Y			Y	Y		Y	Y	Y	Y	Y	Y		Y		
Interphase		Y	Y	Y		Y	Y	Y	Y	Y		Y	Y	Y	Y	Y	Y			Y		Y	Y		Y		
Newbridge	Vivid	Y	Y	Y		Y	Y	Y				Y			Y	Y				Y		Y	Y	Y			
Olicom	RapidFire ATM 155	Y										Y	Y							Y		Y	Y				
Sun				Y				Y														Y	Y	Y	Y		

Some ATM switch vendors also make their own adapters [7], as shown in the list above. This is a great way to avoid interoperability problems between the workstation's adapter card and the switch. Fore Systems leads the market in ATM NIC shipments at the time of publication.

16.4.1.5 Computer Interface Design Impact on Performance

Most applications inherently deal with variable-length data units, and hence don't use individual cells. Witness the fact that no current application makes optimal use of the ATM cell directly. One cell is usually too small to contain a data packet. For example, as described in Chapter 19 even a single TCP/IP acknowledgment enveloped by the IETF's LLC/SNAP multiprotocol encapsulation doesn't fit in one cell. However, a cell is too large to transport voice without delay that may require echo cancellation as described in the next chapter. Undoubtedly, some applications will eventually make direct use of the ATM cell; however, most applications currently require one or more layers of adaptation to make ATM usable.

Current end systems map variable-length data units to a string of fixed length cells using an ATM Adaptation Layer (AAL) as described in Chapter 13. Not all AALs are created equal — some are easier and more efficient than others in support of application software and hardware. In particular, AAL5 began as the Simple Efficient Adaptation Layer (SEAL), originally proposed by computer manufacturers to the telephony-oriented standards organizations. As described in Chapter 13, the vast majority of chip manufacturers and higher-layer protocols utilize AAL5 exclusively.

Some first generation NICs couldn't transmit at the ATM line rate of 25, 51, or 155-Mbps. ATM cards, like any high-speed NIC, share the central processor bus of the workstation to transfer packets or cells into computer memory. Therefore, an ATM NIC impacts overall workstation performance. More recent designs, utilizing processors on-board the NIC minimize the load on the workstation processor. Let's look into this subject in more detail.

The manner in which the NIC and memory access interact with a processor performing other operating system and application tasks significantly affects the achievable communications throughput. The following discussion summarizes two extremes of an end system ATM communication adapter design based upon information derived from References 8 and 9 to illustrate this point. Figure 16.12 illustrates a simple and complex end system ATM/AAL adapter. Each adapter utilizes hardware for interface to the Physical (PHY) layer, ATM layer, and ATM Adaptation Layer (AAL). Typically, the AAL Segmentation And Reassembly (SAR) function employs an integrated circuit to achieve high performance. The Direct Memory Access (DMA) system receives an entire packet from the SAR chip so that the processor is never involved in the PHY, ATM, or AAL functions.

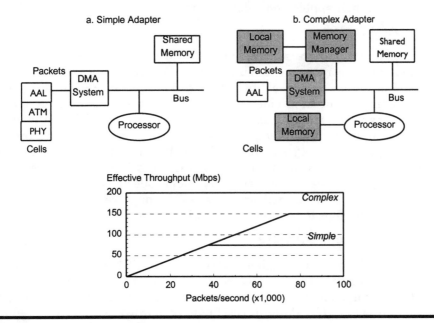

Figure 16.12 ATM Adapter Performance Comparison

The simple adapter shown in Figure 16.12a is typical of many first generation ATM NICs. The card interrupts the processor upon receipt of a packet, causing a process switch to the communications task, which then sets up the DMA transfer. The DMA system then transfers the packet data across the bus to the shared memory, during which time the processor must remain idle. The processor also performs memory management functions, such as allocating buffers for packet data and examining the header to place the packet in the logical queue corresponding to the target higher-layer protocol. Finally, the processor performs another task switch and return to operating system or application software execution until another packet arrives.

Figure 16.12b shows the additional or modified elements in complex adapter as shaded boxes. An intelligent DMA system that does not require the processor to set up data transfers and a separate memory management unit to handle the housekeeping tasks frees up processor cycles. A key benefit of the intelligent DMA system is that the processor and memory management units have their own local memory. Thus, they can still function even when the shared memory is being accessed over the bus, for example, when DMA transfers occur.

The graph at the bottom of Figure 16.12 plots the effective throughput versus the packet arrival rate for 256-byte packets. The increase in effective throughput achieved by the more complex design is nearly 100 percent! More complex adapters operate at close to the ATM line rate, assuming that the workstation processor is fast enough. Of course, ask for benchmark testing results from the manufacturer for your hardware, operating system, higher-layer protocols, and application before you buy.

16.4.1.6 ATM Servers

Modern high-performance servers employ ATM technology. One example of an ATM-enabled server is the Virata Store Server produced by Advanced Telecommunications Modules Ltd. This server has an 8-gigabyte storage capacity optimized for ATM video transmission, guaranteeing picture quality even when handling multiple simultaneous video streams. Thus, the device can handle up to 4 feature-length films and stream up to 25 simultaneous audio clips and video images on its 155-Mbps ATM interface. This server can be directly attached via an ATM adapter card to a Virata ATM workgroup switch — making it one of the first such systems offering a smooth migration path from Ethernet while delivering on ATM's promise of 25-Mbps to the desktop and 155-Mbps to the workgroup switch.

16.4.2 End System Software

ATM software involves the higher layers of the B-ISDN protocol cube in the user, control, and management planes. As described in the previous section, most devices implement the lower physical, ATM, and AAL common part layers in hardware. If ATM is successful, mass hardware production reduces the cost of implementation. In this scenario, the hardware becomes a commodity, and software becomes the value-added, differentiating feature of ATM equipment. Of course, this same effect will occur with the lower layers of ATM-based software in the future after operational experience refines the standards.

Focus now turns, in large part, to the software that exists in the end system equipment, and not on the system software that exists in the switches, bridges, routers, or hubs covered earlier in this chapter. A critical need exists to align the connection-oriented ATM paradigm with the fundamentally connectionless nature of distributed data communications. As a rule, applications don't normally set up a connection before attempting to perform a function; they just request that the function be done and expect a response indicating that either it was done completely or not at all. Most IP and LAN over ATM approaches are similar in the sense that they attempt to (quickly) set up a connection after the first packet arrives or after a routing protocol determines the need to augment the underlying layer two network topology.

16.4.2.1 Application Programming Interfaces (APIs)

An Application Programming Interface (API) allows an application program to gain access to well-defined ATM-based capabilities. An API is a set of software libraries/interfaces that enable applications to use their native language to access lower-level functional modules — such as operating systems, graphical user interfaces, and communication protocols. ATM-enabled APIs are critical to close the gap between the ATM-based capabilities and the application in order for ATM to succeed in the desktop arena. API standards define a framework for developing common interfaces between applications and operating systems (OSs) running on diverse end systems.

Towards this goal, the ATM Forum defined a semantic structure for Native ATM Service (NAS) in 1996, which this section describes first. The Microsoft Winsock and Xopen consortium then began work on the actual API specifications. Study group 8 of the ITU-T is studying a Programming Communication Interface (PCI) for terminal systems. The text then summarizes Microsoft's plans for a native ATM API in Windows 98 and NT 5.0.

16.4.2.2 Native ATM Service (NAS) Semantics

The ATM Forum defines a set of services available to higher-layer applications in its Native ATM Services (NAS) Semantic Description [10]. This document articulates services available to higher-level application programs independent of any programming language or operating system environment. Hence, the ATM Forum's document is not an Applications Programming Interface (API) per se. Instead, the ATM Forum's stated objective is to influence the development of API interfaces, such as Microsoft's WinSock version 2.0 and the X/Open Transport Interface (XTI) protocol for UNIX and Apple machines.

Figure 16.13 illustrates the ATM Forum NAS reference model. The model applies between the API interfaces and the ATM UNI services lines indicated in the figure. These could be implemented in an operating systems kernel, a PBX, or an ATM switch. Native ATM Services include the following:

- Data transfer, supporting both reliable and unreliable delivery modes employing both ATM and ATM adaptation layers
- Means for specifying the parameters and setting up Switched Virtual Circuits (SVC)

- Means for specifying the parameters and setting up permanent virtual connections (PVC)
- Means for supporting traffic management capabilities, such as different ATM layer service categories supporting quality of service guarantees
- A multiplexing/demultiplexing function for distribution of connections and associated data to the associated application — for example, using the higher- and lower-layer information elements in signaling messages
- Means for participation in ATM network management, such as the Integrated Local Management Interface (ILMI) as well as Operations and Maintenance (OAM) functions

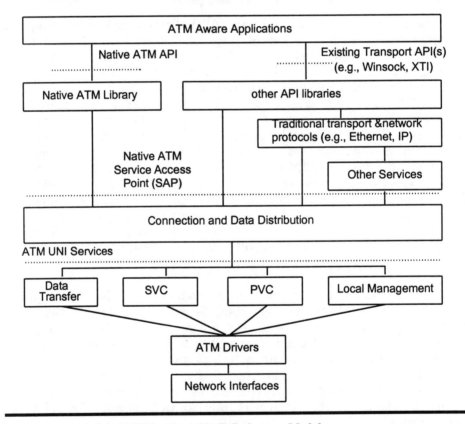

Figure 16.13   Native ATM Services (NAS) Reference Model

Figure 16.14 visually depicts the NAS API interfaces described above. The API provides access to the ILMI information by supporting the capability to send messages to the ILMI Management Entity (IME) on the other end of the UNI. Also, the API provides an interface to query or set the local status. API access to the F4 (for VPC) and F5 (for VCC) Operations And Maintenance (OAM) flows, which carry connectivity and testing information on both a segment and end-to-end basis is a requirement.

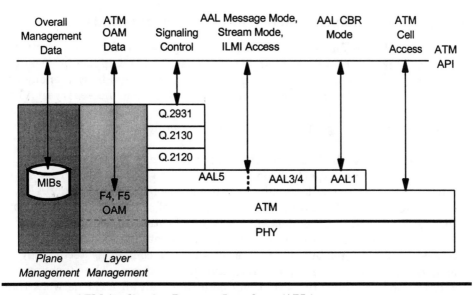

Figure 16.14   ATM Application Program Interfaces (APIs)

Access to the signaling layer allows an application to set up and release a call in a very simple manner, hiding the complexities of the signaling protocol. The API supports point-to-point and point-to-multipoint connection types. The capability to add or release a party in a point-to-multipoint connection is essential to higher-layer applications. See Reference 11 for more coverage on the NAS model for ATM SVCs.

The API provides direct access to the ATM Adaptation Layer (AAL), effectively emulating the AAL Service Access Point (SAP) described in Chapter 13. Data transmission from the application to the AAL occurs in either *message* or *stream* mode. Message mode takes an entire packet and completely fills cells prior to transmission, making it very efficient, but possibly incurring delay in waiting for the next packet before sending the previous packet. Stream mode immediately sends the cells, minimizing delay while incurring a possible loss in efficiency. Thus, the mode used depends on the application's data transfer requirements: efficiency, or minimum delay. The API supports a service where it periodically requests data from the application in support of a Continuous Bit-Rate (CBR) service. For example, this type of interface must transfer clock to the application in support of video transfer.

The API should provide direct access to the ATM cell flow, where cells are transmitted and received without processing or error checking. This interface is useful in a test and development environment, or provides a high-performance interface for applications needing direct access to the power of ATM.

16.4.3 ATM API Implementation Issues, Standards and Plans

The device driver for an ATM API ideally presents the same software interface, therefore requiring no application change. If an application wanted to take advantage of additional capabilities, the API should be rich enough to support this. Many current APIs do not perform functions necessary in ATM, such as point-to-multipoint, specification of traffic parameters, support for connection-oriented signaling, and support for multiple Qualities of Service (QoS).

Something of a chicken and the egg situation exists for ATM-aware applications. Some experts report that since no applications require or support ATM, that users don't see compelling reasons for networking ATM to the desktop. Furthermore, the emergence of cost-effective 100-Mbps Ethernet LAN solutions makes business justification of a more expensive ATM solution difficult without an enabling application. Nonetheless, many early adopters selected ATM networking since no competing technology on the horizon supports real-time interactive applications. The linchpin will likely be popular applications that run better over an all-ATM network than over networks of routers and LAN switches supporting other protocols.

Additionally, legacy equipment not built for ATM may require upgrades or even replacement to utilize native ATM capabilities. This presents a dilemma: unless the underlying Operating System (OS) provides access to native ATM capabilities, then application developers cannot develop applications to take advantage of ATM. The speeds easily supported by ATM may tax software and drivers on adapters and workstations. Furthermore, there may be timing problems in operating system software in supporting synchronous traffic. A next generation of workstations with hardware assists may be necessary to fulfill the vision of true, end-to-end multimedia networked applications.

Good news for native ATM APIs is just over the horizon, however [12]. WinSock version 2 provides support for much of that required to utilize ATM features, and the XTI protocol standard used by UNIX and Apple will provide QoS support. The next enabler for ATM is for independent software vendors to use these APIs as the basis for mainstream applications.

16.4.3.1 Application Programming Interfaces (APIs) - Current Status

The earliest example for support of an API, a UNIX socket call, was the Fore Systems proprietary API. Apple and Fore Systems announced in 1996 that they planned to develop a native ATM API based upon the XTI protocol targeting the film industry. The ForeRunner series of ATM adapters adapters supported both the Winsock 2.0 and the XTI API in 1997 [13]. These APIs enable programmers to take advantage of ATM QoS and point-to-multipoint communications for applications like video-conferencing, video-on-demand, security monitoring, and medical imaging. X-Open XTI specification creates the bridge from ATM to Macintosh, IRIX, and Solaris operating systems [14].

Application programming interfaces (APIs) form the bridge between the operating system (such as Windows NT) and the ATM Network Interface Card (NIC). One example of a fully ATM aware API is Microsoft's Network Driver Interface Specification (NDIS) 5.0, planned for incorporation in Windows 98 and Windows NT 5.0 [15]. Microsoft realizes the importance of ATM in the future of networking, citing ATM's support for guaranteed bandwidth and quality of service as the key drivers for the development of NDIS 5.0 support for ATM. Furthermore, NDIS 5.0 supports LAN Emulation (LANE) and Classical IP over ATM as a viable migration path for legacy data networks. Additionally, ATM's simultaneous support for voice, video and other data networking makes it an enabler for an entirely new spectrum of applications.

Winsock 2.0 API

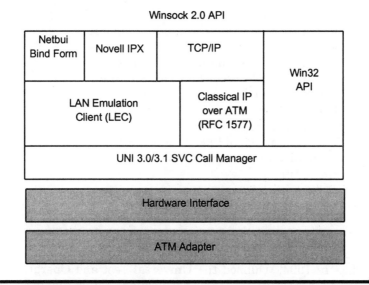

Figure 16.15   Microsoft Network Driver Interface Services (NDIS) 5.0 for ATM

Superseding an earlier developer's kit driver (NDIS 4.1) shipped in November, 1996, NDIS 5.0 allows Winsock 2.0 access to native ATM, TCP/IP, Novell IPX and Netbeui Bind Form (NBF) as illustrated in Figure 16.15. The LAN interfaces (Netbeui and Novell) interface via the LAN Emulation Client (LEC) protocol described in Chapter 19. The native TCP/IP interface uses either the LEC, or Classical IP over ATM as described in Chapter 20. Both the LEC and Classical IP over ATM software drivers utilize an ATM UNI 3.0/3.1 call manager, which then interfaces to the ATM NIC hardware interface as indicated in the figure. The Win32 API interface provides direct access to ATM UNI SVCs via the call manager. Microsoft plans to support this interface in Winsock 2.0 and Active Movie-based Win32 API applications using NDIS 5.0. See Reference 11 for a description on Winsock 2.0 programming extensions and a mapping to the ATM Forum's Native ATM Semantic (NAS) specification. An example application is Optivision's MPEG video over native ATM supporting multimedia applications like distance learning, telemedicine, and videoconferencing.

16.5 ATM CHIP SETS AND PROCESSORS

Most major chip manufacturers provide ATM chip sets that many of the ATM device described in this chapter employ. Want to know more about the latest developments in ATM chipsets and processors? Check out the ATM chip web page at www.infotech.tu-chemnitz.de/~paetz/atm, which has a listing of over 50 ATM chip vendors supporting the following types of functions:

- PHYsical Layer Controllers
- Single chip Network Interface Card (NIC) support
- ICs for 25-Mbps PHY layer
- Line drivers and fiber optic interface support
- Asymmetric Digital Subscriber Line (ADSL) support
- Policing and traffic shaping
- Clock and data recovery
- CBR and AAL1 support
- AAL 3/4,5 Segmentation And Reassembly (SAR)
- Switching elements for building ATM devices
- Interworking units
- Programmable logic
- Application Specific Integrated Circuits (ASICs)

With so many different manufacturers building such a wide range of functional components, the ATM Forum decided to define a standard chip-to-chip interface in 1994. Dubbed the Universal Test and Operations Physical Interface for ATM (UTOPIA) , these specifications define the link between the physical layer (PHY) and the ATM layer, as well as other management entities. The concept is that device manufacturers can employ chips from different suppliers in ATM switches, multiplexers, routers, bridges, and ATM network interface cards. Despite the ephemeral acronym, many chip manufacturers support UTOPIA in their products, thus driving down the cost of ATM hardware.

The ATM Forum has released two UTOPIA specifications: Level 1 and Level 2. Figure 16.16a depicts the UTOPIA Level 1 reference model [16], which defines the interface between a single instance of the ATM layer and one physical interface, for example on an ATM network interface card. Figure 16.16b shows the UTOPIA Level 2 reference model [17], which defines an interface between a single ATM layer instance and two or more physical layer interface, for example multiplexing one high-speed ATM interface to multiple lower speed physical interfaces. The UTOPIA Level 2 specification also defines operation on 16 parallel traces running at 50 MHz, for support of speeds up to 622-Mbps.

a. UTOPIA Level 1 Reference Model   b. UTOPIA Level 2 Reference Model

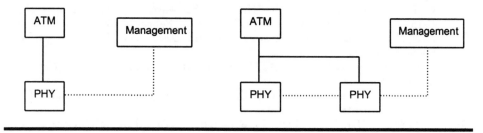

Figure 16.16   ATM Forum UTOPIA Reference Models

16.6 REVIEW

ATM hardware and switching systems take many forms. The chapter began with a discussion of switch network models — the building blocks of ATM hardware. The text then broke down ATM hardware into backbone and edge carrier switches, followed by enterprise switches which often form the core access for many private ATM networks and access to many public ATM networks. The discussion then explored the LAN and campus backbone and native ATM workgroup environment. The chapter then summarized other local ATM-capable systems, such as routers, bridges, hubs, multiplexers, bridges, and CSU/DSUs. We provided a representative, but not exhaustive, list of manufacturers and their products in each of these areas. The chapter then covered hardware and software support for ATM in the PC and workstation, focusing on the ATM Network Interface Card (NIC). The last category of ATM devices discussed were the ATM chip sets, which are essential in ushering in an era of cost effective ATM.

16.7 REFERENCES

[1]   Y. Yeh, M. Hluchyj, A. Acampora, "The Knockout Switch: A Simple, Modular Architecture for High-Performance Packet Switching," *IEEE Journal on Selected Areas in Communication,* October 1987.

[2]   P. Newman, "ATM Technology for Corporate Networks," *IEEE Communications Magazine*, April 1992.

[3]   R. Awdeh, H. Mouftah, "Survey of ATM switch architectures," *Computer Networks and ISDN Systems*, Number 27, 1995.

[4]   J. Turner, "Design of an Integrated Services Packet Network," *IEEE Transactions on Communications*, November 1986.

[5]   D. McDysan, D. Spohn, *Hands On ATM*, McGraw-Hill, 1997.

[6]   R. Bellman, "An enterprise view of ATM," *Business Communications Review,* April 1995.

[7]   K. Cholewka, , "Affordable ATM," *Data Communications*, December 96

[8]   H. E. Meleis, D. N. Serpanos, "Designing Communication Subsystems for High-Speed Networks," *IEEE Network*, July 1992

[9]   C. B. S. Traw, J. M. Smith, "Hardware/Software Organization of a High-Performance ATM Host Interface," *IEEE JSAC*, February 1993.

[10]   ATM Forum, *Native ATM Services: Semantic Description, Version 1.0*, af-saa-0048-000, February 1996.

[11]   J. Chiong, *Internetwokring ATM for the Internet and Enterprise Networks*, McGraw-Hill, 1997.

[12]   R. Jeffries, "Three roads to quality of service: ATM, RSVP, and CIF," Telecommunications, April 1996.

[13]   R. Borden, "FORE Introduces ForeThought™ ATM APIs for Winsock 2.0 and XTI," http://www.fore.com/press/archive/1996/PR610_8.html.

[14]   E. Roberts, "The ATM-ready desktop is just an API away," Communications Week, Issue: 555, May 1995.

[15]   Microsoft, *NDIS 5.0 Extensions and ATM Support in Windows*, http://www.microsoft.com/hwdev/devdes/ndis4_atm.htm., January 16, 1998. http://www.telecoms-mag.com/marketing/articles/aug95/langlois.html.

[16]   ATM Forum, *Utopia Level 1 Specification, v2.01*, af-phy-0017.000, March 1994.

[17]   ATM Forum, *Utopia Level 2 Specification, v1.0*, af-phy-0039.000, June 1995.

ATM Interworking with Voice and Video

This chapter covers the protocols supporting voice and video applications over ATM. This is probably one of the most critical areas to the success of ATM. The text describes the carriage of voice and video over ATM using the set of protocols illustrated in Figure 17.1.

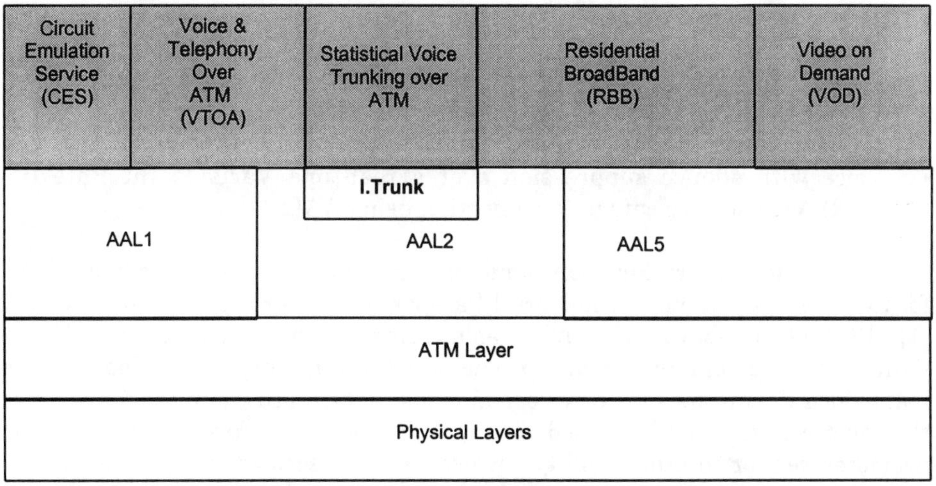

Circuit Emulation Service (CES)	Voice & Telephony Over ATM (VTOA)	Statistical Voice Trunking over ATM	Residential BroadBand (RBB)	Video on Demand (VOD)
AAL1		I.Trunk		
		AAL2	AAL5	
ATM Layer				
Physical Layers				

Figure 17.1   ATM Protocol Support for Voice and Video

The story begins with an overview of voiceband performance and the tradeoffs involved with ATM. Next, the text summarizes the key characteristics of emulating synchronous TDM circuits over an asynchronous ATM network. Next, we review how B-ISDN's predecessor, Narrowband-ISDN (N-ISDN), interworks with ATM-based networks. The chapter then summarizes the important standards created by the ATM Forum's Voice and Telephony Over ATM (VTOA) initiative. We then move on to explore the status of ATM support for video service, summarizing the state of digital video coding, along with the ATM Forum's Residential Broadband (RBB) effort and its Video On Demand (VOD) specification operating over AAL5.

17.1 VOICE OVER ATM

One of the primary benefits of end-to-end ATM is the capability to integrate voice and data traffic. This section reviews integration design alternatives, first looking at the endpoint aggregation methods, and then at ATM service category selection.

17.1.1 ATM Service Categories and AALs Supporting Voice

Voiceband traffic fits into the following categories:

- Human speech
- Fax data
- Modem data
- Recorded audio (e.g., music, sounds from nature, etc.)

The standards-based options today for transporting voiceband traffic over ATM are:

- Circuit emulation over a Constant Bit Rate (CBR) ATM service category connection using AAL1
- Voice with silence suppression over a real-time Variable Bit Rate (rt-VBR) ATM service category connection using AAL2

Some vendor proprietary solutions also exist to transport voice over ATM, for example, as is done in the Nortel Passport multi-service enterprise switch [1]. Each of the above techniques has advantages and disadvantages. While CBR guarantees bandwidth and provides the best quality, it does not free up bandwidth during periods of voiceband inactivity. CBR support for voice is the most common method used today, since the ATM Forum's specification promotes vendor interoperability. However, as described in Chapter 13, the recent adoption of the AAL2 standard by the ITU-T will likely change the situation. In fact, one vendor, General Datacomm, already supports 32-kbps ADPCM voice with silence suppression using the AAL2 standard [2]. Use of the rt-VBR service category offers a more cost effective alternative for human voice, since the ATM network doesn't use any bandwidth during periods of silence. The unused bandwidth during these inactive intervals is then available to other, lower priority ATM service categories.

17.1.1.1 Constant Bit Rate (CBR) Service Category

Table 17.1 shows five standard methods for transmitting voice as CBR service category traffic [3]. These methods use circuit emulation operating over AAL1. Each of these methods operates over either PVCs or SVCs.

Table 17.1   Voice as CBR Service Category Traffic

Voice Format	Standard
64 kbps pulse code modulation (PCM)	ITU-T G.711
64 kbps adaptive differential pulse code modulation (ADPCM)	ITU-T G.722/G.725
16/24/32/40 kbps ADPCM	ITU-T G.726
9.6/12.8/16 kbps linear prediction	ITU-T G.728
1-8 kbps voice for 20 ms packets	IS-54/95 wireless voice

Note that 64-kbps ADPCM supports signals with frequency pass bands up to 7-kHz, whereas 64-kbps PCM only supports a 3.1-kHz frequency pass band. The difference is heard in the upper harmonics of a person's voice — 3.1-kHz will sound flatter and less recognizable than 7-kHz, especially for female voices, which are usually higher in pitch. Support for standard PCM was the first effort of the ATM Forum's Voice and Telephony Over ATM (VTOA) group. Fax and modem modulation are best supported by the CBR service category, or by transcoding to the native bit rate using methods employed by Digital Circuit Multiplication Equipment (DCME).

17.1.1.2 Real-time Variable Bit Rate (rt-VBR) Service Category

Voice support using the real-time VBR (rt-VBR) ATM service category improves efficiency by not transmitting cells during periods of silence [3, 4]. The protocol converting voice to ATM cells listens for silent periods during a call and doesn't generate any cells during these intervals. An integrated voice/data multiplexer sends lower priority data traffic during these silent intervals. Typically, the gaps in normal human conversation last several seconds. Telephone engineers have a great deal of experience on how to make silence suppression yield high fidelity speech transmission from work on undersea cable and satellite communication systems. These gaps in human speech constitute over 60 percent of the average conversation, allowing for approximately a 50 percent bandwidth savings when using rt-VBR compared with CBR. Indeed, the draft ITU-T Service Specific Convergence Sublayer (SSCS) for trunking voice over ATM, called I.Trunk [5], uses exactly this method to suppress silent periods in 64-kbps PCM streams.

17.1.1.3 Comparison of CBR and rt-VBR Transport of Voice over ATM

Table 17.2 compares the advantages and disadvantages of placing voice traffic over a CBR VCC versus a rt-VBR VCC. With the statistical multiplexing inherent in rt-VBR, there is a possibility of loss. As a rule of thumb, between 20 to 50 voice connections can be multiplexed onto a single ATM port of DS1 speed or higher with approximately a 50 percent reduction in required

VCC bandwidth (i.e., SCR) and suffer acceptable loss. Chapter 26 provides more details on the statistical multiplexing of voice over ATM. Loss of voice signal due to cell loss may sound like a click, or, occasionally may render a single syllable (or even an entire word) unrecognizable. In summary, rt-VBR is more efficient than CBR, but at the expense of a small degradation in quality.

Table 17.2   Comparison of CBR to rt-VBR Voice Traffic

Attribute	CBR	rt-VBR
Cost of VC	Highest	Lower
Bandwidth Usage	Less efficient	More efficient
Voice Quality	Effectively lossless	Possibility of loss

17.1.2 Voice over ATM Delay Impact on Echo Control

The ITU-T looked at the basic tradeoff between efficiency and packetization delay versus cell size, eventually deciding on the 53-byte cell in 1990 as described in Chapter 11. This section describes the scenario where echo occurs when a voice-over-ATM-capable workstation interworks with a legacy voice network as illustrated in Figure 17.2. Recall that the packetization delay is the amount of time required to fill the ATM cell payload at the voice encoding rate. When using AAL1 and the standard 64-kbps rate for Pulse Code Modulation (PCM), the resulting packetization delay is approximately 6 ms as indicated in the upper left-hand side of the figure. The ATM network transports the cells containing the voice over AAL1 to a N-ISDN interworking function, which inserts a delay of between 1 and 2 ms to account for variation in the arrival time of the ATM cells. For connection to a standard analog telephone line, most telephone networks pass the signals on the two pairs of wires (i.e., 4 wires) derived from digital telephony into an analog 4-wire to 2-wire directional transformer for transmission over the twisted pair plant to the end telephone subscriber.

If any impedance mismatch exists between the 4 wires on the analog side of the N-ISDN interworking unit and the 2 wires leading to the subscriber's legacy telephone, then an echo of the talker's signal at a reduced level returns. The large arrow in the figure indicates the echo return path. Called *talker echo* [6], this phenomenon is quite annoying for the speaker to hear a delayed version of their own voice if the round-trip delay exceeds approximately 50 ms [7]. Since the same 6 ms packetization delay occurs when converting the narrowband ISDN signal to ATM, along with the 1 to 2 ms playback buffer in the Voice over ATM capable workstation, the round-trip delay in this configuration ranges between 14 to 16 ms. Therefore, only in situations where the echo return loss is quite low (i.e., the talker echo is quite loud) is echo cancellation required.

Therefore, ATM devices supporting a single voice channel over ATM interworking with telephone networks should employ echo cancellation. This costs more than classical telephony where the propagation delay dominates the overall equation, meaning that echo cancellation is not required for calls between points less than 600 miles apart. On the other hand, echo cancelers on a chip help keep costs down. Note that voice over IP and frame relay networking have a similar problem with echo control. One way of avoiding this problem would be to partially fill ATM cells using AAL1 to reduce the packetization delay to a small value. The disadvantage to this approach is reduced efficiency due to partial fill of the ATM cell payload.

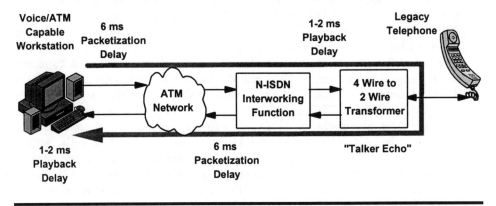

Figure 17.2   Source of Voiceband Echo when Interworking ATM and Legacy Telephony

17.1.3 Applications of Voice Over ATM

Multimedia workstations offer the perfect platform on which to combine voice communications and visual data, such as desktop video-conferencing. Integration of the phone into the desktop workstation and thus LAN interface is one of the benefits of using ATM to the desktop. The next most common method is to provide a direct connection of a telephone, fax machine, or modem to an ATM network switch. Another method of placing voice over ATM is through a direct connection of a private branch exchange (PBX) voice switch to an ATM network switch. The main drivers for placing voice over ATM in a corporate environment include:

- Strong requirement for desktop multimedia
- Volume of voice traffic is lower than data traffic
- Greater volume of voice traffic is intranet, versus extranet to the public switched voice network

17.2 CIRCUIT EMULATION SERVICES (CES)

The ATM Forum specification for Circuit Emulation Service (CES) [8, 9] defines the means for ATM-based networks to employ AAL1 to emulate, or simulate, synchronous TDM circuits over the asynchronous infrastructure of ATM networks. CES defines support for two types of emulated circuits:

- Structured DS1/E1 supporting Nx64-kbps (i.e., fractional DS1/E1)
- Unstructured DS1/E1 (1.544/2.048-Mbps)

17.2.1 Generic Reference Model

Figure 17.3 illustrates the generic CES reference model. On each end, TDM equipment such as a PBX, T1 multiplexer, or a CSU/DSU (not shown) connects to the CES Interworking Function (IWF) via either a standard TDM-formatted DS1 four-wire or E1 coaxial connector and protocols. The IWF implements the AAL1 SAR and CS sublayer functions defined in Chapter 13. The CES specification also defines a logical circuit emulation capability which has no physical DS1 or E1 interfaces. This technique enables an efficient electronic interface to a TDM cross-connect function inside a piece of equipment. CES mandates use of the Constant Bit Rate (CBR) ATM service category and associated quality of service for the Virtual Channel Connection (VCC) that interconnects CES IWFs. This choice means that the playback buffers in the IWF underflow or overflow infrequently. The specification also details handling for error cases, such as lost cells, buffer underflow and overflow, as well as TDM alarm processing and performance monitoring. CES also defines the necessary parameters to set up ATM VCC SVCs to support circuit emulation. The ATM Forum's specification also defines a MIB supporting circuit emulation.

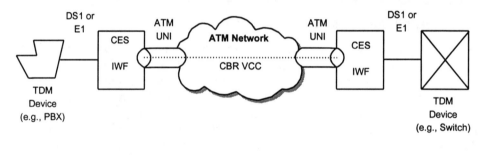

Figure 17.3   Circuit Emulation Service (CES) Reference Model

17.2.2 Structured Mode Circuit Emulation

Figure 17.4 illustrates an example application and key functions of the structured CES IWF, which always provides timing to the connected TDM equipment. As described in Chapter 13, an accurate clock must be present at each interworking function interface for proper structured mode CES operation. The structured capability supports combinations of Nx64-kbps bearer channels, where N ranges from 1 to 24 for DS1s and from 1 to 31 for an E1. Optionally, several emulated groups of Nx64-kbps circuits may occupy the same DS1, E1, or logical interface as illustrated in the figure by the two instances of AAL1, one for each CBR VCC in the example. In this example, the CES function performs a mapping of the 64-kbps time slots from the DS1/E1 TDM transmission pipe into separate AAL1 Segmentation And Reassembly (SAR) functions. Structured CES also specifies support for *Channel Associated Signaling* (CAS), commonly used by PBXs to indicate off-hook and on-hook conditions. Since the structured mapping of the individual time slots does not convey TDM framing information end-to-end, the CAS information is encoded and transported separately, requiring additional overhead.

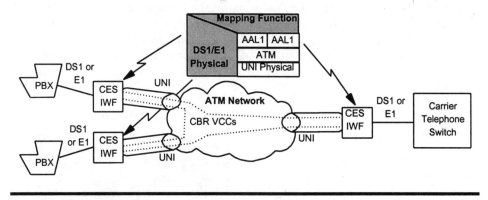

Figure 17.4   Structured Mode CES Interworking Example and Functions

17.2.3 Unstructured Mode Circuit Emulation

Figure 17.5 illustrates an example application and key functions of the unstructured mode CES InterWorking Function (IWF). It provides timing to the TDM equipment in a synchronous mode, or accepts timing in an asynchronous mode as described in Chapter 13. Asynchronous timing uses either the Synchronous Residual Time Stamp (SRTS) method, or adaptive clock recovery to transfer timing across the ATM network. Timing transfer is critical for many legacy TDM networks, specifically T1 multiplexer networks. The unstructured service provides a clear channel pipe at a bit rate of 1.544-Mbps for DS1 or 2.048-Mbps for an E1. This means that the CES IWF supports bit streams with non-standard framing, such as that used by some video codecs and encryptors, in addition to standard, framed DS1 and E1 signals used by multiplexers and PBXs. One disadvantage of devices that don't use standard framing is that the CES IWF cannot support standard DS1 and E1 performance monitoring or alarm handling.

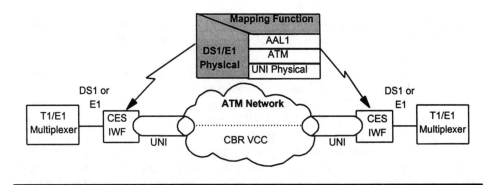

Figure 17.5   Unstructured Mode CES Interworking Function

17.2.4 Dynamic Bandwidth Circuit Emulation Service (DBCES)

The ATM Forum also specified a dynamic bandwidth allocation of circuit emulation service (DBCES) [10]. Although DBCES operates over AAL1 and the CBR ATM service category, it uses a bit mask to indicate when certain 64-kbps timeslots in a structured CES connection are active. Figure 17.6 depicts the reference configuration for an ATM device capable of supporting DBCES.

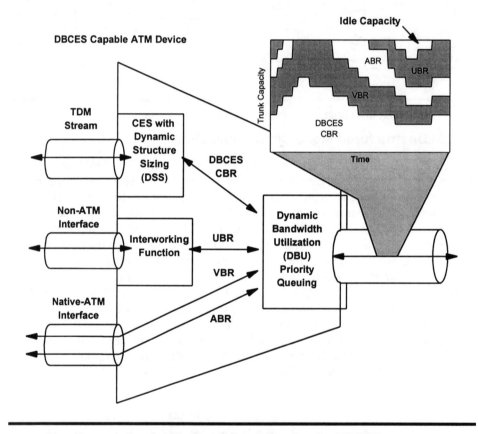

Figure 17.6 Dynamic Bandwidth Circuit Emulation Service (DBCES)

The TDM voice over ATM interface provides structured CES over AAL1 support with or without Channel Associated Signaling (CAS) or Common Channel Signaling (CCS) as described above. The DBCES ATM device may detect voice activity using CAS or CCS on a call-by-call basis, or silence detection directly from the digitized voice samples. The ATM Forum specification does not specify the precise time slot activity detection mechanism, although an informative Annex describes the means to use standard idle codes, repeated bit patterns, or CAS/CCS signaling to do so. The DBCES ATM device utilizes the Dynamic Structure Sizing (DSS) bit mask to indicate whenever a change occurs in the active DS0 time slots in the structured mode CES bundle. The specification also covers the case where all time slots are inactive. Upon reception, the DBCES device maps the received samples for the active time slots onto the TDM interface and inserts a pre-configured idle pattern into the inactive time slots. The ATM VCC carrying the DBCES traffic may be either a PVC, or an SVC. The specification defines the information elements necessary to dynamically establish a SVC between DBCES capable devices.

Another key capability required of the DBCES device is that of Dynamic Bandwidth Utilization (DBU) illustrated in Figure 17.6. Although the allocated CBR bandwidth must still support the largest number of active 64-kbps timeslots, the DSS bit mask makes the unused bandwidth available to other lower priority ATM connections over an ATM network through the DBU capability. For example, as shown in the figure, the DBU capability in the DBCES device makes the unused CBR bandwidth dynamically available to lower priority VBR, UBR and ABR ATM connections using priority queuing. The plot in the upper right-hand corner of the figure illustrates the result of the priority queuing action making the bandwidth unused by DBCES CBR dynamically available to other ATM connections with lower priority. Note that during some time intervals the demand generated by all connections may be less than the ATM trunk capacity. Chapter 25 describes the operation of priority queuing and the performance implications in more detail. Systems using AAL2 will achieve many of the same benefits of DBCES type techniques to suppress silent periods in normal voice conversations. Chapter 26 analyzes the statistical multiplex gain for systems implementing this type of capability.

17.3 INTERWORKING WITH NARROWBAND ISDN (N-ISDN)

ATM's ancestor in the ITU-T is the Narrowband Integrated Services Digital Network (N-ISDN). Hence, a key focus of this international standards group is to provide seamless interworking between the future Broadband ISDN (B-ISDN) and the legacy N-ISDN. Good business reasons drive these needs; for example, many networks want new B-ISDN users to be able to interface to many existing users of the older N-ISDN service. Indeed, many manufacturers of existing switches owned by carriers provide an upgrade path to the new ATM-based B-ISDN technology for this very reason. While Recommendation Q.2931 specifies the signaling protocol required for interworking between B-ISDN and N-ISDN networks, Recommendation I.580 gives the overall view of interworking between these generations of integrated networking technologies.

Figure 17.7 depicts the scope of ITU-T Recommendation I.580 [11] in terms of three interworking scenarios:

I. Covers interworking of 64-kbps-based N-ISDN user circuit, frame, and packet mode services with a B-ISDN user

II. Covers the case where 64-kbps-based N-ISDN user's circuit, frame, and packet mode services transparently trunk over a B-ISDN to another N-ISDN user

III. Covers the case where only 64-kbps-based ISDN circuit, frame, and packet mode services are provided between B-ISDN users

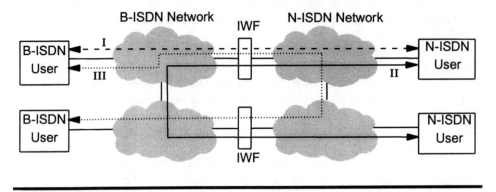

Figure 17.7   Scope of Recommendation I.580 B-ISDN/N-ISDN Interworking Standard

Recommendation I.580 covers an number of potential interworking scenarios and requirements that define mappings between N-ISDN and B-ISDN services as either one-to-one, or a many-to-one mapping suitable for the trunking of scenario II above. It defines the use of AAL1 for the circuit mode bearer service in the InterWorking Function (IWF), an unspecified AAL for frame mode and FR-SSCS, plus AAL5 as one option for packet mode in the IWF as illustrated in Figure 17.8.

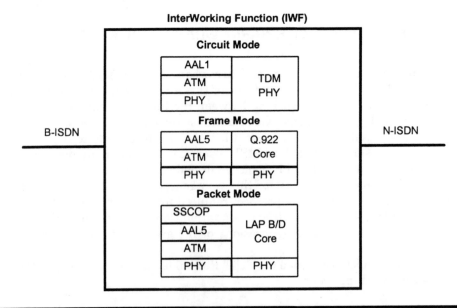

Figure 17.8   Example of I.580 InterWorking Function (IWF) Protocol Mappings

As described in Chapter 7, Q.922 is the ITU-T Recommendation that specifies the frame relay protocol. Link Access Procedure B-channel or D-Channel (LAP B/D) is the link layer protocol of the X.25 packet standard. An IWF may implement one or more of these protocol mapping functions between B-ISDN and N-ISDN. The standard presents several options for performing some of these functions. Therefore, the standards are incomplete at this point, and additional specifications must first be defined in order to achieve full interoperability.

17.4 VOICE AND TELEPHONY OVER ATM (VTOA)

Key to realization of one of ATM's goals is standards support for voice over ATM. Major business drivers and requirements for transporting voice over ATM include [12]:

- Access and/or wide area network transmission and switching consolidation
- Preserve investments in existing telephone and PBX equipment
- Maintain traditional levels of quality and availability in voice services
- Maintain existing line features, such as Calling Line Identification
- Coexistence of desktop voice over ATM with LAN applications

Specifications produced by the ATM Forum's Voice and Telephony over ATM (called VTOA) group include:

- Trunking for Narrowband Services over ATM, also called *land line trunking* [13]
- Voice Services at a Native ATM Terminal, also called *ATM to the desktop* [14]

17.4.1 Trunking for Narrowband Services over ATM

Figure 17.9 illustrates the reference configuration for ATM trunking for Narrowband services. The roman numerals in the figure illustrate the following narrowband network trunking via ATM network(s):

I.   Private to private Narrowband trunked via private ATM
II. Private to private Narrowband trunked via public ATM
III. Public to public Narrowband trunked via public ATM
IV. Private to public Narrowband trunked via private and public ATM
V.   Private to private Narrowband trunked via private and public ATM
VI. Public to private Narrowband trunked via public ATM

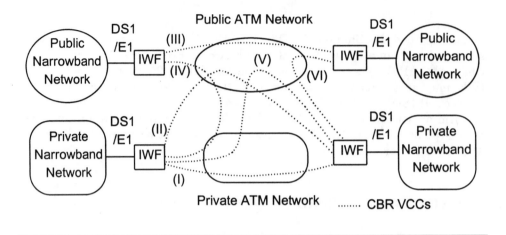

Figure 17.9   ATM Trunking for Narrowband Services Reference Configuration

Note that all ATM connections are between InterWorking Functions (IWFs), never directly to an end user. This means that legacy TDM systems can utilize these capabilities without change or additional investment, an important migration consideration. The IWF provides a DS1/E1 physical circuit with N-ISDN signaling.

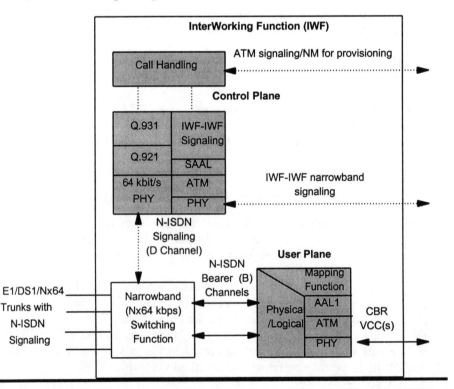

Figure 17.10   ATM Trunking for Narrowband Services InterWorking Function (IWF)

Figure 17.10 illustrates the functions within the ATM Trunking for Narrowband Services InterWorking Function (IWF). Note the similarity of interworking for Nx64-kbps circuit emulation described earlier in this chapter to the functional block in the lower right-hand corner. The lower left-hand functional block switches Nx64-kbps connections between DS1 or E1 interfaces — a function commonly performed by telephone switches and N-ISDN PBXs. The majority of the new function and complexity in this specification is covered in the shaded areas, which define interworking between N-ISDN and B-ISDN signaling and the conversion between TDM circuits and ATM's AAL1. The specification also defines a Management Information Based (MIB) for managing the IWF.

17.4.2 Voice over ATM to the Desktop

Figure 17.11 illustrates the ATM to the B-ISDN desktop interworking function (IWF) reference configuration utilized by the ATM Forum [14]. The interworking configurations are derived from scenario B, case 1 of Annex A of ITU-T Recommendation I.580. Basically, the specification defines how a voice-enabled ATM desktop device accesses either a public or private ISDN network. Connections between B-ISDN and N-ISDN private networks may be either via PNNI or an ATM UNI, as illustrated in cases a and b in the figure. The third scenario represented in Figure 17.11c utilizes different signaling protocols optimized for private networking as detailed below.

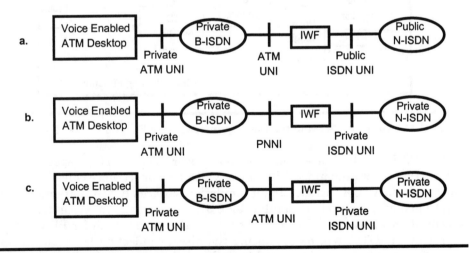

Figure 17.11   Voice over ATM to the Desktop Interworking Reference Configurations

Figure 17.12 illustrates the details of the VTOA ATM to the desktop protocol specification. On the right-hand side, the Narrowband ISDN (N-ISDN) physical interface connects to the IWF — a Primary Rate Interface (PRI) at either the DS1 or E1 rate in this example. A simple TDM multiplexer separates the Bearer (B) channels from the signaling Data (D) channel and provides these streams to user and control plane protocol stacks and mapping functions as illustrated in the left-hand lower part of the figure.

The user plane protocol and mapping function is similar to that used for structured mode circuit emulation described earlier in this chapter. The protocols in the control plane depend upon the reference configuration that applies as illustrated in the table in the upper right-hand corner of Figure 17.12. Digital Subscriber Signaling System number 2 (DSS2) is specified by the ITU-T Q.2931 standard. The acronym PSS1 stands for Private integrated services Signaling System number 1. A separate signaling VCC and a CBR VCC for the voice connects the IWF with the private ATM network via either the UNI or PNNI protocol stack as shown on the left-hand side of the figure.

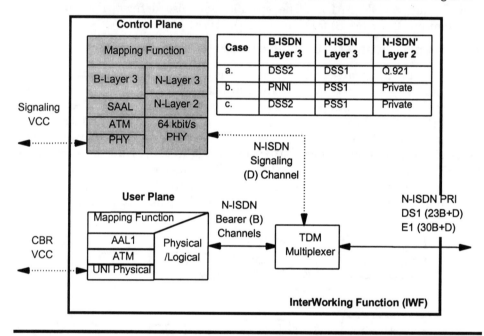

Figure 17.12   Voice over ATM to the Desktop Interworking Function (IWF) Detail

The ATM Forum Voice and Telephony over ATM (VTOA) specification defines the use of predefined PVCs over which to connect voice calls. The two efforts within VTOA include circuit emulation services (CES) that emulate a point-to-point private line similar to CBR service, and a voice trunking specification where the ATM network acts as the voice switch for PBXs. The real gain for VTOA will be enhancing the direct connection between a PBX and ATM switch — providing call-by-call switching — establishing an SVC for each voice call. This group is also working on a silence suppression specification.

Future voice over ATM applications will ride SVCs instead of PVCs [15]. SVCs free up unused bandwidth when not in use that PVCs reserve 100 percent of the time. Once service providers offer SVCs for CBR and rt-VBR service categories, we expect to see more voice and circuit emulation over ATM demand. SVCs have been a challenge for service providers, as they are hard to administer and manage, require significant switch processing time and signaling overhead, along with complex pricing and billing schemes. But clearly SVCs are the way to go for voice over ATM. One benefit of SVCs is freeing up capacity for other traffic types as described in section 17.2.4.

17.5 RESIDENTIAL BROADBAND (RBB)

The ATM Forum established the Residential Broadband (RBB) working group in April, 1995 to define a complete end-to-end, full service ATM capability both to the home, as well as within the home [16, 17]. The RBB group maintains a close liaison with other standards bodies working on various aspects of residential broadband networks. The scope of RBB includes the Video on Demand capability described in the next section. This group envisions a Home ATM network (HAN), called a "toaster net" by some, that interconnects a wide range of household devices, such as PCs, televisions, appliances, security systems and, of course, ATM capable toasters.

The large number of choices for physical UNI interfaces impedes interoperability in the cost-conscious residential marketplace. Although carriers may deploy different access UNI technologies based upon regulatory, demographic or business considerations, the group targets a technology independent interface usable in any home ATM network, possibly through use of an appropriate adapter.

Figure 17.13 illustrates the RBB home ATM network architecture. Starting from the right-hand side of the figure, a backbone (or core) ATM network interconnects a number of distribution and access subnetworks, as well as provides access to ATM-based service providers, such as video on demand.

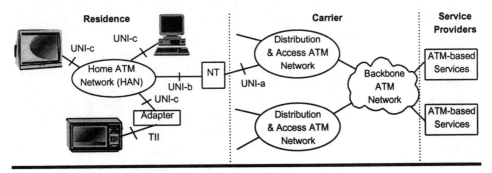

Figure 17.13   Residential Broadband Home Area Network Architecture

The access ATM network uses a particular UNI protocol, labeled UNI-a in the figure, to extend ATM to the residence. ATM over ADSL as detailed in Chapter 32 would be an example of the UNI-a protocol. Often a carrier deploys a Network Termination (NT) device in the residence, potentially providing a different UNI interface to the home ATM network, labeled UNI-b in the figure. For example, the NT may provide a 155-Mbps multi-mode fiber optic ATM interface. The home ATM network operates on another UNI protocol, shown as UNI-c in the figure. For example, the HAN may use 25-Mbps over twisted pair as UNI-c to minimize cost. The RBB group is defining the Technology Independent Interface (TII) so that manufacturers can develop a single ATM interface for their products, a microwave oven (capable of browning toast, of course) in this example. An adapter that converts between UNI-c and the TII is, of course, required.

The end-to-end ATM architecture envisioned for the RBB application has several advantages [17]. First, none of the competing technologies provide the guaranteed performance required for integrated voice and video applications. Secondly, ATM provides a universal infrastructure and a single platform for all applications. Furthermore, the rich signaling capabilities of ATM SVCs support the many different connections types required by a wide range of applications. Additionally, ATM efficiently multiplexes multiple traffic types onto a single physical media, which is a key requirement for ATM over ADSL using existing twisted pair cabling to residences. Finally, ATM is a scalable, manageable solution that supports not only residential services, but meets the needs of large corporations, government agencies, and meets the requirements of a carrier infrastructure as well.

17.6 AUDIO/VISUAL MULTIMEDIA SERVICES (AMS)

This section covers Audio/Visual Multimedia Service (AMS) market demand, an overview of video coding standards, and the ATM Forum Video On Demand (VOD) specification.

17.6.1 Video over ATM Market Demand

A study conducted by IDC Consulting on behalf of the ATM Forum analyzed residential broadband market demand in 1996 [18]. Figure 17.14 shows the breakdown of market opportunity for Residential Broadband (RBB) by application from the survey. Entertainment was the largest segment, which prompted the marketing driven ATM Forum SAA committee to produce the Video On Demand (VOD) specification first.

This report predicted that over the next fifteen years nearly 60 million households are likely to adopt ATM-enabled applications. Furthermore, the strongest demand for ATM-enabled applications will emerge from families with children 10 to 18 years old, home office workers, telecommuters, and World Wide Web users: a population of over 19 million households in the United States alone!

The most promising opportunity to entice leading edge users to invest in ATM hardware, software, and services involves offering entertainment-on-demand and home office applications. Applications such as home shopping and distance learning are also attractive, but appeal to a smaller subset of the residential population.

The survey identified several barriers to adoption, including: price, fear of technology, fear of lost control, security, and privacy. These barriers, along with resolving the various issues that vendors have regarding infrastructure, standards, and economic viability, must be overcome before ATM to the home becomes a reality. To create demand for ATM-enabled applications RBB must deliver the following features:

- Graphical User Interface (GUI)/ interactive controller
- Clear consistent audio/video quality
- Voice-activated control of services
- Integrity of file transfers
- Full motion video
- High-speed connectivity to various data sources
- Intelligent agents

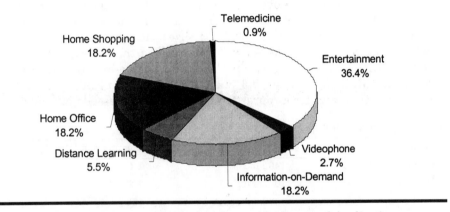

Figure 17.14 RBB Adoption Rate by Early Adopters for Surveyed Applications

The industry is working on standards and implementation agreements to realize these capabilities in response to market demand. For example, one major focus of the current work is achieving efficient coding of full motion video.

17.6.2 Video over ATM Coding Standards

One of ATM's tenets is support of simultaneous data, voice, and video transmission. While the predominant standard for video-conferencing today is TDM based, ATM's higher bandwidth and controlled latency and jitter enable many video applications, including the Motion Photographic Experts Group (MPEG) and MPEG2 video standards.

High-end video applications require accurate bandwidth and delay control, which includes both end-to-end latency and jitter control. Therefore, precise quantification of the tolerable amount of latency and jitter is a key ATM switch and network design consideration. Achievable end-to-end requirements for Video over ATM are:

- Cell loss ratio across the network on the order of one in a billion
- Cross network cell transfer delay on the order of the speed of light propagation delay

- Cell delay variation on the order of a few milliseconds

These are key attributes to look for when selecting an ATM switch or service provider for video applications. Most modern switches meet these requirements; however, some older switches do not. Network designers must pay close attention to the aggregation of these impairments in larger networks to achieve these objectives.

MPEG and MPEG2 coding removes jitter introduced by ATM networks through the use of internal buffers in the coder/decoders (commonly called codecs). Additionally, the clock signal embedded in MPEG/MPEG2 streams enables recovery of timing through the use of a phase-locked loop at the decoder, even in the presence of jitter in the ATM cell interarrival times. MPEG protection and recovery techniques include structured packing of encoded video into macroblocks, which enables rapid re-synchronization by discarding all data until the decoder recognizes the next macroblock.

A principal reason that the ATM Forum created the real-time Variable Bit Rate (rt-VBR) service category is to enable variable-bit-rate video such as MPEG and MPEG2. Table 17.3 summarizes the bandwidth requirements and compression ratio for a representative set of compressed video encodings.

Table 17.3   Bandwidth Requirements for Compressed Video Images

Standard/Format	Bandwidth	Compression Ratio [1]
Uncompressed NTSC	140 Mbps	1:1
Motion JPEG	10-20 Mbps	7-27:1
MPEG-1	1.2-2.0 Mbps[2]	100:1
H.261	64 kbps-2 Mbps	24:1
DVI	1.2-1.5 Mbps	160:1
CDI	1.2-1.5 Mbps	100:1
MPEG-2	4-60 Mbps[3]	30-100:1
CCIR 723	32-45 Mbps	3-5:1
U.S. commercial systems	45 Mbps	3-5:1
Vendor methods	0.1-1.5 Mbps	100:1
Software compression	1-2 Mbps	6:1

[1] Compared with broadcast quality video
[2] Baseline standard; other compression ratios feasible

Although image quality improves only slightly at transmission rates above 8- to 10-Mbps, this is crucial for applications like production video editing and broadcast quality program distribution.

17.6.3 ATM Forum Video On Demand (VOD) Specification

The ATM Forum's Audiovisual Multimedia Services Video on Demand (VOD) version 1.0 [19] specification defines the carriage of audio, video, and data over ATM. Figure 17.15 illustrates the reference configuration of a client (e.g., a set-top box), a server (e.g., a video content provider), an ATM network, and the logical services in control of its connections and the VOD session. The ATM network may be hybrid fiber/coax (HFC), ADSL, xDSL, cable modems, or a digital fiber network. Users would make an ATM connection to view a particular video in a session. Users have VOD session level controls such as "pause," "fast-forward," and "rewind:" similar to that available on Video Cassette Recorders (VCRs) today. In order for users to perceive their actions as interactive (i.e., similar to pressing buttons on their VCR), the network response time should be on the order of 100 ms [17]. This functional requirement places a stringent bound on ATM SVC call setup delay in the ATM Connection Control component and the VOD Session control protocol.

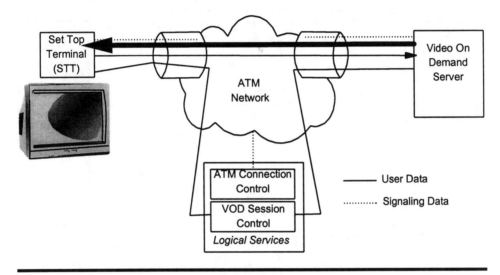

Figure 17.15 Video On Demand Reference Configuration

The VOD specification defines how connection control, video, audio, data and user control streams are encoded and multiplexed over up to three ATM VCCs using AAL5 as illustrated in Figure 17.16. ITU-T Recommendation H.222.1 [20] defines the means for multiplexing the combined audio, video, private data, and user-to-user control over a single transport stream. ITU-T Recommendation H.222 [21] defines the characteristics of the overall multimedia transport stream. The specification defines standard user signaling for connection control based upon the UNI 4.0 signaling specification as shown in the left-hand side of the figure. Specific audio, video and private data coding is outside the scope of the ATM Forum's VOD specification. Example coding standards mentioned in the specification are H.262 [22], H.245 [23], or MPEG [24]. See Reference 25 for an in-depth discussion of supporting MPEG2 over the VOD architecture. This analysis states that the underlying ATM network must provide a bit error rate on the order of 10^{-10}. Furthermore, the impact of jitter on MPEG2 clock recovery will be a significant design challenge. The user-to-user and session control could be a web browser operating over HTTP and TCP/IP, for example. The application interface to the native ATM signaling stack and AAL5 capabilities (indicated in the shaded portion of the figure) conform to the ATM Forum's Native ATM Services (NAS) API model defined in Chapter 16 as indicated by the solid line in the figure.

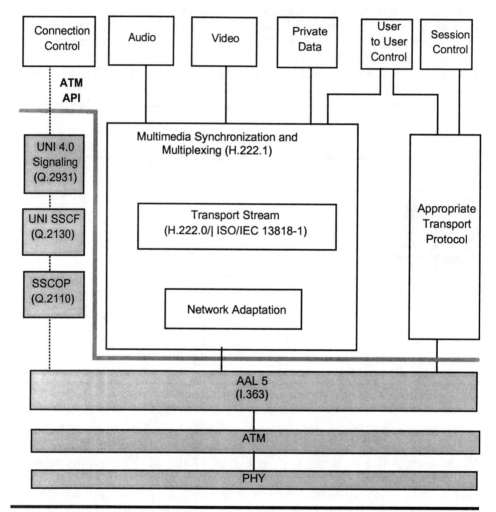

Figure 17.16 Video on Demand (VoD) Protocol Reference Model

MPEG2 uses a Single Program Transport Stream (SPTS) over a single VCC using AAL5 segmentation and reassembly as shown in Figure 17.17. This mapping employs a null SSCS sublayer. The information (movies, commercials, etc.) stored in the MPEG2 SPTS format results in packets that are 188 octets long as indicated in the figure. Note that since the video and audio information are compressed and formatted as an MPEG2 SPTS, the server doesn't require an encoder or multiplexer. The mandatory support of N=2 SPTS packets per AAL5 CPCS-SDU results in a 100 percent efficient packing of the MPEG stream into the payload of 8 ATM cells as shown in the figure.

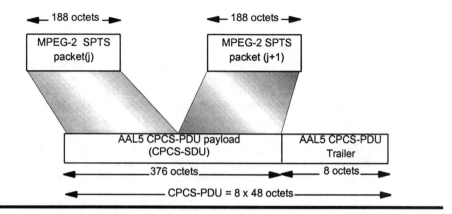

Figure 17.17 Mandatory MPEG2 SPTS over AAL5 Mapping

17.7 REVIEW

This chapter covered higher-layer ATM protocol support for voice and video. First, the text described the basic drivers and context for voice over ATM. The discussion then covered the Circuit Emulation Service (CES) in support of Nx64-kbps TDM circuits using structured mode and support of standard DS1 and E1 signals in the unstructured mode. We then summarized how the ITU-T focuses on interworking between the next generation Broadband ISDN (B-ISDN) and the legacy Narrowband ISDN (N-ISDN) by summarizing the user and control plane functions. The chapter then moved on to the specific protocols developed in the ATM Forum's Voice & Telephony over ATM (VTOA) working group in support of voice trunking over ATM and ATM to the desktop. The coverage then moved on to the ATM Forum's Residential BroadBand (RBB) effort. The chapter concluded with a survey of the business drivers for video over ATM, focusing on the ATM Forum's Video On Demand (VOD) specification.

17.8 REFERENCES

[1]   Nortel, *Voice Networking over ATM with Magellan Passport — White Paper*, 1996.

[2]   M. McLoughlin, K. Mumford, J. Birch, *APEX Voice Service Module Product Overview*, http://www.gdc.com/.

[3]   D. Wright, "Voice over ATM: An Evaluation of Network Architecture Alternatives," *IEEE Network*, September/October 1996.

[4]   M. McLoughlin, K. Mumford, *Adapting Voice for ATM Networks: A Comparison of AAL1 versus AAL2*, http://www.gdc.com/.

[5]   ITU-T, *Working text for planned Recommendation I.Trunk — AAL2 SSCS for Trunking*, September 1997.

[6]   R. Vickers, H. Shimung, "Voice on ATM – Issues and Potential Solutions," Proceedings of the International Conference on Communication Technology, ICCT '96, Beijing, China.

[7]   ITU-T, Control of Talker Echo, *Recommendation G.131*, August 1996.

[8]   ATM Forum, *Circuit Emulation Interoperability Specification*, af-saa-0032.000, September, 1995

[9]   ATM Forum, *Circuit Emulation Service 2.0*, af-vtoa-0078.000, January, 1997.

[10]   ATM Forum, *Dynamic Bandwith Utilization in 64 Kbps Time Slot Trunking Over ATM - Using CES*, af-vtoa-0085.000, July 1997.

[11]   ITU-T, *General Arrangements for Interworking between B-ISDN and 64 kbit/s Based ISDN*, Recommendation I.580, November 1995.

[12]   G. Onyszchuk, "Will Voice Take Over ATM?," *ATM Forum 53 Bytes*, October 1995.

[13]   ATM Forum, *ATM Trunking Using AAL1 for Narrow Band Services v1.0*, af-vtoa-0089.000, July 1997.

[14]   ATM Forum, *Voice and Telephony Over ATM to the Desktop*, af-vtoa-0083.000, May 1997.

[15]   ATM Forum, "Voice Networking in the WAN," http://www.atmforum.com/atmforum/library/vtoa.html.

[16]   S. Ooi, "The Rhyme and Reason of Residential Broadband," *ATM Forum 53 Bytes*, Volume 4, Issue 1, 1996.

[17]   T. Kwok, " A Vision for Residential Broadband Services: ATM-to-the-Home," *IEEE Network*, September/October 1995.

[18]   ATM Forum, "Families Lead the Way in Demand for ATM Applications," ATM Forum Press Release, September 3, 1996.

[19]   ATM Forum, *Audiovisual Multimedia Services: Video on Demand Specification 1.0*, af-saa-0049.000, December 1995.

[20]   ITU-T, *Multimedia Multiplex and Synchronization for Audiovisual communication in ATM environments*, Recommendation H.222.1.

[21]   ITU-T, ISO/IEC, *Information Technology - Generic Coding of Moving Pictures and Associated Audio - Part 1: Systems*, ISO/IEC IS 13818-1, ITU-T Recommendation H.222.0.

[22]   ISO/IEC, *Information Technology - Generic Coding of Moving Pictures and Associated Audio - Part 2: Video*, ISO/IEC IS 13818-2/ ITU-T Recommendation H.262.

[23]   ITU-T, *Line Transmission of Non-Telephone Signals - Control Protocol For Multimedia Communication*, Recommendation H.245.

[24]   ISO/IEC, *Information Technology - Generic Coding of Moving Pictures and Associated Audio - Part 3: Audio*, IS 13818-3.

[25]   S. Dixit, P. Skelly, "MPEG-2 over ATM for Video Dial Tone Networks: Issues and Strategies, *IEEE Network*, September/October 1995.

18

ATM Interworking with Frame Relay, SMDS, DXI, and FUNI

This chapter covers the application of ATM to interworking with legacy wide area data networking protocols, a critical area to the success of ATM. The text does this by describing ATM's support for WAN data protocols, specifically Frame Relay (FR) and SMDS over ATM as shown in Figure 18.1. The chapter begins by defining the general concepts of interworking, logical access, trunking and physical access. The text then provides specific examples of interworking Frame Relay with ATM, trunking Frame Relay over ATM, and logical access to SMDS via ATM. The chapter concludes with a review of low-speed, frame-based ATM protocols, namely, the ATM Data eXchange Interface (DXI) and the Frame-based UNI (FUNI).

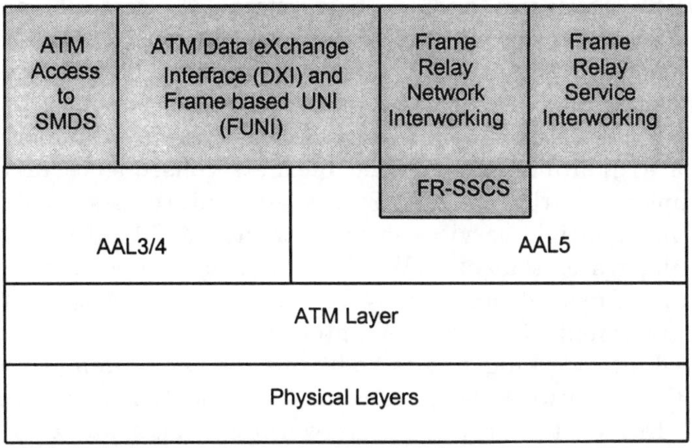

ATM Access to SMDS	ATM Data eXchange Interface (DXI) and Frame based UNI (FUNI)	Frame Relay Network Interworking	Frame Relay Service Interworking
		FR-SSCS	
AAL3/4		AAL5	
ATM Layer			
Physical Layers			

Figure 18.1 ATM Protocol Support for Frame Relay, SMDS, DXI and FUNI

18.1 INTERWORKING, ACCESS, AND TRUNKING

This section looks at how the commonly used data communication protocols interact. Basically, types of interactions fall into three categories: direct interworking, access to one protocol via another, and the carriage (or trunking) of one protocol over another. Figure 18.2 illustrates the relationships defined in standards between the major WAN protocols: X.25, Frame Relay, ATM, SMDS and IP. The notation of A and B defined at the bottom of the figure connected via a particular line or arrow style defines the relationship between protocols A and B. The following narrative speaks to each of these relationships. Figure 18.2a illustrates true protocol interworking with a thick, solid line. Note that standards define true interworking only between X.25 and Frame Relay, and between ATM and Frame Relay. The analogy of protocol interworking with human languages occurs when each has exactly the same semantics, or meaning. Thus, as in languages, true interworking only occurs when protocols have similar semantics. For some protocols, as is the case in some languages, some concepts simply don't translate. X.25/FR and FR/ATM interwork because they are all connection-oriented protocols possessing similar status indication methods, as well as connection establishment and release procedures as studied in Part 2. On the other hand, IP and SMDS are connectionless, and hence cannot directly interwork with a connection-oriented protocol since the basic semantics differ. A similar situation occurs in human languages where different cultural concepts don't translate. Figure 18.2a also illustrates a lesser form of interworking, called logical access, via directed arrows. Many standards define how to access one protocol via another. For example, a user connected to an ATM network can access an IP network. The terminology used in the figure is that the user accesses IP via ATM. Note from the figure that standards exist to access IP via all of these WAN protocols.

Figure 18.2b illustrates another key concept of protocols, namely, that almost every one can be trunked over another. The directed dashed arrows in the figure indicate the cases where one protocol trunks over another. Although ATM trunks any of these protocols; ATM cannot be trunked over these other WAN protocols since they do not currently support quality of service in a standard manner. The business consequence of this fact of protocol trunking is that many carriers and enterprises have announced plans to trunk multiple services over a common ATM infrastructure. The figure also illustrates that each WAN protocol has separate physical access requirements. Physical access may be either dedicated or switched, with different requirements for each WAN protocol.

The limited interworking, access and trunking rules summarized in Figure 18.2 mean that a network designer cannot arbitrarily interconnect protocols over ATM. In fact, the only true interworking design would be to provide access via various protocols to IP and trunk over ATM. Beware of other texts or papers that claim ATM provides the ultimate translation capability between all protocols — it is simply an incorrect statement at this point in time. The Internet Protocol (IP) is the only standard protocol that meets this need today since it can be accessed via FR, X.25, SMDS and ATM, as well as served by dedicated or dial access as shown in Figure 18.2b. The next sections cover the interworking, trunking and access protocols for Frame Relay and SMDS. Chapter 20 covers the subjects of access via ATM to IP and IP trunking over ATM.

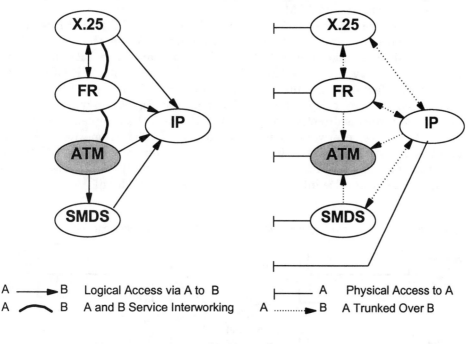

a) Logical Access and Service Interworking   b) Physical Access and Trunking

A ——▶ B   Logical Access via A to B   ⊢——— A   Physical Access to A
A ⌒ B   A and B Service Interworking   A ·······▶ B   A Trunked Over B

Figure 18.2   Logical Access, Interworking, Physical Access, and Trunking

Another commonly used name for the operation of one protocol over another (either in the context of access or trunking as described above) is that of *overlay networking*. This means that one protocol operates more or less transparently over another. For example, a user may access IP over a frame relay network; the IP network may then be trunked over IP to the destination router. The target IP user may then access IP via SMDS. Although a violation of the traditional OSI layered model, the use of such overlay networking is common in real-world networks.

18.2 FRAME RELAY/ATM INTERWORKING

The frame relay service market in the United States will be over five billion dollars in 1999. However, many frame relay users are moving to higher-speed connections and applications that require multiple service categories.

Typically, many customer networks need to migrate their largest locations to ATM first, which requires interworking between a few large ATM sites and many smaller frame-relay-connected sites. Furthermore, many carrier networks and large enterprises plan to trunk multiple services over ATM, including frame relay and IP. In response to these business needs, ITU-T Recommendation I.555, the ATM Forum B-ICI specification, and two Frame Relay Forum (FRF) Implementation Agreements (IAs) specify interworking between frame relay and ATM.

ITU-T Recommendation I.555 [1] labels these types of FR/ATM interworking as scenario one and two as shown in Figure 18.3. In scenario 1, FR is interworked (or trunked) *over* ATM, while in scenario 2, a FR end system interworks directly *with* an ATM end system. In scenario 1, FR CPE either directly interfaces to an Interworking Function (IWF) via a FR UNI, or connects via a frame relay network. The access configuration where an ATM end system connects directly via an ATM UNI which then connects to an IWF applies to service interworking in scenario 2 only. The Frame Relay Forum details these scenarios in two implementation agreements: FR/ATM Network Interworking [2], and FR/ATM Service Interworking [3]. Let's now explore the network and service interworking scenarios in more detail.

a. Interworking Scenario 1: Frame Relay *Over* ATM (Network Interworking)

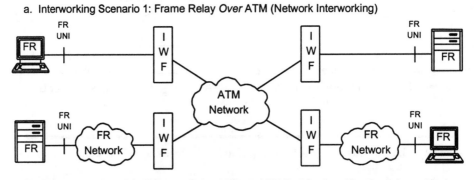

b. Interworking Scenario 2: Frame Relay *With* an ATM End System (Service Interworking)

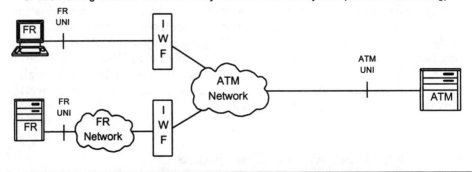

Figure 18.3   FR/ATM Interworking Scenarios and Access Configurations

18.3 FRAME RELAY/ATM NETWORK INTERWORKING

Figure 18.4 illustrates further details of the FR/ATM network interworking protocol. The FR to ATM network Interworking Function (IWF) converts between the basic Frame Relay functions and the FR Service-Specific Convergence Sublayer (FR-SSCS) defined in ITU-T Recommendation I.365.1 [4], operating over the Common Part of AAL5 (see Chapter 13). The network IWF also conveys the FR status signaling for a FR UNI port across the one or more VCCs corresponding to the frame relay Data Link Connection Identifiers (DLCIs). In the FR/ATM network interworking scenario, the upper layer protocols must be compatible.

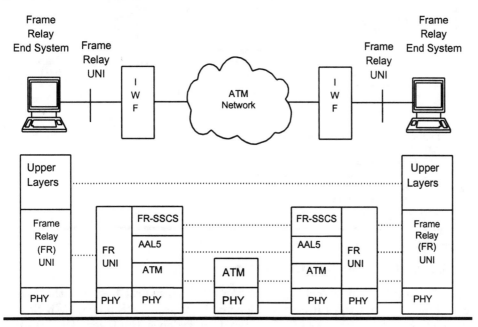

Figure 18.4 FR/ATM Network Interworking

18.3.1 FR Service Specific Convergence Sublayer (FR-SSCS)

Figure 18.5 depicts the Frame Relay Service Specific Convergence Sublayer (FR-SSCS) PDU format (essentially the FR frame summarized in Chapter 7) with inserted zeroes and the trailing CRC both removed. Frame relay supports either 2- , 3- , or 4-octet addressing as indicated in the figure. The FR-SSCS supports multiplexing through the use of the DLCI field, with the ATM layer supporting connection multiplexing using the VPI/VCI fields in the cell header. There are two methods of multiplexing FR connections over ATM: many-to-one and one-to-one. Many-to-one multiplexing maps many FR logical connections identified by the FR DLCIs over a single ATM Virtual Channel Connection (VCC). One-to-one multiplexing maps each FR logical connection identified by DLCI to a single ATM VCC via VPI/VCIs at the ATM layer. Many-to-one multiplexing is best suited for efficiently trunking FR over ATM, since this method efficiently carries the status signaling between FR networks. However, the many-to-one mapping is optional in the standard, hence many vendors don't implement it.

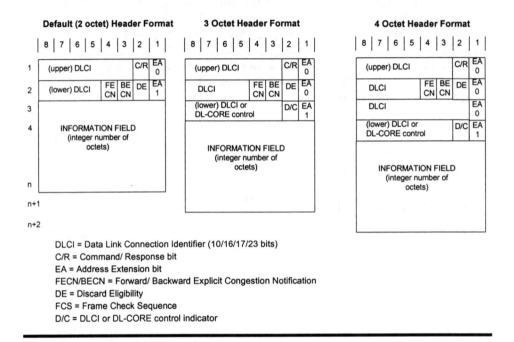

DLCI = Data Link Connection Identifier (10/16/17/23 bits)
C/R = Command/ Response bit
EA = Address Extension bit
FECN/BECN = Forward/ Backward Explicit Congestion Notification
DE = Discard Eligibility
FCS = Frame Check Sequence
D/C = DLCI or DL-CORE control indicator

Figure 18.5   FR-SSCS PDU Formats

18.3.2 Status Signaling Conversion

The FR to ATM InterWorking Function (IWF) converts between the Q.922 core functions and the FR Service-Specific Convergence Sublayer (FR-SSCS) defined in I.365.1, the AAL5 Common Convergence Sublayer (CPCS), and Segmentation And Reassembly (SAR) sublayers from I.363 as shown in the left-hand side of Figure 18.6a. The IWF must also convert between the Q.933 Annex A PVC status signaling for a single, physical FR UNI port and the one or more VCCs that correspond to the DLCIs. The FR-SSCS Protocol Data Unit (PDU) is the CPCS SDU of the AAL5 Common Part as described in Chapter 13. Figure 18.6b illustrates the FR/ATM interworking protocol of an ATM end system. This function is identical to the right-hand side of the FR/ATM IWF. The ATM end system must support Q.933 Annex A frame relay status signaling for each FR/ATM network interworking VCC as indicated in the figure. If the FR/ATM IWF and the end system both support many-to-one multiplexing, then the encapsulated FR status signaling channel contains the status for more than one DLCI. In the case of one-to-one multiplexing, each VCC carries the user DLCI and the status signaling DLCI, which reports on the status of a single DLCI.

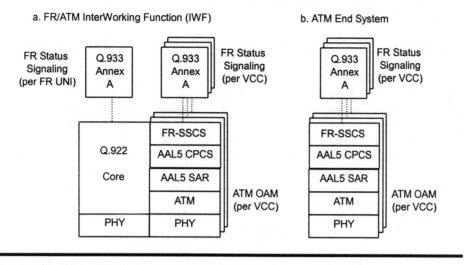

Figure 18.6   FR to ATM Interworking Control Plane Protocol Stacks

18.3.3 Congestion Control and Traffic Parameter Mapping

The InterWorking Function (IWF) either maps or encapsulates control and addressing functions along with the associated data. For example, in the FR to ATM direction the IWF encapsulates the DLCI, DE, FECN, and BECN fields in the FR-SSCS PDU. The IWF maps the FR DE bit to the ATM CLP bit, and also maps the FR FECN bit to the ATM EFCI bit. The IWF encapsulates the FR BECN bit in the FR-SSCS. The IWF may also map an EFCI indication in the last cell of a frame reassembled in the ATM-to-frame relay direction to the BECN bit. The IWF maps the frame relay FCS (that is, it replaces it by) the AAL5 CRC function.

In the ATM-to-frame relay direction the CLP bit may be logically ORed with the DE bit as a configuration option on a per-DLCI basis. The IWF checks the AAL5 CRC, and recomputes the FR FCS for delivery to a frame relay UNI. The FR-SSCS PDU carries the encapsulated FECN, BECN, and DE bits intact.

As described in Chapter 7, the FR traffic parameters include access line rate (AR), committed burst size (Bc), excess burst size (Be), and measurement interval (T). These FR traffic parameters define a Committed Information Rate (CIR) and an Excess Information Rate (EIR) as described in Appendix A of the ATM Forum B-ICI specification [5], which maps these FR traffic parameters to the ATM traffic parameter in terms of Peak Cell Rate (PCR), Sustainable Cell Rate (SCR), and Maximum Burst Size (MBS). The ATM Quality of Service (QoS) class and service category for the ATM VCC must be compatible. Usually, frame relay utilizes the ATM nrt-VBR service category; however, other categories may also be used in conjunction with prioritized FR service.

18.4 FRAME RELAY/ATM SERVICE INTERWORKING

The ATM Forum worked closely with the Frame Relay Forum to develop a FR/ATM service interworking specification [2]. Figure 18.7 illustrates the user plane protocol stacks for FR/ATM service interworking. Note that a Frame Relay end system directly communicates with an ATM end system in this scneario. The figure illustrates three possible implementations of the InterWorking Function (IWF). Case a in the figure illustrates the situation the ATM network implements the IWF, while case c shows the IWF implemented in the frame relay network. Case b illustrates a design employing a separate interworking function. A fourth case (not illustrated in the figure) is also possible where the same multi-service switches implement both the frame relay and ATM protocols.

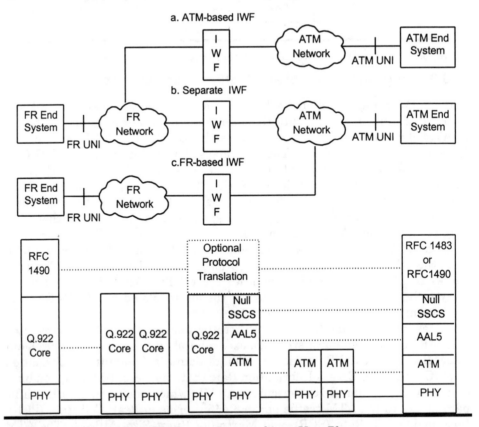

Figure 18.7   Frame Relay/ATM Service Interworking - User Plane

Mapping between different multiprotocol encapsulations standards for FR and ATM, as specified in RFCs 1490 [6] and 1483 [7], respectively, is an optional protocol translation function. Unfortunately, the multiprotocol encapsulation formats for frame relay and ATM differ. RFC 1490 specifies a Network Level Protocol ID (NLPID), SubNetwork Attachment Point (SNAP) format for frame relay, while RFC 1483 specifies a LAN-compatible Logical Link Control (LLC) SNAP format. RFC 1483 handles the case where the IWF does not provide a protocol translation function by requiring that the ATM end system use the NLPID/SNAP multiprotocol encapsulation specified in RFC 1490.

18.4.1 Status Signaling Interworking

Figure 18.8 illustrates the mapping between FR status signaling and the ATM fault and status indications. If the frame relay status signaling active bit indicates a failed DLCI, then the IWF generates an ATM OAM cell indicating an Alarm Indication Signal (AIS) fault as defined in Chapter 28. The frame relay new bit causes the IWF to generate an ATM ILMI status change trap. This mapping communicates the semantics of end-to-end DLCI and VCC status to the FR and ATM end systems, respectively.

FR/ATM service interworking interconnects end users, and was not designed as a trunking protocol. FR/ATM network interworking was designed for carrier interconnection of FR services over ATM [5]. The mapping of status signaling to ATM OAM cells and ILMI traps is less efficient than the many-to-one multiplexing mode of FR/ATM network interworking.

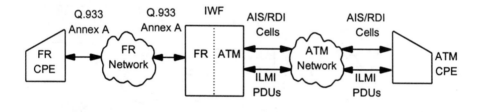

Figure 18.8   Frame Relay/ATM Status Signaling Interworking

18.4.2 Address Resolution Protocol Interworking

The Frame Relay Forum's FR/ATM Service Interworking specification also specifies procedures for mapping the Address Resolution Protocol (ARP) [8] and Inverse ARP (InARP) [9] between their Frame Relay [6] and ATM counterparts (i.e., the PVC portions of RFC 1577 detailed in Chapter 20) when operating in the optional translation mode. The use of these encapsulation mappings allows the interworking function to recognize and perform special processing on these ARP and InARP packets. As described in Chapter 9, IP devices, such as routers or workstations, automatically determine the link layer address using the ARP protocol. ARP encapsulation mapping allows IP end systems to automatically determine the correct FR DLCI and ATM VPI/VCI corresponding to a FR/ATM Service Interworking PVC on which to transfer IP packets. In order to perform the ARP and InARP mapping functions, the interworking function contains a table with the following information statically configured in each row for each of the four FR/ATM Service Interworking PVCs as illustrated in Figure 18.9:

- IWF frame relay port number (P1)
- Frame Relay DLCI number on the Frame Relay Port (aa, bb, cc, dd)
- ATM Port number on the IWF corresponding to the FR PVC (P1, P2)
- VPI/VCI on ATM Port corresponding to the FR PVC (rr, ss, tt, uu)

We now give an example of the ARP protocol with reference to Figure 18.9. Note that a full mesh of PVCs interconnect the four CPE devices in the example. Furthermore, the IP addresses are all on the same subnet as defined in Chapter 9. Therefore, when FR-CPE with IP address IPa wants to send a packet to the device IP address IPx, but doesn't know the correct link layer address, it sends out an ARP packet on all the PVCs. The FR/ATM IWF receives the ARP packet on DLCIs aa and bb on Port P1, converts it into the correct ATM format and sends it out on ATM Port P2 on VPI/VCIs rr and ss. The ATM-CPE with IP address IPx responds to the ARP request with anARP reply in the ATM format on VPI/VCI rr on Port P2, which the FR/ATM IWF converts into the frame relay format and transmits on Frame Relay Port P1 on DLCI aa. Now, FR-CPE with address IPa sends IP packets to address IPx on DLCI aa, which the FR/ATM IWF transmits to the ATM-CPE with address IPx on VPI/VCI rr on ATM Port P2.

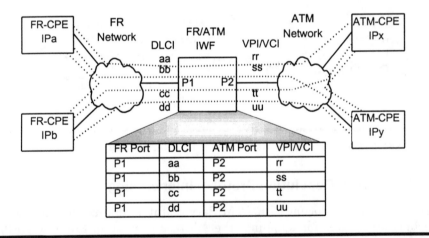

Figure 18.9   FR/ATM Address Resolution Protocol Example

The Frame Relay Forum's FR/ATM Service Interworking specification [3] also specifies SVC addresses as an optional, yet desirable addition to above mapping in anticipation of FR/ATM SVC Interworking.

18.4.3 FR/ATM Interworking Applied

Large enterprise networks typically have a few large locations that serve as major traffic sources and sinks. Typical applications are large computer centers, large office complexes with many information workers, campuses requiring high-tech communication, server farms, data or image repositories, and large-volume data or image sources. These large locations have a significant community of interest among them; however, the enterprise usually also requires a relatively large number of smaller locations needing at least partial, lower performance access to this same information. The smaller locations have fewer users, and generally cannot justify the higher cost of equipment or networking facilities. Generally, cost increases as performance, number of features, and flexibility increase.

Many users turn to hybrid networking solutions using high-speed ATM at the larger sites and interworking with lower-speed frame relay at the many smaller locations. These lower-speed access sites require more efficient access rather than high performance, and thus frame relay access through low-end routing and bridging products is often more cost effective than ATM. This is because the cost per bit per second generally decreases as the public network access speed increases. For example, the approximate ratio of DS1/DS0 and DS3/DS1 tariffs is approximately 10:1, while the speed difference is approximately 25:1. This means that a higher-speed interface can be operated at 40 percent efficiency at the same cost per bit per second. Conversely, the lower-speed interface costs 3.5 times as much per bit per second, and therefore efficiency use of capacity is more important.

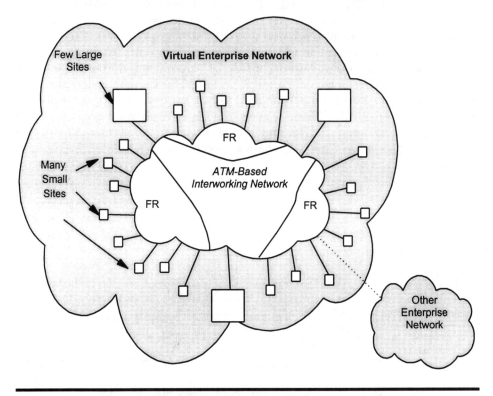

Figure 18.10   Typical FR/ATM Based Enterprise Network

What does the virtual enterprise network look like for the beginning of the next century? Figure 18.10 illustrates an ATM-based interworking network cloud connecting a few large ATM sites to many smaller Frame Relay sites.

Such a network has many smaller sites and few larger sites, which is typical of large enterprises, such as corporations, governments, and other organizations. Principal needs that drive the need for ATM are multiple levels of service characterized by parameters such as throughput, quality, and usage-based billing.

An issue with FR/ATM interworking arises if the application requires end-to-end QoS support. Since the ATM designers started with QoS as a fundamental requirement, while frame relay was initially conceived as a data only service; FR/ATM interworking doesn't necessarily guarantee end-to-end QoS [10]. Therefore, if QoS is critical to your application (for example, voice or video), check to see that the frame relay service provider and equipment supplier supports your needs.

18.5 SMDS ACCESS TO ATM

Bellcore created the Switched Multimegabit Data Service (SMDS) to bridge the interval until ATM matured. A number of users utilize SMDS, particularly in Europe where the service is called Connectionless Broadband Data Service (CBDS). Many of these users want to move to ATM, but don't want to write off their investment in SMDS. Also, carriers want to trunk multiple services over ATM. The ATM Forum B-ICI 1.1 specification first defined how to transport SMDS between carriers. The ATM Forum, SMDS Interest Group (SIG), and European SIG (E-SIG) jointly specified how a user accessed SMDS across an ATM UNI to support customers who want to share ATM access lines between SMDS and access to other ATM and ATM-based interworking services. This was the last specification generated by the SIG before it shut down.

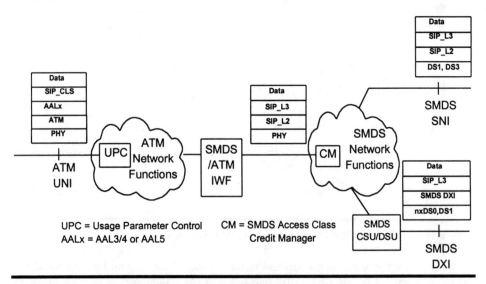

Figure 18.11 Logical Configuration for SMDS Access over ATM

Figure 18.11 depicts the access configuration and logical placement of function for accessing SMDS features over an ATM User-to-Network Interface (UNI) [11]. An ATM end system accessing SMDS over ATM must format either an AAL3/4 or AAL5 CPCS PDU containing the SMDS Connectionless Service (SIP_CLS) PDU as shown on the left-hand side of the figure. The ATM network performs a Usage Parameter Control (UPC) function to emulate the SMDS access class. The ATM network relays the cells to an SMDS Interworking Function (IWF), which may be implemented in a centralized, regionalized, or distributed manner. The SMDS/ATM IWF converts the AAL stream into the SMDS Level 2 and 3 protocol stack and passes this to an SMDS network, which implements the SMDS service features, including access class enforcement via the Credit Manager (CM) . The SMDS network can interface to a subscriber using the SMDS Subscriber Network Interface (SNI) [12] or the SMDS Data eXchange Interface (DXI) [13] as shown in the figure. See [14] for more information on ATM access to SMDS. Several of the functions which the SMDS/ATM IWF performs include:

☞ Conversion between SIP_L3 and the SIP_CLS PDU carried via either AAL3/4 or AAL5

☞ Conversion between 802.6 Level 2 PDUs (slots) and either the ATM AAL3/4 or AAL5 SAR

☞ Multiplexing of 802.6 Multiplex IDs (MIDs) into a single ATM VCC

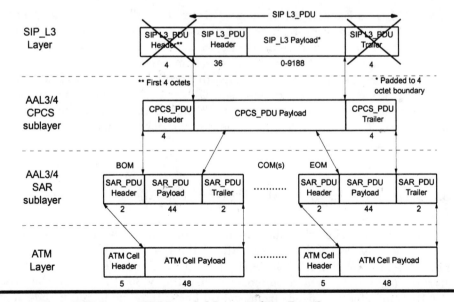

Figure 18.12 SMDS over ATM Protocol Interworking Detail

Figure 18.12 illustrates how the SMDS Interface Protocol (SIP) is carried over ATM using the InterWorking Function (IWF). The IWF removes the 4-

octet header and 4-octet trailer of the SMDS SIP_L3 PDU prior to inserting the remaining fields in the AAL3/4 CPCS PDU (see Chapter 8 for details). Padding of the truncated SIP_L3 PDU must be performed to retain 4 octet alignment in the AAL3/4 CPCS PDU. The figure illustrates the operation of AAL3/4 Segmentation And Reassembly (SAR) using Beginning, Continuation, and End Of Message (BOM, COM, and EOM) and the corresponding mapping to ATM cells.

The mapping between the SMDS access class credit manager parameters and the ATM traffic parameters is a key function. Section B.2 of Appendix B of the ATM Forum version 3.1 UNI specification describes a means for doing this using the peak and sustainable cell rates. Or, the IWF may use only the peak rate as described in appendix A of the SIG/E-SIG/ATM Forum specification [11].

18.6 FRAME-BASED INTERFACES SUPPORTING ATM

Initially, the ITU-T and the ATM Forum focused on the specification of high-speed interfaces for ATM. However, the expense of high-speed lines in carrier networks and the costs of early ATM devices drove the need for lower-speed interfaces, or means to utilize existing hardware via software changes only. In response to this need, the ATM Forum defined two types of low-speed, frame-based ATM interfaces to support such connections to ATM networks.

The ATM Data Exchange Interface (DXI) protocol allowed early adopters to utilize ATM with existing routers and data equipment with serial interfaces using a separate piece of equipment to convert between the ATM DXI protocol and an ATM UNI, similar to early implementations of SMDS. The ATM Forum then adapted the ATM DXI protocol to a WAN interface, calling this specification the ATM Frame-based UNI (FUNI). A main advantage of FUNI is that it eliminates the external CSU/DSU required in the ATM DXI specification. This section describes the context, operation, and application of these two important protocols.

18.7 ATM DATA EXCHANGE INTERFACE (DXI)

Many users have asked the following question: what if I want the capabilities of ATM over the WAN, but I can't afford the cost of a DS3 or OC-3 access line? The answer could be the ATM Forum specified ATM Data eXchange Interface (DXI) [15], which supports either the V.35, RS449, or the HSSI DTE-DCE interface at speeds from several kbps up to and including 50-Mbps.

ATM DXI specifies the interface between a DTE, such as a router, and a DCE, usually called an ATM CSU/DSU, which provides the conversion to an ATM UNI, as illustrated in Figure 18.13. Like the SMDS DXI on which the Forum patterned the ATM DXI specification, the context is a limited distance

DTE-DCE interface. The DTE manages the ATM DXI interface through a Local Management Interface (LMI), while the CSU/DSU passes the ATM UNI Interim Local Management Interface (ILMI) Simple Network Management Protocol (SNMP) messages through to the DTE.

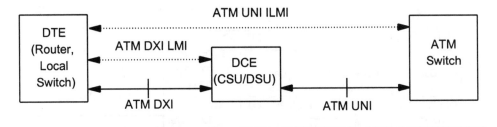

Figure 18.13 ATM Data eXchange Interface (DXI) Configuration

Table 18.1 summarizes the key attributes of the various modes of the ATM DXI specification. All modes support AAL5, while some modes also support AAL3/4. The maximum number of Virtual Channel Connections (VCCs) differs in the various modes, as well as the maximum frame size due to the address bits in the DXI header and the length of the Frame Check Sequence (FCS). Mode 1a is the most widely used ATM DXI interface.

Table 18.1   Summary of ATM DXI Mode Characteristics

Characteristic	Mode 1a	Mode 1b	Mode 2
Maximum number of VCCs	1023	1023	16,777,215
AAL5 Support	Yes	Yes	Yes
AAL3/4 Support	No	Yes	Yes
Maximum DTE SDU Length			
AAL5	9232	9232	65,535
AAL3/4	N/A	9224	65,535
Bits in FCS	16	16	32

18.7.1 ATM DXI — Mode 1a and Mode 1b

ATM DXI Mode 1 supports two implementations. Both mode 1a and 1b define DCE support for AAL5 as shown in Figure 18.14. The protocol encapsulates the DTE Service Data Unit (SDU) in the AAL5 CPCS, and then segments it into ATM cells using the AAL5 Common Part Convergence Sublayer (CPCS) and Segmentation And Reassembly (SAR) sublayer functions defined in Chapter 13. The 2-octet DXI header defined later in this section prefixes the DTE SDU. The 2-octet Frame Check Sequence (FCS) is the same as that used in frame relay and HDLC and hence much existing DTE hardware supports mode 1 through software changes only.

Mode 1b adds support for the AAL3/4 CPCS and SAR on a per-VCC basis as shown in Figure 18.15. The DTE must know that the DCE is operating in mode 1b AAL3/4 since it must add the 4 octets for both the CPCS PDU header and trailer as indicated in the figure. This decreases the maximum-length DTE SDU by 8 octets. The same 2-octet DXI header used for the AAL5 VCC operation is employed.

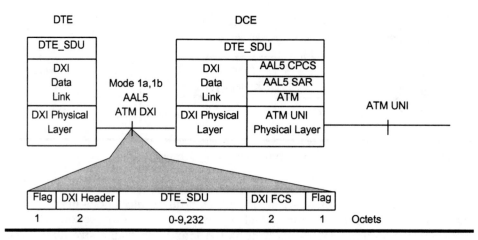

Figure 18.14   Modes 1a and 1b DXI Using AAL5

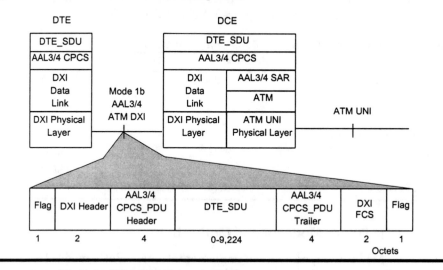

Figure 18.15   Mode 1b ATM DXI Using AAL3/4

18.7.2 ATM DXI — Mode 2

Mode 2 uses the same interface between DTE and DCE regardless of whether the VCC is configured for AAL5 or AAL3/4, as shown in Figure 18.16. The DTE must place the DTE SDU inside the AAL3/4 CPCS header and trailer, and then the DCE performs the appropriate function depending upon whether the VCC is configured for AAL3/4 or AAL5. The DCE operates the same as in mode 1b for a VCC configured for AAL3/4, performing the AAL3/4 SAR on the AAL3/4 CPCS_PDU as shown in the top part of Figure 18.16.

The DCE must first extract the DTE_SDU from the AAL3/4 CPCS_PDU for a VCC configured to operate in AAL5, as shown in the bottom half of the figure. The net effect of these two transformations is that a mode 2 DCE can interoperate with a mode 1 DCE. The mode 2 DXI frame has a 4-octet header and a 4-octet FCS, which usually requires new hardware. Because the FCS is longer, the maximum DTE_SDU length can be larger. The 32-bit FCS used in the DXI is the same as that used for FDDI and AAL5.

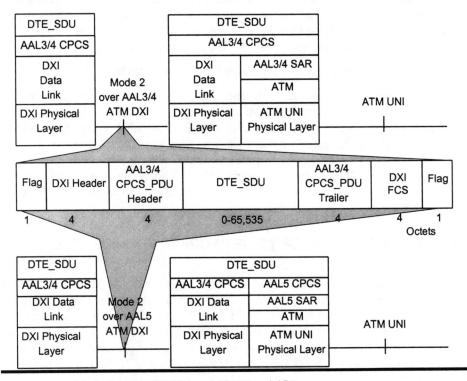

Figure 18.16   Mode 2 ATM DXI Using AAL3/4 or AAL5

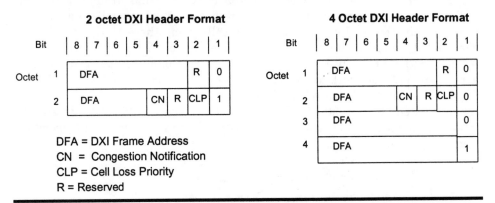

Figure 18.17   Two- and Four-Octet ATM DXI Header Formats

18.7.3 ATM DXI Header Formats

Figure 18.17 illustrates the details of the 2- and 4-octet DXI header structure. The DCE maps the DXI Frame Address (DFA) into the low-order bits of the VPI/VCI. The protocol maps the Congestion Notification (CN) bit from the last ATM cell of the PDU's Payload Type (PT) congestion indication (see Chapter 12). The DTE can set the CLP bit so that the DCE will in turn set the CLP bit in the ATM cell header with the same value, thus allowing the user to mark the cells from selected PDUs as a low loss priority. A great deal of similarity exists between these DXI formats and the frame relay Service-Specific Convergence Sublayer (SSCS) formats covered earlier in this chapter.

18.7.4 Local Management Interface (LMI) Summarized

The DXI Local Management Interface (LMI) defines a protocol for the exchange of SNMP GetRequest, GetNextRequest, SetRequest, Response, and Trap messages between the DTE and the DCE. The LMI allows the DTE to set or query (Get) the mode of the DXI interface as either 1a, 1b, or 2. The LMI also allows the DTE to set or query the AAL assigned on a per-VCC basis as indexed by the DXI Frame Address (DFA). A shortcoming of the current LMI is that the DCE does not communicate the ATM UNI status to the DTE.

18.8 FRAME-BASED USER-TO-NETWORK INTERFACE (FUNI)

The ATM Forum specified the Frame-based User-Network Interface (FUNI) version 1.0 specification [16] so that CPE without ATM hardware, such as many currently deployed routers, could interface to ATM networks with only minor software changes. Most importantly, the ATM FUNI specification obviated the need for an expensive external ATM DXI converter. FUNI provides low-speed, WAN ATM access protocol rates of Nx64-kbps, DS1, and E1.

Figure 18.18 illustrates how frame-based CPE sends frames using the FUNI data link protocol to a network-based ATM switch, which then segments the frames into standard ATM cells using the AAL5 protocol. The same ATM switch reassembles cells transmitted by ATM CPE and delivers frames to the FUNI user. Thus, FUNI users communicate transparently across an ATM network with either other FUNI users (FUNI-to-FUNI) or ATM UNI users (FUNI-to-ATM UNI).

An updated version 2.0 FUNI data link layer specification [17] details the following functions and clarifies some ambiguities in the version 1.0 specification:

- VPI/VCI multiplexing
- SVC Signaling
- Network management
- Traffic policing
- ATM Operations, Administration, and Maintenance (OAM) functions

- VBR and Unspecified Bit Rate (UBR) traffic

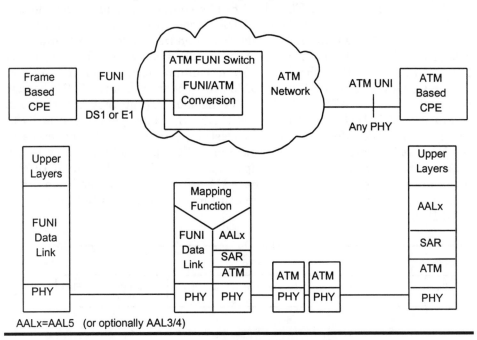

Figure 18.18 ATM Frame-based User-Network Interface (FUNI)

18.8.1 FUNI Frame Formats

The FUNI specification requires operation over AAL5 in DXI Mode 1a, and makes operation over AAL3/4 in Mode 1b an option. It doesn't use Mode 2, but instead defines two additional modes numbered 3 and 4. Table 18.2. summarize the key attributes of the various FUNI modes. These include whether interoperable implementations must implement particular modes, the AAL employed, the number of octets comprising the header, the maximum payload length, the number of Cyclic Redundancy Check (CRC) octets in the trailer, the maximum number of user definable connections, and the target interface speed for the specified mode. In the interest of brevity, this section covers only the details of the required Mode 1a mapping.

Table 18.2 Frame-based UNI (FUNI) Modes

Mode	Implementation	AAL	Header Octets	Maximum Payload Octets	CRC Octets	Maximum User VCCs	Speed Range Required
1a	Required	5	2	4096	2	512	≤DS1/E1
1b	Optional	3/4	2	4096	2	512	≤DS1/E1
3	Required	5	2	9232	4	512	≥DS1/E1
4	Optional	5	4	9232	4	16,777,216	≥DS1/E1

A FUNI PDU has a header and trailer delimited by HDLC flags as illustrated in Figure 18.19. The FUNI header contains a 10-bit frame address in the same bit positions as the ATM DXI Mode 1a header. Note, however, that the the ATM DXI and ATM FUNI frame address fields map to different sets of VPI/VCI bits in the ATM cell header. The two Frame IDentifier (FID) bits determine whether the FUNI frame contains either user information (i.e., data, signaling or ILMI) or an OAM cell. The format of the FUNI header is identical to the ATM DXI Mode 1a header, except that the two bits reserved in ATM DXI are used by FUNI. The other bits operate identically to those in ATM DXI Mode 1a. The Congestion Notification (CN) bit maps to the Explicit Forward Congestion Indication (EFCI) payload type value in the ATM cell header as described in Chapter 12. The Cell Loss Priority (CLP) bit maps to the corresponding bit in the ATM cell header. The additional two bits (0 and 1) serve an address extension function analogous to that used in ATM DXI.

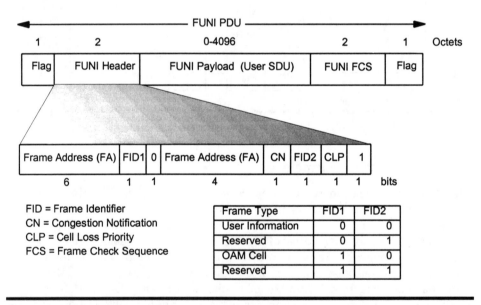

Figure 18.19   FUNI Mode 1a Data Link Frame Format

The mapping of the FUNI frame address fields into the VPI/VCI bits in the ATM cell header allows FUNI to support a limited form of Virtual Path Connection Service as illustrated in Figure 18.20. Thus, this mapping supports at most 16 VPCs or 1,024 VCCs. The FUNI specification further limits the total required number of connections to 512 for Mode 1a operation.

18.8.2 FUNI versus ATM DXI and Frame Relay

Two key functional differences separate FUNI and DXI. FUNI provides improved access line utilization compared with cell-based access of an ATM DXI CSU/DSU. For example, at a typical packet size of 300 bytes FUNI is 15 to 20 percent more efficient than ATM. The second difference is that FUNI supports Nx64-kbps rates, while the lowest speed supported by ATM DXI is DS1/E1.

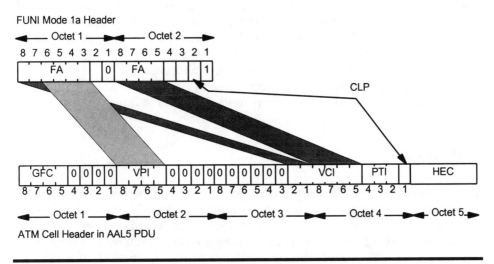

Figure 18.20   FUNI Header to ATM Cell Header Bit Mapping

A paper published by the Frame Relay Forum [18] analyzes the relative advantages and disadvantages of FUNI versus Frame Relay, since these technologies overlap. The paper asserts that the following business considerations drive a choice between FR and FUNI:

- Existing installed base of user and network equipment
- Availability of equipment and service from multiple suppliers
- Cost of network and user equipment
- Recurring cost of service
- Cost of managing the network
- Bandwidth utilization efficiency
- Service classes provided by the network
- Performance (interface speed, delay, throughput)
- Interoperability between vendors

Experts in discussions on the Frame Relay mail exploder believe that wide availability of the FUNI interface won't occur in user equipment until public service providers offer FUNI access service at a cost comparable to frame relay. Otherwise, there is little market demand on the end system (CPE) vendors to implement it.   Even if this situation occurs, users must incur a migration cost to switch their FR sites to FUNI to connect to presumably higher bandwidth ATM sites for reasons such as increased peak rates, QoS support, or seamless networking as described in Section 18.4.3. Some users believe that FUNI offers a number of capabilities that FR does not at lower speeds, using frames that do not incur the ATM cell and AAL5 overhead. Capabilities that FUNI provides, but FR does not, are:

- Access to richer, better defined QoS support in ATM
- Eliminate the need for complex FR/ATM signaling interworking
- Eliminates LLC/NLPID conversion required at FR/ATM boundary
- AAL5 supports longer frames (64K) versus FR (4K), eliminating fragmentation at FR/ATM boundary (Note: This requires the optional 32-bit CRC in FUNI)
- Direct Support for ATM's scalable NSAP addressing
- Direct participation of FUNI interfaces in PNNI networks
- Eliminate traffic descriptor conversion in FR/ATM interworking

18.9 REVIEW

This chapter covered ATM protocol support for cost-effective wide area data networking. The coverage began by defining the basic concepts of physical and logical access, interworking, and trunking and introduced the concept of overlay networking. The text then covered specific examples of frame relay trunking over ATM using the network interworking protocol, true service interworking between frame relay and ATM, and access to SMDS via ATM to illustrate these concepts. Finally, the chapter covered the ATM Forum's specifications on the frame-based interfaces defined by the ATM Data eXchange Interface (DXI) and Frame-based UNI (FUNI) protocols. ATM DXI and FUNI make ATM available to existing CPE cost effectively via external converters and a software defined protocol, respectively.

18.10 REFERENCES

[1]   ITU-T, *Frame relaying bearer service interworking*, Recommendation I.555, November, 1993.
[2]   Frame Relay Forum, *Frame Relay/ATM PVC Network Interworking Implementation Agreement*, FRF.5, December 20, 1994.
[3]   Frame Relay Forum, *Frame Relay / ATM PVC Service Interworking Implementation Agreement*, FRF.8, April 14, 1995.
[4]   ITU-T, *Frame relaying service specific convergence sublayer (FR-SSCS*, Recommendation I.365.1, November 1993.
[5]   ATM Forum, *BISDN Inter-Carrier Interface (B-ICI) Specification*, Version 2.0, af-bici-0013.003, December 1995.
[6]   T. Bradley, C. Brown, A. Malis, *Multiprotocol Interconnect over Frame Relay*, IETF RFC 1490, July 1993.
[7]   J. Heinanen, *Multiprotocol Encapsulation over ATM Adaptation Layer 5*, IETF RFC 1483, July 1993
[8]   D. Plummer, *Ethernet Address Resolution Protocol: Or converting network protocol addresses to 48.bit Ethernet address for transmission on Ethernet hardware*, RFC826, November 1982.
[9]   T. Bradley, C. Brown, *Inverse Address Resolution Protocol*, IETF RFC 1293, January 1992.

[10]   R. Sullebarger, "ATM and Frame Relay: Making the Connection," *Data Communications*, http://www.data.com/tutorials/connection.html, June 1997.

[11]   SMDS Interest Group, *Protocol Interface Specification for Implementation of SMDS over and ATM-based Public UNI, Revision 2.0*, October 31, 1996.

[12]   Bellcore, *Generic System Requirements in support of Switched Multi-Megabit Data Service*, TR-TSV-000772 Issue 1, May 1991.

[13]   SMDS Interest Group, *SMDS Data Exchange Interface Protocol, Revision 3.2*, SIG-TS-001/1991, October 1991.

[14]   M. Figueroa, S. Hansen, "Technology Interworking for SMDS: From the DXI/SNI to the ATM UNI," *IEEE Communications Magazine*, June 1996.

[15]   ATM Forum, *Data Exchange Interface Version 1.0*, af-dxi-0014.000, August 1993

[16]   ATM Forum, *Frame-based User-to-Network Interface (FUNI) Specification*, af-saa-0031.000, September 1995

[17]   ATM Forum, *Frame Based User-To-Network Interface (FUNI) Specification v2.0*, af-saa-0088.000, July 1997.

[18]   Frame Relay Forum, "Frame Relay and Frame-Based ATM: A Comparison of Technologies," http://www.frforum.com/4000/fratm/fratm.toc.html

19

ATM in the Local Area Network

Despite many pronouncements to the contrary, ATM is not dead in the LAN. This chapter covers ATM's support for Local Area Networks (LANs). Most major enterprises embraced LANs in the 1980s, and now even some residences have them. The ATM industry developed LAN Emulation (LANE to provide a means to seamlessly interconnect legacy LANs with new, high-performance local area ATM networks. The LANE protocol empowers ATM host and server applications to interwork with devices already resident on existing, legacy LANs. We show how the LANE protocol supports seamless ATM host and server interconnection with legacy Ethernet, Token Ring, and FDDI LANs. The chapter then covers the Cells in Frames (CIF) concept where ATM cells are encapsulated in native LAN frames. Finally, the chapter covers the IETF RFC 1483 Multiprotocol Encapsulation over ATM standard that defines a means to multiplex multiple protocols over a single ATM VCC.

19.1 LAN EMULATION (LANE)

LAN Emulation (LANE) provides an interoperable transition from legacy Ethernet and Token Ring networks to ATM. Prior to LANE, legacy LAN and WAN protocols (Ethernet, Token Ring, FDDI, etc.) required proprietary conversion devices to benefit from ATM. Although some pundits criticized the ATM Forum LANE effort for focusing on backward compatibility instead of designing for the future, most experts applauded the business savvy of the approach. LANE empowers LAN designers to reap the benefits of high-capacity, scalable, bandwidth controlled, guaranteed quality ATM networking while preserving the best elements of their legacy LAN infrastructure.

This section introduces LANE by first describing the hardware and software involved in an emulated LAN. Next, we cover the components and connections involved in LANE. The text then walks through the step-by-step procedure used by a LAN Emulation Client (LEC) in automatic configuration, initialization, joining a virtual network, registration, broadcast, and data transfer. We then discuss some key implementation considerations and optional LANE capabilities. Finally, the text describes some highlights of the version 2.0 LANE specification and planned work within the ATM Forum.

19.1.1 Hardware and Software Components in an Emulated LAN

Recall from Chapter 8 the key characteristics of a Local Area Network (LAN): high-speed, broadcast capable, connectionless service, along with plug and play operation. Since ATM targeted operation over fiber optics and the fixed cell size is well suited to hardware design, it meets or exceeds the speed requirement of most LANs. A key challenge for LANE is resolving the fundamental difference between ATM's connection-oriented, point-to-point protocol and the inherently connectionless, shared-medium broadcast capable LANs (e.g., Ethernet and Token Ring). Hence, a key function in LANE is the emulation of a broadcast medium. Note that the amount of broadcast traffic limits the overall capacity of an emulated LAN to that of the slowest interface. Broadcast limits overall size of an ELAN. Generally, it is not a good idea to do a lot of LANE over the wide area because of broadcast traffic. For example, the IP ARP procedure described in Chapter 8 generates a smaller amount of broadcast traffic than other network layer protocols do. For example, if 5 percent of the traffic is broadcast, than an OC3 worth of traffic offers almost 8-Mbps of traffic to a single 10-Mbps Ethernet station. Hence, ELANs should have no more than 500 to 2,000 stations. As an existence proof, actual LANE networks with 2,000 users were in operation in 1998.

However, seldom content to just copy an older design, we show how the LANE designers improved upon the overall throughput of shared-medium LANs by combining a broadcast server and fast connection switching. Finally, the LANE specification defines how a workstation with an ATM NIC card and LANE software automatically joins an existing LAN. Thereby, LANE achieves plug and play operation by supporting the standard Ethernet spanning tree or Token Ring source route bridging described in Chapter 9. Here again, the ATM design goes one step further by supporting virtual LANs, which allow a network administrator to flexibly assign users to different virtual workgroups. It is not only what you know, but who knows it, that determines the final value of any information.

LANE enables computers running the same applications on legacy LANs like Ethernet to directly connect to ATM-enabled systems via high-performance ATM networks. Figure 19.1 illustrates how LANE enables legacy applications to run essentially unchanged using existing software device driver interfaces on a workstation with an ATM NIC. Starting on the left-hand side, a host workstation with an ATM Network Interface Card (NIC) runs a set of applications on a specific Operating System (OS) which has a particular set of LAN driver software, for example, the Microsoft Network Driver Interface Specification (NDIS). The NIC card (or the OS) provides LAN Emulation software that interfaces to the ATM NIC card via NIC driver software in the host. The cells from the ATM NIC traverse an ATM network to an ATM capable bridge/router, which maps to a legacy LAN Medium Access Control (MAC) and physical layer. The bridge/router interfaces to the host on the right-hand side of the figure which has a native LAN NIC connected to a legacy LAN; for example, a 10-Mbps Ethernet. The LAN NIC in this workstation interfaces to LAN NIC driver software in the host, which provides an identical interface to the operating system's LAN driver software. Hence, applications running on the ATM-NIC and Native LAN-NIC networked computers see no difference in terms of software functions. Note that the ATM-NIC enabled workstation with LANE software will likely perform better when communicating with other ATM-NIC enabled hosts and servers.

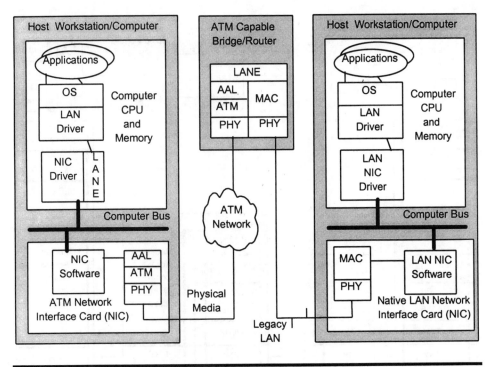

Figure 19.1   Seamless Interworking of ATM LAN Emulation and Legacy LAN Devices

Let's take a deeper look inside the operation of the LANE protocol with reference to Figure 19.2. The figure shows the two types of LANE 1.0 protocol data flows: a signaling connection shown by the dashed line and a data path connection shown by the solid line. The signaling connection sets up a SVC for the data direct path between the ATM client or server and the LAN bridge or switch for each extended data flow between computers. Starting on the left-hand side, an ATM host or server runs an existing application and networking protocol (such as IP or IPX) that interfaces using the Logical Link Control (LLC) protocol implemented in the host's LAN driver software. Note how the LANE software provides the same LLC interface to the network layer that Ethernet attached hosts and servers on the right-hand side of the figure do. Moving to the right, an ATM network interconnects ATM clients and servers. In the user plane, the ATM switch operates only on the ATM cells. As indicated by the dashed line, the ATM switch operates in the control plane to dynamically establish and release connections. In the upper middle portion of the figure is a LAN router, bridge and/or switch with an ATM interface and a LAN interface. This device terminates both the user and control plane components of the ATM-based LANE protocol and converts the frames to the MAC sublayer for transmission over a legacy LAN like Ethernet. On the right-hand side of the figure, the applications running on the native LAN workstation sees the same LLC and layer 3 networking protocol that the ATM-empowered workstation on the left does.

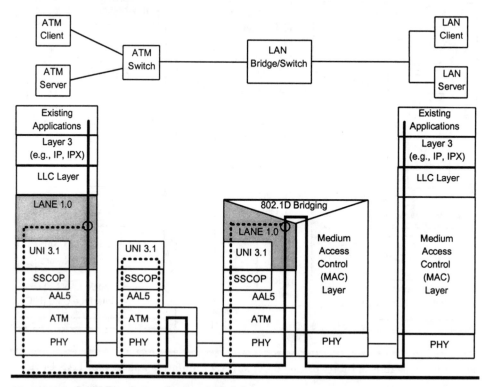

Figure 19.2 LAN Emulation Protocol Model

The LANE 1.0 specification defines a software interface for network layer protocols identical to that of existing LANs that encapsulates user data in either an Ethernet or Token Ring MAC frame. LANE does not emulate the actual media access control protocol of a particular LAN concerned (i.e., CSMA/CD for Ethernet or token passing for 802.5). Instead, LANE defines three servers that clients accesses over a number of ATM connections designated for specific control and data transfer purposes. LANE 1.0 does not directly define support for FDDI; however, devices readily map FDDI packets into either Ethernet or Token Ring using existing translation bridging techniques. Since the two new 100-Mbps LAN standards — Fast Ethernet (100Base-T) and 802.12 (100VG-AnyLAN) — use existing MAC packet formats, they map directly into LANE Ethernet or Token Ring formats and procedures. As described above, LANE literally bridges ATM and LANs by interworking at the Media Access Control (MAC) layer, which provides device-driver interfaces such as Open Data-Link Interface (ODI) and Network Driver Interface Specification (NDIS) to higher-level applications.

The ATM Forum's LANE 1.0 specification [1, 2] defines operation over the ATM best effort, or UBR service class similar to existing LANs. The LANE 2.0 [3] specification adds QoS guarantees, giving ATM based LAN Emulation a distinguishing characteristic over most other LAN protocols.

19.1.2 LANE Components and Connection Types

Figure 19.3 illustrates how virtual channel connections interconnect the following four logical components in the LANE specification:

- LAN Emulation Client (LEC)
- LAN Emulation Configuration Server (LECS)
- LAN Emulation Server (LES)
- Broadcast and Unknown Server (BUS)

The LEC runs in every ATM end and intermediate system (e.g., a host, server, bridge, or router) that provides a standard LAN service interface to higher-layer interfaces. An LEC performs data forwarding, address resolution, and other control functions in this role. ATM Network Interface Card (NICs) in hosts and servers, as well as ports on switches, bridges, and routers are examples of LEC implementations. A unique ATM address identifies each LEC, which the LANE protocol associates with one or more MAC addresses reachable through its ATM UNI. A LAN switch or bridge implementing an LEC dynamically associates all the MAC addresses reachable through its LAN ports to a single ATM address.

An LEC joins an emulated LAN by first connecting to the LECS using an ATM SVC. A unique ATM address identifies the LECS. The LEC address is either pre-configured into the LEC, or dynamically discovered via the Integrated Local Management Interface (ILMI) protocol. The LECS assigns individual LANE clients to particular emulated LANs by directing them to the controlling LES. Logically, one LECS serves all clients within an administrative domain. Vendors often use the LECS to implement virtual LANs, a capability where an administrator controls the allowed interconnection of sets of ATM and LAN users.

A single LES implements the address registry and address resolution server for a particular emulated LAN. An LEC joining an emulated LAN sets up an SVC to the LES and registers the association of its MAC address(es) with its ATM address. The LES assigns the LEC the address of a particular BUS. During operation of an emulated LAN, an LEC may query the LES to resolve a particular destination MAC address to its corresponding ATM address. The LES either responds directly to the address resolution query based upon previous registration information, or forwards the query to other LECs that may be able to respond.

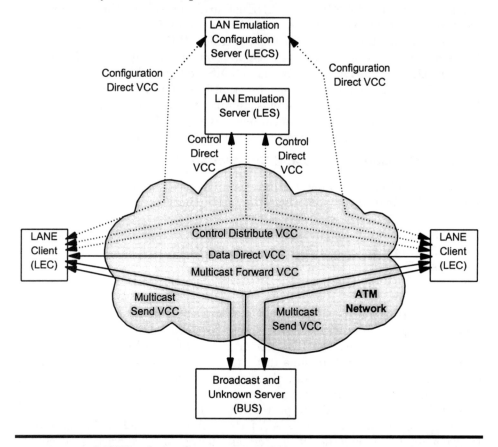

Figure 19.3   LAN Emulation Components and Interconnections

The BUS is a multicast server that floods unknown destination address traffic, and forwards multicast and broadcast traffic to clients within an Emulated LAN (ELAN). An emulated LAN may have multiple BUS's for throughput reasons, but each LEC transmits to only one BUS. Typically, the association of the broadcast MAC address (i.e., "all ones") with the BUS is pre-configured into the LES. As described later, an LEC can send frames to the BUS for delivery to the destination without setting up an SVC. However, the LANE 1.0 specification limits the rate at which an LEC may send frames to the BUS to prevent broadcast storms.

Figure 19.3 also illustrates the control and data virtual channel ATM connections between LANE components via directed arrows. The figure shows the following control connections as dashed lines.

- A bidirectional, point-to-point Configuration Direct VCC set up by the LEC to the LECS.
- A bidirectional, point-to-point Control Direct VCC set up by the LEC to the LES.
- A unidirectional Control Distribute VCC set up from the LES back to the LEC. Typically, this is a point-to-multipoint connection, but may be implemented as a set of unidirectional point-to-point VCCs.

Figure 19.3 shows the following LAN emulation data connections as solid lines.

- A bidirectional, point-to-point Data Direct VCC set up between two LECs to exchange data.
- A bidirectional, point-to-point Multicast Send VCC set up by the LEC to the BUS.
- A unidirectional VCC Multicast Forward VCC set up from the BUS to the LEC. Typically, this is a point-to-multipoint connection with each LEC as a leaf, but may also be a set of unidirectional point-to-point connections from the BUS to each served LEC.

The following sections detail the following basic steps in LAN Emulation using the above components and connections:

- Initialization and Configuration
- Joining and Registration
- Data Transfer Phase
- LANE and Spanning Tree

The remainder of the section then covers implementation considerations, optional capabilities and future directions.

19.1.3 Initialization and Configuration

Upon initialization; for example, a new installation or upon start up, the LEC first obtains its own ATM address. The LEC gets its address through ILMI address registration or pre-configuration. The LEC then determines the LECS address by one of three methods:

- Using the ILMI procedure
- Using a well-known LECS address
- Using a well-known permanent connection to the LECS (e.g., VPI=0, VCI=17)

Next, the LEC sets up a configuration-direct connection control VCC to the LECS. The LECS informs the LEC of information required for entry into its target ELAN through use of a configuration protocol. This information includes the LES's ATM address, the emulated LAN type, maximum frame size, and an ELAN name. Generally, an administrator configures the LECS with this information. Since the LECS may assign particular LECs to different LESs and BUSs, it enables a basic virtual LAN capability.

19.1.4 Joining and Registration

After the LEC obtains the LES address from the LECS, it may clear the configuration-direct VCC to the LECS since it requires no further configuration information. The LEC next sets up a control-direct VCC to the LES using standard signaling procedures. The LES assigns the LEC a unique LEC Identifier (LECID). The LEC registers its MAC address and ATM address with the LES. It may optionally also register other MAC addresses for which it acts as a proxy, for example, other reachable MAC addresses learned by a spanning tree bridge.

The LES then adds the LEC to the point-to-multipoint control-distribute VCC. The LEC employs the control direct and distribute VCCs for the LAN Emulation Address Resolution Protocol (LE_ARP). The LE_ARP response message returns the ATM address corresponding to a particular MAC address. The LES responds to an LE-ARP directly to the LEC if it recognizes this mapping, otherwise it forwards the request on the point-to-multipoint control-distribute VCC to solicit a response from a LEC that recognizes the requested MAC address. LANE uses this procedure because an LES may be unaware of a particular MAC address, because the address is "behind" a MAC bridge that did not register the address.

An LEC may respond to an LE_ARP because it is acting as a proxy for that address on the control direct VCC to the LES. The LES then forwards this response back to either the requesting LEC, or, optionally, on the point-to-multipoint control distribute VCC to all LECs. When the LES sends the LE_ARP response on the control distribute VCC, then all LECs learn and cache the particular address mapping. This aspect of the protocol significantly reduces the transaction load on the LES.

An LEC uses this LE_ARP mechanism to determine the ATM address of the BUS by sending an LE_ARP for the all ones MAC broadcast address to the LES, which responds with the BUS's ATM address. The LEC then uses this address to set up the point-to-point multicast send VCC to the BUS. The BUS, in turn, adds the LEC to the point-to-multipoint multicast forward VCC. Having completed initialization, configuration and registration, the LEC is now ready for the data transfer phase.

19.1.5 Data Transfer Phase

The following text uses Figures 19.4 and 19.5 to illustrate the LANE data transfer phase. To begin, LEC_1 needs to transmit a MAC frame. This occurs in an end system when a higher-layer protocol generates a packet in a NIC, or else when an ATM LANE port in a bridging device receives a frame from another LAN (or LANE) port. For the first frame received with a particular destination MAC address, the LEC doesn't know the ATM address needed to reach the destination. The LEC formulates an LE_ARP packet and sends it to the LES in step 1.

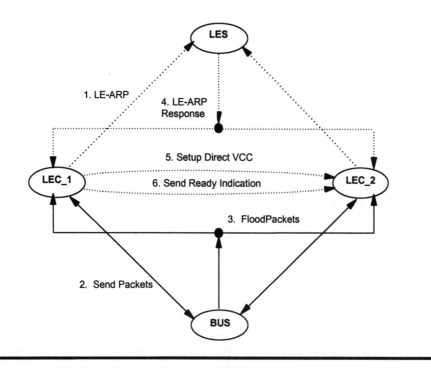

Figure 19.4   LANE Data Transfer Example (1 of 2)

While waiting for a response to the LE_ARP, the LEC may forward packets to the BUS in step 2, which floods the packet to all LECs over the multicast distribute connection in step 3. Alternatively, the LEC could buffer the packets. However, this increases response time to perform the other functions described below. Flooding ensures that no packets are lost, and also reaches MAC addresses "behind" bridging devices that don't register all of their MAC addresses with the LES. Many network protocols are very sensitive to loss, hence the flooding step ensures good performance. However, note that the amount of broadcast traffic limits the size of the emulated LAN. Furthermore, the time required to resolve the LE_ARP request may increase response time unacceptably. As studied in Chapter 9, spanning tree bridges also employ a similar flooding procedure for packets with unknown destination addresses.

When LEC_1 receives an LE_ARP response in step 4 indicating that LEC_2 is the target ATM address, then it sets up a data-direct VCC in step 5. LEC_1 then sends a ready indication frame in step 6. LEC_1 could then use this data direct VCC for data transfer, instead of the BUS multicast distribute VCC. The use of direct VCCs makes efficient use of the underlying ATM network capabilities. Of course, the ATM switches in the network must be capable of processing a large number of SVC requests per second in order for LANE to realize this efficiency. If considering LANE for your enterprise or wide area network, be sure to inquire about the SVC call attempt rates supported by your ATM switch vendor or service provider.

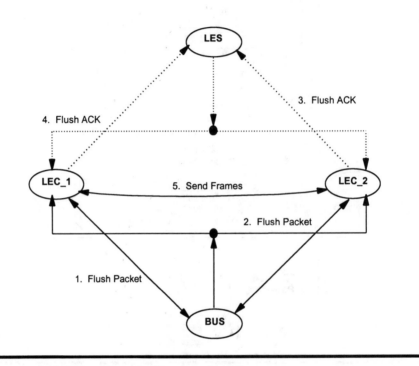

Figure 19.5 LANE Data Transfer Example (2 of 2)

Before LEC_1 uses the direct path, it may choose to utilize the LANE flush procedure to ensure in sequence packet delivery as illustrated in Figure 19.5. Recall from Chapter 8 that most LANs deliver frames in sequence, unlike IP networks. The flush procedure ensures that packets previously sent to the BUS arrive at the destination prior to sending any packets over the data direct VCC. The LEC sends a flush cell over the BUS multicast send VCC in step 1 and receives it back over the point-to-multipoint multicast forward VCC in step 2. LEC_1 buffers any other packets while awaiting a flush acknowledgment (ACK). Once LEC_2 receives the flush packet, it generates a flush acknowledgment on the control-direct VCC to the LES in step 3. The LES distributes the flush ACK to at least LEC_1 in step 4. Now the LECs can exchange frames over the data direct path in step 5.

If a data direct VCC already exists to the destination LEC, the source LEC may choose to use this same data direct connection, thereby efficiently utilizing connection resources and reducing latency. However, a pair of LECs may set up parallel connections to support application with different QoS requirements. If the LEC receives no response to the LE_ARP, then it continues sending packets to the BUS. The LEC periodically re-sends LE_ARPs packets in an attempt to solicit a response. Typically, once the BUS floods a packet through the emulated LAN, another LEC will learn the destination's location, and respond to a subsequent LE_ARP.

Often, an LEC locally caches MAC address to ATM address mappings learned from listening to LE_ARP responses. The LEC first consults a local cache table for the destination MAC address, and uses the cached mapping if a match is found instead of resolving the MAC address to the destination ATM address using the LE_ARP procedure described previously. The LEC ages cached entries over an adjustable interval, usually on the order of minutes, and removes the cached entry if no packets were sent to that MAC address. This aging procedure ensures that invalid mappings eventually leave the cache. The LEC clears data direct VCCs if no activity occurs for the predetermined adjustable interval on the order of minutes. This time-out procedure ensures that ATM network connection resources are used efficiently.

LECs also utilize the BUS for broadcast and multicast packets. Since the BUS sends these packets to all LECs over the multicast forward VCC, the source LEC receives its own broadcast/multicast packet. Because some LAN protocols prohibit this, LANE specifies that the source LEC prefix each packet with its LECID so that it can filter out packets received from itself from the BUS over the multicast forward VCC.

19.1.6 LANE and Spanning Tree

The LANE protocol supports the IEEE spanning tree protocol described in Chapter 9. A number of complex situations result when external networks connect to emulated LANs via LAN switches and bridges that employ the spanning tree protocol. Furthermore, these external switches and bridges may be connected over shared-media LANs, creating the possibility of multiple paths between source and destination, and hence the possibility of fatal bridging loops.

LECs within LAN switches exchange spanning tree bridge packets over the BUS. If a bridging device detects a loop via the spanning tree protocol, then it disables one of the external ports involved in the loop. Since the spanning tree protocol employs bandwidth weighted metrics, it first turns off lower-speed LAN ports prior to disabling any high-speed ATM ports.

Within a complex bridged network using the spanning tree protocol, the reachability of external MAC addresses through a particular LEC changes whenever network conditions change. This dynamic nature of the spanning tree protocol can interact unfavorably with the LANE protocol. For example, the ARP cache could map one or more external MAC addresses to an LEC's ATM address; which is no longer capable of reaching the MAC address as determined by the spanning tree protocol in response to a change in the legacy LAN topology.

The LANE protocol supports LE-Topology-Request messages to minimize the duration of these transient lapses in connectivity. Any LEC implementing the spanning tree protocol that detects a bridged topology change which triggers a BPDU configuration update message should also distribute a LE-Topology-Request via the LES. Upon receipt of the LE Topology Request message, all LECs must reduce the aging period on their cached ARP information. This action flushes out the cached information more rapidly and causes LECs to update mapping information through LE-ARPs. LECs do not disconnect existing data-direct connections; however, the updated cache information will cause inactivity on the data direct VCC, causing it to eventually time out as well. Hence, the emulated LAN heals itself in conformance to the dynamically changed LAN within no more than a few minutes, as controlled by the aforementioned timers.

19.1.7 LANE Implementation Considerations

An LEC resides on every ATM-attached station in an emulated LAN. Although Figure 19.3 depicts all server functions implemented separately, the LANE protocol does not specify the location of any of the server components; any device or devices with ATM connectivity suffice. For the purposes of reliability and performance; however, most vendors implement these server components on networking equipment, such as ATM switches or routers, rather than on a workstation or host.

Each LEC is part of an ATM end station. It represents a set of users identified by their MAC address(es). Communication among LECs and between LECs and the LES occurs over ATM Virtual Channel Connections (VCCs). Each LEC communicates with the LES over control and data VCCs. Emulated LANs operate in any of the following environments: Switched Virtual Connection (SVC), Permanent Virtual Connection (PVC), or mixed SVC/PVC. There are no call setup and release procedures in a PVC-only LAN; instead, layer management sets up and clears connections. However, the large number of PVCs required make all but the smallest emulated LAN networks too complex to manage.

Since MAC addressing is flat, (i.e., no logical hierarchy exists) bridges must flood connectivity data throughout the emulated LAN. LANE makes the same tradeoff that bridges do, achieving plug and play operation at the cost of decreased scalability. A further differentiating advantage of ATM LAN emulation over LAN switching is that it implements LANs at aggregate speeds on the order of gigabits per second. For example, a network of sixteen workstations with 622-Mbps ATM NIC cards operates at an aggregate rate of 10-Gbps (full duplex). Note, however, that when connected to low-speed legacy LANs, like 10-Mbps Ethernet, that the slowest port on the LAN limits the total amount of broadcast traffic, and hence the maximum emulated LAN capacity. Also, emulated LANs make efficient use of WAN bandwidth, avoiding the problems in spanning tree bridging.

LAN designers must use routers to connect the ELANs to routers interconnecting bridged networks as described in Chapter 8. This works because LANE looks exactly like a LAN to a router. The administrator may also assign file servers, printer, Internet gateways, and other resources to the ELAN with the highest traffic usage to minimize the amount of traffic flowing through the router.

Multi-vendor interoperability of LANE NIC cards and bridges/routers is quite good. In this sense, the LANE specification has been very successful. Many vendors have developed LANE protocol implementations in the following physical implementations:

- ATM Network Interface Card (NIC)

- ATM Switches
- LAN Switches
- Routers

ATM NIC cards enable ATM connected workstations to communicate with legacy LAN attached workstations and servers. As seen in Chapter 16, almost all ATM NICs support the LANE protocol. Since LANE protocols use standard ATM signaling procedures, many vendors find ATM switches convenient platforms for implementing LANE server components. The 3Com Corebuilder 7000 is an example of such an implementation [4]. LANE provides a standard way to connect LAN switches, which effectively act as fast multiport bridges. Basically, the LANE protocol resolves MAC addresses into ATM addresses, and hence implements bridging, which makes it a natural extension to current LAN switches. Routers may also use LANE to implement bridged virtual LANs. For example, the Cisco 4000 and 7000 series routers support LANE [5].

A LAN bridge learns about MAC addresses on adjacent LAN segments and doesn't propagate this information. Hence, a bridge shields LAN segments from each other by not propagating broadcast information. Several LANE functions specifically support the operation of LAN bridges.

19.1.8 Optional LANE Capabilities

Although the LANE protocol requires certain aspects for interoperability, many aspects are left open to vendor differentiation. The following are examples of optional LANE capabilities:

- LE_NARP messages
- Intelligent BUS
- Virtual LANs

LANE defines an unsolicited LE_NARP messages that allows LECs to indicate that a particular MAC address is now reachable locally. The LES redistributes the LE_NARP messages to all other LECs which use this information to update their address caches. LE_NARP messages may speed convergence under particular conditions, but are optional since the other methods described above eventually achieve convergence within minutes. If you have a dynamically changing LAN, for example, workers carrying laptops from room to room and plugging in at different LAN segments look for the LE_NARP capability.

An intelligent BUS shares knowledge of MAC address reachability with the LES. The intelligent BUS can then forward packets received from other LECs directly to the appropriate LEC across the bidirectional multicast send VCC, instead of over the point-to-multipoint multicast forward VCC. The intelligent BUS minimizes the complexity of the LEC implementation. The performance, of course, is limited by the forwarding capacity of the intelligent BUS. The LANE protocol, while allowing for intelligent BUS's, does require all LECs to set up data-direct VCCs whenever possible, and also restricts the number of flooded (unicast) packets sent to the BUS over a specified interval.

Vendors utilize LANE as the basis for virtual LAN (VLAN) service over ATM backbones. As mentioned earlier, a vendor may implement a VLAN capability through extensions to the LECS and LES. Employing network management interfaces, a network administrator can establish multiple emulated LANs over a common ATM network by controlling the LES and BUS assignments to end systems, bridges, and LAN switches. VLANs overcome the limitation of many early bridges that associated physical ports on a particular device with a specific LAN segment. VLANs place the control of LAN connectivity within software so that enterprises can dynamically move, add and change users across LAN segments without the requirement for physically moving workstations and servers. This benefit often justifies LANE deployment, especially in a multi-vendor environment. LAN switches which support each user as an individual segment may also implement a VLAN function [6].

Since LANE is a bridging protocol using flat addressing, it is subject to limitations and problems of other bridged protocols. One limitation is scalability, hence LANE is applicable only to smaller private networks. Also, bridged networks are subject to broadcast storms.

Users should plan to utilize routers to interconnect emulated LANs in a manner similar to that commonly in use today to connect smaller bridged networks. LANE allows high-performance ATM routers to be utilized. In an IP-based network, LECs have the same IP subnet address and use LANE procedures to establish data direct VCCs between themselves. If the destination IP address is not on the same subnet, then the LEC uses the MAC address of the default router; which then routes the packet towards its destination using its next hop table. Of course, eventually the capacity of the router becomes the limiting factor in such networks. Chapter 20 describes how the ATM Forum's MultiProtocol Over ATM (MPOA) design solves this problem by operating hand-in-hand with LANE.

19.1.9 LANE Updates and Advances: LUNI v2 and LNNI

The LANE 1.0 protocol specifies only the operation of the LAN Emulation User to Network Interface (LUNI, pronounced "Loony") between a LEC and the network providing the LANE service in a non-redundant mode. The Phase 1 LANE protocols specify only the LUNI operation; furthermore, the Phase 1 LANE protocol does not allow for the standard support of multiple LESs or BUSs within an ELAN. Hence these components represent both single points of failure and potential bottlenecks.

The LANE UNI version 2 (LUNI v2) specification [7] supersedes LUNI 1.0. The LUNI v2 specification defines enhanced capabilities, such as: LLC multiplexing for VCC sharing, support for ABR and other ATM service categories with different Qualities of Service (QoS), enhanced multicast capabilities, along with support for the ATM Forum's Multiprotocol Over ATM (MPOA) standard described in the next Chapter. LANE v2 defines protocol mechanisms in support of locally administered Quality of Service

(QoS) for communication between native ATM endsystems. LANE v2's enhanced multicast capability separates multicast traffic from the general broadcast path, defining a filtering protocol that determine which members of the emulated LAN receive particular multicast frames.

The ATM Forum plans to extend the specification that also adds a LAN emulation NNI (LNNI, pronounced "Lenny") interface, which operates between the server components within a single emulated LAN.

19.2 CELLS IN FRAMES (CIF)

Wouldn't it be nice if you could keep your existing legacy Ethernet NIC cards and LAN devices, yet still get much of ATM's capabilities to the desktop? Enter Cells in Frames (CIF), which uses software to place ATM cells inside LAN frames and forward them to a separate CIF attachment device that interfaces to native ATM networks. An eclectic group of hardware and software vendors, service providers, user organizations, and members of the trade press joined to form the Cells In Frames (CIF) Alliance in early 1996 [8]. Membership in the alliance is free, as the National Science Foundation funds part of the work at Cornell University. The Cells In Frames Alliance seeks to:

- Ease deployment of ATM-based networks
- Make migration to ATM more economical by reusing existing Ethernet hardware
- Establish specifications for carrying ATM over existing LAN infrastructure
- Empower user applications by giving them control over network methods used to provide Quality of Service

Glenville Armitage proposed the concept of carrying cells in legacy LAN frames as early as 1990 as a means to generate test traffic for ATM networks. To accomplish these goals, the CIF Alliance released version 1.0 of a specification in October, 1996 [9] detailing how to carry ATM over Ethernet, 802.3, and Token Ring networks.. Some believe that CIF may be the great enabling API for ATM, in other words, it is ATM for the rest of us outside of those small workgroups with high-end workstations.

Figure 19.6 illustrates the network context of CIF [9, 10]. A legacy LAN network connects to an ATM network via a CIF Attachment Device (CIF-AD) that supports ATM service, in particular multiple classes of service, over an existing LAN Network Interface Card (NIC). CIF utilizes existing LAN framing protocols transparent to applications written to an ATM-compliant Application Programming Interface (API).

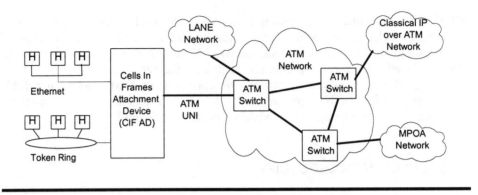

Figure 19.6   Cells in Frames (CIF) Network Architecture

Critical to the CIF concept is the End System (ES) architecture depicted in Figure 19.7. Functions defined in the CIF specification define how a LAN-attached workstation can implement ATM Header Error Checking (HEC), AAL5 Cyclic Redundancy Check (CRC), priority scheduling, leaky bucket traffic shaping, and end-to-end protocol independent flow control using the Available Bit Rate (ABR) service — all in software! The specification actually provides source code examples for many of these features.

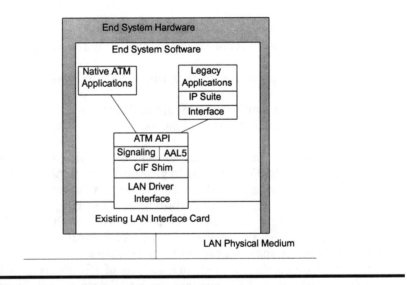

Figure 19.7   Cells in Frames (CIF) End System Architecture

CIF end systems also implement signaling and ILMI modules as part of an ATM processing layer that works with the existing LAN driver. Legacy applications have access to the AAL layers via LAN Emulation (LANE), MultiProtocol Over ATM (MPOA), or classical IP over ATM enabling direct connection with high-end native ATM end systems.

Some ATM vendors view CIF as a way to rapidly extend ATM's reach to the desktop in order to support real-time interactive traffic [11]. The attachment device does most of the segmentation and reassembly work and hence relieves the PC of a low-level, processor-intensive burden. The PC must implement standard ATM signaling and build ATM protocol data units (PDUs) as its part of the implementation. ATM thus uses CIF to bridge the gap between legacy and ATM end systems — without the requirement to change desktop hardware. This allows ATM vendors to capture market share away from legacy LAN system vendors — like routers and LAN switches.

CIF offers one option to the purchase of relatively expensive alternatives like new 25-Mbps or 155-Mbps ATM NIC cards. CIF is available in switching and concentrator devices. CIF bridges the time gap until more ATM APIs can be written. CIF also directly competes with RSVP signaling. See References 11 and 12 for more information on CIF.

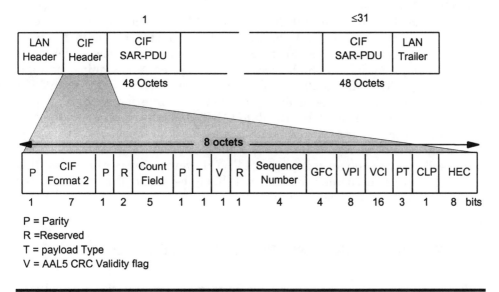

Figure 19.8   Cells in Frames (CIF) PDU Format

Figure 19.8 depicts the CIF frame format used between CIF end systems and the CIF Attachment Device (CIF-AD). A LAN header and trailer envelope the CIF frame comprised of one or more CIF headers and one or more 48-octet CIF SAR-PDUs. The CIF specification limits AAL5 PDUs to 31 cells in length to stay within the frame size restrictions of Ethernet LAN segments. These SAR-PDUs may contain information according to any AAL. Hence, CIF supports all of the AALs described in Chapter 13 — not just AAL5. The 8-octet format 2 CIF header carries data frames. Other CIF format types carry administrative information. Since more than one CIF header may be present in a CIF frame, the count field indicates the number of 48-octet SAR-PDUs associated with the current header. The last CIF header in the CIF frame has a value of zero for the count field. Since an AAL3/4 or AAL5 PDU can be larger than a LAN frame, the Sequence Number field allows CIF to segment a larger AAL-PDU into multiple smaller LAN frames for transfer between the end system and the CIF-AD. In the future of high performance networking, very long frames (with very long IPv6 headers) may become commonplace. The CIF-AD uses the sequence number to reassemble and segment the larger AAL PDUs. For all AALs except AAL5, the CIF-AD takes the 48-octet CIF SAR-PDUs and places them in cells using the cell header template values specified in the remainder of the CIF header, namely: GFC, VPI, VCI, PT, CLP, and HEC as defined in Chapter 12.

The CIF-AD performs some special processing for AAL5. The T bit indicates that the CIF-AD should set the AAL_indication bit in the cell header corresponding to the last SAR-PDU. Recall that AAL5 operates in this manner from Chapter 13. The V bit indicates whether the AAL5 CRC is valid in the last SAR-PDU associated with the current CIF header. If it is invalid, then the CIF-AD computes the AAL5 CRC and inserts it in the last cell of the CPCS-PDU.

CIF advocates [13] claim that their approach delivers high-speed service at a cost lower than ATM since the approach eliminates the need for SAR hardware in the end systems. Furthermore, custom ASICs allow switching of variable length packets at high-speeds in a weighted fair queuing manner. Additionally, CIF is up to 20 percent more efficient than TCP/IP over ATM when used in native frame mode over legacy LAN interfaces. Finally, CIF supports native ATM Available Bit Rate (ABR) flow control on an end-to-end basis, even over legacy Ethernet and Token Ring networks. The vendor community hasn't exactly embraced CIF in their products yet; we found only one vendor with announced plans [14]. However, some aspects of this research and prototype development will likely be adopted in some manner by other standards and specifications.

19.3 MULTIPROTOCOL ENCAPSULATION OVER AAL5

IETF RFC 1483 [15] defines the specific formats for routing or bridging a number of commonly used protocols over ATM Adaptation Layer 5 (AAL5) using either protocol encapsulation or VC multiplexing. *Protocol encapsulation* provides the capability to multiplex many different protocols over a single ATM Virtual Channel (VC) (which is commonly used shorthand name for Virtual Channel Connection (VCC)). The *VC multiplexing* method assumes that each protocol is carried over a separate ATM VC. Both of these encapsulation methods utilize AAL5. This section first covers the protocol encapsulation method and then the VC multiplexing method.

19.3.1 Protocol Encapsulation

Protocol encapsulation operates by prefixing the Protocol Data Unit (PDU) with an IEEE 802.2 Logical Link Control (LLC) SubNetwork Attachment Point (SNAP) header. Hence, RFC 1483 calls this *LLC/SNAP encapsulation*. The LLC/SNAP header identifies the PDU type. As described in Chapter 9, LLC/SNAP allows two hosts to multiplex different protocols over a shared

LAN MAC sublayer. This method was initially designed for public network or wide area network environments where a premises device would send all protocols over a single VCC, because some carrier pricing structures favored a small number of PVCs. In other words, this capability saves money by sending multiple protocols over one VCC, avoiding the expense of ordering a separate VCC for each protocol. Note that all packets get the same QoS since they share a single VCC.

Figure 19.9a illustrates protocol encapsulation by routers, showing a network of three routers each multiplexing separate Ethernet and Token Ring LANs over a single VCC interconnecting the locations. The routers multiplex the Ethernet and Token Ring PDUs onto the same VCC using the encapsulation described below. The drawing in Figure 19. 9b illustrates how bridges multiplex PDUs from an Ethernet and Token Ring interfaces to yield a bridged LAN. The Ethernet bridges only use a spanning tree of the ATM VCCs at any one point in time, which the example illustrates by a dashed line to indicate the unused VC link in the spanning tree where the center bridge assumes the role of the spanning tree root. The token ring bridges may make more efficient use of the ATM VCCs through source routing as described in Chapter 9.

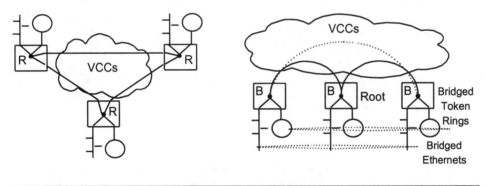

a. LLC Encapsulation - Routing

b. LLC Encapsulation - Bridging

Figure 19.9   Routing and Bridging Use of LLC Encapsulation

19.3.1.1 LLC Encapsulation for Routed Protocols

The following text describes LLC encapsulation for routed protocols with reference to Figure 19.10. Figure 19.10a depicts the LLC-encapsulated routed ISO protocol payload structure. The 3 octets of the LLC header contain three 1-octet fields: a Destination Service Access Point (DSAP), a Source Service Access Point (SSAP), and control. The routed ISO protocol is identified by a 1-octet NLPID field which is included in the protocol data. NLPID fields include, but are not limited to, SNAP, ISO CLNP, ISO ES-IS, ISO IS-IS, and Internet IP. The values of these octets are shown in hexa-decimal notation as X'zz', where z is a hexadecimal digit representing 4 bits. For example X'1' corresponds to a binary '0001' and X'A' corresponds to a binary '1010'. Figure 19.10b depicts the payload for a routed non-ISO protocol, specifically showing the example for an IP PDU. The 3-octet SNAP header follows the LLC and contains a 3-octet Organizationally Unique Identifier (OUI) and the Protocol IDentifier (PID). RFC1483 defines support for other protocols, such as ISO CLNP and IS-IS through the use of different PIDs.

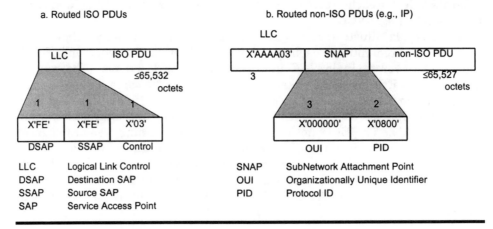

a. Routed ISO PDUs

b. Routed non-ISO PDUs (e.g., IP)

LLC	Logical Link Control		SNAP	SubNetwork Attachment Point
DSAP	Destination SAP		OUI	Organizationally Unique Identifier
SSAP	Source SAP		PID	Protocol ID
SAP	Service Access Point			

Figure 19.10   LLC Encapsulation Using Routed Protocols

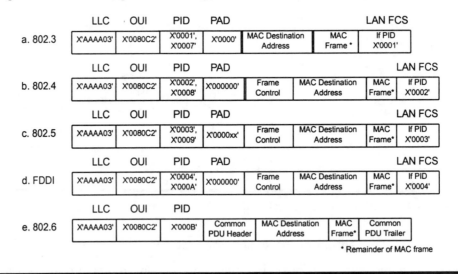

Figure 19.11   LLC Encapsulation Using Bridged Protocols

19.3.1.2 LLC Encapsulation for Bridged Protocols

The LLC encapsulation method supports bridging for LAN and MAN protocols. The SNAP header identifies the type of bridged-medium and whether the original LAN Frame Check Sequence (FCS) is included with the PDU as illustrated in Figure 19.11. The LLC header identifies a non-ISO PDU as before. The OUI field identifies an 802.1 identification code. The

PID field then identifies the actual protocol. The remaining fields are either padding or the actual LAN PDU. Bridgeable protocols include 802.3 Ethernet, 802.4 Token Bus, 802.5 Token Ring, Fiber Distributed Data Interface (FDDI), and 802.6 DQDB, as shown in the figure.

19.3.2 VC-Based Multiplexing

The second method of carrying multiple protocols over ATM is through VC-based multiplexing, which supports a single protocol per virtual connection. In other words, the VCs are multiplexed rather than the protocols themselves as done in the protocol encapsulation method. With this method, different protocols can have different bandwidth allocations and QoS, unlike the multi-protocol encapsulation method.

Figure 19.12a illustrates the VC multiplexing concept for routed protocols, showing a separate VCC connecting the routing point for the attached Ethernet and Token Ring LANs. Figure 19.12b illustrates the same situation for bridged protocols, again requiring twice as many VCCs as for protocol encapsulation. In the example, the source routed bridged Token Ring load balances across all VCCs, while the spanning tree Ethernet does not use the VCC indicated by a dashed line. Comparing this to Figure 19.9, observe that the only difference is the use of one VCC for each protocol that is being routed or bridged versus one VCC between each pair of routers or bridges — Ethernet and Token Ring.

a. VC Multiplexing - Routing   b. VC Multiplexing - Bridging

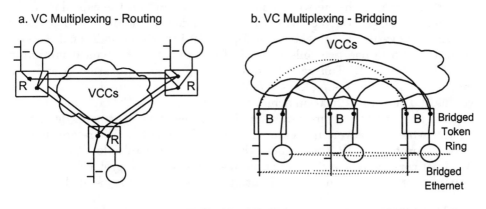

Figure19.12   Routing and Bridging Usage of VC Multiplexing

The bridged PDU payload is devoid of the LLC and SNAP protocol identifiers used in protocol encapsulation, resulting in less overhead, less processing, and higher overall throughput at the expense of the lost routing function. This method is designed for environments where the user can dynamically create and delete large numbers of ATM VCCs in an economical fashion, which occurs in private ATM networks or ATM SVC networks. Routed protocols can make use of the entire 65,535-octet AAL5 CPCS PDU. Bridged protocols have the same format as shown in Figure 19.11 without the

LLC, OUI, or PID fields. The use of LAN FCS in the bridged protocols is implicitly defined by association with the VCC.

19.3.3 Selection of Multiplexing Method

Either of the two types of multiplexing methods, encapsulated or VC multiplexing, can be used with PVCs and SVCs. The method is selected by a configuration option for PVCs. SVCs use information elements in the signaling protocol for the two routers to communicate whether to employ LC/SNAP protocol encapsulation or VC multiplexing. Signaling also indicates, when using VC multiplexing, whether the original LAN FCS is being carried in the PDU.

If your application is IP-based, then the choice of RFC 1483 encapsulation method can have a profound impact on efficiency. Recall from Chapter 8 that the minimum length TCP/IP version 4 packet for a standalone acknowledgment is exactly 40 bytes in length. Although TCP attempts to "piggyback" acknowledgments on packets headed in the reverse direction, the experience in real world networks is that approximately one third of the packets on live TCP/IP networks are exactly 40 bytes long. Recall from Chapter 13 that AAL5 adds 8 octets of overhead, along with an optional padding field to extend the AAL5 PDU to exactly fit within an integer number of cells. If the network designer selects VC multiplexing , then the TCP/IP packet and the AAL5 overhead exactly fit into one cell as illustrated in Figure 19.13a. On the other hand, if the network designer chooses protocol encapsulation for non-routed ISO PDUs (i.e., IP), the router adds an additional 8 bytes of LLC/SNAP overhead. As shown in Figure 19.13b, these additional 8 bytes require AAL5 to use two cells instead of one to carry the typical minimum size TCP/IP packet. This results in an ATM payload utilization of approximately 58 percent (i.e., (48+8+8)/96)) when using protocol encapsulation versus 100 percent utilization achieved when employing VC multiplexing. Of course, the utilization improves for TCP/IP packets of greater length, a subject which we cover in greater depth in Chapter 31. Therefore, if your network protocol is IP, then unless the economics of multiple VCCs in your ATM network overcome the reduction in efficiency due to protocol encapsulation, seriously consider using VC multiplexing over multiple parallel VCCs.

a. VC Multiplexing   b. Protocol Encapsulation

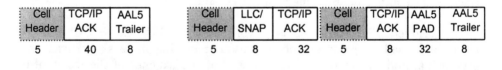

Figure 19.13 TCP/IP Efficiency Using VC Multiplexing and Protocol Encapsulation

19.4 REVIEW

This chapter covered the important topic of ATM in Local Area Networks (LANs) . The text provided an in-depth look at how the ATM Forum's LAN Emulation (LANE) protocol supports seamless interconnection of ATM hosts and servers with their legacy LAN counterparts.   The treatment covered hardware and software aspects, and defined the components and connections involved in the LANE protocol.   The discussion then walked through the steps involved when a LAN Emulation Client (LEC) initializes, registers, resolves addresses and transfers data. The chapter then described optional LANE capabilities, implementation considerations and highlighted key new features in the LANE version 2 specification. Next, the chapter surveyed the Cells In Frames (CIF) proposal to allow end systems to encapsulate ATM cells in native LAN frames using software only implementations. Although not widely adopted yet, aspects of this research effort may yet find their way into standards and implementations. Finally, we covered the multiprotocol over ATM encapsulation standard defined in IETF RFC 1483. The text described the protocol encapsulation and VC multiplexing methods, providing guidelines for selection of the most economical and efficient method for a particular application.

19.5 REFERENCES

[1]   ATM Forum, "LAN Emulation Over ATM: Version 1.0 Specification," af-lane-0021.000, January 1995.

[2]   N. Finn and T. Mason, "ATM LAN Emulation," IEEE Communications, June 1996.

[3]   ATM Forum, "LAN Emulation Over ATM - Version 2 - LUNI Specification," af-lane-0084.000, July 1997.

[4]   B. Klessig, *ATM LAN Emulation*, 3Com, http://www.3com/nsc/500617.html.

[5]   Cisco, "LAN Emulation — Technology Brief," 1995.

[6]   J. Chiong, *Internetwokring ATM for the Internet and Enterprise Networks*, McGraw-Hill, 1997.

[7]   ATM Forum, *LAN Emulation Over ATM Version 2 - LUNI Specification*, AF-LANE-0084.000, July 1997.

[8]   "Cells in Frame Home Page," http://cif.cornell.edu.

[9]   S. Brim, *Cells In Frames Version 1.0: Specification, Analysis, and Discussion*, , Cornell University, http://www.cif.cornell.edu/specs/v1.0/CIF-baseline.html, October 21, 1996.

[10] R. Dixon, "Cells in Frames: A System Overview," *IEEE Network*, July/August 1996.

[11] R. Jeffries, "Three roads to quality of service: ATM, RSVP, and CIF," *Telecommunications*, April 1996.

[12]   R. Cogger, P. Elmer, "Is CIF a viable solution for delivering ATM to the desktop?," Network World, Jun 24, 1996. Vol.13, Issue 26.

[13]   L. Roberts, "CIF: Affordable ATM, At Last," *Data Communications on the web*, http://www.data.com/tutorials/cif.html, April 1997.

[14]   M. Rae McLean, "Madge Rallies Integrated Services," *LAN Times*, http://www.wcmh.com/lantimes/96jul/607b019a.html, July 1996.

[15]   J. Heinanen, , "RFC 1483: Multiprotocol Encapsulation over ATM Adaptation Layer 5," IETF, July 1993.

ATM and
Internetworking

This chapter covers ATM's support for internetworking protocols, such as the popular Internet Protocol (IP). The emergence of the World Wide Web (WWW) as the killer application for data communications established IP as the de facto standard for internetworking to the corporate desktop as well as the home office and residential user. The standards and some proprietary implementations have also made significant progress in addressing the more difficult problem of seamless interoperation of ATM-based networks with existing internets, intranets, and extranets. These designs overcome the fundamental difference between ATM's non-broadcast, connection-oriented protocol and the Internet's broadcast, connectionless orientation via a combination of servers, address resolution and emulation of broadcast capabilities. The text delves into this important subject, showing how industry experts crafted designs involving servers and clever paradigm shifts to emulate broadcast, connectionless LANs and Internets over ATM's non-broadcast, connection-oriented infrastructure.

Figure 20.1 illustrates the higher-layer user plane protocols covered in this chapter that support internetworking protocols, primarily IP, over an ATM infrastructure. As indicated in the figure, several of these protocols build upon the infrastructure of LAN Emulation (LANE) and multi-protocol encapsulation defined in the previous chapter. On top of this infrastructure, we describe support for classical IP subnetworks and emulation of the IP multicast group service. Advanced IP over ATM approaches either extend IP address resolution beyond IP subnet boundaries, or interconnect IP subnets. The first approach directly connects classic IP subnets via Non-Broadcast Multiple Access (NBMA) media, such as ATM, in the *Classical IP over ATM* approach while the second involves IP routing and IP forwarding in the *Next Hop Resolution Protocol* approach. Finally, we introduce several proprietary, leading edge, high-performance, QoS aware designs that support IP networking over an ATM infrastructure. These IP and tag switching techniques defining the IETF's work on a MultiProtocol Label Switching standard offer promise to meet the rapid growth created by the information content of the Web over ATM networks.

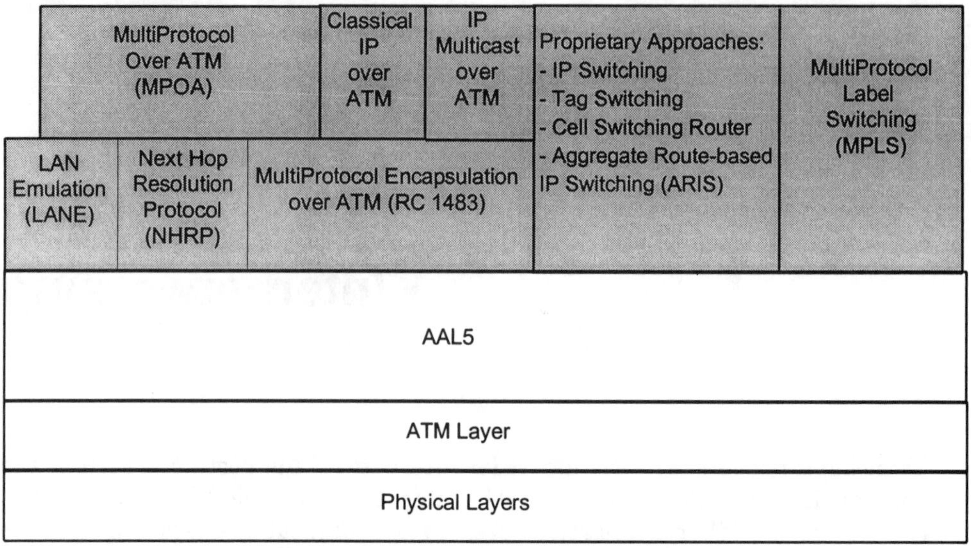

MultiProtocol Over ATM (MPOA)		Classical IP over ATM	IP Multicast over ATM	Proprietary Approaches: - IP Switching - Tag Switching - Cell Switching Router - Aggregate Route-based IP Switching (ARIS)	MultiProtocol Label Switching (MPLS)
LAN Emulation (LANE)	Next Hop Resolution Protocol (NHRP)	MultiProtocol Encapsulation over ATM (RC 1483)			
AAL5					
ATM Layer					
Physical Layers					

Figure 20.1 User Plane Higher-Layer Protocols for Internetworking with ATM

20.1 IP MAXIMUM TRANSFER UNIT (MTU) OVER AAL5

RFC 1626 defines the Maximum Transfer Unit (MTU) negotiation over the Internet over AAL5 [1]. This standard specifies the default MTU size over ATM AAL5 at 9180 bytes, aligning it with the default MTU size for IP over SMDS specified in RFC 1209. Experience shows that networks perform better with larger MTU sizes by avoiding processor intensive fragmentation and reassembly at the IP layer.

The standard also specifies procedures for use with ATM SVCs that allow dynamic negotiation of larger MTU sizes, up to the AAL5 limit of 64 kbytes. Larger MTU sizes are more efficient because they minimize AAL5 overhead and processing overhead. This standard also specifies that all routers utilize the IP Path MTU discovery mechanism defined in RFCs 1191 and 1435 to avoid IP packet fragmentation, resulting in the greatest efficiency. Use of the path MTU discovery mechanism is important because ATM uses a default MTU size significantly different from older subnet technologies such as Ethernet and FDDI. The consequence of choosing an MTU size too large is reduced performance, due to the high overhead process of IP packet fragmentation and reassembly at intermediate routers that convert between ATM and other media with shorter MTU sizes.

20.2 CLASSICAL IP OVER ATM

A basic concept involved in all methods supporting IP over ATM is resolution of an IP address to a corresponding ATM address. End stations must have both IP and ATM addresses in Classical IP over ATM subnetworks. IETF RFC 1577 specifies Classical IP over ATM for the use of ATM as a direct replacement for the "wires" interconnecting IP end stations, LAN segments, and routers in a Logical IP Subnetwork (LIS) [2]. An LIS consists of a group of hosts or routers connected to an ATM network belonging to the same IP subnet, that is, they all have the same IP subnet number and mask as described in Chapter 9. The Classical IP over ATM procedures apply to both PVCs and SVCs. We cover the PVC case first.

20.2.1 Inverse Address Resolution Protocol (InARP)

RFC 1293 [3] defines an *Inverse Address Resolution Protocol (InARP)* as a means for routers to automatically learn the IP address of the router on the far end of an ATM VCC PVC. Basically, it involves a station sending an InARP message containing the sender's IP address over the ATM VCC PVC. This situation occurs in PVC networks upon initialization, or when a router reloads its software because the VCC is known, but the IP address reachable via the VCC is unknown. The station on the other end of the ATM VCC PVC then responds with its IP address, establishing an association between the IP addresses of the pair and the ATM VCC's VPI/VCI on each ATM interface.

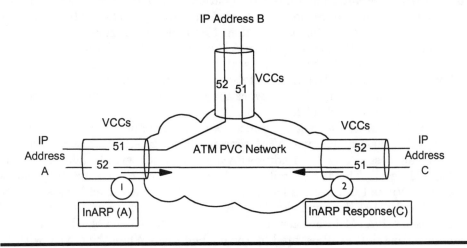

Figure 20.2 Example of Inverse ARP Procedure

Figure 20.2 illustrates the principle of Inverse ARP (InARP) over an ATM network comprised of three VCC PVCs. The router port on the left hand side with IP address *A* sends an inverse ARP over VCC 52 [InARP(*A*)] in the first step. This appears on VCC 51 to the router port with IP address *C* on the right hand side of the figure. The router port with IP address *C* responds with its identity [InARP Response(*C*)] on VCC 51 in the second step. Now the router ports with IP addresses A and C know that they can reach other by transmitting on ATM VCCs 52 and 51, respectively.

20.2.2 ATM Address Resolution Protocol (ATMARP)

Having to manually configure large numbers of PVCs does not scale in larger networks. Therefore, RFC 1577 also specifies an automatic configuration method using SVCs along with an ATM Address Resolution Protocol (ATMARP) server. The standard requires a logical ATMARP server for each LIS, which is a single point of failure in the current standard. The IETF has standardized a Server Cache Synchronization Protocol (SCSP) to support a multi-server environment. A single physical device may implement multiple logical ATMARP servers. The ATMARP server resolves the ARP requests for all IP stations within the LIS. IP stations first register with the ATMARP server, then act as clients to the server when making Address Resolution requests. Hosts must also age their ARP table entries to remove old data.

IP stations register with the ATMARP server by placing an SVC "call" to the ATMARP server. The ATMARP server may either transmit an Inverse ARP request to the newly attached client to determine the node's IP and ATM addresses, or determine this information from ARP requests. In either case, the ATM ARP server stores the association of a node's IP and ATM addresses in its ATMARP table. The ATMARP server time stamps entries and may periodically test that the IP station is still there using an InARP message after approximately 15 minutes of inactivity. A response restarts the time in the ATMARP server. The server ages old entries using this time stamp and eventually removes unresponsive IP stations after periods of 20 minutes or more. The key server function is then to enable IP stations to resolve the association of an IP address with an ATM address. Armed with the destination's ATM address, the originating station dynamically sets up an ATM SVC. After a period of inactivity, the stations take down the SVC to efficiently utilize bandwidth. The ATM network may also release the VCC in response to some failure or overload conditions.

Figure 20.3 illustrates the reference configuration and a simple example of the IETF's classical IP over ATM concept. Two interfaces are shown: one with IP address A and ATM address X and the other with IP address B and ATM address Y. The devices with IP addresses A and B have already made a Switched Virtual Connection (SVC) to the ATMARP server and registered so that the ATMARP server knows the association of their IP and ATM addresses. IP/ATM interfaces have a VCC over which ATM UNI signaling messages are sent as illustrated by the dashed line. When the device with IP address A wishes to send data to IP address B, the first step is to send an Address Resolution Protocol (ARP) message to the ATMARP server. In the second step, the server returns the ATM address Y of the device with IP address B. In the third step, the device with ATM address X originates a SETUP message to ATM address Y. The ATM network switches the SETUP message through to the destination ATM address Y on VCC 5. The switched ATM network makes a connection and requests a SETUP to ATM address Y

using the signaling VCC 5. The SETUP message specifies that data traffic should be sent on VCC 58 as indicated in the figure.

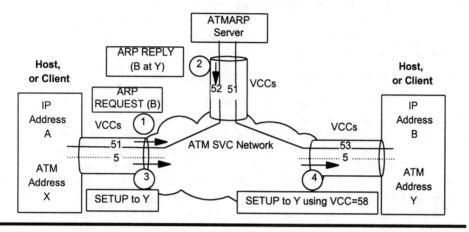

Figure 20.3 Classical IP over ATM — ARP and Setup Steps

Figure 20.4 illustrates the final steps in the classical IP over ATM scenario. In the fifth step, the device with ATM address Y responds with a CONNECT message which the switched ATM network uses to establish the VCC back to the originator, ATM address X. In the sixth step, the ATM network sends a CONNECT message to the device with ATM address X indicating that VCC 54 makes the connection with the device with ATM address Y that is using VCC 58. In the seventh and final step, communication between IP addresses A and B occurs over VCC 54 for IP address A and VCC 58 for IP address B as indicated in the figure. Either ATM address X or Y could release the ATM SVC by issuing a RELEASE message. Classical IP over ATM emulates the connectionless paradigm of datagram forwarding by automatically releasing the SVC call if no packets flow over the VCC for a pre-configured time interval. Typically, these timers are on the order of minutes.

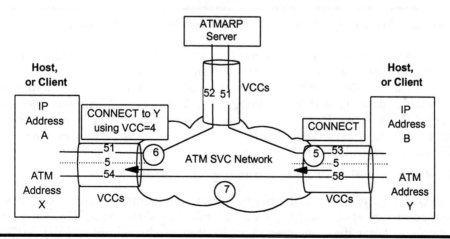

Figure 20.4 Classical IP over ATM — Connect and Data Transfer Steps

RFC 2225 [4] combines RFCs 1577 and 1626 into one document and clarifies several points gained from operational experience in operating real world classical IP over ATM networks. Hence, it defines a stable baseline specification in support of logical IP subnetworks over ATM PVC or SVC networks using the techniques summarized above.

20.2.3 Interconnecting Logical IP Subnetworks

Figure 20.5 illustrates the operation of multiple interconnected LISs. Each of the three LISs operates and communicates independently of all other LISs, even if they are all on the same ATM network. The classical model of RFC 1577 requires that any packet destined outside the source host or routers logical IP subnet (LIS) must be sent to an IP router. For example, a host on LIS-1 must send packets to the router 1 (i.e., the default router for packets not on its LIS) for delivery to hosts or routers on LIS-2 or LIS-3. Routers 1 and 2 are configured as endpoints of the LISs to which they are connected. RFC 1577 notes that this configuration may result in a number of disjoint LISs operating over the same ATM network, but the standard states that hosts on different LIS's must communicate via an intermediate IP router, even though said hosts could open a direct VC between themselves. If the stations did not follow this rule, then routing loops could result.

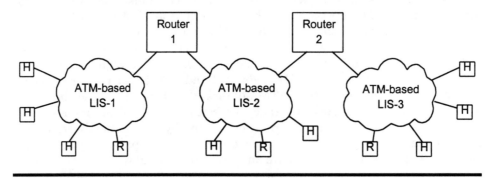

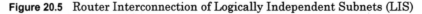

Figure 20.5   Router Interconnection of Logically Independent Subnets (LIS)

RFC 1577 specifies that implementations must support IEEE 802.2 Logical Link Control/SubNetwork Attachment Point (LLC/SNAP) encapsulation as described in RFC1483 covered in Chapter 19. LLC/SNAP encapsulation is the default packet format for IP datagrams. The default Maximum Transfer Unit (MTU) size for IP stations operating over the ATM network is 9180 octets. Adding the LLC/SNAP header of 8 octets, this results in a default ATM AAL5 protocol data unit MTU size of 9188 octets.

20.2.4 Classic IP over ATM Signaling Standards

RFC 1755 [5] specifies the details that hosts and routers require to achieve interoperability when using the SVC capabilities referred to in RFC 1577. In particular, it specifies the precise utilization of the following UNI 3.0/3.1

information elements in the SVC implementation of RFC 1577's classical IP over ATM.

- AAL Parameters IE
  ⇒ AAL5, Max SDU Size 65,535 Bytes, Null SSCS
- Broadband Low Layer Information (B-LLI) IE
  ⇒ Logical Link Control (LLC)
  ⇒ Layer 3 protocol is IP
- ATM Traffic Descriptor IE (See Table 20.1)
- Broadband Bearer Capability IE (See Table 20.1)
- QoS Parameter IE (See Table 20.1)
- Called/Calling Party Address
- Called/Calling Party Subaddress (Optional)
- Transit Network Selection (Optional)

All fields listed are mandatory unless otherwise indicated. For example, end stations use the AAL parameters and B-LLI information elements at call setup time to set up the correct mappings to higher layer protocols. The address and subaddress fields may be either of NSAP format, or E.164 as described in Chapter 15.

Table 20.1 lists the combinations of the Broadband Bearer Capability, Traffic Descriptor, and QoS Parameter information elements required in RFC 1755 in all IP-over-ATM end systems. Other combinations are optional. The first column supports RSVP type systems employing a token-bucket style characterization of the source, such as that defined in the RSVP standards (see RFC 1363 [6] and RFC 1633 [7]). The second column supports virtual ATM "pipes" between two routers. The last column supports a best effort service targeted for end system usage. Blank entries are not applicable, that is, they are null. See RFC 1755 for detailed coding of the UNI 3.0/3.1 signaling messages. An appendix in RFC 1755 lists a much larger set of optional signaling message codings, but it requires only those listed in Table 20.1 for interoperability. Hopefully, this summary gives the reader an appreciation of the level of detail required to achieve interoperability when utilizing ATM SVCs.

RFC 2331 [8] updates these guidelines for the UNI 4.0 signaling specification. This specification defines support for Available Bit Rate (ABR) signaling for point-to-point Calls, traffic parameter negotiation and frame discard. The document also defines a procedure for holding up the ATM SVC as long as the receiver periodically refreshes its reservation using the RSVP protocol as discussed in Chapter 8.

20.2.5 Signaling Support Issues

The two previous examples of supporting IP over ATM rely heavily on ATM Switched Virtual Connection (SVC) capability. Several questions are left unanswered. How rapid must the ARP and call setup process operate for it to be useful to end applications? What should be the holding time duration before the switched connection is released to make efficient use of resources?

What call attempt rate will these applications generate? How many active calls need to be supported? Will the Q.2931 protocol be able to support the required call setup time and attempt rate?

The answers to these questions will be answered through implementation experience and user feedback. Acceptable call setup times and the support for reasonable call attempt rates will be critical factors in the success of the approach of supporting connectionless services by dynamic connection switching. In a typical subnetwork environment, the number of ATM SVC calls per end station is relatively small. Usually, a single user never has more than a connection open to a few file servers at most, an Internet firewall router, and a print server simultaneously. Since these connections are all on the order of minutes to hours, many users will have fewer connections. Hence, the largest call setup rates occur when a server or gateway fails and many stations attempt to reconnect.

Table 20.1 Required RFC 1755 Traffic/QoS Signaling Parameters

Information Element Parameter	RSVP Capable Source	Virtual Router Pipe	Best Effort Host
Broadband Bearer Capability			
Broadband Bearer Class	C	X	X
Traffic Type		CBR	No Indication
Timing Required		Yes	No Indication
Traffic Descriptor			
PCR(CLP=0)			
PCR(CLP=0+1)	Specified	Specified	Specified
SCR(CLP=0)			
SCR(CLP=0+1)	Specified		
MBS(CLP=0)			
MBS(CLP=0+1)	Specified		
Best Effort Indicator			Specified
Tagging	No	No	No
QoS Parameter			
QoS Class	0	0	0

20.3 NEXT HOP RESOLUTION PROTOCOL (NHRP)

Recall that the Classical model for IP over ATM specified in RFC 1577 requires that all communication between different Logically Independent Subnets (LIS) occur through a router. Such networks fail to take advantage of more efficient, ATM-only "short cut" routes enabling direct communication between nodes that are connected to the same ATM network, but members of two different LISs, hence bypassing throughput limiting intermediate router hops. The continued penetration of ATM into internetworks increases the possibility that two stations connect to the same ATM network. Furthermore, ATM better serves voice, video and real-time data applications than legacy connectionless IP routers.

The IETF worked for several years evaluating protocols designed to address these issues, and after considering various approaches, the IETF advanced a solution known as the Next Hop Resolution Protocol (NHRP) in RFC 2332 [9]. NHRP operates over Non-Broadcast, Multi-Access (NBMA) subnetworks (e.g., ATM, FR and SMDS) as described in Chapter 9. NHRP enables different Logically Independent Subnetworks (LIS) present on the same NBMA network to decouple the local versus remote forwarding decision from the addressing convention that defined the LIS's. Thus, NHRP enables systems to directly interconnect over an NBMA network, independently of addressing based upon considerations such as traffic parameters or QoS characteristics.

20.3.1 Operation of NHRP Across Logical IP Subnetworks

NHRP utilizes the concept of administrative domains, where a separate instance of the NHRP applies separately to each domain. These domains may be separate NBMA networks, or a partitioning of a single larger NBMA network into multiple, *logical NBMA subnetworks*. The NHRP standard assumes that no closed user groups or address screening partitions the underlying NBMA network.

NHRP stations (i.e., hosts or routers) utilize the NHRP protocol to determine the internetworking layer address (e.g., IP, IPX and Appletalk) and NBMA subnetwork addresses of the *NBMA next hop* on the path toward a destination station. Stations act as either servers, or clients. Routers act as both a server and as a client for each ATM interface.

Within each administrative domain, *Next Hop Servers (NHS)* implement the NHRP protocol. An NHS is always closely associated with a router; however, the draft RFC assumes that all routers connected to an NBMA may not participate in NHRP. The standard defines the last NHS along the routed path to a client as the *serving NHS*. Next Hop Servers maintain *next-hop resolution* cache tables that map internetworking addresses to NBMA addresses. NHS's construct this cache from *NHRP Register packets, NHRP Resolution Request/Reply* packets, or through manually configured table entries. NHS's automatically populate their interworking address tables through internetworking routing protocols.

Next Hop Clients (NHC's) also maintain a cache mapping internetworking addresses to NBMA addresses, populated through from NHRP Resolution Reply packets or manual configuration. Administrators may configure the NHS NBMA address into stations connected to an NBMA network, the NHS may be the station's default router, or the station may utilize the ATM Anycast address to connect to the closest NHS. Stations register their NBMA address and reachable internetworking addresses with the NHS determined by the above rule using NHRP Register packets.

The following example illustrates operation of NHRP for the same network utilized in the previous section on Classical IP over ATM involving three

LIS's. Figure 20.6 shows the example network which uses IP as the inter-working protocol and ATM SVCs as the NBMA networking protocol. In step 1, the source station with IP address S, receives a packet destined for IP address D. The source station is a client configured to maximize transmissions over the ATM NBMA network, instead of across the current default routed path shown by the dashed line. The source station's NHC issues an NHRP Resolution Request, containing the source IP address (S), the source ATM address (X) and the destination IP address (D), to NHS-A in Router 1 in step 2. NHS-A examines its cache, and finds no entry. It examines its IP forwarding table and determines that Router 2 (and NHS-B) is the next hop on the routed path to IP address D. Therefore, in Step 3, NHS-A sends an NHRP Resolution Request packet on to NHS-B. The station with destination IP address D is registered with NHS-B; therefore, in step 4 NHS-B returns an NHRP Resolution Reply containing station D's ATM address, Y, to NHS-A. NHS-A stores the address mapping between IP address D and ATM address Y in its cache memory and returns the NHRP Resolution Reply to station S using the addressing information contained in the packet in step 5. Now, station S can set up a "short cut" ATM SVC connection using the ATM address returned by the NHRP protocol in step 6. Once station D (with ATM address Y) responds with a CONNECT message in step 7, bi-directional packet exchange occurs over the ATM network in step 8. As long as these stations continue to exchange information the SVC remains active. After a timer controlled period of inactivity ranging from tens of seconds to several minutes, the stations disconnect the SVC.

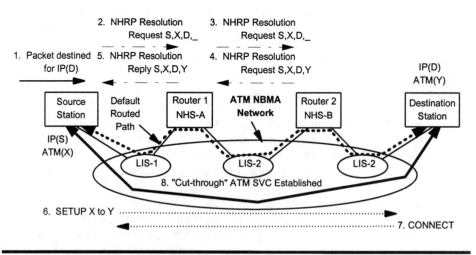

Figure 20.6 Illustration on NHRP "Short cut" for a single NHS

Note that in the preceding example, the source station, S, could have forwarded the packet destined for D over the default routed path while awaiting the NHRP protocol to complete. The NHRP RFC advises against this because forwarding the packet along the default routed path could cause NHS-A (and other intervening NHS's) to generate NHRP Resolution Request packets. The RFC calls this the "domino effect," noting that it may result in additional traffic due to the resolution request packets and/or the establishment of an excessive number of VCCs.

20.3.2 Operation of Interconnected NHRP Domains

If the destination is connected to the same NBMA subnetwork as the originator, then the NBMA next hop is simply the destination station itself. Otherwise, the RFC defines the NBMA next hop as the egress router on the NBMA subnetwork that is closest to the destination station. The example of Figure 20.7 illustrates this concept for a case of two separate NBMA ATM networks interconnected by an IP Router labeled R3. In this example, the source station with IP address S utilizes the same procedure, and ends up setting up a "short cut" SVC to the egress router, R3. Meanwhile, R3 has determined through use of the NHRP protocol with R4 and R5 that a short cut SVC can be set up through the second NBMA network to access destination D through egress router R5. Once the short cut SVCs are established, packets flow from station S to station D over the efficient ATM SVC short cut routes and the LAN connection between R5 and R6.

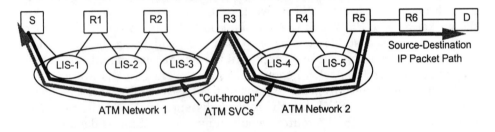

Figure 20.7 Illustration of NHRP Operation for Multiple Domains

As the NHRP Resolution Reply traverses the return path, intermediate NHS's learn and cache the IP-to-ATM mapping so that the NHS can respond directly to subsequent packet flows between these same domains without creating another Resolution request. The draft RFC defines an option where an originating node may request an authoritative mapping option; in which case NHS's must not utilize cached information. The draft RFC identifies how to employ this method to ferret out closed used groups that may be implemented in NBMA networks by requesting a trace of the NHS's along the routed path, and using the ATM address information to determine which NHS's are reachable over the NBMA network.

So far, so good, you say — such is the case with many of the best laid plans. Figure 20.8 adds some traditional LANs and routers to the previous example, a realistic assumption in the hybrid environment expected in many networks adding ATM as part of their infrastructure. To start with, the "short cut" SVCs are set up as in the previous example. The example adds a legacy router, R7, which is interconnected via subnets which preserve the exchange of IP routing information as indicated in the figure. The routing metrics are setup such that the low speed connection between R1 and R7 is undesirable.

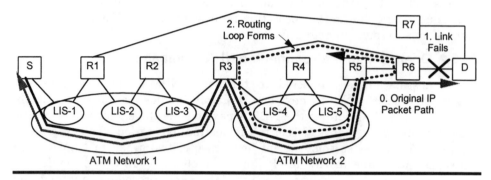

Figure 20.8 Illustration of a Routing Loop in NHRP via a Back Door Path

Now for the important part — *NHRP short cut SVCs do not pass IP routing information*. Let us examine what happens in a failure scenario. While the short cut SVCs are in operation, the link between R6 and destination D fails in step 1. Router R6 determines that the next best path to destination D is through R3. R3 still has the cached association indicating that the short cut SVC through the second ATM network is the best route to D, and forwards the packet back to R5, which then sends it on to R6 and a routing loop forms in step 2. Such a stable routing loop is a very bad thing, because once a packet begins looping, it doesn't stop until the routers decrement the Time To Live (TTL) field in the packet header to zero and eventually discard it.

The "back door" path from R6 to R3 is utilized because R5 cannot inform R3 through the short cut SVC that the routing distance to the destination D has changed, otherwise, the IP routers would converge on the less desirable low-speed path via R1 through R7 to D. The draft RFC gives several guidelines for avoiding routing loops, but leaves the automatic avoidance of routing loops as an area for further work. For example, directly connected stations cannot have this problem, since if they become physically disconnected, then the ATM short cut SVC disconnects as well. Another guideline points out the danger of "back door" routes, as illustrated above.

The draft RFC details other procedures regarding cache timeouts, cache purging, and the handling of error conditions. NHRP also supports optional features, such as route recording to detect loops, and address aggregation which allows the protocol to return a subnet mask and consequently identify a range of reachable addresses instead of a single address.

20.4 MULTIPROTOCOL OVER ATM (MPOA)

The ATM Forum initiated the MultiProtocol Over ATM (MPOA) work in response to a widespread industry consensus to extend ATM to protocols other than IP (e.g., IPX, Apple Talk and DECNET) and to enhance native mode protocols. This section presents a brief history regarding the origins of

MPOA, the current version 1.0 standard, and the challenges and opportunities ahead.

20.4.1 MPOA History and the Roads not Traveled

The early stages of the MPOA group considered three markedly different models for multiprotocol operation over ATM [10]:

- Peer model
- Integrated PNNI as an overlay model
- Distributed (or virtual) routing – separation of forwarding and route computation

Several companies proposed a peer model that employed an algorithmic mapping of all network layer addresses into NSAP-based addresses. This approach had the advantage of allowing PNNI to directly route signaling requests and precluded the need for a separate address resolution protocol. On the other hand, it required different routing protocols in mixed ATM and router networks, resulting in sub-optimal end-to-end routing and concerns about multivendor interoperability. Furthermore, every ATM switch requires address tables large enough for the ATM NSAP formatted addresses and the addresses resulting from the algorithmic mapping of the other address spaces.

The Integrated PNNI model (I-PNNI) proposed that ATM switches and routers universally employ the PNNI protocol. A motivation for this approach is the fact that PNNI is the most powerful and scalable routing protocol defined to date. I-PNNI could support the separate ATM address space, also called an overlay model, defined by the ATM Forum in UNI 3.1, or the peer model with mapped addresses. I-PNNI's principal disadvantage was that all routers would need to implement it, and hence, its introduction would take years and the migration process would likely be extremely complex.

Finally, the ATM Forum agreed on a new vision of virtual LANs as the basis for MPOA utilizing extensions to the LANE protocol and layer 3 switching with distributed routing. In this context, layer 3 switching refers to devices that make forwarding decisions based upon the network layer in the packet header. Advantages of this approach were:

- Elimination of multiple hops through software-based IP routers
- Reduce the impact of broadcast storms seen in bridged networks
- Implement high-performance filtering
- Map ATM's QoS benefits to other protocols

Recognizing that the majority of the cost in current routers is in the high-performance processors and memory for route processing, and not in the hardware involved in packet forwarding, the ATM Forum standardized centralized route processing which disseminated routing information to simpler packet forwarding engines, called edge devices. The proposal allowed vendor differentiation in the internal operation of the layer 3 switches and the routing protocols operating between them. The proposal also centralized the route processors since only the route distribution protocol would be standardized. The ATM Forum also chose to reuse the layer three next hop and address resolution components of the IETF's Next Hop Resolution Protocol (NHRP) described earlier in this section. The design specified a query/response protocol that an MPOA client could use to request the next hop and address resolution information corresponding to a layer three address from an MPOA route server. The hope was that use of NHRP would accelerate MPOA development, since the IETF had worked on NHRP for several years already. The MPOA group adopted this approach for its first version of the specification as detailed later in this section. A number of texts summarize the functional group terminology used in draft MPOA specifications [11, 12], which the version 1.0 MPOA specification does not use.

20.4.2 MPOA Architectural View

Conceptually, MPOA distributed the principal functions of a router — data forwarding, switching and routing control and management — into three networking layers [13] as illustrated in Figure 20.9. The traditional router implements all of these functions on a single machine as shown in Figure 20.9a. Often, the central router processor card is the bottleneck in such designs. MPOA defines the concept of a *virtual router*, which distributes the traditional router functions to different network elements. Figure 20.9b illustrates how MPOA maps the router I/O card forwarding function to MPOA edge devices, distribution of the router backplane to a network of ATM switches, and consolidation of the routing computation at every network node to a smaller number of MPOA route servers.

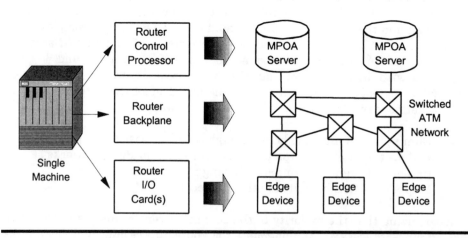

Figure 20.9 MPOA Architectural Separation of Functions — The Virtual Router

MPOA aims to efficiently transfer data between different subnets in an emulated LAN environment. MPOA integrates the LANE and NHRP protocols and adds internetwork layer communication over dynamically established ATM SVCs without requiring routers in the data path. Thus, MPOA synthesizes bridging and routing in an ATM-based network supporting a diverse range of link and network layer protocols. A key architectural aspect of MPOA is the physical separation of the internetwork layer routing protocol and hop-by-hop packet forwarding in a technique known as virtual (or distributed) routing. This separation of routing, switching and forwarding provides the following benefits:

- supports efficient inter-subnet communication via performing only forwarding for the majority of traffic, minimizing router processing
- decreases the number of internetwork layer devices requiring configuration
- increases scalability by reducing the number of nodes participating in internetwork layer routing
- reduces complexity of edge devices by eliminating the need to perform internetwork layer routing

a. Traditional Router Network **b. MPOA Virtual Router Network**

Routing Control Plane Routing Control Plane

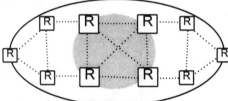

Data Forwarding and Switching Plane Data Forwarding and Switching Plane

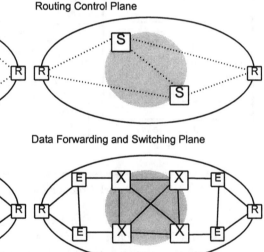

R = Router E = MPOA Edge
X = ATM Switch S = MPOA Server

Figure 20.10 Comparison of Control and Data Forwarding Plane Architectures

A good way to visualize this benefit of the virtual router concept of the MPOA architecture is to compare the distribution of function in separate logical planes for routing control and data forwarding and switching [13]. Figure 20.10a shows how the physical structure of a typical routed network, which often employ edge and backbone routers arranged as a hierarchy to support even moderately sized networks. Every machine processes the network routing protocol in the control plane as well as forwards and switches packets as shown in the data plane. Figure 20.10b illustrates the same physical network topology, but implemented using ATM switches, MPOA servers and MPOA edge devices. Note how traditional routers on the edge of the network peer with the MPOA servers in the routing control plane. Also, more cost-effective ATM switches replace the throughput-limited routers in the core of the network. Let's look further into how the MPOA design achieves these goals.

20.4.3 MPOA Version 1.0

The ATM Forum worked closely with the IETF in developing the MPOA version 1.0 specification. For example, the ATM Forum published drafts of the MPOA specification and contributions, normally accessible to members only, to the public in support of this effort. This summary is based upon the final version of the MPOA specification [14] and other descriptions [15].

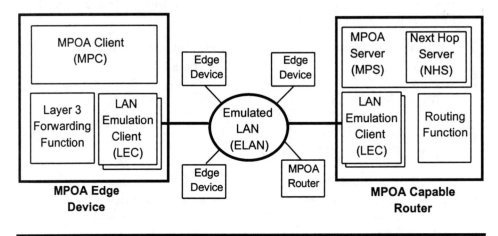

Figure 20.11 MPOA Network Components

Figure 20.11 illustrates the two components in an MPOA network: edge devices, and MPOA capable routers. An emulated LAN (ELAN) connects a network of edge devices and MPOA routers. LAN Emulation Clients (LECs) interconnect MPOA edge devices (also called hosts) and MPOA capable routers. An MPOA edge device also contains an MPOA Client (MPC) and a layer 3 forwarding function. Edge devices reside on the periphery of ATM networks, usually supporting traditional LAN interfaces, such as Ethernet and Token Ring. Edge devices are capable of bridging with other edge devices and hence can be part of virtual LANs. They also have a limited amount of layer 3 processing capability based upon information fed from MPOA route servers. MPOA edge devices do not participate in routing protocols.

On the right hand side of Figure 20.11, an MPOA capable router contains a MPOA Server (MPS), which includes the NHRP Next Hop Server (NHS) function and a routing function. MPOA synthesizes bridging and routing with ATM. LAN Emulation (LANE) performs the bridging function, while extensions of the IETF's Next Hop Resolution Protocol (NHRP) perform the layer 3 forwarding function. MPOA separates switching (or forwarding) from routing in a technique called virtual routing where the MPOA routers compute the values for the next hop forwarding tables and download them to the edge devices. MPOA servers participate in standard routing protocols, like OSPF, IS-IS, and RIP with each other as well as with traditional routers on the periphery of the MPOA network. The routing function in MPOA servers exchange topology information and calculate routes, while the edge devices perform the majority of the layer 3 forwarding. Since the edge devices perform layer 3 forwarding, latency is reduced and higher throughput is achieved more cost effectively than router-based forwarding capabilities. The NHRP Next Hop Server (NHS) distributes layer 3 forwarding information to MPOA clients via NHRP with the addition of a cache management protocol.

MPOA specifies that a single route server may serve multiple edge devices, and that multiple route servers may support a single edge device. This many-to-many mapping provides redundancy for the route server function and eases administration, since MPOA route servers are simply configured to join the virtual LAN for the edge devices they serve. MPCs and MPS's use LANE bridging for communication within their own emulated virtual LANs as illustrated in Figure 20.12. Normally, packets flow over the default routed path until the MPOA client recognizes a long lived flow. Once the MPOA client resolves the ATM address of the shortcut next hop using the NHRP protocol described in the previous section, it places an SVC call to establish a shortcut switched path as shown at the bottom of the figure. For example, if a client requests a 1-Mbyte file from an application server connected to an Ethernet, a routed network would require processing of over 600 packets. The MPOA design instead sets up a VCC, performing routing, address resolution and forwarding decisions only once. The server then rapidly transfers the 20,000 cells corresponding to the 1-Mbyte file to the client over the switched ATM network.

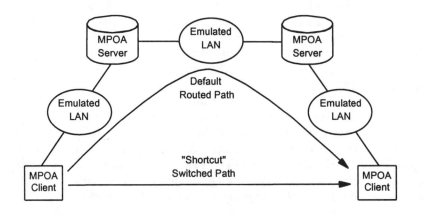

Figure 20.12 MPOA Server and Client Usage of Emulated LANs

20.4.4 Example of MPOA in Action

Figure 20.13 illustrates the operation of the NHRP protocol used in an MPOA network. Each MPC maintains an ingress cache indexed by the MPS MAC address, the MPS ATM address and the Internetwork Layer Destination Address. Hosts attached to edge devices utilize the MPS MAC address as the default address for layer 3 addresses outside their own subnetwork. LANE-based bridging is used for transfer of layer 3 addressed packets within the same subnetwork. Upon the receipt of a layer 3 address outside the MPC1's subnetwork addressed at the MAC layer to MPS1 in step 1, MPC1 creates a new cache entry and begins counting the number of frames sent to that destination layer 3 address. MPC 1 forwards the packets to MPS1 until the counter exceeds a threshold in step 2.

The ingress MPC (MPC1) then sends an MPOA Resolution Request to the ingress MPS (MPS1) in step 3. MPS1 then translates the MPOA Resolution Request to an NHRP Resolution Request, which it then forwards along the routed path to the Internetwork layer destination address using the NHRP Next Hop Server (NHS) function in step 4. MPS2 is the last router on the default path to the destination, and translates the NHRP Resolution Request to an MPOA Cache Imposition Request and sends it to the appropriate Egress MPC in step 5. The egress MPC (MPC2) creates an egress cache entry in step 6 and responds to the Cache Imposition Request with a MPOA Cache Imposition Reply in step 7. MPS2 translates the MPOA Cache Imposition Reply to an NHRP Resolution Reply and forwards the reply back along the routed path to the Ingress MPS address contained in the NHRP Resolution Request in step 8. MPS1 translates the NHRP Resolution Reply to an MPOA Resolution Reply and returns it to the original requester, MPC1 in step 9. Now, MPC1 establishes a MPOA shortcut using the ATM SVC procedure with the ATM address of MPC2 returned by the MPOA resolution process in step 10. Once the SVC is established, MCP1 forwards packets over the shortcut instead of over the default routed path in step 11. The Egress MPC (MPC2) is prepared to receive data over the shortcut because of the egress cache processing performed in step 6. MPC1 continues to forward packets over the shortcut until an inactivity timer expires, or the ingress cache is flushed by management command.

Note that MPOA's augmentation of the NHRP protocol with the Cache Imposition Protocol extends the ARP query all the way to the destination edge device [15]. Normally, NHRP identifies a router serving the destination's LAN, and may not select the router closest to the destination if more than one router serves the destination. This can lead to sub-optimal routing since another emulated LAN hop could be required to reach the destination edge device. In MPOA, the egress MPS ARPs for the destination and uses the response to identify the correct edge device. Therefore, MPOA establishes a shortcut directly from the source edge device to the destination edge device, avoiding any additional routing hops across the ATM network.

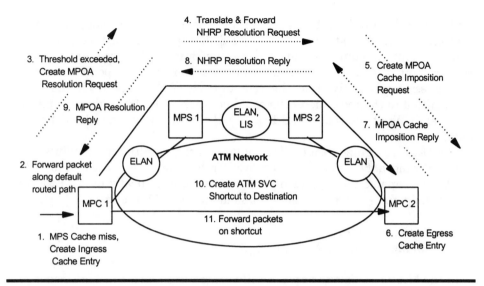

4. Translate & Forward
   NHRP Resolution Request

3. Threshold exceeded,
   Create MPOA
   Resolution Request

8. NHRP Resolution Reply

5. Create MPOA
   Cache Imposition
   Request

9. MPOA Resolution
   Reply

MPS 1   ELAN,
         LIS   MPS 2

7. MPOA Cache
   Imposition Reply

2. Forward packet
   along default
   routed path

ELAN   ATM Network   ELAN

MPC 1

10. Create ATM SVC
    Shortcut to Destination

11. Forward packets
    on shortcut

MPC 2

1. MPS Cache miss,
   Create Ingress
   Cache Entry

6. Create Egress
   Cache Entry

Figure 20.13   MPOA Shortcut Example

20.4.5 Challenges Ahead for MPOA

Users with large emulated LANs supporting thousands of clients, such as McDonald's and Amoco [16], require MPOA as soon as possible. Many manufacturer's have announced MPOA products, many with proprietary extensions. MPOA offers the promise of true multi-vendor interoperability of new, high-performance ATM networks with the legacy multi-protocol networks present in many enterprises today. In response to this demand, a number of manufacturers now support MPOA on their products.

20.5   IP MULTICAST OVER ATM

Current ATM standards do not define a service comparable to the group addressing feature of SMDS. The capability of allowing one address to be able to broadcast to all other addresses in the group effectively emulates a LAN, which is very useful in the exchange of LAN topology updates, Address Resolution Protocol (ARP) messages, and information required in routing protocols.

20.5.1 Basic Components and Operation

IETF RFC 2022 specifies the means to implement multicast over ATM [17] enabling interworking with IP multicast using the Class D address space as specified in RFC 1112 [18]. The IETF's choice of AAL5 for carriage of IP over ATM has important consequences [10] on the design of a multicast service. An inherent property of AAL5 is that *all* cells from a packet must be transmitted sequentially on a single VCC. This stands in sharp contrast to AAL3/4, which has a Multiplex ID (MID) inside each cell, enabling cells from multiple packets to be interleaved on a single VCC. In practice this means that each system must have a separate SAR engine for each VCC, in either point-to-point, or point-to-multipoint configurations. The sequential receipt requirement of AAL 5 also means that in a point-to-multipoint connection, transmission is strictly from the root to each of the leaves. If the leaves transmitted to the root, then cells could be interleaved from multiple packets when arriving at the root, resulting in AAL5 SAR failures and loss of packet data.

The IETF recognized that ATM requires a true multicast capability since most existing data networking protocols rely upon a broadcast, inherently multicast capable facility, such as that occurring naturally on an Ethernet. RFC 2022 defines two methods for implementing ATM multicast:

- Multicast Server
- Full Mesh Point-to-Multipoint

In the Multicast Server approach illustrated in Figure 20.14, all nodes join a particular multicast group by setting up a point-to-point connection with a Multicast Server (MCS). The MCS may have a point-to-multipoint connection as shown in the figure. Alternatively it may emulate the broadcast connection via a transmitting packets in the reverse direction on every point-to-point connection. The MCS receives packets from each of the nodes on the point-to-point connections and then retransmits them on the point-to-multipoint connection. This design ensures that the serialization requirement of AAL5 is met, that is, all cells of an entire packet are transmitted prior to cells from any other packet being sent.

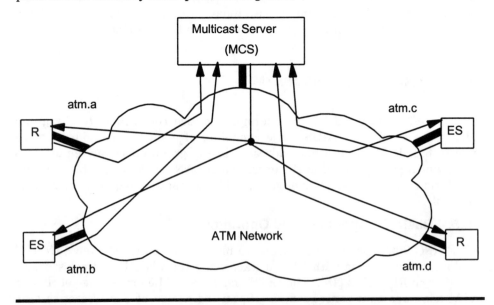

Figure 20.14 MultiCast Server (MCS) ATM Multicast Option

The full mesh point-to-multipoint connection approach, illustrated in Figure 20.15, involves establishment of a point-to-multipoint connection between every node in the multicast group. Hence, as seen from inspection of the figure, every node is able to transmit and receive from every other node. Note the number of connections required even for this relatively simple four node network. Considering the complexity involved in configuring PVCs for such a network, we believe the a point-to-multipoint SVC capability is essential in the practical deployment of a multicast over ATM network.

Each of these approaches for implementing the multicast capability has advantages and disadvantages for a group of N nodes. The point-to-multipoint mechanism requires each node to maintain N connections for each group, while the multicast server mechanism requires at most two connections per node. The point-to-multipoint method places a connection burden on each of the nodes, as well as the ATM network, while the multicast server approach requires that only the server support a large number of connections. Hence, the multicast server mechanism is more scalable in terms of being able to dynamically change multicast group membership, but presents a potential bottleneck and a single point of failure in ATM networks supporting IP multicast.

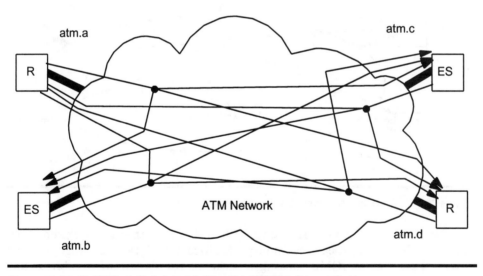

Figure 20.15 Full Mesh Point-to-Multipoint ATM Multicast Option

IP multicast over ATM utilizes the Unspecified Bit Rate (UBR), or best-effort, ATM service class as specified in RFC 1755. The current RFC does not define how to request a guaranteed ATM quality of service.

20.5.2 Multicast Address Resolution Server (MARS)

RFC 2022 supports IP Multicast requirements [18] using a Multicast Address Resolution Server (MARS) connected to a cluster of end system nodes via another point-to-point VCC as shown in Figure 20.16. The separate VCC separates Address Resolution and connection setup messages from the multicast application packets. End systems employing SVCs are configured with the ATM address of the MARS. The MARS keeps a mapping of IP multicast addresses to a list of ATM addresses that are currently members of the particular multicast group. The IP Multicast over ATM protocol allows end systems on LANs and IP networks to seamlessly interoperate with ATM end systems.

RFC 1112 defines IP multicasting as a service where a datagram sent by one host is sent to all other hosts in the multicast group. Hosts dynamically join and leave specific multicast groups by sending Internet Group Management Protocol (IGMP) report messages to the targeted multicast group. IGMP is a required part of the IP protocol and is supported on the specific range of Class D IP addresses from 224.0.0.0 to 239.255.255.255. One of these addresses uniquely identifies a particular multicast group.

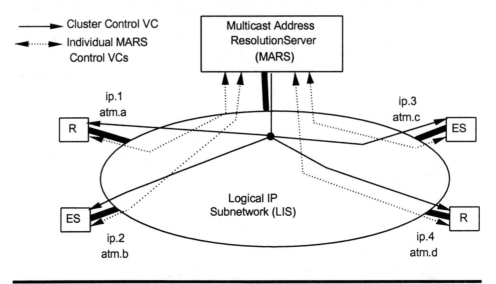

Figure 20.16 Multicast Address Resolution Server (MARS)

Often a router maps the IP multicast to either an Ethernet or Token Ring-hardware-based multicast capability. Multicast routers also form multicast groups based upon IGMP messages received from their own subnets and selectively direct multicast traffic using particular routing techniques, such as Protocol Independent Multicast (PIM) [19]. RFC 1112 implicitly assumes that routers listen on all multicast group addresses for IGMP messages.

All multicast end systems and routers on a subnet are members of a particular group, called the all hosts group, on IP address 224.0.0.1. All IGMP messages are in IP datagrams with the all hosts address, so that all multicast devices on a subnet receive every IGMP message. Routers send IGMP query messages to poll the status of previously active multicast groups once per minute. Any host that is a member of that multicast group waits a random amount of time, and then responds with an IGMP response message, unless some other node responds first. Hence, IGMP avoids congesting the underlying network by efficiently sending join messages once, and polling to detect inactive multicast groups once per minute using only two messages.

The MARS distributes group membership update information over the Cluster Control VC. When the emulated IP multicast service employs Multicast Servers (MCSs), the MARS also establishes a separate point-to-multipoint VC to the set of registered MCSs. For simplicity in explanation, the following treatment does not consider the case employing MCSs.

Any node wishing to join and transmit to any multicast group, for instance, as triggered by an IGMP Report message, must first register with the MARS server using a MARS_JOIN message. The MARS then adds the node as a leaf to its ClusterControl VC. The MARS stores the joining node's address list associated with that multicast cluster. When a node has a packet to send to a particular IP multicast group, and finds that it does not have an open connection to the group, its sends a MARS_REQUEST to the MARS. The MARS responds with either a MARS_MULTI message, identifying the ATM addresses of the other members of the group, or a MARS_NAK if the group is empty. Upon receipt of a MARS_MULTI message, the node with a multicast packet enqueued first sets up a point-to-multipoint connection to each of the ATM addresses identified in the MARS response and then sends the packet.

Because the IETF based RFC 2022 on UNI 3.1 signaling, which defines only root initiated point-to-multipoint SVCs, the MARS emulates a hardware multicast media by broadcasting the JOIN message to the roots of all nodes involved in the multicast group on the Cluster Control VC. All nodes add the joining node to their existing point-to-multipoint connections using add-leaf signaling messages.

Finally, any node wishing to leave a multicast mesh multicast group sends a MARS_LEAVE request to the MARS Server. The MARS removes the node's ATM address from the list of ATM addresses registered for the specific IP multicast address and forwards the message on its Cluster Control VC. The nodes in the particular multicast group remove the departing node from their point-to-multipoint connection using a remove leaf signaling message. Nodes use timers to clear inactive connections caused by lost messages.

Recall that RFC 1112 specifies that multicast routers receive IGMP Report messages on all multicast group addresses. The IP multicast over ATM RFC supports this by requiring such routers to "promiscuously" join all multicast groups through use of a block join message sent to the MARS. Therefore, any node sending a MARS_REQUEST also transmits to the multicast router. Furthermore, all nodes are members of the all hosts multicast group, and hence the IGMP Query/Report protocol for polling multicast group status works as specified in RFC 1112. The result is that end systems that operate on Ethernet, interoperate with ATM end systems implementing the RFC2022 IP/ATM Multicast protocol.

20.6   REINVENTING IP OVER ATM

Recently, network layer switching emerged on the scene, biting at the heels of LAN switching in the continual quest to improve networking price-performance. Network switches operate at the edges of backbone networks with the goal of performing routing and switching decisions only once, and not for every packet that transits the network to the same destination [20]. While the IP and ATM standards bodies were developing the protocols described earlier in this section, manufacturers were building better mousetraps in response to the tremendous growth in IP internetworking. Interestingly, a number of these manufacturers published at least an overview of their approach in IETF informational RFCs, in addition to issuing public proclamations that their approach would be an open, non-proprietary solution. The first company to break with the momentum of the standards bodies was a startup company in Silicon Valley, Ipsilon Networks, whose IP Switching approach proposed placing IP over ATM on a strict protocol efficiency diet. A similar approach from Toshiba proposed Cell Switch Routers for efficiently interconnecting Classical IP over ATM and MPOA networks. The latest entrant onto the scene is Cisco Systems' Tag Switching architecture, which works with not only ATM, but a number of legacy technologies as well. IBM's Aggregate Route-based IP Switching (ARIS) approach differs from the other approaches by a clever means of overlaying data transmission in the reverse direction over the spanning tree rooted in each node determined by most routing protocols.

The IETF formed a Multi-Protocol Label Switching (MPLS) working group to sort this all out and come up with a common standard. The following sections present a brief overview of each approach, concluding with a comparison of the various approaches and the current status of the IETF's MPLS work. We cover the other competing alternatives of switched 100-Mbps Ethernets and Gigabit Ethernets in Part 8.

20.6.1 Ipsilon's IP Switching

Beginning in 1995, Ipsilon Networks introduced a fresh idea into the industry. The company asserted that ATM had already lost the battle for the desktop to Fast Ethernet, that LANE did little more than LAN switching, and that the software burden of the complex ATM protocols made proposed IP over ATM implementations cost prohibitive [21]. Furthermore, their publications questioned the direction of the ATM Forum and the IETF LANE, NHRP, MPOA and entire approach for implementing IP over ATM. They pointed out duplication of function, scaling problems and difficulties in multicast implementations. The answer to all of these problems was a simplified form of IP over ATM that they called IP switching.

Ipsilon published the key aspects of their protocol in Internet RFCs 1953, 1954 and 1987. This bold move made the aspects of the protocol open to all manufacturers, not only were the algorithms available to hosts and routers that would use Ipsilon's devices, but competitors as well. The company even made source code available to the research community free of charge.

Basically, Ipsilon's approach classified traffic into either short or long-lived flows. The new components of the IP switching approach applied only to the longer duration flows, such as FTP, long Telnet sessions, HTTP, and extended web multimedia sessions. IP switching handled short-lived, interactive traffic, such as DNS, E-mail and SNMP in exactly the way IP routers handle it today. Therefore, in order for IP switching to improve performance, a majority of the total traffic must be in long-lived flows. If the majority of the traffic were in short-lived flows, then the performance of IP switching is no better than that of routers. A number of studies published by Ipsilon of corporate networks indicate that although a high percentage of flows are of short duration, these short-lived flows carry a small percentage of the packet traffic. Indeed, these studies report that the small number of long-lived flows carry the majority of packet traffic in corporate networks. A more recent study [22] reported that optimizing 20 percent of the flows on a public Internet backbone optimized 50 percent of the packets traversing the backbone. These studies reinforce the common wisdom that a small portion of the overall flows make up a majority of the traffic.

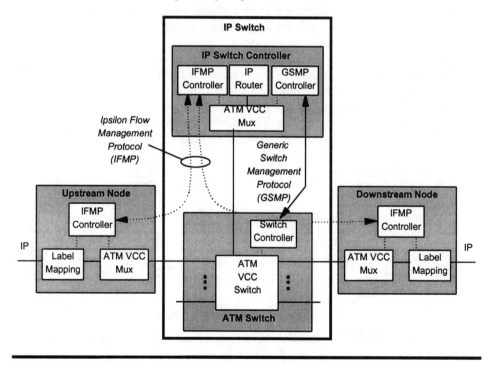

Figure 20.17   Ipsilon Networks IP Switching Architecture

Figure 20.17 illustrates the components, interfaces, and protocols of Ipsilon's IP switching functional architecture [23]. An IP switch has two logical components: an ATM switch and an IP Switch Controller. Any ATM switch with a switch controller capable of making and breaking hundreds of virtual channel connections per second can be part of an IP switch. Ipsilon produces software for IP switch controllers, which connect via an ATM UNI to the ATM switch, and multiplexes ATM cells at the VCC level to an IP router and Ipsilon switch controllers. ATM UNIs connect the IP switch to a number of upstream and downstream nodes as indicated in the figure. Nodes (i.e., hosts or routers) connect to interfaces on the IP switch to transfer IP data packets. These nodes must be capable of adding labels to IP packets, as well as multiplexing various flows of IP packets associated with the labels onto different VCCs. These nodes interface to the IP Switch Controller via Ipsilon's Flow Management Protocol (IFMP) as specified in IETF RFC's 1953 [24] and 1954 [25] as the text describes through an example below. The IP Switch Controller also implements a Generic Switch Management Protocol (GSMP) as specified in RFC 1987 [26] to make and break ATM VCC connections through interaction with the ATM switch's controller.

Figures 20.18 and 20.19 illustrate the operation of Ipsilon's IP switching and protocols through a simple example [21, 23]. Initially, each IP node sets up a default forwarding VCCa to the IP switch controller's IP router in step 1, which forwards the packet to the next hop on VCCb as determined by standard IP routing protocols as shown in Figure 20.18a. Simultaneously, the IP switch controller performs flow classification by detecting long-lived sequences of IP packets sent between a particular source and destination address pair employing the same protocol type (e.g., UDP or TCP), and type of service. Once the IP switch controller identifies a long-lived flow as a candidate for optimization, its IFMP controller requests that the upstream node label the traffic for that flow and use a different VCCx than the default forwarding VCCa in step 2 as shown in Figure 20.18b. If the upstream node's IFMP controller concurs, then the flow begins on the new VCCx, which is still handled by the IP router in the IP switch controller in step 3.

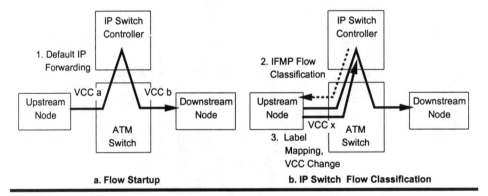

a. Flow Startup   b. IP Switch Flow Classification

Figure 20.18 Ipsilon IP Switching Example (1 of 2)

Independently, the downstream node may request via IFMP that the IP Switch Controller set up an outgoing VCC for the same flow in step 4 as depicted in Figure 20.19a. The IP Switch Controller responds to the downstream node and directs the flow to VCCy, which differs from the default forwarding VCCb in step 5. Finally, in step 6 depicted in Figure 20.19b the IP Switch Controller's GSMP process instructs the ATM switch's controller to connect ATM VCCx to ATM VCCy. This step directly connects the upstream and downstream nodes for the designated flow, removing the IP router from the IP packet data path. After several minutes of inactivity, the IP switch controller instructs the upstream and downstream nodes to disconnect the flow.

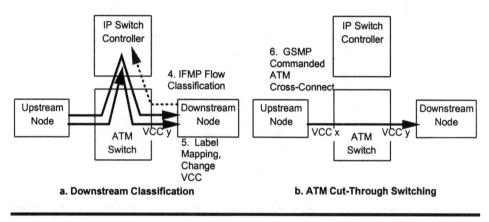

a. Downstream Classification   **b. ATM Cut-Through Switching**

Figure 20.19 Ipsilon IP Switching Example (2 of 2)

Ipsilon claims that first-generation IP Switches support throughputs of over five million packets per second using the above protocols. In contrast, the fastest routers available today have a maximum forwarding capacity of between 500,000 and 3 million packets per second [27]. Flow classification and switching are local, soft-state decisions that time out within seconds unless refreshed. Hence, the system is resilient to failures. Additionally, flow characterization allows IP switches to allocate quality-of-service to different flows in conjunction with ATM switch QoS capabilities. Also, IP switches support IP multicast transparently.

20.6.2 Toshiba's Cell Switching Router (CSR)

RFC 2098 [28] describes another vendor proprietary proposal for handing IP over ATM networks. A white paper by Toshiba provides further insights to this approach [29]. A Cell Switch Router (CSR) has ATM cell switching capabilities in addition to conventional IP datagram routing and forwarding as illustrated in Figure 20.20. Note that this architecture is very similar to Ipsilon's IP Switch at this functional block diagram level.

The routing function in the CSR normally forwards IP datagrams along hop-by-hop paths via a routing function, exactly as in the IP switching approach. The routing function automatically recognizes long-lived flows and either assigns, or establishes efficient short cut ATM paths, similar to IP switching again. But, CSR adds several new concepts. First, it proposes to handle more than the IP protocol. Secondly, it allows short cut connections to be pre-configured, or established via interaction with RSVP. CSR also proposes setting up short cut routes that may bypass several routers. CSRs interact using a Flow Attribute Notification Protocol (FANP) as indicated in the figure. CSRs also implement standard IP routing protocols, the ATM Forum PNNI protocol, and ATM signaling.

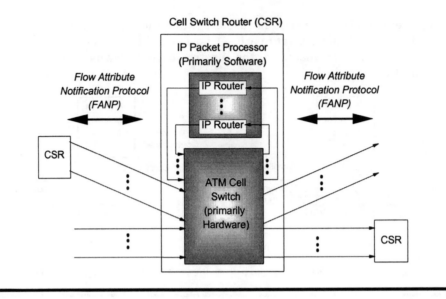

Figure 20.20 Toshiba Cell Switch Router (CSR) Architecture

The CSR proposal also focuses strongly on meeting applications specific QoS and bandwidth requirements, even over non-ATM networks. Furthermore, CSR purports to interconnect Classical IP over ATM, MPOA and switched IP networks [29] as illustrated in Figure 20.21. One of CSR's goals is to internetwork these various flavors of IP over ATM at much higher throughput than the current interconnection methods, as indicated by the thicker lines in the figure. The current RFC only gives an overview of the CSR approach, and further details must be specified before interoperable implementations can be specified.

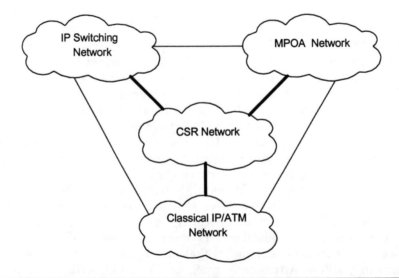

Figure 20.21 Toshiba CSR Internetworking Capability

20.6.3 Cisco's Tag Switching

Cisco announced its tag switching architecture in September, 1996. A white paper on Cisco's web page [30], an informational RFC 2105 [31] and a detailed white paper [32] give an overview of the architecture and protocols involved in tag switching. Figure 20.22 illustrates the basic components and interfaces of Cisco's tag switching architecture in an ATM network environment. Tag edge routers at the boundaries of an ATM network provide network layer services and apply tags to packets. Tag switches/routers at the core of the network switch tagged packets or cells based upon tags determined via information piggybacked onto standard routing protocols, or via Cisco's Tag distribution protocol (TDP). Tag switch/routers and Tag Edge Routers implement standard network layer routing protcols, such as OSPF and BGP, as shown in the figure. Additionally, they implement TDP in conjunction with standard network layer routing protocols to distribute tag information.

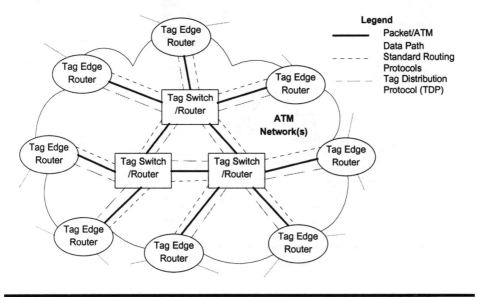

Figure 20.22   Cisco Systems' Tag Switching Architecture

Tag switching is a high-performance packet forwarding technique based on the concept of *label swapping*. A label is a generic name for a header. By swapping labels at intermediate nodes, an end-to-end connection results. Since ATM VCC switching directly implements a special case of the general label swapping using the VPI/VCI fields in the cell header, the switch/routers know whether to switch cells, or assemble the cells and route the resultant packets based upon information derived from TDP.

Tag Edge Routers run standard routing protocols and populate their next hop tables with the most desirable routes based upon the routing criteria, such as the destination address prefix. Tag routers and switches utilize these next hop tables and distribute VCC tag information via TDP. Tag edge routers examine network layer headers of received packets, perform network services (e.g., filtering), select a next hop route, and then apply a tag. In other words, they perform traditional routing. For example, a tag edge router may apply a VCC tag such that several intermediate Tag Switch/Routers may switch the cells directly through to the destination Tag Edge Router without performing any routing! Thus, this design replaces the complex processing of each packet header by the simpler processing of only the label. At the destination edge router the tag is removed, that is the packet is reassembled, and forwarded to the destination.

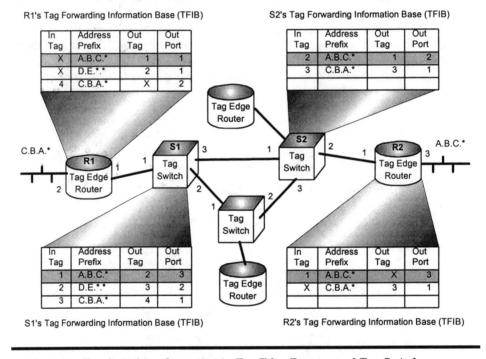

Figure 20.23 Tag Switching Operation in Tag Edge Routers and Tag Switches

Figure 20.23 illustrates the operation of tag switching for a simple network. Tag Edge Router R1 on the lefthand side serves a Class C IP address subnet designated by C.B.A.*, while Tag Edge Router R2 on the far right hand side of the figure serves IP subnet A.B.C.*. As shown in the Figure, each device in a tag switched network has a Tag Forwarding Information Base (TFIB), which contains the incoming tag (In Tag), address prefix obtained from the internetworking routing protocol, an outgoing tag (Out Tag), and an outgoing interface number (Out Port). Let's trace a packet destined for IP address A.B.C.55 from port 2 on the LAN connected to Tag Edge Router R1 through the tag switched network to the destination port 3 on router R2. The shaded entry in R1's TFIB instructs the Tag Edge Router to prefix packets with IP address prefix A.B.C.* with an outgoing tag of 1 and transmits the resulting tagged packet on port 1. Tag switch S1 receives this packet, and consults its TFIB. The shaded entry indicates that S1 changes the tag to a value of 2 and transmits the result on port 3. Tag switch S2 receives the packet and sees from the shaded entry in its TFIB that it should swap the tag to a value of 1 and transmit the tagged packet on port 2. Note that tag switches S1 nor S2 only examined the tag values to make a switching decision. Finally, Tag Edge Router R2 receives the packet with tag value 1, removes the tag (indicated by an X in the TFIB in the figure), and transmits it onto the destination LAN with IP addresses of the form A.B.C.*.

The TFIBs in Figure 20.23 also show the corresponding swapping of tag labels at each node in the reverse direction from Tag Edge Router R2 to the IP subnet C.B.A.* on Tag Edge Router R1. Note that in this example a tag can never be reused in the input tag or outgoing tag columns in any device's TFIB, otherwise an ambiguity would result.

Since this destination-based forwarding approach is topology driven, rather than traffic driven, tag switching does not require high call setup rates, nor depend on the longevity of flows to achieve increased throughput. Tag switching makes ATM switches peers of other routers since they participate in standard network layer routing protocols with edge routers.

20.6.4 IBM's Aggregate Route-based IP Switching (ARIS)

IBM's Aggregate Route-based IP Switching (ARIS) [33, 34] defines a route as a multicast distribution tree rooted at the egress point, traversed in reverse. The egress point is specified by a unique identifier; for example, an IP address prefix, an egress router IP address, or a multicast source and group address pair. Recall from Chapter 9 how the result of the Djikstra algorithm computation the least cost paths from a root node to every other node in the network results in a minimum spanning tree. ARIS effectively uses this tree to determine the forwarding path as the reverse direction along such a minimum spanning tree as shown in Figure 20.24. The thick solid line originating from the egress point rates the minimum spanning tree. The arrows show the data forwarding path in the opposite direction toward the root of the spanning tree (i.e., the egress router).

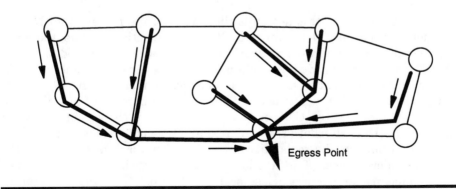

Figure 20.24 Example of IBM's Aggregate Route-Based IP Switching (ARIS)

Note that flows from the leaves of the spanning tree back toward the root merge at several points. ARIS operates over either frame switched networks or cell switched networks. For use over cell switched networks, ARIS requires ATM switches capable of VC-merging in order to support larger networks. An ATM switch capable of VC-merging transmits all cells from an individual AAL5 PDU received on a particular input branch onto the merged VCC prior to transmitting cells received on another branch from another AAL5 PDU as illustrated in Figure 20.25. Thus, VC-merging allows ARIS Integrated Switch Routers (ISR) to group cells from individual AAL5 PDUs from different inputs and switch them onto a shared VCC on the path back to a common egress identifier. Note that the VC merge function need not reassemble the entire AAL5 PDU, but only need ensure that the sequence of cells belong to one AAL5 PDU remains intact as shown in the figure. The standards call this configuration of forwarding in the reverse direction *multipoint-to-point*.

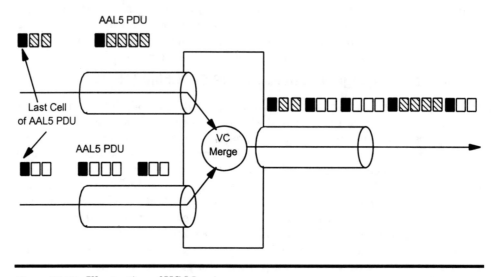

Figure 20.25 Illustration of VC Merging

A key consequence of this design is that every egress router in an N node ARIS network has only N-1 point-to-multipoint connections! Thus, the ARIS design significantly improves network scalability when compared with the N(N-1)/2 connection required in a full-mesh network. In the absence of a VC-merging capability, ARIS defines the means to merge Virtual Circuits through the use of Virtual Paths (VP); however, the number of bits allocated to the VPI, either 8 at the UNI, or 12 at the NNI, limit the size of a network supported by this alternative to approximately 4,000 nodes. The other decision that impacts the scalability of the ARIS approach is selection of the egress identifier. Choice of an IP address prefix results in a larger number of egress endpoints, and hence a less scalable design. Choosing a router IP address makes the network scale in terms of the number of routers. Although not a solution to a large IP backbone in itself, ARIS is well suited to size of many corporate IP networks.

ARIS defines an Integrated Switch Router (ISR) supporting standard IP routing protocols. It also implements an IP forwarding table that includes a reference to the point-to-multipoint switched path determined by an explicitly specified egress point on the IP network. Switched paths terminate in neighboring ISRs, or may traverse a number of ISRs along the best path to the egress ISR. By the nature of its design, ARIS guarantees that switched paths are loop-free. Similar to IP tag switching, ISRs forward datagrams at hardware speeds. The ARIS specification defines a protocol and message exchange emanating at the egress ISR to broadcast its existence and eventually establish the reverse merged paths. ARIS switched paths are soft state, maintained only as long as ARIS keep alive messages are exchanged. A stable ARIS network has N multipoint-to-point trees rooted in each egress node. Note that only control traffic occurs in the point-to-multipoint direction, all data traffic transfer occurs in the multipoint-to-point direction.

20.6.5 IETF's MultiProtocol Label Switching (MPLS)

In response to the unsolicited proposals for building scalable Internet backbones, the IETF established the MultiProtocol Label Switching (MPLS) group in 1997 to come up with one common specification. The current framework document [35] combines a number of functional requirements and design decisions from the proprietary approaches described above. We list some of the key requirements from this framework document and their origins below.

- All of the approaches use a short, fixed length layer 2 switching label to achieve lower cost, higher performance packet forwarding than traditional router techniques.
- From IBM's ARIS work, scalability on the order of N streams for best effort traffic.
- MPLS must support unicast as well as multicast.
- Must support RSVP and flow recognition from Ipsilon's IP switching and Toshiba's CSR proposals.
- Must support topology driven protocols, such as those defined by Cisco's tag switching.
- Retain compatibility with existing and legacy IP routing protocols as well as coexist with devices not capable of supporting MPLS.
- Prevent, or rapidly detect and remove, routing loops.
- Must define operation in a hierarchical network and be capable of supporting large-scale internetworks. The framework document defines the concept of stacking multiple labels in front of one packet to achieve this goal.
- Must be independent of any specific data link technologies. Specific optimizations for particular data link networks, such as frame relay and ATM as well as new data link technologies optimized for MPLS, may be considered.

The framework document goes on to compare the MPLS approach with the two existing approaches for constructing large internetworks: directly interconnected routers, and routers connected by a full mesh of virtual channel connections over a core ATM network. Table 20.2 summarizes the key components of this comparison. Note that MPLS does not claim to support QoS better than ATM networks can. Furthermore, MPLS plans to use explicit routes to support specific bandwidth and QoS guarantees. In essence, MPLS is a different means for setting up connections in a layer 2 network. Much has already happened in the area of reinventing IP over ATM. Inevitably, the final MPLS standard will differ from any of the proprietary approaches since it aims to meet such a broad range of requirements. In any event, the influence of ATM experience can be seen in many of the proposals.

Table 20.2 Advantages of MPLS over Traditional IP and IP/ATM Internetworks

Traditional Directly Connected IP Router Network	Router Network Connected via an ATM Core Network
Forwarding Capacity: Processing short, fixed MPLS layer 2 labels more efficient than longer, variable length layer 3 headers.	Scalability: MPLS avoids the N(N-1)/2 connections required in a full mesh network.
Explicit Routing: More efficient since entire path transferred only once, and not every packet.	Common operation over multiple media: MPLS specifications operate over frame or cell based link layer networks.
Traffic Engineering: More efficient than adjusting administrative routing weights in traditional routing protocols	Common route management and control: ATM network connections must be administered in conjunction with IP routing to achieve efficient operation.
QoS Routing: Supported by MPLS by setting up explicit routes.	Simplicity: MPLS eliminates the need for short cut based routing used in NHRP.
Functional Partitioning: MPLS allows simpler nodes to perform transit forwarding.	
Reuseable Paradigm: It may be possible for MPLS and ATM to operate on the same devices.	

What the heck is a label you ask? A draft document from the MPLS group [36] defines the proposed contents of a four-byte MPLS packet *label* as shown in Figure 20.26. The contents of the label give some insight into the eventual capabilities of the anticipated MPLS standard. The twenty bit label value defines the index used into the forwarding table. The three bit Class of Service (COS) field may be a function of the IP Type Of Service (TOS) field, the packet's input interface, or the flow type. The Stack (S) bit indicates the last label prior to the packet header. The MPLS device only processes the label at the top of the stack (i.e., last before the packet header). MPLS supports hierarchical networks by prefixing packets with multiple labels. See [36] for an example of how this is done. Finally, the eight bit Time To Live (TTL) parameter provides a means for loop detection and removal.

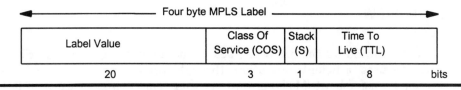

Figure 20.26 Proposed MultiProtocol Label Switching Four-Byte Label

20.6.6 Comparison of Alternative IP over ATM Approaches

Table 20.3 compares some key attributes of these alternative vendor proposals in support of IP (and other network layer internetworking protocols) over ATM. Some of the assessments are subjective since details for some of these protocols are still being defined. The throughput measure reflects the total network capacity and its sensitivity to traffic patterns. The flow mapping row describes the routing decision the scheme makes when mapping a particular flow. The connection churn rate describes the relative rate of dynamically established connections required by the technique.

Table 20.3   Comparison of Alternative IP/ATM Protocols

Attribute	Ipsilon's IP Switching	Toshiba's Cell Switch Router (CSR)	Cisco Systems' Tag Switching	IBM's Aggregate Route--Based IP Switching (ARIS)
Throughput	Low to Medium	Low to Medium	High	Medium to High
Throughput Sensitivity to Traffic	Degrades for short-lived flows	Degrades for short-lived flows	None	None
Flow Mapping	Link by Link	Link or Subnetwork	Edge to Edge	Egress Point
Hierarchy	None	None	Standard Routing	None
Mapping Basis	Long-Lived Flows	Long-Lived Flows or Pre-Configured	Derived from routing	Full mesh
State Persistence	Soft, Refresh Required	Soft, Refresh required	Hard, tied to network routing protocol	Soft, Refresh required
Connection Churn Rate	High	High	Low	Very Low
Complexity	Low	High	Medium	Medium

Note that all of theses approaches are proprietary until the IETF approves the MPLS RFC. Therefore, enter into these solutions with either a commitment to the proprietary solution, or else select a supplier with a firm commitment to implement the MPLS standard once it is finalized.

20.7   REVIEW

This chapter moved up the protocol stack to the network layer by describing how IP ensures efficient operation through proper negotiation of the Maximum Transfer Unit (MTU) size in networks employing a variety of link layers supporting different maximum frame sizes. The text then covered the simplest design supporting a single Logical IP subnetwork (LIS) over ATM employing the Classical IP over ATM protocol. As background we then introduced the Next Hop Resolution Protocol (NHRP) designed by the IETF to support larger internetworks involving multiple IP subnets. We then reviewed the latest output from the IETF and the ATM Forum, the MultiProtocol Over ATM (MPOA) standard, which promises to provide mult-vendor, interoperable, high-performance internets using an ATM-based infrastructure. The text also presented an overview of how the inherently non-broadcast ATM infrastructure supports IP multicast over ATM.

The chapter then reviewed and compared recent industry innovations supporting IP over ATM, in particular, Ipsilon's IP switching, Toshiba's Cell Switching Router (CSR), Cisco's Tag Switching, and IBM's Aggregate Route-based IP Switching (ARIS). We then summarized the current direction of the IETF's MultiProtocol Label Switching (MPLS) group. Unlike the standards based implementations, these proprietary schemes require the same equipment throughout the network. Whatever the future brings, it can only benefit from the cross fertilization afforded by the exchange of ideas between the ATM Forum, the IETF and the manufacturers of routers and switches.

20.8   REFERENCES

[1]   R. Atkinson, *RFC 1626: Default IP MTU for use over ATM AAL 5*, IETF, May 1994.
[2]   M. Laubach, *RFC 1577: Classical IP and ARP over ATM*, IETF, January 1994.
[3]   T. Bradley, C. Brown, *RFC 1293: Inverse Address Resolution Protocol*, IETF, January 1992.
[4]   M. Laubach, J. Halpern, *RFC 2225: Classical IP and ARP over ATM*, April 1998
[5]   M. Perez, F. Liaw, A. Mankin, E. Hoffman, D. Grossman, A. Malis, *RFC 1755: ATM Signaling Support for IP over ATM*, IETF, February 1995.
[6]   C. Partridge, *RFC 1363: A Proposed Flow Specification*, IETF, September 1992
[7]   R. Braden, D. Clark and S. Shenker, *RFC 1633: Integrated Service in the Internet Architecture: An Overview*, IETF, July 1994.
[8]   M. Maher, *RFC 2331: ATM Signalling Support for IP over ATM - UNI Signalling 4.0 Update*, April 1998.
[9]   J. Luciani, D Katz, D. Piscitello, B. Cole, N. Doraswamy, *RFC 2332: NBMA Next Hop Resolution Protocol (NHRP)*, April 1998.
[10] A. Alles, "ATM Interworking," http://cio.cisco.com/warp/public/614/12.html.
[11] J. Chiong, *Internetworking ATM for the Internet and Enterprise Networks*, McGraw-Hill, 1997.

[12] G. Sackett, C. Metz, *ATM and Multiprotocol Internetworking*, McGraw-Hill, 1996.

[13] E. Riley, *MPOA: MultiProtocol Over ATM*, www.techguide.com, 1997.

[14] ATM Forum, *MultiProtocol Over ATM (MPOA), Version 1.0*, af-mpoa-0087.000, July 1997.

[15] G. Swallow, "MPOA, VLANS and Distributed Routers," *ATM Forum 53 Bytes*, Volume 4, Issue 3, 1996

[16] E. Nolley, "MPOA Goes Under the Microscope at User Panel Discussion," *ATM Forum 53 Bytes*, Volume 4, Issue 3, 1996

[17] G. Armitage, *RFC 2022: Support for Multicast over UNI 3.0/3.1 based ATM Networks*, IETF, November 1996.

[18] S. Deering, *RFC 1112: Host Extensions for IP Multicasting*, IETF, August 1992.

[19] S. Deering et al, *Protocol Independent Multicast (PIM): Motivation and Architecture*, IETF Internet Draft, January 1995.

[20] D. Passmore, "Route Once, Switch Many," *Business Communications Review*, April 1997.

[21] P. Newman, T. Lyon and G. Minshall, "Flow Labelled IP: Connectionless ATM Under IP," Networld + Interop Presentation, April, 1996, ttp://www.ipsilon.com/staff/pn/presentations/interop96.

[22] K. Thompson, G. Miller, R. Wilder, "Wide-Area Internet Traffic Patterns and Characteristics," IEEE Network, November/December 1997.

[23] Ipsilon Networks, "IP Switching: The Intelligence of Routing, the Performance of Switching," February 1996.

[24] P. Newman, W. Edwards, R. Hinden, E. Hoffman, F. Liaw, T. Lyon, G. Minshall, *RFC1953, Ipsilon Flow Management Protocol*, IETF, May 1996.

[25] P. Newman, W. Edwards, R. Hinden, E. Hoffman, F. Liaw, T. Lyon, G. Minshall, *RFC1954, The Transmission of Flow Labelled IPv4 on ATM Data Links*, IETF, May 1996.

[26] P. Newman, W. Edwards, R. Hinden, E. Hoffman, F. Liaw, T. Lyon, G. Minshall, *RFC 1987, Ipsilon General Switch Management Protocol*, IETF, August 1996.

[27] E. Roberts, "IP on Speed," *Data Communications on the Web*, http://www.data.com/roundups/ip_speed.html, March 1997.

[28] Y. Katsube, K. Nagami, H. Esaki, "RFC 2098 : "Toshiba's Router Architecture Extensions for ATM : Overview," IETF, February 1997.

[29] H. Esaki, Y. Katsube, S. Matsuzawa, A. Mogi, K. Nagami and T. Jinmei, "Cell Switch Router - Version 1.0," Toshiba, November 1996.

[30] Cisco Systems, "Scaling the Internet with Tag Switching," http://www.cisco.com/warp/public/732/tag/pjtag_wp.html, 2/4/97.

[31] Y. Rekhter, B. Davie, D. Katz, E. Rosen, G. Swallow, "RFC 2105 : Cisco Systems' Tag Switching Architecture Overview," IETF, February 1997.

[32] B. Halabi, J. Lawrence, "Tag Switching in Service Provider ATM Networks," *Cisco White Paper*, 1997.

[33] A. Viswanathan, N. Feldman, R. Boivie, R. Woundy, *ARIS: Aggregate Route-based IP Switching*, IETF draft-viswanathan-aris-overview-00.txt, March 1997.

[34] N. Feldman, A. Viswanathan, *ARIS specification*, IETF, draft-feldman-aris-spec-00.txt, March 1997.

[35] R. Callon, G. Swallow, N. Feldman, A. Viswanathan, P. Doolan, A. Fredette, *A Framework for MultiProtocol Label Switching*, IETF draft ftp://ietf.org/internet-drafts/draft-ietf-mpls-framework-02.txt, November 26, 1997.

[36] D. Farinacci, Tony Li, A. Conta, Y Rekhter, E. Rosen, *MPLS Label Stack Encoding*, IETF Draft ftp://ietf.org/internet-drafts/draft-ietf-mpls-label-encaps-00.txt, November 18, 1997.

5

ATM Traffic Management and Congestion Control

This section provides the reader with an application-oriented view of the ATM traffic contract, Quality of Service, ATM service categories, traffic control, adaptive flow control, measures of congestion, and congestion control. First, Chapter 21 summarizes the basic ATM proposition: a network guarantees bandwidth with a specified QoS for that portion of the user's offered traffic conforming to a precisely specified set of traffic parameters. Together, these conditions form a traffic contract between the user and network. We define the basic QoS parameters involving errors, loss, and delay through precise definitions complemented by practical examples. Chapter 22 first introduces the basic concepts of traffic and congestion control. We then detail the Usage Parameter Control (UPC), or traffic policing, function using some simple examples followed by the formal definition. This chapter also describes the Available Bit Rate (ABR) closed loop flow control service category and protocol in some depth. Next, the text describes how users can employ traffic shaping to ensure that their traffic conforms to the traffic contract. Chapter 23 addresses the topic of congestion control with phases of response ranging from management, to avoidance, and as a last resort, recovery. This chapter also gives several examples of how popular flow control protocols react in the presence of congestion. The text highlights key standards when defining specific traffic management terminology. We explain each concept using analogies and several different viewpoints in an attempt to simplify a subject usually viewed as very complex. Furthermore, the text cites numerous references to more detailed discussions for the more technically oriented reader.

21

The Traffic Contract

This chapter explains the formal concept of a traffic contract from ITU-T Recommendation I.371 and the ATM Forum version 3.1 and 4.0 specifications in an application-oriented manner. This agreement exists between a user and a network. It embodies a philosophy and a paradigm defined by a specific set of terminology engineering to deliver guaranteed QoS at a traffic level defined by a set of traffic parameters. This chapter defines and illustrates the following specific terms and concepts: Quality of Service (QoS), traffic parameters, and conformance checking (commonly called traffic policing). Finally, we put all of these detailed terms and concepts back together using the ATM service category terminology introduced in Chapter 12. The text also gives some guidelines for choosing traffic parameters and tolerances.

21.1 THE TRAFFIC CONTRACT

In essence, the network establishes a separate traffic contract with the user of each ATM Virtual Path Connection (VPC) or Virtual Channel Connection (VCC). This traffic contract is an agreement between a user and a network at a specific User-Network Interface (UNI) regarding the following interrelated aspects of any VPC or VCC ATM cell flow:

- ☞ A set of QoS parameters for each direction of the connection
- ☞ A set of traffic parameters that specify characteristics of the cell flow
- ☞ The conformance checking rule used to interpret the traffic parameters
- ☞ The network's definition of a compliant connection

The definition of a compliant connection allows the network some latitude in the realization of checking conformance of the user's cell flow. Unquestionably, a connection that conforms precisely to the traffic parameters is compliant. A network may treat a connection with some degree of non-conformance as compliant; however, this definition is up to the network. In this case, the QoS guarantees apply to only the conforming cells. Finally, the network need not guarantee QoS to non-compliant connections, even the conforming cells.

The next sections of this chapter define the reference configuration for the traffic contract, Quality of Service (QoS) parameters, traffic parameters, and the leaky bucket conformance checking rule. We provide practical guidelines for use of these traffic parameters and how tolerances should be allocated for them through examples and discussion in the text.

21.2 REFERENCE MODELS

The basis of the traffic contract is a reference configuration, which the standards call an *equivalent terminal*. Figure 21.1 illustrates the components and terminology of this reference configuration.

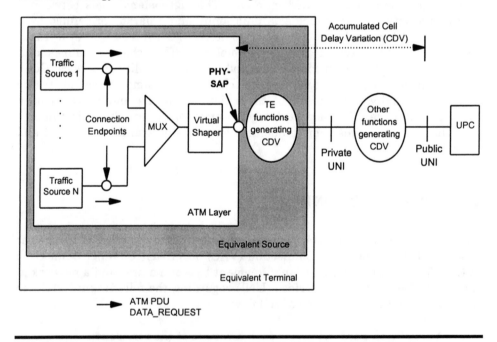

Figure 21.1   Equivalent Terminal Reference Model

An equivalent terminal need not be a real device, and indeed may be a collection of devices, such as an ATM workgroup, an ATM-capable router, bridge, or hub. Inside the equivalent terminal a number of traffic sources at the ATM layer; for example, applications running on a workstation with an ATM NIC, generate cells as ATM Protocol Data Unit (PDU) transmit requests to a specific a VPC or VCC connection endpoint. The ATM layer inside the terminal multiplexes these sources together, for example using a switching backplane or fabric in a local ATM switch, router, or hub. Associated with the multiplexing function is a virtual traffic shaper that spaces out the cells from each connection to ensure that the cell stream emitted to the physical layer Service Access Point (PHY SAP) conforms to the set of traffic parameters defined in the traffic contract with the network.

After the shaper function, other functions within the Terminal Equipment (TE) may create variation in the spacing of cells actually transmitted over a physical private ATM UNI interface so that it no longer conforms to the traffic parameters. Standards call this impairment *Cell Delay Variation* (CDV), a concept detailed in the next section. This ATM cell stream may then traverse other ATM devices in the private network, such as a collapsed ATM backbone, before arriving at the public ATM UNI, potentially accumulating even more CDV. Hence, the network provider defines a tolerance for this accumulated CDV in the traffic parameters, called *Cell Delay Variation Tolerance*, CDVT. Although these terms are similar, they mean different things: CDV specifies a variation in cell delay arrival times, while CDVT specifies a slack parameter for the peak inter-cell spacing when the network polices the user's cell stream. These terms are frequently confused, so remember that CDV is a QoS parameter as defined in 21.3.2 and CDVT is a traffic parameter as described in section 21.4.1.

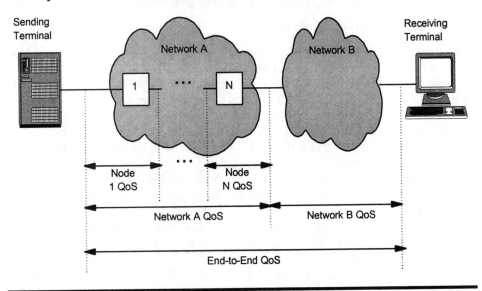

Figure 21.2 End-to-End QoS Reference Model

The end-to-end QoS reference model usually contains one or more intervening networks, each potentially with multiple nodes as depicted in Figure 21.2. Each of these intervening networks may introduce additional fluctuations in the cell flow due to multiplexing and switching, thereby impacting

QoS. Furthermore, statistical variations in the offered traffic may result in loss within congested network nodes. Of course, a network can also implement shaping between nodes, or between networks to minimize the accumulation of cell spacing variations and loss. In principle, the user should not know about the intervening networks and their characteristics, as long as the connection delivers the end-to-end QoS as specified by the traffic parameters.

As described later in this chapter, ITU-T Recommendation I.356 takes the approach of defining a worst case concatenation of networks and devices for specifying the worst possible QoS. Therefore, as long as connections across networks stay within these bounds, users experience a consistent level of QoS performance.

21.3 QUALITY OF SERVICE

The ATM Forum's UNI 4.0 Traffic Management Specification [1] defines specific Quality of Service (QoS) performance parameters for cells conforming to the parameters of the traffic contract. The UNI 3.1 specification tried to simplify a user's request for a certain QoS by defining a small number of specific classes. As described later in this chapter, UNI 4.0 further refines this concept by combining the concepts of specified QoS and traffic parameters into a small set of ATM layer service categories.

21.3.1 Cell Transfer Outcomes

Standards define QoS on an end-to-end basis — the perspective most relevant to an end user. An end user may be either a workstation, a customer premises network, a private ATM UNI, or a public ATM UNI. The following cell transfer outcomes define the various QoS parameters.

➢ A **Transmitted Cell** by the originating user enters the network.
➢ A **Successfully Transferred Cell** delivered by the network to the destination user.
➢ A **Lost Cell** which does not reach the destination user.
➢ An **Errored Cell** which arrives at the destination, but has errors in the payload.
➢ A **Misinserted Cell** which arrives at the destination but was not sent by the originator. This can occur due to an undetected cell header error or a configuration error.
➢ A **Severely-Errored Cell Block** where M or more lost, errored or misinserted cells occur within a received block of N cells.

See the discussion in Chapter 29 regarding ATM performance measurement for further details on cell transfer outcomes.

21.3.2   QoS Parameter Definitions

Table 21.1 lists the ATM layer QoS parameters defined in this section along with their commonly used acronyms. The last column provides an indication of whether the ATM Forum's TM 4.0, UNI 4.0 Signaling and PNNI 1.0 specifications define a means for the user to negotiate the QoS parameter with a network.

Table 21.1   Quality of Service (QoS) Parameter Terminology

QoS Acronym	QoS Parameter Name	Negotiated?
peak-to-peak CDV	Cell Delay Variation	Yes
maxCTD	maximum CTD	Yes
CLR	Cell Loss Ratio	Yes
CER	Cell Error Ratio	No
SECBR	Severely Errored Cell Block Ratio	No
CMR	Cell Misinsertion Rate	No

Propagation delay dominates the fixed delay component in wide area networks, while queuing behavior contributes to delay variations in heavily loaded networks. The effects of queuing strategy and buffer sizes dominate loss and delay variation performance in congested networks. A large single, shared buffer results in lower loss, but greater average delay and delay variation. Multiple buffers offer more flexibility in the tradeoff between delay variation and loss as explored in Chapter 22. The transmission network error rate defines a lower bound on loss. Transmission network characteristics are the leading cause for errors, and hence CER is a QoS parameter common to all QoS classes. Undetected errors in the cell header or configuration errors are the principal cause for misinserted cells. Burst of errors or intermittent failures are likely causes of errored blocks. See Chapter 12 for further discussion regarding the sources of impairments impacting the above QoS parameters.

The following formulae define the CLR, CER, SECBR, and CMR in terms of the cell transfer outcomes defined in section 21.3.1.

$$\text{Cell Loss Ratio} = \frac{\text{Lost Cells}}{\text{Total Transmitted Cells}}$$

$$\text{Cell Error Ratio} = \frac{\text{Errored Cells}}{\text{Successfully Transferred Cells} + \text{Errored Cells}}$$

$$\frac{\text{Severely Errored}}{\text{Cell Block Ratio}} = \frac{\text{Severely Errored Cell Blocks}}{\text{Total Transmitted Cell Blocks}}$$

$$\text{Cell} \quad \text{Misinsertion} \quad \text{Rate} \quad = \quad \frac{\text{Misinserted} \quad \text{Cells}}{\text{Time} \quad \text{Interval}}$$

21.3.3   Cell Transfer Delay and Cell Delay Variation

Two key components of ATM QoS are the Cell Transfer Delay (CTD) and Cell Delay Variation (CDV). Various components within ATM devices contribute to the statistics of delay within ATM network as illustrated in Figure 21.3. In general, fixed and variable delays occur on the sending and receiving sides of the end terminal, in intermediate ATM nodes, as well as on the transmission links connecting ATM nodes.

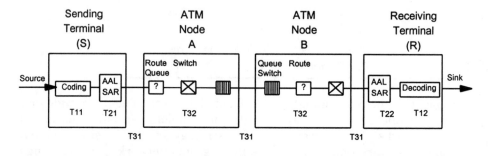

Figure 21.3 Illustration of Sources of Delay

The detailed delay components indicated in Figure 21.3 are:

> ➤ T1 = Coding and Decoding Delay:
>   ◦ T11   = Coding delay
>   ◦ T12   = Decoding delay
> ➤ T2 = Segmentation and Reassembly Delay:
>   ◦ T21   = Sending-side AAL segmentation delay
>   ◦ T22   = Receiving-side AAL reassembly/smoothing delay
> ➤ T3 = Cell Transfer Delay (End-to-End):
>   ◦ T31   = Inter-ATM node transmission propagation delay
>   ◦ T32   = Total ATM node processing delay (due to
>   queueing, switching, routing, etc.)

The principal statistical fluctuations in delay occur due the random component of queuing embodied in the T32 variable. Other terms contribute to a fixed delay, such as the terminal coding/decoding delays T11 and T12, along with the propagation delay T31. Part 6 covers the effects of statistical cell arrivals, queuing and switching techniques on the loss and delay parameters. The interaction of the sources of random and fixed delay depicted in Figure 21.3 result in a probabilistic representation of the likelihood that a particular cell experiences a specific delay. Mathematicians call such a plot a *probability density function*, with an example related to the particulars of the ATM Forum's QoS parameters covering delay and delay variation given in Figure 21.4. Of course, no cells arrive sooner than the fixed delay component, but cells arriving after the peak-to-peak Cell Delay Variation (peak-to-peak CDV) interval are considered late. The user may discard cells received later than this interval, hence the Cell Loss Ratio (CLR) QoS parameter bounds this area under the probability density curve as indicated in the figure. The maximum Cell Transfer Delay (maxCTD) is the sum of the fixed delay and peak-to-peak CDV delay components as indicated at the bottom of the Figure.

Standards currently define Cell Delay Variation (CDV) as a measure of cell clumping. Standards define CDV at either a single point against the nominal intercell spacing, or as the variability in the pattern measured between an entry point and an exit point. ITU-T Recommendation I.356 [2] and the ATM Forum UNI specifications 3.1 [3] and 4.0 [1] cover details on computing CDV and its interpretation. Cell clumping is of concern because if too many cells arrive too closely together, then cell buffers may overflow. Cell dispersion occurs if the network creates too great of a gap between cells, in which case the playback buffer would underrun. Chapter 26 illustrates the cases of playback buffer overflow and underrun via a simple example. An upper bound on the peak-to-peak CDV at a single queuing point is the buffer size available to the particular QoS class or connection. This bound results by observing the fact that the worst case (i.e., peak-to-peak) variation in delay occurs between the buffer empty and buffer full conditions. The worst case end-to-end CDV is the aggregation of the individual bounds across a network.

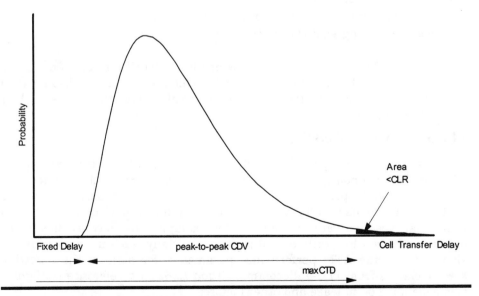

Figure 21.4   Cell Transfer Delay Probability Density Model

21.3.4 ATM Forum QoS Classes

In order to make things simpler on users, the ATM Forum UNI 3.1 specification defined the five QoS classes listed in Table 21.2. Each network specified particular values for the QoS parameters for each class. The applications indicated in the table are representative; other associations of the same application with another QoS class are possible. For example, video could employ QoS class 1. The ATM Forum TM 4.0 specification states that all implementations must support these QoS classes. Optionally, an implementation may support individually negotiated QoS parameters on a per connection basis as specified in the ATM Forum UNI 4.0 signaling and PNNI 1.0 specifications.

Table 21.2 ATM Forum QoS Classes

QoS Class	QoS Parameters	Application
0	Unspecified	"Best Effort" "At Risk"
1	Specified	Constant Bit Rate Circuit Emulation
2	Specified	Variable Bit Rate Video and Audio
3	Specified	Connection-Oriented Data
4	Specified	Connectionless Data

A QoS class defines at least the following parameters:

- Cell loss ratio for the CLP=0 flow
- Cell loss ratio for the CLP=1 flow
- Cell delay variation for the aggregate CLP=0+1 flow
- Delay for the aggregate CLP=0+1 flow

The CLP=0 flow refers to only cells which have the CLP header field set to 0, while the CLP=1 flow refers to only cells which have the CLP header field set to 1. The aggregate CLP=0+1 flow refers to all cells in the virtual path or channel connection.

21.3.5 Specified QoS Classes

A specified QoS class provides performance to an ATM virtual connection (VCC or VPC) defined by a subset of QoS parameters. For each specified QoS class, the network specifies an objective value for each QoS parameter. Note that a particular parameter may be essentially unspecified depending upon the actual objective value assigned – for example, any network meets the objective of a cell loss ratio of 100 percent. Initially, each network provider should define the ATM performance parameters for at least the following service classes from ITU-T Recommendation I.362 in a reference configuration depending on mileage and other factors:

- Service Class A: circuit emulation, constant bit rate video
- Service Class B: variable bit rate audio and video
- Service Class C: connection-oriented data transfer
- Service Class D: connectionless data transfer

Chapter 13 defined how these service classes relate to AALs. In the future, the ITU-T or the ATM Forum may define more QoS classes for a given AAL service class. The ATM Forum UNI 3.1 and TM 4.0 specifications define the following specified QoS Classes:

Specified QoS Class 1 supports a QoS that meets service class A performance requirements. This class should yield performance comparable to current digital private line performance.

Specified QoS Class 2 supports a QoS that meets service class B performance requirements. This class is intended for packetized video and audio in teleconferencing and multimedia applications.

Specified QoS Class 3 supports a QoS that meets service class C performance requirements. This class is intended for interoperation of connection-oriented protocols, such as Frame Relay.

Specified QoS Class 4 supports a QoS that meets service class D performance requirements. This class is intended for interoperation of connectionless protocols, such as IP or SMDS.

In general, an ATM virtual connection may use any one of these QoS classes. However, some higher layer protocols won't operate very well if the QoS provided by the network is too poor. For example, circuit emulation generally requires QoS class 1 for proper operation.

21.3.6   Unspecified QoS

In the unspecified QoS class, the network operator need not specify any the QoS parameters; however, the operator may set internal network performance objectives. Services using the unspecified QoS class must state a peak cell rate traffic parameter. Note that this traffic is "at risk" since the network makes no performance guarantees. In this case, the network admits this traffic and allows it to utilize capacity unused by connections with specific traffic parameters and a QoS class associated with performance guarantees. Connections utilizing the best effort capability should be able to dynamically determine the available capacity on the route allocated by the network; for example, as the TCP slow start procedure does (see Chapter 8).

An example application of the unspecified QoS class is the support of *best effort* service, where the user effectively specifies no traffic parameters and does not expect a performance commitment from the network. Most local area networks support only best effort service.

21.3.7   QoS in International ATM Networks

ITUT Recommendation I.356 [2] defines QoS objectives for a reference configuration spanning multiple carrier networks in a related, yet different set of QoS classes. The reference configuration considers transmission distances of up to 27,500 km, which creates a propagation delay of approximately 170 ms. The informative guidelines of Appendix II describe reference configurations traversing up to five networks containing up to 17 VP switches or 25 VC switches. The analysis assumes that most of the inter-switch connections occur at 155-Mbps or higher. However, the objectives apply even with a number of 34/45-Mbps circuits. The I.356 objectives do not include the delay contributed by geostationary satellites.

Table 21.3 shows the numerical values cited in ITU-T Recommendation I.356 for three QoS classes. These are provisional values and not firm requirements, since operational experience may indicate that better (or worse) performance is realistic. Even with this caveat, having quantitative numbers for QoS parameters on a global scale is a significant step forward. QoS class 1 meets the stringent requirements of constant bit rate traffic, while tolerant QoS class 2 addresses applications that do not differentiate between CLP=0 and CLP=1 loss. The bi-level QoS class 3 targets applications that expect guaranteed performance on CLP=0 cells, but don't expect guarantees on CLP=1 cells. For example, QoS class 3 addresses applications where the network uses the CLP bit to tag non-conforming cells. Recommendation I.356 also defines an unspecified class similar to the ATM Forum's definition.

Table 21.3   Recommendation I.356 QoS Class Objectives

QoS Parameter	Notes	QoS Class 1	QoS Class 2	QoS Class 3
CTD	Mean value	400 ms	Unspecified	Unspecified
CDV	2 point 10^{-8} quantile	3 ms	Unspecified	Unspecified
CLR (0+1)	Applies to CLP=0+1	3×10^{-7}	10^{-5}	Unspecified
CLR(0)	Applies to CLP=0	N/A	Unspecified	10^{-5}
CER	Upper Bound	4×10^{-6}	4×10^{-6}	4×10^{-6}
CMR	Upper Bound	Once per day	Once per day	Once per day
SECBR	Upper Bound	10^{-4}	10^{-4}	10^{-4}

21.3.8   Application QoS Requirements

Although the concept of QoS classes simplified matters somewhat, some applications have unique QoS requirements, and hence don't readily fit into a rigid set of QoS classes. Figure 21.5 [4, 5] illustrates several examples of application level QoS requirements for the loss and CDV QoS parameters. Based upon this market need, the ATM Forum designed the means for applications to request specific QoS parameters on a per connection basis. Furthermore, absolute delay also plays a critical role in many applications as discussed below.

Engineers know many requirements for voice based upon almost forty years of experience with digital telephony. If voice has greater than 50 ms of round-trip delay, then echo cancellation is required as detailed later in Chapter 17. Furthermore, a loss of 0.5 percent of normal speech is not objectionable to most listeners. Newer applications do not have such a basis, or well-defined requirements; however, there we summarize some general requirements in the following paragraphs.

Video application requirements depend upon several factors, including the video coding algorithm, the degree of motion required in the image sequence, and the resolution required in the image. Loss generally causes some image degradation, ranging from distorted portions of an image to loss of an entire frame, depending upon the extent of the loss and the sensitivity of the video coding algorithm. Also, variations in delay of greater than 20 to 40 ms can cause perceivable jerkiness in interactive video applications. Some video coding schemes require much tighter CDV as described in Chapter 17.

Combined video and audio is very sensitive to differential delays. Human perception is highly attuned to the correct correlation of audio and video, as is readily apparent in some foreign language dubbed films. File transfer applications are also sensitive to loss and variations in delay, which result in retransmissions and consequent reduction in usable throughput. The sensitivity to loss occurs as a result of the time-out and loss identification algorithm, as well as the retransmission strategy of the higher-layer protocol.

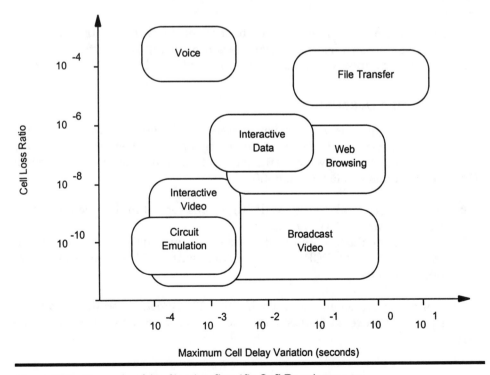

Figure 21.5   Example of Application Specific QoS Requirements

On the other hand, broadcast video can tolerate much longer delays since a large playback buffer compensates for variations in delay. However, the loss rate must be small since lost or distorted frames degrade playback quality. One way to compensate for loss is via error correction using AAL1 as described in Chapter 13. A good benchmark for broadcast video transmission over ATM is the quality of cable television transmission systems.

Users of interactive applications are also sensitive to loss and variations in delay due to retransmissions. For example, inconsistent response time can decrease productivity. Consistent response time (or the lack thereof) affects how users perceive data service quality.

Distributed computing and database applications are also sensitive to absolute delay, loss, and variations in delay. The ideal for these types of applications is infinite bandwidth with latency close to that of the speed of light in fiber. A practical model is that of performance comparable to a locally attached disk drive or CD-ROM, which has an access time ranging from 10 to 100 ms. In other words, the goal of high-performance networking is to make the remote resource appear is if it were locally attached to the user's workstation.

21.4 TRAFFIC PARAMETERS AND CONFORMANCE CHECKING

This section defines two more key components of the traffic contract between the user and the network: the parameters comprising the traffic descriptor and the conformance checking rule, commonly called the leaky bucket algorithm.

21.4.1 Traffic Descriptor

The traffic descriptor is a list of parameters which captures intrinsic source traffic characteristics. It must be understandable and enforceable. This section describes the following traffic parameters defined by the ATM Forum UNI specifications version 3.1 and 4.0:

- A mandatory Peak Cell Rate (PCR) in cells/second in conjunction with a CDV Tolerance (CDVT) in seconds
- An optional Sustainable Cell Rate (SCR) in cells/second (always less than or equal to PCR) in conjunction with a Maximum Burst Size (MBS) in cells

The following statements summarize the ATM Forum's traffic parameter specification. Figure 21.5 illustrates the following traffic contract parameters:

◎   Peak Cell Rate (PCR) = 1/T in units of cells/second, where T is the minimum intercell spacing in seconds (i.e., the time interval from the first bit of one cell to the first bit of the next cell).

◎   Cell Delay Variation Tolerance (CDVT) = τ in seconds. This traffic parameter normally cannot be specified by the user, but is set instead by the network. The number of cells that can be sent back-to-back at the physical interface rate $R=\delta^{-1}$ is $\tau/(T-\delta)+1$.

◎   Sustainable Cell Rate (SCR) = $1/T_s$ is the rate that a bursty, on-off traffic source can send. The worst case is a source sending MBS cells at the peak rate for the burst duration T_b as depicted in Figure 21.5.

◎   Maximum Burst Size (MBS) is the maximum number of consecutive cells that a source can send at the peak rate. A Burst Tolerance (BT) parameter, formally called τ_s (in units of seconds), defines MBS in conjunction with the PCR and SCR cell rate parameters as detailed in section 21.0.

Peak Cell Rate (PCR) + Cell Delay Variation Tolerance (CDVT)

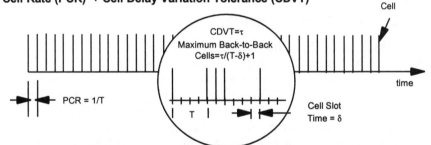

PCR + CDVT + Sustainable Cell Rate (SCR) + Maximum Burst Size (MBS)

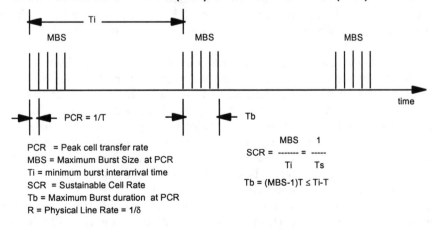

PCR = Peak cell transfer rate
MBS = Maximum Burst Size at PCR
Ti = minimum burst interarrival time
SCR = Sustainable Cell Rate
Tb = Maximum Burst duration at PCR
R = Physical Line Rate = 1/δ

$$SCR = \frac{MBS}{Ti} = \frac{1}{Ts}$$

$$Tb = (MBS-1)T \leq Ti-T$$

Figure 21.5 Illustration of Principal ATM Traffic Parameters

Figure 21.5 also depicts the minimum burst interarrival time as Ti, which relates to SCR and MBS according to the equations at the bottom of the figure. Specifically, SCR is equivalent to sending MBS cells within an interval of Ti seconds. The maximum burst duration in seconds is given by Tb, which an equation at the bottom of the figure also defines. These definitions are only intended to help the reader in understanding the traffic parameters — they are not part of the formal traffic contract. See the ATM Forum TM 4.0 specification [1] for the detailed definitions. The ATM Forum UNI 3.0 and 3.1 specifications first defined the sustainable cell rate and maximum burst size in a manner patterned after the PCR definition to better model bursty data traffic. In 1996, ITU-T Recommendation I.371 [6] added a specification for the sustainable cell rate to the previous standard which defined operation at only the peak cell rate. Modeling of the peak, average, and burst length characteristics enables ATM networks to achieve statistical multiplex gain with a specified loss rate as detailed in Chapter 25.

Figure 21.5, however, does not represent a rigorous definition of the traffic parameters. Standards define a formal, rigorous definition called the Generic Cell Rate Algorithm (GCRA). The next chapter defines this formal algorithm (which has the informal name of a leaky bucket). But first, we illustrate the basic concepts through simple analogies and examples.

21.4.2   Leaky Bucket Algorithm

The leaky bucket algorithm is key to defining the meaning of conformance. The leaky bucket analogy refers to a bucket with a hole in the bottom that causes it to "leak" at a certain rate corresponding to a traffic cell rate parameter (e.g., PCR or SCR). The "depth" of the bucket corresponds to a tolerance parameter (e.g., CDVT or BT). A subsequent section details these tolerance and traffic parameters. Each cell arrival creates a "cup" of fluid flow "poured" into one or more buckets for use in conformance checking. The Cell Loss Priority (CLP) bit in the ATM cell header determines which bucket(s) the cell arrival fluid pours into. First, the text describes the case of a single bucket via an example, then extends the model to the case of two buckets.

In the leaky bucket analogy, the cells do not actually flow through the bucket; only the check for conformance to the contract does. Note that one implementation of traffic shaping is to actually have the cells flow through the bucket as we cover in the next chapter. The operation of the leaky bucket is described with reference to the following figures for examples of a conforming and non-conforming cell flow. In all of the following examples, the nominal cell interarrival time is four cell times, which is the bucket increment, and the bucket depth is six cell times. Many of our examples employ the notion commonly employed in queuing theory of a fictional "gremlin" performing some action or making an observation. Real devices don't have gremlins (even though certain users disgruntled with early ATM devices may disagree here), but we believe that this treatment gives an intuitive insight into the operation of ATM traffic controls.

For each cell arrival, the gremlin checks to see if adding the increment for a cell to the current bucket contents would create overflow. If the bucket would not overflow, then the cell is *conforming*; otherwise it is *nonconforming*. The gremlin pours the fluid for nonconforming cells on the floor. Fluid from a cell arrival is added to the bucket only if the cell is conforming, otherwise accumulated fluid from nonconforming cells might cause later cells to be identified as nonconforming. The hole in the bucket drains one increment each cell time. Each cell arrival adds a number of units specified by the increment parameter. The fluid added by a cell arrival completely drains out after a number of cell times given by the leaky bucket increment. We now look at detailed examples of conforming and nonconforming cell flows.

For the conforming cell flow example shown in Figure 21.6, the first cell arrival finds an empty bucket, and hence conforms to the traffic contract. Thus, the first cell arrival fills the bucket with four units, the bucket increment in our example. The gremlin indicates that this cell is conforming, or okay. At the third cell time two units have drained from the bucket, and another cell arrives. The gremlin determines that the fluid from this cell would fill the bucket to the brim (i.e., to a depth of six); therefore, it also conforms to the traffic contract so that the gremlin adds its increment to the bucket, filling it completely in time slot 3. Now the next earliest conforming cell arrival time would be four cell times later (i.e., cell time 7), since 4 increments must be drained from the bucket in order for a cell arrival to not cause the bucket of depth equal to 6 units to overflow. This worst case arrival of conforming cells continues in cell times 15 and 19 in the example.

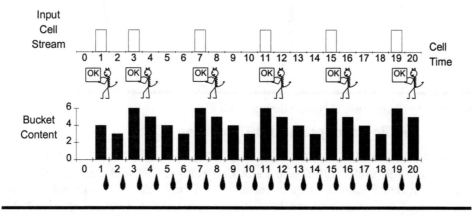

Figure 21.6 Illustration of a Conforming Cell Flow

In the nonconforming cell flow example of Figure 21.7, the first cell arrival at cell time 1 finds an empty bucket, is therefore conforming, and fills the bucket to a depth of four units. Over the next four cell times the bucket drains completely – one unit per cell time. At the fifth cell time, another cell arrives and fills the empty bucket with four increments of fluid. At the sixth cell time a cell arrives, and the gremlin determines that adding the arriving cell's fluid would overflow the bucket. Therefore, this cell is nonconforming and the gremlin pours the fluid for this cell onto the floor, bypassing the bucket. Since the gremlin did not pour the fluid for the non-conforming cell into the bucket, the next conforming cell arrives at cell time 7, completely filling the bucket to the level of 6 units. The next cell arrives at cell time 13, and fills the empty bucket with four units. The next cell arrival at cell time 15 fills the bucket to the brim. Finally the cell arrival at cell time 17 would cause the bucket to overflow; hence it is by definition nonconforming, and the gremlin pours the non-conforming cell's fluid on the floor, bypassing the bucket again.

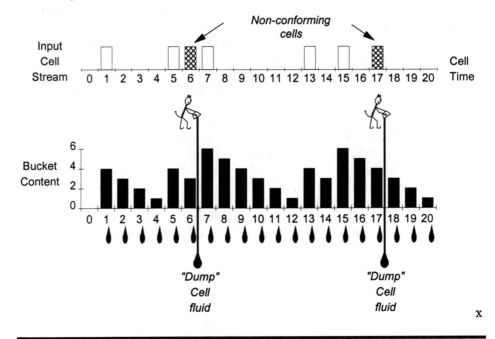

Figure 21.7   Illustration of Nonconforming Cell Flow

21.4.3   Traffic Descriptors And Tolerances

ITU-T Recommendation I.371 and the ATM Forum UNI 3.1 and TM 4.0 specification formally define the Peak Cell Rate (PCR) and Sustainable Cell Rate (SCR) traffic parameters in terms of a virtual scheduling algorithm and the equivalent leaky bucket algorithm representation detailed in the next chapter. The user specifies these traffic parameters either in a signaling message or at PVC subscription time, or else the network implicitly defines these parameters according to default rules.

The Peak Cell Rate (PCR) is modeled as a leaky bucket drain rate, and the Cell Delay Variation Tolerance (CDVT) defines the bucket depth as CDVT+T for Peak Rate conformance checking on either the CLP=0 or the combined CLP=0+1 flows.

The Sustainable Cell Rate (SCR) is modeled as a leaky bucket drain rate, and BT+T_s+$CDVT_s$ defines the bucket depth for sustainable rate conformance checking on either the CLP=0, CLP=1, or CLP=0+1 flows. CDVT$_s$ is the tolerance parameter associated with the SCR. The burst tolerance defines the SCR bucket depth by the following formula from Appendix C of the ATM Forum TM 4.0 specification:

$$\text{Burst Tolerance (BT)} = \tau_s = (\text{MBS}-1)\left(\frac{1}{\text{SCR}} - \frac{1}{\text{PCR}}\right)$$

The burst tolerance, or bucket depth, for the SCR is not simply the MBS because the sustainable rate bucket drains at the rate SCR. The formula above calculates the bucket depth required for MBS cells arriving a rate equal to PCR, while draining out at the rate equal to SCR.

21.4.4 Allocation of Tolerances

There are several considerations involved in setting the leaky bucket depths (i.e., traffic parameter tolerances). These differ for the peak rate and the sustainable rate. For the peak rate the bucket depth should not be much greater than that of a few cells; otherwise cells may arrive too closely together. Recommendation I.371 defines the minimum CDVT at a public UNI. For a single bucket depth of $CDVT$ cells and a nominal cell interarrival spacing T, note that approximately $CDVT/T$ cells can arrive back-to-back.

For the sustainable rate, the burst tolerance (or equivalently MBS) should be set to a value greater than the longest burst generated by the user. Of course, the source should shape its traffic to this parameter. The burst length is at least as long as the number of cells corresponding to the longest higher layer Protocol Data Unit (PDU) originated by the user. Furthermore, some transport protocols, such as TCP, may generate bursts that are many PDUs in length; however, the network may not guarantee transfer of extremely large bursts. Chapter 25 describes the relationship between burst length and loss for statistically multiplexed sources. Also, some additional tolerance should be added to allow for the effect of multiplexing and intermediate networks prior to the point which checks traffic conformance.

21.5 ATM SERVICE CATEGORIES

Are you confused now that we've defined the QoS parameters, the traffic parameters and the conformance checking rule? Don't worry, help is on the way from the ATM Forum's definition of service categories that puts all of these acronyms together in a manner meaningful to end user applications. Each service category definition uses the QoS and traffic parameter terms defined earlier in this Chapter. The ATM Forum Traffic Management 4.0 specification [1] defines the following ATM layer service categories:

- **CBR**   Constant Bit Rate

- **rt-VBR**   real-time Variable Bit Rate
- **nrt-VBR**   non-real-time Variable Bit Rate
- **UBR**   Unspecified Bit Rate
- **ABR**   Available Bit Rate

The TM 4.0 specification advanced the concept of service categories begun in UNI 3.1 by adding a new category, ABR; splitting the VBR category into real time (rt) and non-real time (nrt) components; and better defined the UBR (also known as best effort) service. Video service requirements largely drove the distinction between rt-VBR and nrt-VBR. The ABR service category defines a closed loop flow controlled service which uses feedback to achieve a specified loss objective if the user conforms to a traffic parameter defined by the Allowed Cell Rate (ACR). The next chapter details ABR.

Table 21.4   ATM Layer Service Category Attributes

Attribute	ATM Layer Service Category				
	CBR	rt-VBR	nrt-VBR	UBR	ABR
Traffic Parameters:					
PCR, $CDVT_4$	specified			specified$_2$	specified$_3$
SCR, MBS, $CDVT_{4,5}$	n/a	specified		n/a	
MCR_4	n/a			n/a	specified
QoS Parameters:					
peak-to-peak CDV	specified	specified	unspecified	unspecified	unspecified
maximum CTD	specified	specified	unspecified	unspecified	unspecified
CLR_4	specified			unspecified	See Note 1
Other Attributes:					
Feedback	unspecified			unspecified	specified

Notes:
1. CLR is not signaled for ABR because it has a low value.
2. Should not be subject to CAC and UPC procedures.
3. Represents maximum source rate.
4. Parameters are either explicitly or implicitly specified for PVCs or SVCs.
5. CDVT is not signaled.

Table 21.4 summarizes the QoS, traffic parameter, and feedback attributes for these service categories. Many world-renowned traffic engineering experts participated in the definition and specification of ATM-layer traffic management and service classification taxonomies in the ITU-T, the ATM Forum, and standards bodies around the world. The definition of ATM traffic management and congestion control is now stable within the ITU-T, as defined in conjunction with the ATM Forum, which now enables interoperability in multivendor implementations. The most recent definitions in the ITU-T's Recommendation I.371 and the ATM Forum's Traffic Management Specification Version 4.0 are closely aligned in many areas, but also have some key differences of which practicing engineers should be aware.

Table 21.5 [7] compares the ATM Forum's TM 4.0 service category terminology with the analogous ATM Transfer Capability from ITU-T Recommendation I.371 The third column lists characteristics of a typical application of the category, or capability. Although these groups chose different names, the close relationships established between the groups resulted in compatible definitions in most major areas.

Table 21.5   Mapping of ATM Forum and ITU-T Traffic Management Terminology

ATM Forum Service Category	ITU-T ATM Transfer Capability	Representative Application
Constant Bit Rate (CBR)	Deterministic Bit Rate (DBR)	Circuit Emulation
real-time Variable Bit Rate (rt-VBR)	(for further study)	Video on Demand
non-real-time Variable Bit Rate (nrt-VBR)	Statistical Bit Rate (SBR)	Packet traffic
Available Bit Rate (ABR)	Available Bit Rate (ABR)	Adaptable rate sources
Unspecified Bit Rate (UBR)	(no equivalent)	Best effort LAN traffic
(no equivalent)	ATM Block Transfer (ABT)	Burst level feedback control

Table 21.6 illustrates the mapping of service category/transfer capability to the resource allocation strategy, traffic parameter characterization, and generic QoS requirements.

Table 21.6   Mapping of ATM Forum and ITU-T Traffic Management Terminology

ATM Service Category	ATM Transfer Capability	Resource Allocation	Traffic Parameters	QoS Requirement
CBR	DBR	Constant	Peak, or Maximum Rate	Low CDV, Low Loss
rt-VBR		Statistical	Peak , Average Rates and Burst Size	Moderate CDV, Low Loss
nrt-VBR	SBR	Statistical	Peak , Average Rates and Burst Size	Moderate Loss
ABR	ABR	Dynamic	Minimum Rate	Low Loss
UBR		None	Peak	No Guarantees
	ABT	Per block, or burst	Block Size, Burst Rate	Low Loss

An ATM Service Category or, interchangeably, ATM-layer Transfer Capability, represents a class of ATM connections with similar characteristics. Resource allocation defines how much bandwidth a network must allocate to the service category/capability. Possible strategies are assignment of a fixed amount to each connection; statistical multiplexing of connections; or a dynamic, also called elastic, allocation based upon the current state of network utilization.

21.6 CONFORMANCE DEFINITIONS

The ATM Forum TM 4.0 specification defines conformance definitions for combinations of leaky buckets. This specification defines the associated traffic parameters; use of the Cell Loss Priority (CLP) bit for tagging non-conforming cells; and as a declaration of the cell flow to which the Cell Loss Ratio (CLR) QoS parameter applies. As noted previously, the terminology of "0" means that the parameter applies to only the CLP=0 flow, while the terminology "0+1" means that the parameter applies to the combined CLP=0+1 flow. Table 21.7 summarizes these conformance definitions.

Table 21.7   ATM Forum TM 4.0 Conformance Definitions

Conformance Definition	PCR Leaky Bucket	SCR Leaky Bucket	CLP Tagging Option?	CLR Specification
CBR.1	0+1	Not specified	No	0+1
VBR.1	0+1	0+1	No	0+1
VBR.2	0+1	0	No	0
VBR.3	0+1	0	Yes	0
UBR.1	0+1	Not specified	No	Unspecified
UBR.2	0+1	Not specified	Yes*	Unspecified

Note: The network may overwrite the CLP bit for any cell of the connection in the UBR.2 conformance definition.

Figure 21.8 illustrates these interoperable ATM Forum Leaky Bucket configurations. When there are two buckets, the conformance checking rule pours an equal amount of cell "fluid" into both buckets when the figures show a diagonal line and a two-headed arrow. The analogy for the diagonal line which the fluid pours over into the buckets is a "rough board" which creates enough turbulence in the fluid such that a single cup of fluid from an arriving cell fills both buckets to the same depth as if the fluid from a single cell was smoothly poured into one bucket.

Tagging is the action of changing the CLP=0 cell header field to CLP=1 when the ATM network detects a nonconforming cell. This controls how a bucket acting on CLP=0 only interacts with a bucket that operates on both CLP=0 and CLP=1 flows, referred to as CLP=0+1. We cover tagging in more detail in Chapters 22 and 23.

Note that all of these configurations contain the Peak Cell Rate (PCR) on the aggregate CLP=0+1 cell flow to achieve interoperability with the minimum requirement from ITU-T Recommendation I.371.

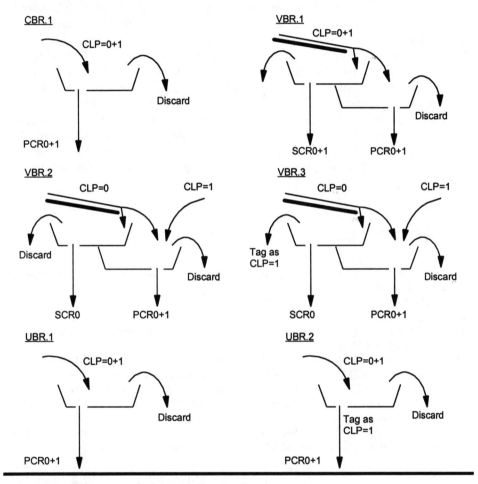

Figure 21.8   ATM Forum Leaky Bucket Configurations

21.7 REVIEW

This chapter introduced several key concepts and some basic terminology used in the remainder of Part 5. A traffic contract is an agreement between a user and a network regarding the Quality of Service (QoS) that the network guarantees to a cell flow — if the cell flow conforms to a set of traffic parameters defined by the leaky bucket conformance checking rule. The principal QoS parameters are: maximum Cell Transfer Delay (maxCTD), Cell Delay Variation (CDV), and Cell Loss Ratio (CLR). The traffic parameters define at least the Peak Cell Rate (PCR), but may optionally define a

Sustainable Cell Rate (SCR) and Maximum Burst Size (MBS) for specific cell flows identified by the Cell Loss Priority (CLP) bit in the ATM cell header. The conformance rule also associates a CDV Tolerance (CDVT) parameter with the peak rate, which the network specifies. A leaky bucket algorithm in the network checks conformance of a cell flow from the user by pouring a cup of fluid for each cell into a set of buckets leaking at rates corresponding to the PCR, and optionally the SCR. If the addition of any cup of cell fluid would cause a bucket to overflow, then the cell arrival is considered *nonconforming*. The algorithm discards fluid for nonconforming cells. The text also described additional considerations in setting the depth of the leaky buckets to account for tolerances in traffic parameters. The text then defined the ATM Forum service categories and ITU-T transfer capabilities in terms of these QoS parameters and traffic parameters. Finally, we summarized the interoperable set of leaky bucket conformance checking combinations defined by the ATM Forum.

21.8 REFERENCES

[1]   ATM Forum, *ATM Forum Traffic Management Specification, Version 4.0*, af-tm-0056.000, April 1996.

[2]   ITU-T, *B-ISDN ATM layer cell transfer performance*, Recommendation I.356, October 1996.

[3]   ATM Forum, *User-Network Interface Signaling Specification, Version 3.1*, af-uni-0010.002, September 1994.

[4]   G. Woodruff, R. Kositpaiboon, "Multimedia Traffic Management Principles for Guaranteed ATM Network Performance," *IEEE JSAC*, April 1990

[5]   Z. Dziong, *ATM Network Resource Management*, McGraw-Hill, 1997.

[6]   ITU-T, *Traffic control and congestion control in B-ISDN*, Recommendation I.371, 1996.

[7]   L. Lambarelli, "ATM Service Categories: The Benefits to the User," http://www.atmforum.com/atmforum/service_categories.html

ATM Traffic Control and Available Bit Rate (ABR)

This chapter introduces, compares, and contrasts some basic concepts from standards groups and the industry regarding traffic control. Traffic control provides the means that allow a user to ensure that the offered cell flows meet the rate specified in the traffic contract, as well as the means for networks to enforce the traffic contract in order to deliver the negotiated QoS performance across all users. Toward this end, we describe several types of Usage Parameter Control (UPC), commonly known as policing, functions based upon the leaky bucket algorithm, windowing techniques, and the Generic Cell Rate Algorithm (GCRA) formally defined in the standards. The text then covers the ATM Forum and ITU-T's Available Bit Rate (ABR) adaptive, closed loop traffic control algorithm. Like citizens using radar detectors, we then describe how the user can meet these stringent network requirements using traffic shaping. The text then covers the concepts of delay priority and loss thresholds using a simple equivalent queuing model. The chapter concludes with an overview of Generic Flow Control (GFC), Connection Admission Control (CAC), and virtual path based resource management.

22.1 TRAFFIC AND CONGESTION CONTROL

This section provides an overview of traffic control as an introduction to the topics covered in this Chapter. A generic reference model derived from I.371 as shown in Figure 22.1 illustrates the placement of various traffic and congestion control functions. Starting from the left hand side, user terminal equipment and private network switches may shape cell flows to conform with traffic parameters. The ingress port of the network then checks conformance of this cell flow to traffic parameters with a Usage Parameter

Control (UPC) function at the public User Network Interface (UNI). In a similar manner networks may check the arriving cell flows from a prior network using Network Parameter Control (NPC) at a public Network-Node Interface (NNI). Networks may employ additional traffic control functions such as Connection Admission Control (CAC), resource management, priority control, ABR flow control, or traffic shaping as indicated in the figure.

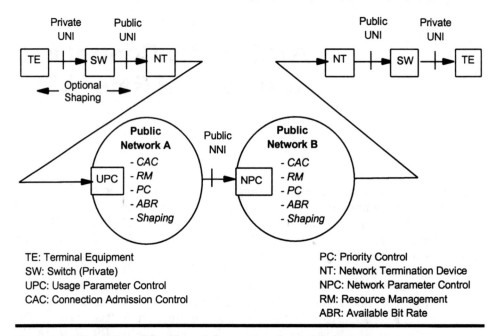

Figure 22.1   Overview of Traffic and Congestion Control Functions

   In traffic and congestion control the time scale over which a particular control is applicable is important. Figure 22.2 illustrates the time scales of various traffic and congestion control methods. The minimum time scale that traffic or congestion control acts on is a cell time (which is approximately 10 μs on a DS3, and 3 μs on an OC3/STS-3c). For example, UPC traffic control (commonly called policing) and traffic shaping act on this time scale. Also, selective cell discard and priority queue servicing as described in the next chapter act at the cell time level. Other functions operate at the AAL PDU time scale, such as frame level policing or selective frame discard. The next time scale that makes sense is that of round-trip propagation time, as covered in the section on ABR flow control later in this chapter. The next major event that changes over time is the arrival of either PVC provisioning or SVC signaling that creates requests to either establish or relinquish virtual connections. Typically, this time scale is much greater than the round-trip delay. Finally, long-term network engineering and network management procedures operate on much longer time scales, on the order of hours to days, extending to months and years for long-range network planning.

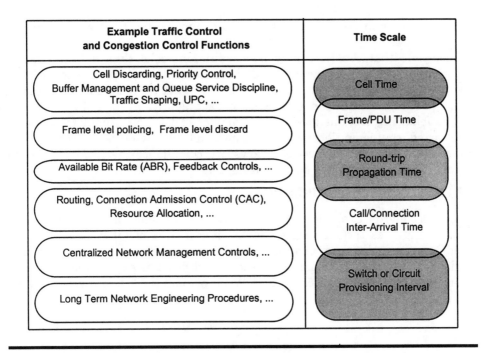

Example Traffic Control and Congestion Control Functions	Time Scale
Cell Discarding, Priority Control, Buffer Management and Queue Service Discipline, Traffic Shaping, UPC, ...	Cell Time
Frame level policing, Frame level discard	Frame/PDU Time
Available Bit Rate (ABR), Feedback Controls, ...	Round-trip Propagation Time
Routing, Connection Admission Control (CAC), Resource Allocation, ...	Call/Connection Inter-Arrival Time
Centralized Network Management Controls, ...	Switch or Circuit Provisioning Interval
Long Term Network Engineering Procedures, ...	

Figure 22.2 Time Scales of Traffic and Congestion Control Functions

22.2 USAGE AND NETWORK PARAMETER CONTROL

Networks employ Usage Parameter Control (UPC) and Network Parameter Control (NPC) to check conformance of cell flows from a user, or another network, against negotiated traffic parameters, respectively.   Another commonly used name for UPC/NPC is *policing* [1]. This is a good analogy, because UPC and NPC perform a role similar to the police in society. Police enforce the law, ensuring fair treatment for all people. The ATM UPC/NPC algorithms enforce traffic contracts. Connection Admission Control (CAC) fairly allocates bandwidth and buffering resources among the users. Without UPC/NPC and CAC, unfair situations where a single user "hogs" resources can occur. Most of the functions and discussion in the following sections apply equally to UPC and NPC, with any differences identified explicitly.

Standards do not specify the precise implementation of UPC and NPC functions; instead they bound the performance of any UPC/NPC implementation in relation to a Generic Cell Rate Algorithm (GCRA), which is essentially a fancy name for the leaky bucket algorithm. Indeed, the compliant connection definition part of the traffic contract identifies how much nonconforming traffic a network will still provide a QoS guarantee. The other requirement is that the UPC should not take a policing action (i.e., tag or discard) on more than the fraction of cells which are nonconforming according to the leaky bucket rule; or in other words, the UPC cannot be too tight and overpolice user cell flows.

Also note that the UPC may police different cells than a leaky bucket algorithm does due to inevitable differences in initialization, the latitude defined for a compliant connection, or the fact that the UPC implementation of a a particular device is not the leaky bucket algorithm.

The following sections give three examples of UPC/NPC implementations; one using the leaky bucket algorithm, and two windowing schemes to illustrate differences in how cell flows are compliance-checked by different algorithms using the same traffic parameters.

22.2.1 Leaky Bucket UPC/NPC

The leaky bucket example of Figure 22.3 uses the same nonconforming cell flow as input as the shaping example later in this chapter. In all of the following examples, the nominal cell interarrival time is four cell times, which is the bucket increment, and the bucket depth is six cell times. Three more gremlins: "Tag," "Discard," and "Monitor" join the "Dump" gremlin from Chapter 21 in this example to illustrate the range of UPC actions.

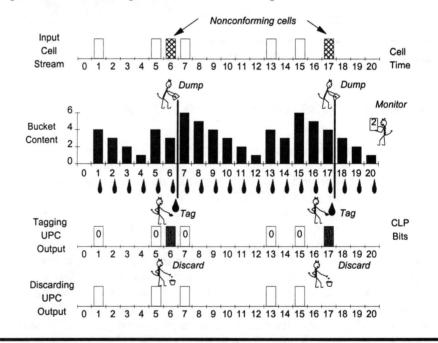

Figure 22.3 UPC via Leaky Bucket Example

Cell arrivals occur along the horizontal axis at the top of Figure 22.3, with the "Dump" gremlin pouring the fluid from nonconforming cells (indicated by cross-hatching in the figure) past the leaky buckets. The other gremlins: "Tag," "Discard," and "Monitor" all operate in conjunction with "Dump's" identification of a nonconforming cell. "Tag" sets the CLP bit to 1 (regardless of its input value) and lets the cell pass through, with its position unchanged.

"Discard" simply blocks the cell from transmitting. "Monitor" simply keeps track of how many cells were nonconforming on his notepad. There is a fourth possible UPC action in the standard, namely, do nothing, which corresponds to all of the gremlins being out to lunch in this example.

22.2.2 Sliding and Jumping Window UPC

This section compares two windowing UPC mechanisms to the leaky bucket mechanism to illustrate how different UPC implementations, designed to police the same proportion of nonconforming cells, yield different results. Some older ATM implementation used these techniques; however, most modern devices use the leaky bucket for reasons demonstrated in this section. Observe from Figure 22.4 that the worst-case conforming cell flow can have at most 3 cells in 10 cell times. For any set of single leaky bucket parameters, an apparently equivalent relationship is at most M cells in a window of N cell times. In the following examples, M=3 and N=10. Two types of windowing UPCs are considered: a sliding window and a jumping window method. We use two more gremlins to describe these UPCs: "Slide" and "Jump."

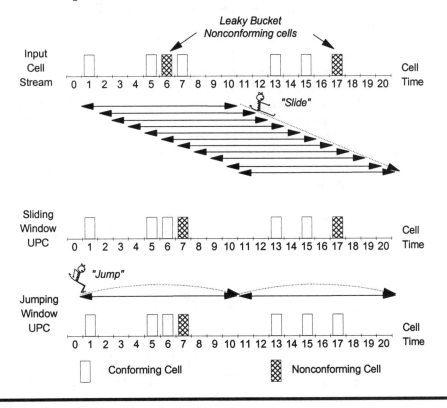

Figure 22.4   UPC via Windowing Example

For reference, the top of Figure 22.4 shows the same nonconforming cell flow input to the leaky bucket in the previous example. In the sliding window protocol the gremlin "Slide" moves a window covering N cell times to the right each cell time. If no more than M cells exist in the window covering N cell times, then no UPC action occurs. However, if by sliding the window one unit to the right, more than M cells exist within the window, then UPC takes action on the cells that violate the M out of N rule. "Slide" repeats this process each cell time. The sliding window and leaky bucket UPCs act upon the same number of cells; however, they act on different cells! In the jumping window scheme, "Jump" moves the window N units to the right every N cell times as shown at the bottom of the figure. "Jump" applies the same M out of N count rule, but detects only one nonconforming cell as compared with two in the leaky bucket and sliding window UPC examples.

In the example of Figure 22.4, the jumping window UPC was looser than the leaky bucket and sliding window UPC methods. The fact that different UPC algorithms, or even the same algorithm, may police different cells is called measurement skew. This is the reason that standards state UPC performance in terms of the fraction of conforming cells according to a specific reference algorithm. However, in some cases the sliding and jumping window UPC algorithms may discard a larger proportion of cells than the leaky bucket algorithm.

Figure 22.5 illustrates a pathological case where the difference between a leaky bucket, sliding window, and jumping window UPC is even more pronounced. Each algorithm has parameters chosen to admit one cell every three cell times on the average, and allow at most two back-to-back cells. The leaky bucket UPC has an increment of three cells and a bucket depth of five cells, while the sliding and jumping window algorithms have parameters M=2 and N=6. Figure 22.5 illustrates the arrival of 10 cells. The leaky bucket UPC identifies 20 percent of the cells as nonconforming, the sliding window identifies 40 percent as nonconforming, and the jumping window identifies 30 percent as being nonconforming.

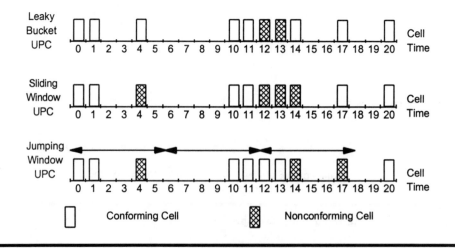

Figure 22.5   Pathological Example of UPC Differences

22.2.3 Generic Cell Rate Algorithm (GCRA)

As seen from the previous example, because different UPC implementations result in markedly different proportions of policed cells, the ATM Forum specified a formal algorithm as a reference model in the UNI 3.1 specification in 1994 [2]. Figure 22.6 illustrates the two equivalent interpretations of the ATM Forum and the ITU-T Recommendation I.371 [3] Generic Cell Rate Algorithm (GCRA): the virtual scheduling algorithm and the leaky bucket algorithm. Each of these algorithms utilizes the two traffic parameters that define either the peak rate or the sustainable rate and the associated tolerance parameters: an Increment (I) and a Limit (L), both expressed in units of time. As stated in the standards, these representations are equivalent. The virtual scheduling representation appeals to time sequence oriented people, while the leaky bucket method appeals to the mathematical and accounting types among us.

a. Virtual Scheduling Algorithm

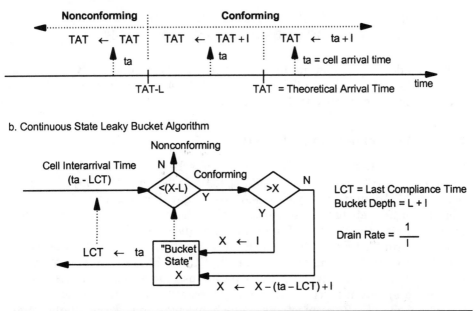

b. Continuous State Leaky Bucket Algorithm

Figure 22.6 Virtual Scheduling and Leaky Bucket Algorithms for the GCRA

Figure 22.6a illustrates the virtual scheduling algorithm which utilizes the concept of a Theoretical Arrival Time (TAT) for the next conforming cell. If a cell arrives more than the tolerance limit, L, earlier than the TAT, then it is nonconforming as shown in the figure. Cells arriving within the tolerance limit are conforming, and update the TAT for the next cell by the increment, I, as indicated in the figure. Cells arriving after the TAT make the tolerance available to subsequent cells, while cells arriving within the tolerance interval reduce the tolerance available to subsequent cells.

The flowchart in Figure 22.6b shows the detailed workings of the leaky bucket algorithm. Here, the GCRA parameters have the interpretation that the bucket depth is L+I and the bucket drain rate is 1/I. The leaky bucket algorithm uses the concepts of a Last Compliance Time (LCT) variable to hold the arrival time of the last conforming cell as well as a bucket fill state variable, X. The algorithm compares the difference between a cell arrival time and LCT to the bucket state minus the tolerance limit (i.e., X-L) to determine if the cell arrival would overflow the bucket. If overflow would occur, then the cell is deemed nonconforming. Otherwise, the flowchart checks to see if the bucket completely drained since the LCT before updating the bucket state variable X accordingly. The algorithm then substitutes for the current values of the LCT and bucket state X in preparation for processing the next cell arrival.

The ATM Forum TM 4.0 specification uses the GCRA(I,L) rule to formally define the conformance checks for the peak and sustainable rate conformance checks as follows:

- Peak Rate: GCRA(T, τ), where PCR=1/T and CDVT=τ
- Sustainable Rate: GCAR(T_s, τ_s), where SCR=1/T_s and BT=τ_s

The reader interested in more details on the formal GCRA algorithm should download the ATM Forum's TM 4.0 specification [4] from www.atmforum.com or consult normative Annex A of ITU-T Recommendation I.371.

22.2.4 Guaranteed Frame Rate (GFR)

Beginning in 1997, the ATM Forum embarked upon specifying another service category targeted to provide specified QoS to higher layer protocols using ATM Adaptation Layer 5 (AAL5). What distinguishes this category from the other cell based categories is that the service either accepts entire frames, or else rejects entire frames at every queuing point.

The Guaranteed Frame Rate (GFR) service targets users who either cannot specify ATM traffic parameters (e.g., PCR, SCR, MBS), or cannot comply with the implied source behavior. Many existing user devices (e.g., routers, hosts, and servers) fall into this category. Currently, these users access ATM networks via UBR connections, which provide no service guarantees. Therefore, many users with legacy devices have little incentive to migrate to ATM technology.

The GFR service aims to provide ATM performance and service guarantees to support such non-real-time applications. The GFR service requires that the user data cells employ AAL5 to segment frames into cells as described in Chapter 13. The GFR service provides the user with a Minimum Cell Rate (MCR) guarantee under the assumption of a given maximum frame size (MFS) and a given Maximum Burst Size (MBS). MFS and MBS are both expressed in units of cells. GFR provides a minimum service rate (determined by MCR) during network congestion, while the user may be able to send at a higher rate during uncongested intervals. Basically, GFR is UBR service with MCR and MBS traffic parameters that implements the intelligent frame discard feature.

If the user sends frames with less than MFS cells in a burst, which is also less than the MBS, then the network should deliver the associated frames minimal loss. GFR allows a user to send traffic exceeding MCR and the associated MBS; however, the network provides no performance guarantees for this excess traffic. Furthermore, the network should service excess traffic using available resources fairly across all GFR users in proportion to the traffic contract. Furthermore, the user may mark cells for entire frames using the CLP bit. The MCR traffic parameter applies only to unmarked frames. Similar to VBR services, the network may tag cells in unmarked frames only if the user requests the tagging option, either via signaling (for SVCs) or via subscription (for PVCs). Currently, the GFR service only applies to virtual channel connections since this is the level at which AAL5 delineates frames. The ATM Forum Traffic Management 5.0 specification will detail the GFR capability.

22.3 AVAILABLE BIT RATE (ABR) CLOSED-LOOP FLOW CONTROL

The ABR specification from the ATM Forum 4.0 specification [4] and ITU-T Recommendations I.371 [3] and I.371.1 [5] share a common goal: to make unused bandwidth available to cooperating end users in a fair, timely manner. Therefore, ABR targets the many existing applications with the ability to reduce their information transfer rate, such as TCP. As such, this objective goes well beyond that of today's best-effort LANs, where a single selfish, or "highly motivated," user can paralyze a shared network. If sources conform to the rules of ABR, then the network guarantees a Minimum Cell Rate (MCR) with minimal cell loss. The network may optionally penalize nonconforming sources, or enforce a policy that users not having transmitted recently lose any rights to access bandwidth at rates greater than MCR during periods of congestion.

On the other hand, TCP congestion control implicitly assumes everyone follows the same rules. For example, policing of TCP/IP conformance is limited to a professor's failing the graduate student who hacks the OS kernel to eliminate TCP's adaptive flow control and achieves markedly improved throughput at the expense of other TCP users sharing the same IP network. ABR's congestion control overcomes this problem by providing the means for an ATM network switch to police users that do not conform to the mandated source behaviors. Hence, a rogue user who sends at a greater rate than that determined by the ABR protocol receives poorer performance than those ABR users that followed the rules. In the previous TCP scenario, the single rogue user achieves good performance while all the other honest users suffer.

Although TCP sounds more like our daily lives and ABR seems to strive for a lofty, utopian goal, we now describe how ABR achieves fairly arbitrated congestion control. Let's begin by considering some simple analogies from everyday life that involve adaptive flow control.

ABR's binary mode is like the green/red lights at the entrance to congested freeways. The light turns green and another car (cell) enters the freeway (interface). Downstream sensors detect congestion and meter cars entering the freeway so that reasonable progress occurs. A police officer tickets vehicles that do not conform to this rule. By analogy, ATM switches indicate congestion using the EFCI bit in the ATM cell header in the forward direction. The destination end system employs a Congestion Indication (CI) field in a Resource Management (RM) cell to communicate the presence of congestion back to the source, which makes green-light/red-light type decisions regarding the source's current transmission rate. The recipient of the CI bit does not completely stop sending, but instead reduces its transmission rate by a fraction of the currently available cell rate.

The Explicit Rate (ER) mode of ABR operation adds another degree of complexity. Similar to the manner in which air traffic controllers control the speed of multiple airplanes (cells) converging on a crowded airport, ATM switches along the path of the end-to-end connection communicate an explicit rate for each source. In this way, controller (RM cells) throttle back fast planes (user cells) during periods of congestion to yield a regular arrival pattern at the congested airport (interface).

The Virtual Source/Virtual Destination (VS/VD) mode is analogous to air traffic controllers and airline carriers coordinating the speed and route of aircraft (cells) as they approach airports (interfaces), making connections that maximize the number of seats filled — as well as customer satisfaction. Each airport-to-airport (virtual source-destination pair) route makes decisions with an awareness of congestion at other airports (other VS/VD pairs).

Of course the details of ABR are more complex than these simple analogies; however, keeping these insights in mind may help the reader to grasp the essence of ABR as the following treatment proceeds into more abstract details. In a WAN environment, a key issue that a network ABR switch addresses is how to set the explicit rates to control the sources.

22.3.1 The Great Rate versus Credit Debate

During the initial stages of the definition of ABR at the ATM Forum in late 1993, two candidate algorithms emerged [6, 7]: a rate-based scheme and a credit-based scheme. The rate-based scheme [8] touted simplicity at the price of reduced efficiency in some cases. Initially, the credit-based scheme [9] was more complex and required larger buffers, but claimed theoretically ideal efficiency. The debate about which scheme the ATM Forum should follow raged until the fall of 1994, making the popular communications press on a regular basis, until the Forum's technical committee voted in the rate-based scheme by a wide margin. Leading up to the final Traffic Management 4.0 (TM 4.0) specification in the spring of 1996 [4], the rate-based scheme gained considerable complexity in response to credit-based skeptics, but addressed the majority of the efficiency drawbacks of earlier, simpler rate-based proposals. Some members from the credit-based camp went away and started the Quantum Flow Control (QFC) consortium [10] .

A key criticism of the credit-based approach was the requirement to dedicate buffer capacity to each connection proportional to the round-trip delay. Hence, for a LAN environment, many agreed that the credit-based approach was optimal, and simpler than the rate-based approach. However, in the quest for a single approach that scaled across both LAN and WAN environments, the ATM Forum chose a rate-based scheme where the source generated Resource Management (RM) cells periodically which the destination in turn looped back. The destination, or switches in the return path, either indicate detected congestion, or explicitly signaled the maximum source rate they could currently support in these looped back RM cells. Closing the flow control loop, the ATM Forum TM 4.0 specification detailed the behaviors conforming sources must follow. Despite earlier proposals to specify network enforcement, the final specification left network policing of ABR sources as an implementation option. Informative Appendix III of TM 4.0 describes a Dynamic GCRA as a possible conformance checking rule for use at the network edge to police ABR sources. As a consequence of these decisions, the VS/VD became the de facto means for networks to dynamically shape user traffic to ensure conformance to the specified traffic contract.

22.3.2 Three Flavors of ABR

ABR specifies a rate-based, closed-loop, flow control mechanism. A key objective of ABR is to fairly distribute the unused, or available, bandwidth to subscribing users while simultaneously achieving a low cell loss rate for all conforming ABR connections. All user data cells on ABR connections must have the CLP bit set to zero. The bandwidth allocated by the network to an ABR connection ranges between the Minimum Cell Rate (MCR) negotiated at connection establishment time and the Peak Cell Rate (PCR) depending upon network congestion and policy.

Users must conform to feedback provided via RM cells according to rules detailed in the ATM Forum's TM 4.0 specification [11]. ABR flow control occurs between a sending end system, called a *source*, and a receiving end system, called the *destination* connected via a bidirectional, point-to-point connection. Each of the terminals is both a source and a destination for each direction of an ABR connection. ABR specifies an information flow from the source to the destination composed of two RM flows, one in the forward direction and one in the backward direction, that make up a closed flow control loop. The forward direction is the flow from the source to the destination, and the backward direction is the flow from the destination to the source. In the following sections, we describe the information flow from the source to the destination and its associated RM flows for a single direction of an ABR connection; the procedures in the opposite directional are symmetrical, but possibly with different parameter values.

22.3.2.1 Binary Mode ABR

The binary mode, shown in Figure 22.7, involves ATM switching nodes setting EFCI in the forward direction so that the destination end station can set the CI field in a returned RM cell to control the flow of the sending end station. The binary mode ensures interoperability with older ATM switches which can only set the EFCI bit in the forward direction in response to congestion. This is the simplest mode; however, it experiences higher loss rates in certain situations, such as those where congestion occurs at multiple points in the network. Furthermore, unless the network elements perform per connection queuing, unfairness may result when sharing a single buffer.

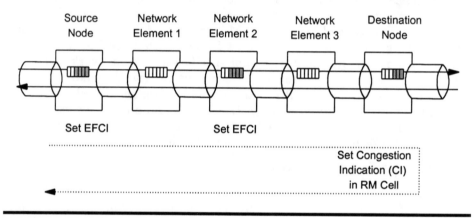

Figure 22.7 Binary-Mode ABR

The complexity in the end system rate control procedures compensates for the simple operation in the network in Binary mode. An end system must tune over a dozen parameters to achieve good performance in this mode. The ATM Forum specification also defines a relative rate marking scheme using the Congestion Indication (CI) and No Increase (NI) bits in the RM cell to provide more granular adjustments of the source systems transmission rate.

22.3.2.2 Explicit Rate (ER) ABR

Figure 22.8 illustrates the ER mode where each Network Element (NE) explicitly sets the maximum allowed rate in RM cells looped back by the destination as they progress backward along the path to the source. The ATM Forum TM 4.0 specification gives examples of how switches may set the explicit rate in the feedback path in Informative Appendix I, leaving the implementation as a vendor-specific decision. Therefore, a key issue that a network ABR switch must address is how it to set the explicit rates to control the sources. The goal is that each user receive a fair allocation of available bandwidth and buffer resources in proportion to their traffic contracts in a responsive manner. Simultaneously, the ABR service should operate at high utilization with negligible loss. This mode requires tuning of far fewer parameters, and hence is the preferred method in networks that are capable of supporting explicit rate ABR.

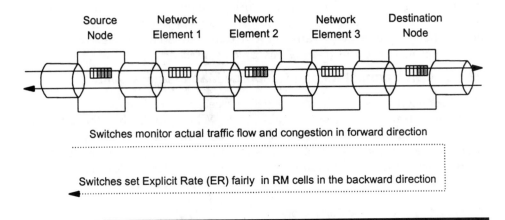

Figure 22.8 Explicit Rate ABR

22.3.2.3 Virtual Source/Virtual Destination (VS/VD) ABR

Figure 22.9 illustrates an ABR virtual connection that incorporates segmentation using the concept of mated pairs of virtual sources and destinations. These sources and destinations form closed flow control loops across the sequence of network elements involved in an ATM ABR connection as shown in the figure.

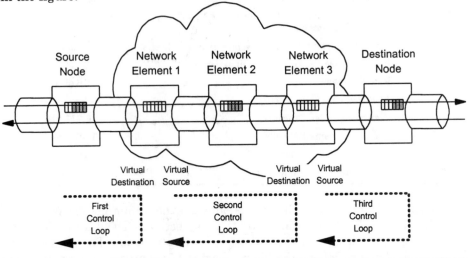

Figure 22.9 Virtual Source/Virtual Destination (VS/VD) ABR Control Loops

A configuration where every switch on the path is a virtual source/destination is also called *hop-by-hop* flow control. The example in the figure shows the network elements using VS/VD mode to isolate the source and destination node control loops. The VS/VD scheme also provides a means for one network to isolate itself from nonconforming ABR behavior occurring in another network. On any network element, the virtual destination terminates the ABR control network for its corresponding source. The same device then originates the traffic on a virtual source for the next control loop that conforms to the end-to-end ABR traffic contract.

22.3.3 ABR Parameters and Resource Management Cells

Upon connection establishment, ABR sources request and/or negotiate the operating parameters shown in Table 22.1. Where applicable, the table also gives the parameter's default value. The user negotiates these parameters with the network via information elements in signaling messages for SVCs, or network management interfaces for PVCs. The following narrative describes each of these parameters and the context in which ABR uses them.

Table 22.1   ABR Service Parameters

Acronym	Meaning	Default Value
PCR	Source's Peak Cell Rate policed by the network	-
MCR	Minimum Cell Rate guaranteed by the network	0
ACR	Currently Allowed Cell Rate by the network	-
ICR	Initial Cell Rate used by source prior to feedback or after an idle period	PCR
TCR	Tagged Cell Rate for "out-of-rate" RM cells	10 cps
Nrm	Number of cells between forward RM cells	32
Mrm	Control on number of RM cells between forward and backward directions	2
Trm	Upper bound on interval between forward RM cells	100 ms
RIF	Rate Increase Factor used in binary mode	1/16 of PCR
RDF	Rate Decrease Factor used in binary mode	1/16 of ACR
ADTF	ACR Decrease Time Factor	0.5 s
TBE	Transient Buffer Exposure	16,777,215
CRM	Count of Missing RM cells	TBE/Nrm
CDF	Cutoff Decrease Factor	1/16
FRTT	Fixed Round Trip Time	

The Allowed Cell Rate (ACR) varies between the minimum and peak rates (i.e., MCR and PCR) negotiated for the connection. The Initial Cell Rate (ICR) applies to either the case where a source transmits for the very first time, or to the case where the source begins transmitting after a long idle period. The ABR specification defines a means to calculate ICR based upon the TBE, Nrm and FRTT parameters. The other parameters control the source, destination and network behavior in conjunction with the RM cell contents as described below.

In order for the ABR algorithms to operate responsively, feedback must occur. Figure 22.10 shows how RM cells sent periodically by the source probe the forward path, while the destination assumes the responsibility for turning around these RM cells by changing the DIRection bit in the RM cell and sending the RM cell in the opposite direction.

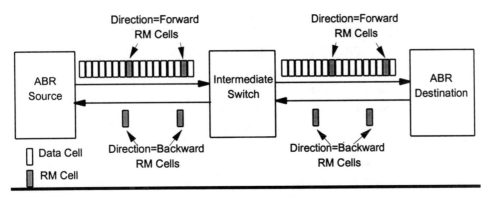

Figure 22.10   ABR RM Cells Insertion and Feedback

The ABR specification allows the destination to return fewer RM cells than it receives. This design handles the case where the MCR of the backward connection cannot sustain feedback at a rate corresponding to one RM cell for each of Nrm cells received from the source in the forward direction. This occurs in practice in ATM over ADSL applications where the bandwidth differs by an order of magnitude in the forward and backward directions. *In-rate* RM cells count toward the Allowed Cell Rate (ACR) of an ABR connection in each direction. A source must send an RM cell at least once every Trm milliseconds. If the source does not receive a backward RM cell within CRM cell times, then it reduces ACR by the CDF factor, unless this would cause ACR to become less than MCR. End systems may optionally send *out-of-rate* RM cells at a lower priority by setting CLP = 1 at a rate limited by the TCR parameter to make the ABR algorithm more responsive. However, since these cells have a lower priority, the network may discard them during periods of congestion.

Table 22.2 depicts the content of the ATM Forum 4.0 ABR Resource Management (RM) cell. The fields are aligned with the ITU-T ABR specification in ITU-T Recommendation I.371. We briefly summarize how the various ABR operating modes specify source and network behaviors using these fields in RM cells. See Reference 11 for more information on source behaviors.

The destination or the network may optionally set the Backward Explicit Congestion Notification (BECN) field in RM-cells in the backward direction in order to command the source to reduce its rate immediately. The source always sets CI=0 in forward RM cells. Either the destination or an intermediate switch sets the Congestion Indication (CI=1) bit to cause the source to decrease its ACR. When operating in binary mode, sources utilize the Rate Increase and Decrease Factors (RIF and RDF) to calculate a new ACR.

The destination or an intermediate switch sets the No Increase (NI) bit to prevent a source from increasing its ACR. The intent of the NI bit is to address detection of impending congestion in networks with long delay paths. Sources set NI=0 in the forward direction and switches cannot change NI from 1 to 0 in the backward direction.

Table 22.2   ATM Forum Resource Management (RM) Cell Contents

Field	Description	Size
Header	ATM RM Cell: VPC VCI=6, PTI=110, VCC: PTI=110	5 bytes
ID	Protocol Identifier = '0000001'	1 byte
	Message Type Field	
DIR	Direction: Forward = 0 and Backward = 1	1 bit
BECN	Backward Explicit Congestion Notification	1 bit
CI	Congestion Indication	1 bit
NI	No Increase	1 bit
RA*	Request/Acknowledge	1 bit
Reserved	Reserved for future use	3 bytes
ER	Explicit Rate	2 bytes
CCR	Current Cell Rate	2 bytes
MCR	Minimum Cell Rate	2 bytes
QL*	Queue Length	4 bytes
SN*	Sequence Number	4 bytes
Reserved	Reserved for future use	246 bits
CRC-10	Cyclic Redundancy Check	10 bits

* Fields defined in ITU-T I.371, but not used in ATM Forum 4.0

TM 4.0 encodes all rates (i.e., ACR, CCR, MCR) using a 16-bit floating point format which represents values of over 4 trillion cells per second. A practical limitation is the 24-bit representation of rate values utilized in the signaling messages, which limits the maximum rates to approximately 16 million cells per second, or approximately 7-Gbps.

In the ER and VS/VD modes, the source sets the Current Cell Rate (CCR) field to its current ACR in forward RM-cells. The destination and intermediate switches may use the CCR field to calculate ER in the backward direction, but they cannot modify it. The destination and/or any intermediate switch uses the Explicit Rate (ER) field to set the source's Allowed Cell Rate (ACR) to a specific value. Typically, the destination initially sets ACR to the maximum possible rate, that is PCR. Subsequently, any network element in the backward path may reduce the ER to communicate the supportable fair share for the connection. According to the rules of ABR, no network element in the backward path can increase ER.

The source sets the Minimum Cell Rate to the negotiated value for the connection. The destination and intermediate switches may use the MCR field to calculate ER. If the MCR value in the RM cell differs from the one signaled, then a switch may correct it.

22.3.4 ABR Conformance Checking - Dynamic GCRA

Informative Appendix III of the ATM Forum's TM 4.0 specification defines a technique called Dynamic GCRA that meets the requirements for checking conformance of an ATM cell flow against the ABR traffic contract in terms of a sequence of Explicit Rate (ER) fields transmitted by the network within RM cells in the backward direction. The principal difference between the dynamic GCRA and the static GCRA defined in section 22.2.3 is that the Increment (I) may change over time. A network may use dynamic GCRA to police ABR users to ensure fairness and conformance to the ABR traffic contract.

22.3.5 Point-to-Multipoint ABR

As shown in Chapters 19 and 20, ATM makes extensive use of point-to-multipoint connections in LAN Emulation (LANE) as well as the protocol supporting IP multicast over ATM defined in RFC 2022. Therefore, the ATM Forum described a framework in support of an optional capability to provide ABR service over point-to-multipoint connections. Recall that an ATM point-to-multipoint connection eminates from a root node and branches at a number of intermediate nodes to each of the leaves as illustrated in Figure 22.11.

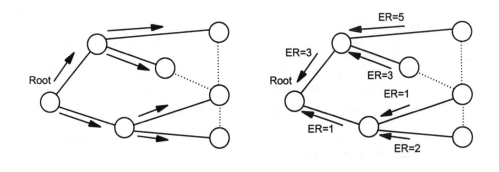

Figure 22.11 Illustration of ABR Support for a Point-to-Multipoint ATM Connection

As required in point-to-multipoint connections, the network must return cells from each of the leaves back to the root. The branch nodes implementing ABR in a point-to-multipoint connection perform two important functions as illustrated in the figure: replicating data and RM cells in the direction from the root to the leaves, and merging RM cells in the direction from the leaves back to the root. Figure 22.11a illustrates the replication of both data and RM cells from a root node out through two branching points to leaves in a small network. In the backward direction (i.e., from the leaves to the root) , the branching nodes may operate in different modes. In one mode of operation, a branching node queues RM cells and consolidates the feedback; for example, taking the minimum Explicit Rate (ER) value and returning it back toward the root as shown in Figure 22.11b. Note that this mode of operation reduces the throughput of the entire point-to-multipoint connection to that of the slowest receiver, or bottleneck at a branching point. Although this may seem bad, it isn't if the goal is for effectively lossless broadcast communication to all parties in the point-to-multipoint connection. As stated in the TM 4.0 specification, branching nodes may also implement the virtual source and destination closed loop protocol between themselves.

22.4 TRAFFIC SHAPING

For a user to derive maximum benefit from the guaranteed QoS from ATM, then the device connecting to the network should ensure that the cells sent to the network conform with the parameters in the traffic contract. Standards call the method to achive this goal *traffic shaping*. In other words, the user equipment processes the source cell stream such that the resultant output toward the network conforms to the traffic parameters according to the leaky bucket algorithm. Although the standards make traffic shaping optional, recall that the network definition of a compliant connection need not guarantee QoS performance for nonconforming cells. Therefore, a user wanting guaranteed QOS must shape traffic to ensure conformance to the traffic parameters in the contract. A network may employ shaping when transferring a cell flow to another network in order to meet the conditions of a network-to-network traffic contract, or in order to ensure that the receiving user application operates in an acceptable way. A real-life example of the need for traffic shaping occurs when a receiving device or user application accepts only a certain maximum rate of cells before its buffers overflow.

22.4.1 Overview of Possible Shaping Methods

Various papers in the literature as well as the standards propose the following traffic shaping implementations:

- Leaky Bucket and Buffering
- Spacing
- Peak Cell Rate Reduction
- Scheduling
- Burst Length Limiting
- Source Rate Limitation
- Framing

We give a brief summary for each of the proposals listed above. The text gives detailed examples for leaky bucket buffering, spacing, and framing.

Buffering operates in conjunction with the leaky bucket algorithm to ensure that cells don't violate the traffic parameters of the contract by buffering cells until the leaky bucket would admit them.

Spacing involves the end terminal holding cells from multiple virtual connections in a queue, scheduling their departures according to the traffic parameters while simultaneously minimizing delay variation. Spacing discards cells if the delay variation would be too great.

Peak cell rate reduction involves operating the sending terminal at a peak rate less than that in the traffic contract, reducing the possibility of conformance violation.

Burst length limiting constrains the transmitted burst length to a value less than the Maximum Burst Size (MBS) in the traffic contract.

Source rate limitation is an implicit form of shaping that occurs when the actual source rate is limited in some other way; for example, in DS1 circuit emulation the source rate is inherently limited by the TDM clock rate.

Framing superimposes a TDM-like structure onto the sequences of ATM cells, and utilizes this frame structure to schedule those cell streams that require controlled delay variation into the next frame time.

22.4.2 Leaky Bucket Buffering

This section gives an example of traffic shaping using buffering and a leaky bucket implementation to transform a nonconforming cell flow into a conforming cell flow. This example uses the same notation for cell arrivals over time along the horizontal axis, the same nominal interarrival time of 4 cell times and the same leaky bucket depth of 6 as the previous example. We're pleased to introduce two new gremlins: "Stop" and "Go," to illustrate the buffering and scheduling operation.

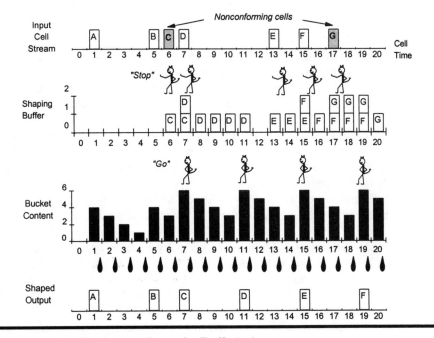

Figure 22.12   Traffic Shaping Example (Buffering)

The gremlin "Stop" replaces "Dump" in the nonconforming example. "Stop" commands the ATM hardware genie to buffer the cell if its fluid flow would cause bucket overflow, and "Go" allows a cell transmission as soon as the bucket drains far enough to admit the latest cell. When "Stop" and "Go" are out of synch, then cells build up in the shaping buffer as shown in Figure 22.12. The figure illustrates this operation with the individual cells labeled A through G, and the nonconforming cells from the previous examples indicated by shading. Cell arrivals A and B are conforming, and leave the bucket in a state such that arrival C at cell time 6 is nonconforming, and hence "Stop" stores cell C in the shaping buffer. Cell D arrives immediately after C, so the gremlin "Stop" also buffers D. In the same cell time the bucket empties enough so that "Stop's" partner "Go" transmits cell C and adds its flow to the bucket. At cell time 11, "Go" sends cell D and fills the bucket. At cell time 13 cell arrival E would cause the bucket to overflow; hence "Stop" buffers it. Cells F and G are similarly buffered by "Stop" and transmitted at the earliest conforming time by "Go" as illustrated in the figure. For the reader wishing to continue the example, cell G would be transmitted at cell time 23 (not shown). Note that the output cell flow from this process is conforming, as can be checked from the conformance test of the leaky bucket defined in the previous chapter.

The leaky bucket shaper smoothes out the input stream and does not drop any cells unless its buffer overflows. A leaky bucket policing algorithm would find this shaped output stream conforming.

22.4.3 Spacing

Another proposal for a UPC/NPC traffic control function is that of spacing [12]. In this shaping implementation the resultant output never violates the nominal intercell spacing, but may discard additional cells. We employ yet another gremlin called "Space" to describe the function of this traffic shaping algorithm. "Space" implements a virtual scheduling algorithm which computes a Theoretical Reemission Time (TRT) such that the output never violates the nominal cell spacing, and discards any input bursts that cannot be spaced out within a tolerance specified in a manner analogous to the leaky bucket. Figure 22.13 illustrates the spacing function with the same nonconforming cell flow example used above. Cells arrive as shown in the upper axis, with nonconforming cells shown shaded as before. The gremlin "Space" appears wherever a cell is rescheduled, or spaced, showing the TRT on the card in his hands.

Note that the output of the spacer is very regular, and never violates the nominal cell emission interval. This has advantages in terms of controlling the delay variation observed by subsequent networks or the destination device. Such regular arrivals are important to video, audio, and circuit emulation applications which expect to receive cells at a regular rate. A disadvantage of spacing is that the algorithm discards cells from a burst, disrupting higher-layer functions as illustrated for cell C in the example. However, the spacer implementation could tag the cells using the CLP bit instead of discarding them.

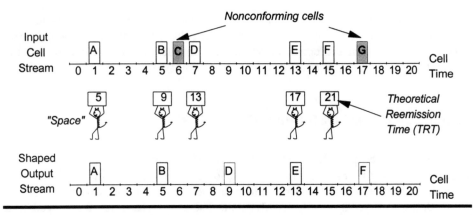

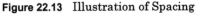

Figure 22.13 Illustration of Spacing

22.4.4 Framing

Figure 22.14 illustrates the concept of framing, which is a proposed method of controlling delay variation for delay-sensitive virtual connections [13].

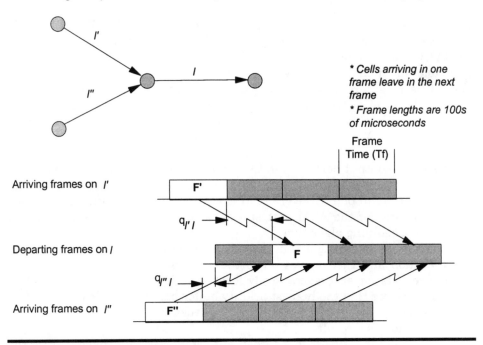

Figure 22.14 Illustration of Framing

Framing basically overlays a synchronous structure of frame boundaries on the asynchronous cell stream. Cells from flows that require a tightly controlled delay variation are scheduled to depart from a switching node in the next frame in a priority manner, similar to catching the next train. If the arrival time between frames is small enough, then the variation can be controlled with a limit equal to the number of nodes times the frame length. Framing is well suited to AAL1 circuit emulation over ATM, where the receiving connection endpoints have only a limited amount of cell buffer for absorption of the jitter accumulated across the network.

22.5 DELAY AND LOSS PRIORITY CONTROL

Priority control helps achieve the full range of QoS loss and delay parameters required by a diverse set of high-performance applications. Generally, prioritization operates to control delay or loss using one of two mechanisms. First, delay priority basically involves the use of separate queues served in a prioritized manner. Secondly, loss priority typically involves thresholds within each buffer for different traffic types and CLP values. Operating in concert, these two mechanisms ensure that ATM devices meet the delay and loss QoS parameters specified in the traffic contract.

22.5.1 Priority Queuing

Priority queuing, service scheduling, or fair queuing [14, 15, 16] all basically implement multiple queues in the switch, such that traffic on certain VPC/VCCs that are not tolerant of delay can "jump ahead" of those that are more tolerant of delay. ATM switches employ *priority queuing* between VPCs and VCCs in different QoS classes or service categories to meet different delay and loss priorities simultaneously. Some switches even implement a queue for each connection thus meeting QoS requirements at the most granular level. We describe priority queuing with reference to the switch block diagram of Figure 22.15. In our example the priority queuing function resides on the output port of an ATM switch. The switch takes arriving cell streams from multiple input ports, looks up an internal priority value, and directs the cells to one of several corresponding queues on the output port. The output side of the ATM port serves each of the queues according to a particular scheduling algorithm.

A simple scheduling algorithm serves the highest priority nonempty queue to exhaustion and then moves on to the next lower priority queue. This process repeats for each successively lower priority queue. This scheduling function ensures that the highest priority buffer has the least loss, delay, and delay variation. However, lower priority queues may experience significant delay variation as a result.

Other scheduling algorithms spread out the variation in delay across the multiple queues. For example, a Weighted Fair Queuing (WFQ) scheduler sends out cells just before reaching the maximum delay variation value for cells in the higher priority queues. This action decreases delay variation in the lower priority queues.

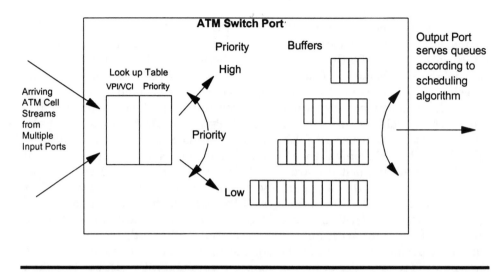

Figure 22.15   Illustration of Priority Queuing Operation

Actual switch designs may dedicate a set of buffers to each output port or split the buffers between the input and output ports. Some switches share memory between multiple ports and/or the multiple priorities. Switches employing a shared memory approach usually limit the individual queue size for each port, service class or in some cases individual virtual connections.

22.5.2 Discard Thresholds

Figure 22.16 illustrates the operation of discard thresholds within a single buffer.

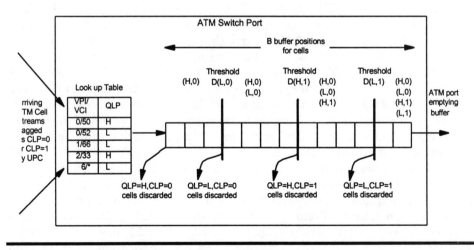

Figure 22.16 Illustration of Discard Threshold Operation

On the switch port, each delay priority queue may be segmented into several regions via *discard thresholds*. Arriving cell streams have the CLP bit set by either the end user or the UPC/NPC at ingress to the network. The switch port consults a lookup table to determine the Queue Loss Priority (QLP) with a value of either (H) or Low (L) based upon the VPI/VCI combination in the cell header. The single buffer with B cell positions has four thresholds in our example. Starting from the right hand side, the four thresholds determine the priority order for cell discard based upon buffer occupancy as follows; discard QLP=L, CLP=1 cells first; discard QLP=H, CLP=1 cells second; discard QLP=L, CLP=0 cells next; and finally discard QLP=H, CLP=0 cells only if the entire buffer is full. Note that the buffer full condition is the implicit discard threshold for QLP=H, CLP=0 cells in our example. Note that the selection of discard thresholds in our example first give preference to CLP=0 traffic over CLP=1 traffic, then give preference to QLP=H traffic over QLP=L traffic. Other choices of discard thresholds are possible. Thus, cells with any combination of (QLP, CLP) may occupy the rightmost portion of the buffer before the first discard threshold D(L,1) as shown in the figure. For buffer occupancy greater than D(L,1), but less than D(H,1), all cells except those of type QLP=L, CLP=1 may occupy the buffer. This partitioning of the buffer continues as we move past successive thresholds toward the left until only cells with QLP=0 and CLP=0 may occupy the leftmost portion of the buffer above the D(L,0) threshold.

22.5.3 Performance Implications of Priority Control

Combinations of discard thresholds operating in each of several separate priority queues result in the capability to support a range of loss and delay QoS parameters. Simultaneously, the CLP discard thresholds provide a measure of congestion control as described in the next chapter. Figure 22.18 shows the combined effects of priority queuing and discard thresholds on the Cell Delay Variation (CDV) and Cell Loss Ratio (CLR) QoS parameters. The figure shows the resultant QoS values for a switch having four delay priority queues as shown in Figure 22.17, each with four discard thresholds as shown in Figure 22.16. The values in this example are only a representative example of the type of CLR versus CDV plot resulting for such a system. The statistics and mix of the offered traffic for each delay and loss priority, the actual values of the discard thresholds, and the queue servicing algorithm all play roles in determining the actual QoS parameters; however, plots similar to the one in our example often result in devices employing combinations of priority queuing and priority discard.

Note how the priority queuing supports differentiated QoS performance for applications that tolerate both increased CLR and CDV. In other words, lower priority queues experience both higher loss rates and greater delay variation. Comparing Figure 22.18 with the plot of typical applications against the same CDV versus CLR scale in Chapter 21, observe that most applications require priority queuing. Fewer applications require the differentiation offered by discard thresholds which offers increased CDV with decreased CLR. In general, many applications tolerate both CDV and CLR impairments equally well and cannot trade one off against the other.

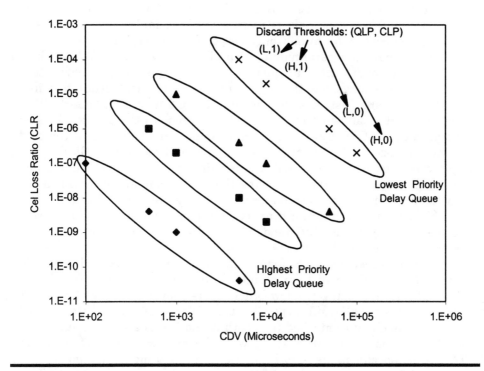

Figure 22.18   Effects of Priority Queuing and Discard Thresholds on QoS

22.6 GENERIC FLOW CONTROL

The concept of Generic Flow Control (GFC) has a long history in the standardization process. Initially, experts viewed the GFC as a means to implement a function similar to the Distributed Queue Dual Bus (DQDB) protocol on a shared access medium. What was standardized in 1995 in ITU-T Recommendation I.361 [17] is a point-to-point configuration that allows a multiplexer to control contention for a shared trunk resource through use of traffic-type selective controls.

The cell header allocates four bits for Generic Flow Control (GFC) at the ATM UNI. Since the GFC is part of the cell header, a multiplexer requires no additional bandwidth or VPI/VCI allocations to control terminals. The GFC bits have different meanings depending upon the direction of cell transmission. The four bits of GFC represent almost 1 percent of the available ATM cell payload rate, and are therefore an important resource. Figure 22.19 illustrates the multiplexer configuration for GFC standardized in I.361 showing the usage of the GFC bits in each direction between the multiplexer and two types of terminals. The default coding of the GFC is *null*, or all zeroes, which indicates that the interface is not under GFC, which is called an *uncontrolled* mode. ATM terminals have either one or two queues, called connection groups A and B as shown in the figure. The protocol between the multiplexer and the terminals is asymmetric; the multiplexer controls the terminals, while the terminals only respond or convey information. The multiplexer commands terminals to stop all traffic via the HALT bit based upon the state of its internal queues. For example, if the shared trunk resource becomes congested, the GFC multiplexer may halt certain terminals. The SET command instructs the terminal to load a credit counter with an initial "Go" value, allowing the terminal to decrement the counter for every cell transmitted. GFC controlled terminals may send cells only if the credit counter is non-zero. Therefore, as long as the multiplexer periodically sends a SET command to the terminal, the terminal may continue sending data. The NULL meaning of the bit means that the terminal cannot reload the credit counter for that connection group. For example, the GFC multiplexer may SET the credit counter for connection group A, but leave the B bit set to zero (i.e., NULL) in the GFC field. The addressed terminal can then only send connection group A traffic.

The terminal responds to commands from the multiplexer, indicating that it understands GFC, which in standards terminology means that it is a *controlled* terminal using the low order bit as shown in Figure 22.19. The terminal indicates the traffic type for a particular VPI/VCI in the second and third bits, namely, uncontrolled (i.e., both A and B bits are zero), queue A, or queue B. Although standardized, few implementations support GFC. The ATM Forum specifications require only uncontrolled mode (i.e., no GFC) for interoperability. Also note that GFC operates only across an ATM UNI, not between ATM devices.

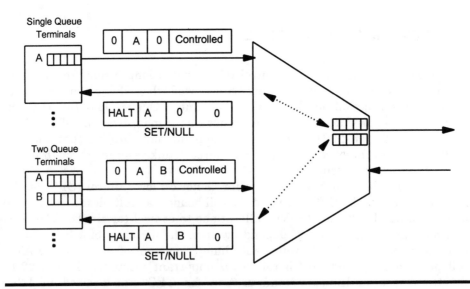

Figure 22.19   Illustration of GFC Configuration and Function

GFC may also be employed in a unidirectional ring using this protocol. There is also a possibility that further information could be multiplexed into the GFC field for more sophisticated controls.

22.7 CONNECTION ADMISSION CONTROL (CAC)

Connection Admission Control (CAC) is a function in ATM devices that determines whether to admit or reject connection requests. Many devices implement CAC in software. Recall from Chapter 21 that a connection request defines the source traffic parameters and either the requested QoS Class, or the user specified QoS parameters. ATM devices use CAC to determine whether admitting the connection request at PVC provisioning time or SVC call origination time would violate the QoS already guaranteed to active connections. CAC admits the request only if the network can still guarantee QoS for all existing connections after accepting the request. Frequently, each node performs CAC for SVCs and SPVCs for performance reasons. A centralized system may perform CAC for PVCs. For accepted requests, CAC determines UPC/NPC parameters, routing decisions, and resource allocation. Resources allocated include trunk bandwidth, buffer space, and internal switch resources.

CAC must be simple and rapid to achieve high SVC call establishment rates. On the other hand, CAC must be accurate to achieve maximum utilization while still guaranteeing QoS. CAC complexity is related to the traffic descriptor, the switch queuing architecture, and the statistical traffic model.

The simplest CAC algorithm is peak rate allocation, where the ATM device simply keeps a running total of the peak rate of all admitted connections. Peak rate allocation CAC denies a connection request if adding the peak rate of the candidate connection exceeds the trunk bandwidth.

Figure 22.20 illustrates peak rate allocation. Starting in the upper left hand corner, the ATM device receives a request for peak cell rate R to the CAC logic. The trunk bandwidth is P, of which a certain portion A is already assigned according to peak rate allocation. If the request R exceeds the available bandwidth (P-A), then CAC denies the request; otherwise, CAC accepts the connection request, incrementing the allocated bandwidth by R. Actually, the admission threshold may be somewhat less than (P-A) due to the slack implied by the compliant connection definition, the CDV tolerance parameter, and the buffer size available for a certain cell loss and delay QoS objective.

CAC implementations may also permit a certain amount of resource oversubscription in order to achieve statistical multiplex gain. CAC algorithms may also use a concept called equivalent bandwidth to allocate bandwidth based upon a combination of the PCR, SCR and MBS as covered in Chapter 25.

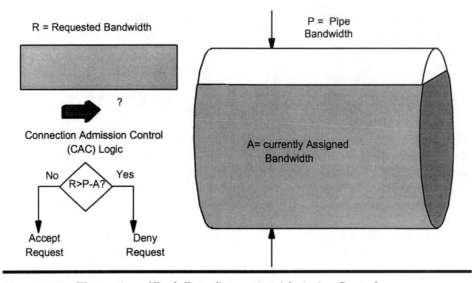

R = Requested Bandwidth

P = Pipe
Bandwidth

?

Connection Admission Control
(CAC) Logic

No ___ R>P-A? ___ Yes

A= currently Assigned
Bandwidth

Accept
Request

Deny
Request

Figure 22.20   Illustration of Peak Rate Connection Admission Control

22.8 RESOURCE MANAGEMENT USING VIRTUAL PATHS

There is a need to manage critical resources in the nodes of an ATM network. Two critical resources are buffer space and trunk bandwidth. One way of simplifying the management of the trunk bandwidth is through the use of Virtual Paths (VPs). Recall from Chapters 11 and 12 that a VP contains many VCs, and that VP cell relaying only operates on the VPI portion of the cell header. If every node in a network is interconnected to every other node by a VPC, then only the total available entry-to-exit VPC bandwidth need be considered in CAC decisions. A VPC is easier to manage as a larger aggregate than multiple, individual VCCs. The complexity and number of changes required to implement routing, restoration, and measurement also are reduced by VPCs as compared to VCCs.

Note that QoS is determined by the VCC with the most stringent QoS requirement in a VPC. One could envision a network of nodes interconnected by a VPC for each QoS class; however, this could quickly exhaust the VPI address space if there are more than a few QoS classes. Unfortunately, a full mesh design does not scale well. Even in partial mesh networks, allocating VPC capacity efficiently is a challenge. The principal issue is the static nature of VPC allocation in current ATM standards.

22.9 OTHER TRAFFIC CONTROL METHODS

Some methods covered in the next chapter on congestion control apply to traffic control purposes as well. For example, the use of selective cell discard based upon the Cell Loss Priority (CLP) bit allows traffic control for lower priority cells.

22.10 REVIEW

This chapter introduced the topics of traffic and congestion control, clearly indicating where these functions reside in real world user equipment and networks. The text organized traffic and congestion control schemes based upon the time scale over which the control operates. We then described the Usage/Network Parameter Control (UPC/NPC) policing functions using Cell Loss Priority (CLP) bit tagging, cell discarding, and monitoring employing the same leaky bucket example introduced in the last Chapter. The text then compared two different windowing-based UPC implementations to the leaky bucket method; showing how the different implementations police disparate proportions of cells. We summarized the formal Generic Cell Rate Algorithm (GCRA) which precisely defines the operation of leaky bucket policing and shaping. Next, the coverage moved on to adaptive traffic control, highlighting the ABR closed-loop flow control capability developed to provide high-performance, fairly allocated service to data applications.

The chapter then covered the means by which user equipment conforms to the traffic contract by "shaping" the cell flow using several methods. The discussion then introduced the concept of priority queuing and discard thresholds, showing how these aspects of ATM devices support a range of QoS parameters suited to the applications introduced in Chapter 21. The text then summarized the use of the Generic Flow Control (GFC) field in the ATM cell header to implement prioritized multiplexing. Next, we discussed how Connection Admission Control (CAC) operates at the call level time scale to meet QoS guarantees. The treatment then moved to the role of resource management, and in particular the use of virtual paths in simplifying this problem.

22.11 REFERENCES

[1]   E. Rathgeb, "Modeling and Performance Comparison of Policing Mechanisms for ATM Networks," *IEEE JSAC*, April 1991.

[2]   ATM Forum, *User-Network Interface Signaling Specification, Version 3.1*, af-uni-0010.002, September 1994.

[3]   ITU-T, *Traffic control and congestion control in B-ISDN*, Recommendation I.371, 1996.

[4]   ATM Forum, *ATM Forum Traffic Management Specification, Version 4.0*, af-tm-0056.000, April 1996.

[5]   ITU-T, *Traffic control and congestion control in B-ISDN: conformance definitions for ABT and ABR*, Recommendation I.371.1, June 1997.

[6]   K.K. Ramakrishnan, P. Newman, "ATM Flow Control: Inside the Great Debate," *Data Communications on the Web*, http://www.data.com/Tutorials/ATM_Flow_Control.html, June 1995.

[7]   K. Fendick, "Evolution of Controls for the Available Bit Rate Service," *IEEE Communications*, November 1996.

[8]   F. Bonomi, K. Fendick, "The Rate-Based Flow Control Framework for the Available Bit Rate ATM Service," *IEEE Network*, March/April 1995.

[9]   H.T. Kung, R. Morris, "Credit-Based Flow Control for ATM Networks," *IEEE Network*, March/April 1995.

[10] M. Gaddis, W. Kelt (Editors), "Quantum Flow Control Version 2.0," Ascom Nexion, July 25, 1995.

[11] R. Jain et al, "Source Behavior for ATM ABR Traffic Management: An Explanation," *IEEE Communications*, November 1996.

[12] P. Boyer, F. Guillemin, M. Servel, J-P. Coudreuse, "Spacing Cells Protects and Enhances Utilization of ATM Network Links," *IEEE Network*, September 1992

[13] L. Trajkovic', S. Golestani, "Congestion Control for Multimedia Services," *IEEE Network*, September 1992.

[14] D. Hong, T. Suda, "Congestion Control and Prevention in ATM Networks," IEEE Network, July 1991.

[15] M. Katevenis, S. Sidiropoulos, C. Courcoubetis, "Weighted Round-Robin Cell Multiplexing in a General-Purpose ATM Switch Chip", *IEEE JSAC*, October 1991.

[16] H. Kröner, G. Hébuterne, P. Boyer, A. Gravey, "Priority Management in ATM Switching Nodes," *IEEE JSAC*, April 1991.

[17] ITU-T, *B-ISDN ATM Layer Specification*, Recommendation I.361, November 1995.

23

Congestion Indication and Control

Webster's New World Dictionary defines *congestion* as "filled to excess, or overcrowded; for example, highway congestion." Although the best solution to congestion is to simply avoid situations where and when congestion is likely to occur, this strategy isn't always possible. Unfortunately, congestion occurs in many real world networking environments because there is always a bottleneck of some sort — a slow computer, a low-speed link, or an intermediate switch with low throughput. Therefore, depending upon the severity and duration of congested intervals, ATM devices, networks and operators can take different levels of response.

First, this chapter introduces the levels of congestion that occur, along with the types of responses, or controls, that exist. The options for congestion indication and control range from the proactive, to cooperative, and when all else fails, reactive. Basically, proactive congestion management involves good long-term planning, traffic measurement and network engineering. The next step involves congestion avoidance by indicating congestion, simply blocking new connection requests, and the operation of adaptive flow control. As an example, we study how the TCP, ABR, and QFC flow control methods respond to congestion. Finally, we describe the last recourse: reactive congestion recovery procedures. If network congestion reaches the need for recovery, then all of these techniques discard traffic in some way. The difference between them is how they choose the traffic to discard, and what the eventual impact on the higher layers becomes.

23.1 WHAT IS CONGESTION CONTROL?

This section introduces the definition of congestion through everyday examples, and then presents terminology more specific to ATM. Next, we define the factors that categorize congestion control schemes. The treatment then moves to a definition of the metrics commonly used in comparing the relative performance of congestion control schemes. Finally, the text introduces the classes of congestion control schemes that define the outline for the remainder of this chapter.

23.1.1 Congestion Defined

Many of us experience congestion daily in the form of traffic jams, long checkout lines at stores, movie ticket lines, or just waiting for some form of service. Congestion is the condition reached when the demand for resources exceeds the available resources for a specified interval of time. Take the real-life example of a vehicular traffic jam. Congestion occurs because the number of vehicles wishing to use a road (demand) exceeds the number of vehicles that can travel on that road (available resources) during a rush hour (a time interval).

More specific to ATM, congestion is the condition where the offered load (demand) from the user to the network approaches, or even exceeds, the network design limits for guaranteeing the Quality of Service (QoS) specified in the traffic contract. This demand may exceed the resource design limit because the network incorrectly oversubscribed resources, because of failures within the network, or because of operational errors.

In ATM networks the congestable resources include buffers, ATM Adaptation Layer (AAL) Segmentation And Reassembly (SAR) devices, and Connection Admission Control (CAC) processors. We call the resource where demand exceeds capacity the *bottleneck*, congestion point, or constraint.

23.1.2 Determining the Impact of Congestion

A number of application characteristics determine the impact of congestion, such as: connection mode, retransmission policy, acknowledgment policy, responsiveness, and higher-layer flow control. In concert with the application characteristics, certain network characteristics also determine the response to congestion, such as: queuing strategy, service scheduling policy, discard strategy, route selection, propagation delay, processing delay, and connection mode.

Congestion occurs on several time scales as discussed in the last chapter; either at the cell level, the burst (or packet) level, or the call level. See References 1, 2, and 3 for more details on categorization of congestion time scales. The detection of congestion as a prelude to subsequent action is *congestion indication*, feedback, or notification. Traffic forecasts, utilization trends, buffer fill statistics, cell transmission statistics, or loss counters are all indications of congestion.

The reaction to indicated congestion occurs in either time or space. In *time,* reactive controls operate on either a cell-by-cell basis, on a packet (or burst) time scale, or at the call level. In *space*, the reaction can be at a single node, at the source, at the receiver, or at multiple nodes.

Definition of the congestion control problem is difficult because of the large number of potential combinations of application and network characteristics, with the various levels of congestion indication and reaction. One congestion control scheme that works well for certain applications and network characteristics at a certain level may work poorly for different characteristics or a different level. Congestion control in broadband networks has been the subject of intensive research and many papers.

23.1.3 Congestion Control Performance

Two basic measurements help us better understand congestion — useful throughput and effective delay. *Useful throughput* is the data transfer rate actually achieved by the end application. For example, if file transfer using TCP loses a packet, then it retransmits at least that packet, and frequently in many implementations, all of the packets sent after it! Although the ATM network transferred at least some of the cells corresponding to the lost packet (and possibly those after it), the end application ends up retransmitting all of the cells corresponding to those packets. Obviously, we wish to avoid such unproductive use of resources whenever possible. In a similar manner, the effective delay was not the delay required to send the packet unsuccessfully the first time, but the interval from the first transmission until the final successful reception at the destination after one (or more) retransmission(s). Note that the useful throughput and effective delay for some applications is identical to that of the underlying ATM network. For example, voice or video coded to operate acceptably under loss conditions is not retransmitted, and hence experiences the usable throughput and effective delay as the underlying ATM layer. In practice, voice and video coding only accept loss or delay up to a critical value; after which point the application performance, subjective perception of the image, or audio playback, becomes unacceptable.

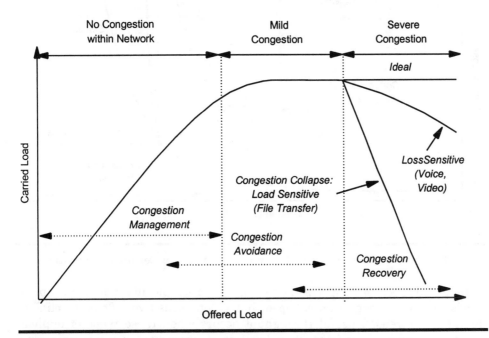

Figure 23.1 Illustration of Congestion Regions and Collapse

These two examples, file transfer and voice/video traffic, represent extremes of application sensitivity to loss, a fact which we use throughout the remainder of this chapter to illustrate the relative performance of various congestion control schemes.

When congestion occurs, a critical phenomenon called *congestion collapse* can occur as illustrated in Figure 23.1. As offered load increases into the mild congestion region the actual carried load increases to a maximum value since the bandwidth and buffering resources limit throughput. As offered load increases further into the severe congestion region, the carried load can actually *decrease* markedly due to user application retransmissions caused by loss or excessive delay. We call these types of applications *load sensitive*. Applications like voice and video which don't perform retransmissions may experience degraded performance due to loss or delay during periods of severe congestion. We call these types of applications *loss sensitive*. The degree to which carried load decreases in the severe congestion region is known as the *congestion collapse* phenomenon. Together, the application and network characteristics determine the degree of congestion collapse.

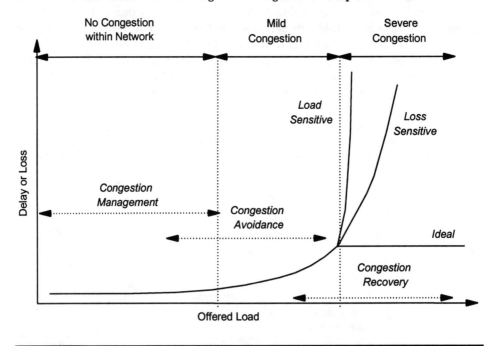

Figure 23.2   Effect of Congestion on Delay on Load and Loss Sensitive Applications

In the absence of adaptive flow control, the simple file transfer example tends to have a congestion collapse that is load-sensitive. Thus, a "cliff" of markedly decreased useful throughput occurs when these types of applications encounter severe congestion because of unproductive retransmissions, which in turn actually increases the offered load — making the congestion even worse. As shown later in this chapter, adaptive flow control protocols like TCP avoid the congestion collapse phenomenon. The other example of voice or video coding is more robust to loss and sometimes has a shallower drop-off in useful carried load. In other words, the network simply transfers less productive information. An *ideal* congestion control scheme is one where no congestion collapse occurs, with the carried load increasing up to the available capacity of the bottleneck resource, and then staying constant. As we show later in this chapter, a number of adaptive flow control protocols approach this ideal.

In the severe congestion region, the principle QoS degradation is either markedly increased delay or loss. A measure of the effectiveness of a particular congestion control scheme is how much delay or loss occurs under the scenario of an offered load in excess of the design limit. Figure 23.2 illustrates this concept. A load-sensitive application has delay and loss that increase markedly as severe congestion occurs, for example, file transfer due to retransmissions. A loss-limited application, like voice or video, still achieves acceptable performance until crossing the threshold between mild and severe congestion. The ideal congestion-controlled application has bounded delay and loss at all values of offered load. Figure 23.2 illustrates only the general trend of delay and loss versus load. In general, the curves look different when plotting delay on a linear scale and loss on a logarithmic scale as shown in examples of Chapter 25.

23.1.4 Categories of Congestion Control

The terminology for the categorization of traffic and congestion control varies. For other categorizations of congestion see References 1, 3, and 4. We chose the following categories as a means to structure the presentation of congestion control techniques in this chapter. The categories of response to congestion are: management, avoidance, and recovery. Each of these may operate at the cell level, the burst (or packet) level, or the call level as illustrated in Table 23.1, which is essentially the road map to the remaining sections in this chapter.

Table 23.1 Congestion Control Categories and Levels

Category	Cell Level	Burst Level	Call Level
Management	UPC Discard	Resource Allocation	Network Engineering
Avoidance	EFCI, UPC Tagging	Window, Rate, or Credit Flow Control	Oversubscribed CAC, Call Blocking
Recovery	Selective Cell Discard, Dynamic UPC	Loss Feedback, EPD/PPD	Call Disconnection, Operations Procedures

Congestion management operates in the region of no congestion with the objective of never entering the congested regions as illustrated in Figures 23.1 and 23.2. This includes allocation of resources, discarding Usage Parameter Control (UPC), oversubscription policy, Connection Admission Control (CAC), and network engineering.

Congestion avoidance is a set of real-time mechanisms designed to prevent entering a severely congested interval during periods of coincident peak traffic demands or transient network overloads. One example of its use is when nodes and/or links have failed. Congestion avoidance procedures usually operate around the point between no congestion and mild congestion, and throughout the mildly congested region as illustrated in Figures 23.1 and 23.2. Congestion avoidance includes Explicit Forward Congestion Indication (EFCI), Usage Parameter Control (UPC) tagging using the Cell Loss Priority (CLP) bit, oversubscribed Connection Admission Control (CAC), SVC call blocking, and either window-, rate-, or credit-based flow control at either the ATM layer or higher layers.

ATM networks and devices invoke congestion recovery procedures to prevent severely degrading end user perceived Quality of Service (QoS) delivered by the network. Typically, networks utilize these procedures only after loss or markedly increased delay occurs due to sustained congestion. Congestion Recovery includes selective cell discard, dynamic setting of UPC parameters, Early or Partial Packet Discard (EPD/PPD), disconnection of existing connections, and operational procedures.

23.2 CONGESTION MANAGEMENT

Congestion management attempts to ensure that the network never experiences congestion. For example, one form of congestion management is to avoid travel during rush hour traffic, or to wait until there are short lines for a particular service. This section covers the following congestion management methods:

☞ Resource allocation
☞ Connection Admission Control (CAC)
☞ Usage Parameter Control (UPC) discard
☞ Proper Network Engineering

23.2.1 Resource Allocation

Of course, one way to control congestion is to avoid it entirely. One way that successful network designers accomplish this is by proper resource allocation. Resources subject to allocation and parameters controlling resource allocation include:

☞ Physical trunk capacity
☞ Allocated Buffer Space
☞ UPC/NPC Parameters and Tolerances
☞ Traffic Shaping Parameters
☞ Virtual Path Connection (VPC) Bandwidth

The manner in which a network allocates resources to meet a balance between economic implementation cost and the degree of guaranteed QoS is, of course, a network decision. For example, networks may optionally allocate resources to CLP=1 flows, although most don't. Many network providers oversubscribe Variable Bit Rate (VBR) traffic to achieve economies of statistical multiplexing across multiple bursty data streams generated by a large number of users. Some network providers offer best effort, or Unspecified Bit Rate (UBR) service to fill in capacity unused by higher-priority services. In any event, the network engineer is responsible for allocating sufficient resources to meet the performance requirements for the expected traffic mix. Toward this end, the Connection Admission Control (CAC) function makes a connection-by-connection decision (or in the case of SVCs call-by-call decision) on whether to admit, or reject, a request based upon available resources and network policy. Also, the resource allocation policy and implementation may vary. For example, the resources for all QoS classes may be in a single shared pool, or placed in separate pools in order to achieve isolation between QoS classes. First, we describe a foolproof approach to resource allocation: peak rate CAC.

23.2.2 Peak Rate Allocated Connection Admission Control (CAC)

Observe that if a network sets the UPC (i.e., policing) action to discard cells in excess of the peak rate, and the switches allocate all trunk bandwidth and buffer resources for the peak rate, then congestion simply cannot occur. The worst that can happen is that arrivals of various streams will randomly clump at highly utilized ATM network nodes, creating some Cell Delay Variation (CDV). A modest amount of buffering keeps the probability of loss quite small in most applications of peak rate allocation. Extending this design to reserve enough capacity to handle likely failure scenarios is straightforward.

Although this approach avoids congestion completely, the resulting utilization of the network may be quite low, making this a potentially expensive proposition. Note that loose resource allocation policies can make sense in a local area where transmission and ports are relatively inexpensive. The literature refers to the practice of consistently allocating more resource than required as "overengineering." In other words, the network designer allocates more than enough capacity to the problem at every potential bottleneck point. As we said before, this approach always works — if you can afford it.

In general, a network uses the peak cell rate, sustainable cell rate, and maximum burst size for the two types of CLP flows (0 and 1) as defined in the traffic contract to allocate the buffer, trunk, and switch resources. Peak rate allocation ensures that even if all sources send the worst-case, conforming cell streams, the network still achieves the specified Quality of Service (QoS). Similar CAC algorithms using the SCR and MBS parameters also achieve lossless multiplexing. Of course, in order for peak rate allocation to work, the Usage Parameter Control (UPC) must enforce the traffic contract by discarding non-conforming traffic. Networks that oversubscribe certain service categories usually employ some form of congestion avoidance or recovery procedures.

23.2.3 Usage Parameter Control (UPC) Controls

As described in Chapter 22, Usage Parameter Control (UPC) acts as the traffic cop at network ingress. Network Parameter Control (NPC) performs the analogous function, acting as the border patrol at interconnection points between networks. UPC/NPC with a discard capability ensures that congestion cannot occur if the network fully allocates resources at the peak rate. The ingress node at the edge of an ATM network implements UPC/NPC. Hence, UPC/NPC discards non-conforming traffic before it even enters the network, proactively preventing congestion in downstream nodes and trunks serving other users. In this case, UPC/NPC provides a measure of fairness by isolating a user attempting to send at greater than the contracted rate from other users who diligently send at no more than their contracted rate. UPC/NPC tagging using the CLP bit for traffic exceeding the Sustainable Cell Rate (SCR) or Maximum Burst Size (MBS) parameters of a VBR traffic contract provides a useful indication to the selective cell discard congestion recovery function studied later in this chapter.

23.2.4 Network Engineering

One method for efficiently allocating resources is to base such decisions upon long-term, historical trending and projections. This is the method used in most large private and public networks today. The decisions involved include designing an overall network, deciding on when and where to install or upgrade switches, homing users to switching devices, and installing transmission capacity. Many network operators collect various statistical measurements and actual performance data to accurately model the offered traffic. They use this as input into network planning tools that provide specific answers to "what if" questions regarding candidate network upgrades or changes. Essentially no standardization exists in this area, and hence it is a network provider decision. We discuss several basic traffic source models, resource models, and traffic engineering methods in Chapter 25.

23.3 CONGESTION AVOIDANCE

Congestion avoidance attempts just that — to avoid severe congestion — while simultaneously keeping the offered load at the lower edge of the mildly congested region. In other words, it attempts to operate at the "knee" of the throughput versus load curve. This is analogous to life in the fast-paced modern world where we try to travel either just before, or just after, rush hour. Or when we try to arrive at the airport with as little time to spare as possible. This section covers the following congestion avoidance methods:

- ☺ Explicit Forward Congestion Indication (EFCI)
- ☺ Usage Parameter Control (UPC) Tagging
- ☺ Connection Admission Control (CAC) Oversubscription
- ☺ SVC Call Blocking

The next section covers how various generic types of flow control protocols respond to congestion. Specifically, we compare how well the TCP window-based protocol, the ABR rate-based protocol, and the credit-based QFC protocol avoid congestion situations.

23.3.1 Explicit Forward Congestion Indication (EFCI)

A network element in a congested state may set the Explicit Forward Congestion Indication (EFCI) payload type codepoint in the cell header for use by other network nodes or the destination equipment to avoid prolonged congestion. Typically, ATM switches set the EFCI bit when the number of cells queued in a buffer exceeds a threshold. A network element in an uncongested state should not modify EFCI, since intermediate nodes use EFCI to communicate the existence of congestion to any downstream node. As described in Chapter 22, the ABR service category interworks with ATM devices that set EFCI in binary mode.

As introduced in Chapter 7, Frame Relay (FR) has a similar congestion indication called Forward Explicit Congestion Notification (FECN). Additionally, frame relay also has a Backward Explicit Congestion Notification (BECN), which ABR supports using the Congestion Indication (CI) bit in the Resource Management (RM) cell. One reason that backward congestion notification wasn't included in the ATM cell header (as it was in the frame relay header), was that experts believed the destination application protocol should communicate to the source destination protocol the command to slow down transmissions when experiencing network congestion. To a large extent, the analogous congestion indications from FR (FECN and BECN) have not been widely utilized by end systems or higher-layer protocols such as Transmission Contol Protocol (TCP). An improved situation exists for ATM since a number of end and intermediate systems support the ABR capability to react to the EFCI bit. Intermediate equipment, such as routers, have a fundamental issue with utilizing congestion indication information. If they slow down and the source application protocol does not, then loss will occur anyway in the intermediate equipment. Therefore, most routers don't do anything with frame relay or ATM congestion indications regarding user traffic. Some routers do give priority to routing traffic in response to congestion indication messages to ensure that layer 3 routing protocols remain stable. However, many routers do collect statistics on the number of congestion messages received, which is useful network planning information.

Some early networks used ATM EFCI along with a proprietary backward congestion notification in a feedback loop, such as that illustrated in Figure 23.3. Basically, if congestion is detected anywhere along the route, including congestion for the outgoing link as shown in the figure, the destination sends a feedback message to the originating node. The originating node may service the queue for that connection more slowly, selectively discard cells, or a combination of both. In either case, the network nodes throttle back source(s) responsible for congestion, hence avoiding the severely congested state and congestion collapse resulting from an uncontrolled flow. The application of control results in useful throughput that avoids congestion collapse as indicated in the graph in the figure. See References 5 and 6 for further examples on this early use of feedback control and simulation results.

This method only works if the congestion interval is substantially greater than the round-trip delay; otherwise the congestion abates before any feedback control can act. The worst-case scenario for such a feedback scheme would be that of periodic input traffic, with a period approximately equal to the round-trip time. A realistic scenario that can result in long-term overload would be that of major trunk and/or nodal failures in a network during a busy interval. This will likely result in congestion that persists for the duration of the failure, in which case feedback control can be an effective technique for avoiding congestion and splitting the impairment fairly across different sources. Subsequent sections compare the performance of several closed-loop flow control techniques.

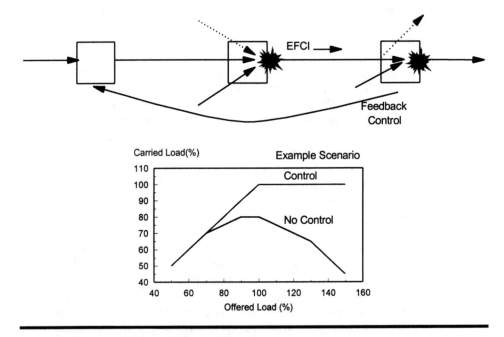

Figure 23.3 Simple Feedback Control Example

23.3.2 Usage Parameter Control (UPC) Tagging

Usage Parameter Control (UPC) applies primarily to traffic control; but also helps avoid congestion. One example is the UPC tagging cells by changing the Cell Loss Priority (CLP) bit to indicate the nonconforming cells (CLP=1). This action allows admission of traffic in excess of the traffic parameters to the network, which may cause congestion to occur. If a network employs UPC tagging for congestion avoidance, then it must also implement a corresponding technique, such as selective cell discard or dynamic UPC, to recover from intervals of severe congestion .

23.3.3 Connection Admission Control (CAC) Oversubscription

A more aggressive form of Connection Admission Control (CAC) than that of peak rate allocation described earlier is where the network allows a certain degree of oversubscription. Users frequently request traffic parameters, such as the sustainable cell rate, that exceed their typical usage patterns. Hence, if a network takes these traffic parameters at face value, it allocates too much capacity and creates call blocking congestion. Furthermore, when a large number of these connections share a common resource, it is unlikely that they all simultaneously use the resource at their peak demand level. Therefore, the network may admit more connections than the traffic parameters indicate could be supported and still achieve the specified Quality of Service (QoS) due to the statistical properties of averaging traffic from many users. These statistical QoS guarantees achieve a good balance between efficiency and quality in well run networks. Chapter 25 describes the equivalent bandwidth model and its use in predicting statistical multiplexing gain.

As in UPC with tagging, the use of CAC with oversubscription must employ a corresponding congestion recovery mechanism, such as dynamic UPC or disconnection of existing connections, to recover from periods of severe congestion.

23.3.4 Call Blocking

Before the network becomes severely congested, Connection Admission Control (CAC) can simply block any new PVC connection requests or SVC call attempts. A good example of this type of congestion avoidance is that which occurs in the telephone network — if there is blockage in the network, you will get a fast busy signal, effectively blocking your call attempt. Another example derived from telephone networks is *call gapping*. If a network detects high blockage levels for calls to the same destination number or area, the network places gaps between attempts destined for the congested number or area. In other words, the network only processes a fraction of the attempts to the congested destination, returning a busy signal or message to the gapped call attempts. These approaches help networks avoid or even recover from severe congestion for SVC services, but do little to help a PVC network avoid congestion.

23.4 CONGESTION AVOIDANCE VIA FLOW CONTROL

The class of adaptive flow control algorithms implementing congestion avoidance algorithms is so important to stable data network operation that this section covers it in some detail. First, we describe some motivation and general principles, followed by the definition of three classes of flow control methods: window based, rate based, and credit based. We then provide an example of each, illustrating key properties of the flow control technique.

Many data communications applications hungrily utilize as much available bandwidth as possible, thereby creating the potential for congestion. The basic idea of congestion avoidance is to back off offered load just before any loss occurs in the network, thus maximizing usable throughput. Furthermore, the network should fairly dole out bandwidth to contending users. In other words, no one user should get all of the available bandwidth of a bottleneck resource if several users equally contend for it. Additionally, conforming users should be isolated from the effects of non-conforming or abusive users.

The generic name given to this balancing act is *adaptive flow control*. In essence, the objective is to control traffic to achieve a throughput close to that of the maximum resource capacity, with very low loss. Such protocols require close cooperation between users and the network. For example, when the network notifies users of congestion in a timely fashion; the user's application reduces its traffic accordingly. On the other hand, when the network has available capacity; users transmit as much as they wish.

This section summarizes three examples of flow control methods: window-based, rate-based, and credit-based. The Internet's Transmission Control Protocol (TCP) employs an adaptive window-based flow control. TCP limits the amount of data a source may transmit by defining a dynamically sized transmit window size based upon detected loss and timeouts. The Available Bit Rate (ABR) rate-based flow control dynamically adapts the source transmit rate in response to explicit feedback from the network. The Quantum Flow Control (QFC) credit-based flow control scheme involves receivers transmitting permission to send (called credits) to sources and intermediate nodes. Each of these methods strives to meet the common goal of controlling the flow from the sender to maximize throughput, yet minimize loss, hence avoiding the region of severe network congestion. As we shall see, each of the methods differs in the way it detects congestion indication as well the response taken to avoid congestion.

In comparing each of these approaches, the text uses the example of two sources (workstations or clients) attempting to continuously send a large file to a single output link (to a server) using the same metropolitan area network ATM switch as shown in Figure 23.4. Both the switch and the server are collocated while the clients are remote. The speeds of the two access source links (workstations or clients) and the single output link (server) are equal in the example, shown as a local 25-Mbps ATM connections in the figure. The two sources are separated from the switch by a Round-Trip Time (RTT) of 16 cell times. This means that the clients are approximately 25 miles away from the server. The RTT is a measure of the delay from a source to the switch (and collocated server) and back to the same source. Each client transmits independently of, hence simultaneously, with respect to the other.

We chose the parameters in each of the following examples to result at least 90 percent of the maximum throughput. Each example covers 512 cell times, or 32 round-trip times. For each example, the text plots the following: the parameter being controlled at the source (i.e., window, rate, or credit), the transmitted cell stream, and the buffer occupancy at the switch.

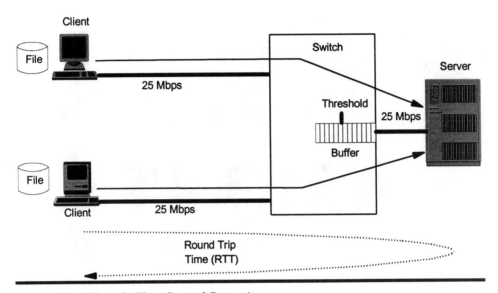

Figure 23.4 Example Flow Control Scenario

23.4.1 Window-Based Flow Control

Figure 23.5 illustrates an example of window-based flow control. As shown at the top of the figure, each source has a dynamically changing transmit window that determines how many cells it can transmit during each successive Round-Trip time (RTT). The switch has a shared buffer that has a capacity of 20 cells in this example. The source continues increasing its transmit window if no cells were lost within the last RTT. Once the buffer overflows, the source throttles back its transmission via a multiplicative decrease. The increasing width of the transmit cells versus time graph corresponds to increasingly longer bursts of cells in Figure 23.5. Most window flow-controlled protocols exhibit this effect of increasing throughput and buffer utilization (terminated in TCP by loss instead of network feedback) followed by a period of rapid backoff.

The delay encountered by the traffic is directly related to the number of cells contained in the switch buffers. The number of cells in the switch buffer is shown plotted at the bottom of Figure 23.5. Once the number of cells exceeds the shared buffer capacity, loss occurs even though a large number of cells are already in transit. Therefore, since the feedback encounters a round-trip time delay, the system loses a number of cells. Window-based flow control is very simple. It was the first type of flow control implemented in data communications. The Internet Transmission Control Protocol (TCP) works in a similar manner if cells are replaced with packets in the example. As stated earlier, this is a worst-case example where the sources are exactly in phase. TCP contains several refinements to increase throughput as well as adapt better to a wide range of network conditions as detailed in Chapter 8.

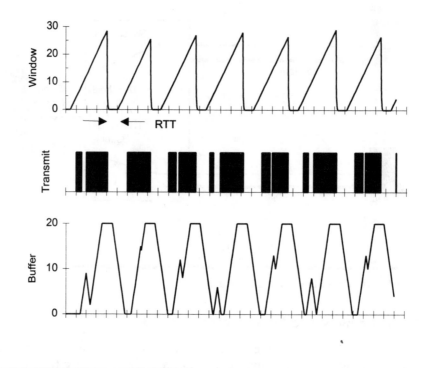

Figure 23.5   Illustration of Window-Based Flow Control

23.4.2 Rate-Based Flow Control

The next examples are similar to the first example; therefore, we cover them in somewhat less detail, highlighting only key differences. Figure 23.6 illustrates an example of rate-based flow control, such as that used in the ABR service. In this case the feedback loop controls the transmit *rate* of the sources instead of the window size, as shown plotted in the upper portion of the figure. The units of the transmit rate is cells per Round-Trip Time (RTT). In this example, the transmit rate is initially zero. Every round-trip time, the switch provides feedback on whether the source should either increase or decrease its rate. Many simulations and analyses have shown that rate-controlled schemes perform better if the feedback uses the rate of buffer growth instead of an absolute threshold. Therefore, the switch measures the fill rate of the shared buffer and compares this to a threshold every RTT. If the input rate is less than the threshold, then the switch provides feedback to each source allowing it to *increase* its transmit rate by an additive increase. If the buffer fill rate exceeds the threshold, then the switch provides feedback to the sources indicating that they should decrease their transmit rate by a multiplicative decrease.

The net result of the rate-based flow control method is that the cell transmission is more evenly spaced, as shown in Figure 23.6 by the nearly continuous transmission of cells in the plot of transmitted cells versus time shown in the middle of the figure. The overall rate ramps up to the thresholded value and oscillates around this as seen in the figure. The effect upon buffering is shown at the bottom of this same figure. In general, properly tuned rate-based flow control achieves a higher throughput than window-based flow control and also results in small delay variation as indicated in the buffer fill trace versus time. A disadvantage of rate control is that the sources must be capable of controlling their transmit rate using some form of traffic shaping.

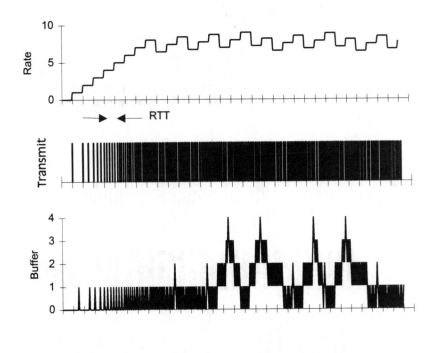

Figure 23.6   Illustration of Rate-Based Flow Control

23.4.3 Credit-Based Flow Control

Figure 23.6 illustrates an example of credit-based flow control, for example that specified by the Quantum Flow Control (QFC) consortium. In this method of flow control, the parameter adjusted by the switch is the *credit* available for transmitting cells as shown at the top of the figure. A source may continue to send cells, decrementing its credit counter by one for each transmitted cell, as long as the credit counter is greater than zero. The algorithm starts with the credit counter initially set to zero. Each Round-Trip

Time (RTT) the switch sends a feedback message indicating the credit counter value for each source, which may vary over time. The credit control scheme requires that the switch dedicate a set of buffers to each virtual connection. The switch computes the credit as the number of remaining cells in the buffer for each virtual connection. In this example the buffer is 10 cells for each source, achieving almost 100 percent throughput.

The credit flow control protocol results in a bursty, yet regular transmission of cells as seen in the transmitted cells versus time plot in the middle of Figure 23.7. Because the switch sends the credit value every RTT, each source bursts that number of cells, and then stops until the next credit is received. This results in a total buffer contents plot that is very jagged as shown as shown at the bottom of the figure. However, the algorithm keeps the buffer contents full much of the time. Credit-based flow control inherently operates in a region that keeps the buffers relatively full. For further details and references on credit-based flow control see Reference 7. Note that the total buffer required in this scheme (2 times 10) is approximately equal to the window- and rate-based examples in the previous sections. Thus, when the round-trip times are small (hence the dedicated buffer capacity is small) such as commonly encountered in LANs, credit flow control performs well.

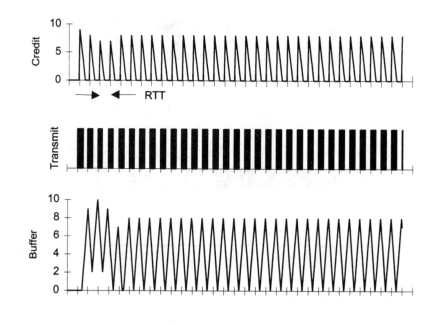

Figure 23.7   Credit-Based Flow Control

In general, the credit flow control algorithm dedicates buffer space proportional to the link delay and maximum VC bandwidth product, in which case throughput approaching the theoretical maximum is readily achieved as illustrated in the previous example. Furthermore, credit control isolates all virtual connections from each other and causes congestion to back up through the network, on a per-virtual-connection basis. Since basic credit-based flow control dedicates buffers to each VC, other VCs don't experience any impact from congestion. Disadvantages of this method are the complexity in the switch and source of implementing the credit control logic, the relatively large amount of storage required for longer propagation delays, and the usage of approximately 10 percent of the link bandwidth for per-VC credit messages.

23.5 CONGESTION RECOVERY

ATM devices initiate congestion recovery procedures after entering the severely congested region. As a real-life example, if you were caught in a particularly bad traffic jam, you might take an alternate route or pull off the road and do something else while waiting for congestion to abate. Another example is when we attempt to access a particular service, and find that the line is too long — we may leave and decide to try again later if conditions allow us to do so. This section covers the following examples of congestion recovery:

- ⊗ Selective Cell Discard
- ⊗ Early/Partial Packet Discard (EPD/PPD)
- ⊗ Dynamic UPC
- ⊗ Loss Feedback
- ⊗ Disconnection
- ⊗ Operations Procedures

23.5.1 Selective Cell Discard

Standards define selective cell discard as the mechanism where the network may discard CLP=1 cell flow while meeting Quality of Service (QoS) on both the CLP=0 and CLP=1 flows. Recall that the Cell Loss Priority (CLP) bit in the ATM cell header indicates whether a cell is of high priority (CLP=0) or low priority (CLP=1). Selective cell discard gives preferential treatment to CLP=0 cells over CLP=1 cells during periods of congestion.

Selective cell discard in ATM is a key, standardized network equipment function for recovering from severe congestion. The network may use selective cell discard to ensure that the CLP=0 cell flow receives a guaranteed QoS. If the network is not congested, then the application may achieve higher throughput by also transferring CLP=1 cells, but never less than the requested amount for the CLP=0 flow. Of course, the network's UPC must implement tagging using the CLP bit as described earlier in this chapter.

The user may also tag cells as CLP=1 if it considers them to be of a lower priority. However, user-tagged cells create an ambiguity because intermediate network nodes have no way to discern whether the user set the CLP bit, or the network's UPC set it as a result of tagging non-conforming cells. If the user sets the CLP bit, and the network does tagging, then it may not be possible to guarantee a cell-loss ratio for the CLP=1 cell flow. In practice, this isn't much of a problem because few applications utilize the CLP bit.

Figure 23.8 shows an example implementation of the selective cell discard mechanism. The switch fills a single buffer having B cell positions with arriving cells from the left. The physical layer empties the buffer from the right. Since arrivals may occur simultaneously from multiple inputs from other switch ports, the buffer can become congested. One simple way of implementing selective cell discard is to set a threshold above which the switch port discards any incoming CLP=1 cells, but still admits CLP=0 cells. Note that CLP=0 cells may occupy any buffer position, while CLP=1 cells may only occupy the rightmost portion of the buffer below the threshold as indicated in the figure. Therefore, by controlling the buffer threshold, the network controls CLP=1 loss performance. A refinement of this idea involves flushing out CLP=1 cells when another threshold is passed [8]. We analyze the delay and loss performance for selective cell discard in Chapter 25.

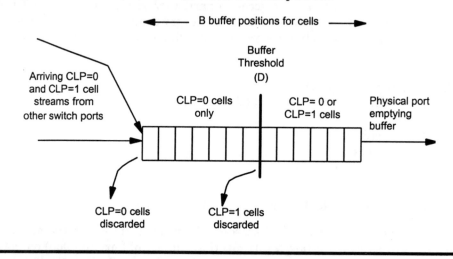

Figure 23.8   Illustration of Selective Cell Discard Function

Figure 23.9 gives an example of how hierarchical video coding makes use of selective cell discard congestion avoidance. Hierarchical video coding encodes the critical information required to contruct the major parts of the video image sequence as the higher priority CLP=0 marked cells, while encoding the remaining, detailed minor change information as a separate stream of lower-priority CLP=1 marked cells. Thus, when there is a scene change, the video coder generates CLP=0 cells at an increased rate for a period of time as indicated by the slowly varying solid line in the figure. The video application sends the detail and minor adjustments within the scene as CLP=1 cells as shown by the jagged line in the figure. When a switch multiplexes several such video sources together and utilizes selective cell discard as shown in Figure 23.9, this congestion recovery scheme ensures that the critical scene change information gets through, even if some of the detail is momentarily lost during transient intervals of congestion.

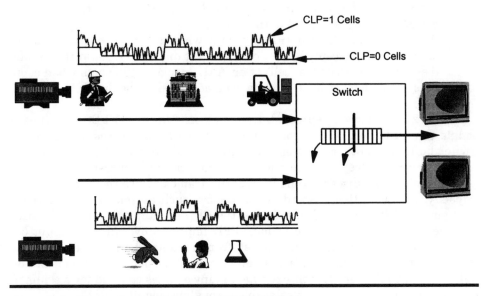

Figure 23.9   Illustration of Hierarchical Video Coding

23.5.2 Early/Partial Packet Discard (EPD/PPD)

The ATM Forum TM 4.0 specification also specifies an intelligent frame discard function as an optional congestion recovery procedure. For AAL5, a number of studies and tests show that a more effective reaction to congestion is to discard at the frame level rather than at the cell level [9, 10]. The situation where intelligent frame discard helps the most occurs when many sources congest a particular resource, such as an output queue on an ATM switch serving a heavily utilized link. Note that a network element that discards at the cell level may discard portions of many packets. Usually, the objective of intelligent frame discard is to maximize the number of complete packets transferred. However, other objectives like maximizing the number of bytes transferred (in complete large frames) are also possible. Intelligent frame level discard helps networks recover from the phenomenon of congestion collapse during periods of severe overload.

An ATM device may treat user data as frames only if the user indicates so using the broadband ATM traffic descriptor IE in an SVC message, or by setting the value in the ILMI MIB for a PVC at subscription time. Once the user negotiates frame level discard service with the network using these means, the ATM switches use the Payload Type Indicator (PTI) in the ATM cells to detect the last cell of an AAL5 PDU as detailed in Chapter 13. The commonly used industry terminology for the intelligent frame discard capability is the following two actions :

- Early Packet Discard (EPD) occurs when a device in a congested state discards every cell from an AAL5 PDU. EPD prevents cells from entering the buffer, reserving remaining buffer capacity for the cells from packets already admitted to the buffer.
- Partial Packet Discard (PPD) occurs when a device discards all remaining cells, except the last one, when it discards a cell in the middle (i.e., not the first or last cell) of an AAL5 packet. PPD acts when some cells from a packet have already been admitted to the buffer.

Figure 23.10 shows a representative implementation of EPD/PPD using a single shared buffer. Cells from multiple frame level sources arrive at the switch buffer from other ports on the switch as indicated on the left hand side of the figure. Once the buffer level exceeds the EPD threshold, the EPD gremlin selectively discards cells from entire frames. Once the buffer reaches capacity, or some other discard action occurs, such as selective cell discard dropping a CLP=1 cell within a frame, the PPD gremlin takes over, discarding the remaining cells of the frame. PPD improves overall throughput because if the network loses even one cell from a frame, the end user loses the entire frame. For example, a maximum size Ethernet data frame has 30 cells. When applied in the context where many sources contend for a common resource EPD/PPD can improve useable throughput significantly.

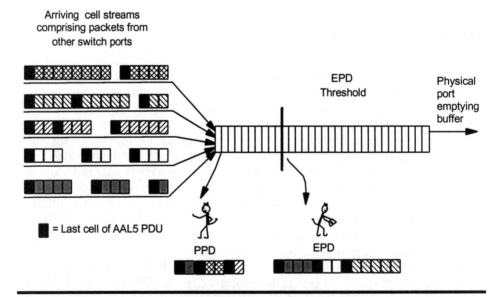

Figure 23.10 Example Early /Partial Packet Discard (EPD/PPD) Implementation

23.5.3 Dynamic Usage Parameter Control (UPC)

Another way to recover from congestion is to dynamically reconfigure the UPC parameters. This could be done by renegotiation with the user [4], or unilaterally by the network for certain types of connections. This technique is related to the dynamic GCRA defined for ABR conformance checking in Chapter 22.

23.5.4 Loss Feedback

Another method of congestion control is for a higher-layer protocol (such as TCP) at the end system to infer that congestion has occurred, and throttle back the end user's application throughput at the end system itself. This approach has the advantage of the end system reducing the offered load, thereby recovering from congestion. There are several possible disadvantages: the response time to congestion is at least the round-trip time through any intervening networks plus any time spent in the end systems, and the possibility that all end systems do not infer congestion and back off in the same manner, creating unfairness. One noteworthy implicit congestion control scheme is that of the Internet Transmission Control Protocol (TCP), which Chapter 26 analyzes in detail.

23.5.5 Disconnection

Another rather drastic response that provides recovery from congestion is to disconnect some connections if and when severe congestion persists. For example, some connections may be preemptible. The U.S. government levies such a requirement upon carriers to support national defense traffic or local community emergency services at the highest priority, such that all other traffic is subject to disconnection if these priority connections require capacity.

23.5.6 Operational Procedures

If all of the automatic methods fail, then human operators can intervene and manually disconnect certain connections, reroute traffic, or patch in additional resources. Network management actions for controlling the automated reroutes are not standardized, and are therefore a proprietary network implementation. These procedures must be carefully coordinated, especially if multiple networks are involved.

23.6 REVIEW

This chapter defined congestion as demand in excess of resource capacity. The degree of congestion impacts contention for resources, which can reduce throughput and increase delay, as occurs in vehicular traffic jams. Congestion occurs at multiple levels in time and space. In time, congestion occurs at the cell level, the burst (or packet) level, or the call level. In space, congestion occurs at a single node, multiple nodes, or across networks. This chapter categorized congestion control schemes in terms of time scale and their general philosophy. First, in the network planning time scale of weeks to months, congestion management attempts to ensure that congestion never occurs, which may be done at the expense of reduced efficiency. Next, acting in the region in the time scale of calls, bursts and cells; congestion avoidance schemes attempt to operate on the verge of mild congestion to achieve higher utilization at nearly optimal performance. The text then compared how well popular flow control schemes using the window-, rate- and credit-based paradigms avoid congestion. Finally, we described congestion recovery techniques that move the network out of a severely congested state in the event that the previous two philosophies fail, sometimes using rather drastic measures such as selective discard, or even disconnection of some users.

23.7 REFERENCES

[1]   J. Hui, "Resource Allocation for Broadband Networks," *IEEE JSAC*, December 1988.
[2]   G. Awater, F. Schoute, "Optimal Queueing Policies for Fast Packet Switching of Mixed Traffic," *IEEE JSAC*, April 1991.
[3]   D. Hong, T. Suda, "Congestion Control and Prevention in ATM Networks," *IEEE Network Magazine*, July 1991.
[4]   A. Eckberg, "B-ISDN/ATM Traffic and Congestion Control," *IEEE Network*, September 1992.
[5]   M. Wernik, O. Aboul-Magd, H. Gilbert, "Traffic Management for B-ISDN Services," *IEEE Network*, September 1992.
[6]   A. Eckberg, B. Doshi, R. Zoccolillo, "Controlling Congestion in B-ISDN/ATM: Issues and Strategies," *IEEE Communications Magazine*, September 91.
[7]   H. T. Kung, R. Morris, T. Charuhas, D. Lin, "Use of Link-by-Link Flow Control in Maximizing ATM Networks Performance: Simulation Results," *Proc. IEEE Hot Interconnects Symposium '93*, August 1993.
[8]   H. Kröner, G. Hébuterne, P. Boyer, A. Gravey, "Priority Management in ATM Switching Nodes," *IEEE JSAC*, April 1991.
[9]   Fore Systems, "ForeThought Bandwidth Management," http://lina.vis.com.tr/fore/FTBMWP.html.
[10]   N.E.T., "Advanced Traffic Management for Multiservice ATM Networks," http://www.net.com/techtop/adtmatm_wp/white.html.

6

Communications Engineering, Traffic Engineering, and Design Considerations

This part provides the reader with an application-oriented view of the communications engineering, traffic engineering, and design considerations applied to ATM. First, Chapter 24 defines and applies communications and information theory terminology. This includes the concepts of signal structure, frequency passband, noisy communication channels, bit errors, channel capacity, and error correcting codes. The text applies the theory to determine the undetected error rates of ATM HEC and AAL5. Next, Chapter 25 covers the important topic of traffic engineering. The treatment begins by introducing random processes and basic models from queuing theory. The analysis applies these results to ATM switch design tradeoffs, performance of CBR and VBR traffic types, equivalent capacity, statistical multiplexing, and priority queuing. Finally, Chapter 26 discusses additional design considerations involved in ATM networks. These include a look at the impact of delay, loss, and delay variation on applications. The text also describes the performance of TCP over various types of ATM services, focusing on the impact of buffer size, packet discard technique, and congestion scenarios. This chapter also analyzes the statistical multiplex gain achieved by AAL2 and the savings achieved by integrating voice and data. Throughout this part, we apply theoretical approaches to the real-world business problems defined in the previous sections. A spreadsheet can implement most formulas presented in this part.

24

Basic Communications Engineering

This chapter deals with some useful approximations for modeling and estimating ATM network performance. The text introduces the concepts of probability theory, followed by an overview of digital signals and their frequency spectra. Next, we analyze the effect of errors on the achievable capacity over a given communications channel. The chapter concludes with an overview of error correcting codes and data compression. In particular, the text gives a simple method for evaluating the probability of undetected error when using the ubiquitous Cylic Redundancy Code (CRC) technique. All formulas used in these examples are simple enough for spreadsheet computation so that the reader can readily evaluate a specific network situation. The chapter cites references to more detailed treatments and extensions of these concepts for the interested reader.

24.1 PHILOSOPHY

This section discusses several dimensions of communications engineering philosophy. First, we cover the basic notion of a communication channel and how factors such as noise, frequency pass band and errors impact reliable communications. Next, the text touches on the different roles for deterministic and random models in communication links. Finally, the section concludes with some practical approaches to basic communications engineering.

24.1.1 Communications Channel Model

Figure 24.1 illustrates the fundamental elements of any communications system [1, 2]. Starting from the left-hand side, a source generates digital information. Next, a source encoder transforms this information into a series

663

of binary digits. A source encoder may also perform data compression to decrease the required number of binary digits. The source encoder may also add other overhead at protocol layers below the application layer. The digital signal transmitter may add error detection and/or error correction prior to sending modulated digital waveforms onto a noisy communication channel. The transmitter may encode multiple bits as a single baud, or channel symbol. A digital signal receiver takes waveforms received over the channel, which now contain the original signal plus noise, and attempts to re-create the same sequence of binary digits sent by the transmitter. The source decoder takes this received stream of binary digits from the receiver and applies the higher-layer protocols prior to passing the information on to the information sink on the far right-hand side. For example, the source decoder may perform data decompression.

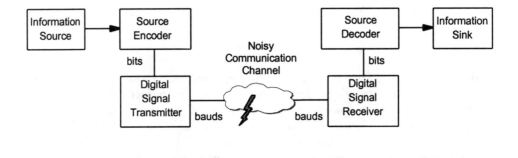

Figure 24.1   Basic Communications Channel Model

24.1.2 Deterministic versus Random Modeling

The discipline of communications engineering utilizes both deterministic and random models depending upon the situation. Deterministic models describe signal structures and their frequency spectra. Random models approximate impairments on real-world communication channels. Although the actual phenomena that cause errors on real channels are very complex, often a simplifying random model allows an engineer to design applications for the noisy channel. Of course, the available frequency spectrum and signal to noise ratio place a fundamental limit on the rate of communication, known as Shannon's channel capacity as described later in this chapter.

24.2 PROBABILITY THEORY

This chapter introduces some basic probability theory, with parameters chosen to model some reasonable communications channels encountered in ATM networking.

24.2.1 Randomness in Communications Networks

Probability theory is used in communication engineering in two principal areas: modeling source behaviors and modeling the effects of noisy communication channels. Central to the theory is the notion of sets, or groupings of experimental outcomes. A familiar experiment with two equally likely outcomes is a fair coin toss: it comes up either heads or tails. A particular type of binary communications channel examined in more detail later exhibits a similar dual outcome: either the received bit is received correctly, or in error. The term Bit Error Rate (BER) is the probability that the bit is received in error. The same random model also applies to a source generating random data.

Often, however, the outcome of one trial is not independent of preceding trials. Such is the case for communication channels that experience bursts of errors, for example twisted copper pairs, protection switching in fiber optic systems, and many radio channels. Information sources also tend to have bursty behavior, a subject which the discussion on equivalent bandwidth addresses in the next chapter.

24.2.2 Random Trials and Bernoulli Processes

A Bernoulli process is the result of N independent "coin flips" in an experiment where the probabilities of heads and tails are unequal: p being the probability of "heads" occurs and (1-p) being the probability that "tails" occurs as the result of each coin flip. The probability that k heads occur, and hence (N minus k) tails also occur, as a result of N repeated Bernoulli trials ("coin flips") is called the *binomial distribution* as given by:

$$\Pr[k \text{ "heads" in N "flips"}] \; = \; b(N,k,p) \; = \; \binom{N}{k} \; p^k \; (1-p)^{N-k}$$

$$\text{where } \binom{N}{k} \; \equiv \; \frac{N!}{(N-k)!\,k!}.$$

Microsoft Excel implements b(N,k,p) in the BINOMDIST(k, N, p, FALSE) function.

24.2.3 The Normal/Gaussian Distribution

A consequence of the deMoivre-Laplace and Central Limit theorems states that the Gaussian, or Normal, distribution is a good approximation to the binomial distribution when Np is a large number in the Np(1-p) region about the mean [3]. Figure 24.2 compares the binomial and Gaussian distributions for an example where N=100 and p=0.1 to illustrate this point. The distributions have basically the same shape, and for large values of Np, in the Np(1-p) region about Np, the Gaussian distribution is a good approximation to the binomial distribution.

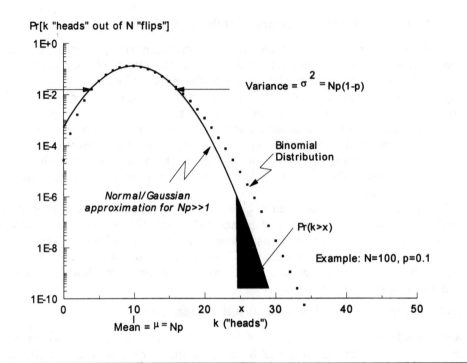

Figure 24.2 Normal Approximation to Binomial Distribution

This is helpful in analyzing relative performance in that the probability area under the tail of the Gaussian, or normal, distribution is widely tabulated and implemented in many spreadsheets and mathematical programming systems. The cumulative distribution of the normal density, defined as $Q(\alpha)$ below, is a good approximation to cumulative sum of the tail of the binomial distribution.

$$\text{Prob}[k > x] \approx Q\left(\frac{x - \mu}{\sigma}\right) = Q(\alpha) \approx \frac{1}{\sqrt{2\pi}} \, e^{-\alpha^2/2}$$

$$\text{where } Q(\alpha) \equiv \frac{1}{\sqrt{2\pi}} \int_{\alpha}^{\infty} e^{-x^2/2} \, dx$$

This text uses this approximation in several contexts to estimate loss probability, statistical multiplex gain, and delay variation in subsequent chapters. Microsoft Excel implements the commonly used function $Q((x-\mu)/\sigma)$ as 1-NORMDIST(x,μ,σ,TRUE) for a normal random variable with mean μ and standard deviation σ. Note that from the above approximation for $Q(\alpha)$, a closed form expression for $\varepsilon \approx \text{Pr}[k > x]$ is

$$\alpha \approx \sqrt{-2 \ \ell n(\varepsilon) \ - \ 2\ell n(2\pi)} \ .$$

Note that Microsoft Excel implements the function to determine α more precisely as $\alpha = -NORMINV(\varepsilon,0,1)$. The above approximation overestimates the required capacity; therefore, use the accurate formula if your software provides it.

As discussed in the next section, the normal distribution is also an excellent model for noise generated on electrical and optical communication channels. The same probability density and distribution functions defined above are widely used by communication engineers to estimate the error rate of physical layer connecting switching nodes. The normal distribution is widely used in other branches of engineering, for example, in approximating buffer overflow and statistical multiplexing gain in queuing theory studied in the next chapter.

24.3 COMMON DIGITAL SIGNALS AND THEIR SPECTRA

This section summarizes key concepts of digital modulation to achieve efficient utilization of the available frequency range for a bandwidth-limited channel. This notion of restricted bandwidth is key to the maximum information rate achievable over a noisy channel. Since frequency limitations and noise are key impairments in many real communication channels, a background in these areas is important in designing physical links in ATM-based networks. Electrical and optical digital communications systems use different types of modulation schemes to physically transfer information from a transmitter to a receiver. The text begins with the simplest scheme, basically a synchronous telegraph signal and finish with the scheme that is the basis of modulation techniques employed by modern high-speed modems.

24.3.1 The Telegraph Pulse — Binary On/Off Keying

The simplest signal is a pulse, similar to that in nature to the telegraph. The transmitter emits a pulse of electromagnetic energy (e.g., an optical pulse, a current or a radiofrequency wave) into the physical medium (e.g., an optical fiber, a pair of wires or the atmosphere), which the receiver detects. Figure 24.3a shows the time domain representation of a random telegraph signal mapping the input sequence of ones and zeros into either a transmitted pulse, or the absence of a pulse, respectively once every T seconds. The information rate of this system is R=1/T bits per second. The baud rate of this system is also R.

The power spectral density, or spectrum for short, is the Fourier transform of the time domain signal, which operates in the frequency domain. Communications engineers define frequencies in a measure defined in terms of the number of complete sinusoidal cycles of an electromagnetic wave in a 1-s interval. The unit of Hertz (abbreviated Hz) corresponds to one sinusoidal cycle per second. Common Alternating Current (AC) power systems in the United States operate at 60 Hz. Although this book frequently uses the term "bandwidth" to refer to the digital bit rate of a transmission link, please note that many communication engineers use the term bandwidth to refer to the *frequency passband* of a particular signal.

Figure 24.3b illustrates the frequency spectrum for the binary on/off signal. Binary on/off keying does not make efficient use of the available frequency passband. For example, at three times the source bit rate, the energy is still over 5 percent (-13 dB) below the peak signal power level. For a particular absolute signal to noise ratio, x; the SNR in decibels is computed as SNR=10 log(x). For example, x=2 corresponds to roughly 3 dB, x=4 corresponds to approximately 6 dB and x=10 corresponds to 10 dB. The frequency *sidelobes* from one signal's spectrum can interfere with signals in adjacent frequency bands. Hence, shrewd designers use more spectrally contained signals in band-limited circumstances. However, on optical fiber systems available spectrum is not a concern in some networks, since each fiber has over 1 trillion Hertz of available bandwidth. Furthermore, on/off keying is the cheapest means to pulse a laser or light-emitting diode (LED) in an optical communications system.

a. Binary On/Off Keying (OOK) Time Domain Signal

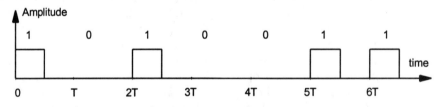

b. Binary On/Off Keying (OOK) Frequency Domain Spectrum

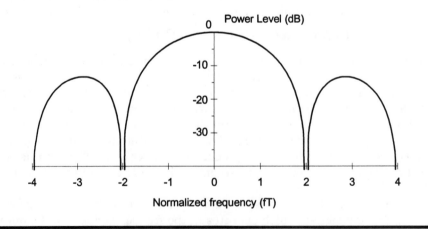

Figure 24.3 Binary On/Off Keying Signal Trace and Spectrum Plot

24.3.2 A Better Way — Pulse Shaping

As seen in the previous section, an unfortunate result of the telegraph pulse is that it doesn't utilize the frequency passband very efficiently. As communication engineers strove to develop better ways to transfer information at increasingly higher rates, they invented more spectrally efficient waveforms. As you might expect, making the pulse smoother could improve the spectrum characteristics, and this is indeed what communications engineers discovered. One of the "smoothest" pulses is the raised cosine shape illustrated in Figure 24.4a. This smoother pulse makes a significant difference in the frequency spectrum as shown in Figure 24.4b, shown plotted alongside the binary OOK spectrum for comparison purposes. The plot normalizes the power for these two signaling methods. The raised cosine pulse must have approximately a 20 percent higher amplitude than the rectangular pulse in order to convey the same energy. The sidelobes are down over 30 dB, (i.e., a level of only 0.1 percent) from the main spectral lobe. Furthermore, the side lobe occurs much closer to the main lobe than it does it binary OOK. This means that more channels fit within a band-limited channel without interfering with each other. Electrical and radio systems can readily implement this technique, as well as some optical systems.

a. Raised Cosine On/Off Keying (OOK) Time Domain Signal

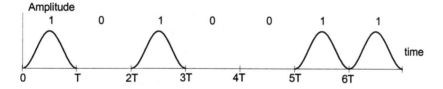

b. Raised Cosine On/Off Keying (OOK) Frequency Domain Spectrum

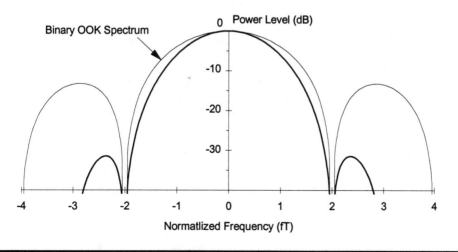

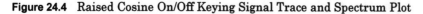

Figure 24.4 Raised Cosine On/Off Keying Signal Trace and Spectrum Plot

24.3.3 Pushing the Envelope — Quadrature Amplitude Modulation

If you're a communications engineer at heart, your thoughts should be racing now, wondering how much better the frequency passband can be utilized. This quest rapidly becomes a very complex and specialized subject; however, we briefly introduce one more example of signal modulation techniques, in particular the one used in the popular standard V.32 modem signaling at 9.6-kbps over most telephone networks in the world. To start with, real electromagnetic physical media actually support two carrier frequencies called the in-phase and quadrature channels. This fact is due to the mathematics of the complex sinusoids. Specifically, when correlated over a full period, the cosine and sine functions cancel each other out. Communications engineers say that these signals are orthogonal, or that they are in quadrature since the sine and cosine functions are 90 degrees out of phase. Additionally, the amplitude of the pulse can also be modulated. Hence, combining the notions of sinusoidal phase and amplitude, the signal space makes up a two-dimensional constellation for Quadrature Amplitude modulation from the V.32 modem standard for non-redundant 9.6-kbps coding shown in Figure 24.5. Since the constellation uses four values of amplitude and phase independently, the sixteen resulting amplitude and phase combinations represent all possible four-bit patterns as shown in the figure.

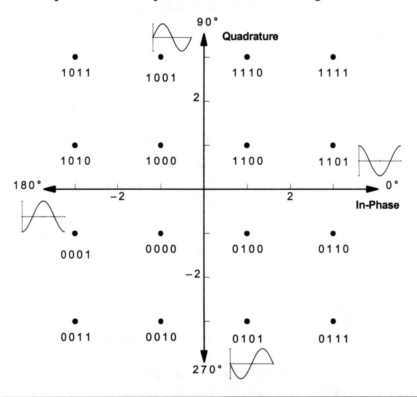

Figure 24.5   Quadrature Amplitude Modulation (QAM) Signal Constellation

The time domain representation of QAM is straightforward, yet somewhat too involved for our coverage here. See References 2, 4, or 5 for more details. Figure 24.6 illustrates an example of the time domain characteristics of QAM signals. The top of the figure shows the input binary sequence which modulates the in-phase and quadrature channels corresponding to the x and y axes on the signal constellation plot of Figure 24.5. Below the in-phase and quadrature input levels is the resulting waveform modulated by a sine and cosine wave, respectively. The trace at the bottom of the figure is the composite sum of the modulated in-phase and quadrature channels. In this case, the information rate is 4/T bits per second. Communication engineers call the signal transmitted every T seconds a baud, or they refer to the channel symbol rate in units of bauds per second. For QAM, the information bit rate is four times the baud rate.

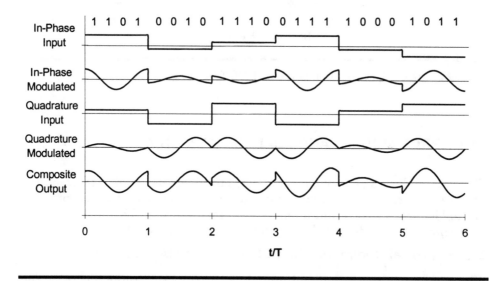

Figure 24.6   QAM In-Phase, Quadrature and Composite Time Domain Signals

What did this additional processing gain? It improves the spectral efficiency markedly as shown in Figure 24.7. The figure shows the spectrum for on/off keying for comparison purposes. Note that the spectral width for the main lobe of QAM is half that of OOK. Furthermore, the system transmits four times as much information per channel symbol time (T), that is, the information rate of the channel is R=4/T bits per second. Furthermore, the sidelobes are significantly lower, reducing interference seen by adjacent channels. Input level pulse shaping, similar to the raised cosine shape discussed earlier, drives these sidelobes down to very small levels [4, 6]. The shaped QAM signal uses spectrum very efficiently. As we shall see, these types of techniques form the basis for higher-speed modems and xDSL transmission systems to deliver higher information rates over real-world, band-limited noisy physical-layer channels.

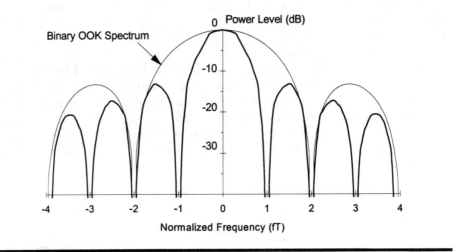

Figure 24.7   QAM Spectrum Compared with OOK Spectrum

24.4 ERROR MODELS AND CHANNEL CAPACITY

This section describes how basic physics, machines and human intervention create errors on modern digital communication systems. But, despite the inevitable errors, communication engineers approach the ideal limits of performance for band-limited, noisy communication channels.

24.4.1 Typical Communication Channel Error Models

What causes errors on communication links? Why can't we just increase the transmit power? What causes noise anyway? This section answers these questions for a number of commonly utilized digital transmission systems. Over the years, communications engineers grouped channels into various types. The text covers three common cases, explaining how various channel characteristics manifest themselves to channel modems.

24.4.1.1 Additive White Gaussian Noise

You'll often see it abbreviated as AWGN in textbooks, and it has the distribution covered in section 24.2.3. This model covers electromagnetic channels (i.e., radio, satellite and cellular), as well as thermal noise occurring in the electronics of all communication transmission systems, including electro-optical receivers in many fiber optic based systems. The term "white" refers to the fact that the power spectrum of AWGN is flat across the signal passband. A corollary of this flat spectrum is the fact that the noise is uncorrelated from instant to instant. An example of AWGN is the sound heard when the TV is receiving no input, or the noise heard on an analog telephone line when no one is speaking.

24.4.1.2 Binary Symmetric Channel

The Binary Symmetric Channel (BSC) is a good model for the low-level errors on fiber optic communication links. Typically, the random residual error rate on fiber optic links is on the order of 10^{-12} or less. This simple channel has a certain probability of bit error, p, for each bit independent of all other preceding bits. As shown in Figure 24.8, the probability of receiving a bit correctly is 1-p. The channel error probability is also independent of whether the input was a one or a zero. The study of error correcting codes later in this chapter uses this simple model.

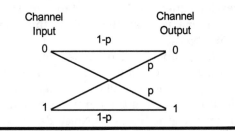

Figure 24.8   Binary Symmetric Channel Error Model

24.4.1.3 Burst Error Channel Model

The basic burst error channel model is where a string of bits received over a communications channel has an error in a starting and ending position, with an arbitrary number of other bit errors in between. Often, analysis assumes that every bit in the string between the first and last errors is randomly in error. Intermittent periods of high noise energy cause these types of errors on many atmospheric communication channels. Unshielded electrical transmission systems, like the twisted pair of copper wires leading to most residences, also pick up noise from lightning strikes, faulty vehicle ignitions, and power tools. Although less prone to bursts of errors, fiber optic transmission systems also experience bursts of errors due to protection switching in the equipment, maintenance activities that flex the fibers, and noise within the receiver electronics. Another commonly used model for bursts of errors is the Poisson arrival process defined in the next chapter.

24.4.2 Shannon's Channel Capacity

In 1948, Claude Shannon derived an upper bound on the maximum error-free information transfer rate that any communication system could achieve over a noisy communication channel with a particular frequency passband of W Hz [1, 5, 7]. The disciplines of information and communication theory show that the performance of communications systems with appropriate source encoding, channel encoding, and signal selection approach this bound. The simple formula for this maximum rate, C, called *Channel Capacity* is

$$C \ = \ W \ \log_2 \ (1 + \text{SNR}),$$

where SNR is the Signal to Noise Ratio of the communication channel. The model used by Shannon assumed additive white Gaussian noise.

Applying this theory to a familiar situation, a typical telephone line has a frequency passband of 3,000 Hz and a SNR of approximately 30 dB (i.e., the signal power is one thousand times greater than the noise power), note that channel capacity is approximately 30-kbps from Shannon's formula above. This is the maximum speed of the highest-speed telephone modems. Note that channel capacity refers to the information-carrying bit rate, and that many commercial telephone-grade modems also employ data compression of the source information to achieve even higher effective user data transfer rates. Data compression squeezes out redundancy in data using sophisticated coding schemes as summarized at the end of this chapter. This is how modems operate at 56-kbps.

A basic lesson from Shannon's theorem is that in order to increase the information transfer rate, a communications engineer must increase the frequency range (i.e., bandwidth) and/or the signal to noise ratio. Figure 24.9 plots this tradeoff by depicting the normalized channel capacity, C/W, in units of bps per hertz versus the Signal to Noise Ratio (SNR). Another way of looking at this result is to note that for a line transmission rate of R=1/T bits per second, each symbol conveyed over the link must carry N= 2WT bits of information in order to approach channel capacity. In other words, the baud rate must be 2W times the bit rate to achieve channel capacity.

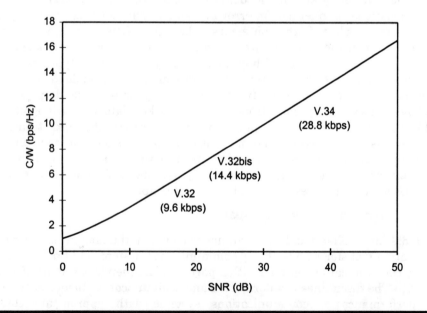

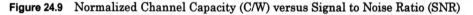

Figure 24.9   Normalized Channel Capacity (C/W) versus Signal to Noise Ratio (SNR)

We've indicated several popular modem standards on the chart; V.32 (9.6-kbps), V.33 (14.4-kbps) and V.34 (28.8-kbps). This chart also illustrates the reason for modems automatically negotiating down to a lower speed if the telephone connection is too noisy. Higher-speed modems over telephone lines also use several other techniques besides sending multiple bits per channel symbol. One that xDSL also employs is that of echo cancellation, where the transmitter is able to effectively cancel out its own signal. This means that both directions of communication can utilize the frequency passband simultaneously.

Shannon's theorem gave an important clue to modulator-demodulator designers to cram as many bits into the transmission of each discrete digital signal over the channel as possible. The evidence of their craft is the ever increasing higher performance of modems for use over the telephone network to access the Web, high-performance radio and satellite communications and the promise of ubiquitous high-speed Digital Subscriber Line (xDSL) modems operating at over ten times the speeds of the fastest telephone-grade modems or ISDN lines. As described in Chapter 32, xDSL achieves this tremendous gain by using a much larger frequency passband.

24.4.3 Error Performance of Common Modulation Methods

Figure 24.10 plots the Bit Error Rate (BER) performance of the modulation schemes studied in the previous section versus the signal to noise ratio per bit (i.e., SNR/N). Note that the total signal power is N times that plotted below, where N is the number of bits per channel symbol (i.e., N=1 for OOK and N=4 for QAM). Although the simple OOK return to zero on/off pulse scheme makes slightly more efficient use of signal power, beating the QAM scheme by approximately 1 dB; remember that the QAM system is sending four bits for every one conveyed by the OOK system.

Furthermore, as we saw in the last section the QAM system also makes more efficient use of the available frequency spectrum. The increasing sophistication of digital electronics has enabled designers to apply these more sophisticated techniques on channels where frequency spectrum is scarce, such as telephone lines, terrestrial radio and satellite transmission systems. Where frequency spectrum is not scarce, for example, on fiber optic cables, a simpler and cheaper scheme such as on/off keying works fine.

All right, the communication depicted in the plot above isn't error free as Shannon's theorem promises, and for lower signal to noise ratios, the performance is marginal. For example, for QAM a 30 dB channel SNR equates to a signal to noise ratio per bit of approximately 8 dB, meaning that a significant number of bits in every hundred would be in error. How do communication engineers achieve a nearly lossless channel? What if my application can't tolerate the delay required to retransmit errored information. Communication engineers use two techniques to further improve the error rate. One is through the use of trellis coding that makes symbol transitions to adjacent places in the symbol constellation impossible. This improves the "distance" between signal points in the constellation and hence

markedly reduces the error rate. See References 2 and 4 for more information on trellis coding. The second technique is through the use of error
correcting codes, as covered in the next section.

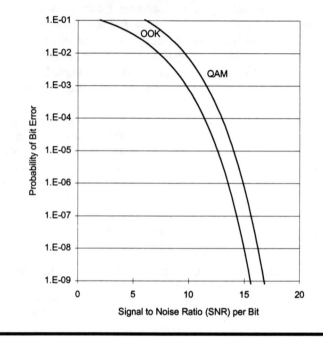

24.11   Bit Error Rate (BER) Performance of Various Modulation Schemes

24.5 ERROR DETECTING AND CORRECTING CODES

Now that we've learned that signals can be transmitted over a channel
frequency passband within a certain error rate, how do communication
engineers achieve the effectively lossless communication essential to
applications? The answer is through error detecting and correcting codes.
This section reviews the performance of the two basic schemes in use today:
the simple parity check typically done in software and the cyclical redundancy check frequently performed in hardware. The text then evaluates the
performance of IP running over HDLC and ATM in terms of undetected error
rate.

24.5.1 Simple Parity Check Schemes

The concept of parity checking originated in the era of digital computing in
the 1950s. Since electronic circuits used for storage were quite unreliable,
computer engineers designed a simple scheme in computer logic to compute a
parity check. The required circuit was a modulo 2 adder (i.e., exclusive or
logical function), implemented with a total of three electronic logic gates. The
direct result of this calculation is called even parity, which means that the

parity bit is zero if there are an even number of ones. A key concept defined during this period is the Hamming distance of a particular code. The distance of a code is the number of bit positions in which two code words differ. A single parity bit added to a bit string results in a code with a distance of two — that is, at least two bit errors must occur to change one valid code word into another. The parity check bit on 7-bit ASCII characters is an example of a simple parity scheme.

A means to do bit-wise parity checking easily in software involves bit-wise exclusive or of subsequent segments of the message. Parity checking is commonly called a *checksum*. Both TCP and IP use this technique. The following simple example illustrates such a bit-wise checksum performed across three bytes of binary data.

	0	0	1	1	1	0	1	1
	0	1	1	0	1	1	0	1
	1	1	0	1	0	1	1	1
Bit-wise Checksum	1	1	0	0	0	0	0	1

A checksum encoder generates the checksum at the transmitter and sends it as part of the message to the receiver, which computes the checksum again on the received string of bits. This simple parity checksum detects any single bit error in the message. If two bit errors (as well as all other even-numbered multiple bit errors) occur in the same position of the checksum columns, then the checksum will not detect the errors. For channels where the bit errors are random, hence the likelihood of a single error is by far the largest, then a simple parity check works reasonably well, for example, within computer systems and over some local area network connections. However, for channels subject to bursts of errors, or non-negligible error rates, then engineers include more powerful error detection schemes in data communication networks. Since these noisy channels constitute the majority of the physical layer in real-world networks, we are fortunate indeed that communications engineers developed efficient and effective coding techniques that integrated circuits implement so inexpensively.

24.5.2 Cyclic Redundancy Check (CRC) Codes

The magic that enabled reliable error detection over noisy communication channels for the early packet-switched networks has a rather complex mathematical basis, yet a relatively simple hardware implementation. The text takes a brief look at the mathematics followed by a description and example of the hardware implementation.

Coding theory represents strings of bits as polynomials with binary coefficients, for example, the bit string 1 0 0 1 would be represented as $x^3 + x^0$. Coding theorists say that a bit string of length n is represented by a polynomial of degree n-1. Standards define specific CRC codes via a generator polynomial, G(x), of degree k. An encoder adds k zeros to the end of a message polynomial M(x) to yield a new polynomial $x^k M(x)$ and divides it modulo 2 by G(x) to yield a remainder R(x). The encoder then computes

$T(x)=x^kM(x)-R(x)$. The receiver demodulates the bit pattern $T(x)+E(x)$, where $E(x)$ is a polynomial with ones representing the positions of bit errors. By the manner in which $T(x)$ was computed, $T(x)/G(x)=0$. Hence, the receiver examines $E(x)$ to see if it contains any ones, which indicate errors on the channel. This mathematical treatment is highly simplified, see References 8 and 9 for further details.

The generator polynomial for the 8-bit CRC used in the ATM Header Error Control (HEC) octet of every ATM cell [10] is

$$G(x) = x^8 + x^2 + x + 1.$$

Fortunately, the hardware implementation is easier to understand, as well as implement (it's also straightforward to implement on a spreadsheet or in a software lookup table). Figure 24.12 depicts an example of the ATM CRC code with generator polynomial of degree 8 protecting an ATM cell header for a user cell on VPI=0, VCI=33 without any congestion or AAL indications. Starting at the top of the figure, the pre-multiplication by x^8 means that the header is appended with eight zeroes. During the first twenty bit positions corresponding to the ATM cell header GFC, VPI, VCI, PT and CLP, the gate feeds back the modulo two sum of the input and the leftmost shift register stage to the rightmost stages. The plus circled symbols indicate modulo two addition. For the last eight bit positions, the gate breaks the feed back loop and shifts out the contents of the shift register, which are the HEC sequence "0110 0101."

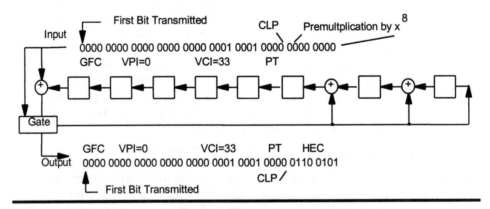

Figure 24.12   Illustration of ATM Cyclic Redundancy Code (CRC) Operation

The I.432 standard also recommends that ATM transmitters perform a bit-wise exclusive or with the pattern "0101 0101" in order to improve cell delineation in the event of bit-slips. This "exclusive or" setting also ensures that if the entire header is all zeros, then the input to the physical link scrambler is non-zero. An ATM receiver first subtracts this pattern from the HEC octet prior to shifting the received bits into the above shift register. After shifting all of the received bits into the register, the pattern remaining is called the *syndrome*. If no errors occurred, then the syndrome is identical to the received HEC field. If errors occur, then the syndrome indicates the single bit error correction required, or the presence of an uncorrectable double bit error.

There are several different ways to use cyclic codes to avoid all zero bit patterns. Some standards, for example HDLC and AAL5, initialize the shift register to all ones so that an all zero information packet results in a non-zero CRC. The basic operation is then the same as that described above using a shift register with taps corresponding to the non-zero coefficients of the code's generator polynomial.

24.5.3 Performance of ATM's HEC

This section gives some handy, practical tips for selecting and evaluating the performance of particular codes. When comparing the performance of different CRC standards for your application, note that the error detection distance (i.e., the number of bit positions in which packets differ) is no more than the number of non-zero coefficients in the generator polynomial [9]. For example, the number of non-zero coefficients in the ATM HEC generator polynomial is four, and in fact, the minimum error detection distance for this code is also four.

CRC codes also have excellent error burst detection properties. Specifically, a CRC code with degree n (i.e., the highest numbered coefficient in the generator polynomial) detects all burst of errors up to length n bits. A burst of errors is defined as an error in the first and last positions with an arbitrary number of other bit errors in between. Thus, CRC codes work well on both random and burst error channels.

The standards give two options for using the HEC: error correction and error detection. In the presence of random errors, the probability of x errors in a 40-bit cell header is given by the binomial distribution $b(40,x)$ from section 24.2.2 with p set equal to the probability of bit error. With header error detection, the probability that an errored cell header is falsely mistaken as a valid header is determined by the probability of randomly matching a valid codeword (1/256) if three or more bit errors occur. Therefore, the probability of falsely passing an errored header given that the receiver performs HEC error detection is:

$$P[\text{False} \mid \text{Detection}] = \frac{1 - b(40,0) - b(40,1) - b(40,2)}{256}$$

When using HEC detection, the receiver discards the cell if the decoder detects any errors. Of course, the receiver cannot detect a falsely matched valid cell header, hence the probability of discard given header detection is:

$$Pr[\text{Discard} \mid \text{Detection}] = (1 - b(40,0))\,(1 - P[\text{False} \mid \text{Detection}])$$

In an implementation employs header error correction, then a false match occurs if three or more bit errors cause an exact codeword match — or, if the errors result in a codeword which appears to have only one bit error and is then inadvertenly corrected into a valid codeword. Actual testing uncovered this unexpected result. In fact, the probability of invalid cells is approximately forty times greater than that when the receiver employs HEC detection as given by the following formula:

$$P[\text{False} \mid \text{Correction}] = \frac{41(1 - b(40,0) - b(40,1) - b(40,2))}{256}.$$

When using HEC correction, the receiver corrects all single bit errors and detects all other bit errors that don't result in a falsely matched valid cell header. Therefore, the probability of discard given that header correction is performed is

$$\Pr[\text{Discard} \mid \text{Correction}] = (1 - b(40,0) - b(40,1)) (1 - P[\text{False} \mid \text{Correction}]).$$

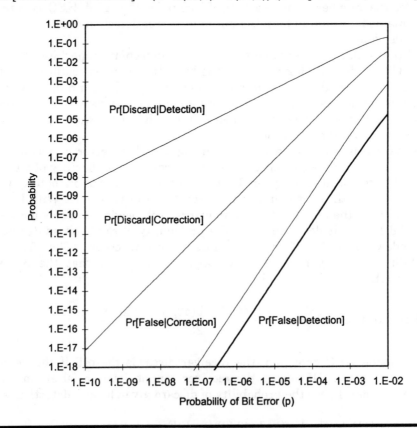

Figure 24.13   Undetected Cell Header Error for HEC Correction and Detection

Figure 24.13 plots the results of these calculations. The values of Pr[Discard | Correction] and P[False | Correction] from the above formulas correspond closely to those of Figure A.1/I.432 [10]. The figure also plots Pr[Discard | Detection] and Pr[False | Detection] for comparison purposes. Note that the use of error detection results in a false cell matching probability somewhat less than when using correction, in fact they differ by a factor of forty as shown above. The probability of discarding a cell due to detected errors; however, is much greater when using HEC detection when compared with HEC correction.

If your application can tolerate a few false cells via further error checking using higher-layer protocols, then use HEC correction. If your application doesn't have a higher-layer protocol, for example, video over ATM, you may want to use error detection instead.

A useful computational trick when computing small numbers (such as the ones in this example) in a spreadsheet is to raise 10 to the power of the sum of the base 10 logarithms of product terms. If you don't do this, then you'll find that you're getting a result of one or zero instead.

24.5.4 Undetected Error Performance of HDLC and AAL5

Despite the good error performance of CRC coding, the bit stuffing mechanism of HDLC is susceptible to random and burst errors [11]. The bit stuffing mechanism used to eliminate the occurrence of a flag sequence within an HDLC frame as described in Chapter 7 causes this flaw. One or two bit errors in the wrong place in a received frame creates a valid flag field, which truncates the frame. Also, the trailing flag field may be obliterated by bit errors. The resulting probability of undetected error for HDLC is well approximated by the following formula

$$\Pr[\text{Undetected Error} \mid \text{HDLC}] \approx \left(1.36 \; kp + \left(\frac{m}{k} \right) p^4 \right) 2^{-16},$$

where the variable k represents the number of bytes in the HDLC frame, p is the random bit error probability, and m is the average HDLC frame length (in bits) after bit stuffing given by the following formula:

$$m = 8 \left(\frac{64}{63} \; k + 2 \right).$$

For comparison purposes, the undetected AAL5 PDU performance does not have the bit stuffing problem. Since the AAL5 32 bit CRC generator polynomial has fifteen non-zero coefficients, its Hamming distance is also 15. Hence, a particular pattern of exactly 15 random bit errors must occur and match a valid codeword with probabilty 2^{-32}. Hence, the undetected frame level error performance of AAL5 for an information field of k octets is

$$\Pr[\text{Undetected Error | AAL5}] \approx b\left(8\left\lceil\frac{k+8}{48}\right\rceil,\ 15,\ p\right)2^{-32}$$

where b(n,x,p) is the value from the binomial distribution defined in section 24.2.2 for bit error probability p with $\lceil x \rceil$ denoting the smallest integer greater than x, commonly known as the "ceiling" function, implemented in Microsoft Excel as CEILING(x,1).

Figure 24.14 depicts the undetected frame error rate for HDLC and ATM's AAL5 versus random bit error probability p. Note that the undetected error performance approaches one in ten thousand for high error rate channels for HDLC. This is the reason why higher-level protocols like TCP/IP add a length field and additional parity checks. On the other hand, the undetected error probability for AAL5 is negligible, even at high channel error rates. The practical implication of this analysis is that ATM AAL5 is much better suited for noisy, error prone channels than HDLC is when the objective is reliable error free operation.

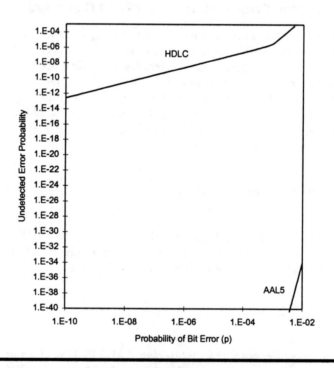

Figure 24.14   Undetected Error Rate for HDLC and AAL5

24.6 DATA COMPRESSION

Most real-world sources generate data sequences which contain redundancy. For example, in ordinary text, some letters are much more likely to occur than others. For example, the letter "e" is hundreds to thousands of times more likely to occur in common text than the letter "z." The basic idea of data compression exploits this difference in likelihood of particular inputs by coding frequently occurring source symbols with fewer bits and coding rarely occurring source symbols with longer sequences of bits. The net result is a reduction in the average number of bits required to transmit a long message. Another simple data compression method used in facsimile transmission is called run-length coding. Since most facsimile input consists of characters or lines, a run-length coder scans the input pages one scan line at a time from left to right. If the input page is dark in the scanned dot, then the source encodes a one, otherwise it encodes a zero. If the source detects a long run of ones or zeroes, then it inserts a special source symbol indicating that a specified number of subsequent scan positions have the same value. The receiver reverses the process by printing out a series of dots for a long run of ones, or leaving the fax output blank if a long run of zeroes is received. Data compression requires additional software processing, or even custom hardware to operate a higher speeds. If bandwidth is scarce, then investigate the use of data compression to see if the additional processing costs less than the incremental cost required to increase the available bandwidth.

24.7 REVIEW

This chapter discussed several key aspects of communications engineering philosophy: probability theory, performance measurement, digital communications, error rates, channel capacity, error detecting and correcting codes, and data compression. The communications system model includes source coding (data compression), channel coding and modulation on the transmitter side along with corresponding functions at a receiver in order to achieve reliable digital transmission. We then covered some basics of probability theory using the discrete binomial distribution and the continuous normal (or Gaussian) distribution as background. The text then summarized some concepts of modulated signals and their frequency spectra. We then introduced the seminal concept of channel capacity, and reviewed the bit error performance of some common modulation schemes. The chapter then introduced the techniques of parity checksums and cyclic redundancy codes used to detect channel errors. The text applied this theory in a comparison of undetetected error rates for ATM HEC, HDLC, and AAL5. Finally, the chapter concluded with a brief overview of data compression techniques.

24.8 REFERENCES

[1]   R. Gallagher, *Information Theory*, John Wiley & Sons, 1968.

[2]   J. Proakis, *Digital Communications*, McGraw-Hill, 1983.

[3]   A. Papoulis, *Probability, Random Variables and Stochastic Processes*, McGraw-Hill, 1965.

[4]   J. Bellamy, *Digital Telephony*, John Wiley, 1982.

[5]   G. Held, R. S arch, "Data Communications," McGraw-Hill, 1995.

[6]   I. Korn, *Digital Communications*, Van Nostrand Reinhold, 1985.

[7]   C. Shannon, "A Mathematical Theory of Communication," Bell System Technical Journal, vol 27, October, 1948.

[8]   A. Tannenbaum, "Computer Communications, Third Edition," Prentice-Hall, 1996.

[9]   F. Peterson, F. Weldon, "Error-Correcting Codes," MIT Press, 1972.

[10] ITU-T, "Recommendation I.432, B-ISDN User-Network Interface - Physical Layer Specification," 1991.

[11] J. Selga, J. Rivera, "HDLC Reliability and the FRBS method to Improve It," Proceedings of the 7th Data Communications Symposium, Mexico City, 1981.

25

Traffic Engineering

This chapter deals with the modeling of traffic sources and switch perform-ance. It covers many useful approximations for estimating performance. The text begins with a discussion of traffic modeling philosophies as an introduc-tion to an application-oriented approach. We introduce several simple models for source traffic to illustrate key concepts. The treatment then introduces the traditional telephone system call arrival, blocking and queuing models, and applies them to B-ISDN. The text analyzes the performance of the buffering schemes described in Chapter 16. Next, the analysis continues with an evaluation of the performance resulting from multiplexing many Constant Bit-Rate (CBR) and Variable Bit-Rate (VBR) traffic sources. An important consideration in the economics of integrated voice, video, and data networks. The coverage then moves on to define the concept of equivalent bandwidth. The text defines statistical multiplexing gain and the traffic characteristics for which it is attractive. Finally, the chapter concludes with an analysis of priority queuing traffic control performance. All formulas used in these examples are simple enough for spreadsheet computation so that the reader can use them to evaluate a particular switching machine or network configuration. Throughout the chapter, the text gives references to more sophisticated models for the interested reader.

25.1 PHILOSOPHY

This section discusses several dimensions of traffic engineering philosophy: source model traffic parameter characterization, performance specification and measurement, and modeling accuracy.

25.1.1 Source Model Traffic Parameter Characteristics

There are two basic philosophies for characterizing source traffic parameters: deterministic and random.   These approaches often embody a tradeoff between accuracy and simplicity.

Deterministic parameters use the traffic contract outlined in Chapters 21 and 22, with conformance verifiable on a cell-by-cell basis using the leaky bucket algorithm. Thus, either the user or the network may unambiguously measure the fraction of cells conforming to the traffic contract.   This deterministic traffic model clearly defines the source characteristics as understood by the user and the network.

The other philosophy for modeling source behavior utilizes random (also called probabilistic or stochastic) models for traffic parameters.   Usually, these parameters correspond to measurable long-term averages. However, a random model describes the short-term behavior.   Since the method and interval for averaging differ, conformance testing should define the details of the measurement method.   In addition to specifying the parameters, the random model defines the short-term traffic source statistics.   With these additional assumptions the user and network can agree on performance for a certain level of traffic throughput.   While these statistical methods are not standardized, they are useful approximations to the deterministic traffic contract behavior.   These methods are useful in analysis when employing a simple statistical model, as shown in this chapter.

25.1.2 Modeling Accuracy

A key aspect of traffic engineering philosophy relates to the required accuracy of the model. As expected, the more complicated the model, the more difficult the results are to understand and calculate. A good guideline is to make the accuracy of the switch and network model comparable to the accuracy of the source model traffic. If you only know approximate, high-level information about the source, then an approximate, simple switch and network model is appropriate.   If you know a great deal of accurate information about the source traffic, then an investment in an accurate switch and network model, such as a detailed simulation, is appropriate.

While theoretically optimal, detailed source modeling can be very complex and usually requires computer-based simulation. Often this level of detail is not available for the source traffic.   Using source traffic details and an accurate switch and network model will result in the most realistic results.

When either traffic or switch and network details are not available, approximations are the only avenue that remains. Approximate modeling is usually simpler, and can often be done using only analytical methods. One advantage of the analytical method is that insight into relative behavior and tradeoffs is much clearer. The analytical method is the approach used in this book. One word of caution remains, however; these simplified models may yield overly optimistic or pessimistic results, depending upon the relationship of the simplifying assumptions to a specific real world network. Modeling should be an ongoing process. As you obtain more information about the source characteristics, device performance, and quality expectations, feed these back into the modeling effort. For this reason, modeling has a close relationship to the performance measurement aspects of network management.

25.2 OVERVIEW OF QUEUING THEORY

The use of a particular source model, and its accurate representation of real traffic, is a hotly debated topic and the subject of intense on-going research and publication. This section defines some basic probability and queuing theory concepts utilized within this chapter to model commonly encountered traffic engineering problems.

25.2.1 General Source Model Parameters

This section defines some general source model parameters used throughout the remainder of this chapter and in many papers and publications which are listed at the end of the chapter.

Burstiness is a commonly used measure of how infrequently a source sends traffic. Traffic engineers call a source that infrequently sends traffic bursty, while a source that always sends at the same rate is nonbursty. The formula that defines burstiness in terms of the peak cell rate and the average cell rate is:

$$\text{Burstiness} = \frac{\text{Peak Rate}}{\text{Average Rate}}$$

The *source activity probability* is a measure of how frequently the source sends, defined by the following formula:

$$\text{Source Activity Probability} = \frac{1}{\text{Burstiness}} = \frac{\text{AverageRate}}{\text{PeakRate}}$$

25.2.2 Poisson Arrivals and Markov Processes

This section describes the Poisson (or Markov) random arrival processes with reference to Figure 25.1. Interestingly, Poisson developed these statistical models based upon his experience in Napoleon's army regarding the likelihood of soldiers dying after being kicked in the head by their horses. Like many random events (most not quite so morbid), Poisson arrivals occur such that for each increment of time (T), no matter how large or small, the probability of arrival is independent of any previous history. These events may be either individual cells, a burst of cells, cell or packet service completions, or other arbitrary events.

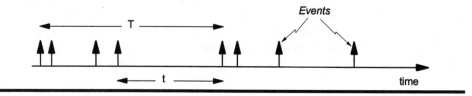

Figure 25.1 Illustration of an Arrival Process

The probability that the interarrival time between events t, as shown in Figure 25.1, has a certain value is called the *interarrival time probability density*. The following formula gives the resulting probability that the interarrival time t is equal to some value x when the average arrival rate is λ events per second:

$$\text{Prob}(t = x) = \lambda e^{-\lambda x}$$

Queuing theorists call Poisson arrivals a *memoryless process*, because the probability that the interarrival time will be x seconds is independent of the *memory* of how much time has already expired. This fact greatly simplifies the analysis of random processes since no past history, or memory, affects the computation regarding the statistics of the next arrival time. These types of processes are commonly known as *Markov processes*, named after the Russian mathematician of the nineteenth century.

The probability that n independent arrivals occur in T seconds is given by the famous *Poisson distribution*:

$$\text{Prob}(n,T) = \frac{(\lambda T)^n}{n!} e^{-\lambda T}$$

We combine these two thoughts in a commonly used model called the Markov Modulated Poisson Process (MMPP). There are two basic types of this process: the *discrete* that corresponds to ATM cells, and the *continuous* that corresponds to higher-layer Protocol Data Units (PDUs) that generate bursts of cells. The next two figures give an example for the discrete and continuous Markov process models.

The labels on the arrows of Figure 25.2 show the probability that the source transitions between active and inactive bursting states, or else remains in the same state for each cell time. In other words, during each cell time the source makes a state transition, either to the other state, or back to itself, with the probability for either action indicated by the arrows in the diagram.

The burstiness, or peak-to-average ratio, of the *discrete* source model is:

$$b = \frac{\alpha + \beta}{\beta},$$

where α is the average number of bursts arriving per second, and β is the average rate of burst completions.

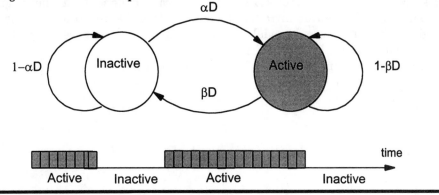

Figure 25.2   Discrete Time Markov Process Model

Often we think in terms of β^{-1}, which has units of the average number of seconds per burst. We define D as the cell quantization time having units of seconds per cell. Therefore, αD defines the probability that a burst begins in a particular cell time, and βD defines the probability that a burst ends in a particular cell time. The average burst duration d (in cells) is the mean of the standard geometric series:

$$d = \frac{1}{\beta D}.$$

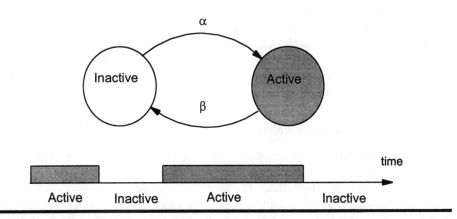

Figure 25.3   Continuous Time Markov Process Model

Figure 25.3 illustrates the *continuous* time Markov process, which models the statistics for the time duration of bursts instead of modeling the individual cells as the discrete model does. Since this model eliminates the effects of quantization inherent in segmentation and reassembly, it is inherently less accurate. However, its analysis is somewhat simpler; therefore, this book relies on the simplicity it provides. Queuing theorists call the diagram depicted in Figure 25.3 a *state transition rate diagram* [1] since the variables associated with the arrows refer to the rate exponent in the negative exponential distribution introduced earlier in this chapter. Both the discrete and continuous Markov models yield equivalent results except for the cell quantization factor D.

The corresponding burstiness b for the continuous process is:

$$b \ = \ \frac{\alpha \ + \ \beta}{\beta},$$

and the average burst duration (in seconds) is:

$$d \ = \ \frac{1}{\beta}$$

Note how these formulas are identical to the discrete case except for the absence of the discrete cell time D in the denominator of the equation for the average burst duration in the continuous model.

Another distribution sometimes used to model extremely bursty traffic is that of the hyperexponential, which is effectively the weighted sum of a number of negative exponential arrivals. This turns out to be a more pessimistic model than Poisson traffic because bursts and burst arrivals are more closely clumped together. For further information on this distribution see Reference 1.

Recent work based upon actual LAN traffic measurements indicates that these traditional traffic models may be overly optimistic. For further information on this work see References 2 and 3. These results show that the LAN traffic measured at Bellcore is *self-similar*, which means that the traffic has similar properties regardless of the time scale on which it is observed. This is in sharp contrast to the Poisson and Markovian models, where the traffic tends to become smoother and more predictable when considering longer and longer time averages.

25.2.3 Queuing System Models

Figure 25.4 depicts a widely used notation employed to categorize queuing systems. This chapter makes use of this shorthand terminology, and it is widely used in the technical literature. Hence, readers doing further research should become familiar with this notation originally attributed to Kendall. The notation designates arrival processes (denoted as A and B) as either M for Markovian, as described above, G for General, or D for Deterministic. The required parameter s defines the number of "servers" in the queuing system; for example, in the case of communications networks the transmission path a single server. The optional w and p parameters specify the waiting room for unserved arrivals and the source population, respectively.

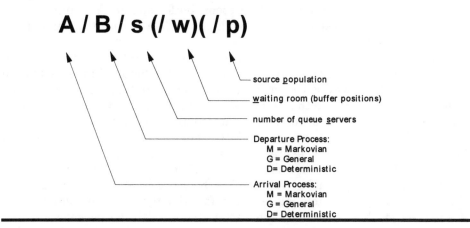

Figure 25.4   Queuing System Notation

Figure 25.5 illustrates two particular examples of queuing systems applied to ATM traffic, namely, the M/D/1 and M/M/1 systems. Each of these queuing systems has Markovian arrivals at a rate of λ bursts per second. The M/D/1 system has constant length bursts as shown in Figure 25.5a, while the M/M/1 system has random length bursts with a negative exponential distribution (Markov) as shown in Figure 25.5b.

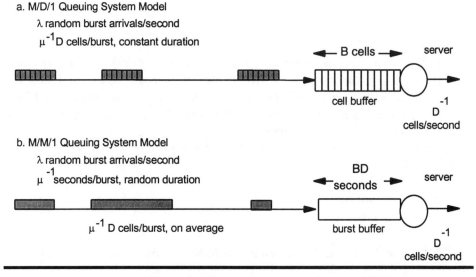

Figure 25.5   Application of M/D/1 and M/M/1 Queuing Systems to ATM

The parameter μ^{-1} defines how may seconds the transmission link requires to send each burst. For the M/D/1 system this is the constant or fixed length of every burst, while in the M/M/1 system this is an exponentially distributed random number with average length μ^{-1}. Both systems also have a single server (i.e., physical transmission link) and an infinite population (number of potential bursts) and infinite waiting room (buffer space). The units of the buffer in the M/D/1 model are cells, while in the M/M/1 case the units of the buffer are bursts.

This is a good example of the tradeoffs encountered in modeling. The M/D/1 system accurately represents the fact that the buffers in the switch are in units of cells; however, the model also assumes that bursts are all of the same fixed length. While this is a good model for voice, video and some data sources, it is a bad model for many data sources like file transfers or web surfing. On the other hand, the M/M/1 system does not model the switch buffers accurately since it expresses waiting room in units of burst-seconds and not cells; however, the modeling of random burst lengths is more appropriate to many traffic sources. Furthermore, the M/M/1 model is the simplest queuing model to analyze. Therefore this text uses the M/M/1 model extensively to illustrate general characteristic of ATM and packet switching systems. In general, if the traffic is more deterministic than the M/M/1 model (for example, more like the M/D/1 model), then the M/M/1 model is pessimistic (there will actually be less queuing and less delay in the modeled network). If the traffic is more bursty than the M/M/1 model, then the M/M/1 results will be optimistic (there will actually be more queuing and more delay in the modeled network).

In many of the following results the system delay and loss performance will use the concept of offered load ρ as defined by the following formula:

$$\rho = \frac{\lambda}{\mu}$$

Recalling that λ is the average number of arriving bursts per second, and that μ^{-1} is the average number of seconds per burst required by the transmission link, observer that the offered load ρ is unitless. Thus, the offered load has the interpretation of the average fraction of the resource capacity the source would use if no loss occurred.

The service rate μ is computed as follows for a burst of B bytes at a line rate of R bits per second:

$$\mu = \frac{8 \; B}{R} \left(\frac{bursts}{second} \right)$$

The probability that n bursts are in the M/M/1 queue awaiting service is:

$$\text{Prob[n burst in M / M / 1 queue]} = \rho^n \; (1 - \rho)$$

The average queuing delay (i.e., waiting time) expressed in seconds for the M/M/1 system is:

$$\text{Avg}[M \, / \, M \, / \, 1 \text{ queuing delay}] \ = \ \frac{\rho \, / \, \mu}{1 - \rho}$$

M/D/1 queuing predicts better performance than M/M/1. Indeed the average delay of M/D/1 queuing is exactly one-half of the M/M/1 delay. The probability for the number of cells in the M/D/1 queue is much more complicated, which is one reason the text uses the M/M/1 model in many of the following examples. The CBR model in section 25.5 gives an approximation to the M/D/1 distribution.

25.3 CALL ATTEMPT RATES, BLOCKING AND QUEUING

This section looks at traditional traffic modeling derived from over a century of experience with telephone networks, largely attributed to work published by the Danish mathematician A. K. Erlang in 1917. Since B-ISDN uses the connection oriented paradigm, and the fact that ATM SVCs are similar in nature to telephone calls, these older call models still apply. One significant difference, however, is that unlike phone calls which are always the same speed, ATM call attempts may have vastly different bandwidths. Furthermore, ATM SVCs may request different service categories and QoS parameters.

25.3.1 Statistical Model for Call Attempts

Through extensive measurements, traffic engineering experts know that the Markov process is a good model for that telephone call attempts. The primary parameters are the call arrival rate λ, usually expressed in terms of Busy Hour Call Attempts (BHCA), and the average call holding time (i.e., call duration) T. Without any blocking, the average number of calls in progress during the busy hour in such a system is λT. By convention, telephony engineers assign the units of offered *Erlangs* to this quantity, although in fact the measure is unitless. In the following sections we refer to the average offered traffic load as:

$a = \lambda \, T$ average offered load (in units of Erlangs)

In a real telephone network, blocking occurs with a certain probability B. Therefore, telephone engineers say that the system carries a (1-B)a Erlangs. Erlang also modeled systems that queue calls instead of blocking them. For example, a call answering system that places calling users on hold listening to music if all the operators are busy. We study these two types of systems — called blocked calls cleared and blocked calls held — in the next sections.

Implicit in the Erlang model is the assumption that the switching system under consideration supports a large number of telephone users. Furthermore, the model assumes that only a fraction of these users are active during the busy hour. When these assumptions are not valid, use a more complicated model developed by one of Erlang's contemporaries, Engset instead [1].

Let's look at a simple numerical example with reference to Figure 25.6. A group of 1,000 subscriber lines originates an average of λ=2,000 call attempts during the busy hour. Each completed call lasts for 3 minutes on average, or T=0.05 hours. Thus, the offered load to the trunk on the telephone switch is a=λT=(2,000)(0.05)=100 Erlangs. The interpretation of this model is that without any call blocking, on average only 100 of the subscribers are actually on the phone at any point in time. In the next sections we examine the performance when the switching system blocks or queues calls that exceed the trunk capacity.

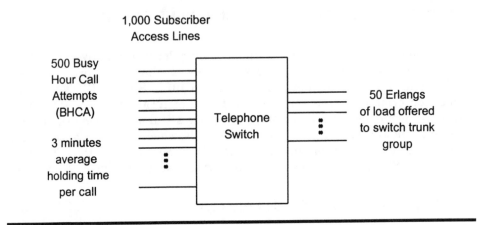

Figure 25.6   Telephone System Call Arrival Model

Another note on historical terminology for telephone call attempt rates that you may encounter is that of the Call Century Seconds (CCS). The name CCS derived from the operation where a camera took a picture of the call peg counters on electromechanical switches once every 100 seconds for billing purposes. If the counter advanced between photographs for a particular subscriber line, then the carrier assumed that the call was active for 100 (a century) seconds, which explains the name CCS. Prior to extended dial-up sessions on the web, a typical residential phone line carried 3 to 6 CCS on average, that is, its average load was between 8 percent and 16 percent. Now, many residential lines carry 9 to 12 CCS.

25.3.2 Erlang's Blocked Calls Cleared Formula

A blocked calls cleared switching system has Markovian call arrivals, Markovian call durations and n servers (trunks), and n waiting positions. Therefore, a M/M/n/n queuing system model defines the blocking performance. For an average offered load of a Erlangs and n trunks, the following formula gives the probability that the system blocks a call attempt,

$$B(n,a) = \frac{a^n / n!}{\sum_{k=0}^{n} a^k / k!}$$

where $k! = k(k-1)(k-2)...(3)(2)(1)$.

Typically, texts call this formula the *Erlang-B* formula [1, 4]. Many older books contain tables for the values of the Erlang-B (lost calls cleared) probability. However, you can easily compute your own using the following simple recursion [4] in a spreadsheet,

$$B(n+1,a) = \frac{aB(n,a)}{n + 1 + aB(n,a)}$$

where $B(0,a)=1$. This is a useful result to solve the commonly encountered problem of determining the number of trunks (n), given an offered load (a) to meet an objective blocking probability $B(n,a)$. You can either write a spreadsheet macro, or define two columns of cells in the spreadsheet. The first column contains the values of n, starting at zero. The second column contains the value 1 in the first row corresponding to $B(0,a)$. All subsequent rows contain the above formula for $B(n+1,a)$ coded to use the preceding row's result, $B(n,a)$, and the average offered load a.

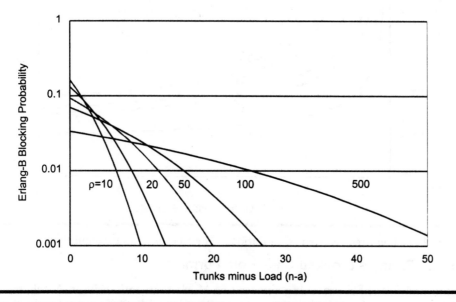

Figure 25.7   Erlang-B Blocking Probability versus Offered Load

Figure 25.7 illustrates the result of using this recursive method of calculation for some values of the Erlang-B blocking probability for various trunk group sizes (n) versus the difference between the number of trunks and the offered load (n-a). Note how the blocking probability decreases more rapidly when adding trunks to serve smaller loads than it does when adding trunks to serve larger loads. Intuitively, we expect this result since larger systems achieve an economy of scale.

One way of expressing the economy of scale is to compare the percentage of additional trunking required to achieve a fixed blocking probability across a range of offered loads. We define this measure as the *overtrunking ratio*, which is simply the required trunks divided by the average offered load, namely n/a. For an overtrunking ratio of 100 percent, the number of trunks exactly equals the average offered load yielding a particular blocking probability. For typical blocking probabilities in the range of 0.1 percent to 1 percent, the overtrunking ratio ranges between 110 percent and 120 percent for large values of offered load. Figure 25.8 illustrates the overtrunking ratio versus the average offered load for a set of representative values of blocking probabilities. Many commercial voice networks are engineered for a blocking probability in the range of ½ to 1 percent.

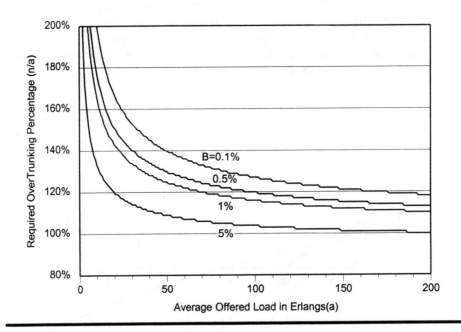

Figure 25.8 Required Overtrunking Ratio for Various Blocking Probabilities

25.3.3 Erlang's Blocked Calls Held Formula

A blocked calls held switching system has Markovian call arrivals, Markovian call duration and n servers (e.g., operators) and an infinite number of waiting positions. Therefore, a M/M/n queuing system model defines the performance. For an average offered load of ρ Erlangs and n servers, the following formula gives the probability that the system queues a call attempt,

$$C(n,a) = \frac{\omega}{\displaystyle\sum_{k=0}^{n-1} \frac{(na)^k}{k!} + \omega}$$

$$\text{where } \omega = \left(\frac{(na)^n}{n!}\right)\left(\frac{1}{1-\rho}\right).$$

Typically, texts call this formula the *Erlang-C* formula [1, 4]. Many older books contain tables and plots for the values of the Erlang-C (lost calls held) probability. However, you can easily compute your own using the following simple formula [4] utilizing the Erlang-B recursion defined in the previous section,

$$C(n,a) = \frac{nB(n,a)}{n - a[1-B(n,a)]}.$$

Figure 25.9 illustrates some values of the Erlang-C formula for various server group sizes as a function of offered load. A similar trend exists to that of blocking probability when comparing the queuing probability of a system with smaller offered load to a system with a larger offered load. The fact that the queuing probability decreases much more slowly with each additional server for larger systems than it does for smaller ones results in an economy of scale similar to that of blocking systems.

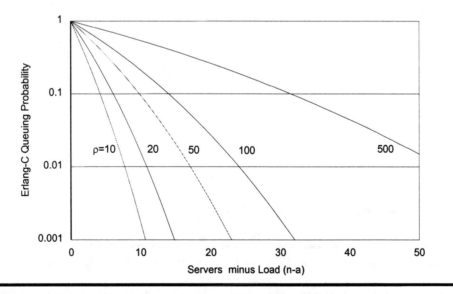

Figure 25.9   Erlang-C Queuing Probability versus Offered Load

25.4 PERFORMANCE OF BUFFERING METHODS

This section analyzes several simple models of switch delay and loss performance determined by the switch buffer architecture. For simplicity, the text assumes Poisson arrivals and negative exponential service times as the traffic model.

25.4.1 Input versus Output Queuing Performance

Recall from Chapter 16 the basic types of switch buffering: input, internal, and output. Output queuing delay performance behaves as a classical M/M/1 system. However, input queuing incurs a problem known as Head Of Line (HOL) blocking. HOL blocking occurs when the cell at the head of the input queue cannot enter the switch matrix because the cell at the head of another queue is traversing the matrix.

For uniformly distributed traffic with random message lengths, the maximum supportable offered load for input queuing is limited to 50 percent [5], while fixed message lengths increase the supportable offered load to only about 58 percent [6] as shown in the figure. On the other hand, output queuing is *not* limited by utilization as in input queuing.

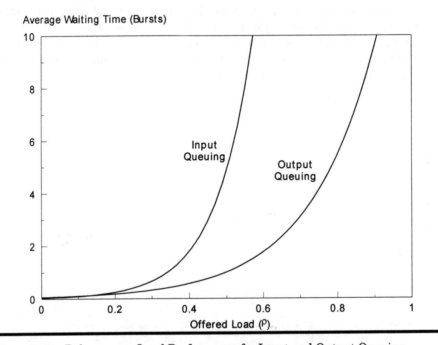

Figure 25.10   Delay versus Load Performance for Input and Output Queuing

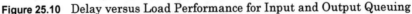

Figure 25.10 illustrates this result by plotting average delay versus throughput for input and output queuing. For a more detailed analysis of input versus output queuing see Reference 7 which shows that these simple types of models are valid for switches with a large number of ports.

The consequence of this result is that almost all ATM switches have some form of output buffering. Modern switches use input buffering in conjunction with some means to address Head Of Line (HOL) blocking. Examples of methods to address HOL blocking involve schemes where other cells pass an HOL-blocked cell. For example, a switch may have one input queue for each destination port, one for each destination port and service category, or even one queue per virtual connection.

25.4.2 Output Buffer Overflow Probability

This example gives a simple, useful approximation for the output buffer overflow probability. For simplicity, the analysis assumes an M/M/1 queuing system which has an infinite buffer, instead of a M/M/1/B system which has a finite buffer of B cell positions. The overflow probability for a buffer of size B cells with input bursts containing P cells on average is approximately the probability B/P bursts exist in the queuing system with infinite capacity. Comparison with simulation results and exact analysis shows this to be a reasonable approximation [8]. The approximate buffer overflow probability is:

$$\text{Prob[Overflow]} \approx \rho^{B/P+1}$$

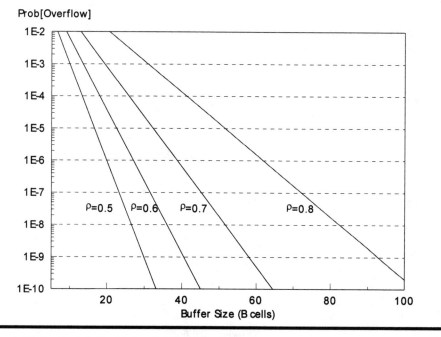

Figure 25.11 Switch Buffering Performance

Figure 25.11 plots this approximate buffer overflow probability versus buffer size for various levels of throughput ρ assuming a Protocol Data Unit

(PDU) size of P=1 cells. The performance for other burst sizes can be read from this chart by multiplying the x axis by the average PDU burst size P.

Typically, a network design requires a specific overflow probability objective. A curve of overflow probability versus buffer size varies as a function of the offered load as shown in Figure 25.11. For a fixed buffer size B, a particular value of offered load, ρ, generates a curve that exactly equals the desired overflow probability. This is the maximum offered load that the queuing system operates at while still meeting a specified overflow objective.

This formula also provides a simple guideline for allocating buffer sizes in an ATM device system with software definable queue thresholds. Solving the above equation for overflow probability in terms of the required buffer size B to achieve an objective Cell Loss Ratio (CLR). The result is the following:

$$B \approx P \; \frac{\log(\text{CLR})}{\log(\rho)}$$

Figure 25.12 shows a plot of the required number of cells in the buffer to achieve an objective Cell Loss Ratio (CLR) for a PDU size P=1 from this equation. Note that as the offered load increases, the required number of buffers increases nonlinearly.

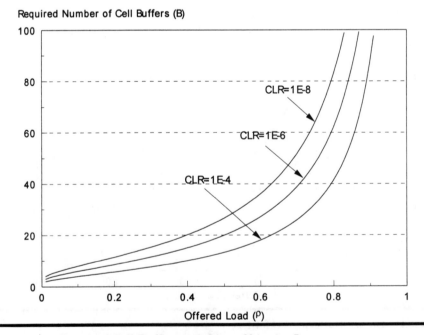

Figure 25.12   Required Cell Buffers to Achieve Objective Loss

Figure 25.13 illustrates the impact of higher-layer PDU size P on buffer overflow performance for various values of output buffer size B. As expected, the buffer overflow probability increases as the higher-layer PDU size P increases. When the PDU size approaches the buffer size, note that the loss rate is almost 100 percent.

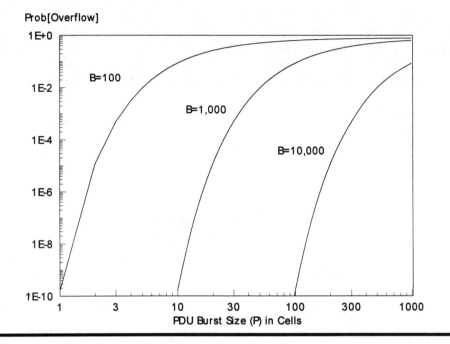

Figure 25.13 Overflow Probability versus PDU Burst Size

25.4.3 Shared Buffer Performance

The shared buffer scheme exhibits a marked improvement on buffer overflow performance. Since it is unlikely that all ports are congested at the same time; sharing a single, larger buffer between multiple ports is more efficient than statically allocating a portion of the buffer to each port.

The exact analysis of shared buffer performance is somewhat complicated [7]; therefore, we present a simple approximation based on the normal distribution. In the shared-buffer architecture, N switch ports share the common buffer, each having the M/M/1 probability distribution requirement on buffer space as defined in section 25.4.2. The sum of the individual port demands determines the shared-buffer probability distribution. The normal distribution approximates a sum of such random variables for larger values of N. The mean and variance of the normal approximation are then given by the following:

$$\text{Mean} = \frac{N\rho}{1-\rho} \quad , \quad \text{Variance} = \frac{2N\rho^2}{(1-\rho)^2}$$

Figure 25.14 shows a plot of the overflow probability versus the equivalent buffer size per port for shared buffers on switches of increasing port size (N), along with the dedicated output buffer performance for large N from Figure 25.11 for comparison purposes. The offered load is $\rho=0.8$, or 80 percent average occupancy. The total buffer capacity on a shared buffer switch is N times the buffer capacity on the x axis. Note that as N increases, the capacity required per port approaches a constant value. This illustrates the theoretical efficiency of shared buffering. Typically, the access speed of the shared memory limits the maximum size of the switch to the range of 10 to 100 Gbps.

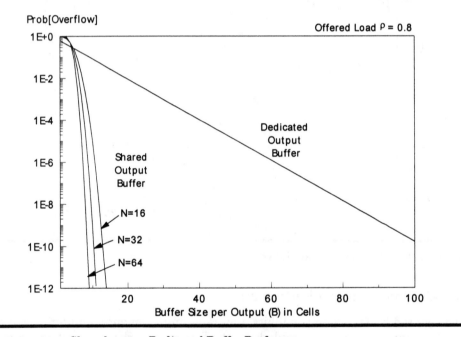

Figure 25.14   Shared versus Dedicated Buffer Performance

25.5 DETERMINISTIC CBR PERFORMANCE

The accurate loss performance measure of another very important traffic type, Constant Bit Rate (CBR), turns out to be relatively easy to calculate. Figure 25.15 illustrates the basic traffic source model. N identical sources emit a cell once every T seconds, each beginning transmission at some random phase in the interval (0,T). Thus, even if the network allocates capacity according to the peak rate, the random phasing of the CBR sources can still create overflow within the buffers in devices in an ATM network.

A good approximation for the cell loss rate for such a randomly phased CBR traffic input is [9];

$$CLR = \text{Prob[Cell Loss]} \approx \exp\left[-2B^2/n - 2B(1-\rho)\right],$$

where n is the number of CBR connections, B is the buffer capacity (in cells), and ρ =nT is the offered load.

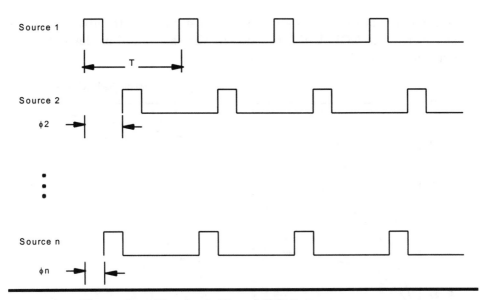

Figure 25.15 Illustration of Randomly Phased CBR Sources

A closed-form solution for the number of buffers required to achieve a specified loss probability results by solving the above formula for the minimum buffer size as follows [10]:

$$ B \approx \frac{\sqrt{[n(1-\rho)]^2 - 2n \ \ln(CLR)} - n(1-\rho)}{2} $$

Figure 25.16 illustrates the results of this calculation by plotting the required buffers B versus the number of CBR connections n for various levels of overall throughput, ρ. If the ATM switch implements priority queuing, this performance measure can be applied independently of other performance measures. This means that in the example of Figure 25.16, that only the fraction ρ of the switches overall capacity is used for CBR. The remaining capacity is available for other ATM service classes.

The actual CDV, measured in absolute time, encountered in the switch is dependent upon several parameters. First, the link rate, R is a key factor: slower links generate more CDV in inverse proportion to the link speed since the cell duration is D=384/R. The CDV for each switch (expressed in microseconds) for a CBR Cell Buffer Capacity, B, selected to achieve a specified CLR as described above is the difference between the CBR Buffer Capacity and the average number of CBR cells in the buffer [9] divided by the link rate (in cells per second) is:

$$CDV(B,R) = \frac{384}{R} \left\{ B - \frac{\sqrt{n}}{2} \frac{1 - \Phi[(1-\rho)\sqrt{n}]}{\varphi[(1-\rho)\sqrt{n}]} \right\}$$

where φ and Φ are the standard normal density and distribution functions, respectively. The NORMDIST function in Microsoft Excel implements the φ and Φ functions in the above formula.

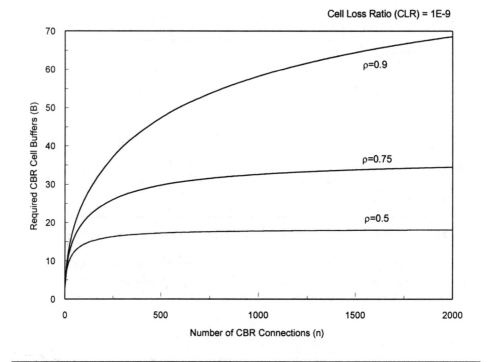

Figure 25.16   Required Cell Buffer Capacity for Deterministic CBR Traffic

Figure 25.17 plots the single switch CDV to achieve a CLR of 10^{-9} versus the number of connections with the overall utilization of the output link as the parameter. The next chapter provides a simpler estimate of end-to-end CDV across an ATM network.

25.6 RANDOM VBR PERFORMANCE

A simple model for Variable Bit Rate (VBR) traffic distribution uses a normal approximation to the binomial distribution for V VBR sources with an average activity per source of p, using the following parameters:

Mean = Vp , Variance = Vp(1 – p)

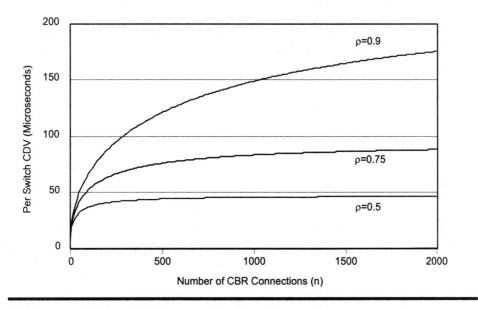

Figure 25.17 CDV per Switch versus Number of Connections

Using these parameters and a specification for the loss rate results in the following solution for the required number of buffers:

$$B \approx Vp + \alpha\sqrt{Vp(1-p)}$$

where the parameter α is determined by equating the Cell Loss Ratio (CLR) to the tail of the normal distribution, namely, $Q(\alpha)=CLR$ as described in Chapter 24.

Figure 25.18 illustrates the required number of buffers to achieve a specified cell loss objective for various values of source activity p. Note that the curves are nearly linear for a large number of sources V.

25.7 EQUIVALENT CAPACITY

Equivalent capacity is widely used analytical technique to model ATM bandwidth requirements based upon work by Guérin [11]. This model approximates the exact solution by combining two separate models, applicable for different regions of operation. For systems with a small number of sources, a fluid flow model considers each source in isolation, allocating a fixed amount of capacity to each source. Unfortunately, for larger systems the fluid flow model overestimates the required capacity because it fails to account for statistical multiplex gain. In the case of larger systems, a model based upon the normal distribution accurately models the statistical benefits of capacity sharing by a large number of sources. The equivalent capacity model is then the minimum of the capacity determined by the fluid flow model and the statistical multiplex gain model. This section first covers the fluid flow model, followed by the normal distribution based statistical multiplexing gain model, concluding with the equivalent capacity model.

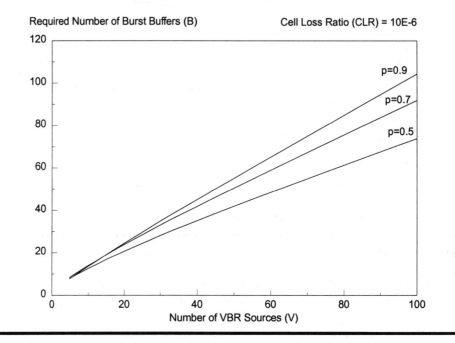

Figure 25.18 Required Buffer Capacity for Random VBR Traffic

The model used in this section assumes N identical traffic sources characterized by:

- PCR = Peak Cell Rate (cells/second)
- SCR = Sustainable Cell Rate (cells/second)
- ABS = Average Burst Size (cells)

The model is readily extended to a mix of different source types using the techniques described in References 11 and 12.

25.7.1 Fluid Flow Approximation

The fluid flow model treats each source independently, reserving bandwidth and buffer capacity to meet a specified Cell Loss Ratio (CLR). Conceptually, the model treats each source as a continuous time two state Markov process, each being in either an active or an idle state. In the active state, the source generates a burst of cells at a rate of PCR cells per second with each burst containing an average number of cells equal to ABS. The model assumes that the ATM device allocates B cells in a buffer for each source. Figure 25.19 illustrates the concept of the fluid flow model:

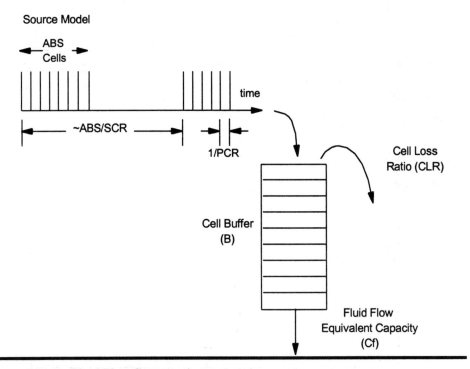

Figure 25.19   Fluid Flow Capacity Approximation

The following formula for the fluid flow equivalent capacity depends largely upon the utilization of the source, ρ=SCR/PCR and the ratio of the average burst size to the buffer capacity,

$$Cf = PCR \; \frac{z - 1 + \sqrt{(z-1)^2 + 4\rho z}}{2z}$$

where $z = -\ell n \; (CLR) \; (1-\rho) \; \dfrac{ABS}{B}$.

Figure 25.20 plots the fluid flow capacity, Cf, from the above equation normalized by dividing Cf by the value of PCR versus the ratio of average burst size to buffer capacity, ABS/B, with the average source utilization, ρ=SCR/PCR as the parameter. This normalized fluid flow equivalent capacity is then the fraction of the actual source peak rate required to achieve the required cell loss ratio. Note that when the burst is very small with respect to the buffer (i.e., ABS/B=0.001) that the required equivalent capacity approaches the average source utilization, ρ. However, as the size of the burst increases with respect to the buffer capacity (i.e., ABS/B>1) then the required equivalent capacity approaches 100 percent of the peak rate. Thus, if your application can tolerate wide variations in delay caused by large buffers, then you can run a network quite efficiently. But if you're integrating voice and video along with this data, your network needs to do more than buffering.

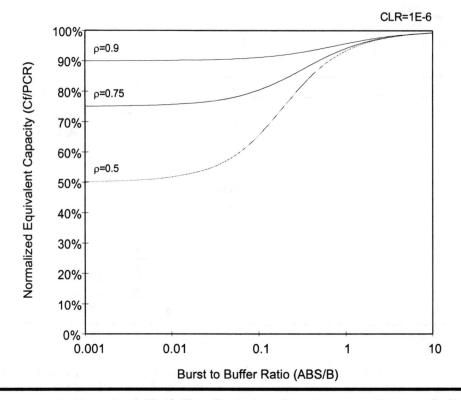

Figure 25.20   Normalized Fluid Flow Equivalent Capacity versus Burst to Buffer Ratio

25.7.2 Statistical Multiplex Gain Model

One key capability that ATM enables is that of statistical multiplexing, which attempts to exploit the on-off, bursty nature of many source types as illustrated in Figure 25.21. In the upper part of the figure, several devices generate and receive bursts of ATM cells: a video display, a server, a camera, and a monitor. The lower trace in the figure shows a plot of the sum of these cell bursts. In this simple example, the aggregate traffic requires only two channels at any point in time. As more and more sources are multiplexed together, the statistics of this composite sum become increasingly more predictable. The overall philosophy of this model stands in sharp contrast to the fluid flow model which considers each source separately.

Statistical multiplexing gain G is the ratio of channels supported to the number of required channels as indicated in the following formula:

$$G = \frac{\text{Number of Sources Supported}}{\text{Required Number of Channels}}$$

Each Source uses an entire channel at the peak rate only a fraction of the time

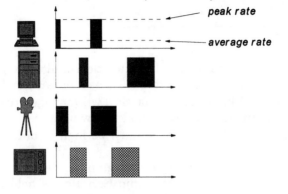

The source outputs can be combined to statistically share channels

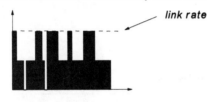

Figure 25.21   Illustration of Statistical Multiplex Gain

Note that the gain is never less than one, and is greater than one only if the number of sources supported greatly exceeds the number of channels required by individual sources. The statistical multiplex gain G for N identical sources defined exactly by the binomial distribution [13], and estimated from the normal distribution with the following parameters:

Mean = N/b   Variance = N/b(1-1/b)

where   N is the number of sources
          b is the burstiness (peak/average rate)

The required number of channels for this Gaussian approximation, expressed in units of the number of peak rate sources, to achieve an objective Cell Loss Ratio (CLR) of $Q(\alpha)$ is:

$$Cg \approx N/b + \alpha\sqrt{N(b-1)}/b$$

The parameter η=PCR/L defines the peak source-rate-to-link-rate ratio, which means that the link capacity is $1/\eta$. Therefore, the statistical multiplex gain reduces to G=N/Cg=Nη. Setting Cg in the above equation equal to the link capacity per source $1/\eta$ and solving for N using the quadratic formula yields the result:

$$G \approx \frac{\eta \left(\sqrt{\alpha^2(b-1) + 4b/\eta} - \alpha\sqrt{b-1} \right)^2}{4}$$

Figure 25.22 plots the achievable statistical multiplex gain G versus the peak-to-link rate ratio η with burstiness b as a parameter for a cell loss ratio of 10^{-6}. This figure illustrates the classical wisdom of statistical multiplexing: the rate of any individual source should be low with respect to the link rate η, and the burstiness of the sources b must be high in order to achieve a high statistical multiplex gain G.

Note that the statistical multiplex gain G never exceeds the source burstiness, b. That is, the relationship $G \leq b$ always holds. Another way of looking at the performance is to consider the link utilization,

$$U = \frac{G}{b}.$$

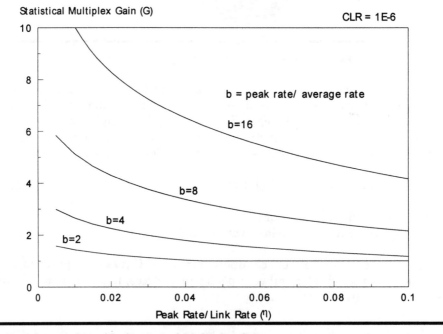

Figure 25.22   Achievable Statistical Multiplex Gain

Figure 25.23 plots the utilization achieved for the same parameters as in Figure 25.22 above. Note that the system achieves high link utilization only when the peak to link ratio, η=PCR/L is very small. The reduction in statistical multiplex gain results in a correspondingly lower utilization as peak to link ratio increases.

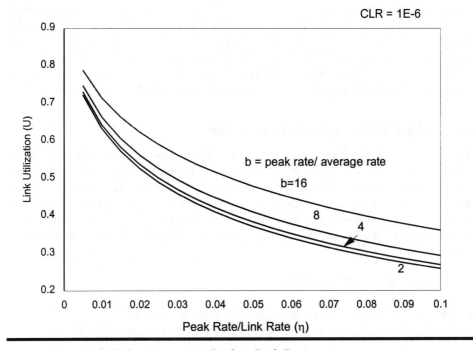

Figure 25.23   Link Utilization versus Peak to Link Ratio

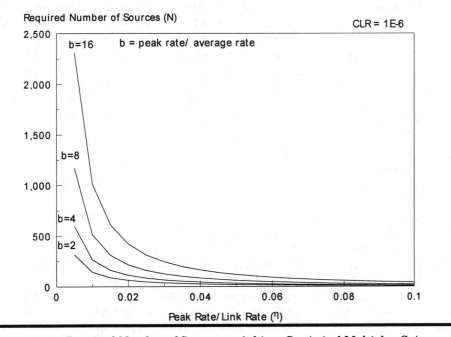

Figure 25.24   Required Number of Sources to Achieve Statistical Multiplex Gain

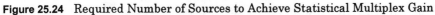

Figure 25.24 illustrates the number of sources that must be multiplexed together to achieve the statistical multiplex gain and utilization predicted in the previous charts. This confirms the applicability of the key statistical multiplex gain assumption that a large number of sources N, with low average rate 1/b, along with a modest peak rate to the link rate ration η, must be multiplexed together in order to achieve a high statistical multiplex gain G.

25.7.3 Equivalent Capacity Approximation

We saw in the preceding sections that the fluid flow approximation predicts efficient operation when the burst size is much less than the buffer size and that the statistical multiplexing model predicts higher efficiency when the source peak rate is much smaller than the link rate for a large number of sources. The approximate equivalent capacity model combines these two models to yield the required equivalent capacity Ce according to the following formula [11],

$$Ce = Min\{ N\ Cf,\ Cg\ PCR\}$$

where   Cf = fluid flow equivalent capacity from section 25.7.1
Cg = statistical multiplex gain capacity from section 25.7.2

Although the equivalent capacity approximation overestimates the required capacity, its computational simplicity makes it an attractive model to understand how the parameters of ATM traffic sources affect overall utilization and efficiency. Let's take a look at how these two approximations combine versus the key parameters of peak to link ratio and burst to buffer ratio through several examples. All of these examples numerically compute the maximum number of identical sources that yield a cell loss ratio of 10^{-6}.

Figure 25.25 plots Utilization U versus the ratio of source peak rate to link speed η=PCR/L for an average source burst size to switch buffer size ratio ABS/B of 10 percent with burstiness b=PCR/SCR as the parameter. This figure illustrates the point that the source peak rate should be much less than the link rate when traffic with long bursts is multiplexed together. As the peak to link ratio increases, the asymptotic statistical multiplexing gain dominates performance.

Figure 25.26 plots Utilization U versus the ratio of source peak rate to link speed for the same parameters, but now the average source burst size to switch buffer size ratio ABS/B is 1 percent. In other words, the buffers employed in the ATM device are ten times larger than those utilized in the previous chart. Note that the utilization is now independent of the source peak to link ratio, since short-term fluctuations in source activity are smoothed out by the larger buffer. Adaptive flow control, such as ATM's Available Bit Rate (ABR), would also achieve effects similar to the large buffer modeled in this example. This analysis stresses the importance of large buffers and/or small burst sizes in order to achieve high link utilization.

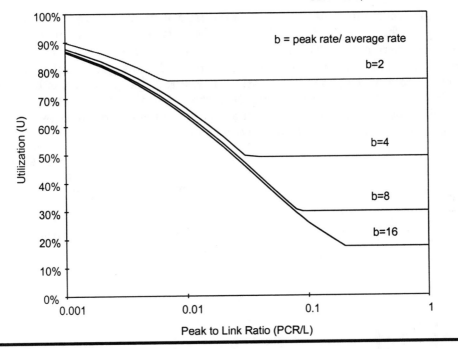

Figure 25.25   Utilization versus Peak to Link Ratio, Burst to Buffer Ratio = 10 percent

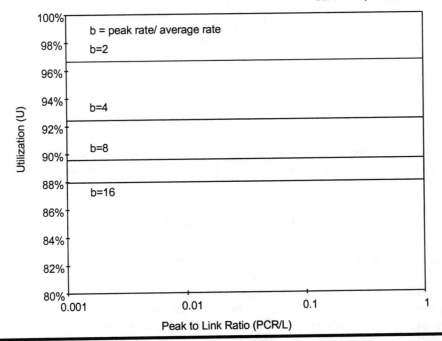

Figure 25.26   Utilization versus Peak to Link Ratio, Burst to Buffer Ratio = 1 percent

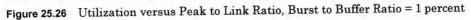

Figure 25.27 illustrates the effect of the ratio of average source burst size to switch buffer size ABS/B on utilization. The PCR divided by the link rate η is now the parameter. The average source activity ρ is 50 percent in this example. For smaller values of burst to buffer ratio, the utilization is independent of the peak to link ratio. However, for larger values of burst to buffer ratio, the utilization approaches that of peak rate allocation. This graphic clearly illustrates the benefits of larger buffers, particularly when the peak to link ratio is larger than a few percent.

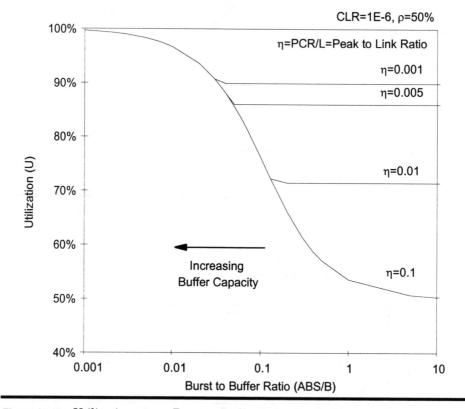

Figure 25.27   Utilization versus Burst to Buffer Ratio, Load =50 percent

25.8 PRIORITY QUEUING PERFORMANCE

This section illustrates the capability of priority queuing to provide different delay performance for two classes of traffic. Figure 25.28 illustrates cell traffic originating from two virtual connections multiplexed onto the same transmission path, the higher priority traffic numbered 1, and the lower priority traffic numbered 2. The system services priority 1 cells first, and priority 2 cells only when no priority 1 cells require service. In the event that more cells arrive in a servicing interval than the device can accommodate, the priority 2 cells are either delayed or discarded as illustrated at the top of the figure.

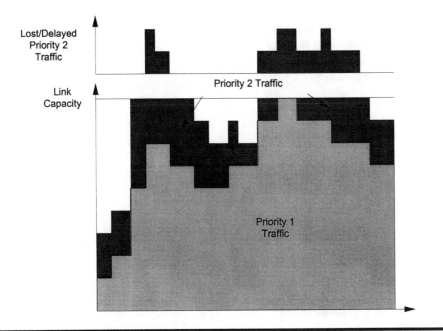

Figure 25.28   Illustration of Priority Queuing

The example assumes that both priority 1 and 2 cells have Poisson arrivals and negative exponential service. Furthermore, the system delays priority 2 cells in an infinite buffer while servicing priority 1 cells. The key to priority queuing is that the priority 1 traffic observes a delay as if the priority 2 traffic did not even exist. On the other hand, the priority 2 traffic sees delay as if the transmission capacity were reduced by the average utilization taken by the priority 1 traffic. The formulas for the average priority 1 and priority 2 queuing delays are [14]:

$$\text{Avg[Queuing Delay for Priority 1]} = \frac{\rho / \mu}{(1 - \rho_1)}$$

$$\text{Avg[Queuing Delay for Priority 2]} = \frac{\rho / \mu}{(1 - \rho)(1 - \rho_1)}$$

where ρ_1 and ρ_2 are the offered loads for priority 1 and 2 traffic, respectively, and $\rho = \rho_1 + \rho_2$ is the total offered load.

Figure 25.29 illustrates the effect of priority queuing by showing the average delay that would be seen by a single priority system according to the M/M/1 model of Section 25.2.3, the priority 1 cell delay, and the priority 2 cell delay. This example assumes that 25 percent of the traffic is priority 1, that is, $\rho_1/\rho=0.25$. Observe that the priority 1 performance is markedly better than the single priority system, while priority 2 performance degrades only slightly.

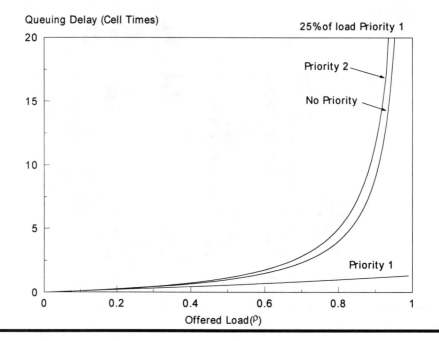

Figure 25.29   Priority Queuing Performance Example

25.9 REVIEW

This chapter first discussed several key aspects of traffic engineering philosophy: source modeling, performance measurement, and switch performance modeling. These affect the accuracy and complexity of the traffic engineering calculations. In general, the more accurate the model, the more complicated the calculation. This book opts for simplicity in modeling and introduces only a few of the popular, simple source models. Next, the text covers key aspects of modeling switch performance: call blocking and queuing performance, a comparison of buffering methods, Continuous Bit-Rate (CBR) loss performance, Variable Bit-Rate (VBR) loss performance, statistical multiplexing, and priority queuing. We demonstrated the superiority of output buffering over input buffering and the increased efficiency of shared buffering over dedicated buffering. The coverage then presented simple formulas to evaluate loss formulas for CBR and VBR traffic.

The text then introduced the equivalent capacity model comprised of a fluid flow and stationary approximation. The fluid flow model considers a single source in isolation, computing the smoothing effect of buffering on transient overloads. The stationary statistical multiplex gain model demonstrates the classic wisdom, that in order to achieve gain, there must be a large number of bursty sources, each with a peak rate much less than the link rate. The equivalent capacity model combines these two models, demonstrating key tradeoffs in the relationship of burst to buffer size and peak to link ratio over a wide range of network operating parameters. Finally, the text showed how priority queuing yields markedly improved delay performance for the high-priority traffic over that of a single priority system.

25.10 REFERENCES

[1]   L. Kleinrock, *Queuing Systems Volume I: Theory*, Wiley, 1975.

[2]   Fowler, Leland, "Local Area Network Traffic Characteristics, with Implications for Broadband Network Congestion Management," *IEEE JSAC*, September 1991.

[3]   Leland, Willinger, Taqqu, Wilson, "On the Self-Similar Nature of Ethernet Traffic," *ACM Sigcomm '93*, September 1993.

[4]   R. Cooper, "Introduction to Queuing Theory, Second Edition," North-Holland, 1981.

[5]   D. McDysan, "Performance Analysis of Queuing System Models for Resource Allocation in Distributed Computer Networks," *D.Sc. Dissertation*, George Washington University, 1989.

[6]   Hui, Arthurs, "A Broadband Packet Switch for Integrated Transport," *IEEE JSAC*, December 1988

[7]   M. Hluchyj, M. Karol, "Queuing in High-Performance Packet Switching," *IEEE JSAC*, December 1988.

[8]   M. Schwartz, *Computer Communication Design and Analysis*, Addison Wesley, 1977.

[9]   L. Dron, Ramamurthy, Sengupta, "Delay Analysis of Continuous Bit Rate Traffic over an ATM Network," *IEEE JSAC*, April 1991.

[10]   M. Wernik, Aboul-Magd, Gilbert, "Traffic Management for B-ISDN Services," *IEEE Network*, September 1992.

[11]   R. Guerin, H. Ahmadi, M. Naghshineh, "Equivalent Capacity and Bandwidth Allocation," *IEEE JSAC*, Sept 1991.

[12]   H. Ahmadi, P. Chimento, R. Guérin, L. Gün, R. Onvural, T. Tedjianto, "NBBS Traffic Management Overview," *IBM Systems Journal*, Vol 34, No. 4, 1995.

[13]   Rasmusen, Sorenson, Kvols, Jacobsen, "Source-Independent Call Acceptance Procedures in ATM Networks," *IEEE JSAC*, April 1991.

[14]   Gross, Harris, *Fundamentals of Queuing Theory*, Wiley, 1985.

26

Design Considerations

This chapter explores important network and application design considerations. First, the text analyzes how delay and loss impact an application performance. The chapter then covers how the accumulation of delay variation in multiple hop networks impacts delay-variation-sensitive applications, such as video, audio, and real-time interactive application traffic. Next, the coverage continues with an in-depth analysis of the performance of TCP over ATM. The text analyzes the operation of TCP over ABR and UBR ATM service, and develops a simple model for TCP performance over ATM. The results presented include: comparisons with measurements and simulations, the impact of buffer size on performance, and the operating characteristics as a function of the number of TCP sources. The treatment then analyzes the statistical multiplex gain achievable for voice traffic, along with the savings occurring by integrating voice and data onto a single transmission facility.

The text then describes some practical implications involved in the decision regarding use of a dedicated private network, a shared public network, or a hybrid private/public network. We compare and contrast the factors involved in the decision between a fixed or usage-based service in a public network service. The chapter closes by describing several open issues relevant to any network design. Furthermore, a good designer considers the reaction of the network response to overloads or significant failures. Finally, the text identifies practical applications of source modeling and traffic engineering.

26.1 IMPACT OF DELAY, LOSS, AND DELAY VARIATION

This section reviews the impact of delay, loss, and delay variation on applications. Delay impacts the achievable throughput for data applications.

Loss impacts the usable throughput for most network and transport layer protocols. Delay variation impacts the performance for applications requiring a guaranteed playback rate, such as voice and video.

26.1.1 Impact of Delay

Two situations occur when a source sends a burst of data at a certain transmission rate (or bandwidth) across a network with a certain delay, or latency: we call these bandwidth limited and latency limited. A *bandwidth-limited application* occurs when the receiver begins receiving data before the transmitter completes sending the entire burst. A *latency-limited application* occurs when the transmitter finishes sending the burst of data before the receiver begins receiving any data.

Figure 26.1 illustrates the consequence of sending a burst of length b equal to 100,000 bits (100 kb) at a peak rate of R Mbps across the domestic United States with a one-way propagation delay τ of 30 ms. In other words, it takes 30 ms for the bit stream to propagate from the originating station to the receiving station across approximately 4,000 miles of fiber since the speed of light in fiber is less than that in free space, and fiber is usually not routed along the most direct path. When the peak rate between originator and destination is 1 Mbps, and after 30 ms, only about one-third of the burst is in the transmission medium, and the remainder is still buffered in the transmitting device. We call this a bandwidth-limited application since the lower transmission rate limits the transmitter from releasing the entire message before the sender begins receiving the burst.

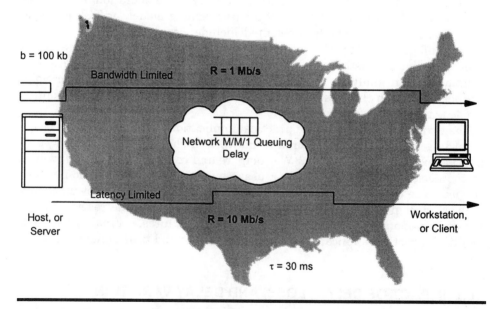

Figure 26.1 Propagation Delay, Burst Length, and Peak Rate

Now let's look at the *latency limited* case where the transmitter sends the *entire* transmission before the receiver gets any data. Increasing the peak rate to 10 Mbps significantly changes the situation — the workstation sends the entire burst before even the first bit reaches the destination. Indeed, the burst occupies only about one-third of the bits on the fly in the fiber transmission system as indicated in the lower part of Figure 26.1. If the sending terminal must receive a response before sending the next burst, then observe that a significant reduction in throughput results. We call this situation *latency limited*, because the latency of the response from the receiver limits additional transmission of information by the sender awaiting an acknowledgment.

Now let's apply the basic M/M/1 queuing theory from Chapter 25 as an additional element of end-to-end delay that increases nonlinearly with increasing load, and becomes a better model of a real world application. The average M/M/1 queuing plus transmission delay in the network is

$$b/R/(1-\rho),$$

where ρ is the average trunk utilization in the network. The point where the time to transfer the burst (i.e., the transmission plus queuing time) exactly equals the propagation delay is called the latency/bandwidth crossover point as illustrated in Figure 26.2.

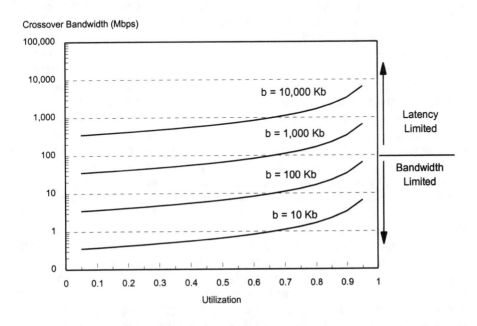

Figure 26.2 Latency/Bandwidth Crossover Point

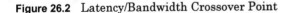

In the previous example, for a file size of b=100 kb the crossover point is 3.33 Mbps for zero utilization, and increases to 10 Mbps for 66 percent utilization. See Reference 1 for a more detailed discussion on this subject.

If the roundtrip time is long with respect to the application window size, then the achievable throughput is markedly reduced. This is a basic aspect of all flow control methods, as seen in Chapter 23 on congestion control. The amount of buffering required in the network to achieve maximum throughput is proportional to the delay, bandwidth product. In our example of Figure 26.1, the round-trip delay-bandwidth product is 60,000 bits (or approximately 8 kbytes) for a 1-Mbps link (30 ms × 1 Mbps × 2) and 600,000 bits for a 10-Mbps link.

This situation is analogous to what occurred in early data communications over satellites, where the data rates were low, but the propagation delay was very high. The delay-bandwidth product is high in satellite communications because the propagation delay is large, while in B-ISDN and ATM communications the delay-bandwidth product becomes large because the transmission speeds are high.

26.1.2 Impact of Loss

Loss is another enemy of data communications applications. For many applications, the loss of a single cell results in the loss of an entire packet (or an entire protocol data unit) because the SAR sublayer in the AAL fails during the reassembly process. Loss (or even excessive delay) results in a time-out or negative acknowledgment in a higher-layer protocol, typically the transport layer, for example in the Internet Protocol's Transmission Control Protocol (TCP) studied in Chapter 8.

Higher-layer protocols recover from detected errors, or time-outs, by one of two basic methods: either they retransmit all information sent after the packet that resulted in the detected error or time-out, or they selectively resend only the information that actually in error or timed out. Resending all of the information means that if the transmitter sent N packets after the packet that caused the detected error or time-out, then the transmitter resends these same N packets again. You'll see this scheme called a *Go-Back-N* retransmission or a cumulative acknowledgment strategy in the technical literature. Obviously, resending the same information twice reduces the usable throughput. The second method requires a higher-layer protocol that explicitly identifies the packet in error. Then, the transmitter only resends the errored packet, improving efficiency when retransmissions occur. The usable throughput increases because only the errored or timed-out information is retransmitted. Standards call this a selective reject, selective retransmission, or selective acknowledgment strategy. However, this type of protocol is more complex to implement. The IETF defines optional selective acknowledgment strategy as extensions to TCP [2] in support of communication over long delay paths; however, few implementations support it.

We now present a simple model of the performance of these two retransmission strategies to illustrate the impact of loss on higher-layer protocols.

The transmission rate R, the packet size p (in bytes), and the propagation delay τ determine the number of cells in the retransmission window W as follows:

$$W \;=\; \left\lceil \frac{2\tau R}{8p} \right\rceil$$

The probability that the receiver loses an individual packet π due to a specific Cell Loss Ratio (CLR) in the ATM network is approximately:

$$\pi \;\approx\; \left\lceil \frac{p}{48} \right\rceil \; CLR$$

In the Go-Back-N strategy, if a single packet is in error at the beginning of a window of W packets, then the sender must retransmit the entire window of W packets. For the Go-Back-N retransmission strategy, the usable throughput η(Go-Back-N) is approximately the inverse of the average number of times the entire window must be sent, which is approximately [3, 4]:

$$\eta(\text{Go Back N}) \;\approx\; \frac{1-\pi}{1+\pi W}$$

In the selective reject retransmission strategy, if a single packet is in error, then the sender retransmits only that packet. For the selective reject retransmission strategy, the usable throughput η(Selective Reject) is approximately the inverse of the average number of times any individual packet must be sent, which is [3, 4]:

$$\eta(\text{Selective Reject}) \;\approx\; (1-\pi)$$

This formula is valid for the case in which only one packet needs to be transmitted within the round-trip delay window. It also applies to more sophisticated protocols that can retransmit multiple packets within a round-trip delay interval, such as ATM's Service Specific Connection Oriented Protocol (SSCOP) as described in Chapter 15.

Figure 26.3 plots the usable throughput (or "goodput") for Go-Back-N and selective reject retransmission strategies for a DS3 cell rate R of 40 Mbps, a packet size p of 200 bytes, and a propagation delay of 30 ms. The resultant window size W is 1500 packets. The retransmission protocols have nearly 100 percent usable throughput up to a Cell Loss Ratio (CLR) of 10^{-6}. As the CLR increases, the usable throughput of the Go-Back-N protocol decreases markedly because the probability that an individual window (of 7500 cells) is error free decreases markedly. As the CLR increases toward 10^{-2}, the probability of an individual packet having a lost cell becomes significant, and even the selective reject protocol's usable throughput degrades.

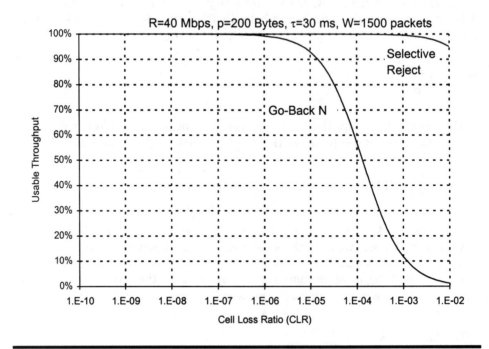

Figure 26.3 Usable Throughput versus Cell Loss Ratio

These examples illustrate the importance of selecting a QoS class with loss performance that meets the application requirements. For example, commonly deployed implementations of the Transmission Control Protocol (TCP) use a Go-Back-N type of protocol and hence work best with low loss rates.

Loss is relatively easy to estimate in an ATM network. Assuming that loss is independent on each switch and link, then the end-to-end loss is 1 minus the product of successful delivery for each element. In practice, a good approximation when the loss probabilities are small is to simply add up the individual loss probabilities.

26.1.3 Impact of Cell Delay Variation

When designing a network of multiple switches connected by trunks, there are several things to consider. This section focuses on aspects common to both private and public networks, pointing out unique considerations that occur in each. Recall the Cell Delay Variation (CDV) QoS parameter, which measures the clumping that can occur when cells are multiplexed together from multiple sources, either in the end system or at any switch, multiplexer, or intermediate system in the network. The resulting affect accumulates when traversing multiple switches in a network. This becomes critical when transporting delay-sensitive applications such as video, audio, and interactive data over ATM. In order to support these applications, the destination device must buffer the received cell stream to absorb the jitter introduced in an end-to-end network.

Jitter refers to the rate of clumping or dispersion that occurs to cells that were nominally spaced prior to transfer across an ATM network. The accumulated jitter must be accommodated by a playback buffer as described with reference to Figures 26.4 and 26.5. The playback buffer ensures that underrun or overrun events occur infrequently enough, according to the CDV (clumping) and cell dispersion that accrues across a network. In our examples the nominal cell spacing is 4 cell times, the playback buffer is 4 cells in length, and begins operation at a centered position (i.e., holding two cells).

In the overrun scenario of Figure 26.4, cells arrive too closely clumped together, until finally a cell arrives and there is no space in the playback buffer; thus cells are lost. This is a serious event for a video coded signal because an entire frame may be lost due to the loss of one overrun cell.

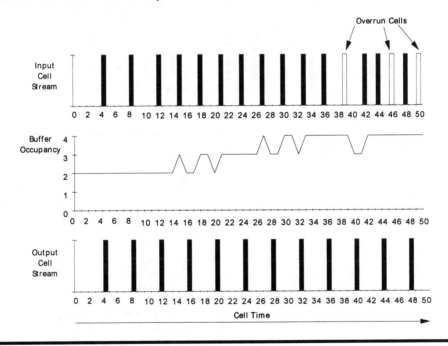

Figure 26.4 Illustration of Playback Buffer Overrun Scenario

In the under-run scenario of Figure 26.5, cells arrive too dispersed in time, such that when the time arrives for the next cell to leave from the playback buffer, the buffer is empty. This too has a negative consequence on a video application because the continuity of motion or even the timing may be disrupted.

A simple approximation for the accumulation of CDV across multiple switching nodes is the square root rule from the ATM Forum B-ICI specification. This states that the end-to-end CDV is approximately equal to the CDV for an individual switch times the square root of the number of switching nodes in the end-to-end connection. The reason that we don't just add the variation per node is that the extremes in variation are unlikely to occur simultaneously, and tend to cancel each other out somewhat. For example, while the variation is high in one node, it may be low in another node.

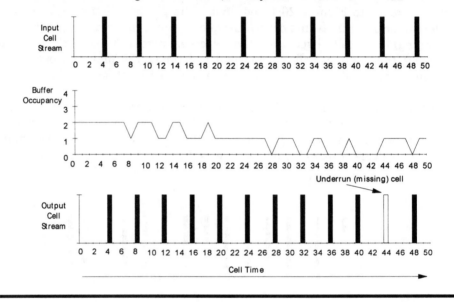

Figure 26.5 Illustration of Playback Buffer Underrun Scenario

Figure 26.6 illustrates the concept of accumulated delay variation by showing the probability distribution for delay at various points in a network.

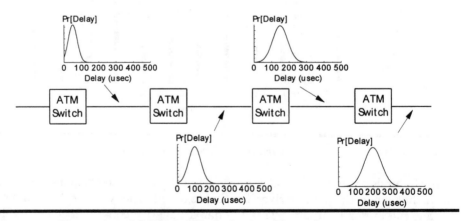

Figure 26.6   Illustration of Cell Delay Variation in a Network

The assumed delay distribution has a fixed delay of 25 μs per ATM switch. The ATM switch adds a normally distributed delay with mean and standard deviation also equal to 25 μs. Therefore, the average delay added per node is the sum of the fixed and mean delays, or 50 μs. Starting from the left-hand side at node A, the traffic has no variation. The first ATM switch adds a fixed delay plus a random delay resulting in a modified distribution. The next ATM switch adds the same constant delay and an independent random delay. The next two ATM switches add the same constant and random delay characteristics independently. The resulting delay distribution after traversing four nodes is markedly different as can be seen from the plots — not four times worse but only approximately twice as bad. Furthermore, correlation between the traffic traversing multiple nodes makes this model overly optimistic as described in informative Appendix V of the ATM Forum's Traffic Management 4.0 specification [5].

Figure 26.7 plots the probability that the delay exceeds a certain value x after traversing N nodes. The random delays are additive at each node; however, in the normal distribution the standard deviation of the sum of normal variables only grows in proportion to the square root of the sum. This is the basis for the square root rule in the ATM Forum B-ICI specification. The sum of the fixed and average delays is 200 μs, and therefore the additional variation is due to the random delay introduced at each node.

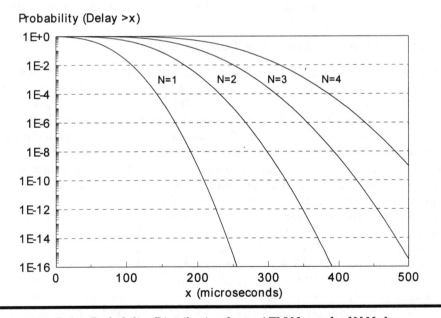

Figure 26.7 Delay Probability Distribution for an ATM Network of N Nodes

The equation for the plots of Figure 26.7 for an average plus fixed delay given by α (50 μs in this example) and a standard deviation σ (25 μs in this example) for the probability that the delay is greater than x after passing through N ATM switches is:

$$\text{Prob[Delay} \geq \text{x]} \approx Q\left(\frac{x - N\alpha}{\sqrt{N}\sigma}\right)$$

Chapter 24 defined the expression for $Q(x)$ for the normal distribution. The presence of CDV in ATM networks means that the end equipment must absorb the jitter introduced by intermediate networks. One option to balance this tradeoff is the capability to reduce CDV within switches in the network via shaping, spacing, or framing as described in Chapter 22. The calculation of average delay is simple — add up all of the components to delay for each switch and transmission link. Be sure to ask switch vendors and service providers about average delays as well as variations in delay.

26.2 TCP PERFORMANCE OVER ATM

This section describes several things to consider when tuning TCP to perform well over ATM networks. These include the effect of window size on performance, the interaction with ATM traffic and congestion control, and the impacts of ATM device buffer size.

26.2.1 TCP Window Size Impact on Throughput

Network congestion resulting in loss affects end-to-end transport layer (e.g., TCP), and ultimately application layer (e.g., HTTP) performance. A key design issue for ATM networks is the selection of service categories and traffic parameters that minimizes loss, and hence maximizes usable throughput performance. Add to this the fact that standard TCP implementations only support a 64Kbyte window size p×W, and you'll see that applications running over TCP can't sustain transmissions at 10Mbps Ethernet speeds across one-way WAN distances greater than approximately 2,500 miles. For applications capable of transmitting at 100 Mbps, the standard TCP window size limits the distance over which maximum throughput is achievable to 250 miles. To achieve higher throughput over networks with large bandwidth-delay products, look for devices that implement TCP window scaling extensions defined in IETF RFC 1323 [6]. A general rule of thumb is that the total throughput of TCP (in kbps or Mbps) increases with the window size in kilobytes linearly up to the point where either the maximum window size is reached, or the window equals the product of the round-trip delay and the bottleneck bandwidth. See Chapters 8 and 23 for further details and examples of the TCP windowing protocol.

26.2.2 TCP over ATM: UBR and ABR

This section looks at a few more issues on placing TCP traffic over the ATM service categories called the Unspecified and Available Bit Rate (UBR, and ABR respectively) as detailed in Chapter 21. TCP/IP is the most common data traffic running over ATM networks today. TCP's adaptive windowing protocol dynamically acts to fill the available bandwidth, if user demand is high enough. Thus, ATM is an ideal network infrastructure to support TCP's ability to fill in bandwidth unused by higher-priority multimedia traffic like voice and video. TCP can use either the ABR or UBR service category, with advantages and disadvantages as summarized below.

Many UBR implementations use a simple cell discard threshold in a buffer based upon the Cell Loss Priority (CLP) bit in the ATM cell header. Once the buffer fills beyond the threshold, the switch discards lower priority cells (i.e., CLP=1 tagged UBR cells). ATM switches should have large buffers, sized to be on the order of the product of the round-trip delay and bandwidth bottleneck when supporting TCP over UBR.

During congestion conditions and subsequent loss of cells, the ATM network device does not notify the sender that retransmission is required — instead higher-layer protocols, like TCP, must notice the loss via a time-out and retransmit the missing packets. Not only does one cell loss cause the missing packet to be retransmitted, but all packets after it up to the end of the transmit window. Excessive packet discards within a TCP window can degrade the recovery process and cause host time-outs — causing interruptions on the order of many seconds to minutes. Loss also touches off TCP's slow start adaptive windowing mechanism, further reducing throughput. If you plan to operate TCP/IP over UBR, be sure that your ATM switches or service provider support Early/Partial Packet Discard (EPD/PPD) as defined in Chapter 23. The EPD/PPD functions ensure that the switch discards entire packets during periods of congestion. This is especially important when a relatively large number of TCP sources contend for a particular bottleneck.

Possibly, the best ATM service category for TCP traffic is ABR, which employs a closed-loop rate-based mechanism for congestion control using explicit feedback. For ABR traffic, the TCP source must control its transmission rate. ABR defines a minimum cell rate (MCR) for each virtual connection, which defines the lowest acceptable value of bandwidth. Note that the MCR may be zero. Operating in an ATM network, the ABR protocol utilizes Resource Management (RM) cells to control the input rate of each source (and thus each connection) based upon the current level of congestion in the switches along the route carrying that connection's traffic. In ABR, the switch buffer size requirements are similar to that in the UBR case.

The network may police ABR connections to ensure that they conform to the traffic contract. The standards allow networks to do this so that an unregulated user on a single connection cannot affect the service of all other users sharing bandwidth with that connection [7, 8].

26.2.3 A Simple TCP/IP over ATM Performance Model

This section defines a simple model for the performance of a number of greedy TCP/IP over ATM sessions contending for a common buffer in an ATM device. The following treatment analyzes the effective TCP throughput for the case where the buffer admission and service logic implements Early Packet Discard/Partial Packet Discard (EPD/PPD), as well as the case where the device does not implement EPD/PPD. The case with EPD/PPD models the performance of a packet switch or router.

Figure 26.8 illustrates the scenario where N TCP clients simultaneous transfer data to a single server attached to a switch egress port and buffer. All hosts connect to the switch via transmission lines running at R bits per second. A single trunk line also running at R bps empties a buffer which has capacity B packets. All clients have the same Round Trip Delay RTD. Since this configuration is completely symmetric, congestion causes all sources to begin retransmission simultaneously. This phenomenon is called "phasing" in the study of TCP performance.

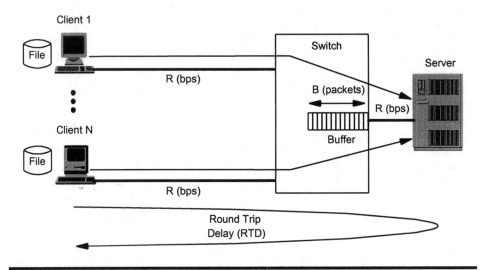

Figure 26.8   TCP Dynamic Windowing Flow Control Contention Scenario

Figure 26.9 illustrates the effect of two (N=2) contending, identical sources on the TCP window size (measured in packets here) and the buffer fill versus the horizontal axis measured in units of Round Trip Delay (RTD). The figure shows the window size by squares on the figure every RTD time. Recall from Chapter 8 that the TCP slow start protocol increases the window size by multiplying by two each RTD until it reaches a value of one half the previous maximum window size. After passing the one half value, TCP linearly increases the window size linearly by one packet for each RTD interval. The figure indicates the number of packets sent in each RTD interval by a horizontal line. Once the buffer overflows, TCP fails to receive an ACKnowledgement. In our example, we assume that TCP times-out after one RTD interval. When the time-out occurs, TCP decreases the window size to one. The buffer fill plot in the figure depicts the worst case situation where each TCP source generates packets according to this same function over time. Notice how the buffer fill returns to zero with the very short window sizes at the beginning of the start-up interval, but gradually increases as the window size increases in the linear increase region. In this example the buffer capacity is B=32 packets, which overflows on the 14th, 15th and 16th RTD intervals as shown in the figure. The maximum number of packets "on the Fly" (F) is 20 in this example.

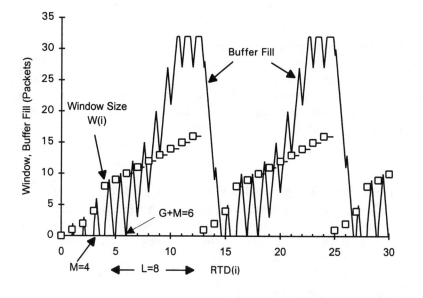

Figure 26.9   TCP Dynamic Windowed Flow Control Window Size and Buffer Fill

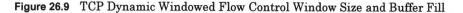

The net effect of this protocol is that each TCP session on average gets about half of the egress port bandwidth. This sawtooth type of windowing pattern is typical for TCP on other types of networks, such as those with routers containing large buffers.

The following simple analytical model approximates TCP slow start performance using the following variables:

M   = Number of RTDs in the Multiplicative increase interval of TCP
L   = Number of RTDs in the Linear increase interval of the TCP
P   = Packet size (Bytes)
R   = Link Rate for each sender (bits/second)
N   = Number of sources
B   = Buffer Size (in cells)
RTD = Round Trip Delay (seconds)

The rate from each transmitter determines the number of packets on the "fly" during a Round-Trip Delay (RTD) time as follows:

$$F \; = \; \frac{48}{53} \; \frac{R}{P} \; \frac{RTD}{8}$$

Translating the TCP slow start algorithm into a formal expression, the window size (in packets) in the ith RTD interval is:

$$W(i) = \begin{cases} 2^{i-1} & 1 \le i \le M \\ L+i & M < i < M+L+1 \end{cases}$$

Since TCP transitions from exponential to linear increase when the window size crosses half of the previous maximum value, we infer that $L \le 2^{(M-1)}$. In the example of Figure 26.9, $M=4$ and $L=8=2^{(M-1)}$. Note that unlimited buffer growth occurs at RTD interval $G+M$ in the linear region when the packets from all sources exceed the capability of the bottleneck link to empty the buffer in an RTD (which is F packets), namely

$$N(G+L) > F.$$

This implies that the point where unlimited growth occurs is $G \approx F/N - L$. Buffer overflow occurs as growth accumulates in the linear increase region after the Gth RTD interval until the $(L+M)$ interval. The total packets arriving minus those transmitted by the link serving the buffer determines the following balance equation:

$$N \sum_{i=G+M}^{L+M} W(i) - (L-G+1)F = B$$

Solving the above equation for buffer overflow for the linear increase interval comprised of $L - G + 1$ RTD time segments yields:

$$L \approx \frac{F}{2N} + \sqrt{\frac{B+F}{2N}}$$

The effective TCP throughput η, or "goodput," is the ratio of good frames transferred to the total frames that could have been sent in a sequence of $(M+L)$ round-trip delay intervals as follows:

$$\eta \approx \frac{N \sum_{i=1}^{M+L} W(i)}{F(M+L)} = \frac{N \; (3L^2/2 + 4L - \Delta)}{F(\log_2 L + L)}$$

The Δ term models the cases where the ATM device employs EPD/PPD and where it does not. When the device implements EPD/PPD, $\Delta = 2L$ while when it does not, $\Delta = 4L$. This is true because in the best case the TCP sources must lose at least $N \times L$ packets to cause the retransmission time-out plus $N \times L$ unproductive packet transmissions that must be resent. In the case where the device does not implement EPD/PPD, the destination fails to receive even more packets because the device discards one or more cells per packet.

This model made a number of approximation to yield the relatively simple expression above. When the buffer size exceeds the packets on the fly number (F), the model may return values greater than 100 percent due to the approximations.

26.2.4 Buffer Size Impact on TCP/IP Performance

Figure 26.10 illustrates the results of the above model compared with simulation results and an actual measurement over an ATM network [9]. The figure plots the effective TCP throughput with and without EPD/PPD. For this example, N=2 sources, R=40 Mbps corresponding to operation over DS3 PLCP, the RTD is 70 ms, and the packet size P=4096 bytes. For these parameters, M is 5 and L is approximately 25. This means that a complete TCP slow start dynamic window cycle takes over 2 s to complete. The ATM devices in the measurement and the simulation did not implement EPD/PPD. The correspondence between the analytical approximation and the measured and simulated results is quite good for this set of parameters.

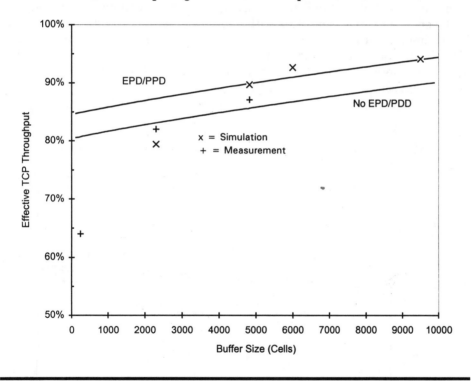

Figure 26.10   TCP Effective Throughput versus Buffer Size

The simple analytical model does not address a situation that occurs in ATM devices with very small buffers, but which the measurements in Figure 26.10 illustrate. When the buffer size of the ATM switch is less than the packet size, then when two packets arrive nearly simultaneously, one or more cells from both packets are lost and both must be retransmitted, resulting in very low throughput as shown by the measurements in the figure. When the buffer size is increased so that two packets can arrive simultaneously, throughput increases markedly. As the buffer size increases further, the throughput increases, but not nearly as dramatically. Similar results occur in a local area for switches with buffers that are small with respect to the packet size as reported in [10].

Figure 26.11 illustrates another example where R=150 Mbps corresponding to operation over an OC3/STS-3c, the RTD is 40 ms, and the packet size P=1500 bytes. Approximately 14,000 cells are on the fly for these parameters. The figure plots the effective TCP throughput versus buffer size B (in cells) as before for the two cases where the ATM device does and does not implement EPD/PPD. However, the figure plots the throughput for two values of the number of sources, namely: N=2 and N=32. For the case where N=2, M is 7 and L is approximately 100. This means that a complete TCP slow start dynamic window cycle takes over 5 s to complete. On the other hand, for N=32 M is 4 and L is approximately 10, meaning that a complete TCP congestion cycle completes within less than half a second. Note how the throughput improves for a larger number of sources N. Even without EPD/PPD, the performance improves because of the sources spending relatively more time in the productive linear increase interval. Note as before that the impact of EPD/PPD is small when N=2. The next section addresses the impact of the number of sources N on TCP/IP throughput.

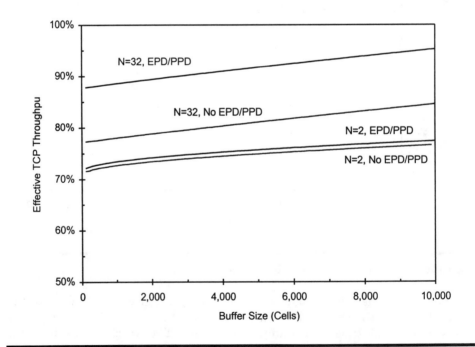

Figure 26.11   TCP Throughput versus Buffer Size for Different Numbers of Sources

26.2.5 Number of Sources Impact on TCP/IP Performance

Figure 26.12 illustrates the same example where R=150 Mbps corresponding to operation over an OC3/STS-3c, the RTD is 40 ms, and the packet size P=1500 bytes as covered in the previous section.   This figure plots the effective TCP throughput versus the number of sources N for a fixed buffer size of 10,000 cells for an ATM device with and without EPD/PPD. This plot clearly shows the benefit of EPD/PPD as the number of sources increases. This effect occurs because EPD/PPD spreads the loss evenly over the sources by discarding the largest possible portions of single packets as described in Chapter 23.   In LAN environments the performance improvement of EPD/PPD is even more marked. However, the approximations in our simple model are not accurate for small values of TCP multiplicative increase interval M typically encountered in LAN environments.

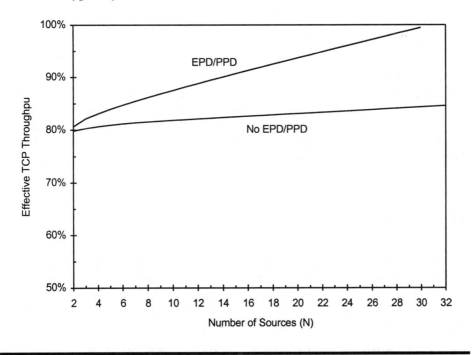

Figure 26.12   TCP Throughput versus Number of Sources N

26.3 VOICE AND DATA INTEGRATION

This section gives a time-tested model for voice transmission statistics within an individual call for use in a simple statistical multiplex gain model. Next, we use this model to estimate the savings resulting from integrating voice and data on the same transmission facility.

26.3.1 Voice Traffic Model

Most people don't speak continuously in normal conversation; natural pauses and gaps create an on-off pattern of speech activity. On average, people speak only 35 to 40 percent of the time during a typical phone call. We call this the speech activity probability p. Furthermore, the patterns of speech activity are independent from one conversation to the next. Therefore, the binomial distribution introduced in Chapter 24 is a good model for the probability that k people are speaking out of a total set of N conversations using the shared facility as follows:

$$\Pr[k \text{ out of N speakers active }] \;=\; b(N,k,p) \;=\; \binom{N}{k} \; p^k \; (1-p)^{N-k}$$

$$\text{where } \binom{N}{k} \;\equiv\; \frac{N!}{(N-k)!\,k!}.$$

The results of many studies performed since the introduction of digital telephony in the late 1950s show that most listeners don't object to a loss of approximately 0.5 percent of the received speech. Of course, the speech encoding and decoding mechanism determines the resulting quality in the event of loss — a very sensitive decoder can magnify small amounts of loss. The parameter of interest is then the fraction of speech lost, commonly called the *freezeout fraction* FF [11] defined by the following formula:

$$FF(N,C,p) \;=\; \frac{1}{Np} \; \sum_{k=C}^{L}(k-C) \; \binom{N}{k} \; p^k \; (1-p)^{N-k}$$

This expression has the interpretation equivalent to the fraction of speech lost by an individual listener. What we need in the following analysis is a function which determines the capacity required C for a given number of speakers N, each with a source activity probability p. The subsequent analysis denotes this function as C(N, p, FF).

26.3.2 Statistically Multiplexing Voice Conversations

Satellite communication and undersea cable communication of voice has long used statistical multiplexing of many voice conversations to reduce costs. Often experts refer to the devices performing this function as Digital Circuit Multiplication Equipment (DCME), since statistical multiplexing effectively packs multiple conversations into a single equivalent voice channel. ATM Adaptation Layer 2 (AAL2) will support the next generation of DCME, as well as integrated voice and data access. However, this gain occurs only a system multiplexes enough conversations together. Unfortunately, the statistical multiplexing of voice reaches a point of diminishing returns after reaching a critical mass. Let's look at an example to see why this is true.

First, the statistical multiplex gain G(N,p,FF) for N conversations with voice activity probability p and a freezeout fraction objective FF is:

Voice Stat Mux Gain: $G(N,p,FF) = \dfrac{N}{C(N,p,FF)}$,

where C(N,p,FF) is the required number of channels defined in section 26.3.1. The following example assumes that the voice activity probability p=0.35 and FF=0.5 percent for all sources. Figure 26.13 plots the results of the required capacity function C(N,p,FF) and the resulting statistical multiplex gain versus the number of sources N.   The curve is not regular because the required capacity function returns integer values. This analysis illustrates several key points. First, until the number of sources reaches 5, there is no gain. After this point, the gain increases slowly and reaches a maximum value of 2. Indeed, for 128 sources the gain increases to only a factor of 2.2.

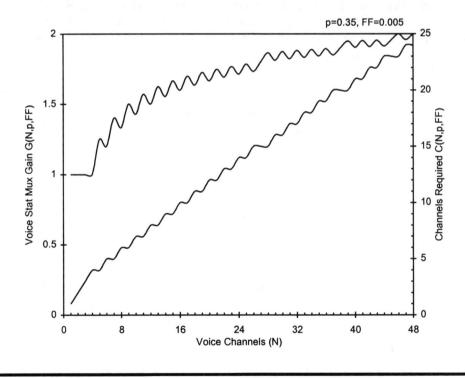

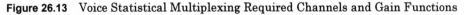

Figure 26.13   Voice Statistical Multiplexing Required Channels and Gain Functions

26.3.3 Voice/Data Integration Savings

The curious reader may now be thinking what gains remain for integrated voice and data transmission.   The answer lies in the observation that although the voice multiplexing system reserves capacity C(N,p,FF) for transmission of speech, the speech utilizes an average capacity equal to Np, the mean of the binomial distribution. Therefore, the percentage of additional data traffic carried by an integrated voice/data system when compared with separate voice and data systems is [12]:

$$\text{Data Integration Savings: } D(N,p,FF) = \frac{C(N,p,FF) - Np}{C(N,p,FF)}$$

Figure 26.14 plots the fraction of additional data traffic supportable D(N,p,FF) versus the number of voice sources N for the same voice activity probability p=0.35 and FF=0.5 percent used in the previous example. For relatively small systems, integrating voice and data makes great sense, and ATM is the only standard way to implement it. Here again, however, a point of diminishing returns occurs as the number of voice sources increases above 30, where the additional capacity available for data is approximately 25 percent. For a system with 128 voice sources, the savings decreases to 11 percent. Hence, don't expect an economy of scale when multiplexing large amounts of data and voice traffic. This result indicates that the most benefit accrues when doing integrated voice/data multiplexing on access lines ranging from nxDS0 up to nxDS1.

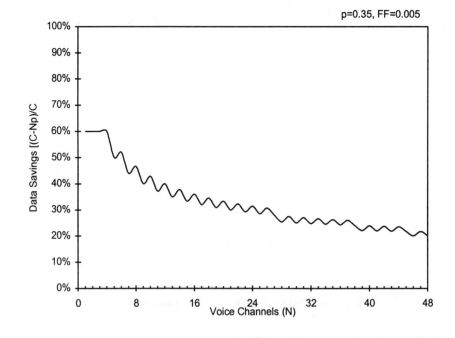

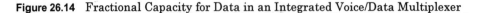

Figure 26.14   Fractional Capacity for Data in an Integrated Voice/Data Multiplexer

This means that the maximum benefits accrue when multiplexing voice and data together on access lines operating at DS1 or E1 rates. This offers the greatest benefits to medium-sized business locations because of reduced access line charges. Another benefit for carriers is better utilization of expensive transoceanic cables, satellite communications, and radiofrequency networks via integrated voice and data multiplexing.

26.4 PRIVATE VERSUS PUBLIC NETWORKING

Is a dedicated private network or a shared public network better for your suite of protocols and applications? Or would a hybrid private/public network be better? This section gives some objective criteria to assist the reader in making this sometimes difficult decision.

Usually, a key element is the overall cost. The economic considerations include not only equipment or services, but also planning, design, implementation, support, maintenance, and ongoing enhancements. Beware that these other indirect activities require the dedication of significant resources in a private network, but are often part of public network services.

The current network likely consists of mainframes, minicomputers, or LANs interconnected by bridges and routers. Their interconnections may use private lines, or an existing private or public data service. Identify if there are some applications, or concentration of traffic, at certain locations that require connectivity with other locations. Try to construct a traffic matrix of what throughput is required between major locations. You can then build a trial network design for a private network to achieve the delay and loss performance that your application requires. In a private network, it is important to estimate the capital and ongoing costs accurately. It is a common mistake to overlook, or underestimate, the planning, support, and upgrade costs of a private network. For a more detailed discussion on the topics of traffic estimation, network design and other practical considerations, see Reference 13.

If you have the site locations and estimated traffic between sites, a carrier will often be able to respond with a fixed cost and recurring cost proposal. These proposals often offer both fixed and usage pricing options. The service provider may guarantee performance of the public service, called a service level agreement, but in a private network the network planning, design, implementation and operations staff determine the resulting network quality. Public networks share expensive switches and trunks across multiple customers, achieving economies of scale difficult to attain in all but the largest private networks. Carriers often implement a shared trunk speed higher than any access line speed, consequently achieving lower delay and loss than a private network does. This occurs because of the economy of scale inherent in the statistical multiplexing gain when combining many smaller sources demonstrated in Chapter 25. Thus, carriers selling smaller portions of larger networks to multiple users, decreases the cost for most users according to a standard set of performance objectives suitable for most applications. If your application has unique performance requirements not met by any of the carrier's offerings, then you may be forced to build a private network, regardless of cost. Be confident however that your current and future application requirements justify this decision and the associated expense.

For comparable performance, the decision then becomes a matter of choosing which costs less, and assessing any risk. Of course, the least costly (and least risky) design might be a hybrid private and public network. In the very general sense, large volumes of high utilization point-to-point traffic may be better handled via a private network or circuit switching, while diverse, time-varying connectivity is usually supported better in public packet-based networks. Chapter 32 considers this tradeoff in more detail. If you are uncertain about your traffic patterns, then a usage-based billing public packet network may be the most economical choice.

If you select a public network service, then a key question is whether fixed-rate or usage-based billing. Not all carriers offer these billing options, so if your application can be more economically supported by choice of billing option, then this should be an important factor in your selection process. A fixed-rate option would charge a fixed monthly price for a certain traffic contract, which specifies a guaranteed throughput and QoS. Examples would be frame relay where the CIR frame delivery rate is guaranteed, or in ATM where the Quality of Service (QoS) is expressed in terms of cell loss and delay as defined in Chapter 21. A usage-based option would specify the guaranteed throughput and QoS, but the actual bill should be based upon the units of data that were successfully received by the destination station. Usage billing is usually subject to a minimum and maximum charging amount.

Most users who are unfamiliar with their traffic patterns choose a fixed-rate option with a guaranteed QoS level. As they become more familiar with their traffic patterns and volumes, they naturally migrate to usage rates, understanding that they still have the minimums and maximums to protect them from unplanned outages or times of excess traffic. Accurate traffic forecasts are difficult, and due to the accelerating bandwidth principle, we believe that the difficulty in future traffic forecasting is increasing. Check to see what the minimum charge is for usage-based billing, and if there is a cap on the usage-based option in relation to the fixed-rate option. If the minimum is low and the cap is reasonable, then usage-based billing may be your best choice.

If the public service provides traffic measurement capabilities, you can use them to obtain more accurate traffic data and make a better informed decision in the future. These products provide the user with detailed daily, weekly, and monthly traffic reports on a physical port and virtual connection basis — ideal for sizing your network. Be sure and check how frequently you can change the public service traffic contract and billing options when selecting a public carrier. It is also very important to understand how frequently the access options can be changed, how long it takes to change them, and what charges are associated with such changes.

26.5 NETWORK PERFORMANCE MODELING

The concepts of modeling traffic sources and buffering strategies at the single node level were introduced in Chapter 25. This is a good way to begin approaching the problem; however, most users are concerned with modeling the performance of a network. As discussed in Chapter 25, there are two basic modeling approaches: simulation and analysis. A simulation is usually much more accurate, but can become a formidable computational task when trying to simulate the performance of a large network. Analysis can be less computationally intensive, but is often inaccurate. What is the best approach? The answer is similar to nodal modeling — it depends on how much information you have and how accurate an answer you require.

Simulation models are very useful in investigating the detailed operation of an ATM system, which can lead to key insights into equipment, network, or application design. However, simulations generally take too long to execute to be used as an effective network design tool for all but the smallest networks.

A good way to bootstrap the analytical method is to simulate the detailed ATM node's performance under the expected mix of traffic inputs. Often an analytical approximation to the empirical simulation results can be developed as input to an analytical tool. An assumption often made in network modeling is that the nodes operate independently, and that the traffic mixes and splits independently and randomly [14]. Analytical models rapidly become intractable without the assumption that nodes are independent of each other.

The inputs and outputs of a network model are similar for any packet switched network design problem. The inputs are the topology, traffic, and routing. The network topology must be defined, usually as a graph with nodes and links. The characteristics of each node and link relevant to the simulation or analytical model must be described. Next the pattern of traffic offered between the nodes must be defined. For point-to-point traffic this is commonly done via a traffic matrix. The routing, or set of links that traffic follows from source to destination, must be defined.

The principal outputs are measures of performance and cost. The principal performance measures of a model are loss and delay statistics. A model will often produce an economic cost to allow the network designer to select an effective price-performance tradeoff [13].

26.6 REACTION TO EXTREME SITUATIONS

Anther important consideration in network design is the desired behavior of the network under extreme situations. The following discussion considers the extreme situations of significant failure, traffic overload, and unexpected peak traffic patterns.

A good network planner considers the impact of significant failures in advance. For example, will a critical trunk outages, or loss of an entire switch do to the network? The general guideline for reliable network design is that any single trunk or switch failure should no isolate another network element. If your private network can survive some loss of connectivity, then you can design it for a lower cost. Usually, public network designs strive for continuous operation for the occurrence of any single failure as well as many multiple failure scenarios. The exception and most vulnerable point is usually access to the public network, in both single-point-of-failure CPE or access circuits, solvable at a price through equipment and access circuit redundancy and diversity. You may have different performance objectives under failure situations than under normal circumstances. Also, you may choose to preempt some traffic during a failure scenario in order to maintain support for mission-critical traffic. A failure effectively reduces either a bandwidth or switching resource, and hence can be a cause of congestion, which we will cover next.

Traffic overloads and unexpected traffic levels cause congestion. For example, offered traffic in excess of the contract may create congestion. Under normal circumstances the network marks excessive traffic for selective cell discard, but usually carries the majority of the excess traffic. Under congestion conditions the network has a first tier of lower priority traffic load to shed. However, if the network oversubscription practice is too aggressive, then selective cell discard may not be enough. As pointed out earlier, congestion can drive a switch or multiple points in a network into overload, reducing overall throughput significantly if congestion collapse occurs. If you expect this situation to occur, then a mechanism to detect congestion, correlate it with its cause, and provide some feedback in order to isolate various traffic sources and achieve the required measure of fairness is desirable. Look for ATM Available Bit Rate (ABR), in particular the Virtual Source/Virtual Destination (VS/VD) variety to handle these types of problems. If you are uncertain as to how long such overloads will persist, then some slow-reacting feedback controls may actually reduce throughput because the reaction may occur after congestion has already abated (such as we saw with higher-layer protocols such as TCP earlier in this section).

In order to support time-varying traffic patterns, two fundamentally different approaches exist: one based upon the telecommunications concept of reserving bandwidth, and the other the data communications concept of fairly shared resources. An ATM Switched Virtual Connection (SVC) capability allows an application to request the network to reserve bandwidth for its exclusive use, similar to circuit switching. In order to use the current SVC protocols, the application must accurately estimate the required bandwidth since it can only be changed by tearing down the call and setting it up again. Future SVC protocols will support dynamically negotiated bandwidth without interrupting the call. Both of these scenarios will require the user to estimate the bandwidth requirement and communicate this to the network.

26.7 TRAFFIC ENGINEERING COMPLEXITY

Realistic source and switch traffic models are not currently amenable to direct analysis, with the results presented in this book providing only approximations under certain circumstances. Beware that these approximate methods may have large inaccuracies, a fact ascertainable only through detailed simulations or actual testing. Simulations are time consuming, and cannot effectively model low cell loss rates since statistics require modeling of an inordinate number of events. For example, to simulate a cell loss rate of one in a billion, a simulation must model at least a trillion cells! Even with today's computers this is a lot of computation to obtain just a single point on a loss curve. Whenever possible, use methods for extrapolating loss rates and estimating the occurrence of unlikely events instead. Also, constantly changing source, switch, and network characteristics creates a moving target for such traffic engineering models.

Traffic engineering has a real-world impact on the Connection Admission Control (CAC) function's complexity. Recall that CAC admits a connection request only if the QoS of existing connections remains within specification after accepting the new request. Implicit in this function is the ability to calculate (or at least estimate) the QoS parameters for a large number of sources, each potentially with different traffic parameters and service categories. Add to this the need to execute CAC in real time to support switching of SVC requests, and it becomes apparent that accurate and complete traffic engineering becomes a key step in the design process.

26.8 REVIEW

This chapter covered several key considerations in the design of networks and switches. It began by reviewing the impact of delay, loss, and delay variation on applications; giving guidelines for the selection of link speed, burst duration, and loss ratio. The text covered the effect of loss on effective throughput for Go-Back-N and selective retransmission. The discussion also showed how Cell Delay Variation (CDV) accumulates nonlinearly across a network of ATM switches. Next, the coverage provided an in-depth look at the factors that affect performance of the Internet's popular Transmission Control Protocol (TCP) over ATM. The topics covered were; effect of ATM service category, impact of buffer size, and performance as a function of the number of network users. Next, the chapter covered the subject of voice statistical multiplexing gain, along with an assessment of the savings achievable in an integrated voice/data network. The chapter then moved on to review some of the key factors involved in the choice between building a private network, purchasing a public network service, or designing a hybrid of the two. The tradeoffs and critical factors include cost, risk, and perform-ance. Part of the network design should consider what will happen under extremes, such as failures, overload, and unexpected peak traffic patterns.

26.9 REFERENCES

[1]   L. Kleinrock, "The Latency/Bandwidth Tradeoff in Gigabit Networks," *IEEE Communications Magazine*, April 1992.

[2]   V. Jacobson, R. Braden, "TCP Extensions for Long-Delay Paths," RFC 1072, October, 1988.

[3]   D. McDysan, "Critical Review of ARPANET, IBM SNA and DEC DNA," George Washington University, April 1985.

[4]   M. Moussavi, "Performance of Link Level Protocols over Satellite," *MELECON '89*, Lisbon, Portugal, April 1989.

[5]   ATM Forum, *ATM Forum Traffic Management Specification, Version 4.0*, af-tm-0056.000, April 1996.

[6]   V. Jacobson,R. Braden,D. Borman, *TCP Extensions for High Performance*, IETF, RFC 1323, May 1992.

[7]   C. Pazos, M. Gerla, V. Signore, "Comparing ATM Controls for TCP Sources," *IEEE*, 1995.

[8]   H. Li, , K-Y Siu, H-Y Tzeng, "A Simulation Study of TCP Performance in ATM Networks with ABR and UBR Services," *IEEE*, 1996.

[9]   L. Boxer, "A Carrier Perspective on ATM," ATM Year 2 Conference, April 1994.

[10]   A. Romanow, "Dynamics of TCP Traffic over ATM Networks," IEEE JSAC, May 1995.

[11]   B. Jabbari, D. McDysan, "Performance of Demand Assignment TDMA and Multicarrier TDMA Satellite Networks," IEEE JSAC, February 1992.

[12]   D. McDysan, "Performance Analysis of Queuing System Models for Resource Allocation in Distributed Computer Networks," *D.Sc. Dissertation*, George Washington University, 1989.

[13]   D. Spohn, *Data Network Design, 2d ed*, McGraw-Hill, 1997.

[14]   L. Kleinrock, *Queuing Systems Volume I: Theory*, Wiley, 1975.

Operations, Network Management, and ATM Layer Management

Now that we've defined the details of ATM networking, this book now takes on the challenge of how to manage ATM networks. Towards this end, Part 7 provides the reader with an overview of operations, network management architectures, network management protocols, object-oriented databases, ATM layer management, and ATM performance measurement. First, Chapter 27 defines the philosophy of Operations, Administration, Maintenance, and Provisioning (OAM&P) to set the stage. Next, the text discusses the state of the solution to the network management problem. The coverage continues with a presentation of network management architectures defined by the standards bodies and industry forums. Chapter 28 then covers the network management protocols developed by standards bodies and industry forums to solve the network management problem. This includes the IETF's Simple Network Management Protocol (SNMP), and the ITU-T's Common Management Interface Protocol (CMIP). We then give a summary of key Management Information Bases (MIBs) defined in support of ATM networks. Finally, Chapter 29 addresses the topics of ATM layer management and performance measurement. The text defines ATM layer Operations And Maintenance (OAM) cell flows and formats. The text first covers fault management, which is the basic determination of whether the ATM service is operating correctly. The chapter concludes with a summary on the use of performance measurement procedures to confirm that the network indeed delivers the specified ATM layer Quality of Service (QoS) objectives.

Operational Philosophy and Network Management Architectures

This chapter covers the important topic of operational philosophy and network management architectures. Starting things off, the text first discusses basic Operations, Administration, Maintenance, and Provisioning (OAM&P) philosophy. We identify generic functions that apply to almost any type of communication network, highlighting advantages that ATM offers. The text summarizes popular network management architectures defined by a number of standards bodies and industry forums. The text covers differences in approach, scope, and protocol usage. The next chapter describes the network management protocols and databases employed by these architectures.

27.1 OAM&P PHILOSOPHY

Network management is about achieving quality. If your network requires quality; then the expense and complexity of comprehensive network management technology is well justified. In order to set the overall stage and context for this part of the book, this section gives a brief definition of each element of Operations, Administration, Maintenance, and Provisioning (OAM&P) and describes how they interrelate as depicted in the flow diagram of Figure 27.1. Each of the major functional blocks performs the following functions.

* Operation involves the day-to-day, and often minute-to-minute, care and feeding of the data network in order to ensure that it is fulfilling its designed purpose.

* Administration involves the set of activities involved with designing the network, processing orders, assigning addresses, tracking usage, and accounting.

* Maintenance involves the inevitable circumstances that arise when everything does not work as planned, or it is necessary to diagnose what went wrong and repair it.

* Provisioning involves installing equipment, setting parameters, verifying that the service is operational, and also de-installation.

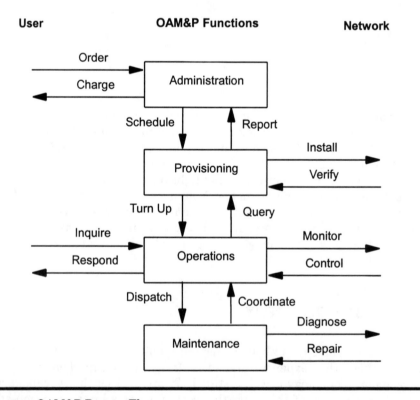

Figure 27.1   OAM&P Process Flow

Despite the order of the acronym, OAM&P, these functions are better understood from the top-down work flow described above. This sequence basically models the life cycle of an element in a network. First, a network planner administratively creates the need to augment or change the network. After the hardware and/or software is purchased, tested and deployed, then operators must provision it. Once in service, operations continuously monitors and controls the network element, dispatching maintenance for diagnosis or repair as required.

27.1.1 Administration

Administration of a network involves people performing planning for the network. This work includes determining the introduction of new elements to grow the network, add feature/function, and the removal of obsolete hardware and software elements. In this context a managed element is either an individual instance of hardware (such as a switch or an interface card), connectivity via underlying transport networks, or a logical design elements (such as address plans or key network operation performance objectives). Administration includes some key management functions and some business functions. First, a network must be designed, either as a private, public, or hybrid network (public and hybrid networks use a service provider for interconnecting remote ATM switches.) Once designed, the elements must then be ordered, along with scheduling installation and associated support. The enterprise must develop an administrative plan for staging the provisioning, operations, and maintenance activities. This often involves automated system support for order entry, order processing, work order management, and trouble ticketing. Orders must be accepted from users and the provisioning process initiated. Address assignments are made where needed.

Once the service is installed, usage data must be collected for traffic analysis and accounting. If departmental charge backs or customer charges are important, an accounting and billing system must have access to the usage data. Based upon forecasts, business requirements, and traffic analysis, changes to the network are required. Network planning frequently makes use of automated tools in the process of administration.

27.1.2 Provisioning

Provisioning of a network involves the day to day activities which actually introduce physical and logical changes in the network to implement the administratively planned growth or change in the network. Ideally, provisioning activity follows set rules and procedures developed in the design phase of a network, or else learned and refined from past provisioning experience. Provisioning functions control commissioning and decommissioning of physical elements as well as the configuration of logical elements. Physical actions associated with provisioning include installation of new or upgraded equipment which may also include updating vendor switch software. Hardware changes require on-site support, while, with proper infrastructure support, software upgrades may be done remotely. Access line or trunk installation and turn-up is also part of provisioning. A key part of processing orders is the establishment of ATM-related service parameters. While this may be done manually for small volumes of orders, automation becomes cost effective when processing a large volume of provisioning work. Once the hardware and software parameters are in place, the final step of the provisioning process ensures that the service performs according to the objectives prior to releasing service to the end user. Verifying performance often involves performing tests in conjunction with operations, such as circuit bit error rate testing, ATM loop back cell testing, or throughput measurements. (Which tests are critical depend on the ATM-related service parameters which were set and the transport network in use.)

27.1.3 Operations

Operating a network involves monitoring the moment to moment fluctuations in the performance of the network and deciding which events require intervention to bring the network into compliance with an ideal performance goal set by designers. Operations provides an organization with a collective memory of the past performance of the network. Included in that memory are activities which identified past service breakdown and the activities which corrected the problem. Some of these corrective actions come from knowledge of ATM and the network design, yet in the real world, some come from practical experience.

Monitoring the network involves watching for faults and invoking corrective commands and/or maintenance actions to repair them. It also involves comparing measured performance against objectives and taking corrective action and/or invoking maintenance. Invoking corrective actions involves operators issuing controls to correct a fault or performance problem, or resolving a customer complaint. A key operational function involves assisting users to resolve troubles and effectively utilizing the ATM network capabilities. Operations coordinates actions between administration, maintenance, and provisioning throughout all phases of the ATM connection's life.

27.1.4 Maintenance

Maintaining a network involves many instances of unplanned changes. Maintenance actions involve changes not instigated via the administrative design or service provisioning process. Examples of maintenance actions are: changing interface cards, replacing failed common equipment, or troubleshooting physical circuit problems. Once operations identifies a problem, it works with maintenance engineers to isolate and diagnose the cause(s) of the problem. Maintenance engineers apply fixes to identified problems in a manner coordinated by operations. Control by operations is critical, since in complex networks like ATM, corrective actions applied incorrectly may result in additional problems. Operations coordination with Maintenance can involve dispatching service personal to the site of the fault, arranging parts delivery, or coordinating repair activities with service suppliers. Besides responding to problem events, an important maintenance activity is performing periodic, routine maintenance so that faults and performance degradations are less likely to occur. Routine maintenance may involve automated test cycles supplemented with application of pre-planned inspections and cyclical service of physical elements.

27.1.5 Unique Challenges Created by ATM

Standards are still being developed in many of these areas, and in a sense they are never done because of the rapid pace of technological change. Usually, standardization of network management occurs later in the technology life cycle, well after the application is introduced. It is commonly believed that only after you have built the network, determined what can go wrong, and discovered what is needed to make it work can you finalize how to operate, administer, maintain, and provision it. However, good planning can provide these OAM&P functions in a much more productive manner soon after the introduction of technology. While there is no substitute for experience, network management is essentially the application of control functions (that is, regulation and feedback) to a system. Therefore, such planning should be a part of the original network design, just as measurements and controls are essential in any engineering endeavor. Experience becomes a feedback mechanism to refine the initial network management and operations designs. The real reasons that network management standards follow after the ATM application standards is historical inertia, "that's the way it has always been done," and the commonly encountered, but shortsighted, business prioritization that getting the network up is more important than controlling its behavior. This thinking has delayed the introduction of some critical ATM OAM functions described in Chapter 29. Truthfully, these OAM standards came after the first wave of detailed ATM networking standards, but now a solid international standard exists for ATM layer management and performance measurement.

In several senses the OAM&P of ATM is similar to that of other private data networking and public data service capabilities. Many of the management functions required for management of an ATM network are identical to a data- or circuit-based network. Yet, ATM network management is complicated by the fact that ATM/B-ISDN is targeted to support older legacy systems as well as to support significant new functions and applications. Also, since ATM changes the underlying, lowest layers of communications through its new paradigm, many tried and true troubleshooting procedures developed for the circuit-oriented paradigm no longer work, or at least require substantial revision. In short, a larger scope of application technology provides a more complex problem for ATM network management to solve.

With ATM the multiplicative factor of logical, or virtual, channels possible on each physical interface is orders of magnitude greater than that of many current data services and more complex than the management of many voice services as well. Simultaneously, ATM includes all of the complexities involved in LAN, MAN, and WAN services along with currently undefined future extendibility. Also, ATM adds some new types of capabilities, such as point-to-multipoint connections. Furthermore, multiservice interworking aspects may require backward compatibility with all of the existing communications networks, and should provide at least the same level of network management, automated provisioning, and billing as done for existing LAN, MAN, and WAN services.

There may be a great advantage to the implementers of ATM in its support of all of these service types [1]. If done well, ATM-based Network Management Systems (NMS) may provide a seamless level of network management — the goal in the interconnection of LAN, MAN, and WAN user interfaces. Great effort has been applied to the standardization of Network Management interfaces (e.g., SNMP or CMIP) to Network Elements (NEs) which, hopefully, will empower management by providing a great deal of visibility and control.

The new features of ATM switching and multiplex equipment will require either enhancements to existing network management systems and/or new systems. The additional protocol parameters of AALs and higher layers will also require management. Monitoring every virtual connection using either OAM cells and/or retrieved counter measurements would result in tremendous volumes of network management data. Sampled or hierarchical data reduction and analysis must be done to reduce this data volume.

The direct inclusion of multiprotocol access cards, which also act to terminate transmission elements, into ATM switches extends the network management domain beyond just ATM. As described in Chapter 16, many ATM-based devices support frame relay, circuit emulation, voice, LAN, and HDLC data ports. Add to this the complexities of behavior introduced between ATM and its physical layer by (sometimes) competing automated restoration and recovery schemes. Together these require a prudent operations unit maintain an integrated network management view of ATM and its supporting physical layer transmission infrastructure. The provision of other protocols carried over ATM, and the peculiarities of behavior sometimes introduced by this, require an integrated view between ATM and the higher services which depend on it. These network interactions place complexity burdens on management application design and may require interfaces between ATM management applications and those managing the physical layer transmission infrastructure. However, when successfully incorporated in the operations view, these integrated management applications provide substantial advances over the uncoordinated monitoring of, for instance, IP over ATM over SONET, all as isolated networks. Standardization efforts have been addressing this problem domain, but are not likely to provide consensus solutions for a few years. Some encouraging standards attempting to solve this are the TINA-C [2] work documented in the ITU-T Recommendations G.803 [3] and G.805 [4]. The prudent OSS deployment will look for practical integration efforts between layers from the OSS designers.

27.1.6 Centralized versus Distributed Network Management?

Returning to the specifics of ATM management, the reader should now understand that when designing an ATM network, it is important to consider how the network management systems impact operational philosophy. For ATM, a key decision is whether to adopt a centralized or distributed Network Management System (NMS) architecture for managing a network of ATM Network Elements (NEs) [5]; Figure 27.2 depicts these two design extremes. The network management subcommittee of the ATM Forum recognized these two possible implementation models and incorporated support for both approaches in their work. Therefore, developers of ATM Forum compliant ATM network management systems provide for management via one or the other of these approaches. Realistically, a network can be managed by a combination of centralized and distributed management connectivity; in this case the management OSS might support local element managers reporting up to a central manager.

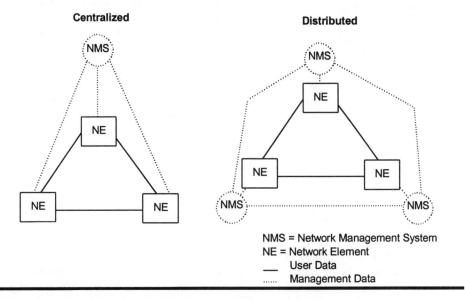

Figure 27.2 Centralized and Distributed OAM&P Architectures

Some will opt for a centralized approach with the expertise concentrated at one location (possibly with a backup site for disaster recovery) with remote site support for only the basic physical actions, such as installing the equipment, making physical connections to interfaces, and replacing cards. In this approach, the software updates, configurations, and troubleshooting can be done by experts at the central site. The centralized approach requires that the requisite network management functions defined in the previous section are well developed, highly available, and effective. Furthermore, a centralized design requires connections from every device back to the centralized management system. Non-real-time functions are often best done in a centralized manner.

Others may have a few management sites. For instance, when riding the "bleeding-edge" of technology, they may want to have expertise at every site. Also, this approach may be required if the network management system is not sophisticated, or the equipment has a number of actions that can only be done at the site. In some cases, the volume of information collected from the network elements drives users towards a distributed processing design. Finally, some lower-level, well-defined, automated functions are best performed in a distributed manner. Real-time functions are often best done in a distributed manner.

There is a performance tradeoff between the centralized and distributed architectures. Transferring large volumes of management data to a central site often requires a large centralized processor. It is all too easy to get in the situation of having too much data and not enough information. On the other hand, defining the interfaces and protocols for a distributed system is complex. One study [6] compares centralized versus distributed Connection Admission Control (CAC). CAC depends upon accurate, timely information about the connections in progress in order to admit, or block, connection attempts. If the information is not current, as it would be in a centralized system, then the throughput carried by the network is less than that of a totally distributed system.

Choosing to adopt either a centralized or distributed Network Management System (NMS) architecture is only one example of the many different and sometimes conflicting design choices that are faced when implementing management of ATM. Choosing network management systems frequently forces the choice of an entire operational philosophy. So, how can you make a cost-effective, technically sound choice?

27.2 THE PROBLEM AND THE STATE OF THE SOLUTION

Since you are reading this book, we assume you either have (or plan to have) an ATM network. Prudence and experience require that you provide for operations and management of that network in order to *keep it functioning* at the desired performance level of your users as well as to *protect your investment*. This means that you must either establish or, via a providers managed service, have access to an organization to carrying out the OAM&P functions described above. Clearly define the charter and interactions of the groups responsible for your network. Operations experts define these via organizational rules, operations plans, and work flows.

27.2.1 Operational Support Systems (OSS)

If you will be running your own network, plan to either purchase or develop Operational Support Systems' (OSS) tools. These tools include vendor-provided or third-party software applications, and the server platforms which host them. These are the tools used by OAM&P groups to perform their tasks. These systems capture requests for new or changed service, aid operators in the configuration of network elements, and implement provisioning changes. An important subgroup of these are the traditional network management systems which monitor the network for fault events and display outages of specific network equipment. Historical examples of vendor products providing this are Sun Microsystems' *SunNetManager* and HP's *Openview Network Node Manager*. Additionally, OSS tools often include the systems which collect statistics and generate performance reports. Included also are the systems which provide control of automated external test equipment.

As you deploy an ATM network, you will find yourself in some combination of a developer, procurer and user of these OSS applications. Under the surface of all OSS are assumptions of how a network should be managed. Network management experts call documentation of these assumptions an *operational model*. When you buy an OSS, it comes with a "built-in" paradigm, that is, an operational model, for how to manage a network. For instance, deploying an application from a vendor which monitors for alarms and displays elements sending alarms as red icons, implies an operations model that alarms equal network faults which require corrective actions in order to fix the network and eventually turn the icon green. However, when you get down to it, there are many different ways to view network management and different applications based upon different operational models.

Much of the body of work of network management literature and network management standards express the evolution of these operational models. As a responsible user, as a developer, and as a procurer of OSS, you must come to grips with this body of experience, these shifting visions of what constitutes a network and how to monitor and control it. Unfortunately, there are many, often conflicting network management models. But together these provide intellectual resources for building network control and feedback systems to meet your networking needs.

Different groups developed these models over the years, documenting their recommendations in published articles, white papers, standards, and applications targeting management of ATM-based networks. None of these resources depicted in Figure 27.3 are completely correct or applicable in every network. There is no ideal way to manage a network for a variety of reasons. Beware of solutions that profess to solve every problem. On the brighter side, together these resources contain the sum of today's knowledge regarding modeling and managing complex networks like ATM. Understanding them and selecting what is right for your network is a matter of looking at how they view what makes up a network (an information model); how a network changes over time (a behavioral model); and what to do to correct problems (an operational model).

Likely, you will have many of these resources in use in different ways, looking at different parts of the ATM management problem. Understanding these different resources helps good designers pick and choose the best solutions to their specific network management problem. It boils down to this: while there is one ATM networking standard, there are several competing ATM management models and standards.

Paradoxically, a great body of information and experience exists to aid in the development of operational models, OSS systems, OAM&P organizations and work flows. The paradox arises because this information is frequently overlapping, contradictory, and unsubstantiated via actual experience in application to ATM. Generally, this collective experience comes from standards groups. Standards come from a process in which experienced engineers publish approaches to network management (and also all the other OSS application types) and use these to build "interface specifications." First, ideas are debated by engineers from many different organizations, on mail lists and in meetings conducted by standards bodies. This material is gathered into documents, reviewed by a technical leadership, and voted upon by voting members of the standards organization. There are many of these organizations with competing and overlapping jurisdictions. This contributes to a muddled scope and to overlapping functions in the various network management standards studied later in this chapter.

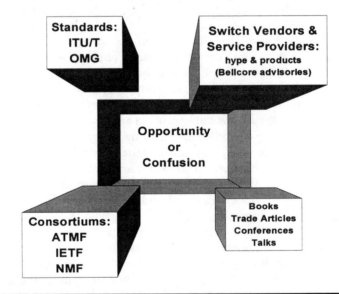

Figure 27.3   Resources for Network Management

There are international standards organization such as the International Telecommunications Union (ITU) and domestic standards organization such as American National Standards Institute (ANSI). Other sources widely followed include the consortiums of business or engineering associations. There are consortiums of businesses incorporated to develop interoperable products such as the ATM Forum (ATMF) and the Network Management Forum (NMF). Sometimes it is a consortium of independent engineers such as the Internet Engineering Task Force (IETF). Much of the specific understanding of ATM network management comes from the directed design of network management standards by these consortia. Notable among these are the network management subcommittee of the ATM Forum Technical Committee and the AToM work group of the IETF.

However, there also is a great body of collective experience and knowledge bound inside vendor applications, the manuals for these applications and in vendor white papers. Sometimes the publications of a specific vendor are followed by a wide range of companies, thus becoming important templates for OSS. *Bellcore Advisories* hold this status. Last, there is a developing body of experience available through books and in articles from industry journals.

At the root of all these operational models is a fundamental representation that

- there are identifiable and classifiable elements which provide services
- these elements are separated from other elements which provided similar or complementary services
- between these elements, which must function together, there exists an interface.

This underlying model is common to most descriptions of the network; this becomes the building blocks of OSS and are the commonly held view of reality concerning network design and data communications.

Most management standards are involved in first describing and classifying the elements of a network. This is what Management Information Bases (MIBs) attempt to do. Some of the older standards stop there. Newer standards also try to formally describe the interfaces between these elements. Emerging standards model the actions and resultant change of the network over time; these are the behaviorist modelers.

Of special importance to network management is the interface between the OSS network management systems and the network elements, and sometimes, when designing this interface, network management designers seem to place more importance on the protocol used to transport commands and information over this interface than on what they are trying to manage. In fact this interface between the OSS and the network has been the source of ideological disagreements (in the past between management interface protocols like SNMP versus CMIP, and now between manager-agent models like TMN and object interfaces like JMAPI. These ideological disagreements often confuse and seldom aid in achieving the goal of management: monitoring and fixing networks.

27.2.2 Open versus Proprietary Interfaces

At many points every developer of an OSS makes decisions to incorporate open or proprietary interfaces. Open interfaces are those which employ either a published standard, or a *de facto* standard. Vendors invent proprietary interfaces for specific OSS applications that are frequently closed, thereby preventing interworking with any other Network Management System (NMS). The normal argument is that open interfaces lead to interoperable implementations and that proprietary interfaces prevent interoperability. Often this is shortened to open interfaces are good; and proprietary interfaces are bad. This is not always completely accurate as covered in the next chapter in the section on proprietary vendor management protocols and interfaces. Nevertheless, the majority of the public-domain resources describe open interfaces. *Indeed, standards are all about defining open interfaces.*

Open interfaces aim to level the playing field for developers by enabling anyone to build an interoperable application by conforming to the standards. In practice, this goal is difficult to achieve. Consider, for example the relative success of the OSI reference model and the IP suite of user plane data communications protocols. These lessons teach us that achieving interoperability is not easy and requires years of concerted effort. The reasons for failure are many. Sometimes the standards is too complex and burdensome and discourages developers. Many manufacturers levy this complaint against the ITU-T's CMIP protocol described in the next chapter. Sometimes the specifications omit some key information, or fail to define a crucial element with insufficient rigor resulting in implementations with incompatible behaviors. Hence, the definition of network management standards must be an iterative process. Developers must talk to each other in open standards bodies, learn from past mistakes and eventually achieve interoperable implementations.

27.2.3 Common Object Request Broker Architecture (CORBA)

The next section covers those groups (i.e., the ATM Forum, the ITU-T and the IETF) directly working on ATM standards. However, the work from other groups will likely impact ATM management OSS architectures. One such group is the Object Management Group (OMG). The OMG integration framework is based on the Common Object Request Broker Architecture (CORBA) and standard interfaces defined in OMG Interface Definition Language (IDL). IDL is a language designed for the precise specification of interfaces. CORBA, the usual short hand reference for the entire work of the OMG, defines a distributed, heterogeneous, multi-vendor, multi-platform computing environment. CORBA can be accessed from almost any programming language, operating system, and hardware platform by the implementation of an Object Request Broker (ORB) and object class libraries in the host environment. The protocol has flexible, comprehensive security. OMG specifications extend interoperability into the application itself through components of the Object Management Architecture (OMA). The initial foundation architecture of basic interfaces and services for object life cycle, transactions and security is now being extended with standard objects in telecommunications.

There is growing acceptance of CORBA for use in TMN. CORBA has been embraced by the Network Management Forum (NMF) for use in business and service management layers of TMN; while the ITU-T Study Groups 4 and 15 are working on CORBA as a target technology for TMN. Common infrastructure and standard components provides a benefit to application designers, builders, and users. However, the performance and scalability of past implementation trials of CORBA has been disappointing until recently. New commercial offerings which are release two compliant versions of

CORBA, coupled with current Object Oriented databases now achieve high transaction rates. Therefore, CORBA applications may show up in new OSS developments.

27.2.4 Too Many Good Theories, No Easy Choices

There is not one vision, one architecture for ATM management. Instead there are many fine but overlapping and often inconsistent descriptions of how to manage the network as shown in Figure 27.4. The next section looks at the IETF and the ATMF standards, which provide the most solid management architecture available for ATM today. However, these designs derive from a heritage of many views about network management. Some of these contrasting views include:

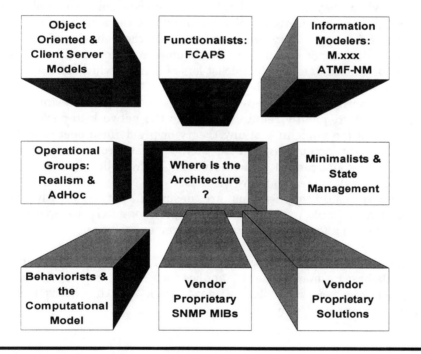

Figure 27.4 Management Views

- CORBA (and to a lesser extent TMN GDMO) provides an object oriented view of how to describe elements and their behavior. CORBA seeks to model reality by observing the behavior of existing systems, TMN GDMO attempts to describe an ideal management architecture and specifies that the network should be designed to support this specification.
- Information modelers go to great pains to describe what makes up a network. Behaviorists claim this is not enough and that the actions of a network must be captured in what they call a computational model.

Work on this is ongoing in the NMF SMART teams and in ITU-T Study Group 4 Question 18.

- OSI FCAPS (described in section 27.3.1) provides a good functional breakdown of management activities. However, the NMF has published a much more comprehensive business model, the Service Management Business Process Model [7], which is more consistent with activities of service providers. Nevertheless this NMF model may be over developed for the operator of a private ATM network.

- Minimalists say a network device should have the least management burden consistent with providing its connection and transmission activities. Device vendors active in the IETF espouse this viewpoint. State management requires that devices be smart enough to determine, on their own, if they are adequately providing their proscribed operation and, when they are not, should communicate this, and only this, to management applications. Service providers request often request this level of function. Between these two falls the behavior of most of the deployed managed communications equipment. Most devices provide lots of unorganized data and broadcast lots of alarms, which OSS applications must then filter and correlate.

- A top priority for any operation group is rapid service restoral. Hence, operations frequently requires access to the network to perform actions outside of the guidelines of any theory or pre-defined operational model. They might accept any *adhoc* method that provides rapid realistic monitoring and control of the network. To front-line operations groups, results speak louder than any theories.

- Standards-based applications are desirable, but frequently the only solution available is either a vendor's private, enterprise specific SNMP MIB, or else a totally proprietary management solution.

Management of ATM networks today is a complex, evolving area. Standards-based management applications are not yet available. Nevertheless, there is a lot of good work on the horizon awaiting standardization and deployment.

27.3 NETWORK MANAGEMENT ARCHITECTURES

This section summarizes the work of various standards bodies and industry forums defining network management architectures, functional models, and protocols. The following covers work done by OSI, NMF, ITU, ATM Forum, and IETF

27.3.1 OSI Network Management Functional Model

OSI defined the following five generic functional areas for network management, commonly abbreviated as "FCAPS" [8]:

- Fault Management
- Configuration Management
- Accounting Management
- Performance Management
- Security Management

The current state of ATM OAM-based management (covered in Chapter 29) primarily covers fault management and performance management. Some aspects of configuration management, connection provisioning, fault management and performance management are covered by ATM Forum Specifications, the ATM Forum Integrated Local Management Interface (ILMI), and IETF defined Management Information Bases (MIBs) as described in Chapter 28. Unfortunately, a great deal of work still remains in the standardization of performance and accounting management. Work in fault determination is the most mature, and meets the minimum requirements for initial ATM deployment. Application of the TMN-based standards from the ATM Forum specification should allow for provisioning and for operations management. SNMP specifications have fallen behind ATM application specifications as this work is currently backlogged in the IETF. Performance management has hopeful possibilities via the ATMF RMON-based test MIB, via the IETF AToM History MIB, and via ITU-T Recommendation I.610. Accounting and billing specifications are in progress in the IETF AToM work group and in the NMF SMART Billing program but vendors have not widely committed to support these.

27.3.2 Network Management Forum (NMF)

The Network Management Forum (NMF) extended the OSI FCAPS model via an extensive requirements gathering process from service providers. This culminated in the publication of a new functional architecture based on these requirements. The NMF calls this the Service Management Business Process Model [7]. It breaks management the management model down into

- Customer Interface management process,
- Customer-care processes (sales, order handling, problem handling, performance reporting, invoicing/collections),
- Service and product development,
- Operations processes (service planning/development, service configuration, service problem resolution, service quality management, rating and discounting),
- Network and systems management processes (network planning/development, network systems administration, installation/maintenance, monitoring/problem resolution, data collection).

The model also formally maps interactions between the above sub-components. This functional model corresponds well to the systems and the organizational structure of service providers. Based on the business process model, the NMF launched a new interoperability specifications program primarily aimed at service providers called Service Management Automation and Reengineering Teams (SMART).

This consortium has made progress in describing the business requirements of service providers for ordering, provisioning, reporting, and trouble management. These provide a clear set of requirements and an organizational structure so that consistency and completeness can be reached when designing and deploying systems [5, 7]. Of direct importance to ATM, the SMART billing team is working on a specification on how to get the massive amounts of accounting data, generated by ATM usage billing, out of ATM switches and into service provider management systems. Until robust usage based accounting can be integrated between switches and service provider OSS applications, customers may not gain the full economic benefit for the multi-service, statistical bandwidth advantages of ATM delivered by usage-based billing.

SMART is also seeking to define trouble-ticket exchange between service providers as well as between providers and customers. This activity plans to extend the ITU-T X.790 trouble ticket model for private line circuits to also include ATM. For the users who contract with ATM service providers for ATM interconnection in the WAN, the trouble ticket management interface will be just as important as the management interfaces for their local switches.

27.3.3 ITU Telecommunications Management Network (TMN)

Figure 27.5 depicts the ITU-T Recommendation M.3010's layered model for the Telecommunications Management Network (TMN) operations functions [9]. A subsequent section defines the interfaces between the layers labeled Q3, and interfaces between the layers and their peers labeled X. This model abstracts lower-level details further up the hierarchy, enabling effective service and resource management.

Starting at the bottom of Figure 27.5, physical network elements are devices, such as ATM switches, LAN bridges, routers or workstations. The element management layer manages network elements either individually, or in groups, to develop an abstraction of the network element functions to higher layer operations functions. Many vendors provide proprietary element managers that control an entire network of their devices. The network management layer addresses functions required across an entire geographic or administrative domain. This layer also addresses network performance by controlling network capabilities and capacity to deliver the required quality of service. The service management layer is responsible for the contractual aspects of services provided to customers by carriers. This includes statistical data reporting, status on interfaces with other carriers, and interactions with other services. The scope of the business management layer is the entire enterprise, encompassing proprietary functions. Since this layer performs proprietary functions, it usually does not provide the standard X interface to a peer NMS layer. Please note that the layers in this model represent functional components, not physical systems.

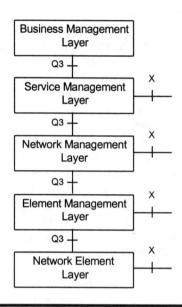

Figure 27.5   The Multi-Layered Reference Architecture for Operations

Of importance to ATM is the strong, current interest among service providers in implementing ATM switch management via the Q3 interface at the interfaces between the network element, element management, and/or network management layers. In a nutshell, the Q3 interface is TMN architecture (ITU-T M.3010), using OSI management (ITU-T X.701). Practically, this becomes the use of GDMO (Guidelines for the Definition of Managed Objects, ITU-T X.722) notation for objects derived from the M.3100 information model, providing CMIS (Common Management Information Service, ITU-T X.710) management services, communicated via CMIP (Common Management Information Protocol, ITU-T X.711) protocol, carried over ROSE (ITU-T X.219 & ITU-T X.229). For ATM, the Q3 interface also generally means the implementation of ATMF M4 (described below) information objects and management operations over this interface. Service providers are investing considerable resources in this approach.

Recommendation M.3010 indicates that other models are valid, so that systems without all of these layers, or systems with different layers, are also acceptable. In fact, at NOMS98 (IEEE Network Operations Management Systems) conference, the ITU secretariat went on record as observing that CORBA was compliant with Recommendation 3010 and therefore it could be used in TMN and the ITU would begin considering it for equal footage with CMIP.

Initial implementations of TMN stressed the Q3 interface. Demand for the X interface was ignored until recently, but now it is skyrocketing. Recent trends in international alliances of service providers have generated the need for a functional interface from service layer accounting systems to matching systems in partner service providers, which can provide 'settlement' of revenue flows by balancing cross customer billing with cross provider leasing of extra-network facilities. Finally, the NMF is actively pursuing a potential specification of the ordering function at the service management to service management interface which would allow carriers to order termination facilities in another providers network. Specifications like PNNI and B-ICI which allow networks to connect have driven similar interface definition needs in the management applications.

Figure 27.6 illustrates several possible implementations of the above reference architecture, showing how the lowest three logical layers may map to physical systems. Figure 27.6a shows separate systems implementing each layer, where element management is performed on a one-for-one basis for each network element. This design could use computers at the element level to convert from a proprietary network element interface to a standard interface, which are then managed by a standard Network Management System (NMS). Figure 27.6b illustrates a system which integrates the network and element level management into a single overall management system. Proprietary vendor management systems often implement this architecture. Figure 27.6c illustrates a system where network management intelligence and standard interfaces are distributed to each network element. Switch vendors who implement all MIB standards and provide open access to their proprietary MIBs follow this model. Finally, Figure 27.6d illustrates a hierarchical system where element management systems manage groups of Network Elements (NEs) and then feed these up into an NMS that manages an entire network. Sometimes this the processing requirements of larger networks dictate this hierarchical structure. The TINA-C subnetwork management architecture spoken of later in this chapter uses this architecture.

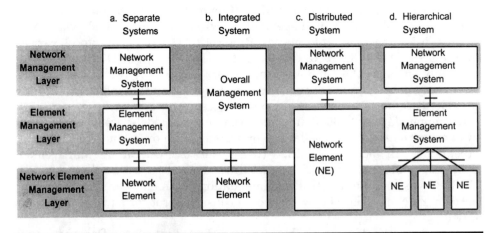

Figure 27.6 Physical Realization Examples of the Multi-Layered Model

Figure 27.7 depicts the ITU-T's vision for the standardized Telecommunications Management Network (TMN) architecture [9, 10]. Starting at the top of the figure, a carrier's Operations Systems (OS) connects to a packet or circuit switched Data Communications Network (DCN) using one of the TMN standard interfaces, denoted by the letters X, F or Q. The X interface supports connections to TMNs in other carriers. For example, the X interface supports coordination of restoration requests between carriers of which the X.790 trouble ticket exchange specification is an example. The F interface allows humans to retrieve and modify management information; for example, via a workstation as shown in the figure. The Q3 interface comprises layers 4 through 7 of the OSI protocol reference model. The ITU-T utilizes the OSI standardized Common Management Information Service Elements (CMISE) and the associated Common Management Information Protocol (CMIP) for the Q3 interface. The Qx interface supports protocols other than the standard Q3 interface; commonly SNMP is called the Qx interface by TMN people. Mediation Devices (MDs), today more often called gateways, convert from these Qx interfaces to the standard Q3 interface. The NMF has been working in this area for several years and has published specifications, mappings and automatic conversion routines for common conversions, such as SNMP to CMIP.

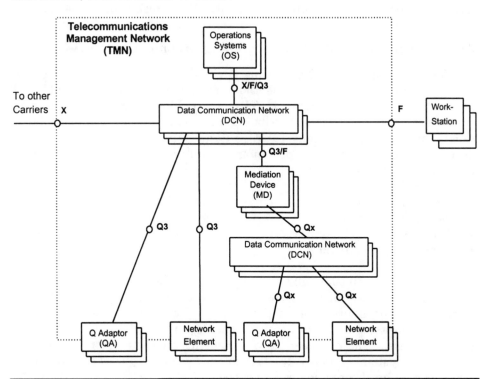

Figure 27.7 Telecommunications Management Network (TMN) Architecture

The software architecture of TMN includes functionally grouped capabilities called *operations systems functions* as follows:

- Business management supports the implementation of policies, strategies and specific services.
- Service management functions that are necessary to support particular services, such as subscriber administration and accounting management.

- Network management supports configuration, performance management, and maintenance.
- Element management support for management of one or more Network Elements (NEs) certainly concerned with maintenance, but also involved in performance management, configuration, and possibly accounting.
- Network element functions at the individual network element or device level that is the source and sink of all network management observations and actions. This includes traffic control, congestion control, ATM layer management, statistics collection, and other ATM related functions.

The mapping of these software functions onto the hardware architecture is an implementation decision. For a detailed description of an example implementation in the European RACE ATM technology testbed, see References 11 and 12.

27.3.4 Telecom Information Networking Architecture (TINA-C)

The Telecom Information Networking Architecture Consortium (TINA-C) is actively engaged in developing methods for integrated management of all parts of a communication network. TINA started by applying principles from software integration, Open Distributed Processing (ODP) and Distributed Communication Environment (DCE) but most importantly object-oriented design. TINA sets forth principles in an overall architecture [2] which is then divided into four components:

- Service Architecture
- Network Architecture
- Management Architecture
- Computing Architecture

TINA also established five viewpoints. The enterprise viewpoint establishes the scope of the system. The information viewpoint gives the semantics of the information in the system. The computational viewpoint broke the systems into logical component parts and defined interfaces between these. The engineering viewpoint describes the infrastructure to support the systems. Finally, the technology viewpoint places the systems in a blueprint of hardware and operating systems.

TINA broke the telecommunications system into three layers: the element layer, the resources layer, and the service layer. The goals of TINA are exciting and derivations of this work are beginning to see limited deployment.

Providers working on this include, but are not limited to, BT, NTT, MCI, and Sprint. But the main force of TINA was felt in its effects on other standards activities and resultant specifications. While the layering of TINA was abstract, the protocol layering of networks was real and concrete.

Inheritors of the concepts of TINA include the ITU-T G.8xx series of documents, specifically Recommendation G.805 covering generic functional architecture of transport networks [4] and Recommendation G.803 covering the architecture of transport networks based on the synchronous digital hierarchy (SDH) [3]. Both of these documents use the generic functional architecture of transport networks to provide a unique way of relating ATM to its transport over SONET. These describe a functional architecture of transport networks in a technology independent way. This generic functional architecture is used as the basis for a closely interworking set of functional architecture recommendations for ATM and SDH transport networks. It provides a basis for a series of recommendations for management, performance analysis and equipment specification.

ITU-T Recommendations G.803 and G.805 are based on the TINA Network Resource Information Model (NRIM) which topically relates switches at an abstraction of their common elements. The SONET port on the ATM trunk card can be related to a logical circuit in the SONET transport network, and to a connected port in the SONET switch. Major ATM service providers are using further documents in this series [13 to 16] as an architectural basis for managing their networks.

These specification provide a client/server perspective for relating one layer of the network to another. Today, in the real world, ATM and SONET systems are managed via different OSS. However, it would be nice to know if the ATM problem showing up with the disruption of hundreds of VPs is actually a failure in the SONET transport infrastructure. These G series Recommendations provide a foundation for building integrated systems which empower the operations groups manage both networks as a single information system. This should provide for more intelligent network design and for better service to customers of full-service providers. The TINA viewpoints are becoming a standard way of systematically designing a telecommunication's system, including OSS. Evolution of these is proceeding in ITU-T Study group 4 Question 18. TINA Consortium itself is now turning to service management and has moved from DCE to use of OMG CORBA where they are customizing CORBA for telecommunications.

27.3.5 ATM Forum Network Management Architecture

Figure 27.8 depicts the ATM Forum's network management reference architecture [17] which identifies five distinct management interfaces. Interfaces M1 and M2 define the interface between a private network management system for one or more customer sites covering private networks and ATM end-stations. The M3 interface allows public network carriers to provide standardized Customer Network Management (CNM) services from their management applications to a private management application. The M4 interface targets standardization of the interface to switches and element managers. M5 provides the management interface between different carrier's network management systems.   Current ATM Forum work efforts have concentrated on the M3 and M4 interface specifications.

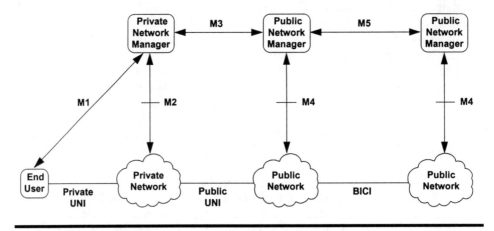

Figure 27.8   The ATM Forum Management Interface Reference Architecture

Although the ATM Forum has not yet standardized the other interfaces, many current standards fit into this framework [18, 19]. Because SNMP is widely deployed by end-users, the M1 and M2 interfaces embrace SNMP-based specifications defined by the IETF. These include relevant standard MIBs for DS-1 (RFC 1406), DS-3 (RFC 1407), and SONET (RFC 1595) physical interfaces and the AToM MIB (RFC 1695) described in the previous section. All of these have been revised and set into SNMPv2 SMI and are currently awaiting   release as standards.   The M3 Customer Network Management (CNM) interface gives a customer a view into their carrier's network.   Several carriers have now deployed CNM offerings where customers dynamically control their services.

The M4 interface provides network wide and single element level views to the carrier's network management system for public ATM networks. Thus, the M4 interface is where the different approaches to management converge, since both the private network manager and the carrier must be able to cooperatively control and monitor ATM service. This situation presents a dilemma. On the one hand, the private network manager wants cost effective public services, but insists on retaining a degree of monitoring and control needed to assure bandwidth and quality of service guarantees. The carrier, on the other hand, wants to monitor and control customer networks to offer end-to-end, value-added network management services.   Business models have been developed which support both these approaches. Today a customer can even outsource management of all their privately owned ATM resources to the service provider providing ATM WAN connectivity. The M5 interface targets the complex area of automatic management connections between carrier network management systems.   The inter-carrier M5 interface is perceived as the most complicated of the management interfaces since it covers all of the TMN X interface at the network management and service management layers, as applied to ATM technology. The ATM Forum has made good progress in connection management at this interface but has not yet published a specification.

The Customer Network Management (CNM) capability offered by the M3 is important for carrier-based services. This includes at least physical port status, VPC/VCC status, order parameters, and selected performance metrics. Delivery of detailed performance counts involves additional complexity and cost. Usage counts from the originating and terminating switch by VPC/VCC may also be used to track performance delivered to customers. Appendix B summarizes the approved specifications and descriptive documents produced by the ATM Forum Network Management group.

Work is in progress in the ATM Forum NM work group on an SNMP MIB for the M4 interface. This work is advanced and should be out soon. From the Service Provider's perspective, it seems to represent a better hope for SNMP management of ATM than does the IETF RCF1695 and AtoM workgroup Supplemental MIB.

27.4 REVIEW

This chapter introduced a model of Operations, Administration, Maintenance, and Provisioning (OAM&P) functions and how they interact to meet the overall needs of network users. It investigated the many schools of management fused into today's ATM management standards. The text then covered specific issues that ATM brings to the area of OAM&P. Next the discussion introduced standards from OSI, the ITU-T, OMG, TINA-C, NMF, ATM Forum, and the IETF. The OSI architecture defines the functions of Fault, Configuration, Accounting, Performance, and Security (FCAPS). The ITU-T defines a physical and logical Telecommunications Management Network (TMN) architecture, which lays out a common structure for managing transmission, voice and data networks. OMG and TINA-C study ATM networks as objects and management from an object-oriented systems point of view. The Network Management Forum (NMF) broadened the OSI functional model to reflect how providers build management systems. The chapter concluded with the ATM Forum's network management architecture, which defines a structure for future MIB definitions.

27.5   REFERENCES

[1]   N. Lippis, "ATM's Biggest Benefit: Better Net Management," *Data Communications*, October 1992.

[2]   TINA-C, "Overall Concepts and Principles of TINA," TB_MDC.018_1.0_94," February 1995.

[3]   ITU-T, *Architecture of transport networks based on the synchronous digital hierarchy (SDH)*, Recommendation G.803.

[4]   ITU-T, *Generic functional architecture of transport networks*, Recommendation G.805, November 1995.

[5]   ATM Forum, *M4 Interface Requirements and Logical MIB*, AF-NM-0020.000, October 1994.

[6]   I. Rubin, T. Cheng, "The Effect of Management Structure on the Performance of Interconnected High-Speed Packet-Switched Networks," GLOBECOM, December 1991.

[7]   NMF, "A Service Management Business Process Model", 1995.

[8]   U. Black, *Network Management Standards*, McGraw-Hill, 1992.

[9]   ITU-T, "Recommendation M.3010 - Principles for Telecommunications Management Network," 1992.

[10]   B. Hebrawi, *OSI Upper Layer Standards and Practices*, McGraw-Hill, 1992.

[11]   G. Schapeler, E. Scharf, S. Manthorpe, J. Appleton, E. Garcia Lopez, T. Koussev, J. Weng, "ATD-specific network management functions and TMN architecture," *Electronics & Communication Engineering Journal*, October 1992.

[12]   J. Callaghan, G. Williamson, "RACE telecommunications management network architecture," *Electronics & Communication Engineering Journal*, October 1992.

[13]   ITU-T, *Management of the transport network - Enterprise viewpoint for simple subnetwork connection management*, Recommendation G.852.01

[14]   ITU-T, *Subnetwork connection management information viewpoint*, Recommendation G.853.02

[15]   ITU-T, *Management of the transport network - Computational interfaces for basic transport network model*, Recommendation G.854.01

[16]   ITU-T, *Common elements of the information viewpoint for the management of a transport network*, Recommendation G.853.01

[17]   ATM Forum, "M4 Network-View Interface Requirements, and Logical MIB," AF-NM-0058.000, March 1996.

[18]   P. Alexander, K. Carpenter, "ATM Net Management: A Status Report," Data Communications, September, 1995.

[19]   IBM, "ATM Network Management Strategy," http://www.networking.ibm.com/atm/atmnman.html#mgt, 1996.

28

Network Management Protocols and Management Information Bases (MIBs)

This chapter summarizes the two major network management protocols defined by the IETF and the ITU-T. We begin with the IETF's Simple Network Management Protocol (SNMP), followed by the ITU-T's Common Management Interface Protocol (CMIP). In network management parlance, databases are Management Information Bases (MIBs). The text gives a summary of the ATM Forum's Integrated Local Management Interface (ILMI) and the IETF ATM MIB (called the AToMMIB) object database structures. These summaries give a good flavor of what network management is available for ATM interfaces, end systems, switches, and networks.

28.1 NETWORK MANAGEMENT PROTOCOLS

This section covers the two major network management protocols used by contemporary ATM systems and networks: the IETF's Simple Network Management Protocol (SNMP) and the ITU-T's Common Management Interface Protocol (CMIP). Workgroup, campus and enterprise networks commonly use SNMP, while larger carrier networks may use SNMP, CMIP or a proprietary network management protocol. Since not all management products use SNMP or CMIP, the text also describes vendor proprietary management systems.

28.1.1 Simple Network Management Protocol

The IETF's network management philosophy, protocol, and database structure (called SNMP for short) is widely used in the data communications industry. This section begins by defining the overall object oriented network management model, summarizes the SNMPv1 and SNMPv2 messaging protocols, and concludes with comments on the usefulness of SNMP-based ATM specific network management standards. The SNMP protocol is part of a management system which is fully described with the addition of the sections below on MIBs and the IETF's AToMMIB work group.

28.1.1.1   Object Model of Network Management

SNMP is the protocol part of the IETF's network management philosophy; however, it alone will not manage your network [1]. The IETF has not invested in the elaborate information models and computational models that the ITU-T specifications lay out. Most of this information was, and still is, passed by tradition and word of mouth between implementers. On the other hand, this works pretty well, because the IETF is a open organization and any engineer can just show up at a meeting or subscribe to a mail list. It also works because the IETF membership devoted to network management is accessible and has a strong mentoring tradition. Today this "folklore" is augmented by some informative books written by the actual specification and MIB designers. For instance, Marshal Rose and Keith McCloghrie, authors of SNMPv1 and SNMPv2, both have good books on SNMP [2, 3]. Furthermore, David Perkins, author of the SMIC SNMP ASN.1 MIB reference compiler has produced the definitive book on MIB design and semantics [4]. Think of the tradition of physics being defined by its great textbooks and the lectures of famous professors. SNMP turns out to be very accessible, and, while perhaps not a simple as the authors originally intended, with due diligence many implementations interoperate.

Figure 28.1 illustrates the key components of an SNMP based NM system. Typically, a single computer system interfaces to a number of network elements. The NM connections may not be physical, and indeed may be carried by the underlying network itself. SNMP is only a basic set of messages for monitoring and controlling the state of the network elements. The intelligence of an NM system is in understanding what the state variables (called Management Information Based (MIB) objects in SNMP parlance) in the network elements actually mean. Here, the collective body of knowledge recorded in RFCs, in the archives of the IETF work group mail lists, in white papers, or vendor documentation is the linchpin to successful understanding. Even so, differences in interpretation of meaning of MIB elements is the greatest problem in interoperability of agents and managers.

The condition of a physical interface — active or inactive — is an example of a state variable modeled by a MIB object. Continuing this example, knowing that two physical interfaces are supposed to be connected together requires intelligence. Unfortunately the defined semantics provided by the Structure of Management Information (the SMI) does not provide for relating an element in one table with a reference in another table; even when they refer to the same thing in the real network. The ATM MIBs show this via indexed tables and references to common indexes in a special MIB called ifIndex (interface index table). Nevertheless, externally supplied configuration information is often mapped by SNMP managers to unchanging MIB references in ifIndex and in cross-connect tables in the MIBs.

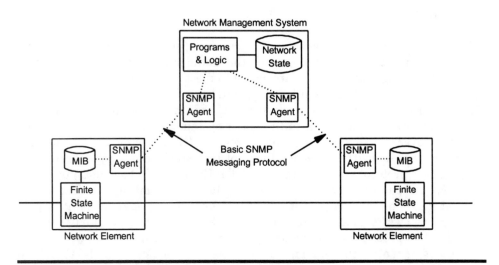

Figure 28. 1   SNMP-based Network Management System

Many network management engineers model more complex network conditions as finite state machines because of the precision and accuracy this provides in capturing the behavior of the elements, under stimulus, over time. In essence, the state machine provides what is called the computational model in the TMN and TINA-C architectures described in the previous chapter.   Another tool often used is rule engines and expert-systems knowledge-bases.   Advanced network management systems capture relationships using these tools and use them to know what variables to compare under what circumstances. These tools are also used to filter SNMP event notifications, called TRAPs, and to associate them with past operational and repair experience.

The MIB objects retrievable and configurable via the basic SNMP messaging protocol reflect and control certain aspects of the network element's state; hence, look for value in the insight conveyed in the vendor's MIB documentation — if the comments aren't detailed and don't reflect specifics of the implementation; beware. Also, when looking for an NMS, we suggest looking at the one provided by the vendor, or else by a third party who is extremely knowledge about the type of equipment being managed.

28.1.1.2   SNMP Message Types

Amazingly, SNMP allows a complex management system to monitor and control a large network of complex device using only the following five simple messages:

- GET
- GET NEXT
- SET

- RESPONSE
- TRAP

The GET message retrieves a particular object, while the GET NEXT message retrieves the next object in the management database structure. The SET message modifies a management object. The RESPONSE message is always paired with a stimulus SET, GET, or GET NEXT message. The TRAP message is very important since it is the unsolicited notification of an unexpected event, such as a failure or a system restart. SNMP normally operates over the User Datagram Protocol (UDP), which then usually operates over IP in the Internet Protocol (IP) stack, but may operate over some other protocol. Note that the UDP/IP protocol does not guarantee delivery of packets, because there is no retransmission or sequence numbering. This means that if a TRAP is lost, then a management system relying on TRAPs alone would miss a failure notification. Real systems resolve this problem by periodically sending out traps for major failure conditions, or else the management system periodically polls the ATM network devices for status. With the increased reliability of modern data network transmission, loss of UDP packets is rare in all but the most congested of networks.

SNMP utilizes a variant of Abstract Syntax Notation 1 (ASN.1) to define a Management Information Base (MIB) as a data structure that can be referenced in SNMP messages. The SNMPv1 SMI (Structure of Management Information) allows MIB definitions of objects in primitives such as strings, integers, and bit maps, using a simple form of indexing. Each object has a name, a syntax, and an encoding. The MIB variables have a textual Object IDentifier (OID) which is commonly used to refer to the objects. The MIB objects are defined as a tree structure that allows organizational ownership of subtrees to be defined. Every registered MIB in the entire world is a member of a single, official registration tree controlled by the IETF IANA. The branches of the tree are identified by a dotted decimal notation. For example, the prefix of the subtree registered to the ATM Forum is 1.3.6.1.4.1.353. Vendor MIBs, called Enterprise MIBs, use a branch of the tree where every vendor registers for a limb and then owns all of the subtending branches and leaves.

28.1.1.3   SNMPv2 and SNMPv3

For several years the IETF has been working on improvements to SNMP to extend its applicability and correct what many have seen as deficiencies. While progress in the standards for SNMP has been bumpy, the IETF eventually advanced a new version of SNMP along the standards track. This version, called SNMPv2, provides two upgrades important to ATM management.

First SNMPv2 (defined in IETF by RFC 1902 [5], RFC 1903 [6], and RFC 1904 [7], RFC 1905 [8], RFC 1906 [9], RFC 1907[10], RFC 1908 [11], RFC 1909 [12] and RFC 1910 [13]) provides a new SMI (read simply as syntax) for defining MIBs. Extensions to the type of OIDs extend MIB features and smooth out problems experienced by SNMPv1 implementers. Another enhancement introduced with SNMPv2 which is important to ATM management is the introduction of a new protocol message:

- GETBULK
- INFORM

GETBULK overcomes the limitations in SNMPv1 which required manager applications to issue many GETNEXT commands to traverse a table. A larger range of MIB values can be returned with one request via GETBULK, increasing the efficiency of retrieving a MIB table by both lowering the number of required transactions. This is important to ATM management because ATM is complex and requires many MIB elements. Furthermore, as seen in Chapter 16, some ATM switches are likely to grow very large, requiring managers to retrieve large volumes of data. Indeed, the size of proposed ATM switches and networks would introduce costly barriers to scaling both managers and switch based agents if these are forced to process thousands of sequential GETNEXTs.

INFORM messages provide unsolicited event notifications. They are similar to SNMPv1 TRAPs, except they expect the receiver to respond to the sender. Thus, INFORMs provide for confirmed alarm notifications, unlike TRAPs which are unacknowledged. Since a manager must respond to an INFORM, probably the most important, most critical events, such as service interrupting events, should be made as INFORMs instead of TRAPs. If the ATM element agent has logic to send a single indication of an problem, and is also intelligent enough to know it need not sent another indication, an INFORM greatly lessens the burden on managers when compared with handling repetitive TRAP notifications (for example, link down notifications every 15 seconds). The issue of per connection TRAPs was hotly debated in the standards groups since a failure could cause a flood of alarms if a devices sends a TRAP for every virtual connection affected by an outage. The proper response is for agents in ATM elements to have enough intelligence to send an alarm notification that covers the range of affected logical elements in a single notification.

Many developers of enterprise MIBs define their MIBs using the SNMPv2 SMI (defined in IETF *SMIv2*: by RFC 1902 [5], RFC 1903 [6], and RFC 1904 [7]), which provides, among other advantages, 64 bit counters. Since ATM switches typically support high speed interfaces, the older SNMPv1 SMI 32 bit counters wrap around much sooner, creating the need for software to sample them more frequently to avoid wraparound. For instance counting cells on OC12, the 32bit counter wraps around in less than an hour. When choosing an SNMP management application for ATM, it is best to look for those which support SNMPv2 or plan to soon.

Much adverse publicity ensued when the SNMPv2 standards development process failed to agree on a mechanism for SNMP security. Since security was a clearly part of the working group's charter, this was a great disappointment. Since it received lots of publicity, many people consider SNMPv2 to be a failure. In fact, the SNMPv2 protocol is fine as far as it goes and relied on the same administrative models (with no security, as is used in SNMPv1.)

After a brief hiatus, a new workgroup attacked the security problem anew. This new team has made excellent progress and a new version of SNMPv3 is on track for IETF approval, complete with new administrative models and security. Specifically, configuration security exists between agents and managers allowing each to uniquely identify the identity of the other and control allowed transactions. Additionally, the SNMPv3 group has set about to clean up a few left over problems such as proxy agents and clear identification of source identity of the agent when a manager receives a TRAP. The general consensus is that they have succeeded in clearing up most of the residual confusion over SNMPv2. At press time, this work is defined by RFC 1905 [8], RFC 1906 [9], RFC 2271 [14], RFC 2272 [15], RFC 2273 [16], RFC 2274 [17], and RFC 2275 [18]. Expect to wait a few years for market availability of SNMPv3, but do not delay deployment and use of SNMPv2 agents and managers.

There is general agreement that SNMP and SNMPv2 are best for CPE and private networks, while CMIP/CMISE is appropriate for carrier interconnection. Although there is still active debate on the usefulness of SNMP for large carrier environments, currently more of these networks using standard's based interfaces employ SNMP than those that either use, or plan to use, CMIP. Many early ATM switches implemented SNMP since it was simpler, and easier to achieve interoperability than with the CMIP protocols. It is likely that SNMP-based management will be the de facto standard for the network and higher layers since these exist primarily in end systems and CPE. Many vendors support hybrid implementations where some functions like statistics capture are performed by SNMP and other functions rely on proprietary interfaces. Where CMIP is deployed, it is being rolled out incrementally; for instance, SNMP TRAPs handle alarms, while CMIP supports complex provisioning actions (since it supports the transaction model required by the multi-step provisioning process.) In general, European PTTs have made more use of CMIP and less of SNMP.

There are more SNMP toolkits, managers, and related products than can be related in this book. Classic managers include SunNetManager, IBM NetView, and HP Openview which together defined the initial SNMP management industry. More recently, Sun has provided a more fully featured manager: Solstice. Cabletron's Spectrum, OSI's NetExpert, Groupe Bull, DEC's TeMIP are all fine managers which have SNMP protocol gateways and support SNMP MIBs. SNMP research and Peer provide SNMP tool kits. Desktalk Systems, Concord, INS, and Frontier provide fine SNMP based performance reporting products. Just about every network element vendor provides an SNMP based management element and some of the larger ATM switch vendors provide SNMP based element managers which often "plug in" to one or the other of these SNMP management systems.

28.1.2 ITU-T's Common Management Interface Protocol (CMIP)

Basically, CMIP operates as the approved protocol for use in ITU-T TMN. Although other protocols, such as CORBA, are being investigated today, CMIP is the established technology. It is sufficiently mature that a wide range of vendor toolkits are available that are essential aids in building element agents and TMN managers. As explained below, CMIP management is not a do-it-yourself activity. The text provides a short list of vendors, because the vendor, and not so much understanding the standards, is the key to gaining a working TMN/CMIP management OSS. DSET, Vertel, NETMANSYS, Open Networks Engineering (ONE), and TCSI specialize in TMN/CMIP tool kits. HP Openview Domain Manager, as do many other CMIP mangers, uses a DSET derived toolkit but extends the features resulting in an environment for strong agent development. IBM/TMN 6000 also provides a strong manager development environment. SUN Solstice is a state of the art, flexible system that adapts well to TMN (and just about every other protocol and information model). Groupe Bull and Objective Systems Integrators (OSI) use TMN principles and provide CMIP protocol gateways in their overall management product.

Initial implementations of CMIP by service providers failed to produce deployable implementations from what was acknowledged to be a very large stack of services in the CMIP specifications. Hence, many doubted TMN's ability to deliver on its promises. Eventually, via efforts of the NMF and some European service trial consortiums, workable subsets of the standard were identified. Most notable of these were the use of ensembles developed in the NMF OMNIpoint program and the NA4 and NA5 information models. Today, CMIP/TMN can be successfully deployed where strong integration and management control is desired over networks and network elements; an area which is key to larger service providers. A great deal of effort still is required to properly subset the features required and to align the information models derived from M.3100 to the target technology. A strong start at this effort for ATM is provided by the publications of the ATM Forum network management work group summarized in Appendix B.

CMIP like SNMP uses a manager/agent model for defining the interface between management OSS and switch elements. The essential difference is that CMIP requires a much more active and function rich role for the TMN manager and the CMIP agent. The manager actually creates object instances in the agent and essentially programs the functions desired into the behavior of the agent. The agent then supplies the information and functions to the manager as packages of information.

CMIP is part of the Q3 interface specification of TMN. There is a fully described data communications stack over which CMIP commands are transported. This stack can include CMISE (ISO 9595 & ISO 9596), ASCE (ITU-T X.217 & ITU-T X.227), ROSE (X.219 & X.229), and ITU-T X.208, ITU-T X.209, ITU-T X.215, ITU-T X.216, ITU-T X.225, ITU-T X.226. In transport and network layers it can also use ISO 8073, and ISO 8473. In ATM, in-band transported CMIP is expected to use AAL5; out-of-band transported CMIP likely will use X.25 or IP. CMIP can use any physical layer transport.

CMIP requires GDMO descriptions of objects and, in practice, only uses objects which inherit from the objects defined in ITU-T M.3100. The information model is therefore much richer in features and vastly more organized than that provided by MIB-II for SNMP. After the initial effort at aligning the details of the objects to be used at the Q3 interface, the management application knows the relationships between all information elements. Management becomes more an engineering exercise and less a subjective art form. Unfortunately, the standard GDMO libraries are even more subject to errors than the existing body of SNMP Enterprise MIBs; probably because they are more feature rich and complex, but also because they are the work of standards, often without benefit of trial implementations before publication. This has to a large degree been corrected by the NMF who now has available a library of "de-bugged" GDMO. Before these, and any ITU-T derived standard, can be used, a license is required from the ITU-T.

CMIP, like SNMP, has a short list of protocol commands; however, the behavior of agent and manager in executing these commands is complex. CMIP acts on GDMO defined objects. GDMO defined objects inherit characteristics from more generalized objects and this is a very powerful tool for designing OSS. The OSI SMI allows the behavior of the manager and agent, when issuing and responding to these commands, to be specified in templates. These templates qualify the characteristics of object classes, that is, these define how managed objects relate and respond to management requests. ACTION and NOTIFICATION templates program an object in regards to the respective CMIP commands (described immediately below.) Generally, CMIP commands are all confirmed, that is, the manager and agent are aware that any specific command is sent, received, and executed. Therefore, CMIP commands have some of the important features of transactions. CMIP command primitives are:

- M-CREATE
- M-DELETE
- M-GET
- M-CANCEL-GET
- M-SET
- M-ACTION
- M-EVENT-REPORT

Brief descriptions of the function of these commands follow. Actual behavior of a command is determined by the content carried with the command and the semantics of the objects on which it works. M-CREATE creates an instance of a managed object. This initializes the specific logical image of an element that management tracks and controls. M-DELETE deletes object instances and all objects instances which inherited from the deleted object. M-GET retrieves the values, that is attributes, of objects specified in the command content. M-CANCEL-GET stops an ongoing M-GET command. M-SET modifies values of attributes in objects specified in the M-SET. M-ACTION invokes the performance of an administrative or operational function. The syntax of the actions which can be invoked are set up in the ACTION template. The above commands are initiated by the manager on the agent. With M-EVENT-REPORT, the managed element communicates an event to the manager. This is generally an alarm and its delivery can be confirmed. The events which are reported are set up in the NOTIFICATION package.

CMIP as a protocol operates in a management environment where the semantics of the objects are well defined. The behavior of the management system and the agents found in the network elements which contain the managed objects should be well known and deterministic, based on the GDMO definitions of the objects. Once a subset of the characteristics of the objects are specified in an ensemble or other interface specification, management stimulus should provide predictable agent response. Therefore, the development of the GDMO interface library is extremely important for CMIP to function to expectations. Vendors are now providing CMIP agents and there are off the shelf CMIP managers available in the market. However, do not expect the plug-and-play behavior of SNMP manager/agent MIBs. Plan to expend considerable effort to turn up a functioning management system. In this respect standards become even more important. Insist on following ATM Forum M4 interface specifications for all but the vendor specific proprietary extensions. Insist that GDMO libraries be certified, bug-free derivations of the standard's specified GDMO. Unless both manager and agent are provided by the same vendor, before buying, have the switch vendor and the manager demonstrate intercommunication in a management trial. Establishing this intercommunication may well involve considerable adaptive effort on the part of both vendors.

28.1.3 Proprietary Network Management Protocols

Often one hears that open interfaces are good and proprietary interfaces are bad. The normal argument is that open interfaces lead to interoperable implementations while proprietary implementations can never interoperate. In fact, business drivers at one time promote open interfaces but at another times drive users to select proprietary solutions. Sometimes it is an accident of circumstance as to whether a specification is open or closed. In actual experience, any specific stack of OSS applications deployed in a big network will contain both standard and proprietary network management protocols.

Some proprietary protocols become open interfaces. If an owner publishes and licenses proprietary interfaces which, by dint of wide spread implementation by different organizations, become *de facto* standards. Microsoft which owns the *de facto* standard MS Windows, has also invented and championed a management protocol which while not as widespread as Windows, does have numerous adherents. Acceptance of the interface is aided by development toolkits and easy integration these into the Windows operating system. Another widely used de facto standard is the API to HP's Network Node Manager (NNM). While HP NNM is an SNMP based application, HP early on published a private API and provided a toolkit so that equipment vendors could talk directly to their devices and represent the results inside NNM or could use HP NNM SNMP to talk to their devices and have NNM call special management applications or equipment maps for their devices. This has been very successful and has seen wide mimicking by other network management vendors. ATM elements managed via the HP API can provide substantial value to a user.

Vendors must invent proprietary interfaces which seems like more work than just following the published standard. What are their motivations?

- Sometimes a particular feature is not defined in the existing open interfaces. Proprietary MIBs may describe proprietary features, valuable new capabilities, not (yet) in the ATM standards. When many people see the advantage of these special features, they often end up being incorporated into later versions of open interfaces.
- Sometimes an important function is awkward to implement using a standard protocol. Provisioning, which requires the notion of a transaction which can be reversed if something goes wrong, is especially awkward to implement in SNMPv1. Vendors selling provisioning applications have often used other methods to guarantee a provisioning transaction.
- Other times vendors invent proprietary interfaces because it is cheaper or easier to do so. The ITU-T CMIP specification was very large and complex and proved very costly in initial implementations by big service providers. Small lean vendors elected to use existing network technology they already sold, like X.25, to interface management systems to their equipment.
- Sometimes a proprietary protocol becomes perpetuated in order to prevent others in the market from writing and selling competitive management systems which would control their ATM switch products. The buyer of these OSS applications and ATM products must then weigh the disadvantages and extra costs of this closed market to the overall strength of the features the vendor provides.

However, even the most successful vendors of proprietary protocols and non standards based OSS applications are under pressure to standardize. For example, Newbridge Networks has long persisted in fielding a proprietary management solution. They argue that the capabilities of their systems are demanded by the buyers. Certainly, the Newbridge solution has performed well and seen wide-scale acceptance. But no vendor is ever deployed alone (except in very small networks); other equipment must be managed. Multivendor integration forces vendors like Newbridge to provide management of other vendors' equipment. This is always costly to do outside of an open interface. Also Newbridge element managers, when deployed in large service provider environments, must interface to many other business applications. These interfaces are often demanded to be standard, open interfaces. Proprietary applications then need complex and costly gateways to their applications; these gateways provide a standards based interface to the outside world.

Today, wide acceptance of toolkits and off-the-shelf management application stacks by special tool vendors for both SNMP and for TMN have made the time to deploy standards as quick and as cost efficient as proprietary methods once were. Keeping old proprietary management code has become very expensive. For all these reasons, expect to see more and more vendors support open interfaces.

28.2   MANAGEMENT INFORMATION BASES (MIB)

Management Information Bases (MIBs) were first widely used in local area network and Internet environments. MIBs have achieved a great degree of interoperability using the SNMP protocol. This section covers the major standardized MIBs available today in many vendor implementations. The usage of MIBs also extends beyond standards. Many vendors utilize SNMP-based MIBs for proprietary extensions.

Note that compiling and loading a MIB into a Manager and using that MIB are different problems. The key to any network management system is understanding what the interplay and changes in the MIB objects mean in terms of network and service performance. Only part of that information is in the semantics of the MIB object definitions. More information is typically carried in the introduction sections of RFC MIBs and, in both standard and Enterprise MIBs, in the comments part of each object definition. Therefore, it is good to have the standard MIBs you will use already 'precompiled' into applications in the SNMP manager you choose. When an Enterprise MIB is not already in a manager, make sure the comments convey information about object and table dependencies and other interactions.

This section summarizes Management Information Bases (MIBs) as examples of the types of information that can be accessed and manipulated in ATM interfaces, end systems, switches, and networks. An example of a Management Information Base (MIB) is the ATM MIB (AToMMIB) defined by the IETF. Of course, other MIBs also support ATM features. Indeed, SNMP based MIBs (formally known as MIB-II) have achieved such a high degree of interoperability that it is highly likely that if you obtain an electronic copy of a vendor's proprietary MIB, it will run on your SNMP network management system (possibly with some slight syntax editing.)

Interoperability does not imply identity; be warned that not all SNMP compilers generate the same results when compiling a MIB; some management application MIB compilers adhere to the standard than more so than others. For example, HP Openview NNM not only supports a wide variety of equipment but accepts less strictly written vendor proprietary MIBs. On the other hand, OSI's NetExpert, needs and demands greater compliance with the SNMP SMI. Generally, stricter compilers are required when a manager supports more management functions.

SNMP MIBs, especially enterprise (private) MIBs, are subject to errors and non traditional syntax constructions. It is best to check a MIB against a reference, diagnostic compiler. Two are available for this purpose and widely used. In the past the publicly available ISO Development Environment (ISODE) MOSY compiler was the reference benchmark and a new version is back in use today. However, today most experts use the SMICng compiler as an aid in checking MIBs, especially during development and assessment. The original SNMP Management Information Compiler (SMIC) is available from Bay Networks. The improved version SMICng is available via Internet web sites or through a CD-ROM included in a (highly recommended) book by the SMICng author [4].

SNMPv2 brought updates to the structure of Managed Information (SMI) which greatly extended the possibilities for MIB definition. New MIBs reference the SNMPv2 SMI via inclusion of RFC 1902 [5] which defines the SMI, the mechanisms used for describing and naming objects for the purpose of management. All MIBs which can respond to SNMPv2 commands issued by the manager refer to RFC 1905 [8] which defines the protocol used for network access to managed objects. All valid MIBs reference the core definition of SNMP MIBs via reference to RFC 1213 which defines MIB-II, the core set of managed objects for the Internet suite of protocols.

28.2.1 ATM Forum Integrated Local Management Interface (ILMI)

When the ATM Forum created the Interim Local Management Interface (ILMI) in 1992, it anticipated that ITU-T and ANSI standards would create a final interface management solution. Four years later, the Forum changed the initial "I" in the acronym to Integrated, since the Integrated Local Management Interface (ILMI) [19] now performs the following critical functions:

- Basic configuration information
- PVC status indication in FR/ATM service interworking   (See Chapter 17)
- ILMI connectivity detection and auto neighbor discovery
- Address registration for SVCs and PNNI (See Chapter 15)
- ABR attribute setting for PVCs
- Auto-configuration of a LAN Emulation Client (LEC) (See Chapter 19)

28.2.1.1 ILMI Configuration

Figure 28.2 illustrates the reference configuration for the ILMI. ATM Interface Management Entities (IMEs) communicate using the ILMI protocol based upon SNMP operating over AAL5, each in turn over physical or virtual links. IMEs may operate in either a user, network, or symmetric mode. Each ATM End System (ES), and every network that implements a Private Network UNI or Public Network UNI, has a ILMI Management Entity (IME) responsible for maintaining the information and responding to SNMP commands received over the ATM UNI. The information in the ILMI MIB can be actually contained on a separate private or public Network

Management System (NMS), or may be accessed over another physical interface. NMSs may also be connected to networks or end systems by other network management interfaces.

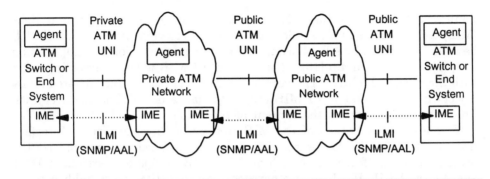

Figure 28.2   Integrated Local Management Interface (ILMI) Context

Initially, the ATM Forum defined an Interim Local Management Interface (ILMI) in advance of a standardized ATM layer management interface. The code used to write SNMP managers and agents was familiar to the authors of ILMI   However, the ILMI departs in several key ways from the SNMP model. SNMP's manager agent administrative model is replaced in ILMI by a symmetric association between the user side and the network side, each of which can SET variables or GET (query) variables in the other's MIB. For instance the network side SETs address prefixes in the user side and the user side SETs addresses in the network side address registration table as described in Chapter 14. Originally, ILMI agents could also double as SNMP agents and also talk to managers. However, some confusion resulted in initial implementations which delayed widespread support for ILMI. These confusions were identified as problems in the early implementations and were resolved in the ATM Forum standards with the work on the ILMI 4.0 specification [19]. A partition of the ILMI MIBs from the SNMP agent MIBs was clearly required. Today a manager wishes to find out information in the ILMI MIB, it must use a "shadow" ILMI MIB implemented in the SNMP agent space of the ATM element. Also clear in the early implementations was that implementations of the ILMI would benefit from using a set of linked state machines at the User and Network sides which carry the true semantics of interaction of command exchanges between the User and Network side.

For the ATM layer management interface, a default value of VPI=0, or VCI=16 for the ILMI, was chosen because CCITT/ITU reserved VCIs 0 through 15 (i.e., the first 16) for future standardization. Alternatively, another VPI/VCI value can be manually configured identically on each side of the UNI for ILMI use. Use of this method is undesirable since it is not automatic and is one more configuration parameter that can be incorrectly set. The ILMI operates over AAL3/4 or AAL5 as a configuration option, with support for a UDP/IP configurable option. Therefore, in order for IMEs to interoperate, the AAL (either 3/4 or 5) and the higher-layer protocol (either UDP/IP or Null) must be chosen.

28.2.1.2 ILMI Management Information Base

Figure 28.3 illustrates the ILMI Interface Management Information Base (MIB) tree structure and its index structure. Three versions of the ILMI MIB have been specified by ATM Forum UNI specification version 2.0 [20], version 3.1 [21], and version 4.0 [19]. The version 3.1 MIB is backward compatible with the version 2.0 MIB. The version 4.0 MIB deprecated (i.e., deleted many objects from the version 3.1 MIB other standards now define these objects. This section summarizes the content of the version 3.1 and 4.0 MIBs. The IME indexes each branch of the MIB tree via a Physical/Virtual Interface (PVI) as indicated in the figure. The following bulleted lists summarize the basic content and function of each of these MIB groups for versions 2.0, 3.1 and 4.0 of the ILMI. Unless otherwise indicated, the object applies to all three ILMI versions.

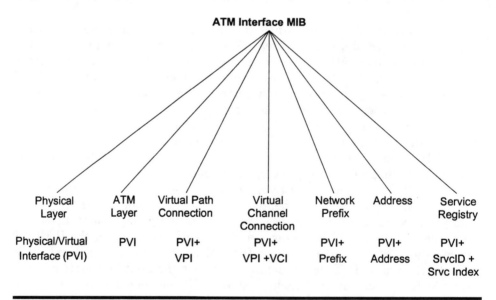

Figure 28.3   ILMI Interface MIB Tree Structure

The Physical Layer MIB information covers:

*   Interface Index
*   Interface Address (2.0 only)
*   Transmission Type (2.0, 3.1)
*   Media Type (2.0, 3.1)
*   Operational Status (2.0, 3.1)
*   Port Specific Information (2.0, 3.1)
*   Adjacency information (4.0)

The ATM Layer MIB information covers:

- Maximum Number of Active VPI and VCI Bits
- Maximum Number of (Allowed) VPCs and VCCs
- Number of Configured VPCs and VCCs
- Maximum SVPC VPI (4.0)
- Maximum SVCC VPI and Minimum SVCC VCI (4.0)
- ATM Public/Private Interface Type Indicator (4.0)
- ATM Device Type (either User or Node) (4.0)
- ILMI Version (4.0)
- UNI Signalling Version (4.0)
- NNI Signalling Version (4.0)

The Virtual Path Connection MIB information covers:

- VPI Value
- Operational Status (either up, down or unknown)
- Transmit Traffic Descriptor (4.0 different from 3.1)
- Receive Traffic Descriptor (4.0 different from 3.1)
- Best Effort Indicator (4.0)
- Service Category (4.0)
- Transmit/Receive QoS Class (3.1)
- ABR Operational Parameters (4.0)

The Virtual Channel Connection MIB information covers:

- VPI and VCI Value
- Operational Status (either up, down or unknown)
- Transmit Traffic Descriptor (4.0 different from 3.1)
- Receive Traffic Descriptor (4.0 different from 3.1)
- Best Effort Indicator (4.0)
- Service Category (4.0)
- Transmit/Receive QoS Class (3.1)
- Transmit Frame Discard Indication (4.0)
- Receive Frame Discard Indication (4.0)
- ABR Operational Parameters (4.0)

The Network Prefix MIB information covers:
- Network Prefix (3.1, 4.0)
- Network Prefix Status (3.1, 4.0)

The Address MIB information covers:
- ATM Address (3.1, 4.0)
- ATM Address Status (3.1, 4.0)
- ATM Address Organizational Scope Indication (4.0)

The ILMI also uses the standard systems group by reference, which supports things such as identification of the system name, and the time that the system has been up. The systems group also provides standard TRAPs, such as when a system is restarted or an interface failure is detected.

As described in Chapter 14, address registration using ILMI is a key component of automatic configuration of Private Network-Network Interface (PNNI) reachability information in the ATM Switched Virtual Connection (SVC) capability.   Basically, address registration allows the network to communicate to the user which address prefixes are valid on the User-Network Interface (UNI). The user can then register the valid remaining portions of the address(es) present locally.   It also provides source authentication for virtual private networks, since the originating switch may screen the calling party information element in the SETUP message against the set of registered addressed prefixes.

The Service Registry MIB information portion of the ILMI provides a general-purpose service registry for locating ATM network services, such as the LAN Emulation Configuration Server (LECS) and the ATM Name Server (ANS).

28.2.2 IETF AToMMIB

IETF RFC 1695 defines an ATM Management Information Base, called the "AToM MIB" [22], which covers the management of ATM PVC-based interfaces, devices, and services. The scope of the AToM MIB covers the management of ATM PVC-based interfaces, devices, and services. This standard defines managed objects for ATM interfaces, ATM VP/VC virtual links, ATM VP/VC cross-connects, AAL5 entities, and AAL5 connections supported by ATM end systems, ATM switches, and ATM networks.

The AToM MIB uses a grouping structure similar to the ILMI described earlier to collect objects referring to related information and provide indexing. The AToM MIB defines the following groups:

- ⊙ ATM interface configuration
- ⊙ ATM interface DS3 PLCP
- ⊙ ATM interface TC Sublayer
- ⊙ ATM interface virtual link (VPL/VCL) configuration
- ⊙ ATM VP/VC cross-connect
- ⊙ AAL5 connection performance statistics

The ATM interface configuration group contains ATM cell layer information and configuration of local ATM interfaces. This includes information such as the port identifier, interface speed, number of transmitted cells, number of received cells, number of cells with uncorrectable HEC errors, physical transmission type, operational status, administrative status, active VPI/VCI fields, and the maximum number of VPCs/VCCs.

The ATM interface DS3 PLCP and the TC sublayer groups provide the physical layer performance statistics for DS3 or SONET transmission paths. This includes statistics on the bit error rate and errored seconds.

The ATM virtual link and cross-connect groups allow management of ATM VP/VC virtual links (VPL/VCL) and VP/VC cross-connects. The virtual link group is implemented on end systems, switches, and networks, while the cross-connect group is implemented on switches and networks only. This includes the operational status, VPI/VCI value, and the physical port identifier of the other end of the cross-connect.

The AAL5 connection performance statistics group is based upon the standard interface MIB for IP packets. It is defined for an end system, switch, or network that terminates the AAL5 protocol. It defines objects such as the number of received octets, number of transmitted octets, number of octets passed to the AAL5 user, number of octets received from the AAL5 user, and number of errored AAL5 CPCS PDUs.

28.2.3 IETF AToM Supplemental MIB(s)

For several years the IETF has been working on extensions to RFC1695 to bring it up to current releases of the ATM Forum standards and to apply implementation experience by various vendors. This work resulted in a set of Internet drafts which are now close to being finalized. These new MIBs include and loosely cover:

- An update of RFC 1695, Definitions of Managed Objects for ATM Management
- The AToM Supplemental MIB, Definitions of Supplemental Managed Objects for ATM Management, draft-ietf-atommib-atm2-12.txt, which updates RFC 1695 to UNI 3.1, partially supports UNI 4.0, provides for SVC support, and facilitates cross connect table reads.
- The ATM Textural Conventions MIB, Definitions of Textual Conventions and OBJECT-IDENTITIES for ATM Management, which explains the SNMPv2 syntax and macro definitions used by these new MIBs.
- The ATM Test MIB, Definitions of Tests for ATM Management, which provides objects for testing ATM interfaces, switches and networks. Of particular interest are objects for controlling ATM loopback tests.
- The ATM Accounting MIB defines Managed Objects for Controlling the Collection and Storage of Accounting Information for Connection-Oriented Networks, draft-ietf-atommib-acct-04.txt, which provides an SNMP administrative interface to control the bulk loading (generally via TFTP or FTP) of statistics from a switch's MIB to a management repository. This MIB might be used to gather data for use in billing or reporting systems.
- The ATM Historical Data MIB, Managed Objects for Recording ATM Performance History Data Based on 15 Minute Intervals, draft-ietf-atommib-atmhist-00.txt, and companion MIB, Textual Conventions for MIB Modules Using Performance History Based on 15 Minute Intervals, draft-ietf-atommib-perfhistTC-01.txt, which provides an RMON like history table for statistics which have been identified as important to ATM operations. This amounts to 24 hours of data stored in ninety-six 15-minute bins, all of which might be collected via a single GETBULK command.

As SNMPv2 and the updates to RFC1695 MIB progressed together through the IETF, the AToM Supplemental MIB used the new SNMPv2 SMI definitions. With the advancement of the SNMPv2 SMI along the standards track, the AtoM Supplemental MIB is also free to progress.

While work on the Supplemental MIB is proceeding, as of draft 12 from March 1998, The MIB provides ATM Switch Support, ATM Service Support, ATM Host Support and importantly, ATM trap support. The following tables are defined. The work group has classified the tables into five groups, depending upon the combination of required support by switches, services, and hosts. These include:

- The ATM Switch Services Host Group, atmSwitchServcHostGroup:
    ATM Interface Signaling Statistics Table, atmSigStatTable
    ATM VPL Statistics Table, atmVplStatTable
    ATM Logical Port Configuration Table, atmVplLogicalPortTable
    ATM VCL Statistics Table, atmVclStatTable
    ATM Interface Configuration Extension Table, atmInterfaceExtTable
    Currently Failing PVPL Table, atmCurrentlyFailingPVplTable
    Currently Failing PVCL Table, atmCurrentlyFailingPVclTable

- The ATM Switch Services Group, atmSwitchServcGroup:
    ATM SVPC Cross-Connect table, atmSvcVpCrossConnectTable
    ATM SVCC Cross-Connect table, atmSvcVcCrossConnectTable
    ATM Interface Signaling Support Table, atmSigSupportTable
    ATM Interface Registered Address Table, atmIfRegisteredAddrTable
    ATM ILMI Service Registry Table, atmIlmiSrvcRegTable
    ILMI Network Prefix Table, atmIlmiNetworkPrefixTable

- The ATM Switch Group, atmSwitchGroup:
    ATM Switch Address Table, atmSwitchAddressTable

- The ATM Services Group, atmServcGroup:
    ATM VP Cross-Connect Extension Table, atmVpCrossConnectXTable
    ATM VC Cross-Connect Extension Table, atmVcCrossConnectXTable

- The ATM Host Group, atmHostGroup:
    ATM Signaling Descriptor Parameter Table, atmSigDescrParamTable
    ATM VPI/VCI to Address Mapping Table, atmVclAddrTable
    ATM Address to VPI/VCI Mapping Table, atmAddrVclTable
    ATM AAL5 per VCC Statistics Table, atmAal5VclStatTable
    ATM VC General Information Table, atmVclGenTable

28.2.4 Other ATM MIBs

The ATM Forum has defined a number of additional MIBs in support of specific functions as summarized in the table below. NMS's using SNMP can utilize these MIBs to manage devices performing these functions.

Table 28.1 Other ATM Related MIBs

Function(s) Supported MIB	Reference(s)
ATM Data eXchange Interface (DXI)	af-dxi-0014.000
ILMI Extensions for LAN Emulation	af-lane-0021.000
Private Network-Network Interface (PNNI)	af-pnni-0055.000
Inverse Multiplexing over ATM (IMA)	af-phy-0086.000
Circuit Emulation Service (CES) MIB	af-saa-0032.000
Physical Interfaces	RFC 1595, 1407, 1406

28.3   REVIEW

This chapter summarized and compared the competing network management protocols. The text first covered the IETF defined Simple Network Management Protocol (SNMP) and Management Interface Base (MIB) which has achieved a high degree of interoperability in the industry. The text then moved on to the ITU-T's Common Management Interface Protocol (CMIP) designed to manage transmission, voice, and data networks. The chapter also highlighted some proprietary network management protocols. We then summarized the ATM Forum's SNMP derived Integrated Local Management Interface (ILMI) for the ATM UNI and the IETF's ATM Management Information Base (AToM MIB) for management of interfaces, end systems, switches and networks.

28.4   REFERENCES

[1]   T. Cikoski, "The Complexities and Future Evolution of SNMP as a Management Protocol," Telecommunications, August 1996.

[2]   M Rose, K. McCloghrie, *How to Manage Your Network Using SNMP: The Networking Management Practicum,* Prentice Hall, January 1995

[3]   M. Rose, *The Simple Book : An Introduction to Networking Management,* Prentice Hall, April 1996.

[4]   D. Perkins, E. McGinnis, *Understanding MSMP MIBs,* Prentice Hall, 1997.

[5]   J. Case, K. McCloghrie, M. Rose & S. Waldbusser, *Structure of Management Information for Version 2 of the Simple Network Management Protocol (SNMPv2),* RFC 1902, IETF, January 1996.

[6]   J. Case, K. McCloghrie, M. Rose & S. Waldbusser, *Textual Conventions for Version 2 of the Simple Network Management Protocol (SNMPv2),* RFC 1903, IETF, January 1996.

[7]   J. Case, K. McCloghrie, M. Rose & S. Waldbusser, *Conformance Statements for Version 2 of the Simple Network Management Protocol (SNMPv2),* RFC 1904, IETF, January 1996.

[8]   J. Case, K. McCloghrie, M. Rose & S. Waldbusser, *Protocol Operations for Version 2 of the Simple Network Management Protocol (SNMPv2)*, RFC 1905, IETF, January 1996.

[9]   J. Case, K. McCloghrie, M. Rose & S. Waldbusser, *Transport Mappings for Version 2 of the Simple Network Management Protocol (SNMPv2)*, RFC 1906, IETF, January 1996.

[10]   J. Case, K. McCloghrie, M. Rose & S. Waldbusser, *Management Information Base for Version 2 of the Simple Network Management Protocol (SNMPv2)*, RFC 1907, IETF, January 1996.

[11]   J. Case, K. McCloghrie, M. Rose & S. Waldbusser, *Coexistence between Version 1 and Version 2 of the Internet-standard Network Management Framework*, RFC 1908, IETF, January 1996.

[12]   K. McCloghrie, *An Administrative Infrastructure for SNMPv2*, RFC 1909, IETF, February 1996.

[13]   G. Waters, *User-based Security Model for SNMPv2*, RFC 1910 , IETF, February 1996.

[14]   D. Harrington, R. Presuhn, B. Wijnen, *An Architecture for Describing SNMP Management Frameworks*, RFC 2271, IETF, January 1998.

[15]   J. Case, D. Harrington, R. Presuhn, B. Wijnen, *Message Processing and Dispatching for the Simple Network Management Protocol (SNMP)*, RFC 2272 , IETF, January 1998.

[16]   D. Levi, P. Meyer, B. Stewart, *SNMPv3 Applications*, RFC 2273, IETF, January 1998.

[17]   U. Blumenthal, B. Wijnen, *User-based Security Model (USM) for version 3 of the Simple Network Management Protocol (SNMPv3)*, RFC 2274, IETF, January 1998.

[18]   B. Wijnen, R. Presuhn, K. McCloghrie, *View-based Access Control Model (VACM) for the Simple Network Management Protocol (SNMP)*, RFC 2275, IETF, January 1998.

[19]   ATM Forum, "Integrated Local Management Interface (ILMI) Specification Version 4.0," af-ilmi-0065.000, September, 1996.

[20]   ATM Forum, "ATM User-Network Interface (UNI) Specification, Version 2.0," June 1992.

[21]   ATM Forum, *User-Network Interface Signaling Specification, Version 3.1*, af-uni-0010.002, September 1994.

[22]   M. Ahmed, K. Tesink, *Definitions of Managed Objects for ATM Management Version 8.0 using SMIv2*, RFC 1695, IETF, August 1994.

29

ATM Layer Management and Performance Measurement

This chapter first introduces the integrated physical and ATM layer Operations And Maintenance (OAM) information flow architecture, details the ATM OAM cell formats, and provides a description of fault management procedures. The next chapter describes performance management. The text begins with a discussion of the usage of OAM cells for fault detection and identification using the Alarm Indication Signal (AIS) and Far End Reporting Failure (FERF) OAM cell types as demonstrated through several examples. Finally, the chapter presents example uses of the loopback capability to verify connectivity and diagnose problems that OAM cells cannot.

This chapter then defines reference configurations for specifying and measuring Network Performance (NP) and user Quality of Service (QoS). Quality of Service is user perception, while Network Performance finds use in network management, OAM, and network design. Next, the text defines OAM cell formats and procedures used to activate and deactivate the performance measurement and continuity check functions. Descriptions and examples illustrate the functions performed by performance measurement OAM cells. The text then gives detailed examples of how the OAM performance measurement cells and procedures estimate each of the QoS parameters from the traffic contract described in Chapter 21. The chapter concludes with a discussion of other topics, including open issues in the estimation of NP/QoS using the current methods, and other elements of performance that require measurement.

29.1 OAM Flow Reference Architecture

Currently, OAM flows are only defined for point-to-point connections. A fundamental part of the infrastructure for network management is that of OAM information. Figure 29.1 shows the reference architecture that illustrates how ATM OAM flows relate to SONET/SDH management flows [1]. The F1 flows are for the regenerator section level (called the Section level in SONET), F2 flows are for the digital section level (called the Line level in SONET), and F3 flows are for the transmission path (call the Path level in SONET). ATM adds F4 flows for Virtual Paths (VPs) and F5 flows for Virtual Channels (VCs). Recall from Chapter 12 that a single VP carries multiple VCs. Each flow either traverses an intermediate connecting point or terminates at an endpoint.

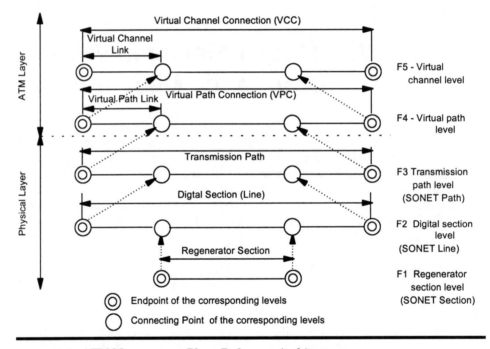

Figure 29.1   ATM Management Plane Reference Architecture

Each of the F4/F5 flows may be either end-to-end or segment-oriented. An end-to-end flow is from one endpoint at the same level to the other endpoint. Only devices that terminate ATM connections receive end-to-end OAM flows.

A segment flow is a concatenation of VP (or VC) links from one connection point to another connection point. Only network nodes receive segment OAM flows. Indeed, network nodes must remove segment flows before they ever reach devices that terminate an ATM (VP or VC) connection. Segment flows cannot overlap.

Recommendation I.610 indicates that OAM flows apply to permanent, semi-permanent and switched virtual ATM connections. The standard does state that procedures for switched connections are for further study. For example, as studied in Chapter 15, the normal response to a trunk or switch failure is to tear down a Switched Virtual Connection (SVC).

Figure 29.2 shows a real-world example of end-to-end OAM flows for an end-to-end ATM Virtual Channel Connection (VCC) connecting two end systems. Starting from the left-hand side, end system 1 connects to Light-wave Terminal Equipment (LTE) 1, which terminates the digital section OAM flow (F2). The transmission path flow (F3) terminates on the Virtual Path (VP) Cell Relaying Function (CRF). The VP flow (F4) passes through the VP CRF, since it is only a connection point; that is, only the Virtual Path Identifier (VPI) value changes in cells that pass through that specific VP. The Virtual Channel Identifier (VCI) value is not changed.

Next, the example traverses a typical transmission path across the wide area network from LTE 2 to LTE 3 through a repeater (indicated by the "bow tie" symbol in Figure 29.2). The regenerator section flow (F1) operates between LTEs 2 and 3 and the repeater, as well as between repeaters. The OAM flow between LTE 2 and LTE 3 is an example of a digital section flow (F2). The transmission path (F3) flow terminates on the VC CRF. The VP flow (F4) also terminates on the VC CRF because in its relaying function it can change the VCI as well as the VPI. A separate digital section OAM flow (F2) then extends from LTE4 to a Customer Premises Equipment (CPE) device (NT 1) as another line flow (F2). The OAM flow to end system 2 from NT 1 is also a digital section level flow (F2). The transmission path flow (F3) extends from VC CRF to end system 2, as does the VP flow (F4) since the VPI cannot change in this portion of the connection. Finally, note that the Virtual Channel (VC) flow (F5) is preserved from end system 1 to end system 2.

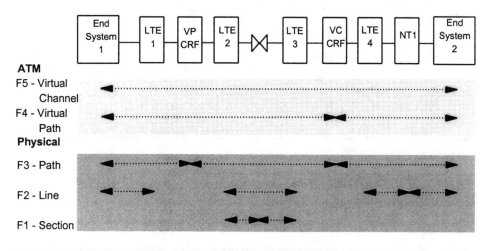

Figure 29.2   Illustrative Example of OAM Flow Layering

29.2 OAM Cell Formats

The I.610 ATM-layer management standard defines cells with a special format, called OAM cells, for VP flows (F4) and VC flows (F5) on either an end-to-end or a switch-to-switch (i.e., segment) basis as described above. Figure 29.3 depicts the format of these special F4 and F5 OAM cells, illustrating the specific coding used to distinguish end-to-end and segment flows within a virtual path or a virtual connection. Note that this use of VCIs within a virtual path and use of Payload Type (PT) within a virtual channel forces OAM cells to implicitly follow the same sequence of switches as user cells. This fact enables many of the ATM OAM functions covered below.

As described in Chapter 12, VP flows (F4) utilize different VCIs to identify whether the flow is either end-to-end (VCI=3) or segment (VCI=4). Recall that the first 16 VCIs are reserved for future standardization. For a VC flow (F5) a specific VCI cannot be used because all VCIs are available to users in the VCC service. Therefore, the Payload Type (PT) differentiates between the end-to-end (PT=100) and segment (PT=101) flows in a VCC.

F4 (VP) OAM Cell Format

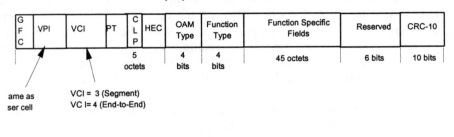

G F C	VPI	VCI	PT	C L P	HEC	OAM Type	Function Type	Function Specific Fields	Reserved	CRC-10
					5 octets	4 bits	4 bits	45 octets	6 bits	10 bits

ame as
ser cell

VCI = 3 (Segment)
VC I= 4 (End-to-End)

F5 (VC) OAM Cell Format

G F C	VPI	VCI	PT	C L P	HEC	OAM Type	Function Type	Function Specific Fields	Reserved	CRC-10
					5 octets	4 bits	4 bits	45 octets	6 bits	10 bits

Same as
user cell

PT =100 (Segment)
PT=101 (End-to-End)

Figure 29.3 ATM OAM Cell Types and Format

Standards define generic types of OAM functions and specific functions within them as follows:

- ❥ Fault management
  - ☞ Alarm indication signal/remote defect indication
  - ☞ Loopback
  - ☞ Continuity check
- ❧ Performance management
  - ☞ Forward reporting
  - ☞ Backward reporting
  - ☞ Monitoring and reporting

Table 29.1 summarizes the OAM type and function type fields in the OAM cells from Figure 29.3. The four OAM types are fault management, performance management, activation/deactivation, and system management as defined in I.610. Each OAM type has further function types with codepoints as identified in Table 29.1. For the fault management OAM type there are Alarm Indication Signal (AIS), Remote Defect Indication (RDI) (also called Far End Reporting Failure (FERF) , and continuity check function types.   For the performance management OAM type there are forward monitoring and backward reporting function types.   The third OAM type defines activation and deactivation of the other OAM types. Currently, there are activation and deactivation function types for performance management and the continuity check.   The final OAM type, system management, has a reserved codepoint, but is currently undefined.

Table 29.1   OAM Types and OAM Function Types

OAM Type		Function Type	
Fault Management	0001	AIS	0000
	0001	RDI/FERF	0001
	0001	Continuity Check	0100
	0001	Loopback	1000
Performance Management	0010	Forward Monitoring	0000
	0010	Backward Reporting	0001
Activation/Deactivation	1000	Performance Monitoring	0000
	1000	Continuity Check	0001
System Management	1111	Not standardized	

Note that there are a significant number of unassigned codepoints in the OAM and function types.   The ATM Forum UNI 3.1 specification allows as much as one second response time for some of these OAM functions.   This means that these functions may be implemented in software.

The "function-specific" fields of ATM OAM cells defined in subsequent sections are based upon the ITU-T I.610 Recommendation.   Many of these functions were first defined in the ATM Forum's B-ICI specification [2]. Subsequent text describes the functional elements of each type and particular values from the B-ICI specification.   Standards define two function-specific fields for the fault management type, one for Alarm Indication Signal (AIS) and Remote Defect Indication (RDI) (previously called a Far End Reporting Failure (FERF)) and a second for loopback.   The continuity check format as currently has no function-specific fields.   There is only one format for the performance management type and the activation/deactivation type.

29.3 Fault Management

The ATM layer uses an approach based upon the SDH paradigm.   Fault management determines when there is a failure, notifying other elements of the connection regarding the failure and providing the means to diagnose and isolate the failure.   Let's consider an analogy for fault management with vehicular traffic where the road itself actually fails! For example, imagine a divided highway with separate bridges crossing a river. A flash flood may wash out one or both of the bridges. The vehicles crossing one bridge cannot see the other bridge because this is very rugged country. The motorists who just passed over the bridge and saw it collapse will travel to the next police station and report the bridge failure. If both bridges wash out, then the police know they must divert traffic away from the bridge in both directions. If the bridge washes out in one direction, then the failure must be reported in one direction, and another vehicle must travel across the remaining bridge in the other direction in order to divert traffic away from the failed bridge.

29.3.1 AIS and RDI/FERF Theory and Operation

Figure 29.4 illustrates the ATM OAM cell AIS and RDI function-specific fields. The meaning of each field is described below.

- **Defect Type** indicates the type of failure as either: unspecified, VP/VC layer defect, or a lower layer defect.

- **Defect Location** indicates where the failure occurred. This is an optional field. If present, the RDI cell contains the same information as the corresponding AIS cell.

Defect Type*	Defect Location*	Unused*
1	16	28

octets

* Default Coding = '6A'Hex for all octets

Figure 29.4   Function-Specific Fields for AIS and RDI/FERF

The following example illustrates the basic principle of Alarm Indication Signal (AIS) and Remote Defect Indication (RDI) shown in Figure 29.5. We cover two failure cases: (a) the failure occurs in both directions simultaneously, and (b) the failure occurs in only one direction. In both examples there a VP (or VC) connection exists between node 1 and node 4. Figure 29.5a illustrates a typical failure in both directions of the physical layer between nodes 2 and 3 that causes the underlying VPs and VCs to simultaneously fail. The failures in each direction are indicated as "Failure-A" and "Failure-B" in the figure so that the resulting AIS and RDI cells can be traced to the failure location. A node adjacent to the failure generates an AIS signal in the downstream direction to indicate that an upstream failure has occurred, as indicated in the figure. As can be seen from example a, both ends of the connection (nodes 1 and 4) are aware of the failure because of the AIS alarm that they receive. However, by convention, each generates an RDI signal.

Figure 29.5b illustrates the purpose of the RDI signal. In most communications applications the connection should be considered failed even if it fails in only one direction. This is especially true in data communications where each packet often requires acknowledgment in the reverse direction. Example b illustrates the case of a failure that affects only one direction of a full-duplex connection between nodes 2 and 3. Node 3, which is downstream from the failure, generates an AIS alarm, which propagates to the connection end (node 4), which in turn generates the RDI signal. The RDI signal propagates to the other connection end (node 1), which is now aware that the connection has failed. Without the RDI signal, node 1 would not be aware that there was a failure in the connection between nodes 2 and 3. This method will also detect any combination of single-direction failures. Note that the node(s) that generate the AIS signals know exactly where the failure is, and could report this to a centralized network management system or take a distributed rerouting response.

a. Failure in Both Directions

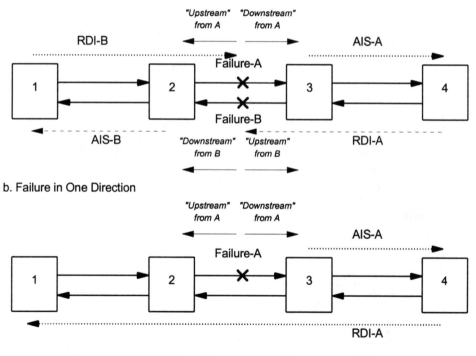

b. Failure in One Direction

Figure 29.5 Illustration of AIS and RDI Theory and Operation

Figure 29.6 illustrates the relationship between the physical layer, VP layer and VC layer failures, AIS, and RDI/FERF indications at a VP and VC level connecting point. Note that a VC connection point is a VP endpoint. The figure shows an exploded view of a VP/VC switch that has PHYsical (PHY) path, Virtual Path (VP), and Virtual Channel (VC) interfaces entering and leaving physical, VP, and VC connection points indicated by the circles. In the top half of the figure the receive PHY path and the ATM VP and VC links enter from the left, terminate on the connection points, and are switched through to the downstream VC, VP, and PHY links. The bottom half of the figure illustrates the upstream physical link and VP link for the VP endpoint. Logic in equipment associated with the respective connection endpoints process the failure, AIS, and FERF indications as indicated in the figure.

At every ATM device (which is always a transmission path termination) either PHY-AIS or physical layer failure is detected (as shown by the "?" in the diamond-shaped decision symbol), resulting in PHY-RDI/FERF being sent upstream. At a VP connection point a detected PHY-AIS causes VP-AIS to be generated in the downstream direction. For a VP endpoint, if VP-AIS is received then VP-RDI/FERF is sent in the upstream direction. If VP-AIS, a physical failure, or PHY-AIS is received, for the VP endpoint corresponding to the VC connection points, then AIS is sent downstream for those VCs on which the AIS feature is activated.

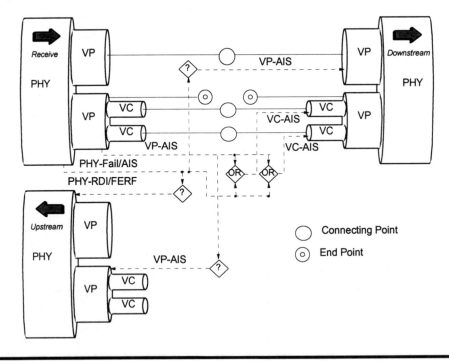

Figure 29.6 Physical/ATM AIS and RDI/FERF Connecting Point Functions

Figure 29.7 illustrates the relationship between the physical layer, VP layer and VC layer failures, AIS, and RDI/FERF indications at a VP and VC endpoint. This would correspond to the actions taken at an ATM end system. Exactly the same as in the switch, if either PHY-AIS or physical layer failure is detected, then PHY-RDI/FERF is sent upstream. For a VP endpoint, if VP-AIS is received, a physical failure is detected, or PHY-AIS is detected, then a

VP-RDI/FERF is returned in the upstream direction. A VP endpoint exists for sets of VC endpoints. For those VC endpoints for which fault management is activated, the receipt of VP-AIS, detection of a physical failure, detection of PHY-AIS, or detection of VC-AIS results in VC-RDI/FERF being returned in the upstream direction.

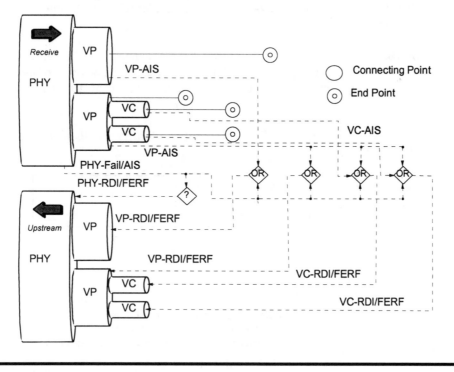

Figure 29.7   Physical/ATM AIS and RDI/FERF Endpoint Functions

If you got the impression that the preceding AIS and RDI/FERF function is complicated, and that a single physical failure can generate many VP and/or VC OAM cells, then you were correct. Once a fault condition is detected, the OAM cell is sent periodically. In order to limit the number of OAM fault management cells, the period for generating OAM cells is on the order of seconds. Furthermore, the ATM Forum UNI does not require VC-AIS and VC-RDI/FERF, while the ATM Forum B-ICI specification specifies that VC-AIS and VC-RDI/FERF is to be used on a selective basis only. These restrictions constrain the amount of OAM cells that are generated (and processed) such that the very useful function of AIS and RDI/FERF is delivered in an efficient manner.

29.3.2 Loopback Operation and Diagnostic Usage

Figure 29.8 illustrates the ATM OAM cell Loopback function-specific fields. A summary of the ATM OAM cell loopback function specific fields is:

↳ **Loopback Indication** is a field that contains '0000 0001' when originated. The loopback point changes it to a value of '0000 0000.' This prevents the cell from looping around the network indefinitely.

↳ **Correlation Tag** is a field defined for use by the OAM cell originator since there may be multiple OAM cells in transit on a particular VPC/VCC. This field allows the sender to identify which one of these has been received.

↳ **Loopback Location ID** is an optional field provided to the sender and receiver for use in segment loopbacks to identify at which connecting point the loopback should occur. The default value of all 1s indicates that the loopback should occur at the end point.

↳ **Source ID** is an optional field provided to identify the loopback source in the OAM cell.

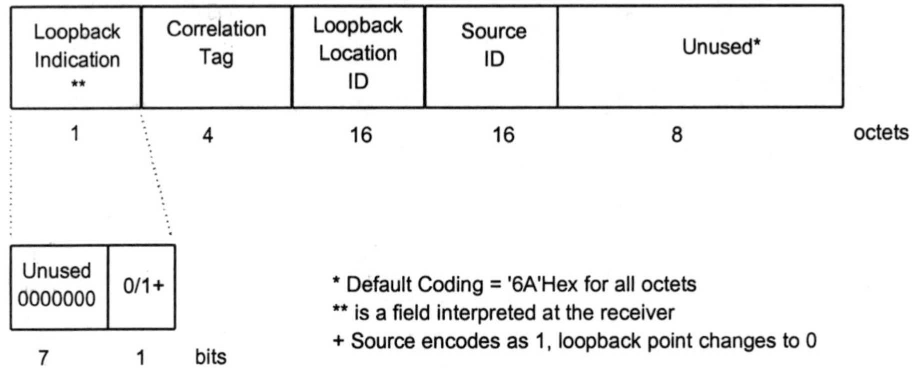

Loopback Indication **	Correlation Tag	Loopback Location ID	Source ID	Unused*
1	4	16	16	8

octets

Unused 0000000	0/1+
7	1

bits

* Default Coding = '6A'Hex for all octets
** is a field interpreted at the receiver
\+ Source encodes as 1, loopback point changes to 0

Figure 29.8   ATM OAM Loopback Function-Specific Fields

As seen in the preceding section, AIS and RDI/FERF are most useful in detecting and identifying to the connection endpoints that a failure has occurred. In some cases there may not be a hard failure, or we may just want to verify continuity. An example of a failure that AIS/FERF does not detect is a misconfiguration of VPI and/or VCI translations in a VPC/VCC service, such that cells do not reach the destination endpoint. The loopback function type OAM cells helps diagnose these types of problems. Let's look at the operation of the loopback function in more detail.

Figure 29.9 illustrates the two basic loopback functions: segment and end-to-end. The OAM cell flow type defined earlier determines whether the loopback cell is either segment or end-to-end. Figure 29.9a illustrates the operation of a segment loopback normally used within a network. The originator of the segment OAM loopback cell at node 1 includes a Loopback Indication of 1, a loopback ID indicating the last connecting point of the segment (node 3), and a correlation tag of "X" so that it can match the received OAM cell. The loopback destination switch (node 3), extracts the loopback cell, changes the Loopback Indication field to 0, and transmits the loopback cell in the opposite direction as shown in the figure. Note that every node is extracting and processing every OAM cell, which results in node 2 transparently conveying the segment loopback cell between nodes 1 and 3. Eventually, node 1 extracts the loopback cell, confirms that the loopback indication changed to 0, and matches the correlation tag. This procedure verifies that ATM continuity exists between the segment of VP (or VC) links from node 1 to node 3. If a physical loopback were in place between nodes 2 and 3, then node 1 would have been able to determine this because the loopback indication would still have a value of 1.

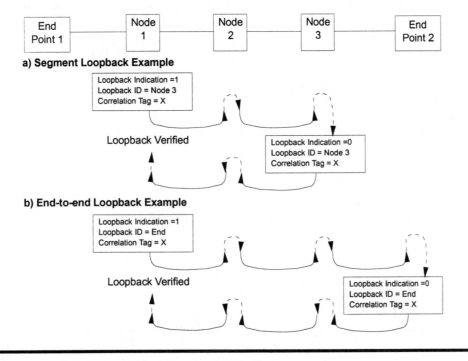

Figure 29.9 Illustration of Basic Loopback Primitives

Figure 29.9b illustrates an end-to-end loopback that could be used by a network switch to verify connectivity with an endpoint, or by an endpoint to verify connectivity with the distant endpoint. In the example, node 1 performs an end-to-end loopback to endpoint 2. Node 1 inserts an end-to-end OAM loopback cell that has a Loopback Indication of 1, a Loopback ID indicating the endpoint (i.e., all ones), and a correlation tag of "X" that it uses to match received OAM cells. The loopback destination, endpoint 2, extracts the loopback cell, changes the Loopback Indication field to 0, and transmits the loopback cell in the opposite direction as shown in the figure. Note that switch nodes and the endpoints are extracting and processing every OAM cell, which results in nodes 2 and 3 transparently conveying the end-to-end loopback cell between node 1 and end point 2. Eventually, node 1 extracts the loopback cell and matches the correlation tag, thus verifying continuity the of VP (or VC) links from node 1 to endpoint 2.

Figure 29.10 illustrates how the segment and end-to-end loopback cells can be used to diagnose a failure that AIS and RDI/FERF cannot. An example of such a failure would be a misconfigured VP or VC Permanent Virtual Connection (PVC). The example shows two endpoints and two intervening networks, each with three nodes. Part a shows the verification of end-to-end continuity via an end-to-end loopback to endpoint 1. If this were to fail, then network 2 could diagnose the problem as follows. Part b shows verification of connectivity between a node in network 2 and endpoint 2 via an end-to-end loopback. If this fails, then the problem is between network 2 and endpoint 2. Part c shows verification of connectivity to endpoint 1 via an end-to-end loopback. If this fails, there is a problem in the link between endpoint 1 and network 1, a problem in network 1, or a problem in the link between networks 1 and 2. Part d shows verification of connectivity across networks 1 and 2 via a segment loopback. If this succeeds, then the problem is the access line from endpoint 1 to network 1. Part e shows verification of connectivity from entry to exit in network 1. If this succeeds, then the problem is in network 1. Verification within any of the networks could also be done using the segment loopback.

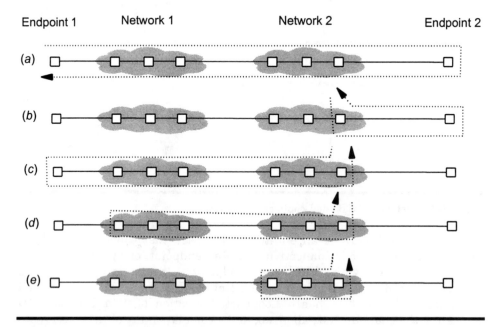

Figure 29.10   Usage of Loopback in Verification/Problem Diagnosis

29.3.3 Continuity Check

The idea behind the continuity check cell is that the endpoint sends a cell periodically at some predetermined interval if no other traffic is sent on the connection so that the connecting points and the other endpoint can distinguish between a connection that is idle and one that has failed. Continuity checking is activated and deactivated by the procedures that we described earlier.

The continuity check cell currently has no standardized function-specific fields. This function is being considered for the Virtual Path (VP) only, and may not be active for all Virtual Paths. Continuity checking is activated and deactivated by the procedures described in the next chapter. The ATM Forum currently does not specify support for the continuity check for interoperability at the ATM User Network Interface.

The continuity check detects failures that AIS cannot, such as an erroneous VP cross-connect change as illustrated in Figure 29.11. Part a shows a VP connection traversing three VP cross-connect nodes with VPI mappings shown in the figure carrying only Continuity Check (CC) cell traffic downstream, interleaved with the user's VP cells. In part b an erroneous cross-connect is made at node 2, shown shaded in the figure, interrupting the flow of CC cells. In part c node 3 detects this continuity failure and generates a VP-RDI/FERF OAM cell in the opposite (upstream) direction.

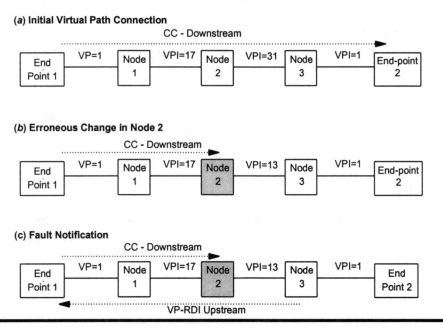

Figure 29.11   Illustration of Continuity Check OAM Cell Usage

29.4 Restoration

The standards currently do not specify what can be done in response to a fault at the ATM layer. There are SONET and SDH standards, however, that define physical layer protection switching on a point-to-point, 1:N redundant basis or a ring configuration. There are also restoration strategies for partial mesh networks. These same concepts could also be applied to restore ATM connections. Restoring Virtual Paths (VPs) that carry a large number of Virtual Channels (VCs) would be an efficient way to perform ATM-level restoration. We briefly discuss these three restoration methods with reference to Figure 29.12.

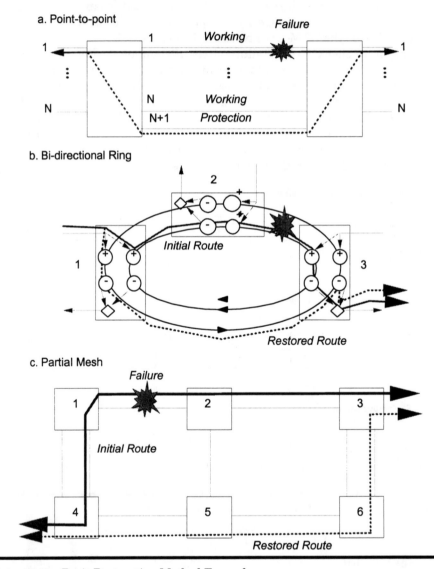

Figure 29.12   Basic Restoration Method Examples

The term 1: N (read as "one for N") redundancy means that there is one bidirectional protection channel for up to N working bi-directional channels, as illustrated in Figure 29.12a. If a working channel fails, its endpoints are switched to the protection channel. If a failure occurs and the protection channel is already in use or unavailable, then the failed working channel cannot be restored.

Figure 29.12b illustrates a bidirectional ring. Traffic from node 1 to node 3 is sent in both directions around the ring. At each node signals may be added as shown by the plus sign inside the circle, or dropped as shown by the minus sign inside the circle. At the receiver (node 3) only one of the signals is selected for output. Upon detection of a failure the receiver will switch to the other, redundant signal, as shown in the example for a failure between nodes 2 and 3. The mirror image of this capability is in place for traffic between nodes 3 and 1. Note that the ring architecture achieves 1:1 redundancy, or in other words, only half of the transmission bandwidth is available to traffic. These types of ring architectures are economically attractive for metropolitan areas and can be further optimized for ATM using Virtual Path level Add Drop Multiplexing (ADM) to improve multiplexing efficiency [3].

Figure 29.12c illustrates a partial mesh network. In the example, traffic between nodes 3 and 4 initially follows the route shown by the solid line. A failure between nodes 1 and 2 impacts the node 3 to 4 traffic on its current route. This can be detected in a centralized, distributed, or hybrid network management system to find a new route shown by the dashed line in the example. Longer distance networks tend to have the type of lattice structure shown in the example and tend to use this type of restoration. In the example of Figure 29.12c, 1:2 redundancy is provided for some routes.

29.5 PERFORMANCE SPECIFICATION AND MEASUREMENT

Quality of Service (QoS) performance is either specified and measured for each individual Virtual Path Connection (VPC) or Virtual Channel Connection (VCC), or measured over the aggregate of many VCCs (or even VPCs). This choice between performance and measurement on an individual or aggregate basis has several implications.

QoS is specified and measured for each connection in the individual case. Devices measure QoS by inserting Operations And Maintenance (OAM) cells on each connection, increasing cell traffic and introducing additional complexity (and cost) for processing these OAM cells. With individual specification and measurement, a situation analogous to a classical digital private line exists, where performance on every line is measured. For example, the error statistics on a DS1 line are estimated from the Extended SuperFrame (ESF) Cyclic Redundancy Check (CRC) for each DS1.

In the aggregate case, QoS is averaged over a large number of connections, which is a more natural interpretation for virtual circuits. Measurement on the aggregate assumes that the performance of all VCCs on a common VPC is identical, which significantly reduces the number of measurement cells that must be transmitted and processed.

Typically, the cost and complexity of individual measurement is justified when it is critical to ensure that the performance of an individual virtual connection is being achieved. Normally, measurement on the aggregate is adequate to ensure that the QoS of a group of virtual connections is being met, and hence the QoS of the individual virtual connection is being met on a statistical basis.

Let's now look at some basic definitions and OAM cells and procedures involved in performance measurement. The complexity of these functions is a commonly cited reason for only measuring QoS in the aggregate case, since individual connection monitoring would be too expensive.

29.6 NP/QoS REFERENCE CONFIGURATION

Network Performance (NP) is observed by the network at various points. Quality of Service (QoS) is observed by the user on an end-to-end basis. The starting point for defining this difference in perception is a reference configuration, against which some observations can be objectively measured and compared.

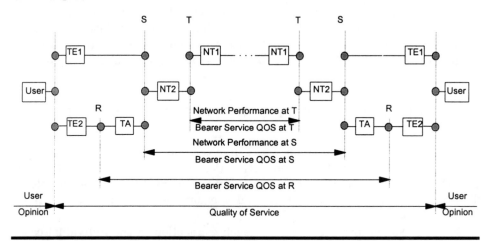

Figure 29.13   NP/QoS Reference Configuration

Some phrases that ITU-T Recommendation I.350 uses to define Quality of Service (QoS) are:

☞ "... parameters that can be directly observed and measured ... by the user ..."

☞ "... may be assured to a user at service access points by the network provider ..."

☞ "... are described in network independent terms ..."

By comparison, I.350 defines Network Performance (NP) using the following phrase:

☞ "... parameters ... meaningful to the network provider ... for ... system design, configuration, operation and maintenance."

Figure 29.13 illustrates these concepts of QoS and Network Performance in a reference configuration derived from ITU-T Recommendation I.350.

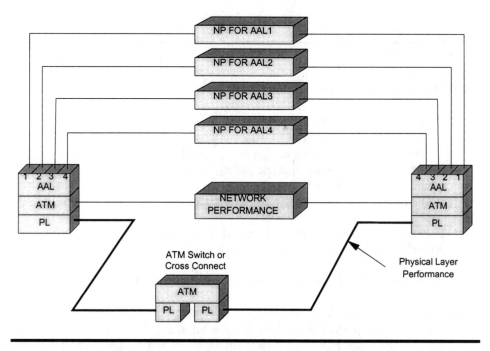

Figure 29.14 QoS/NP for Physical ATM and Adaptation Layers

Figure 29.14 illustrates how Network Performance and QoS apply only to the ATM layer. The NP/QoS for AALs is an item for future study and standardization. This includes common part and service-specific aspects. This will be the QoS that applications actually see.

29.7 ATM NP/QoS MEASUREMENT

This section describes how ATM activates and deactivates Performance Measurement (PM) procedures. We then describe the performance measurement process. A key objective is to measure or estimate these parameters *in-service*; that is, the customer traffic is not impacted by the measurement process.

29.7.1 Activation/ Deactivation Procedures

Figure 29.15 depicts the ATM OAM cell Activation/Deactivation function-specific fields. The following text defines the meaning of each field.

Message ID	Directions of Action	Correlation Tag	PM Block Sizes A-B**	PM Block Sizes B-A**	Unused Octets *
6	2	8	4	4	42x8 bits

* Default coding = '6A' Hex
** Default coding = '0000'

Figure 29.15   ATM Activation/Deactivation OAM Cell Function-Specific Fields

⊠ **Message ID** is defined as follows:

'000001'	Activate (Request)
'000010'	Activation Confirmed
'000011'	Activation Request Denied
'000101'	Deactivate(Request)
'000110'	Deactivation Confirmed
'000111'	Deactivation Request Denied

⊠ **Direction of Activation** coding indicates the A-B direction ('10'), B-A direction ('01'), or a two-way action ('11').

⊠ **Correlation Tag** enables nodes to correlate commands and responses.

⊠ **PM Block Size A-B** identifies the Performance Measurement (PM) block size supported from A to B by a bit mask for sizes of 1024, 512, 256, or 128 from most significant bit (msb) to least significant bit (lsb), respectively. The default value of all zeroes indicates no block size (e.g., when used for activating continuity checking).

⊠ **PM Block Size B-A** identifies the block sizes supported from B to A using the same convention as above. Note that the block sizes may differ in each direction.

Figure 29.16 illustrates the activation/deactivation procedure for performance monitoring or continuity checking. In the example, connection/segment endpoint A generates a De(Activate) request toward B requesting action on either the A-to-B, B-to-A, or both directions. If B can comply with all of the requests, then B returns a (De)Activation Confirmed message. If B cannot comply with the request(s), then it returns a (De)Activation Request Denied message. A denial can result if the endpoint is unable to support the performance management function. If the deactivation refers to a single function that is currently not operating, then the request is also denied. If the deactivation request refers to both directions, yet only one direction is operating, then the request is confirmed, with the reference to the nonoperational function ignored. If the Deactivation refers to a single function that is currently not operating, then the request is also denied. If the Deactivation request refers to both directions, yet only one direction is operating, then the request is confirmed, with the reference to the nonoperational function ignored.

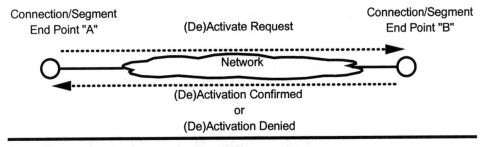

Figure 29.16   Illustration of Activation/Deactivation Flow

Once a performance measurement flow is activated the procedure described in the following section is performed. Activation and Deactivation allow the performance measurement to be performed on selected VPCs and VCCs. This keeps the total processing load required for performance measurements manageable. Continuous monitoring or sampled monitoring uses these Activation and Deactivation procedures.

29.7.2 Performance Measurement Procedure

Figure 29.17 depicts the ATM OAM cell Performance Management (PM) function-specific fields. The following text defines the meaning of each field.

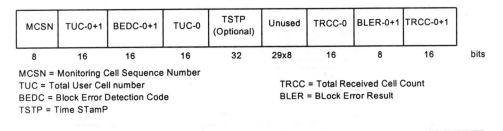

Figure 29.17   Performance Measurement Function Specific Fields

⊠ **Monitoring Cell Sequence Number (MCSN)** is the PM cell number, modulo 256.

⊠ **Total User Cell (TUC)** is the total number of CLP=0+1 cells (TUC-0+1) or CLP=0 cells (TUC-0) containing user data sent since the last PM cell.

⊠ **BEDC-0+1** is a block error detection code computed over all of the CLP=0+1 user cells since the last PM cell. PM procedures employ it for error rate estimation.

⊠ **Time Stamp (TSTP)** is an optional field used in the backward reporting direction to estimate delay.

⊠ **BLock Error Result (BLER-0+1)** is a count of the number of errored parity bits detected across all user CLP=0+1 cells in the last block for backward reporting.

⊠ **Total Received Cell Count (TRCC)** is the number of received CLP=0+1 cells (TRCC-0+1) or received CLP=0 cells (TRCC-0) for use in backward reporting.

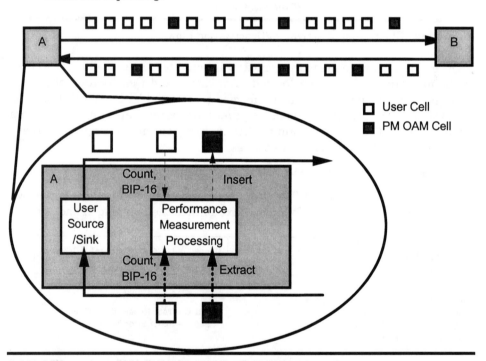

Figure 29.18   ATM OAM Performance Measurement Procedure

Figure 29.18 illustrates the operation of the performance measurement (PM) OAM in cell insertion and processing. The connection or segment endpoints A and B that are involved are determined by the activation/deactivation procedure. In this example the PM cell flow in each direction has different block sizes, every 4 cells from left to right, and every 2 cells from right to left. The functions involved are insertion of OAM cells, counting user cells, and computing the 16-bit parity on the transmit side. At the destination, the receiver extracts OAM cells. The receiver makes the same counts, and recomputes the 16-bit parity for comparison with the value received in the monitoring cell computed by the transmitter. Note that the monitoring cell contains the results for the cells in the preceding block.

Performance monitoring OAM cells detect the following types of impairments on ATM virtual path and channel connections:

☞ Missing or lost cells
☞ Many bit error patterns
☞ Extra or misinserted cells
☞ Delay and delay variation

Higher-level network management systems can then utilize this data to determine if the desired ATM-level Quality of Service (QoS) is being delivered. Calculations based upon the above measurements readily estimate ATM quality of service parameters, such as Cell Loss Ratio (CLR), Cell Error Ratio (CER), and Cell Transfer Delay (CTD). Another means to estimate QoS is to connect ATM test equipment to test connections in the same service category that traverse the same switches and trunks that user cells traverses.

29.8 NP/QoS PARAMETER ESTIMATION

This section defines the Quality of Service (QoS) and Network Performance (NP) parameters in terms of basic cell transfer outcomes. The text also describes a method to estimate these QoS parameters from the Performance Measurement (PM) OAM cells.

29.8.1 ATM Cell Transfer Outcomes

QoS is measured based upon cell entry events at one end and cell exit events at the other end, as illustrated in Figure 29.19. These two events, defined in ITU-T Recommendation I.356 [4], determine the cell transfer outcomes. These outcomes are then used to define the ATM cell transfer performance parameters.

⊕ A *cell exit event* occurs when the first bit of an ATM cell has completed transmission out of an End User Device to a ATM network element across the source UNI Measurement Point 1.
⊕ A *cell entry event* occurs when the last bit of an ATM cell has completed transmission into an End User Device from an ATM network element across the destination UNI Measurement Point 2.

As illustrated in Figure 29.19, a cell that arrives at the exit point without errors and is not too late (i.e., it arrives in less than or equal to a delay of Tmax seconds) is considered successfully transferred. A successfully transferred cell may arrive with the CLP bit set to zero, or one. A succsessfully transferred cell with the CLP bit set to one is called a tagged cell outcome if it entered the network with the CLP bit set to zero. If a cell arrives within a certain delay (Tmax) but has errors, then it is considered an errored outcome. If a cell arrives too late, or never arrives at all, then it is considered lost. There is also a possibility of a misinserted cell, which is defined as the case when a cell arrives at the exit point for which there was no corresponding input cell at the entry point. This can occur due to undetected cell header errors. Depending upon the higher layer protocol, a misinserted cell can be a very serious problem since a misinserted cell cannot be distinguished from a transmitted cell by the receiver. Fortunately, this is a very unlikely event under normal circumstances.

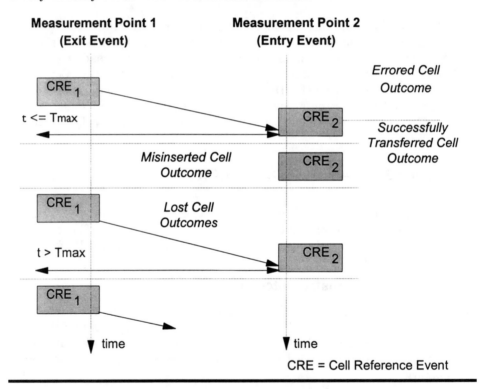

Figure 29.19 Cell Entry Event, Exit Events and Outcomes

The following possible cell transfer outcomes between measurement points for transmitted cells are defined based on ITU-T Recommendation I.356.

☺ **Successful Cell Transfer Outcome:** The cell is received corresponding to the transmitted cell within a specified time Tmax. The binary content of the received cell conforms exactly to the corresponding cell payload and the cell is received with a valid header field after header error control procedures are completed.

☺ **Tagged Cell Transfer Outcome:** This is a successful cell cell transfer outcome where the CLP bit was changed from 0 at measurement point 1 to 1 when received at measurement point 2.

☺ **Errored Cell Outcome:** The cell is received corresponding to the transmitted cell within a specified time Tmax. The binary content of the received cell payload differs from that of the corresponding transmitted cell, or the cell is received with an invalid header field after header error control procedures are completed.

☺ **Lost Cell Outcome:** No cell is received corresponding to the transmitted cell within a specified time Tmax (examples include "never arrived" or "arrived too late").

⊗ **Misinserted Cell Outcome:** A received cell for which there is no corresponding transmitted cell.

⊗ **Severely Errored Cell Block Outcome:** When M or more Lost Cell outcomes, Misinserted Cell Outcomes, or Errored Cell outcomes are observed in a received cell block of N cells transmitted consecutively on a given connection.

29.8.2 ATM Performance Parameters

This section summarizes the set of ATM cell transfer performance parameters defined in I.356. The definitions of these performance parameters use the cell transfer outcomes defined above. This set of ATM cell transfer performance parameters correspond to the generic QoS criteria shown in parentheses as follows:

+ Cell Error Ratio	(Accuracy)
+ Severely Errored Cell Block Ratio	(Accuracy)
+ Cell Loss Ratio	(Dependability)
+ Cell Misinsertion Rate	(Accuracy)
+ Cell Transfer Delay	(Speed)
+ Mean Cell Transfer Delay	(Speed)
+ Cell Delay Variation	(Speed)

The QoS definitions in I.356 summarized here apply to cells conforming to the traffic contract only. Extending the QoS definitions to include nonconforming cells is an area of future standardization. This means that nonconforming cells must be excluded from the cell transfer outcomes. Principally, this affects the cell loss ratio calculation.

The draft ANSI standard [5] assigns the cell counts to all cells, conforming and non-conforming. The spirit of I.356 is still preserved in the ANSI work in that it specifies that counts of the number of cells discarded or tagged due to noncompliance be kept if the monitoring is done in the vicinity of the Usage Parameter Control (UPC) function.

29.8.3 Cell Error Ratio

Cell Error Ratio is defined as follows for one or more connection(s):

$$\text{Cell Error Ratio} = \frac{\text{Errored Cells}}{\text{Successfully Transferred Cells} + \text{Errored Cells}}$$

Successfully Transferred Cells and Errored Cells contained in cell blocks counted as Severely Errored Cell Blocks should be excluded from the population used in calculating the Cell Error Ratio.

Errored Cells (ECs) can only be estimated by counting the number of up to M ($2 \leq M \leq 16$, with a default of 4) parity errors in the BEDC-0+1 code for the block. The successfuly transferred cell count is the Total User Cell number (TUC-0+1) from the PM OAM cell.

29.8.4 Severely Errored Cell Block Ratio

The Severely Errored Cell Block Ratio for one or more connection(s) is defined as:

$$\frac{\text{Severely Errored Cell Block Ratio}}{} = \frac{\text{Severely Errored Cell Blocks}}{\text{Total Transmitted Cell Blocks}}$$

A cell block is a sequence of N cells transmitted consecutively on a given connection. A severely errored cell block outcome occurs when more than a specified number of errored cells, lost cells, or misinserted cells are observed in a received cell block.

An Errored Cell Block (ECB) contains one or more BIP-16 errors, lost cells, or misinserted cells.

A Severely Errored Cell Block (SECB) is a cell block with more than M ($2 \leq M \leq 16$, with a default of 4) BIP-16 errors, or more than K ($2 \leq K \leq M$, with a default of 2) lost or misinserted cells.

29.8.5 Cell Loss Ratio

The Cell Loss Ratio is defined for one or more connection(s) as:

$$\text{Cell Loss Ratio} = \frac{\text{Lost Cells}}{\text{Total Transmitted Cells}}$$

Lost and transmitted cells counted in severely errored cell blocks should be excluded from the cell population in computing cell loss ratio.

The number of lost cells can be estimated as the difference in the past two Total User Counts (TUC) received in the PM OAM cells from the distant end minus the number of cells actually received in a cell block. If this result is negative, then the estimate is that no cells were lost, and cells were misinserted as defined below.

Note that this estimation method would report zero loss and misinsertion if there are an equal number of cell loss and misinsertion outcomes in a cell block.

29.8.6 Cell Misinsertion Rate

The Cell Misinsertion rate for one or more connection(s) is defined as:

$$\text{Cell \ Misinsertion \ Rate} = \frac{\text{Misinserted \ Cells}}{\text{Time \ Interval}}$$

Severely Errored Cell Blocks should be excluded from the population when calculating the cell misinsertion rate. Cell misinsertion on a particular connection is most often caused by an undetected error in the header of a cell being transmitted on a different connection. This performance parameter is defined as a rate (rather than the ratio) since the mechanism producing misinserted cells is independent of the number of transmitted cells received on the corresponding connection.

The number of misinserted cells can be estimated as the number of cells actually received in a cell block minus the difference in the past two Total User Counts (TUC) received in the PM OAM cells from the distant end. If this result is negative, then cell loss has occurred, and the number of misinserted cells is estimated as zero.

29.8.7 Measuring Cell Transfer Delay

The Cell Transfer Delay is defined as the elapsed time between a cell exit event at the measurement point 1 (e.g., at the source UNI) and a corresponding cell entry event at measurement point 2 (e.g., the destination UNI) for a particular connection. The cell transfer delay between two measurement points is the sum of the total inter-ATM node transmission delay and the total ATM node processing delay between measurement point 1 and measurement point 2.

Of course, only the total delay can be estimated using the Performance Measurement (PM) OAM cells, and not the individual components that were described in Chapter 12. The methods currently proposed are optional and utilize either the time stamp function-specific field or a well-defined test signal.

For the time stamp method, tight requirements for Time-Of-Day (TOD) setting accuracy are essential to the measurement of absolute delay, and are currently not standardized. CDV can be estimated by taking differences in time stamps. Any estimation of delay also assumes that the PM OAM cells are processed exclusively in hardware, and not in software as all other OAM cell types could be. For more information on the time stamp estimation method, see the draft ANSI standard [5].

Figure 29.20 illustrates how absolute delay and differential delay can be measured using the time stamp method. The source and destination have highly accurate time stamp clocks which are set to nearly the same time. The source periodically sends OAM Performance Measurement (PM) cells and inserts its time stamp. These cells enter an ATM network and experience varying delays. As an OAM PM cell leaves the network and arrives at its destination, the time stamp is extracted and several operations are performed on it. First, the absolute delay is calculated as the (non-negative) difference between the local time stamp clock and the time stamp received in the OAM PM cell. Next, the value in a memory is subtracted from the absolute delay to yield a differential delay. Finally, the current absolute delay calculation is stored in the memory for use in calculation of the next differential delay.

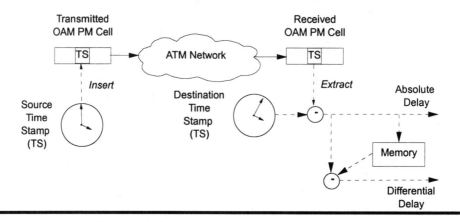

Figure 29.20   Time Stamp Method Delay Estimation

The Mean Cell Transfer Delay is the average of a specified number of absolute cell transfer delay estimates for one or more connections. The above 2-point Cell Delay Variation (CDV) defined in I.356 can be estimated from the differential delays. A histogram of the differential delay can be computed, as well the mean, the variance, or other statistics

ITU-T Recommendation I.356 also define the 1-point CDV as the variability in the pattern of cell arrival events observed at a single measurement point with reference to the negotiated peak rate 1/T as defined in the traffic contract. Figure 29.21 illustrates a method to implement this measurement. A Continuous Bit-Rate (CBR) source emits a cell once every T seconds (note this implies that T is a multiple of the cell slot time in the TDM transmission convergence sublayer). This cell stream, perfectly spaced, is transmitted across an ATM network that introduces variations in delay that we wish to measure. The receiver knows the spacing interval T, and can compute the interarrival times of successive cells and subtract the time T to result in a 1-point CDV estimate. Positive values of the 1-point CDV estimate correspond to cell clumping, while negative values of the 1-point CDV estimate correspond to gaps, or dispersion, in the cell stream. As shown in Figure 29.21 cell clumping occurs for cells that are closer together, while dispersion occurs for cells spaced too far apart. This is important for determining the likelihood of overrun and underrun for CBR services as described in Chapter 16.

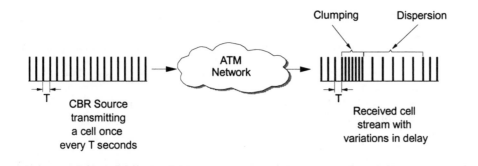

Figure 29.21   CBR Test Source Measurement of CDV

29.9 REVIEW

This chapter defined the reference configuration for ATM Operations And Maintenance (OAM) flows at the physical layer and the ATM Virtual Path (VP) and Virtual Channel (VC) levels. The reference model defines connecting points and endpoints for VPs and VCs at the ATM layer. Coverage moved to definition of the OAM cell format and how VPs and VCs use it on either an end-to-end or segment basis. The discussion then proceeded to the topic of fault management in detail, giving an automotive analogy of a bridge failure as an example. This includes the definition of the Alarm Indication Signal (AIS) and Far End Reporting Failure (FERF), also called a Remote Defect Indication (RDI) in recent standards work. Next, the chapter described the use of loopback and continuity check functions in determining and diagnosing faults where AIS and RDI/FERF does not.

The text then defined the reference configuration for Network Performance (NP) and Quality of Service (QoS) measurement. Basically, QoS is what the user perceives, while NP is what the network uses to make design, operational, or capacity decisions. Next the ATM OAM cell activation/deactivation format and procedure which applies to the continuity check of Chapter 20, as well as, Performance Measurement (PM) was defined. Then decrtiption and examples explained the ATM cell measurement method and the possible outcomes: successfully transferred cell, a lost cell, an errored cell, or a misinserted cell. These outcomes define the NP/QoS parameters of the traffic contract and determine how the OAM PM cells are used estimate them. The chapter closed with a look at the future direction of some key issues in performance measurement.

29.10 REFERENCES

[1]   ITU-T, *B-ISDN Operation And Maintenance Principles And Functions*, Recommendation I.610, November 1995.

[2]   ATM Forum, "BISDN Inter Carrier Interface (B-ICI) Specification, Version 1.0," August 1993.

[3]   T. Wu, "Cost Effective Network Evolution," *IEEE Communications*, September 1991.

[4]   ITU-T, *B-ISDN ATM layer cell transfer performance*, Recommendation I.356, October 1996.

[5]   ANSI, "B-ISDN Operations and Maintenance Principles and Functions (T1S1.5/93-004R2)," January 1994.

Competing Technologies and Future Directions Involving ATM

The telecommunications world continues changing at a mind numbing pace. As Einstein predicted in his famous theory of special relativity in 1905, strange things happen when traveling at high speeds. Strange things are indeed happening in the world of communications networking that will change the way we live and work together in the twenty-first century. As evidence, note the ever increasing combinations of applications operating over communications networks: voice over IP, ATM over wireless, voice over frame relay, facsimile over IP, and even ATM over the existing twisted pair of wires entering your home! Since the key objective of B-ISDN and ATM is to support a broad range of applications, this part compares how well ATM, IP, frame relay, and LANs support voice, video, and data. A recurring theme is how ATM aspires to serve the LAN as well as the WAN; while IP aims to serve the voice domain classically held by telephony and ISDN.

While the first seven parts of the book were largely objective, this last part moves into the more subjective areas of comparison and prediction. In many of these areas the standards are incomplete, hence the analysis proceeds by making assumptions regarding the current state of draft standards. Most of the text does not present the author's opinions; rather, it categorizes other opinions published in the literature or discussed in open forums in a coherent fashion. To the extent possible, we attempt to separate hype and hope from hard facts and reality. Toward this end, the comparison analysis employs objective measures, suitability to application needs, network attrib-

utes, and services provided to end users. The presentation provides numerous references, many accessible on the web, for the reader interested in following up on a particular subject or opinion in more detail.

Chapter 30 surveys technologies competing with ATM in the WAN. The analysis begins with a review of fundamental characteristics of circuit, message, and packet switching. Using a functional model similar to that employed in Chapter 16, the analysis considers the WAN backbone and access components. Next, the chapter focuses on comparing how well ATM, IP and frame relay meet the requirements for TDM circuit emulation, packet data, voice, and video. The objective measure for this analysis is efficiency.

Chapter 31 surveys technologies competing with ATM in the LAN. The LAN analysis covers the desktop as well as the LAN backbone. The text reviews how Gigabit Ethernet burst on to the scene, surrounded by tremendous hype as a contender for the LAN backbone. The analysis surveys how ATM stacks up against the lower speed 10 and 100-Mbps Ethernet technologies along with this new speedster and the traditional FDDI LAN backbone. The text compares IP's bid for QoS: the ReServation Protocol (RSVP) with ATM's traffic management capabilities. Possibly the most important consideration, the chapter concludes with an analysis of connection-oriented versus connectionless networking paradigms.

Chapter 32 gazes into the crystal ball for a glimpse at ATM's potential future. The coverage includes ATM's new frontiers: high-speed Digital Subscriber Lines (xDSL) over existing twisted pairs, wireless ATM, and ATM over satellite. This final chapter also touches on some fundamental issues involved in achieving QoS with frame based networks. The text also highlights the need for high-performance SVCs driven by higher-layer protocols to take best advantage of ATM's connection oriented infrastructure. Finally, the text discusses some key issues and work in progress expected to exhibit a major impact on the future of ATM.

30

Technologies Competing with ATM in the WAN

Earlier, this book studied some of the technologies and services that both compete with, as well as complement, ATM in the Wide Area Network (WAN). The treatment begins with an overview of the basic attributes of circuit, message, and packet switching. We then summarize key WAN networking requirements. The chapter then embarks upon an objective efficiency and performance analysis. The first stop is a comparison ATM with TDM, or STM technology used in traditional voice switching. The next topic is a comparison of voice over ATM, IP and frame relay. Continuing in this theme, the chapter compares how well frame and cell networking techniques carry variable length data packets. Next, the analysis covers the efficiency of video transport over ATM and IP. Finally, the chapter summarizes the efficiency analysis results.

30.1 CIRCUIT, MESSAGE, AND PACKET SWITCHING

This section introduces a comparison of technologies by summarizing the major communications switching techniques developed during the relatively short history of electromagnetic communications. The text then provides motivations for the choice between circuit switching versus packet switching in terms of some simple application examples. Finally, the text covers the still important area of messaging switching that many of us know today through E-mail, but which began the modern era of communications approximately 150 years ago as telegraphy.

30.1.1 A Taxonomy of Communication Methods

Basically, networks employ one of three methods for communicating information between users: circuit switching, message switching, or packet switching. All other categories of communications fit within this taxonomy as illustrated in Figure 30.1.

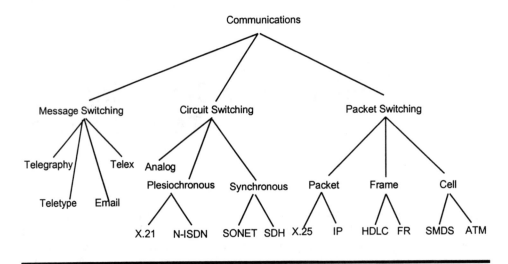

Figure 30.1   A Taxonomy of Data Communication Methods

Message switching started the era of modern communications in 1847 with the invention of the telegraph by Morse and the deployment of message relays connected via overhead wires. Message switching evolved into paper tape teletype relay systems and the modern telex messaging system used for financial transfers and access to remote locations. The most popular message switching system today is electronic mail.

The use of circuit switching in communication networks began over fifty years ago through modems designed for use on analog transmission systems. This evolved into plesiochronous digital transmission systems over the past forty years. The main technologies in use today are digital private lines, X.21 fast circuit switching, and Narrowband ISDN (N-ISDN). Within the past decade, a high-speed transmission network emerged as defined by the ITU-T developed Synchronous Digital Hierarchy (SDH) and a North American SONET variant that provided accurate timing distribution as well as enhanced operations and maintenance support.

Packet switching developed over the past thirty years to overcome the expense and poor performance of transmission systems. Three major classes of packet switching evolved over this period: packet, frame, and cell as indicated as shown in Figure 30.1. The first packet switching systems used by X.25 and IP were designed for extremely poor transmission networks. A simpler frame-based protocol like HDLC sufficed only for local connections where the quality of the links was better. HDLC in turn evolved in the wide area into frame relay. The desire to achieve even higher performance and flexibility led to the development of SMDS; an experience from which ATM designers learned some valuable lessons. Improvements in transmission technology, optics, electronics, and protocol design are key enablers in the continuing evolution of communications networking. The next few sections compare and contrast these three basic networking paradigms.

30.1.2 To Switch or Not to Switch? — An Answer to This Question

This section presents a few generalizations regarding the engineering economics of communications services. First, tradeoffs exist between dedicated and switched communications, as illustrated in Figure 30.2a. Generally, a point occurs where a dedicated facility between two points is more economical than a switched facility for a certain daily usage. This crossover may differ if the individual usage duration is so small that the required switching capability is too expensive. If the point-to-point usage indicates switching is economically desirable, then there are two types of switching to choose from: either connectionless like the Internet Protocol (IP), or connection-oriented like the Narrowband Integrated Services Digital Network family of (ISDN) protocols. As a general rule of thumb, the average transaction duration should be an order of magnitude or greater than the circuit (either physical or logical) setup time, as shown in the setup time regions of Figure 30.2b.

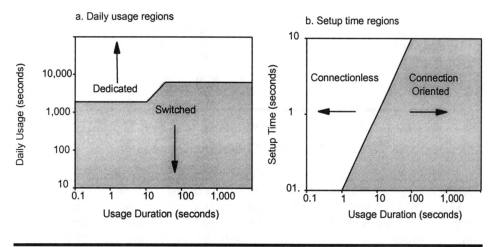

Figure 30.2 Ranges of Data Service Application

Figure 30.3 illustrates the concept of usage duration by showing the time required to transfer an object of a certain size for a range of representative PDH circuit transfer rates. Obviously, a linear relationship exists between transfer time and object size. The figure divides the range of transfer times into two regions based upon the applicability of communications networking technology: connectionless or connection-oriented. Note that the transfer rate is the sustained transfer rate. If the transfer time exceeds more than a few seconds, then connection-oriented service is best. If the transaction takes less than a second, then a connectionless service is best. In some cases, overnight mail is a better solution. For example, backing up a PC disk at 64 kbps would take over one day!

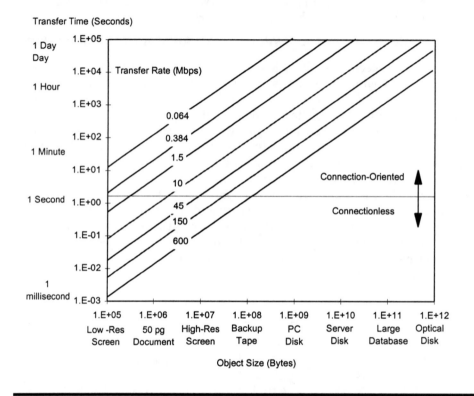

Figure 30.3 Transfer Time versus Transaction Size for Various Transfer Rates

30.1.3 Oh, Where Has Message Switching Gone?

Although digital data communications began with the telegraph, which initially employed human operators to perform the message switching function, it is not the prevalent data communications technology today. The "high-speed" 50-bps teletype run by human operators relaying punched paper tape messages replaced telegraphy. These messaging systems have evolved into the current Telex system of today, without which you could not wire money or reach remote locations.

Why hasn't message switching blossomed in the modern era of data communications? The answer is that messaging has become a higher-layer protocol, and in many cases, an application that operates over packet switching in modern data communication networks. The Simple Mail Transfer Protocol (SMTP) of the Internet drives E-mail around the planet. Packet switching networks allowed non-real-time message applications to coexist with real-time interactive data applications.

The experience gained through this integration of applications with dissimilar performance requirements will enable the networks of tomorrow to carry real-time images, interactive data applications, file transfer, as well as messaging. Let's take a look at some of the key requirements before delving into an efficiency analysis of the WAN protocols competing with ATM.

30.2 KEY WAN REQUIREMENTS

This section introduces some key WAN business and performance require-ments. These include efficiency, support for a wide range of interface speeds, along with quality and throughput tailored to specific applications.

30.2.1 Making Efficient Use of Access and Backbone Bandwidth

Despite what some advocates claim, WAN bandwidth is not free, and is not likely to become significantly less expensive in the foreseeable future. Indeed, the drive to minimize bandwidth charges in the wide area networking environment is a key objective. One way of achieving this is to utilize the most efficient protocol possible to carry information between parties requiring communication services. Hence, the majority of this chapter focuses on efficiency analysis of ATM, frame relay, and IP in support of TDM circuits, voice, packet data, and video information transfer.

WAN bandwidth has two principal components illustrated in Figure 30.4: access and backbone. Access circuits connect large and small business locations, as well as residential subscribers to nodes within an integrated services network. Backbone trunks connect these nodes. Hence, the relative efficiency of various protocols determines the amount of information a particular physical access or backbone circuit can carry. Currently, access charges are the most expensive component of public WAN data services. Hence, protocols that effectively support multiple services on the same access facility are attractive.

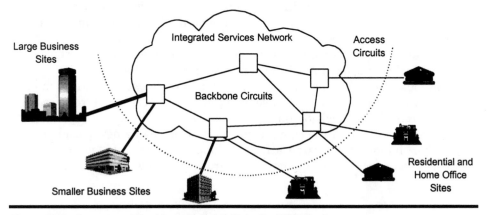

Figure 30.4   Access and Backbone Bandwidth in the WAN Environment

30.2.2 Wide Range of Interface Speeds

A major advantage of ATM in the beginning was that it was developed for high-speed operation; although there is nothing inherent in the protocol itself that prevents operation at lower speeds. In the early 1990s, when frame relay and routers were operating at DS1 and Ethernet speeds, ATM was running at DS3 and OC3 speeds. By the mid-1990s, frame relay and routers operated at DS3 and OC3 speeds. ATM now operates at OC12 speeds in commercial networks while OC48 speeds are available from some vendors. High end routers have also broached the OC12 barrier. Furthermore, operation of frame relay at SONET speeds is also feasible with special purpose hardware. The latest generation of hardware-based routers may reach OC192 (10 Gbps) speeds before ATM does. Therefore, so much overlap exists in terms of supported interface rates that speed *alone* is not a differentiator between various WAN protocols.

Also important in the WAN is support for a wide range of speeds, ranging from analog dial-up for access to the smallest residential users to the highest speeds for backbone trunks. Here, frame relay and IP are on equal footing at the low end, both supporting dial-up access. Conversely, ATM standards currently do not define a means for low-speed dial-up access. However, ATM is the protocol being defined by standards for supporting high-performance Digital Subscriber Line (xDSL) technologies to the residence and home office as described in Chapter 32. Additionally, ATM and frame relay define standard means for support operation at nxDS1 and nxE1 speeds.

30.2.3 Engineering Economic Considerations

Another consideration, secondary only to minimizing bandwidth costs, in WAN networking designs is equipment costs. IP and frame relay already work on most existing data communications hardware platforms. Software support for both of these protocols is also widely available. Few hardware changes are generally required for frame relay at lower speeds, and software is widely available. Upgrading a data communications network to ATM generally requires new hardware and software. Hardware includes new interface cards, and in some cases new switches, hubs, and routers. In many switches and routers, the upgrade to ATM only requires a new interface card. Software includes operating systems, device drivers, and applications.

Many customers have separate voice and data networks. Most modern data networks employ routers to some extent. As these users require more features, they face a decision on whether to replace the routers with higher-performance versions of the same technology, or move to a new paradigm.

Another important consideration is support costs. In more complex networks and applications, support costs are comparable to bandwidth costs.

30.2.4 Quality of Service (QoS) and Throughput

Increasingly, differentiated QoS is becoming a requirement in the WAN, particularly for integrated access. The basic QoS attributes of delay, delay variation and loss defined in Chapter 21 apply to all protocols. Circuit switches have essentially constant nodal delay. On the other hand, any form of packet switching introduces variations in delay. Typically, the lower the speed of the packet switching trunks, the more the variation in delay. Dedicated circuits are best suited to applications that are not bursty and have a strict delay tolerance, such as video, audio, and telemetry data. ATM currently supports the broadest range of burstiness and delay tolerance through the implementation of multiple QoS classes. Some frame relay devices support prioritized queuing; however, a standard means of doing so is not defined yet. As described in Chapter 20, the IP Resource reSerVation Protocol (RSVP) currently defines support for two QoS classes in addition to the current best effort Internet service. The next chapter compares RSVP and ATM QoS support.

Table 30.1 summarizes the approximate QoS and traffic attributes for a number of applications. Typically, real-time applications have low loss requirements since retransmission is not a viable strategy to recover lost information.

TABLE 30.1 Application Traffic Attributes

Application	Burstiness	Loss Tolerance	Response Time	Throughput (Mbps)
Voice	Medium	Low-Medium	Real-Time	.004 to .064
File Transfer	High	High	Batch	.01 to 600
CAD/CAM	High	Medium	Near Real-Time	10 to 100
Transaction Processing	High	Low	Near Real-Time	.064 to 1.544
Channel-Channel	Low	Low	Real-Time	10 to 600
Imaging	High	Medium	Real-Time	.256 to 25
Business Video	Low	Low	Real-Time	.256 to 16
Entertainment Video	Low	Very Low	Near Real Time	1.5 to 50
Isochronous Traffic	Low	Low	Real -Time	.064 to 2.048
LAN-LAN	High	High	Real- Time	1 to 100
Server Access	Avg.	High	Real- Time	1 to 100
Hi-Fi Audio	Low	Low	Real- Time	.128 to 1

Applications require a wide range of peak throughput to achieve acceptable performance. Furthermore, the ratio of peak rate to average rate, defined as burstiness, determines the statistical multiplexing achievable when multiplexing many different sources together on an access line or a backbone trunk as described in Chapter 25. Finally, applications have varying degrees of tolerance for loss and response time. The statistical properties of delay determine the response time as either: real-time (sub-second), near real-time (seconds), or batch (minutes).

30.2.5 Savings from Integrating Multiple Services

Most enterprises and residences have separate voice and data networks, and look to achieve savings through integrating these transmission requirements. As mentioned earlier in this section, a large portion of bandwidth costs accrue on the access side, hence integrated access is a key area of interest. ATM, IP, and frame relay all support transport of voice and data. Currently, however, only IP and ATM define standards in support of video. The need for multimedia and mixed services over a single integrated access line or shared switch port is becoming a key requirement for both the business and residential user. How much savings does this integration achieve? Does it justify the additional cost of new hardware and software?

The following example examines the case of multiple services with different delay requirements sharing a circuit versus establishing separate circuits for each service. The analysis assumes that costs are proportional to the link speed. Chapter 25 presented the basic delay performance model for an individual M/M/1 system and a priority queuing system. This simple example assumes two classes of traffic with different average delay requirements: a real-time class with an average delay of 100 μs and a non-real-time class with average delay of 5 ms. The integrated ATM system uses priority queuing as described in Chapters 22 and 25 for the entire link bandwidth. The separate system uses two separate links, each with half the capacity of the integrated system. The analysis makes a fair comparison by forcing the integrated system real-time load to be equal to the separate system.

Figure 30.5 plots total utilization of the two systems versus link bandwidth. The integrated system has higher utilization because it is carrying much more low-priority traffic than separate packet systems for the high- and low-priority with equal capacity.

Of course, this is only the bandwidth saving for two classes of service. The utilization advantage of an integrated system for more traffic types increases when considering more traffic classes. Further savings due to integration of network management, provisioning, operations, equipment costs, may also be significant. Since operational support costs are a key consideration, look carefully into this area of integration savings.

The next sections compare how well the major WAN technologies competing with ATM meet these application needs in terms of efficiency, QoS, technological maturity and other business considerations.

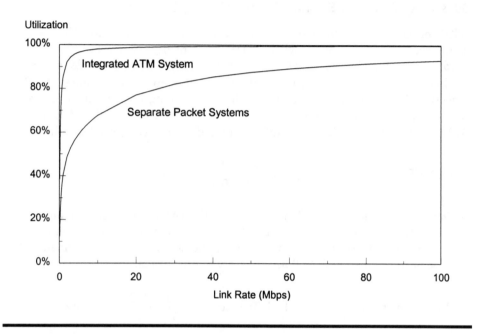

Figure 30.5   Illustration of Savings through Integration

30.3 ATM COMPARED WITH STM

ATM is the only integrated services protocol that supports TDM circuits. This section compares aspects of Synchronous Transfer Mode (STM) with Asynchronous Transfer Mode (ATM). Recall from Chapter 6 the definition of STM as a special case of Time Division Multiplexing (TDM) which controls the clock frequency very accurately. The North American SONET standard and the international Synchronous Digital Hierarchy (SDH) standard (see Chapter 6) are the premier examples of STM today. This chapter uses the terms STM and circuit switching interchangeably.

Private lines offer a single dedicated physical channel between two fixed points, whereas ATM offers each point multiple logical connections to a number of potentially different end points. Furthermore, private lines offer a small number of bandwidth choices dictated by the rigid TDM hierarchy, while ATM provides fine bandwidth granularity and multiple service categories across logical virtual path and channel connections. Private line networks also require one egress port for every ingress port, as contrasted with a single ATM access port logically connected to all other access ports in a virtual private PVC network.

30.3.1 ATM versus STM CBR Multiplexing Efficiency

ATM, of course, has its overhead, as many other protocols do. This section compares ATM efficiency with TDM, specifically SONET. The VT1.5 mapping in SONET uses 27 octets every 125-μs frame inside an STS-1 payload to carry 24 octets of payload (see Chapter 6). Furthermore, the STS-1 section, line and path overhead use 36 octets along with 18 stuff octets per frame out of the total 9×90=810 octets in an STS-1. Thus, only 28 VT1.5s (or 672 DS0s) fit into an STS-1, resulting in an efficiency computed as follows:

$$\text{SONET VT1.5 Efficiency} = \frac{24 \times 28}{9 \times 90} = 83\%$$

Comparing this to the efficiency of ATM's AAL1 unstructured mode (see Chapter 13) uses one extra octet out of the 48 octet payload in the 53 octet cell. Thus, the ATM efficiency in carrying the equivalent user payload in the STS-3c path payload of 9×260 octets (see Chapter 12) is:

$$\text{ATM Unstructured AAL1 Efficiency} = \frac{47}{53} \frac{260}{270} = 85.4\%$$

Furthermore, for a decrease in efficiency of less than 0.3 percent, choosing AAL1 structured mode allows connections of arbitrary multiples of 64 kbps. ATM packs the structured and unstructured connections at one hundred percent efficiency at utilization in excess of 90 percent on any link as analyzed in Chapter 23. The main provision of ATM that achieves this efficiency requires that networks set up connections using the AAL1 circuit emulation protocol using the CBR service category through devices which implement some form of prioritized queue servicing.

Also, use of the structured mode means that no intermediate cross connect stages are required in ATM networks implementing circuit emulation when compared with the classical TDM digital cross connect hierarchy. Even if individual ports on an ATM device performing circuit emulation are more expensive than those on the separate TDM cross connects, the overall cost may be less depending upon the number of ports required to interconnect the levels of the TDM hierarchy.

30.3.2 ATM versus STM Switch Blocking Performance

Our analysis also considers the flexibility of ATM in contrast to systems optimized for circuit switching. The comparison assumes that the blocking probability is 1 percent for any connection rate. A key advantage of ATM over STM is the ability to flexibly, and efficiently, assign multiple Continuous Bit-Rate (CBR) circuit rates on a SONET transmission facility, as shown in Figure 30.6. The upper left-hand corner of the figure illustrates a specific example of STM for the SONET hierarchy carrying STS-1 (51.84 Mbps),

VT1.5 (1.792 Mbps), and H0 (0.384 Mbps) within an OC-3 channel. The STS-1 channel must be dedicated in a circuit-switched network; while an ATM network can allocate it dynamically. The VT1.5 channels are allocated within an STS-1, and the H0 channels are allocated in another set of VT1.5s as shown in the figure.

The same OC-3 channel is allocated as an STS-3c ATM channel as illustrated in the lower left-hand corner of Figure 30.6. Approximately 42 ATM cells fit within each SONET frame (125-μs interval), shown as a series of rectangles in the figure. The STS-1 channel has 16 cells per time slot. The VT1.5 channels require approximately one cell every other time slot. The H0 channels require approximately one cell every eighth time slot. ATM assigns these cells in an arbitrary order, as illustrated in the figure. The inefficiency of the 5 bytes of cell overhead and 4 bytes of AAL overhead results in a maximum utilization of 90.5 percent for ATM (i.e., 48/53) versus the channelized SONET mapping. This is shown by the line labeled STM 100 percent STS-3c and the line labeled ATM.

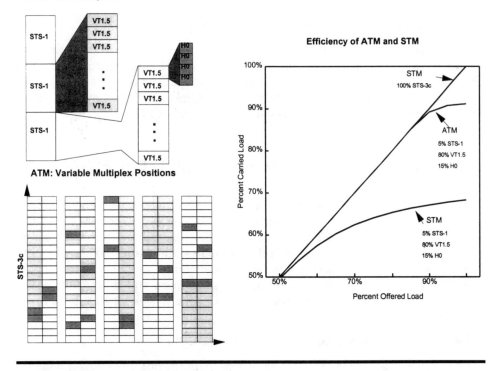

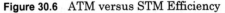

Figure 30.6   ATM versus STM Efficiency

The right-hand side of Figure 30.6 depicts the result of a calculation where 5 percent of the connection traffic is for STS-1, 80 percent for VT1.5, and 15 percent for H0. The figure plots carried load versus offered load. Carried load is proportional to the offered load, minus any blocked load. SONET and

ATM are equivalent at lower levels of offered load, up until the offered load exceeds 50 percent of the trunk capacity. At higher loads, the higher efficiency of ATM becomes evident. For this particular example, the maximum efficiency of STM is about 65 percent, compared to 91 percent for ATM. This represents about a 25 percent improvement in efficiency. When the mix of circuit rates becomes even more varied, or unpredictable, the advantage of ATM is even greater.

30.3.3 Statistical Multiplexing Gain

Another key advantage of ATM over STM is the opportunity to achieve cost savings by statistical multiplexing gain. Chapters 25 and 26 covered the basic mathematics of statistical multiplexing gain for data and voice traffic. The conclusion reached there was that statistical multiplexing gain is higher for bursty traffic which has a peak rate less than the trunk line rate. Recall that burstiness is defined as the ratio of the source's peak rate to the average rate. When a source is very bursty, it has nothing to send most of the time. However, when it does send, it transmits a significant amount of data in a relatively short period of time.

30.3.4 Delay Performance of ATM versus STM

All right, ATM is more efficient than STM in terms of basic encoding, and has better blocking performance and enables statistical multiplexing. Does it have any disadvantages? Yes, it does. As studied in Chapter 17, circuit emulation adds a packetization delay and a playback buffer delay. For lower speed circuits, packetization delay can be a significant factor.

30.4 VOICE OVER ATM, FRAME RELAY, AND THE INTERNET

Voice over ATM, voice over frame relay, and voice over the Internet are all forms of packetized voice [1]. Chapters 13 and 17 covered how ATM's AAL1 and AAL2 support voice. The Frame Relay Forum's FRF.11 implementation agreement describes how frame relay supports the multiplexing of many voice conversations over a single virtual connection. Also called IP telephony, voice over the Internet is a rapidly growing business and a hot item in the trade press at the time of publication. Let's take a look at how packetized voice systems emulate the various aspects of a telephone call.

30.4.1 Packetized Voice Deployment Scenarios

Packetized voice targets several deployment scenarios involving phone-to-phone communications, trunking between PBXs and communication from PBXs to branch office sites, as well as PC-to-phone connections [1, 2, 3]. Figure 30.7 illustrates these representative deployment scenarios. The figure also shows several ways of connecting to the Public Switched Telephone Network (PSTN).

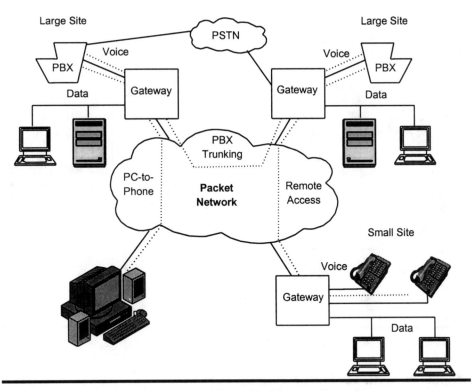

Figure 30.7   Representative Packetized Voice Deployment Scenarios

A packetized voice capability serves as a virtual trunk between the PBXs at the large sites (for example, headquarters or main offices) in the upper left and right-hand corners.   A packetized voice gateway-gateway path also connects the smaller site (for example, a branch office) in the lower right-hand corner with a larger site via a pair of gateways in a remote access configuration.   Finally, a gateway function built into the PC in the lower left-hand corner connects a remote user to the headquarters site in the upper right-hand corner.   This PC-to-phone connection often uses a hardware card in the PC to allow simultaneous voice and data operation by the end user.   Future PC hardware and software will likely offer this as a built in capability.

Let's take a look inside the packetized voice gateway function to better understand its operation.   Figure 30.8 illustrates the basic functions implemented by a gateway: voice packetization, telephony and signal conversion, and the packet network interface.   The voice packetization function converts from analog (or digital) speech to packets according to one of the voice coding standards described in the next section.   Since these coding schemes can't convey DTMF tones, call progress tones, fax, or modem signals; a separate telephony signal processing conversion unit converts these information streams to a separate set of packets as shown in the figure.   The packet network interface function takes the voice and telephony signal packet streams and affixes packet headers for relay across the packet network to the destination gateway. This function also takes other packet data streams and multiplexes them with the voice and signaling packets in a prioritized manner to ensure voice quality.

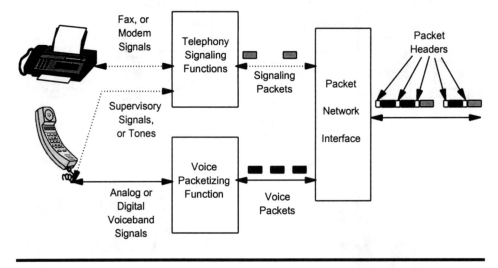

Figure 30.8   Packetized Voice Gateway Functions

30.4.2 Packetized Voice Business Drivers

The principal business driver today for packetized voice is cost reduction. Enterprises reduce costs by: combining voice and data applications, not transmitting the packet samples corresponding to periods of silence, and reducing the transmission rate required for voice through advanced coding schemes.   Combining voice and data effectively requires the network to provide some degree of bandwidth reservation as well as controlled delay and loss for voice packets with respect their data packet counterparts that applications can retransmit if lost.   The analysis of Part 6 on this subject applies to packet voice and data in general, and not just ATM.   The silence supression feature typically reduces the transmission requirement by 50 percent for ten or more conversations multiplexed together as covered in Chapter 26.   Finally, many sophisticated coding schemes convey voice at bit rates much lower than the standard 64-kbps Pulse Code Modulation (PCM) invented over forty years ago.   Table 30.2 shows the major voice coding names, the defining ITU-T standard, the peak transmission bit rate, and the one-way latency [1, 4, 5].   The one-way latency is the sum of the encoding delay, the packetization delay and the look ahead delay [4].   Note that the lower bit rate standards carry human voice only — they do not carry DTMF tones, fax or modem signals. As shown earlier, separate packet streams and coding methods carry these non-voice signals. As seen from the table, the low bit rate G.723.1 coder uses less than 10 percent of the peak transmission bandwidth that the current telephone network does!   However, this isn't the complete story since packet-based voice transmission adds additional overhead.

Table 30.2   Voice Coding Techniques, Standards and Peak Bit Rates

Acronym	Name	ITU-T Standard	Peak Bit Rate (kbps)	One-way Latency (ms)
PCM	Pulse Code Modulation	G.711	64	0.125
ADPCM	Adaptive Delta Pulse Code Modulation	G.726	16, 24, 32, 40	0.125
LD-CELP	Low Delay - Code Excited Linear Prediction	G.728	16	5.0
CS-ACELP	Conjugate Structure - Algebraic CELP	G.729	8	25
MP-MLQ	Multi Pulse - Maximum Likelihood Quantizer	G.723.1	5.3, 6	67.5

Currently, a significant business driver for Internet Telephony is the current economics of Internet access versus local access for long distance service. Surprisingly, the local access charges are the largest components of long distance phone calls [6]. Also, since international telephone charges are also quite high, packetized voice serves this need as well. An enterprise can achieve lower costs because the lower transmission rate allows multiplexing of more conversations over a single access line. Furthermore, bypassing the local (or international) telephone carrier replaces the time based charges with the bandwidth based charging employed by the Internet. In today's best effort Internet, the cost of equivalent bandwidth is less than that charged by international and local carriers.

For example, in the United States, unlimited local calling costs on the order of twenty dollars a month. However, when making a long distance call, the user ends up paying several cents a minute in local access charges. Therefore, making a local call to a voice over IP gateway which carries the call across the Internet to another local gateway within the local calling area of the destination saves several cents per minutes. If you make a lot of long distance calls, this adds up quickly. The savings in international calling are an order of magnitude greater.

So, you ask, if this is so great, why isn't everyone using it already? There are several reasons. One is quality, primarily impacted by the coding fidelity and latency. Secondly, the standards for full interoperation of packetized voice, particularly in the area of signaling, are incomplete. Finally, not all packet voice networks provide for reserved bandwidth and QoS guarantees. Note that IP over Telephony technology is maturing rapidly. For example, Level 3 and Qwest have announced plans to build Internet-based backbones to carry both voice and data.

30.4.3 Packetized Voice Quality

The issue of quality is important in voice. Since packetized voice inserts the algorithmic delay defined in section 30.4.2, and encounters other delays when traversing a packet network; delay is the key quality impairment. Most users can't tell the difference between 64-kbps PCM and the lower bit-rate coded voice signals in subjective tests. The contributions to overall round trip delay include:

- Voice encoding algorithm delay
- Packetization delay (i.e., the time to fill a packet with samples for transmission over the packet network)
- Time required to transmit the packets on the access lines
- Delay encountered by voice packets traversing the packet network
- Propagation delay
- Playback buffer delay

If the total round trip delay exceeds 50 milliseconds, then ITU-T Recommendation G.131 requires echo cancellation [7]. Currently, most packetized voice gateway functions implement echo cancellation. Furthermore, if the one-way delay exceeds 100 to 200 ms, then the conversation is not very interactive. Most IP telephony gateways today have one-way delays that exceed these values, particularly for PC-to-PC packetized voice [4]. This situation is similar to that of satellite communications where people may begin speaking at the same time. Once people get used to this effect, the connection is usable. As an additional benefit, people actually need to be more polite to let the other party finish speaking! As in the day when satellite communications was cheaper than terrestrial microwave transmission, cost conscious users adapt to save money.

With improved voice codecs, lost or delayed voice packets are now the most significant source of voice fidelity impairments. Unlike data applications, voice applications cannot simply retransmit lost packets — it simply takes too long. Therefore, packetized voice gateways account for lost packets in one of two ways [1]. First, the packets contain sequence numbers so that the decoder can interpolate between lost samples. G.723.1 uses this method to operate acceptably when losing 10 percent of the packets while adding 40 ms of delay when compared with the G.729 coding and packetization delays [4]. However, if the decoder does too much interpolation, then fidelity suffers. Another method involves the voice encoder inserting redundancy in the packets. However, this approach increases the bandwidth requirements and also increases delay. Of course, the best solution is to reduce loss to acceptable values. As studied in Part 5, ATM keeps absolute delay and delay variation to minimal values. The Frame Relay Forum Implementation agreement FRF.11 adds the notion of prioritization for voice frames as described in section 30.5.3. The IP protocol adds the RSVP protocol to reserve bandwidth and QoS as described in Chapter 8. Voice over IP also uses the Real Time Protocol (RTP) to sequence number voice packets and aid in the playback process.

Finally, the receiving gateway must handle variations in the packet interarrival times through a playback buffer dimensioned to handle the expected range of packet delay variation. ATM does this by assigning service categories and QOS classes to minimize cell delay variation. This reduces the requirements on playback buffer size as analyzed in Part 6. Many contemporary frame relay and IP networks experience greater variations in packet interarrival times, necessitating larger playback buffers, increasing the end-to-end delay.

30.4.4 Packetized Voice Efficiency

This section compares the efficiency of the various packetized voice techniques described above. Table 30.3 shows the overhead (in bytes) incurred by each approach, the packet size, and the resulting efficiency.

Table 30.3   Voice over Packet Efficiency Analysis

Overhead	Voice over AAL1	Voice over AAL2	Voice over IP Access	Voice over IP Backbone	Voice over FR
AAL	1-2	3-4			
ATM	5	5			
IP			6	40	
UDP				8	
RTP			12	12	
HDLC			6-8	6-8	6-8
FR					3-7
Total Overhead	6-7	8-9	24-26	66-68	9-16
Packet Size	20-47	20-44	20-50	20-50	20-50
Efficiency	38%-89%	38%-85%	37%-59%	23%-41%	67%-71%
Voice Bit Rate (kbps)	6-32	6-32	6-32	6-32	6-32
Peak Bit Rate (kbps)	16-37	16-39	14-49	27-78	9-42

As studied in Chapters 13 and 17, ATM defines two methods for supporting voice: AAL1 and AAL2. The analysis considers cells partially filled to 20 octets for direct comparison with IP and frame relay as well as completely filled cells. Voice over IP and voice over frame relay support a range of coding techniques, hence the packet sizes differ markedly for these techniques. Of course, silence suppression applies in a comparable manner to all of these approaches. The PPP protocol allows voice over IP to compress the IP and UDP headers on an access line to 6 bytes as indicated in the table. Indeed, without this header compression, IP Telephony loses much of its attractive attributes on access lines. Note that the sequence number and time stamp information conveyed by RTP comprise approximately one-half of the overhead. However, the routers in the Internet backbone and the destination gateway require the complete larger header size for packet forwarding and voice conversion at the destination. In conclusion, ATM looks very attractive for transporting voice on xDSL or DS1/E1 access lines from an efficiency point of view, while IP with PPP header compression or frame relay is attractive on dial-up lines or lower speed access circuits.

However, on lower speed access lines, long data frames impose problems in achieving QoS as discussed in Chapter 30.

30.5 CELLS VERSUS FRAMES FOR PACKET SWITCHING

While overall efficiency isn't so important in the LAN — after all, 40 percent utilization on a shared Ethernet is considered acceptable — high utilization of expensive WAN facilities is a key consideration for many network designers. Hence, a great debate has arisen regarding the efficiency of ATM versus alternative frame-based networking techniques in the access and backbone portions of IP networks.

If the decision between IP/ATM versus IP/SONET in the Internet backbone were easy; experts would have made it already. Instead, the IETF continues busily defining support for IP over ATM; while the ATM Forum diligently works on ways to make ATM support IP-based services better. Past experience with communications networking indicates that no one technology ever serves the entire marketplace.

This section compares the efficiency and features of major data protocols used in private and public networks: ATM (using AAL5), frame relay, SMDS, and IP using PPP over HDLC. One thing you should consider is the tradeoff of efficiency versus features. Increasing the number of features often comes at the expense of decreased efficiency.

30.5.1 Average Protocol Efficiency

A major criticism leveled against ATM is its inefficient carriage of varaible length packets when compared with the efficiency of a native frame-based protocol like HDLC. This section models the efficiency of various protocols assuming that all packets have the same average size. Figure 30.9 compares the efficiency of three major data protocols used in private or public networks: frame relay, ATM (using AAL5), and 802.6/SMDS (using AAL3/4). The following text describes the way in which these efficiencies were computed.

Basic frame relay and HDLC use variable-length packets with an overhead of at least 6 bytes per packet (excluding zero stuffing) as described in Chapter 8. Therefore, the protocol efficiency for HDLC is:

$$\text{Efficiency(HDLC)} \quad = \quad \frac{P}{1.03 \quad (P + 6)}$$

where P is the packet size expressed in bytes.

Many analyses covering the efficiency of packet transport over HDLC ignore the increased line bit rate created by zero stuffing. For random data, this results in a 3 percent increase since HDLC stuffs a zero bit for any sequence of five consecutive ones in the data stream to avoid replicating the six ones in an HDLC flag ('01111110'). For random bit patterns, each bit is a one or a zero half the time. Hence, the likelihood of encountering five consecutive ones is one out of thirty-two, or approximately 3.1 percent. For non-random data (for example all binary ones) the overhead of HDLC zero stuffing never exceeds 20 percent, but may be less than 3 percent. IP over SONET as defined in RFC 1619 [10] uses byte stuffing for the flag character, which results in an efficiency loss of approximately 1% (i.e., 1/128) [17].

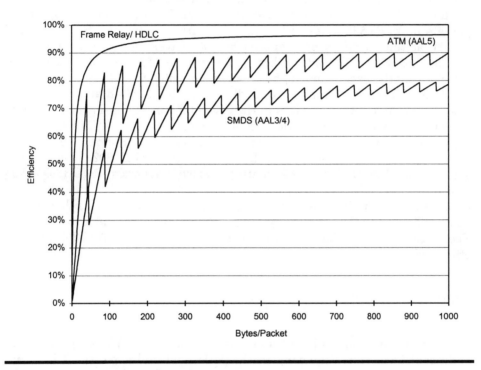

Figure 30.9   Protocol Efficiency versus Packet Size

Frame relay/HDLC is the most efficient protocol with respect to overhead of the three protocols considered in this section. However, frame relay and HDLC don't readily support multiple QoS classes, especially if some frames are very long as discussed later in this section. The longest typical frame size encountered today is approximately 1500 bytes as determined by the Ethernet propagation delay. Frame relay/HDLC approaches approximately 96 percent for long user data packet sizes.

As described in Chapter 13, the 8 bytes in the AAL5 trailer combined with the 5 bytes of ATM cell header overhead and the rounding up to an integer number of cells using the pad field yield a protocol efficiency of

$$\text{Efficiency(AAL5)} \quad = \quad \frac{P}{53} \quad \left\lceil \frac{P+8}{48} \right\rceil^{-1},$$

where $\lceil x \rceil$ is the smallest integer such greater than or equal to x. The formulas for protocol efficiency for variable length packet and fixed cell protocols have the same basic form as shown for HDLC and AAL5. What changes is the amount of overhead and the number of bytes carried by the specific AAL.

ATM using AAL5 provides functions very similar to frame relay and HDLC, while simultaneously providing the additional flexibility of mixing very long packets with other delay sensitive traffic. AAL5 also allows support for packets up to 65,535 bytes in length, which frame relay and SMDS do not. Note that the efficiency only approaches 90 percent for very long packets because the ATM cell header takes up 5 bytes out of every 53. Because the variable-length packet must be segmented into fixed-length cells, the resultant efficiency decreases markedly when this segmentation results in one or a few bytes of packet data in the last cell. For a typical average packet size of 100 to 300 bytes, the difference in efficiency between ATM using AAL5 and FR/HDLC ranges from 15 to 30 percent as seen from Figure 30.9.

As described in Chapter 8, SMDS utilizes the IEEE 802.6 Distributed Queue Dual Bus (DQDB) cell formatting, which is identical to AAL3/4. The first cell in an SMDS Protocol Data Unit (PDU) contains 44 bytes of level 3 information, including the source/destination addresses and other fields, while all of the other cells only utilize 44 bytes of the 48 byte payload. Therefore, the following formula gives the SMDS and AAL3/4 protocol efficiency.

$$\text{Efficiency(AAL3/4)} \;=\; \frac{P}{53} \left\lceil \frac{P + 44}{44} \right\rceil^{-1}$$

Using the additional 4 bytes of overhead in each cell, AAL3/4 provides an additional level of multiplexing in the MID field and a per-cell CRC, consuming an additional 4 bytes per cell. Much of the flexibility and many of the features of SMDS and AAL3/4 derive from this 44-byte header. Packets may be up to 9188 octets long, longer than the maximum in frame relay. This flexibility reduces efficiency, with the best limited to less than 80 percent efficiency for large packets.

Which protocol is best for your application? If you need a feature that only the less efficient protocols support, then the choice is clear; you can't use a more efficient protocol if it doesn't support a critical feature. If your network requires support for multiple QoS classes for different concurrent applications, then ATM is probably a good choice. If you need to connect to other networks via a public service, or require an additional level of multiplexing, then 802.6/SMDS is a possible choice. If raw efficiency is key, and support for multimedia applications is not required, then frame relay or basic HDLC is a good choice. For example, IP operating over HDLC is more efficient than IP operating over any other protocol as analyzed in the next section.

30.5.2 IP/ATM versus IP/SONET Efficiency

The preceding analysis assumed that each packet had the same average size. Real traffic, however, has packets of many different sizes as reported in many recent studies of IP network traffic [8]. Figure 30.10 plots a typical measurement of the relative frequency of occurrence of packets of various on an IP backbone versus the IP packet size in bytes.

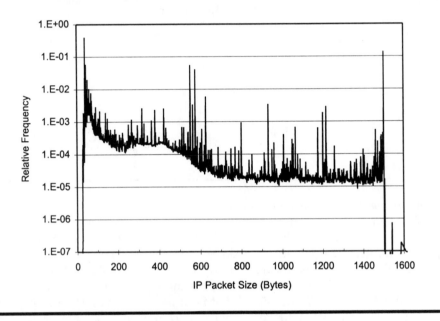

Figure 30.10   Relative Frequency of IP Packets versus Packet Size

Approximately 30 percent to 40 percent of the packets are TCP/IP acknowledgment packets which are exactly 40 bytes long. The majority of packets are less than on equal to the maximum Ethernet frame size of 1500 bytes. Very few packets have lengths greater than 1500 bytes.   Less than one in 10 million packets have lengths greater than 1500 bytes, reflecting the predominant use of Ethernet in end systems.

Table 30.4 illustrates the protocol overhead required when using IP over various frame-based and ATM-based protocol stacks. As pointed out in Chapter 19, the LLC/SNAP multi-protocol encapsulation defined in IETF RFC 1483 adds an extra 8 bytes of overhead when carrying multiple protocols over ATM. Although this standard targeted multiprotocol applications between end systems, many backbone routers use this same form of encapsulation.   This means that the 40 byte IP packets with 8 bytes of AAL5 Common Part (CP) overhead plus the 8 bytes of LLC/SNAP overhead require two cells instead of one if VC multiplexing were used instead. Note that IP operating directly over HDLC adds another 1 (or optionally 2) octets of PPP overhead per RFC 1662 [9] and that RFC 1619 [10] recommends a 32-bit HDLC CRC for IP operating over SONET. The table also includes for comparison purposes IETF work in progress for supporting IP using 4 bytes of overhead for MultiProtocol Label Switching [11]. As discussed later in this part, both MPLS and RSVP are essential for IP to achieve comparable capabilities to ATM's guaranteed QoS and bandwidth on demand capabilities.

Table 30.4   Overhead for Various Protocols Carrying IP Packets

Overhead Field Description	PPP/ SONET	LLC/SNAP over AAL5	VC Multiplexing over AAL5	PPP/ FUNI	MPLS/PPP
HDLC	8 bytes			2 bytes	8 bytes
FUNI				8 bytes	
Zero Stuffing	3.10%			3.10%	3.10%
53 Byte Cell	No	Yes	Yes	No	No
AAL5 CP		8 bytes	8 bytes		
LLC/SNAP		8 bytes			
PPP	1 byte			1 byte	1 byte
MPLS label					4 bytes

Computing the relative efficiency for each of these protocols using the measurements reported in [8] results in the efficiencies shown in Table 30.5. Clearly, IP operating over PPP over SONET is more efficient at the link layer by approximately 15 percent over ATM using AAL5 and RFC 1483 defined LLC/SNAP multiprotocol encapsulation. Other studies [12, 13] using actual packet distributions also measured approximately 80 percent raw efficiency for LLC/SNAP over ATM AAL5. Note that a network using the Frame-based UNI (FUNI) protocol achieves efficiency comparable to native IP over SONET. Operation over either frame relay or the Frame-based UNI (FUNI) alleviates many of the efficiency concerns about ATM when operating over high-speed backbone circuits. Using LLC/SNAP encapsulation in an ISP backbone doesn't make much sense. Why employ a protocol designed for communication between end systems between routers? In fact, if an IP backbone is running only IP, then the network operator incurs approximately 5 percent of unnecessary overhead. An application could use the AAL5 User-User Indication (UUI) byte to multiplex multiple protocols (using PPP for example) and avoid the inefficient carriage of 40-byte TCP/IP acknowledgments caused by LLC/SNAP encapsulation. Note that adding in the overhead for MPLS and RSVP to give capabilities comparable to ATM QoS and bandwidth management is actually less efficient than running IP over PPP over ATM FUNI. Therefore, as discussed in Chapter 31, the real issue is not the link level efficiency, but instead is the choice of routing protocol and the bandwidth management paradigm.

One study [13] shows that the resulting efficiency is relatively independent of the choice of cell size. Therefore, the fundamental difference exists between the choice of a fixed packet size (i.e., a cell) and a variable length packet (i.e., an HDLC frame). Note that when operating over SONET and SDH links that ATM and any other protocol has the same synchronous payload envelope rate listed in Chapter 12.   For example, the actual rate available to any protocol operating over a 155.52-Mbps link is only 149.76 Mbps.   Therefore, SONET itself introduces a 3.7 percent overhead at the OC3/STM-1 rate. At higher speeds, the SONET overhead decreases slightly (see Chapter 12).

Table 30.5   Efficiency of IP Transport over Various Protocols

IP Operating Over the Protocol(s)	Resulting Efficiency
PPP over SONET	96.6%
ATM AAL5 using LLC/SNAP Multiprotocol Encapsulation	80.5%
ATM AAL5 using VC Multiplexing	84.9%
PPP over Frame Relay or ATM Frame-based UNI (FUNI)	96.1%
PPP over MPLS over SONET	95.6%

Nothing is certain but death and taxes. As seen from this analysis, the ATM cell tax can be reduced by operating IP over other protocols, but not eliminated.

30.5.3 Trouble In Achieving QoS with Frames

A fundamental performance issue with store-and-forward packet switching occurs when a long frame gets ahead of a delay-sensitive short one, particularly on a lower speed WAN circuit. You may want to look back at analogy of trains (long packets) holding up urgent voice packets (cars) at a crossing described in Chapter 11 to help you remember this key concept. If an application sensitive to variations in delay, for example voice, generated the short packet, then it may not be able to wait until the long packet finishes transmission. Consider an example where a user's router connects to the network via a DS1 link operating at 1.5 Mbps. Alternatively, this could be the high-speed downstream direction on an ADSL connection to a residential subscriber. Typically, a maximum length Ethernet packet is 1500 bytes, which takes approximately 8 ms to transmit at DS1 speeds. A long Ethernet packet getting ahead of a short voice over IP packet could delay it by up to 8 ms, even if the router performed nonpreemptive priority queuing so that a higher priority voice packet is sent immediately after the lower priority data packet completes transmission. The effect is even more pronounced on lower speed links. For example, transmission of a 1500-byte packet on a 128-kbps N-ISDN BRI link takes almost 100 ms, while a 28.8-kbps modem takes over 400 ms. Therefore, packet voice and data cannot share the same access line unless the user device and the network work together to fragment the large packets.

Voice over IP implementations compensate for these large variations in delay by inserting a large playback buffer at the destination. If this effect occurs at several intermediate DS1s (or worse yet, even slower) links, then the user eventually perceives this delay if the cumulative effect exceeds approximately 100 ms. This effect means that real time gaming, circuit emulation or interactive video conferencing cannot be effectively performed over store and forward packet switching systems operating even at modest link speeds.

One solution to this problem requires a preemptive resume protocol, similar to that used in most multitasking operating systems. In essence, the packet switch must stop transmission of the long packet (i.e., preempt it) by reliably communicating this interrupt to the next node, and then transmit the high priority packet. Once completing service of the higher priority packet, the packet switch can then send the lower priority packet (i.e., resume). Currently there is no standard for preemptive-resume packet switching, although several vendors have announced proprietary implementations. For example, Ascend breaks larger data frames into shorter segments so that voice frames do not experience excessive delay variation [14]. However, if the network experiences congestion, then voice doesn't perform well [15]. The frame relay forum FRF.11 specification [5] defines a means to multiplex several voice conversations between the same endpoints using a technique similar to AAL2. A companion document, FRF.12 [16] outlines how to break up data frames into smaller segments and reassemble them at the destination to avoid the problem of a long frame delaying shorter, urgent voice frames. The specification defines procedures for performing fragmentation across an interface between a user and a network, or on an end-to-end basis. Unfortunately, this approach creates more overhead for every data frame. Of course, IP also provides for fragmentation to overcome this delay variation problem; however, it introduces significant additional processor and packet header overhead when doing so.

On the other hand, ATM inherently implements a preempt-resume protocol in a standard fashion on 53 octet boundaries. Hence, for lower link speeds, ATM is the only standard protocol that enables true support for voice, video and data on the same physical link.

An even simpler solution to this problem in packet networks is to simply increase the link speed and/or decrease the maximum packet size. Increasing the transmission speed is a natural solution in the a packet backbone network, but may be uneconomical for WAN access circuits. As pointed out earlier, the threshold of human perception of delay variations is on the order of 10 ms for auditory and visual stimuli.

Another QoS impacting issue occurring in large Internets occurs in response to frequent state changes in routers, their ports and the links that interconnect them. Whenever a state change occurs, the routing protocol distributes the updated topology information, and the packets may begin following another path. A change in routed path usually results in jittered, lost, misequenced, or in some cases, duplicated packets. These impairments will likely cause annoying interruptions in the video playback. Customers may not be as tolerant of such poor quality when they are paying for video content instead of watching a "free" video clip over the Internet.

If your application requires QoS on low speed lines, then the 10 percent overhead of ATM is worthwhile. If you have legacy TDM, video or native ATM applications mixed with packet data, then ATM provides a superior

WAN solution. Also, applications like interactive video-conferencing work best over ATM SVCs.

30.6 ATM VERSUS IP PACKET VIDEO EFFICIENCY

This section rounds out the efficiency comparisons in this chapter by briefly comparing the efficiency of carrying packet video over ATM and IP. Recall from Chapter 17 that a number of video coding standards generate packetized video at rates ranging from 100-kbps to over 40 Mbps; for example, MPEG1 and MPEG2. An IP network must carry the entire header field across the backbone, while an ATM network can convey information during call establishment and not send information in the header field once the connection is established. The analysis assumed the IPv6 packet header to support the extended addressing capabilities instead of dynamic address assignment as is done today to allow assignment of an address to every subscriber. The 20-octet ATM End System Addresses (AESAs) easily meet this requirement as studied in Chapter 14. Actually, the IPv6 address space is a subset of the ATM address space; hence every IPv6 user has a corresponding ATM address. As covered in the next chapter, a connection-oriented protocol generally provides a more economical solution for long-lived information flows, such as video information transfer. Also, in many case the carriage of the actual video packets is more efficiently done by ATM as well.

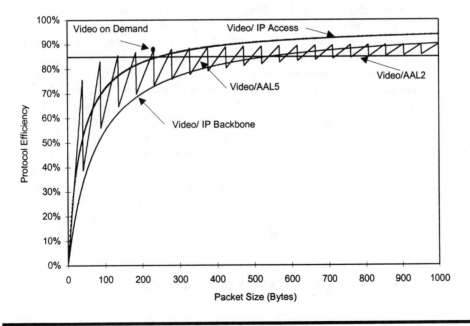

Figure 30.11   Protocol Efficiency for Transport of Video Signals

Figure 30.11 illustrates the protocol efficiency of video carried by AAL5, AAL2, and IP protocols on access and backbone circuits for packet sizes up to 1,000 bytes. The analysis assumed that AAL5 overhead was only 8 bytes because all other parameters were conveyed in the setup message. The AAL2 overhead assumes a constant 8 bytes of overhead per cell, resulting in a constant efficiency of approximately 85 percent. The analysis assumes that on backbone circuits, video over IP uses RTP running over UDP using IPv6 over PPP with HDLC framing for a total of 77 octets of overhead per packet. On access circuits, the figure depicts the resulting efficiency if IPv6 and UDP header compression reduces the overhead by 42 octets. Of course, HDLC zero stuffing adds 3 percent to every packet for video over IP.

Video coders typically send packets on the order of several hundred bytes; for example, MPEG2 uses a packet size of 188 bytes. If the video to ATM mapping accounts for the round off efficiently (as described in Chapter 17 for the ATM Forum's video on demand specification [18]), then video over ATM is more efficient than IP on the backbone by 8 percent and slightly better than the performance achieved using header compression by video over IP on access circuits.

Making the packet size larger to improve video over IP efficiency creates several problems. First, packet lengths of 1,000 bytes create an additional packetization and transmission delay of 20 ms at xDSL speeds (i.e., 1 Mbps). This amount of delay reduces the interactive nature of other services such as voice and interactive network game applications on the lower speed access line. Recall that voice over IP needs to reduce overall delay, so creating more opportunity for delay by using long video packets exacerbates an already marginal level of quality. Secondly, video coders send frames at least 30 to 60 times a second (for example, for basic updates and detailed information). Furthermore, the video session also involves transmission of side information such as audio, closed caption, and other video-related data. Therefore, a video coder sends packets at least one hundred times a second, which results in a packet size of approximately 1,000 bytes for a transmission rate of 1 Mbps. Furthermore, many video coders include some redundancy in case video packets are lost or errored. Finally, many video coding algorithms require some form of timing recovery to synchronize the video, audio, and ancillary information streams for playback purposes. Large gaps or variations in packet arrival times complicate timing recovery at the receiver. On the other hand, ATM readily provides timing recovery to synchronize services; for example, as described in References 18 and 19. Therefore, video coders generally use smaller packets.

30.7 MULTISERVICE EFFICIENCY COMPARISON

All right, now that we've looked at the efficiency comparisons for TDM circuits, voice, packet data, and video; how do they stack up in the big picture? Table 30.6 shows the scorecard of the efficiency analyses from the previous sections. A dash indicates that the protocol does not have a standard means to support the service. As discussed in the previous sections, efficiency alone is not the only attribute; however, it is an objective one that we chose to highlight the advantages and disadvantages of the competing WAN protocols. The next two chapters look at other aspects that differentiate ATM and IP in the LAN environment.

Table 30.6   Relative Protocol Efficiency Comparison Scorecard

Protocol	TDM Circuits	Voice	Packet Data	Video
ATM	1st	1st	3rd	1st
IP	-	3rd	2nd	2nd
Frame Relay	-	2nd	1st	-

The overall measure of efficiency, of course, depends upon the relative volume of each service. However, this table shows some of the rationale behind the decision of many carriers to deploy ATM since it handles multiple traffic types in a relatively efficient manner using a common infrastructure. Many estimates state that data traffic has either already exceeded voice traffic, or will do so by the year 2000. The big unknown is the volume of video traffic, a topic which Chapter 32 addresses. A survey conducted in 1998 regarding why users select ATM provides some insight into this pivotal question [20]. This survey reported that ATM users differ from IP and Gigabit Ethernet advocates in one key respect: they expect delay-sensitive traffic like video to grow significantly.

30.8 REVIEW

This chapter compared data services from an economic, performance, and business point of view. The principal focus was an objective analysis of the efficiency and performance in transferring TDM circuit, voice, packet data, and video information. As described in the text, the principal economic impact of efficiency occurs in the access and backbone circuit costs. The text first categorized data communications based on the method — message, circuit, or packet switching. The trend is that all of the modern methods use packet switching paradigm: frame relay, IP, and ATM. The fundamental difference is that frame relay and IP use variable length packets while ATM employs fixed length cells. The chapter then presented a detailed comparison of ATM with STM, citing many advantages of ATM over STM, but highlighting the fact that ATM adds packetization and playback delay on the order of tens of milliseconds for voice over ATM. Comparing voice over IP with voice over ATM, the analysis concluded that voice over IP not only adds more delay, but is also less efficient. The text then presented an analysis of the relative protocol efficiency performance for transfer of packet data. Depending upon the underlying protocol employed, ATM ends up being in the range of 15 percent less efficient to 2 percent more efficient than packet data transfer data via IP. Use of the ATM Frame-based UNI instead of ATM cells resulted in the 2 percent efficiency gain for ATM over IP for packet data. Finally, the text analyzed the relative efficiency of ATM and IP for transporting high-quality video. The analysis concludes that ATM is the efficiency winner here over IP by a margin of up to 8 percent on the backbone with efficiency on the access line at a comparable level. Finally, the text considered the overall efficiency of a combined voice, video, and data multiservice network. The conclusion was that the answer hinges on the amount of high-quality video traffic required by end users in relation to packet data and voice traffic volumes.

30.9 REFERENCES

[1]   Telogy Networks, "Voice over Packet Tutorial,"
        http://www.webproforum.com/telogy/#topics, 1998.
[2]   Siemens Stromberg-Carlson, "Internet Telephony Tutorial,"
        http://www.webproforum.com/siemens2/, 1998.
[3]   Dialogic, "IP Telephony Basics,"
        http://www.dialogic.com/solution/internet/4070web.htm, 1998.
[4]   T. Kostas, M. Borella, I. Sidhu, G. Schuster, J. Grabiec, J. Mahler, "Real-Time Voice Over Packet Switched Networks," *IEEE Network*, January/February 1998.
[5]   Frame Relay Forum, *Voice over Frame Relay Implementation Agreement*, FRF.11, May 5, 1997.
[6]   Phonezone.com, "Using IP Telephony To Bypass Local Phone Monopolies," http://www.phonezone.com/tutorial/telco-bypass.htm, March, 1998.
[7]   ITU-T, *Control of Talker Echo*, Recommendation G.131, August 1996.
[8]   K. Thompson, G. Miller, R. Wilder, "Wide-Area Internet Traffic Patterns and Characteristics," *IEEE Network*, November/December 1997.
[9]   W. Simpson, *PPP in HDLC-like Framing*, RFC 1662, IETF, July 1994.
[10]   W. Simpson, *PPP over SONET/SDH*, RFC 1619, IETF, May 1994.
[11]   D. Farinacci, T. Li, A. Conta, Y Rekhter, E. Rosen, *MPLS Label Stack Encoding*, IETF Draft ftp://ietf.org/internet-drafts/draft-ietf-mpls-label-encaps-00.txt, November 1997.
[12]   G. Armitage, K. Adams, "How Inefficient is IP over ATM Anyway?" *IEEE Network*, January/February 1995.
[13]   A. Saar, "Efficiency of TCP/IP - Is 53 Bytes for ATM cell a poor choice?" http://194.87.208.92/tech1/1995/ip_atm/tcp_atm.htm#About_document, 1995.
[14]   Cascade White Paper, "Priority Frame - Absolute Quality of Service in a Frame Relay Environment," http://www.casc.com/products/datasheets/008_97.pdf, January 1997.
[15]   S. Mace, "ATM's Shrinking Role," *Byte*, October 1997.
[16]   Frame Relay Forum, *Frame Relay Fragmentation Implementation Agreement*, FRF.12, December 15, 1997.
[17]   J. Manchester, J. Anderson, B. Doshi, S. Dravida, "IP over SONET," IEEE Communications Magazine, May 1998.

[18]   ATM Forum, *Audiovisual Multimedia Services: Video on Demand Specification 1.0*, af-saa-0049.000, December 1995.

[19]   M. DePrycker, *Asynchronous Transfer Mode — Solution for Broadband ISDN*, Prentice Hall, 1995.

[20]   K. Korostoff, "Why They Buy: A Survey of ATM Customers," Business Communications Review," *Business Communications Review*, February 1998.

Chapter

31

Technologies Competing with ATM in the LAN

The claim that ATM is dead in the LAN is somewhat exaggerated, but should teach some key lessons to aspiring ATM LAN and internetworking equipment and network designers. This chapter focuses on comparisons between the combination of IP and Ethernet in the LAN versus ATM-based end systems and local area networks. As in the WAN, the principal focus of the LAN is economical operation. However, in the local area network environment, unlike the WAN, bandwidth is not the most expensive component; instead, equipment and support costs dominate the engineering economic equation. Hence, LAN designers can throw cheap bandwidth at the problem, mitigating the problems in achieving QoS with frames through increased line speeds and greatly reducing the likelihood of congestion at the same time. Given excess bandwidth everywhere in the LAN, the need for QoS and bandwidth guarantees thus diminishes in importance. Therefore, the simpler, connec-tionless IP/LAN paradigm is more cost-effective than the more complex, connection-oriented given the LAN economic parameters employed by most users today. However, an ever increasing number of users with applications requiring bounded delay and guaranteed bandwidth applications select ATM in the LAN.

Additionally, in the real world where increasingly more of the valuable information content doesn't reside on the LAN, integrated LAN/WAN solutions must address the issues of QoS and expensive bandwidth. The text compares how well the IP/LAN and ATM-based LANE and MPOA solutions address this problem in corporate virtual private networks and the Internet backbone.

Furthermore, the choice between connection-oriented and connectionless packet forwarding paradigms is not a simple one. It is unlikely that either approach will meet every need. This somewhat philosophical, and rather deep topic, sets the stage for the final chapter which covers the future direction of ATM and related protocols.

31.1 KEY LAN REQUIREMENTS

This section summarizes key requirements of the local area networking and integrated multi-service applications spanning both the LAN and WAN environment. Users of local area networks seek to exchange data between computers at increasingly higher speeds. The LAN legacy runs strong, starting with a shared media concept from the initial Aloha network, through 1, 10, and 100 Mbps speeds and now soon one billion bits per second – a gigabit per second! All of this happened in less than twenty years.

Chapters 8 and 9 introduced most of the technologies and services that both compete with and complement ATM in the LAN. Competing LAN technologies include 100-Mbps Ethernet, Gigabit Ethernet, and FDDI. Despite ATM's traffic management and quality of service benefits over legacy LAN technologies, like Ethernet and FDDI, ATM has not made significant inroads into the LAN market because applications don't yet require these features. Some LAN backbones have chosen ATM for higher speed as well as anticipated demand for QoS-sensitive applications like video conferencing. The analysis shows that ATM fares more favorably in the LAN backbone environment than in the cost driven desktop and workgroup area networks, but is still up against stiff economic competition from Gigabit Ethernet and Fast Ethernet switching in the LAN backbone.

31.1.1 Keep it Simple to Succeed

The FDDI and Token Ring protocols were too complex, and hence failed to garner significant market share. Furthermore, wiring and cabling costs are significant in many LAN upgrade decisions. ATM gained a great deal of complexity in LANE and MPOA in order to support the LAN. LANs are inherently a broadcast media and naturally provide multicast services. As shown in Chapter 19, ATM's LAN emulation protocol introduces the complexity of additional servers and connections to provide a comparable service. Since 100-Mbps Fast Ethernet and Gigabit Ethernet retain the inherent architectural simplicity of 10-Mbps Ethernet, they are the current favorite to continue domination of the LAN networking marketplace. Some experts place ATM in a niche market for desktop connectivity and LAN backbones [1]. Other users that see switched high-quality video and delay sensitive traffic on the horizon have already selected ATM as their long term LAN strategy [2].

As shown in Figure 31.1, the LAN environment consists primarily of end user computers containing LAN Network Interface Card (NICs) connected to a smaller LAN switch/routers. High-speed backbone LANs interconnect these LAN switch/routers in a hierarchy. In the LAN, minimizing cost centers on the hardware, software and operating expenses associated with the NIC in the user's computer, the cabling connecting to the switch, and the cost per user of the workgroup level LAN switch or router. Typically, although still important, efficient engineering of the LAN backbone is a secondary consideration in the overall cost equation. Another key factor that drives up costs is complexity. Generally, simpler protocols result in less expensive software, more maintainable software, as well as lower operating expenses.

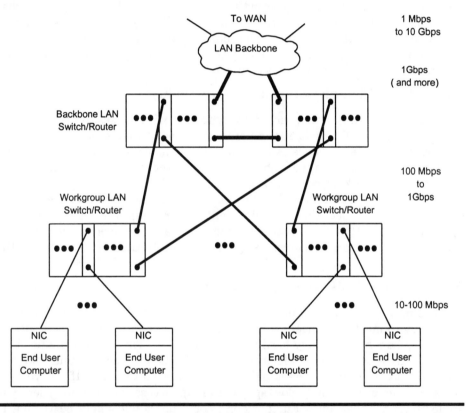

Figure 31.1   Hierarchy of Computers, LAN Switches and Backbones

Currently, ATM interface cards and LAN Emulation (LANE) software have a more complex configuration and installation procedure than their LAN counterparts. ATM-based equipment needs to improve significantly in this area to succeed.

31.1.2 Engineering Economic Considerations

Most existing hardware and software platforms already support IP and LAN protocols. Upgrading a LAN to ATM generally requires the installation of new hardware and software. Hardware includes new interface cards, switches, hubs, and possibly routers. In some cases, only the interface cards require replacement with their ATM-capable counterparts in these devices. Software includes operating systems, device drivers, and applications.

IP, LAN and ATM networks are experiencing phenomenal growth, which indicates the pent-up demand for ubiquitous data communications. Here, we see the accelerating bandwidth principle described in Part 1 in action. Moore's law predicts that the performance of microelectronic devices doubles once every two years. However, the need for communication, particularly Internet traffic growth over the past several years (and frame relay before that), exceeds this processing growth rate by a factor of 2 or more. Higher speed LAN interfaces along with more cost effective WAN access circuits continues fueling this demand. Therefore, a long term networking strategy must provide a potential for scalability that does not depend upon solely upon the increased performance of electronic devices.

31.1.3 Don't Plan on Congestion — Avoid It!

As discussed in Chapter 23, the best way to deal with congestion is to avoid it if at all possible. This is a fundamental premise of high-speed LAN design. However, if past experience is any teacher, then high-speed networks don't eliminate congestion forever, they only delay the inevitable point in time where the network is no longer fast enough. Witness the fact that most people could not comprehend how an individual device could ever send at 10 Mbps and didn't see the collision limited throughput of Ethernet as a problem. Note that 100-Mbps Ethernet was a short-lived savior of LAN backbones, with many network designers looking to Gigabit Ethernet as the next step.

31.1.4 Overall LAN-WAN Performance

Increasingly, LAN users must access the WAN to obtain the information necessary to meet their needs. As discussed earlier in the text, the traffic is changing from being largely local to mostly remote. This creates the need for faster WAN access speeds. Therefore, as identified in the previous chapter, the local access circuit increasingly becomes the bottleneck. If you've ever browsed the web using a low-speed dial up access connection, then you understand the problem. Furthermore, some users will likely be willing to pay more for higher quality Internet service. This is such a key consideration that some experts believe that business users would pay more for superior performance [3].

The wide scale adoption of 100-Mbps Ethernet to the desktop and Gigabit Ethernet will usher in a new era of almost limitless bandwidth in the LAN. But with all things come a price. While these technologies allow LAN environments to scale into the gigabit per second range, WAN access bandwidth of a few 1.5-Mbps DS1 access circuits multiplexed together, a 45-Mbps DS3 or a 155-Mbps OC-3 SONET link pale in comparison if a large portion of the traffic is going someplace else. This impedance mismatch between LAN and WAN imposes a bottleneck to communications between locations and limits true LAN extension over the WAN.

31.2 SURVEY OF COMPETING LAN TECHNOLOGIES

This section surveys and reviews technologies competing with ATM in the Local Area Network (LAN) environment as background to a comparison analysis in the next section. Competing LAN protocols include FDDI, 100-Mbps Fast Ethernet, and Gigabit Ethernet. The text also summarizes key attributes of ATM-based LAN interfaces.

31.2.1 FDDI and FDDI-II

As described in Chapter 9, the Fiber Distributed Data Interface (FDDI) targeted high-performance LANs and campus backbones at a transmission rate of 100-Mbps simplex. FDDI also operates over copper interfaces using the Copper Distributed Data Interface (CDDI) standard. Similar to the IEEE 802.4 token bus standard, a key merit of FDDI is its fault tolerant design. Many process control, manufacturing, health care, and other applications requiring high reliability will continue to use it for years to come. Although the FDDI-II standard addressed the lack of FDDI isochronous service, the vendor community has not widely implemented this feature.

31.2.2 Switched 10-Mbps and 100-Mbps Ethernet

There is no question that Ethernet is the king of the LAN today. A recent estimate [4] placed Ethernet market share at 83 percent. As described in Chapter 9, the IEEE standardized two Ethernet protocols operating at 100 Mbps in response to the accelerating need for more bandwidth. Ethernet is extremely cost-effective: PC network interface cards cost less than $50 for the 10 Mbps variety and less than $100 for the 100 Mbps version. The 10-Mbps limit, or realistically about 4 Mbps when using a medium shared by many stations, was not a limit for many users. LAN administrators accommodate power users and servers by segmenting the LAN to reduce the number of stations sharing a particular medium. A recent trend is to move toward Ethernet switching with only one client or server per Ethernet segment. LANs connect servers via multiple Ethernet segments when traffic demands require more than 100 Mbps of throughput. This allows better throughput and extends the life cycle and investment of the existing LAN and wiring infrastructure.

31.2.3 Gigabit Ethernet

Literally, Gigabit Ethernet has burst onto the networking scene. Just two years after the first IEEE study group meeting, twenty vendors demonstrated interoperable implementations at the Fall 1997 Interop tradeshow [5]. Let's take a closer look under the covers of the latest technology developed to serve the need for higher speeds in the LAN backbone and high-performance applications [4, 6]. Basically, Gigabit Ethernet is almost identical to its lower-speed cousins, but scaled up in speed. After beginning work in Fall 1996, the IEEE 802.3z committee plans to publish the first Gigabit Ethernet standards for full duplex operation over fiber optic cable and short-haul copper by mid 1998. Subsequent standards for half-duplex operation over long-haul copper are planned for early 1999.

The standard employs the same Ethernet frame format and protocol as its 10 and 100-Mbps Ethernet cousins. By preserving 802.3 Ethernet's frame structure and management framework, Gigabit Ethernet promises to seamlessly upgrade existing LAN backbones while preserving existing investments in hardware and software. In essence, Gigabit Ethernet is just a faster Ethernet. On the other hand, the new protocol retains some disadvantages from its ancestors. The maximum frame size is still only 1,5000 bytes, and in some configurations throughput decreases under heavy loads due to collisions.

In half-duplex mode, Gigabit Ethernet utilizes the same collision detection and avoidance mechanism as its lower-speed ancestors, the Carrier Sense Multiple Access/Collision Detection (CSMA/CD) protocol. One key area of difference in the IEEE 802.3z standard defines an enhanced CSMA/CD protocol that achieves a 200-meter collision diameter. As studied in the next section, collisions reduce throughput under heavy loads depending upon the distance and number of stations attached to the shared medium. But, when bandwidth in the LAN is cheap, these considerations aren't as important as they are in the WAN.

Full duplex operation eliminates the CSMA/CD protocol entirely, but inserts a 96-bit interframe gap. Switches and buffered distributors employ full duplex mode, but servers and interface cards do not. Full duplex mode supports a 64-byte minimum frame size.

The standard chose the Fibre Channel signal structure to accelerate product availability. The reach of Gigabit Ethernet over fiber extends from 500 m to over 2 km with multimode and single mode fiber, respectively. It can run up 25 to 100 m over existing twisted pair cabling. More information on Gigabit Ethernet is available on the web from the Gigabit Ethernet Alliance [7].

31.2.4 ATM Interface Speeds Designed for the LAN

The ATM Forum initially defined physical layer standards operating over multi-mode optical fiber at 100 and 155 Mbps. Recognizing the need for lower cost implementations, the Forum embarked upon an effort to define an ATM physical layer standard at lower speeds for use over existing twisted pair wiring. The Forum eventually produced two standards: first one operating at 50 Mbps, and after much debate, a second operating at 25 Mbps. All ATM interfaces defined to date operate in the full-duplex mode. The results from our survey of ATM network interface card manufacturers reported on in Chapter 16 revealed that most manufacturers support only the 25-Mbps variety.

25-Mbps ATM had one key objective — cheap desktop ATM LAN connectivity. Using tried Token Ring based technology, a number of devices now support this recently developed standard. In contrast to cheap 10 or 100-Mbps simplex Ethernet, 25-Mbps ATM offers a full duplex link. Furthermore, as shown later in this chapter, 100-Mbps Fast Ethernet only achieves a maximum of 80 Mbps throughput. Therefore, for fully symmetric traffic, 25-Mbps ATM actually offers 25 percent more usable bandwidth than Fast Ethernet. This more than makes up for the ATM cell tax. Also, every feature offered by ATM at any other speed is also available at 25 Mbps on a seamless, end-to-end basis. For users keen on one or more of ATM's features, this is often the deciding factor.

Perhaps the biggest advantage of 25-Mbps ATM is that it runs over existing Category-3 unshielded twisted pair (UTP) wiring, versus other standards that require installation and upgrade to more expensive Category-5 wiring. The maximum distance is 100 meters per run. Of course, Category-5 cabling is required when upgrading to 155-Mbps ATM. 25-Mbps ATM clearly competes with 100BaseT Fast Ethernet since their costs were within a factor of two in 1998.

31.3 LAN TECHNOLOGY PERFORMANCE COMPARISON

This section compares the key aspects of performance occurring in the LAN. These include the achievable throughput and the limitations incurred by Ethernet's CSMA/CD protocol when operating with shared media 10-Mbps or half-duplex Gigabit Ethernet. The text also analyzes the impact of packet size on efficiency.

31.3.1 Throughput

As workstation power and application bandwidth demands increased; 10-Mbps Ethernet LANs ran out of gas. LAN administrators segmented, and re-segmented LANs, to increase bandwidth allocated to each computer, until in some cases, only one server or workstation remained per Ethernet segment! FDDI served as the high-speed LAN backbone during the early years of LAN growth, but soon became a bottleneck as well. LAN switching then emerged to more efficiently interconnect such highly loaded LAN segments. Next, 100-Mbps Ethernet came on the scene, but still some high-end servers required dedicated 100-Mbps segments to handle the traffic. However, the leading edge of workstation technology makes even 100 MBps a bandwidth constraint. And now Gigabit Ethernet has emerged to once again conquer the LAN bottleneck — but this only moves the bottleneck back to the WAN.

Note that ATM provides the best LAN to WAN single technology migration path (1.5 Mbps to OC-48) in the industry. A major advantage of ATM is this of a broad range of speeds. Ethernet LANs and ATM run at comparable speeds. As described in the previous section, 25-Mbps ATM is equivalent to 100-Mbps Fast Ethernet and using similar reasoning, 622-Mbps ATM is equivalent to simplex Gigabit Ethernet. Thus, a server can have anywhere from a 25-Mbps connection to a giant 622-Mbps pipe – more than the leading servers can currently use. And ATM is already standardizing OC-48, almost 2.5 Gbps of full-duplex bandwidth, over twice that of full-duplex Gigabit Ethernet and four times that of simplex Gigabit Ethernet! Remember that less than a decade ago, most LAN designers never envisioned that users would require more than 10 Mbps.

Remember that what really matters in the LAN is not the efficiency of switching variable length frames or ATM cells, rather the key difference is whether the devices uses hardware or software to perform the switching and routing function. Hardware always beats software in terms of speed.

31.3.2 Loss due to Retransmission

The CSMA/CD protocol employed by Ethernet inherently causes loss under heavy load due to collisions. Because stations must retransmit lost frames, the resulting efficiency declines as collisions increase. Collisions are greater for shorter frame lengths and longer cable runs. The following formula [8] gives the resulting Ethernet channel efficiency for a heavily loaded Ethernet segment:

$$\text{Channel Efficiency} = \frac{A\ F\ c}{A\ F\ c\ +\ 2\ B\ L},$$

where $A = (1\text{-}1/k)^{k\text{-}1}$ = Probability of successful transmission,
k = number of Ethernet stations sharing the media,
F = Ethernet frame size (in bits),
c = speed of signal propagation in Ethernet (10^8 m/s),
B = Ethernet bit rate (bps),
L = distance between stations (m).

The second term in the denominator of the above equation indicates a fundamental scaling limit of the CSMA/CD protocol. Specifically, the term B×L is the bandwidth-distance product. Note that as this term increases, efficiency decreases. However, this is fundamentally at odds with what LAN users require — increased efficiency when going across longer distances at higher speeds. This is why CSMA/CD does not scale beyond the LAN environment. Therefore, network designers use other networking techniques to interconnect LANs. Figure 31.2 plots the resulting 10-Mbps Ethernet channel efficiency as a function of the frame size (F) for k=1, 2 and 4 Ethernet stations connected to the shared media at the maximum distance of L=2,500

m.   Note how the efficiency of even a single station using CSMA/CD, for example a client or server attached to a LAN switch, achieves only 70 percent throughput when operating at the largest allowed Ethernet distance.   This occurs because the transmitting station must receive its own frame before it can send another frame.   Therefore, reducing the distance between stations improves performance as indicated by the above formula.

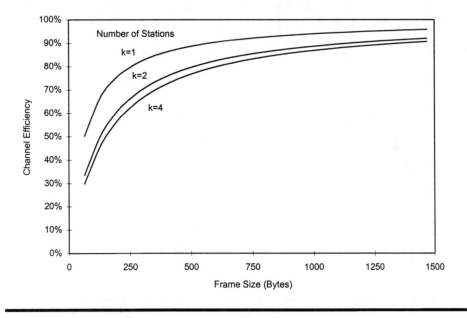

Figure 31.2   10-Mbps Ethernet Channel Efficiency versus Frame Size (F)

31.3.3 Efficiency of Small versus Large Packets

As studied in Chapter 9, the 64-byte minimum frame size comes from the basic physics involved in propagating an Ethernet frame along the maximum distance of 2.5 km on a shared media 10-Mbps LAN.   The Gigabit Ethernet standard increases this to 512 bytes and decreases the maximum distance to 200 m.   Gigabit Ethernet takes on some additional complexity to adress this inefficiency by defining an option called packet bursting.   This allows switches and servers to multiplex multiple 64-byte frames into a the larger 512 byte frame for the sake of efficiency.   Without packet bursting Gigabit Ethernet loses approximately 30 percent efficiency for TCP/IP traffic since as shown in the last chapter, 40 percent of the packets are only 40 bytes long.

On a LAN backbone, efficiency is an important design consideration.   The maximum Ethernet frame size of 1500 bytes also impacts efficiency.   A mismatch with Ethernet, IPv6 supports packets in excess of 65,535 bytes. Recall from Chapter 13 that ATM's AAL5 supports packets of lengths up to 65,535 bytes.   Accounting for the IPv6 overhead in the LAN and the WAN for packet sizes in excess of 1500 bytes decreases the efficiency gap between ATM and frame based protocols in the LAN and WAN.   Figure 31.3 plots the resulting efficiency of carrying 1500 byte IPv6 packets over the Ethernet LAN and an HDLC-based WAN versus single longer packets using ATM's AAL5 using LLC/SNAP encapsulation.   The analysis assumes that IPv6 uses an 8-byte MPLS header and a 1-byte PPP header in the WAN. The IPv6 LAN and WAN figures assume that all packets but the last one contain 1500 bytes of payload.   As seen from the figure, for longer packet sizes, the efficiency difference diminishes to 2 percent between ATM in the WAN and IPv6 in the WAN. In fact, Ethernet is 2 percent more efficient carrying IPv6 in the LAN than IPv6 carried over HDLC, PPP, and MPLS in the WAN.

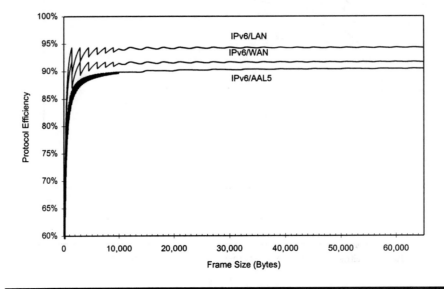

Figure 31.3   IPv6 Efficiency over Ethernet, HDLC in the WAN, and ATM AAL5

31.4 LAN TECHNOLOGY COMPARISON

This section provides the reader with the criteria required to chose a LAN backbone and access technology. The analysis presents technology, business, and cost attributes for   following technologies competing for access to the desktop, connecting the enterprise workgroup, and the LAN backbone environment:

- Legacy LAN   contention-based   protocols   (e.g.,   10/100/1,000-Mbps Ethernet)
- High-speed token-passing protocols (FDDI)
- Low Cost ATM (25 Mbps over twisted pair, 155, and 622 Mbps over multimode fiber)

The general rule when building high-performance workgroups and LAN backbones is switch whenever possible and share media only when necessary. This remains true because switching achieves much greater throughput and mitigates the throughput-robbing contention problems inherent in shared Ethernet media.

31.4.1 The Battle for the Desktop

This section reviews ATM's battle with Ethernet for control of the desktop. ATM did not make its way into the mid-1990s generation of workstation motherboards. There were many reasons for this; first of all, the worksta-tions of the mid-1990s had internal bandwidth constraints of 33-50 MHz buses, where an ATM LAN offered speeds ranging from 25 Mbps up to 600 Mbps. And even when the bus speeds were matched, peripheral speeds were not. New Pentium PC bus speeds operate in excess of 400 MHz and are now capable of providing true multimedia to the desktop.

What makes a user require an ATM interface on their end system? One reason is the requirement for higher bandwidth and guaranteed QoS performance in the LAN. Users typically first try to segment their existing Ethernet LAN into smaller and smaller segments until they reach the threshold of Ethernet switching which provides 10 or 100 Mbps to each user. This is the point where ATM takes over by providing full-duplex 25 and 155-Mbps to the desktop. From an application standpoint, desktop video-conferencing — the integration of the phone, computer, and video — is rapidly becoming a reality for many corporations, along with integrated multimedia applications, all of which require the capabilities of ATM.

For many cost-conscious users, ATM to the desktop is too costly at this point, especially when installation and support costs are added to hardware costs. However, the price for ATM interface cards is dropping rapidly, so that the cost of 100-Mbps Ethernet and ATM adapter cards may soon be close. But support costs for ATM remain high due to its complexity. Although ATM speeds are greater than many current applications, the rapid growth in computer performance and increased need for communications is rapidly changing this situation. When the workstation becomes faster than the current LAN, then an upgrade is in order. An upgrade to ATM is becoming the choice of some.

But is ATM too late? The battle for the desktop is really between IP over legacy LAN protocols like 100-Mbps Ethernet, and ATM. ATM has an uphill battle to fight to become a major contender; let alone attain the role of dominant desktop LAN interface. Beating 100-Mbps switched Ethernet and eventually Gigabit Ethernet is a formidable challenge for ATM.

31.4.2 Ethernet versus FDDI versus ATM in the Workgroup

ATM was designed to handle multimedia. A key protocol difference between ATM and Ethernet is in the segmentation and reassembly of the user protocol data units (PDU). ATM breaks data (the PDU) into fixed-length cells. It then establishes a connection-oriented virtual circuit that guarantees bandwidth and quality to the recipient. In contrast, Ethernet encapsulates user PDUs into frames and transmits them out onto a shared media in a connectionless broadcast fashion, relying on its collision-detection mechanism or switching to manage user contention for the shared bandwidth. If the user PDUs are too large for the shared media, then a higher layer protocol (e.g., IP) must segment and reassemble them.

Despite all the hype surrounding Gigabit Ethernet, ATM remains very much a viable option in the LAN. While ATM remains the future WAN technology of choice, one must look deeper into the economics of placing ATM in the LAN. This is especially true when analyzing the existing legacy LAN environment, whose cost-effective, embedded base of 10/100-Mbps shared and switched Ethernet technology creates a tremendous barrier to entry for ATM. To date, the key for ATM's foray into the LAN environment has ridden in on its extension and integration with a new breed of LAN switches and routers. These devices provide a viable and more cost effective LAN backbone technology than FDDI, as well as a more cost effective per Mbps technology than 10/100-Mbps Ethernet to the desktop.

Table 31.1   Attributes Comparison of FDDI, Fast Ethernet, and ATM

Attributes	FDDI	Switched 100 Mbps Ethernet	1 Gbps Ethernet	ATM
Throughput	100 Mbps Simplex	100 Mbps Simplex	1,000 Mbps Simplex or Duplex	25 Mbps to 622 Mbps Duplex
Backwards Compatibility	Good	Best	Best	Good (via LANE)
Evolution Potential	Little	1 Gbps	None	2.48 Gbps
Reserved Bandwidth	No	Yes	No	Yes
Isochronous Support	No	Yes	No	Yes
Multiple Traffic Classes	No	Priorities in 802.1p	Priorities in 802.1p	Yes
Projected Costs	High	Low	Medium	Medium
Use of Existing Wiring	No	Yes	No	Yes
Scaleable in Speed	No	Yes	No	Yes
Scaleable to WAN	No	No	No	Yes

Table 31.1 compares the key technology and business attributes of FDDI, Switched 100-Mbps Ethernet, Gigabit Ethernet, and native ATM.   This analysis shows that ATM is clearly differentiated from its LAN peers in at least one area. Switched 100-Mbps Ethernet is the closest contender for all

around capability of the group. For this reason, this is why a number of network managers use switched Ethernet to upgrade their existing LAN backbones and high-performance workgroups instead of ATM.

While Ethernet switching provides large amounts of dedicated bandwidth to single users, switched networks still operate in a peer-to-peer, flat layer 2 bridged architecture. This can be a major disadvantage as the size of the network grows, affecting performance, management and scalability. Also, Ethernet is distance limited, whereas ATM networks effectively have no distance limitations. Another major disadvantage of Ethernet technology is its inability to interoperate with dissimilar LAN technologies, such as FDDI, on a single network without the additional expense of translation bridges or routers. However, ATM's LAN Emulation and MPOA protocols achieve interoperability with these legacy LANs. And for FDDI, its high price and lack of widespread support and adoption are still major disadvantages. For all of its benefits, moving to 100-Mbps switched Ethernet and Gigabit Ethernet solutions still only simply throws more bandwidth at the problem. Although simple, less costly and familiar; these solutions don't solve the QoS, bandwidth allocation and WAN extension issues.

31.4.3 ATM versus Gigabit Ethernet in the LAN Backbone

We now turn our attention to a more in-depth comparison of ATM and Gigabit Ethernet, focusing on key requirements in the LAN backbone.

Three key factors drive Gigabit Ethernet to replace ATM in the LAN [9, 10]:

- Equipment availability and pricing favors Ethernet
- Industry forecasts for the hub and switch market clearly show Ethernet dominance for at least the next several years
- The requirement for multimedia, in particular voice and video, has not dominated LAN traffic

Let's look at each of these in more detail. Gigabit Ethernet switching and routing devices soon promise to be only two to three times the cost as 100 Mbps Ethernet full-duplex devices — yet will achieve nearly a 10 to 1 performance advantage over current 100-Mbps internetworking gear. This results in a 3- to 5- fold increase in price/performance. ATM switches are still more expensive than conventional LAN Ethernet switches, but the gap is narrowing. Note that most major equipment vendors are providing both Gigabit Ethernet and ATM products, and are members of both the ATM Forum and the Gigabit Ethernet Alliance. Beware of potential interoperability problems for products developed prior to the final Gigabit Ethernet standard. See our companion book, "Hands On ATM," [11] for more details on the business aspects of Gigabit Ethernet and ATM.

The requirement for multimedia, in particular voice and video, in the LAN has not materialized in force. One example is MPEG2 full-motion video with a 64-kbps voice quality line and data transfer on every workstation. We believe that this lack of adoption is due to the lack of APIs and applications designed to specifically take advantage of ATM's QoS capabilities. The 802.1p prioritized Ethernet Logical Link Control (LLC) standard could change this for native Ethernets. Also, the recent MicroSoft NDIS 5.0 standard described and native ATM LANE 2.0 support could rapidly change this situation.

While Gigabit Ethernet partners well with IP networks, there are still missing pieces: namely, prioritization, switching, and full quality of service support. Table 31.2 provides a comparison of ATM to Gigabit Ethernet.

Table 31.2   ATM Compared with Gigabit Ethernet

Attribute	ATM	Gigabit Ethernet
LAN Protocols Supported	Ethernet, Token Ring, FDDI	Ethernet
WAN Support	Yes	No
Upgrade Path	DS1 through OC-48	10, 100 Mbps Ethernet
Learning Curve	High	Low (given Ethernet experience)
New NIC required	Yes	Yes
Equipment Cost	High	High
QoS Support	Built-in	Planned in 802.1p
Installation Difficulty	High	Medium
Speed	Ranges from DS1 (1.5 Mbps) up to OC-48 (2.5 Gbps)	1 Gbps
Configuration	Some required	Plug and play
Manageability	More complex	Simpler

Thus, Gigabit Ethernet stands a strong chance of displacing ATM in most tactical LAN implementations. But users that choose this route will have to rely heavily on early pre-standard versions of 802.1p and/or RSVP to achieve classes of service and a form of bandwidth reservation. However, until Gigabit Ethernet proves in and collects additional feature support, ATM is currently the best bet for a high-speed LAN backbone.

31.4.4 Cost Comparisons of LAN Protocol Implementations

Of course, in the local area network, cost is frequently the key consideration. Here, the inherent simplicity of Ethernet holds reign over the market. Although ATM adapter sales continue growing at a more rapid pace than their Ethernet counterparts, they would not crossover at the current growth rates for many years. Figure 31.4 illustrates the declining cost per megabit-per-second per port over time [11, 12] for bandwidth in the LAN. This includes the interface card in the workstation and the switch port. Note that this comparison is based upon the cost per Mbps and not the absolute costs.

This means that the absolute cost for 155-Mbps ATM is much greater than the cost for a 10-Mbps Ethernet connection.

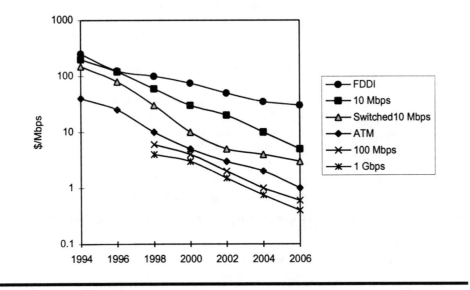

Figure 31.4   Cost Trends per Mbps for ATM, FDDI and Switched Ethernet

From this analysis it is clear that Gigabit Ethernet, 100-Mbps Ethernet and 155/622-Mbps ATM offer the most cost effective bandwidth. Be careful though, the total cost is proportional to bandwidth. In other words, if you don't require 1 Gbps of capacity, a 100-Mbps solution is actually more economical in terms of actual expenditure. Furthermore, note from the previous analysis that ATM offers the additional key capabilities of a seamless LAN/WAN technology, more choices for bandwidths, and mature support for multiple QoS classes.

Obviously, these cost benefits are most relevant when designing a network from scratch. But if an existing network is upgrading to ATM, and has many legacy Ethernet and FDDI LAN NICs and switches in place, moving to an all ATM network may be cost prohibitive. In closing, we list some real questions to ask. Does 1 Gbps of full-duplex bandwidth not seem like all that much? Do your applications require QoS guarantees and reserved bandwidth? Does your application need to scale across the WAN? Yes answers typically point users toward ATM.

31.5 QOS AND BANDWIDTH RESERVATION: RSVP OR ATM?

As discussed in Chapter 16, QoS aware applications face a chicken and egg situation. Application developers don't make use of ATM features since QoS capable networks (the chicken) aren't widely deployed. On the other hand, users aren't deploying Quality of Service (QoS) capable networks since QoS-aware applications have been slow to hatch (the egg).

Do you need to be concerned about QoS? Ask yourself the following three questions, and if you answer yes to any one of them, then you probably do.

1. Do you plan to run video-intensive multimedia applications on the workstation?
2. Do you want to mix voice and/or telephony traffic with data from the desktop across the WAN?
3. Do you want to consolidate voice and data infrastructures (wiring, equipment, maintenance, support staff, etc.)?

This section compares the following technologies that deliver a consistent QoS to applications that require low delay and jitter, such as real-time voice and video:

- IP using RSVP over Ethernet using 802.1p prioritization
- ATM using SVCs and LANE 2.0 or MPOA

31.5.1 RSVP and ATM Approaches to QoS Support

While both RSVP and ATM are signaling protocols, their approaches for supporting communication of traffic parameters and QoS indication are diametrically opposed. RSVP uses receiver-initiated flow indication, ATM employs sender-initiated signaling. RSVP receiving nodes signal through the upstream network back to the transmitting node to indicate acceptable flow characteristics. RSVP duplicates this procedure independently in each direction for a full-duplex application. On the contrary, an ATM sender must signal through the downstream network the traffic and QoS requirements for both directions of a full duplex connection. Usually, the sender doesn't know the characteristics of the receiver before placing the call, which makes deciding on traffic parameters and QoS a challenge for the sender.

Experts often call the RSVP "soft state," since the request and path messages continually communicate the request for service and bandwidth. This stands in contrast to the "hard state" of connection-oriented protocols like X.25, frame relay and ATM. Each of these approaches has advantages and disadvantages. A key difference revolves around the fact that the "hard state" approach enables networks to guarantee QoS and reserve bandwidth by denying connection attempts if admitting the additional traffic would degrade the quality for the existing connections. The ATM network then attempts another route, blocking the service request only if all routes are full. On the contrary, the "soft state" approach cannot perform this function easily since the order of the reservation packets (i.e., Resv packets as defined in Chapter 8) arriving at routers along the path may differ. This effect means that all of the routers along the path may not reserve capacity for the same flows consistently. Thus, ATM provides for a much more predictable network; a critical consideration when offering differentiated services that carry voice, video, and data traffic.

This brings us back to the requirement for end-to-end ubiquitous handling of QoS. Another difference is that RSVP operates in a connectionless IP network; packets transferred between two locations may take different paths, making it a challenge to guarantee bandwidth and QoS. ATM completely avoids this issue since each node along the end-to-end path explicitly verifies through Connection Admission Control that the requested QoS can be met at the specified traffic level in both directions. In other words, in IP reservations and routing operate independently; while in ATM they operate concurrently during the connection establishment interval.

Because of the arguments summarized above, most experts concede that the RSVP model better supports the native IP interface present on many applications, while ATM achieves efficient connection establishment and resource allocation in backbone networks, which is an acknowledged weakness of RSVP [13]. A key issue with both approaches is scalability in the backbone since each protocol effectively operates on a per flow basis. When millions of flows share a large trunk, processing of each flow is simply not practical. As stated in the RSVP RFC, some form of aggregation is essential to building a scalable network. ATM uses virtual path trunking to address this issue as described in Part 5.

31.5.2 RSVP versus ATM: The QoS Battle

One of the key battlegrounds of RSVP and ATM will occur in the Internet. The many claims that the Internet will soon resemble the LA turnpike gridlock during rush hour may be exaggerated. But one thing is for sure — businesses increasingly relying on the Internet to disseminate information, advertise, and conduct mainstream activities will not tolerate productivity impairing traffic jams. This content focused activity causes bandwidth requirements to skyrocket. And the amount of multimedia traffic transiting the Internet is also increasing.

Regarding QoS guarantees, ATM is years ahead of any IP-based approach. Although years in the making, RSVP still has a number of problems and is only an experimental protocol as defined by the IETF. Proponents of ATM in the Internet see ABR service and RSVP competing head on. But ATM has been out for years, and IETF-defined RSVP has yet to hit the street in full production form. QoS and reserved bandwidth is the key battle where ATM is the incumbent over IP [14].

31.5.3 LAN QoS Support

The IEEE 802.1p standard defines a means for Ethernet stations to prioritize traffic. This concept maps directly to the concept of precedence in the Type Of Service (TOS) byte in the IP header as well as the concept of ATM Forum

QoS classes. However, the standard defines no means to enforce the traffic generated by any one station or switch. Instead, the standard recommends the use of priority queuing as described in Chapter 22 to selectively discard lower priority traffic during periods of extended overload. This is only a partial solution since only the highest priority has any hope of guaranteed QoS. Frequently QoS is only a statistical guarantee; however, the following example of priority service without sufficient capacity hopefully helps the reader remember this key point. Although the first class customers on the Titantic had first access to the lifeboats, there simply weren't enough boats (and the service mechanism was inefficient); hence not everyone in this class got the same quality. Therefore, in order to achieve quality of service, the network must have a means to deny requests that it does not have sufficient resources to provide the requested quality. Since differentiated QoS doesn't reserve bandwidth, simple priority queuing schemes like those defined in 802.1p and IP precedence are inferior to ATM's connection-oriented approach. Hence, Ethernet prioritization schemes must be used with a traffic level indication mechanism like RSVP. Furthermore, network devices must enforce the traffic parameters to ensure fairness within a class.

Problems occur on some high speed networks, however, in particular on shared media LANs, even those operating at 100 Mbps. The collisions that still occur on 100 BaseT Ethernets can create significant delay variations during periods or relatively high load [15]. The 100VG-AnyLAN technology overcomes this problem by prohibiting shared media and requiring that intelligent hubs use a demand priority scheme to serve high priority traffic before any low priority traffic. However, this design fails to address the issue of bandwidth reservation as well.

Ethernet is still only a MAC layer, collision-based technology, and must rely on higher layer protocol schemes to assure specific quality of service to users. Thus, QoS planning starts at the workgroup level — because each successive move toward the WAN typically decreases bandwidth. One way to provide provides QoS management between Ethernet segments is using IP's Resource reServation Protocol (RSVP) protocol defined by the Internet Engineering Task Force (IETF). However, applications must be able to signal for QoS support and traffic levels using LANE 2.0 for any of ATM's capabilities to really make a difference to the desktop.

31.5.4 RSVP QoS Support

As described in Chapter 8, IPv4 has only the seldom used precedence field in the header only defines relative priorities. RSVP, on the other hand, provides application developers the capability to request a specific service quality and bandwidth from a IP network. RSVP is a signaling protocol used by a receiver to request QoS and bandwidth from a multicast sender. However, RSVP does not guarantee QoS or reserve bandwidth — every device along the path must do this dynamically.

The RSVP signaling protocol allows an IP user to specify QoS and bandwidth per application flow. IPv4 uses special messages in support of RSVP, while the IPv6 packet header allocates a 24-bit flowspec header designed for RSVP. New APIs like MicroSoft's WinSock specification make QoS available to application developers, making QoS in the application space possible. Currently, RSVP is supported in two ways:

- By most major router vendors in a proprietary way.
- Using an industry standard API like MicroSoft's WinSock 2.0

RSVP capable devices must also implement shaping and policing functions to control bandwidth usage to ensure fair QoS allocation among all users. In comparison, ATM controls and enforces traffic contracts start at the originating edge of the network to manage and guarantee the specified bandwidth and QoS using connection admission control [16]. In contrast, IP does not naturally deny new RSVP requested "connections," since its underlying networking paradigm is connectionless. Instead, routers must examine every packet header to determine if a flow exceeds its requested parameters. Also, routers must implement policy control processing on each RSVP request.

31.5.5 Comparison of RSVP and ATM

Table 31.3 provides a comparison of RSVP and ATM features. The key difference between ATM and RSVP when it comes to QoS is that RSVP only *requests* delivery over fixed bandwidth in a logical broadcast medium, whereas ATM *guarantees* delivery and reserves bandwidth using its connection-oriented paradigm.

Table 31.3   Comparison of RSVP and ATM

Feature	ATM	IP/RSVP
Signaling Initiator	Sender	Receiver
Multicast handling	Add-on	Native
Connection-Orientation	Duplex	Simplex
Traffic Enforcement	Policing	Not specified
Evolution Strategy	Excellent	Poor
End-to-end guaranteed QoS	Yes	Limited
Flow Control	ABR	TCP
Additional workstation software required?	Yes	No

31.5.6 Interworking RSVP and ATM

So, if these protocols are so different, can they ever coexist, let alone work together? The answer appears to be yes. Several IETF drafts address the interoperation of RSVP with ATM networks. One draft [17] maps between IP and ATM services with the goal to achieve end-to-end Quality of Service for

IP networks containing ATM subnetworks. It defines several services and associated parameters that quantify source traffic and QoS requirements. It identifies various features of RSVP's Guaranteed Service and Controlled Load Service, identifying appropriate mechanisms in ATM Virtual Channel Connections (VCCs) which facilitate them. A second draft [18] describes a method for providing IP Integrated Services with RSVP over ATM Switched Virtual Connections (SVCs). It defines a specific method for running over today's ATM networks. It provides guidelines for using ATM VCCs with QoS as part of an Integrated Services Internet. The IETF and the ATM Forum are working jointly on specifying interoperation of RSVP and ATM, so there is hope that these otherwise incompatible protocols will not only be able to coexist, but cooperate.

In the end analysis, ATM and RSVP will probably coexist for some time. Legacy IP networks are growing, not disappearing, and ATM has yet (if ever) to dominate the LAN. These two trends point to a coexistence of the two technologies. In fact, IETF efforts are under way to map ATM's MPOA to RSVP to reconcile RSVP requests to MPOA's virtual routers [19, 20].

31.6 COMPARISON ANALYSIS OF CONNECTION-ORIENTED VERSUS CONNECTIONLESS NETWORKING PARADIGMS

As introduced in Chapter 5, the concepts of connection-oriented and connectionless networking have fundamental differences. As shown in the previous chapter, IP operating over ATM FUNI achieves efficiency greater than IP operating over a protocol stack including RSVP and MultiProtocol Label Switching (MPLS) to achieve comparable feature function. Therefore, the real question comes down to the choice of connection-oriented versus connectionless protocols, and not the efficiency of frames versus cells at the link layer. Fundamentally, frame relay and ATM are connection-oriented protocols, while IP is a connectionless protocol. This section examines the applicability of IP and ATM networking principles to achieve network reliability, scalability and robustness in support an of advanced portfolio of voice, video and data services.

31.6.1 Unassured and Assured Data Delivery

One aspect of packet switching is the service guarantee related to packet delivery. An assured data service guarantees delivery of the packets, while an unassured data service does not. The assured mode is tolerant of errors. The unassured mode is also called a datagram service.

Frame relay, IP and ATM all operate in the unassured mode. ATM offers assured mode service through the Service-Specific Connection-Oriented Protocol (SSCOP) described in Chapter 14. SSCOP is a well defined, solid protocol that works well in congested environments because of its sophisticated, efficient selective retransmission strategy. The IP protocol suite adds assured delivery via the Transmission Control Protcol (TCP) at layer 4 when operating over any lower level network. Chapter 8 introduced the TCP protocol and Chapter 26 analyzed its performance. The nearly ubiquitous deployment and interoperability of TCP testifies to its robustness and inherent simplicity.

The packet switching service may also guarantee that data arrives in the same sequence as it was transmitted, or may reorder the packets upon delivery. A packet service with sequence integrity preserves the order of transmitted packets. Both frame relay and ATM require sequence integrity, while IP does not. Inherent to the nature of a connectionless protocol is the possibility that packets can be delivered in a different order than they were sent. Again, TCP provides sequence integrity at layer 4 for IP-based applications that require this feature.

31.6.2 Addressing Philosophy and Network Scalability

Addressing may be geographically oriented like the telephone numbering plan, service provider oriented like the Internet, or hardware-oriented. The addressing schemes employed include IP, E.164, NSAP, and IEEE MAC, as covered in earlier chapters. The assignment and agreements to interconnect addresses range from monopoly-oriented like the telephone network to cooperative as in the Internet. Agreement on a technical standard for addressing is usually an easy task compared with resolving the political, economic, business, and social issues that arise.

The IPv4 address is 32 bits while the IPv6 address is 16 bytes. Careful design and allocation of the IP address space keeps the number of routing table entries in the global Internet on the order of 100,000. The E.164 addressing plan is 15 Binary Coded Decimal (BCD) digits and currently aggregates well on a global or service provider prefix basis. The number of routing table entries here is on the order of one thousand. MAC addresses are 48 bits and assigned to specific addresses with a manufacturer specific prefix. Hence, MAC addresses have no aggregation potential. The ATM Forum PNNI specification mandates the hierarchical organization of the 20-byte NSAP-based address. ATM SVCs utilize either the E.164 or 20-byte NSAP-based addresses.

Coordinated assignment of the address space is a key enabler to global scalability for both connection-oriented and connectionless protocols since it determines the required size of the routing table. Both IP and ATM NSAP-based addresses support the notions of hierarchy and summarization essential to large networks. LANs have flat addressing, and hence do not scale beyond thousands of addresses. Typically, broadcast traffic limits the size of a LAN since every device must receive the broadcast messages.

31.6.3 A Shared Concept — Guaranteeing QoS and Bandwidth

Semantically, the ATM and IP protocol suites agree that some users require service with guaranteed QoS with reserved bandwidth. Also, they both agree that some users need only a best effort service. In essence, these best effort users are willing to accept lower quality, in exchange for paying a lower price — this is for the most part is what the Internet offered since the mid-1990s.

Therefore, the ATM and IP protocol suites provide a way for users to signal these QoS and bandwidth requests in a way that can be mapped on a one-to-one basis as described in section 31.5.6. You can't get more semantically equivalent than that!

As described throughout this book, ATM is a technology that guarantees QoS and reserves bandwidth. Most industry experts acknowledge this as the key strength of ATM over competing technologies. A key question is then: can a connectionless protocol actually reserve bandwidth and guarantee QoS?

Connection-oriented services use either Permanent Virtual Connections (PVCs) or Switched Virtual Connections (SVCs). A key component of the connection establishment process is reserving bandwidth for only those connections that the network can guarantee QoS. The network tries other possible routes, and eventually blocks connections for which it cannot find a route that guarantees QoS. This design automatically maximizes utilization of expensive WAN bandwidth.

On the other hand, connectionless services utilize dynamic routing protocols to synchronize forwarding tables in a hierarchical routed network. Since the forwarding tables may temporarily be out of synch, connectionless networks must also implement a means of removing looping packets. As analyzed in Chapter 9, contemporary connectionless routing algorithms tend to route all traffic on a minimum cost path without any notion of QoS or bandwidth.

However, the IETF has answers brewing to the problems of QoS guarantees and guaranteed bandwidth: RSVP (see Chapter 8) and MPLS (see Chapter 20). As covered in the previous section, RSVP doesn't perform the job as well as ATM on a network wide basis, at least not yet anyway. Label switching may well provide a solution to the bandwidth management problem for connectionless services. Therefore, to achieve QoS and bandwidth on demand capabilities comparable to ATM, a connectionless-based solution requires at least 4 to 8 bytes of additional overhead for the MPLS label in each IPv4 packet on the backbone. However, full standardization and interoperable implementation of the routing protocol to distribute the labels in response to time-varying demand may take several years prior to ubiquitous deployment across the Internet. Furthermore, as discussed in the previous chapter, frame-based protocols require a standard means for frame-based multi-service access to prevent long frames from delaying short urgent frames.

Not only do the control plane connection-oriented ATM protocols reserve bandwidth and guarantee QoS, but as described in Chapter 29, the management plane measures it as well! When money changes hands for guaranteed QoS and reserved bandwidth, the issue of measurements and service level agreements is soon to follow. The ATM cell-level OAM standards are now largely complete and being implemented by most vendors. Packet switched solutions will require similar management plane capabilities; an area where very little standardization exists today. Of course, these can be defined, but some concepts like packet delay variation will be more complex than their fixed length cell counterparts.

31.6.4 The Inequality of QoS and Bandwidth Reservation

Philosophically, resource reservation creates inequality [21, 22]. Basically, user requests that either ATM's Connection Admission Control (CAC) or IP's RSVP admit get the QoS and bandwidth they require, while those who pay less, or are latecomers, don't.

An alternative to such a blocking system could offer graceful degradation in the event of overflows via adaptive voice and video coding [21]. Currently, voice and video coders have an avalanche performance curve in terms of degradation in the face of increasing loss and delay variation: they degrade gracefully until a certain point where they become unusable. In essence, however, CAC and RSVP could factor in the allowable degradation as part of the admission decision process. Or else the network or server could offer varying grades of service at a different price. ATM and frame relay support the concept of user-specified priority via the Discard Eligble (DE) and Cell Loss Priority (CLP) bits in their header fields, respectively. As described in Chapter 23, some ATM-based video codec designs use the CLP bit to distinguish between the essential information for image replication and the non-essential fine detail. During congested conditions, the network discards only the fine detail information, hence the video and audio is still usable. When the network is not congested, then the user receives higher quality.

The following set of questions and answers illustrates the fundamental inequality of QoS and reserved bandwidth. When you connect to a web site, are you willing to experience only part of its potential audio, visual, video, and interactive capabilties? Possibly. What if the quality was so poor that you had to go back and visit that site at some later time? Sure, everybody does that today. What if you were paying for the visit, as you do when you watch a movie, receive educational training, or receive consultation? No, of course not.

As an illustration of the bandwidth reservation problem encountered in connectionless routing protocols like OSPF, consider the following example of booking a flight reservation from Boston to Los Angeles [23]. First, the agent informs you that the shortest path connects through Chicago and Denver. The agent then tells you that he has good news and bad news. The good news is that a seat is available on the flight from Boston to Chicago. The bad news is that the OSPF reservation system doesn't know if and when seats are available on the Chicago-Denver or Denver-Los Angeles segments. Furthermore, a longer path exists through Dallas and has an available seat on the Boston-Dallas segment, but the status of the Dallas-Los Angeles segment is unknown. This occurs because routers running OSPF only know about the status of the directly connected links. Even if you tried to make a first class reservation (i.e., equivalent to 802.1p priority in a LAN), this doesn't help the situation. A router running OSPF can't tell if there are any "seats" available except on directly connected links.

31.6.5 Scalability Limits of RSVP and ATM SVCs

RSVP and ATM SVCs do not scale to support the current best effort web traffic observed on the Internet. Supporting 100,000 simultaneous flows lasting an average of 30 s [24] each using RSVP or ATM SVCs would require an IP router or ATM switch to process over 3,000 connection setup requests per second for each OC3. Current ATM switches at best process hundreds of SVC request per second per port. RSVP requires that each router along the path make comparable admission decisions several times a minute for each flow. Hence, for long-lived flows, RSVP is less efficient than ATM SVCs. In any event, neither are well suited to supporting the current web-driven traffic flows. Therefore, a hybrid architecture where routers handle best effort traffic while RSVP and/or SVCs handle the flows requesting guaranteed QoS and reserved bandwidth may be the best tradeoff [21, 25]. This has some other advantages because a small portion of the overall flows comprise a majority of the traffic volume seen on the Internet.

Note that 20 percent of the flows comprise 50 percent of the traffic in the measurements reported in [24]. Assuming that the flows are correspondingly longer reduces the peak ATM SVC rate to 600 calls per second, still a large number. Since the RSVP flows are refreshed periodically, this makes no difference to the level of RSVP policy admission control processing levels. Other experts believe that the concept embodied in Ipsilon's version of IP switching of automatically recognizing individual flows and setting up differentiated QoS connections is fundamentally flawed [3]. This analysis warns of the dangers of excessive hype, and identified the fundamental problem as the assignment of a virtual circuit to every flow. The business failure of the approach was failure to recognize that some users are willing to spend more money to achieve a better quality of service.

ATM and frame relay SVCs have a scaling limitation similar to RSVP. Both protocols require control processing of the requests to determine whether to admit, or reject, the new arrival. For long holding time calls lasting at least several minutes, SVCs make sense. Certainly if switched video on demand using ATM over xDSL lines becomes a significant service, then direct SVCs lasting for 1 to 2 hours are much more efficient than video over IP as analyzed in the previous chapter. Connection-oriented service has a relatively high per call processing overhead, but then sets up forwarding tables (which could use either cells or packets). On the other hand, connectionless services perform a fixed amount of processing for each packet, but avoid call processing overhead. Therefore, as illustrated in Figure 31.5 an economic crossover point occurs regarding the overall processing expense for connection-oriented and connectionless services versus the duration of the flow [25]. Ever changing economic figures in the cost of connection switching and connectionless determine the precise numerical value of the crossover point. One analysis [21] placed the crossover point at between 30 and 100 packets. An interesting trend to track would be the duration of the average flow. If it is increasing, then development of connection-oriented solutions is essential to minimize networking costs.

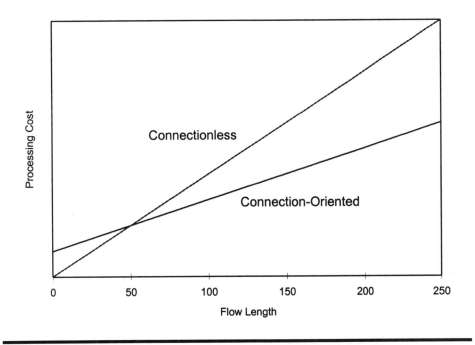

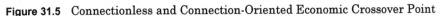

Figure 31.5   Connectionless and Connection-Oriented Economic Crossover Point

31.6.6 Flow Control

A packet service may implement flow control. The flow control method may be window-based, rate-based, or credit-based as described in Chapter 22. Alternatively a packet service may have no flow control whatsoever, for example, UDP used for IP telephony.

TCP provides a universally implemented, adaptive, end-to-end flow control mechanism that is connection-oriented. The difference in scalability here between RSVP and SVCs is that only the end systems participate in the connection, not every tandem node in the path. TCP uses packet loss detection to throttle back transmitters. Unfortunately, the protocol assumes that everyone is a good citizen and follows the rules because it defines no means of enforcement. Frame relay defines backward and forward congestion notification; however, few end systems recognize this feedback. ATM's Available Bit Rate (ABR) feature promises greater efficiency than TCP; however, it is not widely implemented yet. Fundamentally, these protocols must provide feedback layer four to improve efficiency.

31.6.7 Routing Scalability and Efficiency

Another fundamental difference between ATM and IP is the basic routing paradigm. As studied in Chapter 5, ATM uses the connection-oriented paradigm originated by the telephone network, while IP uses a connectionless paradigm where devices use the header within each packet to determine the next hop towards the destination. The connection oriented paradigm must maintain state for every flow, on the order of 100,000 connections per high-speed trunk for today's Internet traffic. On the other hand, connectionless routing protocols require forwarding tables on the order of 100,000 entries for an entire router. Each port uses the same forwarding table. Hence, the connectionless design scales better than the connection-oriented paradigm for Internet traffic. The key difference is aggregation of addresses in the connectionless design. Aggregating flows using dynamically adjusted virtual path trunks in ATM networks would improve the scaling of connection-oriented networking since only the virtual channel switches at the edge of the network would need to perform per-flow processing.

Furthermore, running IP over ATM has other advantages related to bandwidth and traffic management as cited in the IETF's draft MPLS requirements. In fact, the MPLS work draws directly on several key concepts from ATM. The MPLS label and the ATM cell header both have a similar semantic interpretation. Indeed the MPLS draft clearly states that label switching replaces the function of ATM cell header based forwarding. Since the semantics of layer two label switching are identical, ATM connection-oriented signaling and routing protocols can be applied to different hardware. There is no reason that ATM user and control plane protocols can't run over a frame based protocol. For example, the Frame-based User Network Interface (FUNI) protocol described in Chapter 18 supports signaling, QoS, multiple ATM service categories. Although limited by standards to 2.0 Mbps, there is no technical reason that FUNI can't run at 2.5 Gbps (i.e., OC48) or even higher speeds. At backbone speeds greater than DS3 rates, the benefits of the small cell size to preserving QoS diminish in inverse proportion to the trunk rate.

Let's look at a simple example of a four node network depicted in Figure 31.6a to better understand the efficiency of routing and restoration. Each node has an equal amount of traffic destined for every other node, as shown by the two-headed arrows. Figure 31.6b illustrates a network where a four SONET rings connect the nodes. The SONET ring dedicates one physical circuit between nodes for restoration, hence the network operates at 50 percent efficiency while providing for complete restoration. A major advantage such a SONET ring restored network is that restoration after a failure occurs within 50 ms or less. Another advantage is that the router requires fewer ports, an important consideration with the current port-limited router products offered by the major manufacturers.

a. Equally Distributed Offered Load

b. SONET Ring Routing/Restoration

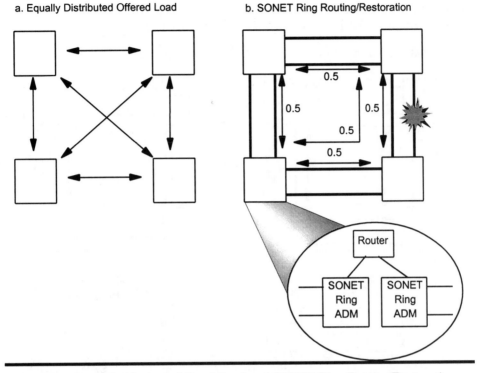

Figure 31.6 Example Network Offered Load and SONET Ring Routing/Restoration

Figure 31.7 illustrates the use of IP directly over SONET and IP operating over an ATM or label switched infrastructure. The IP/SONET design requires the deployment of an emerging technology: Big Fast Routers (BFRs). The example of Figure 31.7a shows the response of a BFR network running the OSPF protocol in response to the same failure. Since OSPF picks the shortest path, traffic rerouting is constrained and still results in only 50 percent efficiency when full restoration is necessary. An advantage of IP/SONET versus SONET ring restoration is that traffic can be statically balanced across equal cost routes as shown in the figure. This means that each link is less loaded, and hence should encounter fewer intervals of congestion. Figure 31.7b illustrates the rerouted traffic loads in response to the same failure scenario for an ATM or label switch based network. Assuming that the ATM or label based routing algorithm is aware of required bandwidth, the design improves efficiency by 50 percent, resulting in 75 percent utilization. ATM's PNNI algorithm won't necessarily yield this performance, but could come close. Note from the figure how the rerouting algorithm splits traffic across both the shortest and longer paths to achieve this goal. Therefore, although IP over SONET achieves improved link level efficiency (due to less overhead), the loss in network level efficiency requires that the routing algorithm changes as well before migrating off of ATM. As described in Chapter 20, the IETF is aggressively defining the MultiProtocol Label Switching (MPLS) standard for this very reason.

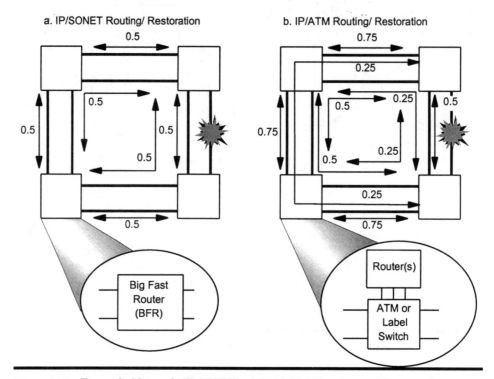

Figure 31.7 Example Network IP/SONET and IP/ATM Routing and Restoration

There is a basic mathematical reason for why the OSPF routing algorithm is well defined and widely implemented while the ATM/label switching routing algorithm is still undefined. Minimizing a single parameter, such as minimum distance has a relatively simple optimal solution that the Djikstra algorithm solves in an optimal amount of time. Adding the constraint of bandwidth makes the problem an NP-complete problem, to which the optimal solution is called multi-commodity flow optimization. This algorithm is much more complex than the Djikstra algorithm, and does not distribute well. Hence, design of suboptimal, yet still relatively efficient, bandwidth-aware routing algorithms is a key challenge for scaling the Internet backbone or any large ATM network.

31.6.8 Reliability and Availability

ATM inherits a penchant for reliable and highly available operation from the telephone network. Many of the manufacturers building ATM equipment have a strong telephony background and use similar design philosophies. The Ethernet LAN technology is also quite mature, but not designed for highly available operation. Although also mature and reliable, FDDI is not widely used.   Tremendous demand and an ever changing set of application and user requirements are straining the Internet technology designed 20 years ago. The Internet needs to come up with a viable guaranteed QoS and reserved bandwidth solution to lose the alias for the acronym frequently assigned to WWW in the trade press: World Wide Wait. Also, you probably wouldn't want to control a nuclear power plant with a best-effort connection-less network [25]. That is, unless you're willing to accept performance at the level Homer provides on the popular show "The Simpsons" aired by the Fox network.

31.7 REVIEW

This chapter compared ATM with other technologies competing in the LAN. The text first described the principal LAN requirements of low cost, simple economical operation, congestion avoidance via over provisioning link speeds, and the increasing need to smoothly integrate LANs across the wide area environment. Next, the chapter provided a brief summary of the competing LAN technologies: FDDI, switched Ethernet, 100-Mbps Fast Ethernet, and Gigabit Ethernet. The coverage then moved on to a comparison of ATM against these technologies at three levels: the desktop, the workgroup and the backbone. Currently, ATM does not fare well in terms of cost and complexity at the desktop or the workgroup except for applications that require guaranteed QoS and reserved bandwidth. The text then moved on to a comparison of IP and ATM resource reservation protocols. Finally, the chapter concluded with an analysis of the scalability, performance, and applicability of connection-oriented protocols like ATM and connectionless protocols like IP to enterprise and backbone network engineering problems.

31.8 REFERENCES

[1]   Essential Communications, "Gigabit Networking Today - Demand, Implementation & Its Future," http://www.esscom.com/whitepapers/gigawhite.html

[2]   K. Korostoff, "Why They Buy: A Survey of ATM Customers," *Business Communications Review*, February 1998.

[3]   T. Nolle, "There's More to Switched IP than IP Switching," Network World, January 12, 1998.

[4]   Gigabit Ethernet Alliance, "Gigabit Ethernet - Accelerating the Standard for Speed," http://www.gigabit-ethernet.org/technology/whitepapers/gige_97/, 1997.

[5]   C. Feltman, "Where will Gigabit Ethernet Fit?," *Business Communications Review*, December 1997.

[6]   3Com, "Gigabit Ethernet on the Horizon: Meeting the Network Challenge," 1997, http://www.3com.com/0files/strategy/600220.html.

[7]   Gigabit Ethernet Alliance, www.gigabit-ethernet.org.

[8] A. Tannenbaum, *Computer Communications, Third Edition*" Prentice-Hall, 1996.

[9] S. Saunders, "Closing the price gap between routers and switches," Data Communications, Feb 1995.

[10] L. Wirbel, "At N plus I, unity--and all hell breakin' loose," Electronic Engineering Times; Manhasset, Sep 16, 1996.

[11] D. McDysan, D. Spohn, *Hands On ATM*, McGraw-Hill, 1997.

[12] U. Black, *ATM: Foundation for Broadband Networks*, Prentice-Hall, 1995.

[13] A. Mankin, Ed., F. Baker, B. Braden, S. Bradner, M . O `Dell, A. Romanow, A. Weinrib, L. Zhang, "RFC 2208 - Resource ReSerVation Protocol (RSVP) -- Version 1 Applicability Statement Some Guidelines on Deployment," September, 1997.

[14] M. Cooney, "Can ATM save the Internet?," *Network World; Framingham,* May 20, 1996.

[15] S. Saunders, *High-Speed LANs Handbook*, McGraw-Hill, 1996.

[16] T. Nolle, "The great QoS debate: IP vs. A TM is only the tip of the iceberg," Network World, Jul 15, 1996.

[17] M. Garrett and M. Borden, "Interoperation of Controlled-Load and Guaranteed-Service with ATM", Internet Draft, <draft-ietf-issll-atm-mapping-01.txt>, 01/03/1997.

[18] S. Berson and L. Berger, "IP Integrated Services with RSVP over ATM", Internet Draft, <draft-ietf-issll-atm-support-02.txt, .ps>, 11/27/1996.

[19] J. Caruso, "Dynamic Duo: RSVP + ATM," Communications Week, June 23, 1997.

[20] IBM, "Whitepaper: Desktop ATM versus Fast Ethernet," 1996.

[21] V. Antonov, "ATM: Another Technological Mirage," Pluris Inc., http://www.pluris.com/ip_vs_atm/, 1996.

[22] C. Lewis, "QoS: Creating Inequality In An Equal World," Network Computing On Line, http://techweb.cmp.com/nc/809/809ws2.html, May 1996.

[23] M. Ladham, "MPOA versus MPLS," ICM L2/L3 Interworking Workshop, April 1998.

[24] K. Thompson, G. Miller, R. Wilder, "Wide-Area Internet Traffic Patterns and Characteristics," *IEEE Network*, November/December 1997.

[25] B. Olsson, "IP Backbone Provisioning Considerations," *Communicator Teleplan AB*, http://www.communicator.se/whitepapers/IP_Backbone.HTM, April 1997.

Future Directions
Involving ATM

The question now is not whether ATM will succeed, rather the question is: how successful will ATM be? How much wide area traffic will ATM carry? Voice traffic? Integrated access traffic? Internet backbone traffic? Will ATM ever penetrate the LAN significantly, or will the ever resurgent spawn of the Ethernet species dominate? Or, will the Internet continue to adopt ATM's best features [1] and hence reduce the need for ATM in end systems, leaving ATM to wither away in its wake? Will ATM become the dominant medium for transport of on-demand video? What will be the future of communications involving ATM?

The roller coaster ride of industry expert predictions about the next killer networking solution must be making a lot of money for indigestion and headache remedy manufacturers. While no one can accurately foretell the future, this chapter refers to other authors who have published their visions of the future. Beginning with a review of emerging ATM technologies, the chapter moves on to explore the range of possible scenarios advocated by various industry experts.

32.1 ATM'S EXPANSION INTO NEW FRONTIERS

ATM continues to expand into brave new areas of communication networking. As analyzed in the case studies in our companion book, *Hands On ATM* [2], even McDonald's is already using ATM — bringing ATM ever closer to the McDonald's invented benchmark of counting the number of millions of units sold. So, you ask: what frontiers remain? One of the latest exciting areas is wireless ATM (WATM), aiming to serve mobile users within the office, a metropolitan area, and even the residence. ATM over satellite and terrestrial radio networks enables many commercial and military applications never before possible at ATM speeds. Another promising area is the developments occurring in the area of high-performance Digital Subscriber Line (DSL) access networking.

32.1.1 Wireless ATM (WATM)

With the advent of wireless LANs, wireless ATM looms on the horizon. Several companies are on the verge of announcing products compliant with today's ATM switches. In the wireless arena, ATM switches set up low latency VCCs among voice switches to provide call and service type signaling. In this manner, ATM switches provide a front end to the wireless services, eliminating the need to connect every wireless switch to every voice switch, and saving lots of money. ATM switching will also play a major role in supporting wireless multimedia services and as Internet gateways for wireless subscribers. Figure 32.1 illustrates these methods of providing broadband wireless subscriber access to wireline services.

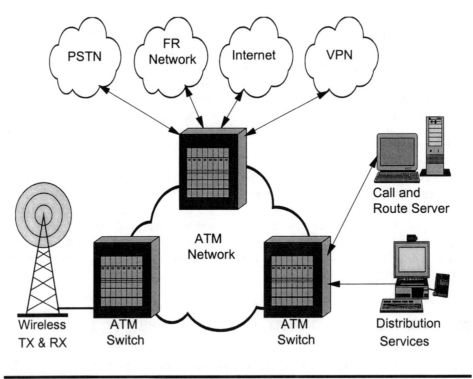

Figure 32.1 Wireless Access to Wireline ATM Switching

Cost-effective wireless ATM could produce a viable access alternative to ATM service providers, especially in hard to reach or downtown areas where constriction and right of way costs are quite high. Wireless also provides an excellent second source of access to improve reliability and disaster recovery.

But many hurdles remain. High error rates inherent in wireless transmission require special modifications to the protocols riding over ATM. Another hurdle is frequency allocation constraints – closely regulated by government agencies both in the US and abroad. The challenges facing wireless ATM include:

- Access Schemes — development of access protocols that can overcome high error rates and noise inherent in radio systems
- Reliability and Availability — coverage areas, fading, temporary outages, error detection and correction
- Service Ubiquity — while wireless can reach hard downtown areas, carriers must acquire licenses, acquire rooftop space, and deploy access stations
- QoS Mobility — how to ensure consistent QoS and handoff as a user roams
- Applications — applications need to be written that understand and overcome the limitations inherent in wireless transmissions.

Most of the major wireless players are active in the ATM Forum's Wireless ATM (WATM) group – having looked at the alternatives and then selected ATM as the technology that best fit the requirements. Mobility is no longer a luxury for ATM: it is a necessity. Responding to this industry consensus, the ATM Forum founded the WATM working group in June, 1996 [3] with the charter to develop a set of specifications applying ATM technology across a broad range of wireless network access scenarios, both private and public. This includes extensions for mobility support within ATM networks and specification of a wireless ATM radio access layer. The WATM specifications target compatibility with ATM equipment adhering to current ATM Forum specifications. The group also is establishing liaison arrangements with other relevant wireless standards bodies. For example, ETSI is specifying the wireless communications layer while the Forum focuses on higher layer protocols. Similar to its role in past endeavors, the ATM Forum is the world-wide coordination point for the specification of wireless ATM.

Several companies in the radio industry spearheaded formation of the group in the hope of utilizing ATM to achieve greater efficiencies in the use of scarce radio spectrum. This effort also aimed to avoid proprietary solutions that could result in another Beta versus VHS shoot-out for a de facto industry standard. The group foresees wireless as an extension to existing ATM signaling, traffic management, and network management specifications, even though the WATM specification will be published as a separate document.

WATM will enable users wireless connectivity to anywhere the ATM network serves, and gain all of its advantages. The initial work items include specification in the following areas as illustrated in Figure 32.2:

- Radio Access Layer protocols including:
  ⇒ Radio's physical layer

⇒ Medium Access Control designed for wireless channel errors
⇒ Data link control for wireless channel errors
⇒ Wireless control protocol for radio resource management
• Mobile ATM protocol extensions including:
⇒ Handoff control (signaling/NNI extensions, etc.)
⇒ Location management for mobile terminals
⇒ Routing considerations for mobile connections
⇒ Traffic/QoS control for mobile connections
⇒ Wireless network management

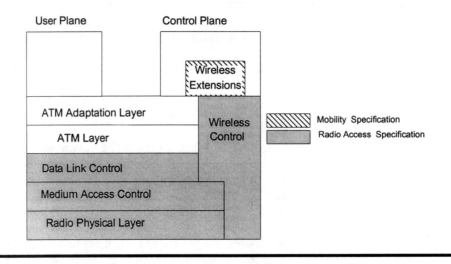

Figure 32.2   Wireless ATM Specification Relationship to Existing Standards

The radio access work will define wireless access to an ATM network from laptop computers or hand held devices inside a building; such as a office, convention center, hotel, or even a residence.   A key challenge for the radio access specification is handling the multipath fading and time varying error rate of the high frequencies targeted by wireless ATM. The work on mobility extensions in the control plane target ATM based access to intelligent network features, such as authentication, roaming, and assured delivery services.

Figure 32.3 illustrates the elements and context of the wireless ATM specification efforts.   The model places the public switching infrastructure and wired private networks at the center of the drawing, surrounded by a number of scenarios envisioned for wireless ATM. These scenarios could be implemented as wireless extensions of the public/private infrastructure, or as mobility-enhanced wireless networks.

Mobile end users communicate directly with the fixed network through wireless access nodes or PCS interworking gateways.   Satellite-based switching could support wireless end users in remote areas (e.g., natural resource exploration) and connectivity to military or emergency vehicles.

Wireless LANs allow users to move within buildings, or between locations. Mobility across locations is a highly valued capability in wireless corporate networks for executives and highly compensated information workers. Wireless ad hoc networks are capable of supporting a group of laptops utilized in a collaborative environment. An attractive feature of wireless ATM is true plug(in)-and-play operation, without having to plug any equipment together! Coupled with VLAN use, this benefit alone can save Information Technology shops a large portion of their desktop support costs.

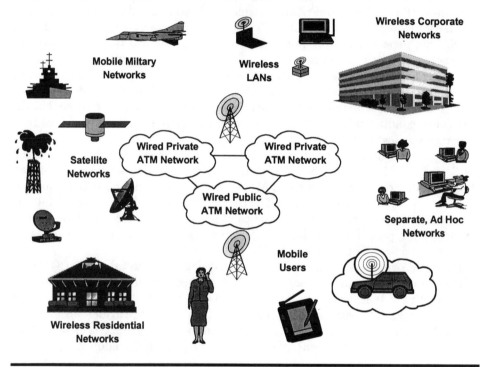

Figure 32.3   Wireless ATM Application Model

There are several competing standards for wireless LANs:

- IEEE 802.11
- HIPERLAN as standardized by ETSI/RES10 in Europe
- IETF Mobile-IP
- ITU AAL class for multiplexed compressed VBR signals for wireless

Each of these standards offer various base station designs and architectures, and the protocols differ markedly.   There are also two efforts under way in Europe to standardize 5 Gbps ATM wireless technology to deliver 10 to 20-Mbps service:

- European Community Wireless ATM Network Demonstrator (Magic WAND)
- Advanced Technology Program's (ATP's) Mobile Information Infrastructure project

32.1.2 ATM over Satellite Trials and Services

A number of trials and demonstration networks have used ATM over satellite [4] since 1994. The American Petroleum Institute's ATM Research and Industrial Enterprise Study (ARIES) project was an early adopter that used a NASA high-speed satellite and terrestrial ATM networks to demonstrate interactive seismic exploration in 1994. In subsequent years, ARIES demonstrations included shipboard exploration and simulated medical emergencies. In the remote parts of Canada, satellites are the only cost-effective way to transmit communications signals, which motivated North-wesTel to conduct trials of ATM over satellite for telemedicine and distance learning applications.

The prize for highest speed performance by ATM over a satellite goes to the Jet Propulsion Laboratory and NASA Lewis Research Center, which achieved TCP/IP data transfers at over 600 Mbps over NASA's Advanced Communication Technology Satellite (ACTS) [5]. In fact, the field of ATM over satellites is an area of active research [6]. Areas of projected advances include higher capacity, support for multimedia and sophisticated on-board processing specifically designed to support ATM.

The ATM over satellite market is now just beginning to take off. COMSAT appears to be the leader in satellite ATM services in the United States. In Europe, the RACE program has spawned the Catalyst Project to provide interconnection of regional ATM services. This section provides an analysis of ATM performance over satellite. Surprisingly, performance is close to that of terrestrial networks, except, of course, for the propagation delay.

32.1.2.1 Benefits of ATM over Satellite

ATM over geostationary satellites (satellites that remain 36,000 km above the earth and have a view of approximately one-half of the earth's surface) provide ATM service to geographically diverse areas, many of which terrestrial fiber cannot economically serve. Thus, ATM over satellite enables nearly ubiquitous worldwide service. Since remote areas may wait years for fiber optic cable installation, ATM satellite provides a fast deployment alternative. Initially, ATM satellite speeds run at fractional DS1 through DS3 speeds. ATM service can even be offered over existing Very Small Aperture Terminal (VSAT) earth stations.

32.1.2.2 TCP Transmission Issues when Using Satellites

As studied in Chapter 26, TCP throughput declines for networks with long end-to-end delays (latency). Furthermore, noise on satellite and wireless links greatly impacts TCP throughput, since many TCP implementations close their transmit window upon detection of a single lost packet. In essence, TCP makes the implicit assumption that all loss is due to congestion. Thus, the round-trip propagation delay of a 1/2 second and occasional bursts of noise bring out the worst characteristics of TCP. Thus, protocol changes within the frame windowing method were required to transport TCP traffic over satellite transmissions. Efforts are under way by BBN Inc. (funded by the National Aeronautics and Space Administration) to improve TCP's performance over networks with so-called high-latency links, such as satellite transmissions. In fact, NASA's Research and Educational Network (NREN) has tested TCP/IP data transfers at speeds of OC-12 over satellite using devices that employ error correction and TCP protocol spoofing.

32.1.2.3 When Is ATM Satellite Cheaper than Terrestrial Fiber?

ATM satellite services can be very cost effective when compared to the fiber land-line alternative depending upon the geographic location of the end-points. This is especially true for cities not near the head-ends of transoceanic communication cables. Thus, ATM over satellite eliminates large amounts of expensive fiber backhaul or high-speed leased lines [7]. Therefore, ATM over satellite is a candidate for smaller, geographically isolated cities in Europe and North America, in addition to remote locations in South America, Asia, and Africa.

32.1.3 Digital Subscriber Line (xDSL) Technologies

Digital Subscriber Line (or DSL for short) technology is a hot topic in communications networking today. First, this section explains this technology by defining the technical parameters and reference configuration. Next, the text summarizes the standards activity in xDS. We then discuss the advantages of xDSL technology for carriers, and conclude with some speculation about the future role of xDSL technologies.

32.1.3.1 xDSL Explained

The term xDSL refers to a family of communication technologies based around the DSL technology, where "x" corresponds to a standard upstream and downstream data rates, defined from the perspective of the end user. The term *upstream* refers to transmission from the user to the service provider; while the term *downstream* refers to transmission from the service provider to the end user. Several groups of manufacturers, carriers and entrepreneurs are actively standardizing and promoting this technology in several forums [8, 9]. One group, the ADSL Forum, has even coined a new acronym for ATM: Another Telecommunications Medium. They've also coined a new set of acronyms, explained in Table 32.1. The table gives the upstream and downstream rates, the number of twisted pairs required, and a list of representative applications.

Table 32.1   The family of Digital Subscriber Line (xDSL) Acronyms Explained

Acronym	Full Name	Twisted Pairs	Upstream Rate	Down stream Rate	Example Applications
DSL	Digital Subscriber Line	1	160 kbps	160 kbps	ISDN service for voice and ISDN modem data
HDSL (DS1)	High data rate Digital Subscriber Line	2	1.544 Mbps	1.544 Mbps	North American T1 service
HDSL (E1)	High data rate Digital Subscriber Line	3	2.048 Mbps	2.048 Mbps	European and International E1
SDSL (DS1)	Single line Digital Subscriber Line	1	1.544 Mbps	1.544 Mbps	North American T1 service
SDSL (E1)	Single line Digital Subscriber Line	1	2.048 Mbps	2.048 Mbps	European and International E1
CDSL	Consumer Digital Subscriber Line	1	64 to 384 kbps	1.5 Mbps	"Splitterless" operation to the Home (g.Lite)
ADSL	Asymmetric Digital Subscriber Line	1	16 to 640 kbps	1.5 to 9 Mbps	Video on Demand, LAN & Internet Access
VDSL	Very high data rate Digital Subscriber Line	1	1.5 to 2.3 Mbps	13 to 52 Mbps	High quality video, high-performance Internet/LAN

Digital Subscriber Line (DSL) technology harnesses the power of state-of-the-art modem design to "supercharge" existing twisted-pair telephone lines into information super-highway on-ramps. For example, ADSL technology enables downstream transmission speeds of over 1 Mbps to a subscriber, while simultaneously supporting transmissions of at least 64 kbps in both directions. xDSL achieves bandwidths orders of magnitude over legacy access technologies, such as ISDN and analog modems, using existing cabling, albeit over shorter distances. Many experts expect xDSL to play a crucial role over the coming decade as small business and consumer demand for video and multimedia information increases, since it uses existing unshielded twisted pair cabling while the installation of new, higher performance cabling (e.g., optical fiber) will take many years.

Figure 32.4 illustrates a typical xDSL configuration. The equipment on the left-hand side is located on the customer premises. Devices which the end user accesses, such as a personal computer, television and N-ISDN video phone connect to a premises distribution network via service modules. These service modules employ either STM (i.e., TDM), ATM or packet transport modes as depicted by the arrows at the bottom of the figure. An existing twisted pair telephone line connects the user's xDSL modem to a corresponding modem in the public network. The xDSL modem creates three information channels — a high speed downstream channel ranging from 1.5 to 52 Mbps, a medium speed duplex channel ranging from 16 kbps to 2.3 Mbps, and a POTS (Plain Old Telephone Service) channel.

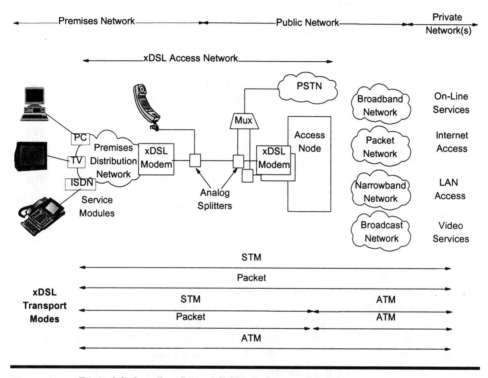

Figure 32.4   Digital Subscriber Line (xDSL) Reference Model

Many xDSL schemes require analog splitters at the user's site to separate the POTS channel from the digital modem. Basically, a splitter is a filter that passes only the lower 4 kHz of spectrum to the local telephone devices. This design guarantees uninterrupted telephone service, even if the xDSL system fails. However, if the telephone company must send a technician to install such a device, the cost of the service increases markedly. The CDSL method targets this problem explicitly by eliminating the need for telephone company installed splitters. Other service providers give users easily installed micro-splitters that attach to each user telephone, fax machine, or other legacy telephone compatible devices.

Moving to the center of Figure 32.4, multiple xDSL modems connect to an access node within the public network. Frequently, these are called DSL Access Multiplexers (DSLAMs). Splitters in the telephone company Central Office (CO) carve off analog voice signals, multiplex them together, digitize them, and deliver the aggregate voice traffic to the Public Switched Telephone Network (PSTN). The access node interfaces to packet, broadband, video or service networks as shown on the right hand side of the figure. Depending upon the transport mode employed by the customer, conversion to/from ATM may be performed as indicated by the lines at the bottom the figure. If the customer employs ATM end-to-end, then cost and administration of such conversions is eliminated. Experts envision these access networks connecting to a variety of service provider's private networks supporting a wide range of applications, such as Internet access, on-line services, remote LAN access, video conferencing and video-on-demand.

Current xDSL modems provide data rates aligned with North American and European digital STM hierarchies at a constant bit rate. Future plans include LAN interfaces and direct incorporation of xDSL modems and service modules directly into TV set top boxes and PC interface cards. The ADSL Forum plans to support ATM at variable rates when the ATM market matures. Since real time signals, such as digital video, can't employ link or network level error control procedures because of the delay they cause, xDSL modems incorporate forward error correction that dramatically reduces errors caused by noise on twisted pair wiring.

The achievable high speed downstream data rate depends upon a number of factors, such as the length of the twisted pair line, the wire gauge, presence of bridged taps from party line days, and cross-coupled interference from other lines. Without bridged taps, ADSL can support a 1-Mbps downstream rate over distances in excess of 3 miles. While the measure varies according to geographic area, studies indicate that ADSL capabilities cover up to 80 percent of a loop plant. Premises outside the 3 mile limit must be served by fiber-based digital loop carrier systems. However, depending upon regional demographics, ADSL may not reach a correspondingly large part of the target market. For example, ADSL typically doesn't reach potential consumers living in far flung affluent suburbs.

From a system viewpoint, xDSL looks simple — high speed data rates over ordinary telephone lines. Of course, several miracles of modern technology work together to achieve this quantum leap in performance over existing POTS analog modems and even the 128-kbps ISDN BRI technology. ADSL modems divide up the 1 MHz telephone line bandwidth using either Frequency Division Multiplexing (FDM) or TDM Echo Cancellation. The FDM method assigns a frequency band for downstream data and a smaller band for upstream data. Echo cancellation assigns the upstream band to overlap the downstream, and separates the two by means of local echo cancellation, as V.32 and V.34 modems do. Echo cancellation uses bandwidth more efficiently, but at the expense of complexity and cost. ADSL multiplexing, interleaving, and error correction coding create up to 20 ms of delay. This process achieves high resistance to impulse noise, for example, that generated by electrical appliances.

Figure 32.5 summarizes speed and distance limitations for typical 24 gauge twisted pair wiring. The double headed arrows are placed vertically at the maximum speed for the DSL technology, with their horizontal dimension indicating the range of feasible operation.

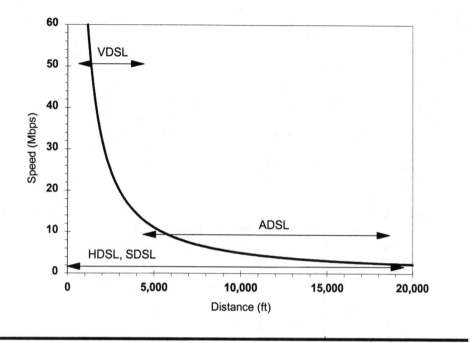

Figure 32.5   Typical xDSL Downstream Rate versus Distance

32.1.3.2 Standards Activity

A coalition of vendors formed the ADSL Forum in December of 1994 to promote the ADSL concept and facilitate development of ADSL system architectures, protocols, and interfaces. The American National Standards Institute (ANSI) approved an ADSL standard at rates up to 6.1 Mbps (ANSI Standard T1.413). The European Technical Standards Institute (ETSI) provided an Annex to T1.413 reflecting European requirements. The ATM Forum and the Data And VIdeo Council (DAVIC) have both recognized ADSL as a physical layer transmission protocol for unshielded twisted pair media.

The ADSL Forum, working in conjunction with the ATM Forum and American National Standards Institute (ANSI), wrote Technical Report TR-002. This pre-standard specifies how to interwork ATM traffic over ADSL modem links. The ATM Forum Residential Broadband Working Group is working on their version of an ADSL/ATM standard that will work with either xDSL or cable modem technologies.

32.1.3.3 ATM and xDSL — Advantages for Carriers

Asymmetric Digital Subscriber Line (ADSL), or for that matter any of the xDSL technologies, are basically access technologies to public network services. ATM is a networking and transport technology that can be offered as a service over digital subscriber line (xDSL) access, where the xDSL line provides the user access at a lower cost per Mbps. This combination is very attractive to service providers, as it leverages the existing copper plant and offers ATM services over it without the requirement to replace existing copper with fiber optic cabling. In fact, ATM offers virtual connections at specified bandwidths with guaranteed QoS [10].

A real advantage exists in the use of ATM over ADSL for the Local Exchange Carriers (LECs). Since LECs already own the (unshielded) twisted pair into almost all residences, as well as many businesses, a combination of ATM service for video (TV channels), data (for Internet or telecommuting access), and voice (for residential dial tone services) could allow them to keep the hold on existing customers for some time to come. LECs could offer lower speed ADSL services, video-on-demand, home shopping and banking, local and long distance phone services, and Internet access more cost effectively than through separate providers over ADSL. The potential benefits are enormous: fewer access lines, less termination equipment, consistent QoS, and more. Of course, the downside for the consumer is the lack of competition from multiple providers.

32.1.3.4 ATM over xDSL - Future Internet Access Method of Choice?

Will ATM over ADSL become the preferred method for high-performance Internet access? While dial access via analog modems are the preferred method of Internet access for consumers today; many telecommuters and remote offices requiring remote LAN access have moved to faster (128 kbps) ISDN BRI access. Most Internet access traffic is small requests upstream from the user to the network, interleaved with much larger data streams coming from the remote web site downstream to the requestng user. Thus, xDSL technologies promise a ten-fold increase in downstream transmission rate in comparison with today's top-of-the-line N-ISDN BRI modems.

Currently, Internet providers are faced with the problem of how to provide different quality of service to different applications over the Internet. ATM offers the best method today for guaranteeing end-to-end QoS across the WAN. Potentially, these facts position ATM over xDSL services as the Internet access method of choice, especially for large corporations with many telecommuters, or remote offices that want to send voice, video, and data traffic over the Internet.

32.1.4 ATM Cable Modems

xDSL services compete with cable modems that can operate at much higher speeds (up to 30 Mbps). Both technologies are optimized for high speed transmission rates from the network service provider toward the user — in other words designed for distribution from server to user, rather than for peer-to-peer, user-to-user traffic patterns.

Typically, cable modems operate at speeds ranging from 10 to 30 Mbps in the shared downstream direction. Yet the bandwidth limited and electrically complex upstream channel in today's cable networks remains a major barrier to complete realization of cable modem access. Also, most office buildings still lack coaxial cable, the necessary transport medium. These cable modem standards can be found in IEEE 802.14.

This higher speed option for running ATM over existing cable plant skips the lower bandwidth ATM over ADSL alternative and goes directly to shared 30-Mbps ATM over cable modems. This technology gives cable companies a competitive potential against the LECs. But if history is an indicator, cable companies may struggle to provide these services as competently as LECs.

32.1.5 Set-Top ATM

CATV providers have had the ability to offer interactive services over their existing infrastructure, such as telephony services and video-on-demand, for some time. Yet they have been unable to successfully offer these types of services. ATM can play a key enabler along with other technologies to deliver these services to devices like set-top boxes within a Home Area Network (HAN) from video servers resident within the ATM service network. Standards groups like the ATM Forum have working groups — Residential Broadband (RBB) — that focus on ATM in the HAN.

32.2 IP AND ATM: HAND IN HAND, OR, IN EACH OTHER'S FACE?

While IP is the current protocol of choice for global end-to-end internetworking, ATM is still a contender to support major portions of the infrastructure and in some cases provide the end user interface via appropriate interworking with IP or LANs as described in Chapter 20. ATM targeted a wide range of services, such as voice, video and data from the outset. The IP protocol suite targeted data communications from the outset. Features to support guaranteed QoS and reserved bandwidth for voice, video, and QOS-sensitive applications are only being added now to IP. It remains to be seen how well this approach will work when compared with ATM's design objective to support multiple service types in a bottom up design. Many in the IP community see ATM as just another communications subnetwork. After all, IP runs over every other communication networking technology ever invented; why should ATM be different? Many in the ATM community have the vision of a ubiquitous ATM network.

32.2.1 Homogeneous versus Heterogeneous Networking

ATM works best in a homogeneous environment: it does not work well in a network with diverse communication technologies. However, IP works well in a heterogeneous environment since it was designed that way from the outset. Presently, IP works best with applications without real-time constraints: it doesn't support a wide range of applications. These are challenges that both IP and ATM must meet for either to succeed in the future. In order for users to reap the QoS and guaranteed bandwidth benefits of ATM, they must replace existing hardware. On the other hand, the IP community must still develop protocols and procedures for supporting QoS and reserved bandwidth on a global scale.

32.2.2 Scalability

Recall that LAN Emulation utilizes bridging as its underlying paradigm. Many experts in the internetworking community view bridging as a foul concept — "It won't scale!" their T-shirts cry out in reaction to the proposal of large bridged networks. Instead, they advocate routing solutions. Indeed, the charter of the IETF is design of a global Internet. However, to an enterprise network, as long as a LAN can be built of a scale to meet the enterprise's needs, then the fact that it won't scale to serve the entire world is not an immediate business concern. Customers are running emulated LANs using the ATM Forum's LANE protocol with between 500 and 2,000 end systems. Nonetheless, for some of those with a vision of a global Internet, ATM still plays a pivotal role. For example, many Internet Service Providers (ISPs) currently utilize ATM-based backbones and interconnect at ATM-based Network Access Points (NAPs).

While some Internet service providers have chosen the IP over ATM path, others pursue high-performance IP over SONET implementations. Beginning in 1995, ATM had an edge over packet switching because it was the only technology capable of operating at 150-Mbps speeds. However, now router manufacturers plan to support IP directly over SONET at 150-Mbps speeds in 1997 and 600-Mbps speeds in 1998.

Another differentiator for ATM was that the relatively simple fixed cell size enables construction of larger scale switching machines than devices utilizing frame-based protocols do. For example, today's routers struggle to get to the 1 Gbps benchmark [11], however, this throughput only qualifies as an ATM enterprise switch in the taxonomy used in Part 3. Carrier edge and backbone switches are already 10 to 100 times larger than the upcoming generation of Gigarouters! This huge difference in scale is due to the fact that many of the parts in ATM switches are now built from commodity integrated circuits. Also, ATM switching avoids the processor or silicon gate intensive layer 3 functions. However, the development of hardware-based routers and layer 3 switches will likely change this in the near future [12, 13].

32.2.3 Beyond Hype to Maturity

ATM is not as exciting as many of the newer technologies, like RSVP, Gigabit Ethernet, switched IP, or Layer 3/4 switching. Part of the reason this happened is that ATM and IP are mature technologies, while many of the newcomers still have a shine and luster in the trade press. Remember that these publications are in business to sell newspapers, magazines, and get their advertisements in front of potential customers. The same criticism applies to many publications on the web as well. Controversy and world impacting predictions sell papers: stability and maturity don't.

It is indeed difficult to predict the future in the rapidly changing world of communications networking. For example, several promising approaches to deliver QoS hyped by the trade press hit the wall only a few years after inception. In 1996 Ipsilon's IP switching appeared ready to take over the world if you believed some industry pundits. However, less than three years later, many of these same advocates now admit that the approach was fundamentally flawed because every flow required a virtual connection. Also, RSVP, in the oven for an even longer period after its inception in the early 1990s, also falls short of the mark. Even the RFC cautions users regarding the limitations of the state of the protocol.

No fault to the popular press for predicting some successes which turned out to be surprise failures. The uncertainty involved in forecasting communications technology trends during these times makes a meteorologist's job seem trivial. Since technology continues changing at such a rapid rate; making predictions even a few years out is risky business indeed! Recall that ATM encountered much the same heightened expectations in the mid-1990s. Now ATM is here, and reality has set in so that a rapidly growing segment of corporate and enterprise users are getting down to business using ATM-based solutions.

32.2.4 Intelligent or Stupid Networks

So where is the real cost element in data networking in the next century? There will always be an access switching and transmission cost, regardless of the technology or media. However, their unit costs should decline over time. Information will be the new currency of the next century. The valuable elements in future networks will be sought after information content, knowledgeable network management, competent technical support, and effective maintenance. Since price usually follows value, intelligent networks of the future will focus on these areas instead of the raw costs of switching and transmission.

Will the eventual answer be inexpensive transmission capacity? In this case, the notions of reserving bandwidth and guaranteeing bandwidth vanish. Indeed, this is the argument for Gigabit Ethernet and the simple 802.1p priority scheme. Are communications networks repeating the same paradigm shift that occurred in computing a decade ago when microcomputers began significantly displacing the need for minicomputers and mainframes? Will all of the intelligence reside on the clients and servers leaving only basic functions to the network. Some experts believe this to be true [14].

32.2.5 On Demand Services — The Key to Success?

As studied in Part 4, the design of many higher layer protocols running over ATM require Switched Virtual Connections (SVCs) to perform optimally. The dynamic bandwidth allocation and resource sharing of SVCs also make them an economical alternative. In contrast to PVCs, where the bandwidth is dedicated 100 percent of the time, SVCs make bandwidth available on demand — returning this resource when a user completes an activity, or when the network devices detect a period of inactivity. The key protocols described in Part 4 that require high-performance SVCs are:

* Classical IP over ATM Subnetworks
* LAN Emulation (LANE)
* MultiProtocol Over ATM (MPOA)
* IP Multicast over ATM

These protocols create brighter prospects for ATM than frame relay SVCs [15]. The operation of these protocols in larger networks create a challenge for equipment manufacturers to develop equipment that can process high rates of SVC call attempts. A number of ATM switch vendors advertise call set up rates on the order of hundreds to thousands of connections per second per switch.

SVCs bridge the gap created by the fundamentally different paradigms employed by IP and LANs when compared with ATM. IP and LANs utilize connectionless packet-by-packet processing as opposed to ATM's connection-oriented protocol that switches once for an entire stream of packets communicated between endpoints. In essence, the standards defining support for IP and LAN protocols over ATM replace packet by packet forwarding with three stages of protocol: flow recognition, address resolution and connection establishment. Hence, ATM's SVC based support for higher layer protocols is efficient only for long-lived traffic flows. In other words, the combined cost required for address resolution and call setup for the more efficient ATM SVC must be less than the cost saved by the less efficient packet by packet routing. The IETF concluded that for short flows the overhead of ATM SVCs is probably not justified [16].

The ATM SVC protocol, as specified in the ITU-T standard Q.2931, is very complex [17]. As discussed in Chapter 15, in addition to the signaling similar to that needed to set up telephone calls, ATM must also signal a number of additional parameters. The additional information elements in ATM signaling messages convey requests for different service categories, traffic parameters, information about higher layer protocols and topology information, such as point-to-multipoint connections. In essence, the ATM signaling standards carry along with them the evolutionary baggage of interworking with telephony protocols conceived of over a century ago, plus all of these additional information elements. Furthermore, the complexity of emulating connectionless protocols with various forms of registration, address resolution, flow recognition and fast connection setup will require some time to optimize the price/performance of hardened implementations.

Thus, the key challenge for ATM switching to succeed is to achieve high call set up rates and low call set up times cost effectively. Since ATM signaling is computer and memory intensive, it should be able to follow the technology curves for computing and storage to achieve these goals. As noted earlier, data traffic is growing at a rate faster than Moore's law, hence the control processing must be distributed since a single processor cannot keep up with the growth rate. Another linchpin to success is interoperability. ATM signaling must be as interoperable as telephone signaling to succeed. Furthermore, ATM addressing must be assigned logically and consistently so that carriers will interconnect and customers can call any other ATM user [18]. Finally, ATM signaling must seamlessly interwork with N-ISDN and telephony signaling and addressing to support voice over ATM.

Another problem is determining how much bandwidth the application should request in an SVC. An engineer easily can determine how much bandwidth is required for a video call and generate an ATM SVC requesting that bandwidth. However, most data applications have little concept of making a connection prior to transferring data. Furthermore, these data applications have no clue about the bandwidth required on such a connection.

User expectations have been shaped by the types of applications that use whatever bandwidth is available and that provide flow control to adapt for congestion. The astounding increase in computing power and sophistication of applications is fueling the demand for this type of service beyond Ethernet's ability to satisfy it. Thus, a new technology such as ATM will be needed.

On the other hand, carriers have been slow to roll out ATM SVCs. Only one carrier, AT&T, has announced plans to roll out ATM SVC services at the time of publication.

32.3 ATM'S KEY CHALLENGES

ATM equipment sales have been much stronger than ATM service revenue, although ATM service revenues continue to double every year. Some studies predict that ATM services could outstrip private line and frame relay revenues early in the twenty-first century, even subsuming some of the nation's voice traffic and switching. Most studies predict that IP-based traffic will dominate the market place. Hence, a key challenge for ATM is to continually improve support for native IP. ATM could also make a strong penetration into the LAN backbone and high performance workgroups if the vendor implementations of LAN Emulation, MPOA and PNNI become cost effective and QoS-aware applications like videoconferencing catch on. A more pessimistic view places ATM as the WAN transport technology of choice for only carriers and service providers. Many experts now believe that Gigabit Ethernet will dominate over ATM in the LAN backbone, while QoS aware IP protocols will deliver services over switched Ethernet to the desktop and workgroup.

32.3.1 Making ATM More Efficient

One on the main technical criticisms leveled against ATM is that it is inefficient. This criticism occurs in both the user and control planes. As discussed in preceding chapters, several changes could make ATM more efficient in the user plane:

- Cell header suppression is a technique which sends only one cell header for a burst of cells with identical header fields [19].
- Use of the Frame-based UNI on high-speed backbone trunks.
- Use the AAL5 User-User Indication for Multiprotocol encapsulation to improve IP over ATM efficiency by 5 to 6% (see Chapter 30).
- Use of longer packet sizes by new applications

Furthermore, MPLS could control ATM cross connects instead of the heavyweight ATM signaling stack. If the IETF had completely written off ATM at this point, then why does the MPLS draft include both frame and cell based switching in its requirements? A key new function required by MPLS not implemented by many current ATM switches is that of VC merging as described in Chapter 20. Furthermore, the Guaranteed Frame Rate (GFR) service being defined in the ATM Forum will also require new ATM hardware. Both of these changes will make IP work better over ATM networks.

32.3.2 Key Challenges for ATM Technology

What are the challenges facing ATM technology for LAN and WAN applications? Some of the key battlegrounds and deciding factors are:

- Cost effective support for voice traffic over ATM
- Interworking between ATM QoS and IP RSVP
- Lower priced CPE ATM Access Devices
- Lower priced ATM NIC cards
- Cost effective LAN and MPOA solutions
- Further standards for video over ATM
- Effective end-to-end Network Management
- Adoption of ATM or QoS aware APIs by application developers

Users cannot afford high-speed access for all of their locations. ATM must provide a range of performance better matched to customers' business needs. This ranges from the very low end to the high end. Better testing for conformance to standards to ensure interoperability is needed; otherwise users will be locked into a proprietary solution. ATM must be economically justified before the majority of corporate users will adopt it, and this must be documented by published success stories. More training on ATM is needed, along with applications that can effectively use ATM. Users require seamless interworking of their current LANs with ATM-based solutions.

The identification of seamless networking from the LAN across the campus and across the WAN has been consistently identified by users as a key benefit of ATM. Completing delivery of this vision will be a key challenge for the standards bodies, vendors, network designers, and service providers.

Two established network management protocol standards serve as the foundation for ongoing efforts to develop ATM network management capabilities; the IETF's highly interoperable Simple Network Management Protocol (SNMP) and the OSI-based Common Management Interface Protocol (CMIP). IETF and ATM Forum activities center on SNMP, while the ITU-T and ANSI are working to adapt the general-purpose Telecommunications Management Network (TMN) architecture based on CMIP. These dual approaches eventually will converge, with SNMP managing private networks and CMIP used within and between public networks. Users should look to SNMP in the interim until this convergence occurs.

Part 7 summarized the current ATM standards use of special Operations And Maintenance (OAM) ATM cells to provide fault management and performance measurement at the ATM layer. The protocols using these cells can indicate faults, detect faults, and verify continuity via a loopback. They also can be used to estimate the virtual ATM connection's performance in terms of delay, error rate, and loss. Such capabilities are critical in managing an ATM network that supports multiple applications, each of which has different quality requirements. Unfortunately, most users operate at the virtual channel connection (VCC) level, where there still are at least three major limitations:

- There may not be support for the fault-detection capabilities at the VCC level
- Performance measurement probably will be economical only when done across large numbers of VCCs enclosed in an ATM virtual path
- Current standards for delay measurement may be inadequate

Users should consider employing continuity checks and delay measurements in higher layer protocols (such as ICMP Ping) until the ATM standards are widely implemented.

32.3.3 Key Challenges for ATM Service Providers

The greatest challenge to the service providers is to offer all the features of ATM in a cost effective and easy to use manner to the user community. This means that they must keep their internal capital costs down while simultaneously keeping the pricing competitive with existing legacy services like private lines, frame relay, and telephony services. There are many options open to service providers. One option left to carriers is the large scale adoption of ATM as a service offering and consolidation of their transport infrastructure. The hurdle they face is a costly depreciation of the existing voice infrastructure balanced against the cost savings of deploying ATM network wide. Another option is to offer many access options to users that want to deploy ATM access multiplexers. These devices connect via scaleable and flexible nxDS0, DS1, and nxDS1 ATM access circuits. Another option for users is to only migrate sites to ATM that require its benefits and perform interworking with other non-ATM sites [20].

Carriers must also deliver on the following features in order for ATM to be successful:

◊ Widespread availability of cost-effective SVC services
◊ Interconnection of carrier SVC networks to enable ubiquitous access
◊ ATM SVC interworking with frame relay SVCs
◊ True end-to-end service QoS guarantees over concatenated public networks
◊ Customer network management

Look for each of these capabilities to play a key role in the future success of ATM as a service in the WAN.

The introduction of SVCs will require accounting and such statistics as blocking, call attempt, invalid attempt, and busy counts. To improve cost-effectiveness for both ATM PVCs and SVCs, subscribers need usage statistics and accounting reports on a per-virtual-connection basis.

32.3.4 ATM Areas of Strength

The charter of the CCITT for the Broadband Integrated Digital Services Network (B-ISDN) was to use ATM as the switching and multiplexing method for carrying all information streams in the future. A few years ago, some industry experts envisioned that B-ISDN would be the universal network of the future. With the content-driven adoption of the world wide web, most of these experts now acknowledge IP as the universal protocol to the desktop client and the application content delivering server.

The practicality of a universal B-ISDN vision is questioned by many however. Experience with technology indicates that someone will come up with a better idea, eventually. Also, since B-ISDN has been developed in large committees the results are inevitably weakened by compromise in the standards process. But, ATM may be a survivor of the standards process and be the foundation for a successful data communications infrastructure. The prolific volume and scope of specification from the ATM Forum and the modernization of the ITU-T bode well for ATM. Furthermore, the ATM standards groups work closely with the IETF.

Probably, ATM networks will coexist with existing STM based networks for a long time. The era of the universal adoption of ATM will likely never occur for various reasons. However, ATM will likely play a key role in at least the following areas:

• Integrated Access
• Virtual Private Networking
• Protocol Interworking
• High-Performance Workgroup and Server LANs

- Carrier WAN Infrastructure

32.3.4.1 Integrated Access

Circuit emulation capabilities will allow circuit-based voice, video and data to share expensive access line costs with guaranteed QoS and reserved bandwidth. Particularly on lower speed xDSL technologies, ATM is the only solution that meets all of these needs. Voice and video integration will flow from the access line, through the LAN, to the desktop, and all the way to the end application, becoming a (virtual) reality.

32.3.4.2 Virtual Private Networking

Many enterprises will require wide area connections for applications that require the bandwidth, quality, and flexibility of ATM. Carrier-provided virtual networks will be more cost effective than private networks for many users. Carrier network support for Switched Virtual Connections (SVCs) and support for connectionless traffic over ATM will be developed to meet user needs. LANE meets this need in the LAN environment, while MPOA extends the capability across the WAN.

32.3.4.3 Protocol Interworking

ATM-based interworking with other protocols will support enterprise connectivity to the many sites which will not have ATM initially. Virtual public networks will provide this capability to interconnect sites with ATM, frame relay, IP, TDM circuits, voice, and video interfaces via a portfolio of interworking services.

32.3.4.4 High-Performance Workgroup and Server LANs

ATM adapter cards in high-end workstations and servers will see wider use. Workgroups will be connected by local switches, with access to legacy LANs and other users via a collapsed ATM backbone. Servers will need the performance of ATM before clients, which will lead the way for ATM to the desktop. Switched 10-Mbps Ethernet and 100-Mbps Fast Ethernet will be competitors and will be the true market test for acceptance of ATM to the desktop. Gigabit Ethernet may capture a large share of the LAN backbone market if it lives up to predictions. However, OC48 ATM is the next technology step for large LAN backbones.

32.3.4.5 Carrier WAN Infrastructure

Most experts acknowledge that ATM will play a significant role in carrier infrastructure since it is standardized, supports multiple types of traffic and provides a migration path from legacy TDM, voice, and ISDN networks. Whether carriers will continue to use ATM as an Internet backbone is an open issue. Since the IETF's MPLS group plans to implement many capabilities already implemented in ATM, plus several new ones, a new technology optimized for connectionless Internet traffic may emerge.

Standards-based ATM, and public addressing with SVCs will enable ubiquitous communications, like the voice network today for longer-lived flows, like video-on-demand.   Unfortunately, politics, regulations, and competitive posturing may delay inter-carrier connections.

32.3.5 Why Customers Choose ATM

Clearly the need for high-capacity networks drives ATM deployment. In a recent survey, over two-thirds of the respondents indicated they chose ATM to provide increased bandwidth [21]. Following in a three-way tie, one third of the respondents indicated that they chose ATM to support high-volumes of data traffic, future-proof their networks, or support delay sensitive applications. Dedicated bandwidth provides no advantage if it isn't used efficiently. ATM users dynamically share bandwidth resources, because ATM allocates bandwidth fairly based on requested service category and traffic levels. Users can harness bandwidth as needed without worrying about controlling the amount and variability of delay through careful traffic engineering of their communication networks, as network designers must do in LANs supporting integrated services.

32.3.6 Standards Development

The ATM Forum is partly to blame for the limited success of ATM in the LAN and voice networking environments [22]. Focusing almost exclusively on the lower layers, witness the plethora of over 20 basic physical interface types listed in Chapter 12 developed by the Forum in its first 5 years. Furthermore, many of these physical interfaces have incompatible options that further complicate basic ATM connectivity. Fortunately, there is only one ATM layer, and it has the best traffic management capabilities of any protocol every designed. The ATM Forum moved into the key areas of LAN and IP later, and now the LANE and MPOA protocols are good solutions for Virtual Private Networks (VPNs). The Forum's efforts for supporting voice and video must remained focused and deliver interoperable results to meet the challenges above.

32.4 POSSIBLE FUTURE DIRECTIONS

What are the possible directions for ATM? What is the best case? What is the worst case? What is the likely scenario? This section provides some answers to these questions.

32.4.1 Competition and Survival of the Fittest

Will ATM become a dominant networking technology in the twenty first century? Or, will it go the way of ISDN? Will Gigabit Ethernet encroach upon its claim to high performance in the LAN? Will RSVP deliver quality of service and reserved bandwidth more effectively? Will a subsequent generation of IP/LAN hardware eventually unseat ATM from the guaranteed QoS and reserved bandwidth throne? No one can say for certain; however, one thing is clear — ATM will continue to compete in an environment of legacy and new competing technologies. The communications industry appears to operate on a Darwinian principle similar to nature's survival of the fittest, where only the strong survive. But, one key difference in this analogy is that in order to survive, the newcomers must coexist (that is, interwork) with those they intend to replace for extended periods of time. While some forecasters predict that IP will displace most other protocols. within the next decade; others believe that X.25, SNA, Ethernet, Frame Relay and ATM networks will exist ten to twenty years from now. ATM is on the cusp, standing at the brink of the chasm, ready to join the cast of legacy protocols. If ATM continues growing at a rate of 100 percent into the twenty first century, then most would concede that ATM succeeded.

As presented in our companion book, "Hands On ATM," [2] most of the leading U.S. carriers and traditional RBOCs now offer ATM service; while the few who don't plan to offer it soon. Carriers continue to work together to deliver switched ATM services. Furthermore, international ATM service is now becoming available. Despite the competitive pressures in the LAN and WAN covered in the previous chapter, ATM continues to expand into new areas, as well as experience significant growth. Furthermore, ATM technology continues to improve price/performance with respect to other technologies in both the LAN and WAN. Leading edge applications comprise an ever increasing, emerging driver for hybrid public-private ATM networking. Announcements by high-tech, high-content as well as high-dollar enterprises jumping on the ATM broadband wagon continue a systematic increase. For example, see the case studies described in Reference 2.

32.4.2 ISDN — Again?

A similar vision and a significant amount of hype regarding the Time Division Multiplexing (TDM)-based Integrated Services Digital Network (ISDN) becoming the universal solution for all communications occurred in the late 1970s. ISDN would handle all voice and data communication in an integrated manner that would work seamlessly in the office, and around the world. The general consensus is that this has not, and will not, come to pass. The fact is that although initial ISDN standards were released in 1984, only a few parties can make an international ISDN call today. Access to ISDN services is simply not available to most users. The current ITU-T standard has reused this name, calling it Broadband ISDN (B-ISDN) to reflect an initial focus on speeds higher than those supported in the original ISDN, which was renamed Narrowband ISDN (N-ISDN). Is this just an unfortunate choice of names, or a move to avoid admitting defeat? The current ITU-T direction has the same universal vision as summarized previously. Can B-ISDN succeed where N-ISDN did not?

Let's look at what happened to N-ISDN to see if the same fate awaits B-ISDN. While the N-ISDN standards were being developed, Local Area Network (LAN) protocols were developed and being standardized by the IEEE. The most successful LAN data communications protocol today is IEEE 802.3 Ethernet family, with IEEE 802.5 Token Ring a distant second. Both of these technologies were invented first, and then formally standardized. The ARAPNET was also being developed at this time, and it has blossomed into the most rapidly growing public data communication network in the world today — the Internet. After experiencing such tremendous success, the Internet established its own de facto standards body — the Internet Engineering Task Force (IETF).

The signaling protocol developed for ISDN had as an objective to interwork with any telephone system deployed within the last century, as well as provide new features and functions for applications that were not even envisioned yet. The result was a very complicated signaling protocol specified in CCITT Recommendations Q.921 and Q.931. Achieving interoperable implementations between user devices and network switches took years to achieve. These recommendations specified a signaling protocol at a User-Network Interface (UNI). Unfortunately, most of the vendors built the network side of the interface expecting the user side of the interface to be developed. One reason for the failure of ISDN was largely due to never achieving a critical mass of user-side signaling implementations. Another key reason is that Local Exchange Carriers (LECs), IntereXchange Carriers (IXCs) and international Postal Telegraph and Telephone (PTT) operators have not been able to coordinate the offering of a ubiquitous international ISDN service.

However, the situation is different for ATM. As outlined in Chapter 10, the ATM Forum was established with participants from the data communications as well as the telephony carrier sectors — this combination of forces is fundamentally different than what occurred in N-ISDN, and is the brightest ray of hope for ATM. The choice of the name ATM avoids the four-letter word ISDN in many of their documents, even though the carrier and ITU-T influence exists in many places in the documentation. The ATM Forum has chosen to build the signaling protocol on the ISDN protocol as described in Chapters 14 and 15. Herein lies a critical juncture for the success of ATM. If users and networks can build interoperable, high-performance signaling systems, then user needs can be met. The ATM Forum also blazed new trails in areas traditionally outside the scope of standards bodies in the areas of applications, interfaces, protocol interworking, routing, video transport, and LAN emulation. These are also very different from ISDN. The hype that surrounded ATM in the early 1990s now focuses on IP-based protocols. Interestingly, each of these technologies: N-ISDN, ATM, and now IP targeted a similar concept: the universal network. Is this an achievable goal, or is it the elusive holy grail?

32.4.3 A Change in Power?

The era of ATM may also usher in a change for those in power. Some industry analysts have published prophesies that ATM will introduce new carriers and service providers into the marketplace. The ones with the most to lose could be the traditional telephone carriers. There is already some evidence of a trend such as this with companies using ATM to offer transparent LAN connection service. The introduction of a new technology or paradigm causes chaos and creates the opportunity for a significant change in market share. The explosion of multiprotocol routing in data communication in the early 1990s is evidence of a similar phenomenon. The explosive growth of the Internet and the web in the late 1990s is another example. Also, virtual private voice networks have taken a large amount of market from integrated voice/data private network solutions. The adoption of ATM-based technology for video-on-demand and future interactive applications to the home, as described in Chapter 18, also points to a potential shift in power.

32.4.4 Multimedia: Now, or When?

The industry press provides a range of answers to the following dilemma:

Will emerging multimedia applications create demand for networks that support multiple QoS class?

Or, will the availability of networks that support multiple QoS classes stimulate the development of multimedia applications?

The arguments for the first stimulus-response model, applications create demand for networks, include the following. Currently, true multimedia enabled hardware and software is too expensive for use in all but the leading edge high-tech or executive user communities. The localized, limited need for multiple QoS categories in networks drives enterprises to implement ATM work groups, or over-provisioned high speed LANs, to support the limited demand for expensive multimedia applications. Only when PC class hardware and software supports multimedia applications cost effectively will there be demand for networks with multiple QoS classes. This will likely occur around the year 2000.

Supporters of the contrary view, that networking stimulates application development, see things differently. Only lightly loaded (or over-provisioned) LANs effectively support multimedia applications involving bandwidth hungry media, such as full motion high quality video today. This lack of an effective networking infrastructure provides a small market to multimedia application developers. Only when networks, their supporting protocols and APIs capable of supporting multiple data types cost effectively become available will multimedia applications be developed.

32.4.5 Future-proofing Investments

A touted benefit of ATM is that investments today in ATM will not become obsolete as soon as other technologies. However, some recent experience with the ATM Forum runs counter to this argument. Many users found that they must replace the UNI 3.1 compatible hardware they purchased with UNI 4.0 hardware to achieve the benefits of ABR closed loop congestion control. The ATM Forum responded to this problem with the "Anchorage Accord," effectively putting a freeze on any major new specification development for several years so that implementers could catch up to the wave of specifications released by the de facto industry standards body in 1996. The current set of ATM standards promise to be quite stable for many years to come. In fact, the ITU-T I.610 standard defines performance measurement methods that will enable users to accurately measure the QoS their networks actually deliver. No other communications standard has anything remotely like this today. IP and LAN networking is years behind in this area.

32.5 ARCHITECTURES OF TOMORROW

What will be the predominant networking technology five years from now? Based upon past experience, the most likely answer is something not even on the radar screen yet. In the LAN, however, an Ethernet derivative will be the likely winner. In the WAN, IP backbones will be a mix of packet over SONET and ATM depending upon the traffic mix. Many integrated access and QoS aware private and virtual private networks will likely use ATM.

Will the telephony model that brought us ubiquitous voice service across the planet prevail in the twenty first century, or will the radical concepts of the Internet explosion dominate networking for the upcoming decades? Or, will evolution result, taking the best from both? Have you ever wondered what the architecture of tomorrow's data network will look like? This section takes a glimpse into what network architectures may look like in the twenty-first century, and how ATM will play a key role.

Static, predefined private communications networks are migrating to dynamic, virtual networks — networks with ubiquitous access that can interwork past, present, and future protocols — that provide anyone-to-anyone, anywhere-to-anywhere communications. The flexible, extensible capabilities of ATM interworking well with LANs and the Internet protocols in both the local and wide area position it well to be a key vehicle in the rapidly changing world of networking. Virtual private networks implemented over public ATM infrastructures employ sophisticated switching and routing intelligence, in essence making the network the computer. Major corporations increasingly depend upon virtual networking to run their day-to-day businesses, increasingly relying on their partitioning and security features. ATM services already are available worldwide and are now moving out towards mobile and telecommuting users via wireless ATM and xDSL technologies. However, access to the Internet is also critical. The Internet must accommodate different classes of service for users willing to pay for differentiated QoS. This must happen because someone must fund the tremendous growth rate of the Internet backbone. ATM enables the user to win the war with technical obsolescence — offering a network infrastructure that will last well into the next century. Will ATM be the Rosetta stone for protocol interworking in both the LAN and WAN, allowing multiple protocols to be deciphered and understood? Will these ATM networks provide ubiquitous, any-to-any access for all users? Only time will tell.

32.5.1 ATM or IP as the Next Voice Infrastructure?

A great topic to bring up at a technocrat voice provider's dinner party is whether ATM switching or IP routing will replace today's voice switch infrastructure. While on first blush this may appear to be a religious more than a technical argument, there are many real reasons why ATM and/or IP may indeed eventually replace the voice switched network — or at least part of it. Tremendous benefits in support cost aggregation, voice-aware client-server applications, and ATM and IP telephony are already being realized by VPN users that consolidate voice and data traffic on a single corporate network, interconnecting via gateways to the PSTN and public data networks.

One major factor that will heavily influence a service provider's future switching fabric is the projected mix of voice and non-voice traffic. Non-voice traffic volumes continue to grow at a more rapid pace than voice traffic, particularly more so in corporate calling patterns than residential. Non-voice traffic's highly variable, bandwidth intensive traffic patterns are no match for existing voice switches and trunking. One technology that is already pushing this envelope is N-ISDN. Voice service providers that offer N-ISDN must carefully plan their trunking requirements, as each N-ISDN user requires much more bandwidth than the typical phone call (128 kbps versus 64 kbps for voice) and for much longer periods. Data calls are on the average tens to hundreds of minutes long, whereas voice calls are usually only a few minutes in length. Thus, growing non-voice traffic requirements place a much greater strain on existing networks than does voice growth. While today it makes sense to keep these two networks separate, network architects must at least consider whether it makes economic and operational sense to merge the two.

The second major issue is voice services. Take a good look at your local service phone bill, and you will see at least a dozen calling options — from call waiting to message services to call blocking security features to automatic redial when a busy number becomes clear. These custom calling and handling features generate huge amounts of revenue for the voice service providers, yet consume resources in large CO voice switches and the associated computers that provide network intelligence. These services might be provided more cost effectively on ATM or Internet based network servers, with the added bonus of even more service features.

The third major issue is mobility. Voice services have been mobile (i.e., cellular phones) for some time. Yet now users want mobile data services — and definitely don't want to pay for two separate services and access devices. This points to a single access infrastructure. You can already see this today in many cell phones that have small data cards for attaching your laptop for remote dial access to corporate resources. The trend toward personal data assistants (PDAs) adds to the amount of data (voice mail is now joined by email, paging, stock quotes, etc. – you get the idea) on the voice network. Quickly, the actual voice traffic is subsumed by the volume of data, and possibly video traffic after that.

ATM takes on this challenge by offering a front end technology that accepts various access types (POTS, ISDN, xDSL, etc.) and converts them to ATM switching — feeding them to private voice and data networks as required. Calls and services originate and terminate at the edges of the ATM network and then are switched and processed from there.

However, the wholesale replacement of our current PSTN infrastructure is still years away. Efforts like the Advanced Intelligent Network (AIN) are an attempt to bring the PSTN into its next generation of switching and services. AIN is a strategy and framework for the next generation of voice services over legacy voice switches. To compound this movement, the depreciation cost tied up in legacy voice switches alone is enough to make even the more progressive service providers shudder. In many cases, a migration strategy using many of the concepts and technologies defined in this chapter will be a service provider's best bet. This replacement will probably happen much faster in countries who lack an existing infrastructure, whose existing voice infrastructure services are expensive and fragmented, and/or in environments where ATM service are particularly attractive for other reasons.

An ATM approach also allows network managers to pick and chose where they want high quality voice (extranet) or lower quality voice (intranet). This will change over time, as ATM and public voice network service providers merge their backbone architectures into a single ATM infrastructure — allowing for seamless inter-corporate voice and data networking.

The key to ATM service and equipment providers to replace existing TDM networks is in the price of the combined equipment and service over the depreciation period, matching the features and functions that are most important to the customer. ATM networks should also support integrated access, particularly in support of voice. In fact, the integration of voice over IP or ATM is very attractive in European and Pacific Rim countries where inter-country voice communications are extremely expensive.

32.5.2 One Possible Scenario

Internet traffic continues growing at a phenomenal rate. Much of the traffic is currently web traffic, and most of that TCP [23]. What will the traffic mix of future traffic be? What will the impact of ADSL be? What new hardware and software capabilities will affordable home computers and information appliances have? What application will sustain the tremendous growth in data traffic? How quickly can a networking need change when people replace their computing devices once every several years? Let's look at one possible scenario as an answer to these questions.

Statistics show that Americans spend several hours a day watching television, more than any other nation (not that this is something to be proud of). The cable system typically carries up 100 channels of approximately 3 MHz each to 50 million households in the United States. Assuming 1 bps per Hz, this equates to approximately 15 Pbps (a petabit per second (Pbps) is 10^{15} bits per second). The problem often is that even with 100 channels, subscribers find nothing on worth watching all too frequently.

Already, tens of millions of people in America spend several hours a day surfing the web, reading E-mail, in chat groups, or other pursuits. At an average transfer rate of 10 kbps for 10 million users, this equates to a demand of approximately 100 Gbps.

When ADSL delivers video on demand to many small businesses and households, a natural conclusion is that the channel changers will spend at least some of their time away from the television. Assume that one-quarter of the approximately 100 million households in the United States use video on demand at 1 Mbps during periods of peak demand. This represents an aggregate demand of 100 Tbps. This is 1,000 times the current level of Internet traffic!

The payback interval for a movie or television show can be quite lengthy. Many movies do not pay back until after videocassettes are released. Similarly, many TV shows aren't profitable until well into syndication reruns. Many companies own a wealth of video content. Also, video on demand offers tremendous potential for remote learning and education. Much of the expense associated with video is paying the rights to the artists and production companies. Video on demand must be cost competitive with video tape rental chains and pay per view cable offerings in order to succeed.

During the business day, assume that the approximately 10 million enterprises generate and average of 10 Mbps of traffic each, again approximately 100 Tbps. The commercial and residential busy hours will likely be non-coincident. Therefore, the total domestic U.S. network needs to carry on the order of 1 Pbps.

Currently, all of the telephone networks in the United States combined carry a peak of approximately 20 million conversations during the business busy hour and during peak periods, like Mother's day. At 64 kbps this level of traffic is only about 1 Tbps. Therefore, ten years from now the potential data and video traffic is several orders of magnitude greater than the expected voice traffic. Eventually, technology will eclipse even the benchmark of a petabit per second and we will measure network capacity in terms of exabits per second (10^{18} bps).

32.5.3 Betting on a Sure Thing

This book has taken you through the business drivers, technology enablers, a review of communications networking, and a detailed look at ATM, spelling out the competing technologies and the challenges ahead. A great deal of ATM is already defined in sufficient detail for interoperability. The hype for ATM across all segments in the communications industry has died down, and the technology is now viewed by many as mature. This is not necessarily a bad thing because it is an indicator of success. ATM will be successful; the only questions that remains is how successful, and in which specific areas.

32.6 REVIEW

This final chapter reviewed the state of ATM and expected directions. The text began with a review of ATM's latest forays into the frontiers of networking: wireless services, satellite, and the renaissance of high bandwidth to small businesses using xDSL over existing copper pairs or cable modems. The coverage then addressed the potential roles for IP and ATM as the infrastructure for integrated access, the backbone network, and as a replacement for the current N-ISDN and TDM-based voice network infrastructure. The text then highlighted several critical success factors for ATM, including: continual cost reduction, support for SVCs, and effective support for voice and video over integrated access lines. Finally, the chapter concluded with further speculations on the architectures of tommorow's networks. One thing is for certain, the telecommunications world will never be the same. Those that aren't ready to change and adapt will soon become extinct. We hope that this book gave you background in the major competing technologies and analytical methods to better prepare you in your own battle for survival.

32.7 REFERENCES

[1]   J. McQuillan, "Deconstructing ATM," Business Communications Review, March, 1997.
[2]   D. McDysan, D. Spohn, *Hands On ATM*, McGraw-Hill, 1997.
[3]   L. Dellaverson, "Reaching for the New Frontier," ATM Forum 53 Bytes, Volume 4, Issue 3, 1996
[4]   V. Oakland, "Broadband from Above," Telephony, Feb 17, 1997
[5]   A. Rogers, "… While IP Gets Speedy in Space," Communications Week, March 17, 1997
[6]   IEEE, "Broadband via Satellite," IEEE Communications Magazine, July, 1997.
[7]   S. Masud, "TCP overhaul needed for satellites", Computer Reseller News, Iss: 702, Sep. 23, 1996.
[8]   ADSL Forum, "ADSL Forum Home Page," http://www.adsl.com/adsl/.

[9]   Telechoice, "The Telechoice Report on xDSL On-line, "
        http://www.telechoice.com/xdslnewz/.

[10]   J. Caruso, "DSL Goes Hybrid by Adding ATM," *Communications Week*, June
        1997.

[11]   E. Roberts, "IP On Speed," Data Communications, March, 1997.

[12]   S. Keshav, R. Sharma, "Issues and Trends in Router Design," IEEE Communi-
        cations Magazine, May 1998.

[13]   V. Kumar, T. Lakshman, D. Stiliadis, "Beyond Best Effort: Router Architec-
        tures for the Differentiated Services of Tommorow's Internet," IEEE Communi-
        cations Magazine, May 1998.

[14]   D. Isenberg, "Rise of the Stupid Network," *AT&T*, June 4, 1997.

[15]   D. Passmore, "Who Needs SVCs?," *Business Communications Review*, June
        1997

[16]   R. Cole, D. Shur, C. Villamizar, *IP over ATM: A Framework Document*, RFC
        1932, IETF, April 1996.

[17]   N. Shelef, "SVC Signaling: Calling All Nodes," *Data Communications*, June
        1995.

[18]   Cascade White Paper, "Addressing and Numbering Plans for Public ATM
        Networks," November, 1996,
        http://www.casc.com/products/datasheets/008_97.pdf.

[19]   J. Bredeson, D. Pillai, "Efficiency of ATM Networks in Transporting TCP/IP
        Traffic," *Proceedings of the 28th Symposium on Systems Theory*, 1996.

[20]   K. Hodges, J. Hamer, "Carrier implementation of ATM Services," *Telecommu-
        nications* Vol: 29, Iss: 4, April 1995.

[21]   K. Korostoff, "Why They Buy: A Survey of ATM Customers," *Business
        Communications Review*, February 1998.

[22]   T. Nolle, "ATM: Is There Life after Hype?," *Business Communications Review*,
        April 1997.

[23]   K. Thompson, G. Miller, R. Wilder, "Wide-Area Internet Traffic Patterns and
        Characteristics," *IEEE Network*, November/December 1997.

Acronyms and Abbreviations

AAL	ATM Adaptation Layer
ABM	Asynchronous Balanced Mode (HDLC)
ABR	Available Bit Rate
AC	Access Control (IEEE)
ACF	Access Control Field (DQDB)
ACK	Acknowledgment
ACM	Address Complete Message
ACR	Allowed Cell Rate (ATM ABR)
ADM	Add/Drop Multiplexer
ADPCM	Adaptive Differential Pulse Code Modulation
ADSL	Asymmetric Digital Subscriber Line
ADTF	ACR Decrease Time Factor (ATM ABR)
AESA	ATM End System Address
AFI	Authority and Format Identifier (AESA)
AII	Active Input Interface (used in UNI PMD specs)
AIR	Additive Increase Rate
AIS	Alarm Indication Signal (OAM)
AMI	Alternate Mark Inversion
ANS	American National Standard
ANSI	American National Standards Institute
AOI	Active Output Interface (used in UNI PMD specs)
API	Application Programming Interface
APS	Automatic Protection Switching
ARM	Asynchronous Response Mode (HDLC)
ARP	Address Resolution Protocol
ASN.1	Abstract Syntax Notation One
ATM	Asynchronous Transfer Mode
ATMARP	ATM Address Resolution Protocol
ATMF	ATM Forum

*NOTE : Additional comments in parentheses are a clarification or refer to the standard from which the term is derived. Many acronyms are used by multiple standards and only the most prevalent are mentioned.

AU	Access Unit (DQDB)
AUI	Attachment Unit Interface (Ethernet 802.3)
B	Bearer (64-kbps DS0 Channel in N-ISDN)
B8ZS	Bipolar with 8 Zero Substitution
BCC	Block Check Characters
BCD	Binary-Coded Decimal
BCOB	Broadband Connection Oriented Bearer
BECN	Backward Explicit Congestion Notification (FR)
Bellcore	Bell Communications Research
BER	Basic Encoding Rules (ASN.1)
BER	Bit Error Ratio or Rate
BFR	Big Fancy Router
BGP	Border Gateway Protocol
B-HLI	Broadband High-Layer Information
B-ICI	Broadband Intercarrier Interface
B-ISDN	Broadband Integrated Services Digital Network
B-ISSI	Broadband Inter-Switching System Interface
B-ISUP	Broadband ISDN User Part
BIP	Bit Interleaved Parity
BIPV	Bit Interleaved Parity Violation
BIS	Border Intermediate System (ATM Forum, PNNI SWG)
BITS	Building Integrated Timing Supply
B-LLI	Broadband Low-Layer Information
B-NT	Broadband Network Termination
BOC	Bell Operating Company
BOM	Beginning of Message (DQDB)
BOOTP	Bootstrap Protocol
BPDU	Bridge Protocol Data Unit
bps	bits per second
BRI	Basic Rate Interface (ISDN)
BSC	Binary Synchronous Communications protocol (IBM)
BSS	Broadband Switching System
B-TA	Broadband Terminal Adapter (ATM)
B-TE	Broadband Terminal Equipment (ATM)
BO	Bit Oriented (SONET)
BT	Burst Tolerance
BUS	Broadcast Unknown Server (ATM LANE)
C/R	Command/Response Indicator or Bit
CAC	Connection Admission Control
CAD/CAM	Computer-Aided Design/Computer-Aided Manufacturing
CAS	Channel Associated Signaling
CBDS	Connectionless Broadband Data Service
CBR	Constant Bit Rate
CCITT	Consultative Committee on International Telephone and Telegraph (now ITU)
CCS	Common Channel Signaling

CCS7	Common Channel Signaling System 7
CCR	Current Cell Rate (ATM ABR)
CD	CountDown counter (DQDB)
CDF	Cutoff Decrease Factor (ATM ABR)
CDV	Cell Delay Variation (ATM QoS)
CDVT	Cell Delay Variation Tolerance
CE	Connection Endpoint
CEI	Connection Endpoint Identifier
CEPT	Conference on European Post & Telegraph
CER	Cell Error Ratio (ATM QoS)
CES	Circuit Emulation Service
CHAP	Challenge Handshake Authentication Protocol
CI	Congestion Indicator (ATM ABR)
CIR	Committed Information Rate (FR)
CL	ConnectionLess
CLIP	Classical IP
CLLM	Consolidated Link Layer Management (FR)
CLNP	Connectionless Layer Network Protocol
CLNS	Connectionless Network Service (OSI)
CLP	Cell Loss Priority
CLR	Cell Loss Ratio (ATM QoS)
CLS	Connectionless Service
CLSF	Connectionless Server Function (ITU-T)
CMIP	Common Management Interface Protocol (ISO)
CMIS	Common Management Information Service (ISO)
CMISE	CMIS Element (ISO)
CMR	Cell Misinsertion Rate (ATM QoS)
CMT	Connection Management (FDDI)
CNMS	Customer Network Management System
CNR	Complex Node Representation (ATM Forum PNNI)
CO	Central Office
COAM	Customer Owned and Maintained
COCF	Connection-Oriented Convergence Function (DQDB)
Codec	Coder/Decoder
COM	Continuation of Message (DQDB)
CONS	Connection-Oriented Network Service (ISO)
COS	Class Of Service
CPCS	Common Part Convergence Sublayer
CPE	Customer Premises Equipment
CPI	Common Part Indicator
CRC	Cyclic Redundancy Check
CRF	Connection Related Function
CRS	Cell Relay Service
CS	Convergence Sublayer
CSMA/CD	Carrier-Sense Multiple Access with Collision Detection
CSU/DSU	Channel Service Unit/Data Service Unit
CTD	Cell Transfer Delay (ATM QoS)

CVSD	Continuously Variable Slope Delta modulation
DA	Destination Address field
DAC	Dual Attached Concentrator
DACS	Digital Access and Cross-connect System
DAL	Dedicated Access Line
DARPA	Defense Advanced Research Projects Agency
DAS	Dual-Attached Station (FDDI)
DAT	Digital Audio Tape
DCC	Data Country Code (AESA)
DCE	Data Communications Equipment
DCS	Digital Cross-connect System
DDD	Direct Distance Dialing
DDS	Digital Data Service
DE	Discard Eligibility (FR)
DHCP	Dynamic Host Configuration Protocol
DLCI	Data Link Connection Identifier (FR)
DMA	Direct Memory Access
DNS	Domain Name Service
DoD	Department of Defense
DQDB	Distributed Queue Dual Bus (IEEE)
DS0	Digital Signal Level 0
DS1	Digital Signal Level 1
DS3	Digital Signal Level 3
DSAP	Destination Service Access Point (LLC)
DSP	Domain Specific Part (AESA)
DSR	Data Set Ready
DSU	Data Service Unit
DSX	Digital Signal Cross-Connect
DTE	Data Terminal Equipment
DTL	Designated Transit List
DTMF	Dual Tone Multifrequency
DXC	Digital cross (X)-Connect
DXI	Data Exchange Interface (SMDS, ATM)
E&M	Ear & Mouth or Earth & Magnet
E1	European Transmission Level 1
E3	European Transmission Level 3
EA	Extended Address
ECN	Explicit Congestion Notification
ECSA	Exchange Carriers Standards Association
ED	End Delimiter (IEEE 802)
EFCI	Explicit Forward Congestion Indication
EFCN	Explicit Forward Congestion Notification
EGP	External (Exterior) Gateway Protocol
EGRP	Exterior Gateway Routing Protocol
EIA	Electronics Industries Association
EIR	Excess Information Rate (FR)

EISA	Extended Industry Standard Architecture
ELAN	Emulated LAN (ATM LANE)
EMI	ElectroMagnetic Interference
EML	Element Management Layer
EMS	Element Management System
EOM	End of Message
EOT	End of Transmission
EPD	Early Packet Discard
EPRCA	Enhanced Proportional Rate Control Algorithm (ATM ABR)
EPROM	Erasable Programmable Read Only Memory
ER	Explicit Rate
ES	End System (OSI) or Errored Seconds
ESI	End System Identifier (AESA)
ESF	Extended Superframe
ETB	End of Transmission Block
ETSI	European Telecommunications Standards Institute
ETX	End of Text
F	Flag
FCS	Frame Check Sequence
FDDI	Fiber Distributed Data Interface (ANSI)
FDDI-II	Fiber Distributed Data Interface Version II
FDM	Frequency-Division Multiplexing
FEBE	Far End Block Error (SONET)
FEC	Forward Error Correction
FECN	Forward Explicit Congestion Notification (FR)
FERF	Far End Reporting Failure (now called RDI)
FM	Frequency Modulation
fps	Frames per second
FR	Frame Relay
FRAD	Frame Relay Assembler/Disassembler, or Access Device
FR-SSCS	Frame Relay Service Specific Convergence Sublayer
FS	Frame Status field (FDDI)
FT1	Fractional T1
FTP	File Transfer Protocol
FUNI	Frame-based User-to-Network Interface (ATM Forum)
Gb	Gigabits (billions of bits)
Gbps	Gigabits per second (10^9 bps)
GCAC	Generic Connection Admission Control
GCRA	Generic Cell Rate Algorithm
GFC	Generic Flow Control
GFR	Guaranteed Frame Rate (GFR)
GOS	Grade of Service
GOSIP	Government Open System Interconnection Profile
GUI	Graphical User Interface

HCS	Header Check Sequence (DQDB)
HDLC	High-Level Data Link Control (ISO)
HDTV	High-Definition Television
HEC	Header Error Control
HOB	Head of Bus (DQDB)
HPPI	High Performance Parallel Interface
HSSI	High-Speed Serial Interface
HTML	HyperText Markup Language (HTML)
HTTP	HyperText Transfer Protocol (IETF)
Hz	Hertz or cycles per second
ICD	International Code Designator
ICI	InterCarrier Interface
ICR	Initial Cell Rate (ATM ABR)
ICMP	Internet Control Message Protocol
ICP	IMA Control Protocol
IDI	Initial Domain Identifier (AESA)
IDP	Initial Domain Part (AESA)
IDRP	Interdomain Routing Protocol
IDU	Interface Data Unit
IE	Information Element
IEC	Inter-Exchange Carrier
IEEE	Institute of Electrical and Electronics Engineers
IETF	Internet Engineering Task Force
IGMP	Internet Group Management Protocol
IGP	Internal (Interior) Gateway Protocol
IGRP	Internal (Interior) Gateway Routing Protocol
IISP	Interim Interswitch Signaling Protocol (ATM Forum)
ILMI	Integrated Local Management Interface
IMA	Inverse Multiplexing over ATM
IME	ILMI Management Entity
IMPDU	Initial MAC Protocol Data Unit (DQDB)
ION	IP Over Non-broadcast multiple access networks (IETF)
IP	Internet Protocol
IPX	Internetwork Packet Exchange protocol (Novell)
IS	Intermediate System (OSI)
ISDN	Integrated Services Digital Network
ISDU	Isochronous Service Data Unit (DQDB)
IS-IS	Intermediate System–to–Intermediate System (OSI)
ISN	Initial Sequence Number
ISO	International Standards Organization
ISSI	Interswitching System Interface (SMDS)
ITU	International Telecommunications Union
ITU-T	ITU - Telecommuncations standardization sector
IUT	Implementation Under Test
IWF	Inter-Working Function
IXC	IntereXchange Carrier

JPEG	Joint Photographic Experts Group
kb	kilobit (thousands of bits)
kbps	kilobits per second (10^3 bps)
km	kilometers (10^3 meters)
LAN	Local Area Network
LANE	LAN Emulation (ATM Forum)
LAP-B	Link Access Procedure — Balanced (X.25)
LAP-D	Link Access Procedure — D (ISDN/Frame Relay)
LAP-F	Link Access Procedure — Frame Mode
LATA	Local Access Transport Area
LBO	Line Build Out
LCGN	Logical Channel Group Number
LCP	Link Control Protocol
LCT	Last Compliance Time (used in GCRA definition)
LEC	Local Exchange Carrier
LEC	LAN Emulation Client (ATM Forum LANE)
LECS	LAN Emulation Configuration Server
LES	LAN Emulation Server (ATM Forum LANE)
LGN	Logical Group Node (ATM Forum PNNI)
LIJ	Leaf Initiated Join
LIS	Logical IP Subnet (IETF)
LLC	Logical Link Control (IEEE 802.X)
LLC/SNAP	Logical Link Control/Subnetwork Access Protocol
LMI	Local Management Interface
LNNI	LAN Emulation NNI
LOC	Loss of Cell Delineation
LOF	Loss of Frame
LOP	Loss of Pointer
LOS	Loss of Signal (UNI fault management)
LSB	Least Significant Bit
LT	Line Termination
LSAP	Link Service Access Point
LTE	Line Terminating Equipment (SONET)
LU	Logical Unit (SNA)
LUNI	LANE UNI (ATM Forum, LANE)
m	meter
MAC	Media Access Control (IEEE 802.X)
MAN	Metropolitan Area Network (DQDB, FDDI)
MARS	Multicast Address Resolution Service (IETF)
Mb	Megabits (millions of bits)
MBONE	Multi-Media Backbone
Mbps	Megabits per second (10^6 bps)

MBS	Maximum Burst Size
MCR	Minimum Cell Rate (ATM ABR)
MHz	Megahertz
MIB	Management Information Base
MID	Multiplexing IDentifier (ATM), Message IDentifier (DQDB)
MIPS	Millions of Instructions per Second
MJPEG	Motion JPEG
MM	Multi-Mode
MMF	Multimode Fiber
MPEG	Motion Picture Experts Group
MPOA	Multiprotocol over ATM (ATM Forum)
ms	millisecond (one-thousandth of a second, 10^{-3} second)
MSB	Most Significant Bit
MSS	MAN Switching System (SMDS)
MTP	Message Transfer Part
MTU	Maximum Transmission Unit
MUX	Multiplexer
NANP	North American Numbering Plan
NAU	Network Access Unit
NBMA	Non-Broadcast Multiple Access
NCP	Network Control Protocol or Point (SNA)
NDIS	Network Driver Interface Specification
NE	Network Element
NEBS	Network Equipment Building Specification
NetBIOS	Network Basic Input/Output System protocol
NEXT	Near End Crosstalk
NFS	Network File Server
NHRP	Next-Hop Resolution Protocol (IETF)
N-ISDN	Narrowband Integrated Services Digital Network
NIU	Network Interface Unit
NLPID	Network-Layer Protocol Identifier
nm	nanometer (10^{-9} m)
NM	Network Management
NMS	Network Management System
NNI	Network Node Interface, or Network to Network Interface
NOS	Network Operating System
NP	Network Performance
NPA	Numbering Plan Area
NPC	Network Parameter Control
NRM	Normalized Response Mode (ISO)
Nrm	Number of cells between successive resource management cells (ATM ABR)
nrt-VBR	non-real time Variable Bit Rate
NRZ	Non-Return to Zero
NRZI	Non-Return to Zero Inverted
ns	nanosecond (10^{-9} second)

NSAP	Network Service Access Point
NT	Network Termination
NTSC	National Television Standards Committee
OAM	Operations And Maintenance
OAM&P	Operations, Administration, Maintenance, and Provisioning
OC-n	Optical Carrier Level n (SONET)
OCD	Out-of-Cell Delineation (ATM UNI)
OH	Overhead
OID	Object Identifier
ONA	Open Network Architecture
OOF	Out of Frame
OOS	Out of Service
OS	Operating Systems
OSI	Open Systems Interconnection
OSI CLNS	Connectionless Network System (OSI)
OSIRM	OSI Reference Model
OSPF	Open Shortest Path First
OUI	Organizationally Unique Identifier (IEEE)
PABX	Private Automated Branch Exchange
PAD	Packet Assembler/Disassembler (X.25)
PAR	PNNI Augmented Routing
PBX	Private Branch Exchange
PCM	Pulse Code Modulation
PCR	Program Clock Reference (MPEG2)
PCR	Peak Cell Rate
PDH	Plesiochronous Digital Hierarchy
PDU	Protocol Data Unit
PG	Peer Group
PGL	Peer Group Leader
PHY	Physical Layer
PID	Protocol IDentifier
PING	Packet INternet Groper
PLCP	Physical Layer Convergence Protocol
PLL	Phase Locked Loop
PM	Performance Monitoring or Physical Medium
PMD	Physical Medium Dependent
PNNI	Private Network-to-Network Interface or Private Network Node Interface
POH	Path Overhead (SONET)
POI	Path Overhead Indicator
PON	Passive Optical Network
PoP	Point of Presence
POTS	Plain Old Telephone Service
PPD	Partial Packet Discard
PPP	Point-to-Point Protocol (Internet)

pps	packets per second
PRI	Primary Rate Interface (ISDN)
PSTN	Public Switched Telephone Network
PT	Payload Type
PTE	Path-Terminating Equipment (SONET)
PTI	Payload Type Identifier
PTSE	PNNI Topology State Element
PTSP	PNNI Topology State Packet
PTT	Postal, Telegraph & Telephone Ministry/Administration
PVC	Permanent Virtual Connection
PVCC	Permanent Virtual Channel Connection
PVPC	Permanent Virtual Path Connection
QoS	Quality of Service
QPSX	Queued Packet and Synchronous Exchange
RAM	Random Access Memory
RBOC	Regional Bell Operating Company
RDF	Rate Decrease Factor
RDI	Remote Defect Indication
REJ	Reject frame (HDLC)
RFC	Request for Comment (IETF)
RIP	Routing Information Protocol
RISC	Reduced Instruction Set Computing
RM	Resource Management (ABR)
RMON	Remote MONitoring (IETF)
RNR	Receive Not Ready
ROLC	Routing Over Large Clouds (Now IP Over NBMA (ION))
ROM	Read-Only Memory
RR	Receive Ready frame
RSVP	Resource reSerVation Protocol (IETF)
RTT	Round-Trip Time
rt-VBR	real-time Variable Bit Rate
RTP	Real-Time Protocol
s	second
SA	Source Address (IEEE)
SAAL	Signaling ATM Adaptation Layer
SAP	Service Access Point
SAR	Segmentation and Reassembly
SAS	Single-Attached Station connection (FDDI)
SCCP	Signaling Connection Control Part
SCR	Sustainable Cell Rate
SCSI	Small Computer Systems Interface
SD	Start Delimiter
SDH	Synchronous Digital Hierarchy (ITU-T)
SDLC	Synchronous Data Link Control protocol (IBM)
SDU	Service Data Unit

SEAL	Simple and Efficient Adaptation Layer (AAL5)
SECB	Severely Errored Cell Block
SES	Severely Errored Seconds
SEL	Selector Byte (AESA)
SF	Superframe
SFD	Start of Frame Delimiter
SIG	SMDS Interest Group
SIP	SMDS Interface Protocol (SMDS)
SIR	Sustained Information Rate (SMDS)
SLIP	Serial Line IP
SM	Single Mode
SMDS	Switched Multimegabit Data Service
SMF	Single-Mode Fiber
SMT	System Management protocol (FDDI)
SN	Sequence Number
SNA	Systems Network Architecture (IBM)
SNAP	Sub-Network Attachment Point
SNI	Subscriber Network Interface (SMDS)
SNMP	Simple Network Management Protocol (IETF)
SOH	Section Overhead
SONET	Synchronous Optical Network (ANSI)
SPE	Synchronous Payload Envelope (SONET)
SPF	Shortest Path First protocol
SPVCC	Soft Permanent Virtual Channel Connection
SPVPC	Soft Permanent Virtual Path Connection
SR	Source Routing
SREJ	Select Reject frame
SRT	Source Route Transparent protocol
SRTS	Synchronous Residual Time Stamp (AAL1)
SS7	Signaling System Number 7
SSAP	Source Service Access Point (LLC)
SSCF	Service-Specific Coordination Function
SSCOP	Service-Specific Connection-Oriented Protocol
SSCS	Service-Specific Convergence Sublayer (ATM)
STDM	Statistical Time Division Multiplexing
STE	Section Terminating Equipment (SONET)
STM	Synchronous Transfer Mode (SDH)
STM-n	Synchronous Transport Module level n (SDH)
STP	Shielded Twisted Pair
STP	Spanning Tree Protocol (IEEE 802.1d)
STP	Shielded Twisted Pair
STS-N	Synchronous Transport Signal Level N (SONET)
STS-n	Synchronous Transport Signal-n
STS-Nc	Concatenated Synchronous Transport Signal Level N (SONET)
SVA	Signature component of AVA/ATV
SVC	Switched Virtual Connection
SVCC	Switched Virtual Channel Connection

SVPC	Switched Virtual Path Connection
SYN	Synchronous Idle
t	Time
TA	Terminal Adapter
TAT	Theoretical Arrival Time (used in GCRA definition)
TAXI	Transparent Asynchronous Transmitter/Receiver Interface
TC	Transmission Convergence sublayer of PHY Layer
TCAP	Transaction Capabilities Applications Part (SS7)
TCP	Transmission Control Protocol (IETF)
TCP/IP	Transmission Control Protocol/Internet Protocol (IETF)
TDM	Time Division Multiplexing
TDMA	Time Division Multiple Access
TDS	Time Division Switching
TE	Terminal Equipment
TFTP	Trivial File Transfer Protocol
TM	Traffic Management
TNS	Transit Network Selection
TUC	Total User Cell Count
TUCD	Total User Cell Difference
TUNIC	TCP/UDP over Non-existent IP Connections
UBR	Unspecified Bit Rate
UDP	User Datagram Protocol (IETF)
UME	UNI Management Entity (used in ILMI Definition)
UNI	User-to-Network Interface
UPC	Usage Parameter Control
UTOPIA	Universal Test and Operations Interface for ATM
UTP	Unshielded Twisted Pair
VBR	Variable Bit Rate
VC	Virtual Channel (ATM) or Virtual Call (X.25)
VC-n	Virtual Container-n (SDH)
VCC	Virtual Channel Connection
VCI	Virtual Channel Identifier
VCL	Virtual Channel Link
VLAN	Virtual LAN
VLSI	Very Large Scale Integration
VoD	Video on Demand
VP	Virtual Path (ATM)
VPC	Virtual Path Connection
VPCI	Virtual Path Connection Identifier
VPI	Virtual Path Identifier
VPL	Virtual Path Link
VPN	Virtual Private Network
VS/VD	Virtual Source/Virtual Destination (ABR)
VSAT	Very Small Aperture Terminal
VT	Virtual Tributary (SONET)

VTx   VT of size "x" (currently x = 1.5, 2, 3, 6)

WAN   Wide Area Network
WWW   World Wide Web (IETF)

μs   microsecond (10^{-6} seconds)

ATM Standards and Specifications Reference

This appendix provides tables of the approved standards numbers and titles along with brief descriptions for ATM-related standards and specifications produced by the ITU-T, ANSI, ETSI, the ATM Forum and the IETF. See Chapter 10 for background and contact information for these important ATM-related standards and specifications.

ITU-T Standards

The ITU-T defines ATM as a component of an overarching vision called Broadband ISDN (B-ISDN). The ITU-T organizes recommendations covering various aspects of B-ISDN in a set of alphabetically identified series. The following series are relevant to ATM (B-ISDN).

- Series E Telephone network and ISDN
- Series F Non-telephone telecommunication services
- Series G Transmission systems and media
- Series H Transmission of non-telephone signals
- Series I Integrated services digital network
- Series M Maintenance: international transmission systems, telephone circuits, telegraphy, facsimile and leased circuits
- Series Q Switching and signalling

The following sections list the Recommendations related to ATM and B-ISDN. The text also summarizes the content and relationship of many of the key standards.

E Series — Network and Service Operation

The Series E Recommendations cover overall network operation, telephone service, service operation, and human factors. Table B.1 lists the ITU-T E series recommendations numbers and titles related to B-ISDN, along with their date of publication or latest update.

Table B.1 ITU-T B-ISDN E Series Recommendations

Number	Title	Date
E.164	Numbering plan for the ISDN era	08/91
E.177	B-ISDN routing	10/96
E.191	B-ISDN numbering and addressing	10/96
E.716	User demand modelling in Broadband-ISDN	10/96

Recommendation E.164, Numbering Plan for the ISDN era, describes a worldwide numbering plan based upon the pre-ISDN numbering plan, which is the worldwide telephone numbering plan. Therefore, the existing international telephone numbering plan is a subset of the full ISDN numbering plan.

Recommendation E.177, B-ISDN Routing, states routing principles in B-ISDN, in particular, it provides guidance on incorporating service requirements into the network routing process.

Recommendation E.191, B-ISDN Numbering and Addressing, provides gives guidance on the different use of numbers and addresses in B-ISDN.

Recommendation E.716, User demand modelling in Broadband-ISDN, characterizes user demand at the UNI, focusing on B-ISDN specific aspects involving ATM. E.716 points to Recommendation E.711 (user demand modelling in ISDN) for aspects in common with N-ISDN.

F Series — Non-Telephony Telecommunication Services

Series F Recommendations deal with telecommunication services other than telephone-based services. Table B.2 lists the ITU-T F series recommendations numbers and titles related to B-ISDN, along with their date of publication or latest update.

Table B.2 ITU-T B-ISDN F Series Recommendations

Number	Title	Date
F.732	Multimedia conference services in the B-ISDN	10/96
F.811	Broadband connection-oriented bearer service	3/93
F.812	Broadband connectionless data bearer service	3/93
F.813	Virtual path service for reserved and permanent communications	8/95

Recommendation F.732, Multimedia conference services in the B-ISDN, defines characteristics involved in combined voice, video and data media in a conferencing environment.

Recommendation F.811, Broadband connection-oriented bearer service, describes information flows and configurations for point-to-point, multipoint, broadcast, and multicast. It lists the attributes for the following subcategories:

- A - Constant bit-rate (CBR)
- B - Variable bit rate (VBR) with timing
- C - VBR without timing
  - ⇒ C1 - emulation of packet mode bearer service
  - ⇒ C2 - emulation of frame mode bearer services
  - ⇒ C3 - others
- X - ATM adaptation layer defined by user

Recommendation F.812, Broadband connectionless data bearer service, lists the attributes for the connectionless data bearer service, referred to as subcategory D in the terminology of F.811.

Recommendation F.813, Virtual path service for reserved and permanent communications, defines stage one of the broadband Virtual Path service for Reserved and Permanent Communications (VPRPC) for public B-ISDNs.

G Series — Transmission Systems, Networks and Media

Series G Recommendations deal with transmission systems and media, digital systems and networks. Table B.3 lists the ITU-T G series recommendations numbers and titles related to B-ISDN, along with their date of publication or latest update.

Table B.3 ITU-T B-ISDN G Series Recommendations

Number	Title	Date
G.702	Digital Hierarchy Bit Rates	6/90
G.703	Physical/electrical characteristics of hierarchical digital interfaces	9/91
G.704	Synchronous frame structures used at 1544, 6312, 2048, 8488 and 44736 kbit/s hierarchical levels	11/95
G.707	Network node interface for the synchronous digital hierarchy (SDH)	8/93
G.804	ATM cell mapping into plesiochronous digital hierarchy (PDH)	11/93
G.805	Generic functional architecture of transport networks	11/95

Recommendation G.702, Digital Hierarchy Bit Rates, specifies the bit rates in Table B.4 for the Plesiochronous Digital Hierarchy (PDH):

Table B.4 ITU-T Digital Hierarchy Levels

Digital Hierarchy Level	first level of 1544 kbps	first level of 2048 kbps
	64	64
1	1 544	2 048
2	6 312	8 448
3	32 064 and 44 736	34 368
4	97 728	139 264

Recommendation G.703, Physical/electrical characteristics of hierarchical digital interfaces, defines the following physical and electrical characteristics for the G.702 transmission bit rates. This includes the following parameters: bit rate, impedance, peak-to-peak voltage, rise and fall times, jitter and pulse masks.

Recommendation G.704, Synchronous frame structures used at 1544, 6312, 2048, 8488 and 44736 kbit/s hierarchical levels, specifies characteristics of nodal interfaces in integrated digital telephony and ISDN networks It defines framing structures, such as frame length, frame alignment patterns, Cyclic Redundancy Check (CRC) procedures and alarming. The standard also specifies 64 kbps channel mutliplexing and basic telephony signaling support.

Recommendation G.707, Network node interface for the synchronous digital hierarchy (SDH), defines signals in the Synchronous Digital Hierarchy (SDH) at the rates in the following table in terms of bit rate, frames structure, PDH mappings, ATM mappings and functions for the many overhead fields.

Recommendation G.804, ATM Cell mapping into plesiochronous digital hierarchy (PDH), defines procedures for transporting ATM cells over PDH networks operating at the 1544 and 2044 kbps based rate hierarchies defined in G.702. The standard defines ATM cell mapping in terms of cell rate adaptation, header error control, cell delineation, cell header verification and extraction, along with physical layer operation, administration, and maintenance.

Recommendation G.805, Generic functional architecture of transport networks, employs a technology independent viewpoint to define a harmonized set of functional architecture recommendations for transporting ATM, SDH and PDH networks within a common framework for management, performance analysis and equipment specification.

H Series — Non-Telephony Signal Transmission

H Series Recommendations cover line transmission of non-telephone signals. Table B.5 lists the ITU-T H series recommendations numbers and titles related to B-ISDN, along with their date of publication or latest update.

Table B.5 ITU-T B-ISDN H Series Recommendations

Number	Title	Date
H.222.1	Multimedia Multiplex and synchronization for AV communications in ATM environment	3/96
H.245	Control protocol for multimedia communication	3/96
H.310	Broadband and audiovisual communication systems and terminals	3/96
H.321	Adaptation of H.320 visual telephone terminals to B-ISDN environments	3/96

Recommendation H.222.1, Multimedia Multiplex and synchronization for AV communications in ATM environment, specifies a Program and Transport

Stream by selection of coding elements from the generic H.222.0 specifications, adding specific items for use in ATM environments as needed.

Recommendation H.245, Control protocol for multimedia communication, specifies the terminal information message syntax, semantics and procedures for in-band negotiation during connection establishment as well as during a multimedia call. These messages define transmit/receive capabilities, as well as receiving end defined mode preference, logical channel signaling, along with Control and Indication communication. The standard specifies reliable signaling procedures for audiovisual and data communication.

Recommendation H.310, Broadband and audiovisual communication systems and terminals, applies selected aspects of the H.200/AV.100-Series Recommendations to the B-ISDN environment. The standard defines support for both uni-directional and bi-directional broadband audiovisual terminals. H.310 classifies terminals into different types based upon a set of audiovisual, network adaptation, and signaling capabilities that support a wide range of conversational and distributed applications and services.

Recommendation H.321, Adaptation of H.320 visual telephone terminals to B-ISDN environments, specifies adaptation of narrow-band visual telephone terminals defined in Recommendation H.320 to B-ISDN environments. This approach achieves not only B-ISDN visual telephone interoperability, but also enables interoperability with existing H.320 N-ISDN visual telephone terminals.

I Series — Integrated Services Digital Networks (ISDN)

Series I Recommendations cover Integrated Services Digital Networks (ISDN), including B-ISDN. Table B.6 lists the ITU-T I series recommendations numbers and titles related to B-ISDN, along with their date of publication or latest update. There are a large number of I series recommendations, since this is the heart of first B-ISDN and now ATM. Many of these recommendations are older since they state the fundamental principles of B-ISDN and ATM, which have not changed.

Recommendation I.113, Vocabulary for B-ISDN, provides a glossary of terms and acronyms used in B-ISDN and ATM.

Recommendation I.121, Broadband aspects of ISDN, introduces other foundational B-ISDN related recommendations, states basic B-ISDN principles, and outlines an approach for evolving networks from telephony and ISDN to support advanced services and applications.

Recommendation I.150, B-ISDN asynchronous transfer mode functional characteristics, defines the ATM layer in terms of a connection-oriented, cell-based service using a two-level, virtual path and virtual channel identifier in the cell headers. It defines the concept of User-Network Interface (UNI) and the Network-Node interface (NNI). The standard also defines signaling, cell level multiplexing, per virtual connection Quality of Service (QoS) and Generic Flow Control (GFC)

Table B.6 ITU-T B-ISDN I Series Recommendations

Number	Title	Date
I.113	Vocabulary of terms for broadband aspects of ISDN	11/93
I.121	Broadband aspects of ISDN	04/91
I.150	B-ISDN asynchronous transfer mode functional characteristics	11/95
I.211	B-ISDN service aspects	03/93
I.311	B-ISDN general network aspects	03/93
I.321	B-ISDN protocol reference model and its application	04/91
I.326	Functional architecture of transport networks based on ATM	11/95
I.327	B-ISDN functional architecture	03/93
I.331	Numbering plan for the ISDN era (see E.164)	08/91
I.350	General aspects of quality of service and network performance in digital networks, including ISDNs	03/93
I.353	Reference events for defining ISDN and B-ISDN performance parameters	08/96
I.356	B-ISDN ATM layer cell transfer performance	11/93
I.357	B-ISDN semi-permanent connection availability	08/96
I.361	B-ISDN ATM layer specification	11/95
I.362	B-ISDN ATM Adaptation Layer (AAL) functional description	03/93
I.363	B-ISDN ATM adaptation layer (AAL) specification	03/93
I.364	Support of the broadband connectionless data bearer service by the B-ISDN	11/95
I.365.1	Frame relaying service specific convergence sublayer (FR-SSCS)	11/93
I.365.2	B-ISDN ATM adaptation layer sublayers: service specific coordination function to provide the connection oriented network service	11/95
I.365.3	B-ISDN ATM adaptation layer sublayers: service specific coordination function to provide the connection-oriented transport service	11/95
I.371	Traffic control and congestion control in B-ISDN	03/93
I.374	Framework Recommendation on "Network capabilities to support multimedia services"	03/93
I.413	B-ISDN user-network interface	03/93
I.414	Overview of Recommendations on layer 1 for ISDN and B-ISDN customer accesses	03/93
I.432	B-ISDN user-network interface - Physical layer specification	03/93
I.555	Frame relaying bearer service interworking	11/93
I.580	General arrangements for interworking between B-ISDN and 64 kbit/s based ISDN	11/95
I.610	B-ISDN operation and maintenance principles and functions	11/95
I.731	Types and general characteristics of ATM equipment	03/96
I.732	Functional characteristics of ATM equipment	03/96
I.751	Asynchronous transfer mode management of the network element	03/96

Recommendation I.211, B-ISDN service aspects, provides guidelines for the B-ISDN services supporting video, audio, data, graphics, and still images in the following modes: conversational, messaging services, retrieval services, and distribution services. It presents example applications along with attributes such as bit rate, QoS support, service level synchronization, signaling, interworking, video coding, and source characteristics

Recommendation I.311, B-ISDN general network aspects, peels back the multilayered ATM onion and begins detailing key ATM concepts, such as physical and ATM sublayers and a link by link construction of end-to-end Virtual Path (VP) and Virtual Channel (VC) connections. It defines the notions of VC switching and VP cross-connects using several examples. The standard then covers the control and management of B-ISDN, including traffic management, relationships to physical network management architecture and general principles of signaling.

Recommendation I.321, B-ISDN protocol reference model and its application, presents the three-dimensional ISDN protocol cube introduced in Part 1, comprised of the user, control and management planes. It also relates the B-ISDN model to the OSI reference model defined in X.200. The standard introduces the functions of the B-ISDN layers and sublayers including: the physical layer, the ATM layer, the ATM Adaptation Layer (AAL).

Recommendation I.326, Functional architecture of transport networks based on ATM, describes ATM transport assembly using the architecture defined in G.805. ATM transport assembly consists of the VCC layer network, VCC to VPC adaptation, VPC layer network, and the VPC to transmission path adaptation.

Recommendation I.327, B-ISDN functional architecture, complements I.324, ISDN Functional Architecture. First, it envisions a generic B-ISDN connection composed of terminal equipment, customer networks, and a public network before defining relationships to ISDN and connectionless services. It further defines signaling controlled VCC switching and network management controlled VPC cross-connects interconnected by physical links as the basic ATM connection model.

Recommendation I.350, General Aspects of Quality of Service and Network Performance in Digital Networks, including ISDN, defines the terms Quality of Service (QoS) as the user's perception and Network Performance (NP) as the network operator's observation. It defines specific performance parameters in terms of a number of generic categories.

Recommendation I.356, B-ISDN ATM layer cell transfer performance, defines the ATM specific measures of speed, accuracy, and dependability performance parameters defined in I.350. These parameters apply to end-to-end ATM connections and the links from which networks construct them. The standard defines methods to measure these parameters in terms of ATM cell transfer reference events observable at physical interfaces.

Recommendation I.357, B-ISDN semi-permanent connection availability, defines network performance parameters, objectives, and measurement methods for the availability of B-ISDN ATM semi-permanent connections. The specified parameters and objectives apply to an international ATM semi-permanent connection reference configuration. These worst-case values provide guidelines to network planners by bounding network impairments, including congestion, equipment failures, and transmission errors. The standard considers both service and network perspectives in defining a state machine model for availability. It also defines an availability estimation procedure using sampling techniques.

Recommendation I.361, B-ISDN ATM layer specification, defines the detailed bits and bytes of the ATM cell format, what they mean, and how they are used. This includes the detailed cell structure, field coding, and protocol procedures. It details the User-Network Interface (UNI) and the Network-Node Interface (NNI) cell formats. Chapter 3 in Part 1 summarized this information.

Recommendation I.362, B-ISDN ATM adaptation layer (AAL) functional description, defines basic principles and sublayering of the AAL layer. It also defines service classification in terms of constant or variable bit rate, timing transfer requirement, and whether the service is connection-oriented or connectionless. It describes the ATM adaptation layer (AAL) in terms of its sublayers; the segmentation and reassembly sublayer (SAR) and the convergence sublayer (CS).

Recommendation I.363, B-ISDN ATM adaptation layer (AAL) specification, describes the interaction between the AAL layer and the next higher layer, interactions between the AAL and the ATM layer, and the AAL peer-to-peer operations. It defines the following AAL types:

- Type 1 for transfer of Service Data Units (SDU) requiring constant bit rate, transfer of timing, along with an indication of lost or errored information
- Type 2 for transfer of SDUs requiring a variable source bit rate, transfer of timing information, as well as indication of lost or errored information
- Type 3/4 for transfer of SDU from one AAL user to one or more AAL users over the same VCC
- Type 5 for transfer of the AAL Service Data Units (AAL-SDU) from one user to one other AAL user through the ATM network over the same VCC

Note that the ITU-T plans to replace this recommendation when it approves draft recommendations I.363.1, I.353.3, and I.363.5. which will define the details of AAL1, AAL3/4 and AAL5 respectively.

Recommendation I.364, Support of broadband connectionless data service on B-ISDN, describes the connectionless network access protocol (CLNAP) for the UNI and compares it with the MAC sub-layer service described in the ISO/IEC 10039 standard. It employs the Type 3/4 AAL unassured service to support transfer of variable size data units from a source to one or more destinations.

Recommendation I.365.1, Frame relaying service specific convergence sublayer (FR-SSCS), defines the specifics of FR/ATM network interworking

Recommendation I.413, B-ISDN user-network interface, describes the reference configuration using points labeled by letters (e.g., RB, SB, TB and UB) between functional elements. Some examples place the functional blocks and reference points in representative user-network deployments. It describes physical layer information flows, interface functions, and OAM issues in terms of the reference configuration.

Recommendation I.414, Overview of Recommendations on layer 1 for ISDN and B-ISDN customer accesses, provides an overview of other ITU Recommendations applicable to ISDN (narrowband) access and B-ISDN access.

Recommendation I.432.1, B-ISDN User-Network Interface: physical layer specification - General characteristics, defines the detailed mapping of ATM cells into the Synchronous Digital Hierarchy (SDH) structure, details generation of the ATM Header Error Control (HEC), and presents analyses of bit error impacts on undetected header errors and cell delineation time. This recommendation should be used with each of the following recently approved recommendations detail the characteristics for various speeds and media. Each part presents the required functions in the Physical Media Dependent and Transmission Convergence sublayers.

Recommendation I.432.2, B-ISDN User-Network Interface: physical layer specification for 155 520 kbit/s, specifies the physical layer for transport of ATM cells at bit rates of 155.520 and 622.080 Mbps over coaxial cable and optical fiber interfaces at the B-ISDN UNI.

Recommendation I.432.3, B-ISDN User-Network Interface: physical layer specification for 1 544 and 2 048 kbit/s, specifies physical layer characteristics for transporting ATM cells using existing ISDN Primary Rate Interfaces (PRI) operating at 1.544 and 2.048 Mbps at the B-ISDN UNI.

Recommendation I.432.4, B-ISDN User-Network Interface: physical layer specification for 51,840 kbit/s, specifies the physical layer for transport of ATM cells at a bit rate of 51.840 Mbps over category 3 Unshielded Twisted Pair (UTP) cabling, commonly used in existing building wiring, over distances of up to 100 meters at the BISDN UNI.

Recommendation I.580, General arrangements for interworking between B-ISDN and 64 kbit/s based ISDN, describes the service capabilities for the following configurations anticipated during the period as ISDN networks evolve to B-ISDN:

- connection of 64 kbps networks via a B-ISDN
- connection of B-ISDNs via a 64 kbps network
- interworking between a B-ISDN and a 64 kbps network.

The standard defines service characteristics and requirements for each major category of services in the above configurations. It also describes two possible interworking scenarios for providing 64 kbps based services between N-ISDN and B-ISDN users.

Recommendation I.610, B-ISDN operation and maintenance principles and functions, covers principles and functions required for Operation, Administration, and Maintenance (OAM) of primarily the ATM layer for permanent, semi-permanent, reserved and on-demand virtual connections. It defines bi-directional F4 and F5 OAM flows for VPCs and VCCs respectively using ATM cell header fields, so that OAM cells follow the same route as user cells. Furthermore, and F4/F5 OAM flows may be either end-to-end flow for use by ATM end systems only, or a segment flow for use by intermediate ATM systems only. The standard defines the format and procedures using different OAM cell types for fault management, performance monitoring, system management and procedures for activation/deactivation of monitoring functions.

Recommendation I.731, Types and general characteristics of ATM equipment, applies the elements of the B-ISDN protocol cube from I.321 to specify ATM Network Elements (NE) using the modelling methodology in Recommendations G.805 and G.806. Specifically, the standard describes ATM NE functional blocks in terms of User Plane, Control Plane, Layer Management, and Plane Management functions. It provides guidelines for the classification of ATM equipment types through description and example. It lists generic performance requirements, Quality of Service (QoS) aspects, along with timing and synchronization needs applicable to ATM equipment, with reference to relevant I-series Recommendations for further detail.

Recommendation I.732, Functional characteristics of ATM equipment, further details ATM Network Element functional requirements based on the general functional architecture in companion Recommendation I.731. The functional requirements are detailed sufficiently to achieve interoperability between different ATM NEs. This specification details information transfer between elements of the Transfer Layer and the corresponding Layer Management elements, as well as the Coordination function of Plane Management which provides an interface between Layer Management functions and the overall NE System management functions specified in Recommendation I.751.

Recommendation I.751, ATM management of the network element, defines plane management functions in terms of ATM specific managed object classes and properties exchanged across interfaces defined for the envisioned Telecommunications Management Network (TMN) architecture defined in Recommendation in M.3010.

J Series — Television and Sound Program Transmission

J Series Recommendations cover Transmission of sound programs and television signals. Table B.7 lists the ITU-T J series recommendations numbers and titles related to B-ISDN, along with their date of publication or latest update.

Table B.7 ITU-T B-ISDN J Series Recommendations

Number	Title	Date
J.82	Transport of MPEG-2 constant bit rate television signals in B-ISDN	7/96

Recommendation J.82, Transport of MPEG-2 constant bit rate television signals in B-ISDN, defines the transport of television signals using ATM via the MPEG-2 standard conveyed, either at a constant bit rate using AAL1.

M Series — Maintenance and Management Aspects

Series M Recommendations cover maintenance: transmission systems, telephone circuits, telegraphy, facsimile and other management aspects. Table B.8 lists the ITU-T M series recommendations numbers and titles related to B-ISDN, along with their date of publication or latest update.

Table B.8 ITU-T B-ISDN M Series Recommendations

Number	Title	Date
M.3010	Principles for a Telecommunications management network	05/96
M.3207.1	TMN management service: maintenance aspects of B-ISDN management	05/96
M.3610	Principles for applying the TMN concept to the management of B-ISDN	05/96

Recommendation M.3010, Principles of Telecommunications Management Network (TMN), defines a generic model, functions and interfaces necessary to manage general telecommunications networks. The recommendation also describes devices which comprise a TMN along with interfaces each device could support.

Recommendation M.3207.1, TMN Management Service for the maintenance aspects of B-ISDN Management, uses Guideline for the Definition of TMN Management Services (GDMS) to detail operations and maintenance functions for B-ISDN.

Recommendation M.3610, Principles for applying the TMN concept to the management of B-ISDN, presents terminology definitions, maintenance principles for B-ISDN, and reference models for the management of B-ISDN.

Q Series — Switching and Signalling

Series Q Recommendations define switching and signaling (note that the ITU-T uses two "ls" in its spelling of signaling). Table B.9 lists the ITU-T Q series recommendations numbers and titles related to B-ISDN, along with their date of publication or latest update. The B-ISDN Digital Subscriber Signaling System 2 (abbreviated DSS2) is so named because the (narrowband) ISDN signaling system is called DSS1. The ITU-T has laid out development of signaling standards in capability sets according to three releases. The majority of the current standards focus on capability set 1.

Recommendation Q.704, Signaling System No. 7 - Signaling network functions and messages, specifies the NNI Message Transfer Part 3 (MTP3) signaling message formats and procedures.

Recommendation Q.2100, B-ISDN signaling ATM adaptation layer (SAAL) overview description, summarizes the components and functions that comprise the overall signaling AAL (SAAL) and provides a guide to other recommendations.

Table B.9 ITU-T B-ISDN Q Series Recommendations

Number	Title	Date
Q.704	Signalling System No. 7 - Signalling network functions and messages	3/93
Q.2010	Broadband integrated services digital network overview - signalling capability set, release 1	02/95
Q.2100	B-ISDN signalling ATM adaptation layer (SAAL) overview description	07/94
Q.2110	B-ISDN ATM adaptation layer - service specifid connection oriented protocol (SSCOP)	07/94
Q.2120	B-ISDN meta-signalling protocol	02/95
Q.2130	B-ISDN signalling ATM adaptation layer - service specific coordination function for support of signalling at the user network interface (SSFC At UNI)	07/94
Q.2140	B-ISDN ATM adaptation layer - service specific coordination function for signalling at the network node interface (SSCF at NNI)	02/95
Q.2144	B-ISDN Signalling ATM adaptation layer (SAAL) - Layer management for the SAAL at the network node interface (NNI)	10/95
Q.2610	B-ISDN- usage of cause and location in B-ISDN user part and DSS 2	02/95
Q.2650	Broadband-ISDN, interworking between Signalling System No. 7 broadband ISDN user part (B-ISUP) and Digital Subscriber Signalling System No. 2 (DSS 2)	02/95
Q.2660	B-ISDN - interworking between Signalling System No. 7 - Broadband ISDN user part (B-ISUP) and narrow-band ISDN user part (N-ISUP)	02/95
Q.2730	B-ISDN signalling system no. 7 B-ISDN user part (B-ISUP) - supplementary services	02/95
Q.2761	B-ISDN -Functional description of the B-ISDN user part (B-ISUP) of signalling system No. 7	02/95
Q.2762	B-ISDN - General functions of messages and signals of the B-ISDN user part (B-ISUP) of Signalling System No. 7	02/95
Q.2763	B-ISDN - Signaling System No. 7 B-ISDN user part (B-ISUP) - formats and codes	02/95
Q.2764	B-ISDN - Signalling system no. 7 B-ISDN user part (B-ISUP) - Basic call procedures	02/95
Q.2931	B-ISDN - Digital subscriber signalling system no. 2 (DSS 2) - User-network interface (UNI) - Layer 3 specification for basic call control	02/95
Q.2951	(Clauses 1, 2, 3, 4, 5, 6 and 8) - Stage 3 description for number identification supplementary services using B-ISDN Digital Subscriber Signalling System No. 2 (DSS 2) - Basic call	02/95
Q.2957	Stage 3 description for additional information transfer supplementary services using B-ISDN digital subscriber signalling system no. 2 (DSS 2) - Basic call; Clause 1 - User-to-user signalling (UUS)	02/95
Q.2961	B-ISDN - Digital subscriber signalling system no.2 (DSS 2) - additional traffic parameters	10/95
Q.2962	Digital subscriber signalling system No. 2 - Connection characteristics negotiation during call/connection establishment phase	07/96
Q.2963.1	Peak cell rate modification by the connection owner	07/96
Q.2971	B-ISDN - DSS 2 - User-Network Interface Layer 3 Specification – User-Network Interface Layer 3 Specification for Point-To-Multipoint Call/Connection Control	10/95

Recommendation Q.2110, B-ISDN ATM adaptation layer - Service specific connection oriented protocol (SSCOP), details the SSCOP common part of the Service Specific Convergence Sublayer (SSCS) of the SAAL. SSCOP provides a generic, connection oriented, reliable data transfer service for use a the UNI via the upper level part of the SSCS, the Service Specific Coordination Function (SSCF).

Recommendation Q.2119, B-ISDN ATM Adaptation Layer (AAL) protocols - Convergence function for SSCOP above the frame relay core service, specifies a mapping function enabling B-ISDN data communications utilizing SSCOP to utilize an HDLC-based data link capability instead. This standard exploits the fact that HDLC supports the core functions of AAL5.

Recommendation Q.2120, B-ISDN meta-signaling protocol, specifies the protocol and procedures for assignment, status checking and removal of signaling virtual channels. These meta-signaling procedures address the multiple access UNI environment where B-ISDN terminals must dynamically allocate signaling channels to the network. Meta-signaling establishes signaling links so that the signaling protocol can then operate between a user and the network.

Recommendation Q.2130, B-ISDN signaling ATM adaptation layer - Service specific coordination function for support of signaling at the user-network interface (SSCF at UNI), operates on top of SSCOP (I.2110) at the User-to-Network Interface (UNI). This UNI SSCF performs a coordination function between the layer 3 signaling service user (Q.2931) and SSCOP by defining the mapping of layer 3 primitives to and from SSCOP signals.

Recommendation Q.2140, B-ISDN ATM adaptation layer - Service specific coordination function for signaling at the network node interface (SSCF at NNI), operates on top of SSCOP (I.2110) at the Network-Node Interface (NNI). This NNI SSCF performs a coordination function between the layer 3 signaling service user (Q.704) and SSCOP by defining the mapping of layer 3 primitives to and from SSCOP signals.

Recommendation Q.2144, B-ISDN signaling ATM adaptation layer (SAAL) - Layer management for the SAAL at the network node interface (NNI), specifies layer management functions for the SAAL at the Network Node Interface (NNI) which perform a coordination function and error monitoring between the systems management function and the SAAL sublayers.

Recommendation Q.2210, Message Transfer Part Level 3 Functions and Messages Using the Services of ITU-T Recommendation Q.2140, specifies the control of signaling links that provide the services of ITU-T Recommendation Q.2140.

Recommendation Q.2610, Broadband Integrated Services Digital Network (B-ISDN) usage of cause and location in B-ISDN user part and DSS 2, describes the interworking between the DSS 2 access interface protocol and the Broadband ISDN User Part (BISUP) protocol as part of capability set 1.

Recommendation Q.2650, Broadband-ISDN, interworking between Signalling System No. 7 broadband ISDN User Part (B-ISUP) and digital subscriber Signalling System No. 2 (DSS 2), further details interworking between the DSS 2 access interface protocol and the Broadband ISDN User

Part (BISUP) protocol as part of capability set 1. The standard uses mapping tables and diagrams to define interworking between the UNI (DDS2) and NNI (BISUP) protocols for basic call establishment and release.

Recommendation Q.2660, B-ISDN - Interworking between Signalling System No. 7 - Broadband ISDN User Part (B-ISUP) and Narrow-band ISDN User Part (N-ISUP), specifies interworking between B-ISDN and N-ISDN NNI protocols for basic calls as well as supplementary services. Message flow diagrams depict interworking in conjunction with tables that define parameter mappings. The standard focuses on the case where a B-ISDN network acts as a transit network between N-ISDN networks.

Recommendation Q.2721.1 B-ISDN User Part (CS2.1) Overview of the B-ISDN Network Node Interface (NNI) Signalling Capability Set 2, Step 1, summarizes B-ISUP capability set 2 which adds the following:

- Support for point-to-multipoint calls
- Support for additional traffic parameters
- Look-ahead capability
- Negotiation of traffic characteristics during the call set-up
- Modification of traffic characteristics during the active phase of the call
- ATM End System Address (AESA) support
- Call prioritization
- Network call correlation ID
- Frame relay interworking

Recommendation Q.2722.1, Network node interface (NNI) specification for point-to-multipoint call/connection control, defines BISUP extensions to support establishment and release of point-to-multipoint calls by the connection root. This achieves alignment with the ATM Forum UNI 3.1 root controlled point-to-multipoint specification at the NNI.

Recommendation Q.2723.1, B-ISDN Extensions to SS7 B-ISDN User Part - Additional traffic parameters for sustainable cell rate (SCR) and Quality of Service (QOS), specifies BISUP extensions supporting additional traffic and QoS parameters in support of VBR services and the additional capabilities defined in Recommendation I.371

Recommendation Q.2724.1, B-ISDN User Part (CS2.1), Look-ahead without state change for the network node interface (NNI), specifies BISUP extensions that allows a network to look ahead to determine the busy or idle status of the terminating side prior to attempting establishment of a connection across the network.

Recommendation Q.2725, B-ISDN User Part (CS2.1) contains individual clauses as follows:

Q.2725.1, specifies BISUP extensions in support of user negotiation of traffic parameters during connection setup via either a user specified alternative or minimum ATM cell rate.

Q.2725.2, Modification procedures, specifies BISUP extensions allowing modification of connection characteristics (e.g., the peak cell rate) during active calls.

Recommendation Q.2726, B-ISDN User Part (CS2.1), contains individual clauses as follows:

Q.2726.1, ATM end system address (AESA), specifies formats and procedures for carrying calling, called, and connected party AESAs over the BISUP protocol using mapping tables for messages and information elements.

Q.2726.2, Call priority, specifies BISUP extensions in support of the UNI call priority service.

Q.2726.3, Network call correlation identifier, specifies BISUP extensions of a network-generated, unique call correlation identifier that enables networks to correlate records associated with a single call after the call has completed.

Recommendation Q.2727, B-ISDN User Part, Support of Frame Relay, specifies BISUP message and procedure usage supporting switched frame relay connections over ATM VCCs.

Recommendation Q.2730, Broadband integrated services digital network (B-ISDN) - Signalling System No. 7 B-ISDN User Part (B-ISUP) - Supplementary services, supports CS 1 at the NNI.

Recommendation Q.2761, Broadband integrated services digital network (B-ISDN) - Functional description of the B-ISDN user part (B-ISUP) of signaling system No. 7, gives an overview of NNI signaling capabilities supporting basic bearer services and supplementary services for CS 1.

Recommendation Q.2762, Broadband integrated services digital network (B-ISDN) - General Functions of messages and signals of the B-ISDN user part (B-ISUP) of Signalling System No. 7, describes signaling information elements and general functions at the NNI.

Recommendation Q.2763, Broadband integrated services digital network (B-ISDN) - Signalling system No. 7 B-ISDN user part (B-ISUP) - Formats and codes, specifies the detailed formats and parameter coding of NNI signaling messages.

Recommendation Q.2764, B-ISDN - Signalling system No. 7 B-ISDN user part (B-ISUP) - Basic call procedures, details procedures for basic call establishment and release, as well as maintenance procedures at the signaling NNI.

Recommendation Q.2931, B-ISDN - Digital subscriber signaling system No. 2 (DSS 2) - User-network interface (UNI) layer 3 specification for basic call/connection control, specifies procedures for establishing, maintaining, and clearing connections at the B-ISDN user network interface, in terms of messages exchanged. It specifies essential features, procedures and messages required for call and connection control.

Recommendation Q.2932.1 B-ISDN DSS2 Generic functional protocol - Core functions, defines a means for exchanging ROSE components on behalf of signaling application in peer entities for the support of supplementary services or protocol support for other features, such as Look Ahead, Status request or Local/remote interrogation in association with existing calls and bearer capabilities.

Q.2933, B-ISDN DSS2 - Signalling specification for frame relay service, defines the optional DSS2 capability for support of frame relay SVC service at the UNI.   The procedures support only a single step frame relay call/connection control.

Recommendation Q.2951, B-ISDN - Stage 3 description for number identification supplementary services using B-ISDN digital subscriber signaling system No. 2 (DSS 2) - Basic Call, defines UNI support for a number of supplementary services via the following individual clauses.

Q.2951.1 - Direct-Dialling-In (DDI), enables a user to directly call another user on a B-ISDN private branch exchange or other private system, without attendant intervention, based exclusively on the use of the ISDN number.

Q.2951.2 - Multiple Subscriber Number (MSN), allows assignment of multiple ISDN numbers to a single access line

Q.2951.3 - Calling Line Identification Presentation (CLIP), provides the calling party's ISDN number, possibly with sub-address information, to the called party.

Q.2951.4 - Calling Line Identification Restriction (CLIR), allows the calling party to restrict presentation of the calling party's ISDN number and sub-address to the called party.

Q.2951.5 - Connected Line Identification Presentation (COLP), allows the calling party to view the connected party's ISDN number, possibly with sub-address information.

Q.2951.6 - Connected Line Identification Restriction (COLR), enables the called party to restrict presentation of the connected party's ISDN number and sub-address to the calling party.

Q.2951.8 - Sub-addressing (SUB), allows the called user to expand his addressing capacity beyond the one given by the ISDN number.

Recommendation Q.2957, B-ISDN DSS2 - Stage 3 description for additional information transfer supplementary services - Basic call, contains the following individual clauses for each service:

Clause 1 - User-to-user signaling (UUS) allows a B-ISDN user to send and receive a limited amount of information to/from another B-ISDN user over the signaling virtual channel in association with a call/connection to another B-ISDN user.

Recommendation Q.2959, B-ISDN DSS2 - Call Priority, defines the a basic call and connection control at UNI, as a service provider option, that allows for preferential treatment of high priority calls during network congestion based upon the priority level assigned to the call by the user.

Recommendation Q.2961.1, B-ISDN DSS2 - Support of additional parameters, defines additional traffic and QoS parameters used for basic call and connection control at the UNI for a VBR bearer capability.

Recommendation Q.2962, B-ISDN DSS2 - Connection characteristics negotiation during call/connection establishment phase, defines negotiation at the UNI during the call/connection establishment phase using either an alternative set or minimum acceptable set of traffic parameters given in an additional information element in the setup message.

Recommendation Q.2963.1 B-ISDN DSS2 Connection modification - Peak cell rate modification by the connection owner, defines formats and procedures that allow a user to modify the traffic parameters (i.e., the peak cell rates) of an established point-to-point connection

Recommendation Q.2964.1 B-ISDN DSS2 Basic look-ahead, defines the protocol at the UNI that enables a network to determine the busy/idle status of an end user.

Recommendation Q.2971, Broadband integrated services digital network - Digital subscriber signaling system No. 2 - User-network interface layer 3 specification for point-to-multipoint call/connection control, supports point-to-multipoint switched virtual channel connections under control of the root at the UNI. The root may add or remove one party at a time. Additionally, a leaf party may independently initiate withdrawal from the call.

ANSI Standards

The American National Standards Institute (ANSI) has two committees defining ATM-related standards: T1 and X1.   This section lists these standards.

ANSI X1 ATM-Related Standards

ANSI Committee X1 specifies standards for computers and computer networking. This committee produced the following standard related to ATM:

X3.299-1997 : Information Technology - High-Performance Parallel Interface - Mapping to Asynchronous Transfer Mode (HIPPI-ATM)

ANSI T1 ATM-Related Standards

ANSI committee T1 adapts CCITT/ITU-T standards to the competitive telecommunications environment and the unique physical layer transmission requirements of North America. Table B.10 lists the standards approved to date.

Table B.10 ANSI T1 ATM-Related Standards

Standard	Title
T1.627-1993	Telecommunications - Broadband ISDN - ATM Layer Functionality and Specification
T1.511-1994	Telecommunications - B-ISDN ATM Layer Cell Transfer - Performance Parameters
T1.636-1994	Telecommunications - B-ISDN Signaling ATM Adaptation Layer - Overview Description
T1.662-1996	Telecommunications - Broadband ISDN - ATM End System Address for Calling and Called Party
T1.652-1996	Telecommunications - B-ISDN Signaling ATM Adaptation Layer - Layer Management for the SAAL at the NNI
T1.629-1993	Telecommunications - Broadband ISDN - ATM Adaptation Layer 3/4 Common Part Functions and Specification

T1.635-1994	Telecommunications - Broadband ISDN - ATM Adaptation Layer Type 5, Common Part Functions and Specification
T1.637-1994	Telecommunications - B-ISDN ATM Adaptation Layer - Service Specific Connection Oriented Protocol (SSCOP)
T1.630-1993	Telecommunications - Broadband ISDN - ATM Adaptation Layer for Constant Bit Rate Services Functionality and Specification
T1.646-1995	Telecommunications - Broadband ISDN - Physical Layer Specification for User-Network Interfaces Including DS1/ATM
T1.646a-1997	Telecommunications - Broadband ISDN - Physical Layer Specification for User-Network Interfaces Including DS1/ATM
T1.638-1994	Telecommunications - B-ISDN Signaling ATM Adaptation Layer - Service Specific Coordination Function for Support of Signaling at the User-to-Network Interface (SSCF at the UNI)
T1.645-1995	Telecommunications - B-ISDN Signaling ATM Adaptation Layer - Service Specific Coordination Function for Support of Signaling at the Network Node Interface (SSCF at the NNI)
T1.663-1996	Telecommunications - Broadband ISDN - Network Call Correlation Identifier
T1.624-1993	Telecommunications - Broadband ISDN User-Network Interfaces - Rates and Formats Specifications
T1.640-1996	Telecommunications - Broadband ISDN Network Node Interfaces and Inter-Network Interfaces - Rates and Formats Specifications
T1.644-1995	Telecommunications - Broadband ISDN - Meta-Signalling Protocol
T1.649-1995	Telecommunications - B-ISDN Cell Relay Service Description
T1.656-1995	Telecommunications - Broadband ISDN - Interworking between Signalling System No. 7 Broadband ISDN User Part (B-ISUP) and ISDN User Part (ISUP)
T1.657-1996	Telecommunications - Broadband ISDN - Interworking between Signalling System No. 7 Broadband ISDN User Part (B-ISUP) and Digital Subscriber Signalling System No. 2 (DSS2)
T1.664-1997	Telecommunications - Broadband ISDN - Point-to-Multipoint Call/Connection Control
T1.665-1997	Telecommunications - Broadband ISDN - Overview of ANSI B-ISDN NNI Signaling Capability Set 2, Step 1

ETSI Standards

ETSI organizes its work according to a number of Technical Committees (TCs) as follows:

TC SPS - Signalling, Protocols and Switching
TC NA - Network Aspects
TC TM - Transmission and Multiplexing
TC TMN - Telecommunications Management Network

These committees produce the following a number of different types of documents, ranging from full standards to informative reports and regulatory guidelines. We list the following standards document types along with a brief description from ETSI's web page:

- ETSI Standard, ES
- ETSI Technical Specification, TS
- European Standard (telecommunications series), EN (telecommunications series),
- European Telecommunication Standard (ETS):

Table B.11 lists the ETSI standards related to ATM approved to date.

Table B.11 ETSI Standards

ETSI Document	Date	Title
prEN 300 301-1	1997-10	Broadband Integrated Services Digital Network (B-ISDN); Traffic control and congestion control in B-ISDN; Conformance definitions for Available Bit Rate (ABT) and ATM Blocked Transfer (ABR) [ITU-T Recommendation I.371.1 (1997)]
prEN 300 443-1	1997-12	B-ISDN; Digital Subscriber Signalling System No. two (DSS2) protocol; B-ISDN user-network interface layer 3 specification for basic call/bearer control; Part 1: Protocol specification [ITU-T Recommendation Q.2931 (1995), modified]
prEN 301 003-1	1997-12	B-ISDN; Digital Subscriber Signalling System No. two (DSS2) protocol; Connection characteristics; Peak cell rate modification by the connection owner; Part 1: Protocol specification [ITU-T Recommendation Q.2963.1 (1996), modified]
prEN 301 003-2	1997-12	B-ISDN; Digital Subscriber Signalling System No. two (DSS2) protocol; Connection characteristics; Peak cell rate modification by the connection owner; Part 2: Protocol Implementation Conformance Statement (PICS) proforma specification
prEN 301 004-1	1997-12	B-ISDN; Signalling System No.7; Message Transfer Part (MTP) level 3 functions and messages to support international interconnection; Part 1: Protocol specification [ITU-T Recommendation Q.2210 (1996), modified]
prEN 301 067-1	1997-12	B-ISDN; Digital Subscriber Signalling System No. two (DSS2) protocol; Connection characteristics; Negotiation during call/connection establishment phase; Part 1: Protocol specification [ITU-T Recommendation Q.2962 (1996), modified]
prEN 301 068-1	1997-12	B-ISDN; Digital Subscriber Signalling System No. two (DSS2) protocol; Connection characteristics; ATM transfer capability and traffic parameter indication; Part 1: Protocol specification [ITU-T Recommendations Q.2961.1 (1995), Q.2961.2 to Q.2961.4 (1997), modified]
ETS 300 298-1	1996-10	B-ISDN; Asynchronous Transfer Mode (ATM); Part 1: B-ISDN ATM functional characteristics [ITU-T Recommendation I.150 (1995)]
ETS 300 298-2	1996-10	B-ISDN; Asynchronous Transfer Mode (ATM); Part 2: B-ISDN ATM layer specification [ITU-T Recommendation I.361 (1995)]

ETS 300 299	1997-06	B-ISDN; Cell based user network access for 155 520 kbit/s and 622 080 kbit/s; Physical layer interfaces for B-ISDN applications
ETS 300 300	1995-02	Broadband Integrated Services Digital Network (B-ISDN); Synchronous Digital Hierarchy (SDH) based user network access; Physical layer interfaces for B-ISDN applications
ETS 300 300	1997-04	B-ISDN; Synchronous Digital Hierarchy (SDH) based user network access; Physical layer User Network Interfaces (UNI) for 155 520 kbit/s and 622 080 kbit/s Asynchronous Transfer Mode (ATM) B-ISDN applications
ETS 300 349	1995-02	B-ISDN; Asynchronous Transfer Mode (ATM); Adaptation Layer (AAL) specification - type 3/4
ETS 300 354	1995-08	B-ISDN; B-ISDN Protocol Reference Model (PRM)
prETS 300 399-4	1995-04	Frame relay services; Part 4: B-ISDN; Frame relay bearer service; Service definition
ETS 300 404	1997-04	B-ISDN; B-ISDN Operation And Maintenance (OAM) principles and functions
ETS 300 428	1995-08	B-ISDN; Asynchronous Transfer Mode (ATM); Adaptation Layer (AAL) specification - type 5
ETS 300 436-1	1995-11	B-ISDN; Signalling ATM Adaptation Layer (SAAL); Service Specific Connection Oriented Protocol (SSCOP); Part 1: Protocol specification [ITU-T Recommendation Q.2110 (1995), modified]
prETS 300 436-3	1997-11	B-ISDN; Signalling ATM Adaptation Layer (SAAL); Service Specific Connection Oriented Protocol (SSCOP); Part 3: Test Suite Structure and Test Purposes (TSS&TP) specification
ETS 300 437-1	1995-11	B-ISDN; Signalling ATM Adaptation Layer (SAAL); Service Specific Co-ordination Function (SSCF) for support of signalling at the User-Network Interface (UNI); Part 1: Specification of SSCF at UNI [ITU-T Recommendation Q.2130 (1995), modified]
ETS 300 437-2	1996-08	B-ISDN; Signalling ATM Adaptation Layer (SAAL); Service Specific Co-ordination Function (SSCF) for support of signalling at the User-Network Interface (UNI); Part 2: Protocol Implementation Conformance Statement (PICS) proforma specification
ETS 300 438-1	1995-11	B-ISDN; Signalling ATM Adaptation Layer (SAAL); Service Specific Co-ordination Function (SSCF) for support of signalling at the Network Node Interface (NNI); Part 1: Specification of SSCF at NNI [ITU-T Recommendation Q.2140 (1995), modified]
ETS 300 38-1/C1	1996-06	B-ISDN; Signalling ATM Adaptation Layer (SAAL); Service Specific Co-ordination Function (SSCF) for support of signalling at the Network Node Interface (NNI); Part 1: Specification of SSCF at NNI [ITU-T Recommendation Q.2140 (1995), modified]
ETS 300 443-2	1997-05	B-ISDN; Digital Subscriber Signalling System No. two (DSS2) protocol; B-ISDN user-network interface layer 3 specification for basic call/bearer control; Part 2: Protocol Implementation Conformance Statement (PICS) proforma specification
ETS 300 455-1	1995-08	B-ISDN; Broadband Virtual Path Service (BVPS); Part 1: BVPS for Permanent communications (BVPS-P)
ETS 300 455-2	1995-08	B-ISDN; Broadband Virtual Path Service (BVPS); Part 2: BVPS for Reserved communications (BVPS-R)

ETS 300 486-1	1996-08	B-ISDN; Meta-signalling protocol; Part 1: Protocol specification [ITU-T Recommendation Q.2120 (1995), modified]
ETS 300 495	1997-05	B-ISDN; Signalling System No.7; Interworking between Broadband ISDN User Part (B-ISUP) and Digital Subscriber Signalling System No. two (DSS2) [ITU-T Recommendation Q.2650 (1995), modified]
ETS 300 496	1997-05	B-ISDN; Signalling System No.7; Interworking between Broadband ISDN User Part (B-ISUP) and narrowband ISDN User Part (ISUP) [ITU-T Recommendation Q.2660 (1995), modified]
ETS 300 647	1997-01	B-ISDN; Signalling ATM Adaptation Layer (SAAL); Layer Management for the SAAL at the Network Node Interface (NNI) [ITU-T Recommendation Q.2144 (1995), modified]
ETS 300 656	1997-05	B-ISDN; Signalling System No.7; B-ISDN User Part (B-ISUP) Capability Set 1 (CS1); Basic services [ITU-T Recommendations Q.2761 to Q.2764 (1995), modified]
ETS 300 657	1997-05	B-ISDN; Signalling System No.7; B-ISDN User Part (B-ISUP) Capability Set 1 (CS1); Supplementary services [ITU-T Recommendation Q.2730 (1995), modified]
ETS 300 662-1	1996-09	B-ISDN; Digital Subscriber Signalling System No. two (DSS2) protocol; Multiple Subscriber Number (MSN) supplementary service; Part 1: Protocol specification [ITU-T Recommendation Q.2951, clause 2 (1995), modified]
ETS 300 662-2	1996-09	B-ISDN; Digital Subscriber Signalling System No. two (DSS2) protocol; Multiple Subscriber Number (MSN) supplementary service; Part 2: rotocol Implementation Conformance Statement (PICS) proforma specification
ETS 300 663-1	1996-09	B-ISDN; Digital Subscriber Signalling System No. two (DSS2) protocol; Calling Line Identification Presentation (CLIP) supplementary service; Part 1:Protocol specification [ITU-T Recommendation Q.2951, clause 3 (1995), modified]
ETS 300 663-2	1996-09	B-ISDN; Digital Subscriber Signalling System No. two (DSS2) protocol; Calling Line Identification Presentation (CLIP) supplementary service; Part 2: Protocol Implementation Conformance Statement (PICS) proforma specification
ETS 300 664-1	1996-09	B-ISDN; Digital Subscriber Signalling System No. two (DSS2) protocol; Calling Line Identification Restriction (CLIR) supplementary service; Part 1: Protocol specification [ITU-T Recommendation Q.2951, clause 4 (1995), modified]
ETS 300 664-2	1996-09	B-ISDN; Digital Subscriber Signalling System No. two (DSS2) protocol; Calling Line Identification Restriction (CLIR) supplementary service; Part 2: Protocol Implementation Conformance Statement (PICS) proforma specification
ETS 300 665-1	1996-09	B-ISDN; Digital Subscriber Signalling System No. two (DSS2) protocol; Connected Line Identification Presentation (COLP) supplementary service; Part 1: Protocol specification [ITU-T Recommendation Q.2951, clause 5 (1995), modified]
ETS 300 665-2	1996-09	B-ISDN; Digital Subscriber Signalling System No. two (DSS2) protocol; Connected Line Identification Presentation (COLP) supplementary service; Part 2: Protocol Implementation Conformance Statement (PICS) proforma specification
ETS	1996-09	B-ISDN; Digital Subscriber Signalling System No. two (DSS2) protocol; Connected Line Identification Restriction (COLR)

300 666-1		supplementary service; Part 1: Protocol specification [ITU-T Recommendation Q.2951, clause 6 (1995), modified]
ETS 300 666-2	1996- 09	B-ISDN; Digital Subscriber Signalling System No. two (DSS2) protocol; Connected Line Identification Restriction (COLR) supplementary service; Part 2: Protocol Implementation Conformance Statement (PICS) proforma specification
ETS 300 667-1	1996- 09	B-ISDN; Digital Subscriber Signalling System No. two (DSS2) protocol; Subaddressing (SUB) supplementary service; Part 1: Protocol specification [ITU-T Recommendation Q.2951, clause 8 (1995), modified]
ETS 300 667-2	1996- 09	B-ISDN; Digital Subscriber Signalling System No. two (DSS2) protocol; Subaddressing (SUB) supplementary service; Part 2: Protocol Implementation Conformance Statement (PICS) proforma specification
ETS 300 668-1	1996- 09	B-ISDN; Digital Subscriber Signalling System No. two (DSS2) protocol; User-to-User Signalling (UUS) supplementary service; Part 1: Protocol specification [ITU-T Recommendation Q.2957, clause 1 (1995), modified]
ETS 300 668-2	1996- 09	B-ISDN; Digital Subscriber Signalling System No. two (DSS2) protocol; User-to-User Signalling (UUS) supplementary service; Part 2ρ Protocol Implementation Conformance Statement (PICS) proforma specification
ETS 300 669-1	1996- 09	B-ISDN; Digital Subscriber Signalling System No. two (DSS2) protocol; Supplementary service interactions; Part 1ρ Protocol specification
ETS 300 669-2	1996- 09	B-ISDN; Digital Subscriber Signalling System No. two (DSS2) protocol; Supplementary service interactions; Part 2: Protocol Implementation Conformance Statement (PICS) proforma specification
ETS 300 685	1997- 01	B-ISDN; Usage of cause and location in Digital Subscriber Signalling System No. two (DSS2) and Signalling System No.7 B-ISDN User Part (B-ISUP) [ITU-T Recommendation Q.2610 (1995), modified]
ETS 300 771-1	1997- 09	B-ISDN; Digital Subscriber Signalling System No. two (DSS2) protocol; B-ISDN user-network interface layer 3 specification for point-to-multipoint call/bearer control; Part 1: Protocol specification [ITU-T Recommendation Q.2971 (1995), modified]
prETS 300 771-3	1997- 11	B-ISDN; Digital Subscriber Signalling System No. two (DSS2) protocol; B-ISDN user-network interface layer 3 specification for point-to-multipoint call/bearer control; Part 3: Test Suite Structure and Test Purposes (TSS&TP) specification for the user
prETS 300 771-6	1997- 11	B-ISDN; Digital Subscriber Signalling System No. two (DSS2) protocol; B-ISDN user-network interface layer 3 specification for point-to-multipoint call/bearer control; Part 6: Abstract Test Suite (ATS) and partial Protocol Implementation eXtra Information for Testing (PIXIT) proforma specification for the network
ETS 300 780	1997- 12	B-ISDN; Broadband Connection-Oriented Bearer Service (BCOBCS) [ITU-T Recommendation F.811 (1996)]
ETS 300 796-1	1997- 09	B-ISDN; Digital Subscriber Signalling System No. two (DSS2) protocol; Generic functional protocol; Core aspects; Part 1: Protocol specification [ITU-T Recommendation Q.2932.1 (1996), modified]
prETS 300 842	1997- 12	B-ISDN; ATM Adaptation Layer (AAL); Sublayers: Service Specific Co-ordination Function (SSCF) to provide the Connection-Oriented Network Service (CONS) [ITU-T Recommendation I.365.2 (1995)]

prEN 301 029-1	1998-01	B-ISDN; Signalling System No.7; B-ISDN User Part (B-ISUP); Part 1: Overview of the B-ISDN network node interface signalling capability set 2, step 1 [ITU-T Recommendation Q.2721.1 (1996)]
prEN 301 029-2	1998-01	B-ISDN; Signalling System No.7; B-ISDN User Part (B-ISUP); Part 2: Network node interface specification for point-to-multipoint call/connection control [ITU-T Recommendation Q.2722.1 (1996)]
prEN 301 029-4	1998-01 -	B-ISDN; Signalling System No.7; B-ISDN User Part (B-ISUP); Part 4: Look-ahead without state change for the network node interface [ITU-T Recommendation Q.2724.1 (1996)]
prEN 301 029-5	1998-01	B-ISDN; Signalling System No.7; B-ISDN User Part (B-ISUP); Part 5: Support of negotiation during connection setup [ITU-T Recommendation Q.2725.1 (1996)]
prEN 301 029-6	1998-01	B-ISDN; Signalling System No.7; B-ISDN User Part (B-ISUP); Part 6: Modification procedures [ITU-T Recommendation Q.2725.2 (1996)]
prEN 301 029-9	1998-01	B-ISDN; Signalling System No.7; B-ISDN User Part (B-ISUP); Part 9: Network generated session identifier , [ITU-T Recommendation Q.2726.3 (1996)]

ATM Forum Specifications

The ATM Forum uses a set of acronyms in the naming and numbering its documents. Table B.12 lists the acronyms used in the document labels, which also defines the organization of this section. You can download most of these specifications (except for a few of the older ones) for free at the ATM Forum's web site, www.atmforum.com.

Table B.12 ATM Forum Document Naming Convention

Acronym	Meaning
bici	Broadband InterCarrier Interface (B-ICI)
dxi	Data eXchange Interface (DXI)
ilmi	Integrated Local Managment Interface (ILMI)
lane	Local Area Network Emulation (LANE)
mpoa	MultiProtocol Over ATM (MPOA)
nm	Network Management subcommittee documents
phy	Physical layer
pnni	Private Network-Network Interface (PNNI)
saa	Service Aspects and Applications (SAA)
sig	Signaling
test	Testing subcommittee documents
tm	Traffic management subcommittee documents
vtoa	Voice and Telephony Over ATM (VTOA)
uni	User-Network Interface (UNI)

The remainder of this section lists the specifications produced by each working group.

Broadband InterCarrier Interface (B-ICI) Specifications

Table B.13 lists the Broadband InterCarrier Interface (B-ICI) specifications defined by the ATM Forum.

Table B.13 ATM Forum B-ICI Specifications

Specification Name	Number	Publication Date
B-ICI 1.0	af-bici-0013.000	Sep 1993
B-ICI 1.1	af-bici-0013.001	
B-ICI 2.0 (delta spec to B-ICI 1.1)	af-bici-0013.002	Dec 1995
B-ICI 2.0 (integrated specification)	af-bici-0013.003	Dec 1995
B-ICI 2.0 Addendum or 2.1	af-bici-0068.000	Nov 1996

Data eXchange Interface (DXI) Specification

Table B.14 lists the ATM Data eXchange Interface (DXI) specification defined by the ATM Forum. See Chapter 18 for a description of this protocol.

Table B.14 ATM Forum Data eXchange Interface (DXI) Specification

Specification Name	Number	Publication Date
Data Exchange Interface version 1.0	af-dxi-0014.000	Aug 1993

Integrated Local Managment Interface (ILMI) Specification

Table B.15 lists the Integrated Local Managment Interface (ILMI) specification defined by the ATM Forum. See Chapters 14 and 28 for descriptions of ILMI. The ATM Forum's UNI 3.1 specification defines the previous version of the Interim Local Management Interface (ILMI).

Table B.15 ATM Forum ILMI Specifications

Specification Name	Number	Publication Date
UNI 3.1	af-uni-0010.02	1994
ILMI 4.0	af-ilmi-0065.000	Sep 1996

Local Area Network Emulation (LANE) Specifications

Table B.16 lists the Local Area Network Emulation (LANE) specifications defined by the ATM Forum. See Chapter 19 for a detailed description of the LANE protocol.

Table B.16 ATM Forum Local Area Network Emulation (LANE) Specifications

Specification Name	Number	Publication Date
LAN Emulation over ATM 1.0	af-lane-0021.000	Jan 1995
LAN Emulation Client Management Specification	af-lane-0038.000	Sep 1995
LANE 1.0 Addendum	af-lane-0050.000	Dec 1995
LANE Servers Management Spec v1.0	af-lane-0057.000	Mar 1996
LANE v2.0 LUNI Interface	af-lane-0084.000	July 1997

MultiProtocol Over ATM (MPOA) Specification

Table B.17 lists the MultiProtocol Over ATM (MPOA) specification defined by the ATM Forum. See Chapter 20 for a detailed description of the MPOA protocol. MPOA utilizes the ATM Forum's LANE Emulation (LANE) specification and the IETF's Next Hop Resolution Protocol (NHRP).

Table B.17 ATM Forum MPOA Specification

Specification Name	Number	Publication Date
Multi-Protocol Over ATM Specification v1.0	af-mpoa-0087.000	July 1997

Network Management subcommittee Specifications

Table B.18 lists the Network Management specifications defined by the ATM Forum. See Part 7 for information on these specifications.

Table B.18 ATM Forum Network Management Specifications

Specification Name	Number	Publication Date
Customer Network Management (CNM) for ATM Public Network Service	af-nm-0019.000	Oct 1994
M4 Interface Requirements and Logical MIB	af-nm-0020.000	Oct 1994
CMIP Specification for the M4 Interface	af-nm-0027.000	Sep 1995
M4 Public Network view	af-nm-0058.000	Mar 1996
M4 "NE View"	af-nm-0071.000	Jan 1997
Circuit Emulation Service Interworking Requirements, Logical and CMIP MIB	af-nm-0072.000	Jan 1997
M4 Network View CMIP MIB Spec v1.0	af-nm-0073.000	Jan 1997
M4 Network View Requirements & Logical MIB Addendum	af-nm-0074.000	Jan 1997
ATM Remote Monitoring SNMP MIB	af-nm-test-0080.000	July 1997

Physical Layer Specifications

Table B.19 lists the Physical Layer specifications defined by the ATM Forum. See Chapter 12 for more information on the physical layer specifications.

Table B.19 ATM Forum Physical Layer Specifications

Specification Name	Number	Publication Date
Issued as part of UNI 3.1: 44.736 DS3 Mbps Physical Layer 100 Mbps Multimode Fiber Interface Physical Layer 155.52 Mbps SONET STS-3c Physical Layer 155.52 Mbps Physical Layer	af-uni-0010.002	
ATM Physical Medium Dependent Interface Specification for 155 Mb/s over Twisted Pair Cable	af-phy-0015.000	Sep 1994
DS1 Physical Layer Specification	af-phy-0016.000	Sep 1994
Utopia	af-phy-0017.000	Mar 1994
Mid-range Physical Layer Specification for Category 3 UTP	af-phy-0018.000	Sep 1994
6,312 Kbps UNI Specification	af-phy-0029.000	June 1995
E3 UNI	af-phy-0034.000	Aug 1995
Utopia Level 2	af-phy-0039.000	June 1995
Physical Interface Specification for 25.6 Mb/s over Twisted Pair	af-phy-0040.000	Nov 1995
A Cell-based Transmission Convergence Sublayer for Clear Channel Interfaces	af-phy-0043.000	Jan 1996
622.08 Mbps Physical Layer	af-phy-0046.000	Jan 1996
155.52 Mbps Physical Layer Specification for Category 3 UTP (See also UNI 3.1)	af-phy-0047.000	
120 Ohm Addendum to ATM PMD Interface Spec for 155 Mbps over TP	af-phy-0053.000	Jan 1996
DS3 Physical Layer Interface Spec	af-phy-0054.000	Mar 1996
155 Mbps over MMF Short Wave Length Lasers, Addendum to UNI 3.1	af-phy-0062.000	July 1996
WIRE (PMD to TC layers)	af-phy-0063.000	July 1996
E-1 Physical Layer Interface Specification	af-phy-0064.000	Sep 1996
155 Mbps over Plastic Optical Fiber (POF)	af-phy-0079.000	May 1997
Inverse ATM Mux	af-phy-0086.000	July 1997

Private Network-Network Interface (PNNI) Specifications

Table B.20 lists the Private Network-Network Interface (PNNI) specifications defined by the ATM Forum. See Chapter 15 for more information on PNNI.

Table B.20 ATM Forum PNNI Specifications

Specification Name	Number	Publication Date
Interim Inter-Switch Signaling Protocol	af-pnni-0026.000	Dec 1994
P-NNI V1.0	af-pnni-0055.000	Mar 1996
PNNI 1.0 Addendum (soft PVC MIB)	af-pnni-0066.000	Sep 1996
PNNI ABR Addendum	af-pnni-0075.000	Jan 1997
PNNI v1.0 Errata and PICs	af-pnni-0081.000	July 1997

Service Aspects and Applications (SAA) Specifications

Table B.21 lists the Service Aspects and Applications (SAA) specifications defined by the ATM Forum. See Chapter 14 for information on the ATM Name Service (ANS); Chapter 16 for information on the Native ATM services semantic description; Chapter 17 for information on circuit emulation and Video on Demand; and Chapter 18 for FUNI.

Table B.21 ATM Forum Service Aspects and Applications (SAA) Specifications

Specification Name	Number	Publication Date
Frame UNI	af-saa-0031.000	Sep 1995
Circuit Emulation	af-saa-0032.000	Sep 1995
Native ATM Services: Semantic Description	af-saa-0048.000	Feb 1996
Audio/Visual Multimedia Services: Video on Demand v1.0	af-saa-0049.000	Jan 1996
Audio/Visual Multimedia Services: Video on Demand v1.1	af-saa-0049.001	Mar 1997
ATM Name Service	af-saa-0069.000	Nov 1996
FUNI 2.0	af-saa-0088.000	July 1997

Signaling Specifications

Table B.22 lists the signaling specifications defined by the ATM Forum. See Chapters 14 and 15 for the details of signaling.

Table B.22 ATM Forum Signaling Specifications

Specification Name	Number	Publication Date
UNI 3.1	af-uni-0010.002	1994
UNI Signaling 4.0	af-sig-0061.000	July 1996
Signaling ABR Addendum	af-sig-0076.000	Jan 1997

Testing Specifications

Table B.23 lists the testing specifications defined by the ATM Forum.

Table B.23 ATM Forum Testing Specifications

Specification Name	Number	Publication Date
Introduction to ATM Forum Test Specifications	af-test-0022.000	Dec 1994
PICS Proforma for the DS3 Physical Layer Interface	af-test-0023.000	Sep 1994
PICS Proforma for the SONET STS-3c Physical Layer Interface	af-test-0024.000	Sep 1994
PICS Proforma for the 100 Mbps Multimode Fibre Physical Layer Interface	af-test-0025.000	Sep 1994
PICS Proformat for the ATM Layer (UNI 3.0)	af-test-0028.000	Apr 1995
Conformance Abstract Test Suite for the ATM Layer for Intermediate Systems (UNI 3.0)	af-test-0030.000	Sep 1995
Interoperability Test Suite for the ATM Layer (UNI 3.0)	af-test-0035.000	Apr 1995
Interoperability Test Suites for Physical Layer: DS-3, STS-3c, 100 Mbps MMF (TAXI)	af-test-0036.000	Apr 1995
PICS Proforma for the DS1 Physical Layer	af-test-0037.000	Apr 1995
Conformance Abstract Test Suite for the ATM Layer (End Systems) UNI 3.0	af-test-0041.000	Jan 1996
PICS for AAL5 (ITU spec)	af-test-0042.000	Jan 1996
PICS Proforma for the 51.84 Mbps Mid-Range PHY Layer Interface	af-test-0044.000	Jan 1996
Conformance Abstract Test Suite for the ATM Layer of Intermediate Systems (UNI 3.1)	af-test-0045.000	Jan 1996
PICS for the 25.6 Mbps over Twisted Pair Cable (UTP-3) Physical Layer	af-test-0051.000	Mar 1996
Conformance Abstract Test Suite for the ATM Adaptation Layer (AAL) Type 5 Common Part (Part 1)	af-test-0052.000	Mar 1996
PICS for ATM Layer (UNI 3.1)	af-test-0059.000	July 1996
Conformance Abstract Test Suite for the UNI 3.1 ATM Layer of End Systems	af-test-0060.000	June 1996
Conformance Abstract Test Suite for the SSCOP Sub-layer (UNI 3.1)	af-test-0067.000	Sep 1996
PICS for the 155 Mbps over Twisted Pair Cable (UTP-5/STP-5) Physical Layer	af-test-0070.000	Nov 1996
PNNI v1.0 Errata and PICs	af-pnni-0081.000	July 1997
PICS for Direct Mapped DS3	af-test-0082.000	July 1997
Conformance Abstract Test Suite for Signalling (UNI 3.1) for the Network Side	af-test-0090.000	September 1997

Traffic Management (TM) Specifications

Table B.24 lists the Traffic Management (TM) specifications defined by the ATM Forum. See Part 5 for a detailed coverage of these subjects.

Table B.24 ATM Forum Traffic Management (TM) Specifications

Specification Name	Number	Publication Date
UNI 3.1	af-uni-0010.002	1994
Traffic Management 4.0	af-tm-0056.000	Apr 1996
Traffic Management ABR Addendum	af-tm-0077.000	Jan 1997

Voice and Telephony Over ATM (VTOA) Specifications

Table B.25 lists the Voice and Telephony Over ATM (VTOA) specifications defined by the ATM Forum. See Chapter 17 for a review of these VTOA specifications.

Table B.25 ATM Forum Voice and Telephony Over ATM (VTOA) Specifications

Specification Name	Number	Publication Date
Circuit Emulation Service 2.0	af-vtoa-0078.000	Jan1997
Voice and Telephony Over ATM to the Desktop	af-vtoa-0083.000	May 1997
(DBCES) Dynamic Bandwith Utilization in 64 KBPS Time Slot Trunking Over ATM - Using CES	af-vtoa-0085.000	July 1997
ATM Trunking Using AAL1 for Narrow Band Services v1.0	af-vtoa-0089.000	July 1997

User-Network Interface (UNI) Specifications

Table B.26 lists the User-Network Interface (UNI) specifications defined by the ATM Forum. See Chapters 11 and 12 for more information.

Table B.26 ATM Forum User-Network Interface (UNI) Specifications

Specification Name	Number	Publication Date
ATM User-Network Interface Specification V2.0	af-uni-0010.000	June 1992
ATM User-Network Interface Specification V3.0	af-uni-0010.001	Sep 1993
ATM User-Network Interface Specification V3.1	af-uni-0010.002	1994

IETF RFCs Related to ATM

This section summarizes many of the ATM-related efforts standardized by the IETF and covered in the IETF RFCs listed in Table B.27. It also briefly covers several other key areas that are nearing completion of the draft stage, involving QoS support and efficient routing over ATM networks.

Table B.27 List of Approved IETF Request For Comment (RFC) Documents

Number	Title	Date	Category
RFC 1483	Multiprotocol over AAL5	Jul 1993	Stds Track
RFC 1577	Classical IP and ARP over ATM	Jan 1993	Stds Track
RFC 1626	Default IP MTU for use over ATM AAL5	May 1994	Stds Track
RFC 1680	IPng Support for ATM Services	Aug 1994	Informational
RFC 1695	ATM Management Objects	Aug 1994	Stds Track
RFC 1755	ATM Signaling Support for IP over ATM	Feb 1995	Stds Track
RFC 1821	Real-time Service in IP-ATM Networks	Aug 1995	Informational
RFC 1932	IP over ATM: A Framework Document	April 1996	Informational
RFC 1937	Forwarding in Switched Data Link Subnets	May 1996	Informational
RFC 1946	Native ATM Support for ST2+	May 1996	Informational
RFC 1953	IFMP Specification	May 1996	Informational
RFC 1954	Flow Labeled IPv4 on ATM	May 1996	Informational
RFC 1987	GSMP Protocol Specification	Aug 1996	Informational
RFC 2022	Multicast over UNI 3.0/3.1 based ATM	Nov 1996	Stds Track
RFC 2098	Toshiba's Router Architecture Extensions for ATM : Overview	Feb 1997	Informational
RFC 2105	Cisco Systems' Tag Switching Architecture Overview	Feb 1997	Informational
RFC 2121	Issues affecting MARS Cluster Size	March 1997	Informational
RFC 2129	Toshiba's Flow Attribute Notification Protocol (FANP) Specification	April 1997	Informational
RFC 2149	Multicast Server Architectures for MARS-based ATM multicasting	May 1997	Informational
RFC 2170	Application REQuested IP over ATM (AREQUIPA)	July 1997	Informational
RFC 2191	VENUS - Very Extensive Non-Unicast Service	Sept 1997	Informational
RFC 2226	IP Broadcast over ATM Networks	Oct 1997	Stds Track
RFC 2269	Using the MARS Model in non-ATM NBMA Networks	Jan 1998	Informational
RFC 2331	ATM Signalling Support for IP over ATM - UNI Signalling 4.0 Update	April 1998	Stds Track
RFC 2332	NBMA Next Hop Resolution Protocol (NHRP)	April 1998	Stds Track
RFC 2333	NHRP Protocol Applicability Statement	April 1998	Stds Track
RFC 2334	Server Cache Synchronization Protocol (SCSP)	April 1998	Stds Track
RFC 2335	A Distributed NHRP Service Using SCSP	April 1998	Stds Track
RFC 2337	Intra-LIS IP multicast among routers over ATM using Sparse Mode PIM	April 1998	Experimental

Glossary

AAL 1 - Addresses Constant Bit Rate (CBR) traffic such as digital voice and video. It supports applications sensitive to both cell loss and delay variation. It emulates TDM circuits (see CES).

AAL 2 - Supports connection-oriented services with variable rate that have timing transfer and limited delay variation requirements. Used with time-sensitive, variable bit rate traffic such as packetized voice or video.

AAL 3/4 - Intended for both connectionless and connection-oriented variable rate services. Originally standards defined two distinct protocols but later merged then into this single AAL. Used primarily to support SMDS.

AAL 5 - Supports connection-oriented and connectionless variable bit rate data services without timing transfer requirements. Examples are the typical bursty data traffic found in LANs. Most commonly used AAL for data applications.

acknowledgement (ACK) - A message that acknowledges the reception of a transmitted packet. ACKs can be separate packets or piggybacked on reverse traffic packets.

ACR Decrease Time Factor (ADTF) - Time permitted between RM-cells before the source reduces rate rate to Initial Cell Rate (ICR). ADTF range is 0.01 to 10.23 s with a granularity of 10 ms.

Additive Increase Rate (AIR) - An ABR service parameter that controls the rate of cell transmission increase. It is signaled as AIR×Nrm/PCR.

Additive Increase Rate (AIR) - An ABR service parameter, AIR controls the rate at which the cell transmission rate increases.

address - An identifier of a source or destination in a network. Examples of addresses are IPv4, E.164, NSAP, and IPv6.

address prefix - A string of the high-order bits of an address. Used in routing protocols.

address registration - A procedure using the ILMI where a network communicates valid address prefixes to the user, and the user registers specific AESAs with the network. LANE Clients (LECs) provide address information to the LAN Emulation Server (LES) using this protocol.

Address Resolution Protocol (ARP) - Protocol used to resolve a destination address for a lower-layer protocol (e.g., ATM or MAC) from a known address for another higher-layer protocol (e.g., IP).

agent - Software residing in a managed network device that reports MIB variables through SNMP.

Alarm Indication Signal (AIS) - An OAM function used in fault management for physical and ATM layers. A system transmits AIS upstream if it detects a fault. The downstream end system responds with a Remote Defect Indication (RDI).

Allowed Cell Rate (ACR) - An ABR service parameter, ACR is the current rate in cells/s at which a source is allowed to send.

alternate routing - A technique that allows selection of other paths if an attempt to set up a connection along a previously selected path fails.

analog - Continuously variable signals (like voice) that possess an infinite number of values, as contrasted with digital signals that have a discrete number of values.

anycast - A capability defined in the ATM Forum's UNI 4.0 and PNNI 1.0 specifications that supports load-balancing. The network routes calls to the closest ATM end-system which is assigned the anycast address within a specified scope.

Application Programming Interface (API) - A set of functions used by an application program to access to system capabilities (e.g., packet transfer or signaling).

assigned cell - A cell generated by the user, control or management planes which the ATM layer passes to the PHY layer. See unassigned cell.

Asynchronous Transfer Mode (ATM) - A high-speed connection-oriented multiplexing and switching method specified in international standards utilizing fixed-length 53 byte cells. to support multiple types of traffic. It is asynchronous in the sense that cells carrying user data need not be periodic.

asynchronous transmission - The transmission of data through start and stop sequences without the use of a common clock.

ATM Adaptation Layer (AAL) - A set of internationally standardized protocols and formats that define support for circuit emulation, packet video and audio, and connection-oriented and connectionless data services. These protocols translate higher layer user traffic into a stream of ATM cells and convert the received cells back to the original form at the destination. Most AALs consists of two sublayers: the segmentation and reassembly (SAR) sublayer and the convergence sublayer (CS). See AAL1, AAL2, AAL3/4, AAL5.

ATM Address Resolution Protocol (ATMARP) - IETF RFC 1577 defines this means to map IP addresses to ATM hardware addresses. It works in much the same way as conventional ARP works on a LAN when mapping network-layer addresses to MAC addresses.

ATM End System Address (AESA) - A 20 byte long, NSAP-based address. The address has a hierarchical structure that supports scaling to global networks. It has an Initial Domain Part (IDP) assigned by the network and a Domain Specific Part (DSP) assigned by the end user.

ATM service categories - A set of categories that define QoS parameters, traffic parameters and the use of feedback. Currently these include: CBR, rt-VBR, nrt-VBR, UBR and ABR.

AToM Management Information Base (AToM MIB) - Defined in IETF RFC 1695, AToM MIB allows SNMP net management systems to monitor and configure ATM devices. AToM MIB reports on device or traffic status and accepts configuration commands.

Authority and Format Identifier (AFI) - The first byte of an ATM address that defines the number assignment authority and the format of the remainder of the address. The ATM Forum current defines formats for International Code Designator (ICD), Data Country Code (DCC), and E.164. See AESA, IDP.

Available Bit Rate (ABR) - An ATM service category in which the network delivers limited cell loss if the end user responds to flow control feedback. The ABR service does not control cell delay variation. ABR operates in one of three modes: binary mode, Explicit Rate (ER), or Virtual Source/Virtual Destination (VS/VD). ABR employs Resource Management (RM) cells to implement a closed loop flow control loop. A number of parameters control the ABR capability. See ADTF, ACR, CCR, CI, and MCR.

B channel - An ISDN bearer service channel that can carry either voice or data at a speed of 64 kbps.

Backward Explicit Congestion Notification (BECN) - Convention in frame relay for a network device to notify the user source device of network congestion.

Backward Explicit Congestion Notification (BECN) cell - A Resource Management (RM) cell type. Either the network or the destination may generate a BECN RM-cell to throttle the ABR source in binary mode.

bandwidth - The amount of transport resource available to pass information, measured in Hertz for analog transmission (i.e., frequency passband) and bits per second for digital transmission.

Basic Rate Interface (BRI) - An ISDN access interface type composed of two B channels, each at 64 kbps, and one D channel at 16 kbps (2B+D).

baud rate - A baud is the discrete signal sent once per signaling interval at the physical layer. If the baud in each signaling interval represents only one bit, then the baud rate equals the bit rate. Some transmission schemes send multiple bits per signaling interval.

Beginning Of Message (BOM) - A protocol indicator contained in the first cell of an AAL3/4 PDU.

best effort - A Quality of Service (QOS) class in which no traffic parameters are specified and no guarantee given that traffic will be delivered. The Unspecified Bit Rate (UBR) service category is an example of a best-effort services.

Binary Coded Decimal (BCD) - A 4-bit expression for zero ('0000') through nine ('1001').

binary mode ABR - A particular type of Available Bit Rate (ABR) that interworks with older ATM switches that only support the Explicit Forward Congestion Indication (EFCI) function in the ATM cell header. Binary mode uses the Congestion Indication (CI) bit in a Resource Management (RM) cell in the reverse direction to control the rate of the source.

B-ISDN Inter-Carrier Interface (B-ICI) - An ATM Forum defined specification for the interface between public ATM networks in support of user services (e.g., CRS, CES, SMDS, or FR) traversing multiple public carriers.

Bit-Interleaved Parity (BIP) - A parity check that groups all the bits in a block into units (such as a byte) then performs a parity check for each bit position in a group. For example, a BIP-8 created eight-bit (one-byte) groups, then does a parity check for each of the eight bit positions in the byte.

bridge - A device interconnecting local area networks at the OSI Data Link Layer, filtering and forwarding frames according to media access control (MAC) addresses.

broadband - Typically Refers to bandwidths in excess of DS3 (45 Mbps) or E3 (34 Mbps).

Broadband Connection Oriented Bearer (BCOB) - Information in the SETUP message that indicates the bearer service requested by the calling user. Types include A (see CBR) and C (see VBR) which may include the specific AAL and its parameters for use in interworking. BCOB-X defines a service where AAL, traffic type, and timing are transparent to the network.

Broadband High Layer Information (B-HLI) - An information element in a SETUP message used by end systems for compatibility checking.

Broadband Integrated Services Digital Network (B-ISDN) - An ITU-T-developed set of standards supporting the integrated, high-speed transmission, switching and multiplexing of data, audio and video. ATM emerged as a suitable transport standard. B-ISDN has three logical planes: user, control and management: The user plane provides the basic service as determined by the control plane signaling. The management plane coordinates and monitors the function of the various layers, as well as the user and control planes. All three planes use a common ATM layer that operates over a variety of physical layers.

Broadband ISDN User Part (B-ISUP) - A protocol defined in ITU-T Q.2761 used between carrier networks operating between origin, destination and transit nodes. Interoperates with the narrowband ISUP using the same SS7 Message Transfer Part (MTP) protocol.

Broadband Low Layer Information (B-LLI) - Information element in a SETUP message identifying the Layer 2 and Layer 3 protocol used by the application.

Broadband Terminal Equipment or Adapter (B-TE or B-TA) - An equipment category for B-ISDN which includes end systems and intermediate systems. TAs convert from other protocols while Tes operate in a native ATM mode.

broadcast - A transmission to all addresses on the network or subnetwork.

Broadcast and Unknown Server (BUS) - A LAN Emulation component that receives all broadcast and multicast MAC packets as well as MAC packets with an unknown ATM address. The BUS transmits these messages to every member of an Emulated LAN (ELAN).

Burst Tolerance (BT) - Proportional to the Maximum Burst Size (MBS), the burst tolerance measures conformance checking of the Sustainable Cell Rate (SCR) as part of the Usage Parameter Control (UPC) function.

Burstiness - A source traffic characteristic defined as the maximum rate divided by the average rate.

Carrier Sense Multiple Access/ Collision Detection (CSMA/CD) - A protocol employed in Ethernet where stations first listen to the bus and only transmit when the bus is free (CSMA). If the transmitting stations detect a collision, they retransmit the frame after a random time-out.

cell - The basic ATM Protocol Data Unit (PDU) used in transmission, switching and multiplexing. An ATM cell is a fixed-length 53-octet packet comprised of a 5-octet header and a 48-octet payload.

Cell Delay Variation (CDV) - A QoS parameter that measures the difference in delay between successive cell arrivals. Standards define one-point and two-point measurement methods. A one-point measurement applies to CBR sources and determines the deviation from the nominal cell spacing. The two-point method measures the difference in cell spacing at different points for the connection. An example of a two-point method is the peak-to-peak CDV associated with CBR and VBR services determined by the probability α that a cell arrives late.

Cell Delay Variation Tolerance - A tolerance parameter used in the Usage Parameter Control (UPC) specification of the Peak Cell Rate (PCR). CDVT defines the depth of the PCR leaky bucket Generic Cell Rate Algorithm (GCRA). CDVT effectively defines how many back-to-back cells may enter the network for a conforming connection.

Cell Error Ratio (CER) - A QoS parameter that measures the fraction of cells received with errors at the destination.

cell header - A 5-octet header that defines control information used in processing, multiplexing, and switching cells. It contains the following fields: GFC, VPI, VCI, PT, CLP and HEC.

Cell Loss Priority (CLP) - A one-bit field in the ATM cell header that indicates the relative discard priority. The network may discard cells with CLP set to 1 during periods of congestion to preserver the Cell Loss Ratio (CLR) performance of cells with CLP set to 0.

Cell Loss Ratio (CLR) - A negotiated QoS parameter that specifies the ratio of lost (i.e., not delivered) cells to the total cells transmitted.

Cell Misinsertion Rate (CMR) - A QoS parameter defined as the number of misinserted cells (those that were never sent by the source) divided by a specified time interval.

Cell Relay Function (CRF) - A basic function of an ATM device operating at the Virtual Path of Channel (VP or VC) level. A CRF translates the VP and VC Identifiers (VPI and VCI) between an input port and an output port on an ATM device.

Cell Relay Service (CRS) - A service offered to end users by carriers that supports the delivery of ATM cells according to standards and specifications.

Cell Transfer Delay (CTD) - A QoS parameter that specifies the accumulated delay between two measurement points for a specific virtual connection. It is the sum of coding, decoding, segmentation, reassembly, processing and queueing delays along the connection route.

Central Office (CO) - A telephone company switching office providing access to the network and its services.

Channel Associated Signaling (CAS) - A technique used in telephony to associate on-hook and off-hook conditions with a TDM signal representing voice channels. For example, in North America, robbed bit signaling uses the LSB of a 8-bit PCM sample to signal telephony channel state.

Channel Service Unit/ Data Service Unit (CSU/DSU) - A function commonly used to interface CPE to the WAN. This may be a separate device, or built into a switch or multiplexer port card. The DSU function provides a Data Communications Equipment (DCE) interface to a user Data Terminal Equiment (DTE). The CSU functions supports the standard WAN interface.

Circuit Emulation Service (CES) - An ATM Forum specification defining emulation of Time Division Multiplexed (TDM) circuits (e.g., DS1 and E1) over ATM Adaptation Layer 1 (AAL1). CES operates over Virtual Channel Connections (VCCs) which support the transfer of timing information. Typically, CES operates at a Constant Bit Rate (CBR). CES operates in an unstructured (i.e., a 1.544 Mbps DS1) or a structured mode (i.e., Nx64 kbps). Optionally, CES supports Channel Associated Signaling (CAS).

circuit switching - A connection-oriented technique based on either time- or space-division multiplexing and switching providing minimal delay.

Classical IP over ATM - An adaptation of the Address Resolution Protocol (ARP) for operation of ATM over a single Logical IP Subnetwork (LIS) defined by IETF RFCs 1483, 1577, and 1755. It either uses PVCs or dynamically establishes SVCs between systems to transfer IP packets using the ATM address returned in the ARP response.

Clock - The source of timing information used in synchronous transmission systems.

Committed Information Rate (CIR) - A term used in frame relay that defines the rate the network commits to deliver frames. Transmissions exceeding the CIR are subject to Discard Eligible (DE) tagging or discard at ingress.

Common Management Information Protocol (CMIP) - An ITU-T-defined management interface standard supporting administration, maintenance and operation information functions. See OAM&P.

Common Part Convergence Sublayer (CPCS) - The mandatory portion of an AAL Convergence Sublayer (CS) that passes primitives to an optional Service Specific Convergence Sublayer (SSCS). AAL1, AAL3/4, and AAL5 standards have a CPCS function.

Compression - A set of techniques that conserve bandwidth and/or storage by reducing the number of bits required to represent information.

congestion - The condition in which demand exceeds available network resources (i.e., bandwidth or buffer space) for a sustained period of time.

congestion collapse - A condition where re-transmissions ultimately result in markedly reduced effective throughput. ATM networks employing adequate buffering, intelligent packet discard mechanisms, or Available Bit Rate (ABR) flow control usually avoid the congestion collapse phenomenon.

Congestion control - A resource allocation and traffic management mechanism that avoids or recovers from congestion situations. See ABR, selective cell discard, and CAC.

Congestion Indication (CI) - A bit in a Resource Management (RM) cell that indicates congestion. The destination sets this bit in response to cells received with the EFCI bit set in the cell header.

Connection Admission Control (CAC) - An algorithm that determines whether an ATM device accepts or rejects an ATM connection request. Generally, CAC algorithms accept calls only if enough bandwidth and buffer resources exist to establish the connection at the requested quality of service (QOS) without impacting connections already in progress. See GCAC.

Connectionless Network Service (CLNS) - An OSI defined service for the transfer of information between users employing globally unique addresses. Unlike CONS, this service requires no establishment of a connection prior to data transfer. Examples are LAN, IP and SMDS protocols.

Connection-Oriented Network Service (CONS) - A service where the network must first establish a connection between the source and destination prior to data transfer. Examples are N-ISDN, X.25, frame relay, and ATM.

Constant Bit Rate (CBR) - One of several ATM service categories. CBR supports applications like voice and video requiring a constant bit rate, constrained CDV, and low CLR connection. Circuit emulation also employs CBR. The PCR and CDVT traffic parameters define the characteristics of a CBR connection.

Consultative Committee International for Telegraphy and Telephony (CCITT) - Now referred to as the ITU-T.

contention - A condition arising when two or more stations attempt to transmit at the same time using the same transmission channel.

Continuation Of Message (BOM) - A protocol indicator contained in the any cell that is not the first or last cell of an AAL3/4 PDU.

Convergence Sublayer (CS) - The higher layer portion of the AAL that interfaces with the lower layer Segmentation And Reassembly (SAR) portion of the AAL. The CS sublayer interfaces between ATM and non-ATM formats. The CS may be either a Common Part (CPCS), or Service Specific (SSCS).

crankback - A mechanism defined in PNNI for partially releasing a connection establishment attempt which encounters a failure back to the last node that made a routing decision. This mechanism allows PNNI networks to perform alternate routing in a standard way.

Current Cell rate (CCR) - A Resource Management (RM) cell field set by the source indicating its transmission rate. When the transmitting end sends a forward RM cell, it sets the CCR to the current allowed cell rate (ACR). No elements in the network can adjust this value.

Customer Network Management (CNM) - A capability that allows public network users to monitor and manage their services. ATM Forum specifications define CNM interfaces for monitoring physical ports, virtual channels, virtual paths, usage parameters, and QoS parameters.

Customer Premises Equipment (CPE) - Equipment that resides and is operated at a customer site. See also DTE.

Cyclic Redundancy Check (CRC) - A mathematical algorithm commonly implemented as a cyclic shift register that computes a check field for a block of data. The sender transmits this check field along with the data so that the receiver can either detect errors, and in some cases even correct errors.

D channel - The ISDN out-of-band (16 kbps or 64 kbps, depending on BRI or PRI, respectively) signaling channel which carries the ISDN user signals or can be used to carry packet-mode data.

Data - Information represented in digital form, including voice, text, facsimile and video.

Data Communications (or Circuit termination) Equipment (DCE) - Function existing as the access for the user Data Terminal Equipment (DTE) to a network. Typically, a modem, CSU/DSU, or switch port implements this function.

Data Country Code (DCC) - One of three Authority and Format Identifiers (AFI) defined by the ATM Forum. The ISO 3166 standard defines the country in which the NSAP-based address is registered.

Data Exchange Interface (DXI) - A frame-based interface between legacy CPE (e.g., a router) and an external ATM CSU/DSU via a High Speed Serial Interface (HSSI). Initially developed to support early introduction of ATM, the Frame-based UNI (FUNI) provides a similar function without the expense of an external ATM CSU/DSU.

Data Link Connection Identifier (DLCI) - A frame relay address designator locally significant to the physical interface for each virtual connection endpoint.

Data Link Layer (OSI) - Layer 2 of the OSI model Establishes, maintains and releases data-link connections between end points. Provides for error detection and optional correction between adjacent network devices when operating over a noisy physical channel. Examples include frame relay, ATM, and the LAN LLC and MAC sublayers.

Data Terminal Equipment (DTE) - A device transmitting data to, and/or receiving data from, a DCE (for example, a terminal or printer).

Data Terminal Equipment (DTE) - The user side device connection to a DCE. Typically implementations are serial ports on computers, routers, multiplexers and PCs.

datagram - The basic unit of transmission, switching, and multiplexing in a ConnectionLess Network Service (CLNS). Typically, a datagram service does not guarantee delivery. Furthermore, a network may route datagrams along different paths, therefore datagrams may arrive in a different order at the destination than that generated by the source.

Designated Transit List (DTL) - An Information Element (IE) defined in the ATM Forum PNNI signaling protocol used in a SETUP message that conveys information about an end-to-end source route from source to destination. Organized as a push down stack, DTLs support hierarchical source routes where each domain along the path expands the source route. This design supports very large networks.

Destination Address (DA) - A six octet value uniquely identifying an endpoint as defined by the IEEE Media Access Control (MAC) standards.

digital - A system having discrete values, such as binary 0s and 1s.

Digital Access and Cross-connect System (DACS) - A space-division and/or time-slot switching device that electronically interconnects at various

PDH and SDH multiplexing level for TDM signals. See SONET, and SDH. Also called a DCC or DXC.

Digital Cross-Connect System (DCC or DXC) - See DACs.

Digital Service Unit (DSU) - See CSU/DSU.

digital signal - An electrical or optical signal that varies in discrete steps. Electrical signals are coded as voltages, while optical signals are coded as pulses of light.

Digital Signal 0 (DS0) - Physical interface for digital transmission using TDM operating at 64 kbps. PDH standards map 24 DS0s into a DS1 or 32 DS0s into an E1.

Digital Signal 1 (DS1) - North American standard physical transmission interface operating at 1.544 Mbps. The payload rate of 1.536 Mbps may carry 24 DS0 channels, each operating at 64 kbps. Colloquially known as T1.

Digital Signal 3 (DS3) - The North American standard 44.736-Mbps digital transmission physical interface. The payload rate of approximately 44.21 Mbps may be used to carry digital data, or multiplexed to support 28 DS1s.

Digital Subscriber Signaling   number 2 (DSS2) - Generic name for signaling between on a B-ISDN User-Network Interface (UNI) defined in standards by the ITU-T. N-ISDN standards define DSS1 signaling between a user and network.

Discard Eligibility (DE) bit - A bit in the frame relay header that, when set to 1, indicates that the particular frame is eligible for discard during congestion conditions.

Distributed Queue Dual Bus (DQDB) - The IEEE 802.6 MAN standard based upon 53-byte slots that supports connectionless and connection-oriented, isochronous integrated services.   A physical DQDB network typically has two unidirectional buses configured in a physical ring topology.

Domain Specific Part (DSP) - The low-order portion of an ATM End System Address (AESA) assigned by an end user. It contains the End System Identifier (ESI) and the selector (SEL) byte.

Dual Attachment Station (DAS) - A workstation that attaches to both primary and secondary FDDI MAN rings. This configuration enables the self-healing capability of FDDI.

E.164 - An ITU-T definined numbering plan for public voice and data networks containing up to 15 BCD digits. Numbering assignments currently exist for PSTN, N-ISDN, B-ISDN, and SMDS public networks. The ATM Forum 20-byte NSAP-based ATM End Systems Address (AESA) format encapsulates an 8-byte E.164 address as the Domain Specific Part (DSP).

E1 - The European CEPT1 standard digital channel operating at 2.048 Mbps. The payload of 1.92 Mbps may be employed trasnparently, or multiplexed into 32 individual 64 kbps channels The E1 frame structure supports 30 individual 64 kbps digital channels for voice or data, plus a 64 kbps signaling channel and a 64 kbps channel for framing and maintenance.

E3 - European CEPT3 standard for digital transmission service at 34.368 Mbit/s. It supports transmission at a payload rate of 33.92 Mbps, or it carries 16 E1s when used in the PDH multiplexing hierarchy.

E4 - European CEPT4 standard for digital physical interface at 139.264 Mbps. It operates at a payload rate of 138.24 Mbps, or carries four E3s when operating in multiplexing mode.

Early Packet Discard (EPD) - An intelligent packet-discard mechanism congestion control technique that drops all cells from an AAL5 PDU containing a higher layer protocol packet. This action helps prevent the congestion collapse phenomenon when an ATM device buffer overflows and loses portions of multiple packets.

Emulated Local Area Network (ELAN) - A logical network defined by the ATM Forum LAN Emulation (LANE) specification comprising both ATM and legacy attached LAN end stations.

End of Message (EOM) - A protocol indicator used' in AAL 3/4 that identifies the last ATM cell in a packet.

End System Identifier (ESI) - A 6-byte field in the Domain Specific Part (DSP) of an ATM End System Address (AESA).

Ethernet - The IEEE 802.3 and ISO set of standards defining operation of a LAN protocol at 10, 100, and 1,000 Mbps. Typically, Etherent uses the CSMA/CD Media Access Control (MAC) protocol. Ethernet operates over coax, twisted pair and fiber optic physical connections.

European Telecommunications Standards Institute (ETSI) - The primary telecommunications standards organization in Europe.

Explicit Congestion Notification (ECN) - In frame relay, the use of either FECN and BECN messages notifies the source and destination of network congestion, respectively.

Explicit Forward Congestion Indication (EFCI) - A one-bit field in the ATM cell header Payload Type Indicator (PTI). Any ATM device may set the EFCI bit to indicate a congested situation; for example, when buffer fill crosses a pre-set threshold. In the EFCI marking operation of the Available Bit Rate (ABR) binary mode, switches set the EFCI bit in the header of cells in the forward direction to indicate congestion.

Explicit Forward Congestion Indication (EFCI) - One of the congestion feedback modes allowed in the Available Bit Rate (ABR) service. In the EFCI marking mode, switches can set a bit in the headers of forward cells to indicate congestion; these are then turned around at the destination end system and sent back to the source end system. (see ABR)

Explicit Rate (ER) - A mode of Available Bit Rate (ABR) service where switches explicitly indicate within a specific field in backward Resource Management (RM) cells the rate they can support for a particular connection. Devices in the backward path may only reduce the value of ER, resulting in the source receiving the lowest supported rate in the returned RM cell. See ABR, ACR, CCR.

exterior link - A link which crosses the boundary of the PNNI routing domain. The PNNI protocol does not run over an exterior link.

Far-End Reporting Failure (FERF) - Refers to the same condition that Remote Defect Indication (RDI) does, but specifies more information about the failure.

fast packet - A generic term used for advanced packet technologies such as frame relay, DQDB, ATM.

FDDI-II - FDDI standard with the additional capability to carry isochronous traffic (voice/video).

Fiber Distributed Data Interface (FDDI) - ANSI defined Fiber-optic LAN operating at 100 Mbps using a dual-attached, counter-rotating ring topology and a token passing protocol.

fiber optics - Thin filaments of glass or plastic carrying a transmitted light beam generated by a Light Emitting Diode (LED) or a laser.

File Transfer Protocol (FTP) - A capability in the Internet Protocol suite supporting transfer of files using the Transmission Control Protocol (TCP).

Fixed Round-Trip Time (FRTT) - The sum of the fixed and propagation delays from the source to the furthest destination and back used in the ABR Initial Cell Rate (ICR) calculation.

Flag (F) - The bit pattern '0111 1110' signaling the beginning or end of an HDLC frame.

Forward Error Correction (FEC) - An error correction technique which transmits redundant information along with the original data. The receiver uses the redundant FEC data to reconstruct lost or errored portions of the original data. FEC does not employ retransmissions to correct errors.

Forward Explicit Congestion Notification (FECN) - Convention in frame relay for a network device to notify the user (destination) device of network congestion.

Fractional T1 (FT1) - Colloquial reference to the transmission of Nx64 kbps TDM service on a DS1 for 1<1<24. The name derives from the fact that the service provides a fraction of the T1.

frame - A commonly used named for the data link layer PDU.

Frame Check Sequence (FCS) - A field in an X.25, SDLC, or HDLC frame which contains the result of a CRC error detection and correction algorithm.

Frame Relay (FR) - An ANSI and ITU-T defined WAN networking standard for switching frames between end users. It operates at higher speeds with less nodal processing than X.25. ATM borrows many concepts from frame relay.

Frame-based User-Network Interface (FUNI) - A protocol using a modified DXI header defined for access to ATM service at speeds ranging from 64 kbps up to DS1/E1 rates. In addition to DXI capabilities, FUNI adds support for OAM and RM cells.

Frequency Division Multiplexing (FDM) - A method of aggregating multiple simultaneous connections over a single high-speed channel by using individual frequency passbands for each connection.

full duplex - The simultaneous bidirectional transmission of information over a common medium.

Generic Call Admission Control (GCAC) - A parameter controlled CAC procedure defined in PNNI to allow a source to estimate the connection admission decisions along candidate paths to the destination.

Generic Cell Rate Algorithm (GCRA) - A reference model proposed by the ATM Forum for defining cell-rate conformance in terms of certain traffic parameters and tolerances (i.e., PCR, CDVT< SCR, BT, MBS). It is often referred as the leaky bucket algorithm. See UPC.

Generic Flow Control (GFC) - A 4-bit field in the ATM cell header with local significance to allow a multiplexer to control two levels of priority in attached terminals. Few implementations support GFC.

Guaranteed Frame rate (GFR) - Another service category under development by the ATM Forum oriented toward frame-based, instead of cell-based, QoS and traffic paramters.

half duplex - The bidirectional transmission of information over a common medium, where information travels in only one direction at any time as determined by a control protocol.

Header Error Control (HEC) - A 1-octet field in the ATM cell header containing a CRC checksum on the cell header fields. HEC is capable of detecting multiple bit errors or correcting single bit errors.

High Level Data Link Control (HDLC) - A widely used ISO and ITU-T standardized link layer protocol standard for point-to-point and multi-point communications. The frame relay, SNA, X.25, PPP, DXI, and FUNI protocols all use HDLC.

host - An end station in a network.

Hyper Text Transfer Protocol (HTTP) - A protocol in the TCP/IP suite operating over TCP that implements home pages on the World Wide Web (WWW).

IMA Control Protocol (ICP) - The mechanism that enables Inverse

Multiplexing over ATM (IMA) to combine multiple cell streams at the receiving end. ICP cells enable endpoints to negotiate and configure, synchronize, and coordinate multiple physical links.

implicit congestion notification - A congestion indication which is performed by upper-layer protocols (e.g., TCP) rather than network or data link layer protocol conventions.

Information Element (IE) - A variable length field within an signaling message defined in terms of type, length and values according to various standards and specifications.

Initial Cell Rate (ICR) - The rate that a source is allowed to start up at following an idle period. It is established at connection set-up and is between the MCR and the PCR.

Initial Domain Part (IDP) - The high-order portion of an ATM End System Address (AESA) assigned by a network. It contains the Authority and Format Identifier (AFI) and the Initial Domain Identifier (IDI).

Institute of Electrical and Electronics Engineers (IEEE) - A worldwide engineering standards-making body for the electronics industry.

Integrated Local Management Interface (ILMI) - An ATM Forum defined, SNMP-based management protocol for an ATM UNI. It provides interface status and configuration information, and also supports dynamic ATM address registration.

Inter-Exchange Carrier (IXC or IEC) - A public switching network carrier providing connectivity between LATAs.

interface - In OSI, the boundary between two adjacent protocol layers (i.e., link to network). In the ITU-T, a physical connection between devices.

Interim Inter-switch Signaling Protocol (IISP) - An ATM Forum defined protocol employing UNI-based signaling for switch-to-switch communication in private networks. Unlike PNNI, IISP relies on static routing tables and makes support for QoS an alternate routing optional..

International Code Designator (ICD) - One of three Authority and Format Identifiers (AFI) defined for ATM End System Addresses (AESA). A 2-byte field within the 20-octet NSAP-based ATM address, I format that uniquely identifies an international organization.

International Standards Organization (ISO) - An international organization for standardization, based in Geneva, Switzerland, that establishes voluntary standards. Well known for the seven-layer OSI Reference Model (OSIRM).

International Telecommunications Union- (ITU) - The new name for the CCITT, a U.N. treaty organization. The ITU Telecommunications standardization sector (ITU-T) defines standards for N-ISDN, frame relay, B-ISDN, and a variety of other transmission, voice, and data related protocols.

Internet Engineering Task Force (IETF) - Organization responsible for standards and specification development for TCP/IP networking.

Internet Protocol (IP) - A connectionless datagram-oriented network layer (layer 3) protocol containing addressing and control information in the packet header that allows nodes to independently, yet consistently, forward packets toward the destination. Associated routing protocols discover and maintain topology information and for routing packets across a wide range of network topologies and lower layer protocols.

Interoperability - The ability of multiple, dissimilar vendor devices and protocols to operate and communicate using a standard set of rules and protocols.

Inverse Multiplexing over ATM (IMA) - An ATM Forum defined protocol that distributes a stream of cells over several physical circuits (typically DS1s or E1s) in a round-robin fashion at the transmitter. The receiving end then reassembles these streams back into their original order using the IMA Control Protocol (ICP).

isochronous - A capability supporting allows a consistent timed access of network bandwidth for time-sensitive transmission of voice and video traffic.

jitter - The deviation of a transmission signal in time or phase. Accumulated jitter may introduce errors and loss of synchronization.

label switching - A generic technique for relaying data packets through a network by switching to an output port based upon the received label and swapping the label value prior to transmission on the output port. Examples are frame relay, ATM, and MPLS. Also called address switching.

LAN Emulation (LANE) - An ATM Forum specification enabling ATM devices to seamlessly interwork with legacy LAN devices using either the Ethernet or Token Ring protocols. Higher layer protocols run without change over LANE on ATM-based systems. LANE employs a number of functional elements in performing this role (see LEC, LECS, LES, and BUS).

LAN Emulation Client (LEC) - Software residing in every ATM device, such as workstations, routers, and LAN switches. An LEC has an ATM address and performs data forwarding, address resolution, connection control, and other control functions.

LAN Emulation Configuration Server (LECS) - The LANE component involved with configuration of Emulated LANs (ELANs) by assigning LECs to specific LANE Servers (LES). Network administrators control the combination of physical LANs to form VLANs.

LAN Emulation Network Node Interface (LNNI) - Specifies the NNI operation between the LANE servers (LES, LECS, BUS).

LAN Emulation Server (LES) - LECs register their ATM and MAC addresses with the LES. The LES then provides MAC address to ATM translation via the LANE Address Resolution Protocol (LE-ARP).

LAN Emulation User-to-Network Interface (L-UNI) - The ATM Forum standard for LAN Emulation on ATM networks; defines the interface between the LAN Emulation Client (LEC) and the LAN Emulation Server components. (see BUS, LES, LECS)

layer management - Network management functions which provide information about the operations of a given OSI protocol layer.

Layer Management Interface (LMI) - In DQDB, the interface between the LME and network management systems.

Leaf Initiated Join (LIJ) - A procedure defined in the ATM Forum 4.0 signaling specification that allows leaf nodes to dynamically join and leave a point-to-multipoint connection using special messages.

leaky bucket - A colloqiual term used to describe the algorithm used for conformance checking of cell flows against a set of traffic parameters. See GCRA, UPC and NPC. The leak rate of the bucket defines a particular rate, while the bucket depth determines the tolerance for accepting bursts of cells.

Least Significant Bit (LSB) - The lowest order bit in the binary representation of a numerical value.

Line-Terminating Equipment (LTE) - A device which either originates or terminates a SONET OC-N signal.

Local Access and Transport Area (LATA) - Regulatory defined, geographically constrained telecommunication areas, within which a local carrier can provide communications services. See LEC, IXC.

Local Area Network (LAN) - A MAC-level data and computer communications network confined to short geographic distances. Various LAN standards exist, with Ethernet as the most widely used.

Local Exchange Carrier (LEC) - An intra-LATA communication services provider, such as an RBOC or an independent telephone company.

Local Management Interface (LMI) - A set of user device–to–network communications standards used in ATM DXI and frame relay.

Logical Group Node (LGN) - A node that represents a lower level peer group in the PNNI hierarchy.

logical link - An abstract representation of the connectivity between two logical nodes used in PNNI. This includes one or more physical circuits or virtual path connections.

Logical Link Control (LLC) - The sublayer defined by the IEEE 802.2 standard that interfaces with the Medium Access Control (MAC) sublayer of the data link layer in LAN standards. It performs error control, broadcasting, multiplexing, and flow control functions.. LLC provides a common interface for all IEEE 802.X MAC protocols to user systems.

logical node - An abstract representation of a peer group or a switching system as a single point in PNNI identified by a unique string of bits.

Loss of Frame (LOF) - A condition at the receiver or a maintenance signal transmitted in the PHY overhead indicating that the receiving equipment has lost frame delineation. This is used to monitor the performance of the PHY layer.

MAC address - A six octet value uniquely identifying an endpoint and which is sent in an IEEE LAN frame header to indicate either the source or destination of a LAN frame.

Management Information Base - A structured set of objects (e.g., integers, strings, counters, etc.) accessible via a network management protocols like SNMP and CMIP. The objects represent values that can be read or changed. See ILMI, and AToMMIB for examples.

Maximum Burst Size (MBS) - A traffic parameter that specifies the maximum number of cells a that can be transmitted at the connection's Peak Cell Rate (PCR) such that the maximum rate averaged over many bursts is no more than the Sustainable Cell Rate (SCR). See BT, UPC.

medium - The physical means of transmission (e.g., twisted-pair wire, coaxial cable, optical fiber, radio waves, etc.).

Medium Access Control (MAC) - A protocol defines by the IEEE that controls workstation access to a shared transmission medium. The MAC sublayer interfaces to the Logical Link Control (LLC) sublayer and the PHYsical layer. Examples are 802.3 for Ethernet CSMA/CD, and 802.5 for Token Ring.

Message Identifier (MID) - In DQDB and AAL3/4, a value used to identify all slots or cells that make up the same PDU.

Message Transfer Part (MTP) - Level 3 of the SS7 signaling protocol stack used within carrier networks.

Metropolitan Area Network (MAN) - A network which operates over metropolitan area distances and number of subscribers. A MAN can carry voice, video, and data. Examples are DQDB and FDDI.

Minimum cell rate (MCR) - An ABR traffic descriptor that specifies a rate at which the source may always transmit traffic.

modem (modulator-demodulator) - A specific type of Data Communication Equipment (DCE) that converts serial DTE data to a signal suitable for transmission, for example, over an analog telephone line. It also converts received signals into serial digital data for delivery to the DTE.

Most Significant Bit (MSB) - The highest order bit in the binary representation of a numerical value.

multicast - A connection type with the capability to broadcast to multiple destinations on the network.

Multicast Address Resolution Server (MARS) - Comparable to the Address Resolution Protocol (ARP) specified for Classical IP over ATM, MARS resolves IP addresses to a single ATM multicast address that defines a cluster of end points.

multimedia - A means of presenting a combination of information such as text, data, images, video, audio, graphics to the user in an integrated fashion.

Multimode Fiber (MMF) - 50- to 100-µm core diameter optical fiber in which the light propagates along multiple paths or modes. Since the pulses travel different distances for each mode, pulse dispersion limits the transmission distance when compared with single-mode optical fiber.

multiplexing - A technique that combines multiple streams of information to share a physical medium. See TDM, FDM, label switching.

multipoint-to-point - A unidirection connection topology where multiple leaf nodes transmit back toward a root node. The intermediate nodes must provide a means to merge these streams; for example by using separate VCCs on a multipoint-to-point VPC.

multiprotocol encapsulation over ATM - Defined in IETF RFC 1483, it allows multiple higher-layer protocols, such as IP or IPX, to be routed over a single ATM VCC using the LLC/SNAP MAC header.

Multiprotocol Over ATM (MPOA) - An ATM Forum defined means to route ATM traffic between virtual emulated LANs, bypassing traditional routers. MPOA utilizes the Next Hop Resolution Protocol (NHRP) to determine short-cut paths between IP-addressed ATM systems attached to a common ATM SVC network.

Narrowband-Integrated Services Digital Network (N-ISDN) - Predecessor to the B-ISDN, N-ISDN encompasses the standards for operation at the PRI rate and below. N-ISDN interworks with the traditional telephony standards.

network - A system of autonomous devices connected via physical media that provide a means for communications.

network interworking - A method of trunking frame relay connections over an ATM network defined in the Frame Relay Forum's FRF.5 implementation agreement.

network management - The process of managing the operation and status of network resources (e.g., devices, protocols).

Network Management System (NMS) - Equipment and software used to monitor, control, and provision a communications network. See OAM&P.

Network Node Interface (NNI) - An interface between ATM switches defined as the interface between two network nodes.

Network Node Interface (or Network-to-Network Interface) (NNI) - An ITU-T-specified standard interface between nodes in different networks. The ATM Forum distinguishes between two NNI standards, one for private networks called PNNI and another public networks called the B-ICI.

Network Parameter Control - Traffic management mechanism performed at the NNI for traffic received from another network. See UPC.

Network Service Access point (NSAP) - An OSI format defining the octet network address used to define an ATM End System Address (AESA).

Next Hop Resolution Protocol (NHRP) - An IETF defined protcol for routers to learn network-layer addresses over NBMA networks like ATM. A Next Hop Server (NHS) responds to router queries regarding the next best hop toward the destination.

node - A point of interconnection with in a network.

Non-Broadcast Multiple Access (NBMA) - A link layer network that cannot perform broadcasting, yet has multiple access points. Examples are frame relay and ATM.

non-real time Variable Bit Rate (nrt-VBR) - An ATM service category for traffic well-suited to packet data transfers. It may experience significant CDV. The traffic parameters are PCR, CDVT, SCR, and MBS.

Nrm - An ABR service parameter that specifies the maximum number of cells a source may send for each forward RM cell.

octet - The networking term designating a field 8 bits long. Usually synonymous with byte from the world of computing

Open Shortest Path First (OSPF) - A routing protocol defined for IP that uses the Djikstra algorithm to optimally determine the shortest path. OSPF also defines a reliable topology distribution protocol based upon controlled flooding.

Open Systems Interconnection Reference Model (OSIRM) - A seven-layer model defining a suite of protocol standards for data communications. X.25 and CMIP are examples.

Operations And Maintenance (OAM) - A set of diagnostic and alarm reporting mechanisms defined by the ITU-T using special purpose cells. Functions supported include fault management, continuity checking and Performance Measurement (PM).

Operations, Administration, Maintenance and Provisioning (OAM&P) - A set of network management functions and services that interact to provide the necessary network management tools and control.

Optical Carrier (OC-n) - Fundamental unit in the Sonet (Synchronous Optical Network) hierarchy. OC indicates an optical signal and n represents increments of 51.84 Mbit/s. Thus, OC-1, -3, and -12 equal optical signals of

51, 155, and 622 Mbit/s.

Optical Carrier level N (OC-N) - The optical carrier level signal in SONET which results from converting an electrical STS-N signal into optical pulses. The basic transmission speed is an STS-1 at 51.84 Mbps, hence OC-3 is 155.52 Mbps, OC-12 is 622.08 Mbps, and OC-48 is 2.488 Mbps.

Organizationally Unique Identifier (OUI) - A 3-octet field defined in the IEEE 802.1a standard in the SubNetwork Attachment Point (SNAP) header. It identifies administering organization for the two octet Protocol Identifier (PID) field in the SNAP header. Together the OUI and PID identify a particular routed or bridged protocol.

packet - An ordered group of data and control signals transmitted through a network as a subset of a larger message.

Packet Assembler Disassembler (PAD) - A device that converts between serial data and packet formats.

packet switching - A switching technique that segments user data into fixed or variable units called packets prefixed with a header. The network operates on the fields in the header in a store-and-forward manner to reliably transfer the packet across the network to the destination.

Partial Packet Discard (PPD) - A mechanism that discards all remaining cells of an AAL 5 PDU (except the last one) after a device drops one or more cells of the same AAL5 PDU. This technique avoids transmitting cells across the network that the AA5 receiver would fail to reassemble. See Early Packet Discard (EPD).

payload - The 48-octet part of the 53-octet ATM cell, that contains either user information or OAM data as specified by the Payload Type Indicator (PTI) and VPI/VCI fields in the cell header.

Payload Type Indicator (PTI) - A 3-bit field in the ATM cell header that indicates whether the cell contains user data, OAM data, or resource management data. The PTI field contains the AAL indication bit used by AAL5 to identify the last cell in a PDU as well as the Explicit Forward Congestion Indication (EFCI) bit.

Peak Cell Rate (PCR) - A traffic parameter that characterizes the maximum rate at which a source can transmit cells according to the leaky bucket algorithm with a tolerance parameter defined by CDVT.

peer group - A set of nodes within a PNNI network sharing a common address prefix at the same hierarchical level. Nodes in a peer group exchange PTSEs.

Peer Group Leader (PGL) - A node elected by its peers to represent the peer group in the next level of the hierarchy.

Permanent Virtual Channel Connection (PVCC) - A Virtual Channel Connection (VCC) permanently provisioned by a network management function.

Permanent Virtual Connection (PVC) - A Virtual Path or Channel Connection (VPC or VCC) provisioned for indefinite use in an ATM network by a network management system (NMS). See PVPC, SPVPC, PVCC, and SPVCC.

Permanent Virtual Path Connection (PVPC) - A Virtual Path Connection (VPC) permanently provisioned by a network management function.

Physical layer (PHY) - The bottom layer of the ATM protocol stack, which defines the interface between ATM traffic and the physical medium. It consists of two sublayers: the Physical Medium-Dependent (PMD) sublayer and the Transmission Convergence (TC) sublayer.

Physical Layer Convergence Protocol (PLCP) - A protocol specified within the TC sublayer defining a framing structure over DS3 facilities based upon the 802.6 DQDB standard. Supplanted by HEC based cell delineation which makes more efficient use of DS3 bandwidth.

Physical Layer Medium Dependent (PMD) - The lower sublayer of the physical layer that defines how to transmit and receive cells over the transmission medium.

Plesiochronous Digital Hierarchy (PDH) - A Time Division Multiplexing (TDM) technique for carrying digital data over legacy high speed transmission systems. Plesiochronous means nearly synchronous as compared with truly synchronous transmission systems defined by the SONET and SDH standards.

PNNI Topology State Element (PTSE) - PNNI information about available bandwidth, QoS capability, administrative cost, and other factors used by the routing algorithm.

PNNI Topology State Packet (PTSP) - The PDU and flooding procedure used by PNNI to reliably distribute topology information throughout a hierarchical network.

point-to-multipoint - A unidirectional connection topology with one root node transmitting to two or more leaf nodes.

point-to-point - A unidirectional or bidirectional connection topology with only two endpoints.

policing - A term commonly used to refer to Usage Parameter Control (UPC). See UPC.

port - The physical interface on a computer, multiplexer, router, CSU/DSU, switch, or other device.

Primary Rate Interface (PRI) - A N-ISDN access interface type operating on a DS1 physical interface organized as 23 Bearer (B) channels operating at 64 kbps and one Data (D) channel operating at 64 kbps (23B+D). The European version operates on an E1 interface organized as 30 B channels and one D channel.

Private Branch Exchange (PBX) - A circuit switch that connects telephony-based equipment within an enterprise and provides access to the Public Switched Telephone Network (PSTN).

private network - A network providing intra-organizational connectivity only. Typically implemented via interconnecting switches, multiplexers, PBXes, or routers via leased lines. See VPN.

Private Network- Network Interface (NNI) - Also known as a Private Network-Node Interface it aims to provide interoperability in multi-vendor networks. ATM Forum defined signaling and routing protocols dynamically discover the topology and establish SVC connections on demand using source routing. See crankback, DTL, GCAC, peer group, source routing, SPVPC, and SPVCC.

protocol - A formal set of conventions and rules governing the formatting and sequencing of message exchange between two communicating systems.

Protocol Data Unit (PDU) - A unit of data consisting of control information and user data exchanged between peer layer processes. Also known as a message it contains information in the header and trailer fields associated with the particular layer.

Protocol IDentification (PID) - See OUI.

public network - A communications network operated by a service provider that shares resources across multiple users.

Public Switched Telephone Network (PSTN) - The telecommunications network commonly accessed by ordinary telephones, PBXes trunks, modems, and facsimile machines

Pulse Code Modulation (PCM) - A technique for converting an analog signal (such as voice) into a digital stream representing digitized samples of the original analog signal.

QoS class - One of five broad groupings of QoS parameters outlined by the ATM Forum's UNI 3.1 specification. Largely replaced by the notion of ATM service categories and specific QoS parameters.

Quality of Service (QOS) - A set of parameters and measurement procedures defined by the ITU-T and the ATM Forum to quantify loss, errors, delay, and delay variation. See CLR, CER, CMR, SECB, CDV, CTD.

Rate Decrease Factor (RDF) - A factor by which an ABR source should decrease its transmission rate if congestion occurs.

Rate Increase Factor (RIF) - A factor by which an ABR source can increase its transmission rate if the received RM cell indicates no congestion.

real time - Variable Bit Rate (rt-VBR) - An ATM service category that supports traffic requires transfer of timing information via constrained CDV. It is suitable for carrying packetized video and audio. The PCR, CDVT, SCR, and MBS traffic parameters characterize the source.

Real Time Protocol (RTP) - Part of the IP suite, RTP conveys sequence numbers and time-stamps between packet video and audio applications. Widely used in Internet Telephony.

Regional Bell Operating Company (RBOC) - Local service telephone companies that resulted from the break-up of AT&T in 1984.

Remote Defect Indication (RDI) - One of the OAM function types used for fault management. The connection endpoint generates RDI after receiving AIS for a specified interval of time, nominally 3 seconds.

Remote Monitoring (RMON) MIB - A database which allows for remote configuration and data retrieval from a network monitoring device.

Resource Management (RM) - Specific cells that communicate information about the state of the network like bandwidth availability and impending congestion to the source and destination. ABR makes use of RM cells to implement closed loop flow and congestion control.

Resource reSerVation Protocol (RSVP) - A protocol developed by the IETF to supporting different classes of service for IP flows. Current standards define guaranteed QoS and controlled load service classes. The RSVP message communicate traffic parameters that are semantically similar to ATM traffic parameters.

ring - A closed-loop, common bus network topology.

Round-Trip Time (RTT) - The delay incurred in a transmission of a message from a source to a destination, the generation of a response by the destination node, and the subsequent transmission of the response back to the source.

router - A device operating at layers 1 (physical), 2 (data Link), and 3 (network) of the OSI model.

routing - The process of selecting the most efficient circuit path for a message.

section - A transmission facility between a SONET network element and regenerator.

segment - In DQDB, the payload (user data) portion of the slot.

Segmentation and Reassembly (SAR) Sublayer - The lower half of the common part AAL for types 1, 3/4, and 5. Converts PDUs received from the CPCS sublayer into ATM cells for transmission by the PHY layer at the source. The destination SAR extracts cell payloads received from the TC sublayer and assembles them back into PDUs for delivery to CPCS. Each AAL type has its own unique SAR format and procedures.

Selector (SEL) - The low order 1-octet field in the Domain Specific Part (DSP) of a 20-octet ATM End System Address (AESA). The SEL byte is reserved for multiplexing by the end user. The network does not use this field for routing decisions.

self-healing - The ability for a network to reroute traffic around a failed link or network element, thus automatically restoring service.

Service Access Point (SAP) - The point where services of a lower layer are available to the next higher layer. The SAP is named according to the layer providing the services. For example, the interface point between the PHY layer and the ATM layer is called the PHY-SAP.

Service Data Unit (SDU) - A unit of information transferred across a Service Access Point (SAP).

service interworking - Defined in the Frame Relay Forum FRF.8 specification, it provides a method of transparently interworking frame relay and ATM end users. An interworking function converts between semantically equivalent fields in the frame relay and ATM headers.

Service Specific Connection-Oriented Protocol (SSCOP) - Part of the SSCS portion of the Signaling AAL (SAAL). SSCOP is an end-to-end protocol that provides error detection and correction by retransmission, status reporting between the sender and the receiver, and guaranteed delivery integrity. UNI and NNI signaling protocols use the same SSCOP layer.

Service Specific Convergence Sublayer (SSCS) - The portion of the Convergence Sublayer (CS) dependent upon the specific higher layer protocol. Examples are the frame relay service-specific convergence sublayer (FR-SSCS), the Switched Multimegabit Data Service service-specific convergence sublayer (SMDS-SSCS), and the Signaling AAL (SAAL).

Service Specific Coordination Function (SSCF) - A function defined in Signaling ATM Adaptation Layer-(SAAL) that resides between the SSCOP and the signaling application. ITU-T standards define different SSCF functions at the UNI and NNI.

shaping - A mechanism used by the source to modify bursty traffic characteristics to match specified traffic parameters, such as PCR, CDVT, SCR, and MBS. See GCRA.

Shielded Twisted Pair (STP) - A twisted-pair wire with jacket shielding, used for longer distance and/or higher speed transmission. Less sensitive to electrical noise and interference than UTP.

Signaling ATM Adaptation Layer (SAAL) - A set of protocols residing between the ATM layer and the signaling application. See SSCOP and SSCF.

Signaling System Number 7 (SS7) - An ITU-T defined common channel signaling standard used in telephony and ISDN. It provides NNI signaling and network intelligence utilized in telephony.

Simple Efficient Adaptation Layer (SEAL) - The original name for AAL5.

Simple Network Management Protocol (SNMP) - The Internet network management protocol. SNMP provides a means to monitor status and performance as well as set configuration parameters.

simplex - One-way transmission of information on a medium.

Single-Attachment Stations (SAS) - An FDDI station attached to only the primary ring. SAS does not provide self-healing in FDDI.

Single-Mode Fiber (SMF) - 8- to 10-μm core diameter optical fiber with a single propagation path (i.e., mode) for light. Typically used for higher speeds or longer distances as compared with multimode optical fiber.

slot - The basic unit of transmission on a DQDB bus.

SMDS Interface Protocol (SIP) - Three levels of protocol which define the SMDS SNI user information frame structuring, addressing, and error control.

Soft Permanent Virtual Channel Connection (SPVCC) - An ATM Virtual Channel Connection (VCC) provisioned at the end points using the PNNI MIB. The network employs signaling message with a unique SPVCC Information Element (IE) to dynamically establish and restore the intermediate portions of the connection.

Soft Permanent Virtual Path Connection (SPVPC) - An ATM Virtual Path Connection (VPC) provisioned at the end points using the PNNI MIB. The network employs signaling message with a unique SPVPC Information Element (IE) to dynamically establish and restore the intermediate portions of the connection.

source routing - A scheme in which the originator determines the end-to-end route. Examples are Token Ring and PNNI.

station - An addressable logical or physical network entity, capable of transmitting, receiving, or repeating information.

Structured Data Transfer (SDT) - An AAL1 data transfer mode which supports transfer of Nx64 kbps TDM payloads. It requires the availability of an accurate clock at both the source and destination.

Subscriber-Network Interface (SNI) - A DQDB user access point into an SMDS network.

Sustainable Cell Rate (SCR) - A traffic parameter that characterizes a bursty source. It defines the maximum allowable rate for a source in terms of the Peak Cell Rate (PCR) and the Maximum Burst Size (MBS). It is equal to the ratio of the MBS to the minimum burst interarrival time.

switch fabric - The central function of switch, which buffers and routes incoming PDUs to the appropriate output ports.

Switched Multimegabit Data Service (SMDS) - A connectionless MAN service offered using the DQDB protocol or HDLC framing.

Switched Virtual Channel Connection (SVCC) - A VCC established and released dynamically via control plane signaling.

Switched Virtual Connection (SVC) - A logical connection between endpoints established by the ATM network on demand based upon signaling messages received from the end user or another network. See SVCC, SVPC.

Switched Virtual Path Connection (SVPC) - Defined in the ATM Forum 4.0 signaling specification as a VPC established and released dynamically via control plane signaling.

synchronous - Transmission, switching, and multiplexing systems utilizing highly accurate clock sources. See SONET, SDH.

Synchronous Data Link Control (SDLC) - IBM defined protocol for use in SNA environments. SDLC is a bit oriented protocol, similar to HDLC.

Synchronous Digital Hierarchy (SDH) - ITU-T defined standard for the physical layer of high speed optical transmission systems. Similar to SONET in terms of transmission speeds; however, differences in overhead functions make these systems incompatible.

Synchronous Optical Network (SONET) - A Bellcore and ANSI defined North American standard for high-speed fiber-optic transmission systems. Similar to the ITU-T SDH, but incompatible.

Synchronous Residual Time Stamp (SRTS) - A mode defined in AAL1 for transparently transferring timing from a source to a destination. It requires that both the source and destination have accurate clock frequencies.

Synchronous Transfer Mode (STM) - An ITU-T defined communications method that transmits a group of Time Dvision Multiplexed (TDM) streams synchronized to a common reference clock. In contrast to ATM, STM systems reserve bandwidth according to a rigid hierarchy regardless of actual channel usage.

Synchronous Transfer Module (STM-n) - Basic unit of SDH (Synchronous Digital Hierarchy) defined in increments of 155.52 Mbps. The variable n represents multiples of this rate. Commonly values of n are STM-1 at 155.52 Mbps, STM-4 at 622.08 Mbps, and STM-16 at 2.488 Gbps.

Synchronous Transfer Signal (STS-n) - Basic unit of the SONET multiplexing hierarchy defined in increments of 51.84 Mbps. Commonly encountered rates are: STS-3 at 155.52 Mbps, STS-12 at 622.08 Mbps, and STS-48 at 2.488 Gbps. Usually, STS is viewed as an electrical signal to differentiate it from the SONET optical signal level designated as OC-N.

Synchronous Transport Module n concatenated (STM-nc) - SDH standard transmission rates that treat the payload as a single concatenated bit stream. ATM uses this mode of SDH transmission. For example, the payload rate for STM-1c is 149.76 Mbps, and STM-4c is 599.04 Mbps.

Synchronous Transport Signal Level nc (STS-nc) - A SONET mode that concatenates the Synchronous Payload Envelopes (SPE) to deliver a single high-speed bit stream to the higher level user. ATM uses this mode of

SONET transmission. For example, the payload rate for STS-3c is 149.76 Mbps, and STS-12c is 599.04 Mbps.

T1 - A four-wire transmission repeater system for operation at 1.544 Mbps. Commonly used as colloquial name for a DS1 formatted digital signal. See DS1.

T3 - Commonly used as colloquial name for a DS3 formatted digital signal. See DS3.

telecommunications - The transmission, switching, and multiplexing of voice, video, data, and images.

time slot - A set of bits in a serial bit stream determined by framing information using the TDM paradigm. In DS1 and E1 TDM systems, an 8-bit time slot transferred 8,000 times per seconds delivers a 64 kbps channel.

Time-Division Multiplexing (TDM) - The method of aggregating multiple simultaneous transmissions (circuits) over a single high-speed channel by using individual time slots for each circuit.

token - A marker indicating the station's right to transmit that can be held by a station on a Token Ring or FDDI network.

Token Ring - A LAN protocol standardized in IEEE 802.5 that uses a token-passing access method for carrying traffic between network elements. Token Ring LANs operate at either 4 or 16 Mbps.

traffic contract - An agreement between the user and the network regarding the expected QoS provided by the network subject to user compliance with the predetermined traffic parameters (i.e., PCR, CDVT, SCR, MBS).

traffic descriptor - Generic traffic parameters that capture the intrinsic traffic characteristics of a requested ATM connection. See UPC, CAC, PCR, SCR, MBS, and ABR.

Transmission Control Protocol (TCP) - A widely used layer 4 transport protocol to reliably deliver packets to higher layer protocols like FTP and HTTP. It performs sequencing and reliable delivery via retransmission since the IP layer does not perform these functions. The dynamic windowing feature of TCP dynamically maximizes the usage of bandwidth.

Transmission Control Protocol/ Internet Protocol (TCP/IP) - The combination of a network and transport protocol developed over the past decades internetworking. Also known as the Internet Protocol Suite. See IP, FTP, HTTP, routing, RTP, TCP, and UDP.

Transmission Convergence (TC) - A sublayer of the PHY layer in ATM standards responsible for transforming cells into a steady flow of bits for transmission by the Physical Medium Dependent (PMD) sublayer. On transmit, the TC sublayer maps the cells to the frame format, generates the Header Error Check (HEC), and sends idle cells during periods with no ATM traffic. On reception, the TC sublayer delineates individual cells within the received bit stream by using the HEC to detect and correct received errors.

twisted pair - The basic transmission medium, consisting of 22 to 26 American Wire Gauge (AWG) insulated copper wire. TP can be either shielded (STP) or unshielded (UTP).

unassigned cell - A cell inserted by the Transmission Convergence (TC) sublayer when the ATM layer has no cells to transmit. This function decouples the physical transmission rate from the ATM cell rate of the source.

Unshielded Twisted Pair (UTP) - A twisted-pair wire without jacket shielding, used for short distances since it is more subject to electrical noise and interference than Shielded Twisted Pair (STP).

Unspecified Bit Rate (UBR) - An ATM service category where the network makes a best effort to deliver traffic. It has no QoS guarantees with only a single traffic parameter, PCR.

Usage Parameter Control (UPC) - A set of actions taken by the network to monitor and control traffic offered by the end user. Commonly used by public ATM service providers to prevent malicious or unintentional misbehavior, which may adversely affect the performance of other connections. UPC detects violations of negotiated traffic contract parameters according to the Generic Cell Rate Algorithm (GCRA). Non-conforming cells may be tagged with the Cell Loss Priority (CLP) bit in the ATM cell header, or discarded.

User Datagram Protocol (UDP) - A connectionless datagram-oriented transport-layer protocol belonging to the IP suite.

User-to Network Interface (UNI) - The interface point between end users and a network defined in ITU-T and ATM Forum documentation.

Variable Bit Rate (VBR) - A generic term for sources that transmit data intermittently. The ATM Forum divides VBR into a real-time and non-real-time (rt-VBR and nrt-VBR) service categories in terms of support for constrained Cell Delay Variation (CDV) and Cell Transfer Delay (CTD). See nrt-VBR and rt-VBR.

Variance Factor (VF) - A measure used in PNNI GCAC computed in terms of the aggregate cell rate on a link

Video on Demand (VoD) - A service that enables users to remotely select and control the playback of video content.

Virtual Channel (VC) - A shorthand name for a VCC.

Virtual Channel Connection (VCC) - A concatenation of Virtual Channel Links (VCLs) connecting endpoints which access higher-layer protocols. VCCs are unidirectional. Switches and multiplexers operate on both the VPI and VCI fields in the ATM cell header. VCCs may be permanent (PVCC), switched (SVCC), or provisioned via PNNI procedures (SPVCC).

Virtual Channel Identifier (VCI) - The 16-bit value in the ATM cell header which in conjunction with the VPI value identifies the specific virtual channel on the physical circuit.

Virtual Channel Link (VCL) - A reference configuration in ATM networks between two connecting points where the VPI and VCI values are either modified or removed. Typically, every ATM device is a VCL.

Virtual Path (VP) - A shorthand name for a VPC.

Virtual Path Connection (VPC) - A concatenation of Virtual Path Links (VCLs) connecting Virtual Path Termination (VPT) endpoints. VPCs are unidirectional. Switches and multiplexers operate on only the VPI fields in the ATM cell header. A VPC carries multiple VCCs. VPCs may be permanent (PVPC), switched (SVPC), or provisioned via PNNI procedures (SPVPC).

Virtual Path Connection Identifier (VPCI) - An index used in signaling messages to logically associate a real VPI with a particular interface. Employed when external VP multiplexing occurs in a network prior to the interface on an ATM switch in proxy signaling and virtual UNIs as defined in the ATM Forum 4.0 signaling specification.

Virtual Path Identifier (VPI) - The value in the ATM cell header which identifies the specific virtual path on the physical circuit. The VPI field is 8 bits long at the UNI and 12 bits long at the NNI.

Virtual Path Link (VPL) - A reference configuration in ATM networks between two connecting points where the VPI value is either modified or removed. Typically, every ATM device is a VPL.

Virtual Source/ Virtual Destination (VS/VD) - A mode of Available Bit Rate (ABR) service that partitions flow control into a series of concatenated flow control loops, each with a virtual source and destination. This capability enables networks to reshape traffic flows via the virtual source function, thereby ensuring fair operation across all ABR users.

Virtual Tributary (VT) - A lower level time division multiplexing structure used in SONET.

Wide Area Network (WAN) - A network that operates over a large geographic region. Usually, WANs employ carrier facilities and/or services.

window - The concept of establishing an optimum number of frames or packets that can be outstanding (unacknowledged) before the source transmits again. Windowing protocols include X.25, TCP, and HDLC.

Index

ABOUT THE AUTHORS

David E. McDysan has worked in a wide field of data communications ranging from undersea, to satellite, to terrestrial radio and fiber optics. Dr. McDysan received a BSEE degree from Virginia Polytechnic Institute, in Blacksburg, Virginia and the MSEE and Ph.D. degrees from the George Washington University in Washington, D.C. His doctoral dissertation was on the subject of performance analysis of queuing system models for distributed computer communications systems, focusing on statistical multiplexing and ATM switching. Dr. McDysan has published a number of papers on various subjects related to traffic engineering, system design considerations, and performance analysis of data communications systems.

Dr. McDysan has worked primarily in the field of data communications, beginning with Computer Sciences Corporation on a number of government contracts covering undersea, radio, and satellite data communications. Since then he has been with Satellite Business Systems, which was acquired by MCI Communications in 1986. He has held a number of engineering and project leadership positions in the development of protocol enhancements, vendor management, architecture definition, performance evaluation, capacity planning, requirements definition, and feasibility demonstration. These projects included active participation in digital telephony, the intelligent network, network management architecture, a switched DS1 service, frame relay, SMDS, and most recently, ATM. Dr. McDysan was responsible for the architecture and planning of ATM-based services at MCI. He managed the design and deployment of the world's first wide area OC3 and OC12 ATM networks as Senior Manager of ATM Engineering at MCI. He now holds an executive staff position at MCI responsible for planning future data networks.

Dr. McDysan is active in many standards and professional organizations. He is a past Vice President of the ATM Forum board, and past chairperson of the traffic management and Frame-based UNI groups in the ATM Forum. He speaks regularly at conferences and seminars about ATM, traffic management, and real world networking experiences.

Darren L. Spohn is founder and principal of Spohn & Associates. Spohn & Associates is an Austin-based firm founded with the primary goal of providing specialized data network consulting and training to end users; corporations; and venture capital, large consulting, and analysis firms.

Prior to founding Spohn & Associates, Mr. Spohn served as the Chief Technology Officer and Vice President of Engineering and Development at NetSolve Inc. in Austin, TX. There he managed NetSolve's engineering, product and software development, information systems, and business channel development efforts. Prior to his role as CTO, he served two years as Vice President of Operations. NetSolve provides comprehensive WAN and LAN network management out-tasking services.

Prior to joining NetSolve, Darren held a number of critical engineering, marketing, and management positions at MCI Communications. He served as the Manager of Data Product Marketing and Development, after holding the position of Manager of Data Network Engineering and Design, with responsibility for MCI's internal data networks, as well as customer-switched and multiplexed network design. These responsibilities included leading the design efforts of four intelligent national data networks, including an NET intelligent multiplexer network and the frame relay/cell switch network that forms the technology platform for MCI's first generation HyperStream Data Services.

Darren previously worked in various LAN and WAN consulting positions in the Washington, D.C. area. He co-founded Design Consultants, a consulting firm that manufactured and marketed personal computers, as well as provided LAN/WAN consulting. Mr. Spohn started his communications career as a non-commissioned officer in the Pennsylvania Air National Guard. In his eight years of service he spent the majority of his time on international telecommunications projects in both the United States and Germany.

Mr. Spohn holds a B.S. in Telecommunications Engineering from Capital College, and has added to his bachelor's degree an M.S. in Engineering Management from Southern Methodist University, where he has completed work toward a Doctorate in Engineering. Mr. Spohn is author of *Data Network Design,* 2nd edition, and co-author of *ATM: Theory and Application* and *Hands-On ATM* — all published by McGraw-Hill.

Both authors have an edge over many other authors in that they have real-life experience in all subjects they write about. This brings a real life feeling to the text. Their experience to date with cell relay and recent publishing projects have produced a text which is not only readable for the common man through the technical guru, but also serves as a working tool for users and anyone considering the use of ATM technology or services for use well into the next millennium.